Subcellular Biochemistry

Volume 91

Series editor

J. Robin Harris, Institute of Zoology, University of Mainz, Mainz, Germany

More information about this series at http://www.springer.com/series/6515

J. Robin Harris • Viktor I. Korolchuk

Editors

Biochemistry and Cell Biology of Ageing: Part II Clinical Science

 Springer

Editors
J. Robin Harris
Institute of Zoology
University of Mainz
Mainz, Germany

Viktor I. Korolchuk
Newcastle University
Newcastle upon Tyne, UK

ISSN 0306-0225
Subcellular Biochemistry
ISBN 978-981-13-3680-5   ISBN 978-981-13-3681-2   (eBook)
https://doi.org/10.1007/978-981-13-3681-2

Library of Congress Control Number: 2018965214

This Springer imprint is published by the registered company Springer Nature Singapore Pte Ltd.
The registered company address is: 152 Beach Road, #21-01/04 Gateway East, Singapore 189721, Singapore

Foreword

Never has it been more important or timely for new volumes on the science of ageing to be produced. Around the world, continuing gains in life expectancy coupled with declining fertility rates in many countries are producing profound shifts in demographic profiles. A growing fraction of the population is living to advanced old age, bringing with it increased prevalence of a wide range of age-related chronic diseases. Whereas it was once thought that ageing was something that just happened and that was relatively low on the priority list for research, recent decades have seen exciting advances in probing the complex mechanisms through which the ageing process develops.

We have come a long way from the days when it was simply assumed there was some internal biological clock that would allow us an allotted span of "three score years and ten" and then kill us. Few had questioned why ageing should impose this fate upon us. It was loosely supposed that it was nature's way of creating living space for the next generation and securing evolutionary succession. We now know that these old-fashioned concepts have little credence. During our evolution, our genomes evolved impressive systems to try and preserve functional homeostasis in the biochemistry and cell biology of our bodies. The trouble is that there was never the evolutionary pressure to make these systems good enough completely to prevent damage from accumulating. Gradually, and at first unobtrusively, things begin to go wrong, starting from the earliest stages of life. And it is not one thing above all others – many systems experience deterioration at the same time. Herein lies the intriguing challenge of trying to unpick the contributions of the individual mechanisms that are being found to play their part in ageing and then of putting it all together.

Understanding the biochemistry of ageing is among the most complex of problems in the life sciences. On the one hand, we need to be intensively reductionist. We need to identify the fine detail of each one of the many biochemical mechanisms that contribute to functional decline. On the other hand, we need to appreciate that knowing everything there is to know about one particular mechanism may tell us rather little about the ageing process itself. To get the bigger picture, we must acknowledge that it makes little sense to argue the case for this mechanism versus that mechanism and so on. It is not a matter of simple alternatives. Instead of rooting for mechanism A *or* B *or* ... *or* Z, we must learn to appreciate that it is A *and* B *and* ... *and* Z. Whether we call this integrative biology, or systems biology, or some other term of a similar nature, the bottom line is that we need to join forces and learn as much as we can about the different biochemical mechanisms and their often synergistic interactions. In some ways, the science of ageing is the science of life itself. In the traditional school of biochemistry, we learn about how life has evolved the remarkable processes of DNA replication, transcription, translation, turnover, signal transduction, cell division and all the rest. These systems are so beautifully coordinated that we might marvel at first at how well they work. But the underlying molecular interactions are noisy and subject to perturbations of all kinds and at all times. It is this reverse side of the orderliness of biochemical processes that we need to appreciate to understand ageing.

In clinical terms, ageing is equally complex and challenging. Age is much the largest risk factor for a whole spectrum of different diseases, dwarfing the contributions from genetic, lifestyle and environmental risk factors. Furthermore, the fact that so many conditions share ageing as their dominant risk factor means it is no surprise that very old people commonly exhibit extensive multi-morbidity. But is ageing normal, or is it a disease? The answer is that ageing is a normal biological process, but it has the distinctive property that it makes us more vulnerable to diseases of many kinds. So it is a bit of a hybrid – both normal and also the source of pathologies. The old arguments about whether ageing is normal or disease are not particularly helpful. Ageing is driven by the accumulation of damage in our cells and organs, and the same is true of age-related, chronic diseases. Thus, there is a huge overlap. Once we understand the basic mechanisms of ageing itself, we will gain valuable knowledge about the many diseases which may affect us in later life. Thus, the study of the biochemistry and cell biology of ageing should seek to combine biomedical and clinical science. It is to be warmly welcomed, therefore, that J. Robin Harris and Viktor Korolchuk have produced these twin volumes, bringing into intimate juxtaposition a collection of state-of-the-art reviews of the biochemistry of ageing from both perspectives.

Emeritus Professor   Thomas B. L. Kirkwood
Newcastle University Institute for Ageing,
Campus for Ageing and Vitality,
Newcastle upon Tyne, UK

Preface

This book, *Biochemistry and Cell Biology of Ageing: Part II, Clinical Science* (along with *Part I, Biomedical Science*) was conceived following the reading (by JRH) of Lewis Wolpert's controversial yet thoroughly enjoyable 2011 book *You're Looking Very Well: The Surprising Nature of Getting Old*. As a broad discipline, *ageing* has been deemed to fit in well with the diverse content of the Springer Subcellular Biochemistry series; the two books covering biomedical science and clinical science were duly commissioned by the Springer Company.

We have attempted to compile a list of chapters written by authoritative clinical scientists to cover the field as thoroughly as possible. Along the way to production, a few chapters failed to appear! Nevertheless, the remaining 17 chapters provide a good coverage of the subject. To place the available chapters in a logical sequence has defeated us; we have simply presented them here as they appear in our initial list of agreed chapters, at the time of compilation. Each clinical science chapter stands firmly on its own merit, with correlation to the *Biological Science* book chapters in some cases. Over recent decades, *ageing* research has expanded enormously world-wide, responding to the increased importance to and interest from the general population, where there is an obvious desire to retain "quality of life", health and self-sufficiency into the later years.

The Contents list page, immediately following this Preface, shows the range of topics that are included. Without singling out any individual topic and author(s), it is clear that most of the important aspects of ageing research are included. Together they provide an in-depth survey of numerous aspects within the field of ageing research with clinical emphasis. We hope that the book will be of value to undergraduate biomedical science students, medical students, postgraduate researchers, clinicians and academics involved and interested in aspects of ageing research.

Mainz, Germany   J. Robin Harris
Newcastle upon Tyne, UK   Viktor I. Korolchuk
August, 2018

Contents

Chapter 1
Poor Early Growth and Age-Associated Disease

Jane L. Tarry-Adkins and Susan E. Ozanne

Abstract The prevalence of age-associated disease is increasing at a striking rate globally and there is evidence to suggest that the ageing process may actually begin before birth. It has been well-established that the status of both the maternal and early postnatal environments into which an individual is exposed can have huge implications for the risk of developing age-associated disease, including cardiovascular disease (CVD), type-2 diabetes (T2D) and obesity in later life. Therefore, the dissection of underlying molecular mechanisms to explain this phenomenon, known as 'developmental programming' is a highly investigated area of research. This book chapter will examine the epidemiological evidence and the animal models of suboptimal maternal and early postnatal environments and will discuss the progress being made in the development of safe and effective intervention strategies which ultimately could target those 'programmed' individuals who are known to be at-risk of age-associated disease.

Keywords Developmental programming · Ageing · Disease · Mechanisms

Introduction

At the biological level, ageing is associated with accumulated damage to cells that, over time, weakens the immune system, diminishes the body's capacity to repair itself and increases the risk of developing a host of different diseases. United Nations (UN) statistics have estimated that the number of people over the age of 65 will increase from 524 million in 2010 to 1.5 billion in 2050 (United Nations 2015). Whether this is due to increased birth rates or improved medical care for the elderly, or a combination of both factors, this increasingly aged population has consequently

J. L. Tarry-Adkins (✉) · S. E. Ozanne
University of Cambridge Metabolic Research Laboratories and MRC Metabolic Diseases Unit, Wellcome Trust-MRC Institute of Metabolic Science, Addenbrooke's Treatment Centre, Addenbrooke's Hospital, Cambridge, UK
e-mail: seo10@cam.ac.uk

© Springer Nature Singapore Pte Ltd. 2019
J. R. Harris, V. I. Korolchuk (eds.), *Biochemistry and Cell Biology of Ageing: Part II Clinical Science*, Subcellular Biochemistry 91,
https://doi.org/10.1007/978-981-13-3681-2_1

1

triggered a rapid rise in age-associated disease. Specifically, from the age of 50 there has been an exponential rise in burden from many conditions, including cardiovascular disease (CVD), type 2-diabetes (T2D), cancers, dementia, chronic kidney disease (CKD) and liver cirrhosis. This is placing an enormous strain on health and social care systems and therefore has been a focus of much research over the years.

Interestingly, it has been shown that people do not age at the same rate at a biological level, and this divergence in biological ageing rate can be detected in early life. A study in 2015 by Belsky and colleagues (Belsky et al. 2015) measured indices of biological ageing in a New Zealand cohort of young, disease-free adults who were followed from birth to 38 years old. Young individuals of the same chronological age varied in their biological ageing (as defined by a decline in multiple organ systems). Those that demonstrated accelerated biological ageing were less physically able, self-reported worse health, had a decline in cognition and looked older at 38 years of age, compared to those with normal biological ageing. This suggests that mechanisms exist which are causing acceleration of biological ageing in the young and these require further elucidation.

The Thrifty Phenotype: Epidemiological Evidence

There is evidence to suggest that the ageing process may actually begin before birth. This stems from studies by Hales and Barker over 25 years ago. They demonstrated increased risk of the development of impaired glucose tolerance and T2D (Hales et al. 1991) and the metabolic syndrome and risk factors for CVD in 64-year old men in Hertfordshire, UK who had a low-birth weight and poor growth in early life compared to those that had a higher birth-weight (Barker et al. 1993). These findings were also recapitulated in women (Fall et al. 1995). Moreover, these findings have been robustly tested and reiterated in many population cohorts throughout the world, including the USA (Dabelea et al. 1999), Europe (Lithell et al. 1996), India (Fall et al. 1998) and China (Mi et al. 2000). Hales and Barker proposed the associations between low birth-weight and age-associated disease arose due to the response of a growing fetus to in-utero nutrition. They proposed that the fetus would develop a "thrifty phenotype" which would make it vulnerable to the effects of postnatal adequate nutrition/over-nutrition. The Dutch Hunger Winter Famine provided a natural experiment in which to study the effects of a poor maternal nutrition upon offspring health. During WWII, between September 1944 and May 1945, due to a German blockade of supplies to Holland, people were restricted to less than 1000 calories per day. After Holland was liberated by the Allies in June 1945, food supplies were restored and people were received 2000 calories per day. Ravelli and colleagues initially demonstrated perturbed glucose tolerance which was associated with insulin resistance in offspring of women exposed to the famine during pregnancy (Ravelli et al. 1998). More recently, the specific time-windows of this famine exposure and *in-utero* growth restriction (IUGR) have been investigated: Offspring of women who were exposed to famine conditions during early gestation had increased prevalence of coronary heart disease (CHD), a more atherogenic profile, increased stress responsiveness, obesity (Roseboom et al. 2006) and an increased rate of mortality (Ekamper et al. 2014). Mid-gestational exposure to famine resulted in offspring with increased prevalence of microalbuminuria and obstructive airways disease (Roseboom et al. 2006), and those exposed during late gestation had the greatest incidence of glucose intolerance (Roseboom et al. 2006), however mortality rates were unaltered (Ekamper et al. 2014). Glucose intolerance was observed in offspring of mothers exposed to famine during all trimesters (Roseboom et al. 2006; de Rooj et al. 2006). These data gave important insight into how crucial specific gestation time-windows of exposure are in the manifestation of age-associated disease in offspring in later life.

Since the conception of the Thrifty Phenotype, the association between suboptimal maternal/early-life environments and subsequent risk of later-life disease has been demonstrated in animal models of many suboptimal environments; including maternal protein restriction, maternal obesity, placental insufficiency and maternal hypoxia. This concept is therefore referred to as the Developmental Origins of Health and Disease (DoHAD) or "Developmental Programming."

The Environment vs. Genetics: The Evidence

In order to elucidate whether this phenomenon has any genetic basis, studies using monozygotic twins have been used. Poulsen et al. demonstrated increased prevalence of poor glucose tolerance and T2D in the monozygotic twin with the lower birth weight (Poulsen et al. 1996; Poulsen et al. 1999). Others have reported hypertension and impaired endothelial function in the monozygotic twin with poor fetal growth, compared to the normal birth weight twin (Poulsen et al. 2002). Interestingly, low birth weight was also associated with glucose intolerance (Halvorsen et al. 2006), insulin resistance (Grunnet et al. 2007) and insulin secretion (Monrad et al. 2009) in elderly monozygotic twin pairs, suggesting that ageing could play an important role on the influence of an adverse intrauterine environment on glucose intolerance, insulin resistance and low insulin secretion in genetically identical twins. Taken together, this demonstrates that age-associated disease risk in the offspring as a consequent of a poor maternal environment occurs independently of genotype.

Mis-matched *in utero* and Postnatal Environments

It has also become apparent that risk to offspring exposed to a suboptimal maternal environment (which elicits low birth weight) can be further exacerbated if the offspring are also exposed to a postnatal environment which induces rapid growth. Eriksson and colleagues initially studied a large cohort of individuals from Finland including those who were born small and underwent rapid postnatal catch-up growth. Individuals who were born small and crossed growth centiles postnatally, had the highest blood pressure (Eriksson et al. 2000) and T2D incidence (Eriksson 2006) and increased prevalence of CVD (Eriksson et al. 1999). Subsequent epidemiological evidence recapitulated the finding of increased risk of hypertension in children (Hemachandra et al. 2007) and in 22 year old adults (Law et al. 2002) who were small at birth and grew rapidly postnatally. Other phenotypes as a consequence of low birth weight and rapid postnatal growth has included poor cognitive development (Estourgie-van Burk et al. 2009), reduced β-cell numbers and diminished whole body glucose uptake (Crowther et al. 2000) and poor glucose tolerance (Crowther et al. 1998) in later life. In a cohort of small-for-gestational-age (SGA) babies, those randomly assigned to a nutrient-enriched formula had accelerated postnatal growth and increased fat mass in later life compared to those assigned to a control formula (Singhal et al. 2010). The induction of a slow postnatal growth trajectory can also have beneficial effects on future offspring health. Evidence that slow growth during the lactation period is beneficial has been found in several studies on pre-term infants. Pre-term babies who were randomly assigned donated breastmilk compared to enriched formula had lower blood pressure as adolescents (Singhal et al. 2001), had reduced leptin concentrations (Singhal et al. 2002) and reduced markers of atherosclerosis risk, compared to those pre-term infants given formula (Singhal et al. 2004). Similarly, a low level of caloric restriction (a restriction of nutrient intake without causing malnourishment) has been established to increase longevity in a range of organisms from worms and fruit flies to rodents and humans (Calabrese et al. 2011).

Maternal Obesity and Gestational Diabetes Mellitus (GDM)

Thus far, we have discussed epidemiological evidence from suboptimal maternal environments in terms of insufficiency during pregnancy. The impact of overnutrition during pregnancy is also a major factor in programming of offspring health. Incidence of obesity during pregnancy is now at epidemic proportions with 64% of all women of reproductive age classified as either overweight (BMI > 25) or obese (BMI > 30), in both the USA and Europe (Heslehurst et al. 2010). Maternal obesity leads to severe problems for both mother and offspring. The mother's risk of gestational diabetes mellitus (GDM), T2D and CVD complications in later life is increased. For the offspring, fetal outcomes include increased prevalence of stillbirth, neonatal death, congenital anomalies and macrosomic birth. The risk of macrosomia in maternal obesity is due to the association with gestational diabetes mellitus (GDM), in which hyperglycaemia occurs in the mother. As maternal glucose can cross the placenta but maternal insulin cannot, the fetus attempts to regulate its own glucose homeostasis by increasing fetal β-cell insulin production. Insulin is a potent growth factor in fetal life, thus this can result in macrosomic offspring. High birth weight (as well as low birth weight) has been shown to be a risk factor for increased risk of cardio-metabolic disease in later life. The best evidence for this phenomenon comes from a well investigated and documented cohort of Pima Indians in Arizona, USA. This population has very high rates of gestational diabetes mellitus (GDM) and it was demonstrated that the greatest incidence of T2D in the offspring was located at the extreme edges of birth-weight, producing a 'U' shaped curve (Pettitt and Jovanovic 2001). Others have also demonstrated, in a general population cohort, that large for gestational age (LGA) offspring from GDM pregnancies had increased risk of developing metabolic syndrome in childhood (Boney et al. 2005). Additionally, children who were born to obese mothers, without GDM are also at increased risk of metabolic syndrome development in later life (Boney et al. 2005), which may have implications for perpetuating the cycle of obesity, insulin resistance, and their consequences in subsequent generations.

Other deleterious adult-onset effects of being born to an obese mother include CVD incidence. In a cohort analysis of 37,709 people, Reynolds and co-authors demonstrated that maternal obesity is associated with an increased risk hospitalisation from a cardiovascular event and premature death in offspring (Reynolds et al. 2013). Other risk factors which are increased in offspring of obese mothers include impaired glucose tolerance, hypertension and obesity (Tam et al. 2017).

Animal Models of Developmental Programming

Since the early 1990's, strong epidemiological evidence supports the concept of developmental programming. Animal models have also been widely utilised to support epidemiological studies and gain further understanding of the complex molecular mechanisms and interactions underlying this phenomenon, which would otherwise be impossible/unethical to conduct in human populations. These models have included rodents, sheep and primates, however, because if the relative ease of studying them across the life-course, rodent models have been the best characterised of all animal models, and therefore this book chapter will focus most on these models.

Maternal Protein Restriction

Studies involving maternal protein restriction have been the most widely studied of all programming models due to amino acids making up an essential part of growth *in utero*. One of the most studied of all maternal protein restriction models is one utilising an 8% protein diet *in utero* in rats which was first used by Hoet and colleagues (Snoeck et al. 1990). Using this model, offspring of maternally protein restricted dams had a low birth weight, reduced β-cell proliferation and pancreatic islet cell size (Snoeck et al. 1990), reduced islet vascularization (Snoeck et al. 1990; Dahri et al. 1991), and pancreatic insulin content (Dahri et al. 1991) and increased β-cell apoptosis (Petrik et al. 1999). This was accompanied by a reduced pancreatic expression of insulin-like growth factor II (IGFII) (Petrik et al. 1999). In our laboratory, we sought to elucidate the molecular mechanisms underpinning these observations. Offspring of mothers fed a restricted (8%) protein diet were insulin sensitive in early life (3 months of age) (Ozanne et al. 1996), however in old age (17 months), male rats developed T2D (Petry et al. 2002), with female offspring becoming insulin resistant by 22 months of age (Fernandez-Twinn et al. 2005). These data highlights the key role that ageing plays in the development of deleterious phenotypes in 'programmed' individuals. The insulin resistance of these rat offspring was related to impaired insulin action in adipose tissue, via impaired PI-3 kinase activation and in skeletal muscle (Ozanne et al. 2001), via down-regulation of PKC-zeta (PKC-ζ), p85alpha (p85-α), p110beta (p110-β) and glucose transporter-4 GLUT4 (Ozanne et al. 2005). Interestingly, the same insulin signaling molecules were also down-regulated; to the same magnitude, in the muscle and adipose tissue of 19 year old men who were born small, compared to those with a normal birth weight (Ozanne et al. 2001; Ozanne et al. 2005). This illustrates the relevance and importance of using the maternal protein restriction rat model of IUGR to accurately mimic human IUGR. Others have recently demonstrated a fatty liver phenotype in rat offspring of mothers fed a restricted (8%) protein diet during pregnancy (Campisano et al. 2017).

Other studies utilising a rat model of maternal protein restriction (9%) have demonstrated hypertension and reduced nephron and glomeruli number in offspring (Langley-Evans et al. 1994, 1999). The reduction in glomeruli number seems to be associated with increased apoptosis of mesenchymal cells at the start of rat metanephrogenesis (Welham et al. 2002) and may be related to alterations in gene expression associated with metanephrogenesis (Welham et al. 2005). Reductions in nephron number have also been shown in mouse offspring who were exposed to protein restriction *in utero* (9%) (Hoppe et al. 2007). Maternal protein restriction (9%) has also been shown to reduce the number of cardiomyocytes (Corstius et al. 2005), influence vascular remodelling (Dodson et al. 2017) and increase interstitial fibrosis in hearts of rat offspring (Lim et al. 2006).

Maternal Protein Restriction and Altered Postnatal Growth: Animal Mdels

In order to further understand the effects of a mismatched maternal and postnatal milieu that has been observed in epidemiological studies, our laboratory have utilised rat and mouse models of cross-fostering. Mouse offspring of mothers fed a low (8%) protein diet which were suckled by control-fed (20% protein), were born small and underwent rapid postnatal growth (Ozanne and Hales 2004). This rapid growth trajectory impacted upon longevity, with these 'recuperated' offspring living for significantly less time compared to control offspring (Ozanne and Hales 2004). When slow postnatal growth was induced by cross-fostering offspring exposed to a control protein diet *in utero* to mothers fed the low (8%) protein diet, these mouse offspring lived significantly longer compared to control offspring and were protected against the detrimental effects of an obesogenic 'cafeteria' diet (Ozanne and Hales 2004). In addition, rats (Bieswal et al. 2006) and mice (Bol et al. 2009) that were calorie restricted *in utero* and over-fed during the sucking period and underwent rapid postnatal growth, were hyperphagic, hyperinsulinaemic, glucose intolerant, insulin resistant and obese. Moreover, this obesity phenotype may be associated with alterations in genes related to adipose tissue differentiation and function in the mouse (Bol et al. 2009). *In utero* food restriction in rats followed by a post-weaning high-fat diet also resulted in rapid postnatal growth, insulin resistance and pancreatic islet steatosis (Xiao et al. 2017).

Maternal Food Restriction

Many animal models of food restriction ranging from mild (30%) to severe (70%) have been used as a model of IUGR. In a rat model of 50% food restriction (FR) during pregnancy, offspring had reduced body and pancreatic weights. At postnatal day 1 these pups had reduced pancreatic insulin content and decreased β-cell mass (Garafono et al. 1997). This was associated with reduced islet number. By postnatal day 21 the reduction in β-cell mass and pancreatic islet content was not restored (Garafono et al. 1997) and re-nutrition up to the weaning period was unable to reverse the decrease in β-cell mass (Garafono et al. 1998). When the offspring of 50% FR dams were aged to 12 months, this caused glucose intolerance and increased β-cell apoptosis (Garafono et al. 1999). These offspring also had disturbed hypothalamic-pituarity-adrenal axis (HPA) which resulted in increased plasma corticosterone levels (Lesage et al. 2001). Another rat model of 50% FR demonstrated a reduction in the postnatal leptin surge and alterations in hypothalamic POMC mRNA expression, which may sensitise the offspring to an obesity phenotype in later life (Delahaye et al. 2008). In a severe model of gestational FR (70%), rat pups were growth restricted and developed hyperphagia, obesity, hypertension, hyperleptinaemia and hyperinsulinaemia via dysregulation of the adipoinsulinar axis during adult life (Vickers et al. 2001). In baboons, a moderate (30%) maternal FR model resulted in cardiac remodelling which mimicked accelerated ageing (Kuo et al. 2017), insulin resistance (Choi et al. 2011) and increased aggressive behaviours (Huber et al. 2015) in the offspring.

Maternal Hypoxia

Fetal hypoxia is a common pregnancy complication in which the fetus mounts a defence to a short-term episode of hypoxia, by redistributing blood flow away from peripheral circulations towards essential vascular beds, such as those perfusing the brain. Offspring born to hypoxic mothers are IUGR and have cardiovascular dysfunction, including enhanced aortic thickening (Giussani et al. 2012), decreased capillary supply, reduced metabolic efficiency and decreased cardiac output (Hauton et al. 2015) and altered vascular tone (Tang et al. 2015). It has also been shown that vascular endothelin-1 function was altered in aged maternally hypoxic male rat offspring (Bourque et al. 2013). Additionally, in a mouse model of maternal hypoxia, ageing male mice had decreased glomerular number which was associated with albuminurea, glomerular fibrosis and renal fibrosis, which was not observed in females. Both sexes however, were hypertensive in old age (12 months of age) (Walton et al. 2017). The placenta is also affected by maternal hypoxia, with increased fetoplacental vascular resistance and vasoconstrictor reactivity in the offspring which is likely to produce placental hypoperfusion and fetal undernutrition (Jakoubek et al. 2008; Kafka et al. 2016). Recently, it is emerging that the fetal lung is also detrimentally affected by maternal hypoxia. In late gestation, chronic maternal hypoxia increases genes which promote lung maturation which may be an adaptive response to air breathing after birth (McGillick et al. 2017).

Maternal Obesity

Many animal models of maternal obesity exist which recapitulate observations in epidemiological studies. In a mouse model of maternal obesity, offspring born to mothers fed an obesogenic diet during pregnancy and lactation had increased adiposity and were hyperphagic, hypertensive and insulin resistant (Samuelsson et al. 2008; Kirk et al. 2009). These phenotypes were accompanied by an amplified and prolonged neonatal leptin surge and increased *leptin* mRNA expression (Kirk et al. 2009). This was associated with a 50% decrease in pancreatic islet insulin secretion (Zambrano et al. 2016). In 8 week mouse offspring from obese pregnancies, our laboratory has also demonstrated cardiac hypertrophy which was associated with re-expression of cardiac fetal genes. By young adulthood (12 weeks) these offspring developed severe systolic and diastolic dysfunction and cardiac sympathetic dominance (Blackmore et al. 2014). Furthermore, these offspring had increased markers of adipose tissue inflammation (Alfaradhi et al. 2016) and had hall-marks of non-alcoholic fatty liver disease (NAFLD), which included reduced hepatic glycogen and protein content and increased hepatic lipid content, which was associated with increased *ppar*γ triglyceride lipase mRNA expression (Alfaradhi et al. 2014). Importantly, these phenotypes manifested in the absence of increased body weight and adiposity (Blackmore et al. 2014; Alfaradhi et al. 2014, 2016). Furthermore, we have recently demonstrated that dysregulation of insulin signaling in maternally obese offspring is further exacerbated when maternally obese offspring are weaned onto an obesogenic diet (de Almeida Faria et al. 2017). Others have shown that placentas of maternally obese mice had disrupted placental morphology, decreased cell proliferation and increased inflammation (Kim et al. 2014), and offspring of maternally obese mouse mothers were glucose intolerant and had increased nephron number (Hokke et al. 2016).

Placental Insufficiency

Placental insufficiency is one of the key causes of IUGR offspring, with between 4% and 8% of all pregnancies being affected by this complication (Hendrix and Berghella 2008). In pregnancies affected by placental insufficiency, a poorly functioning placenta restricts nutrient supply to the fetus and prevents normal fetal growth. One of the most utilised animal models to mimic placental insufficiency is bilateral uterine ligation. Simmons and colleagues developed a rat model of bilateral uterine ligation in which both uterine arteries were ligated to the same degree as observed in human pregnancies complicated with uteroplacental insufficiency. The authors found that these animals were born small and underwent rapid postnatal growth so that they were obese by 26 weeks of age (Simmons et al. 2001). These animals were also hyperglycemic, glucose intolerant and insulin resistant in early life. They also underwent an age-associated decline in β-cell mass, so that by 26 weeks of age, their β-cells were less than one third the size of the controls (Simmons et al. 2001). Others have demonstrated that bilateral uterine ligation in rats caused a decrease in insulin receptor and reduced expression of enzymes of long-chain fatty acid metabolism in the skeletal muscle (Germani et al. 2008) and altered hepatic fatty acid metabolising enzymes (Lane et al. 2001) in the offspring. Placental insufficiency by bilateral uterine ligation has also had an impact upon the fetal kidney. Reduced nephron number, increased blood pressure (Wlodek et al. 2008; Moritz et al. 2009), and altered renal function (Wlodek et al. 2008) have been reported in male IUGR rats. Female rats also demonstrate a nephron deficit and modest renal insufficiency; however this was not accompanied by hypertension (Moritz et al. 2009), however hypertension was detected in older female offspring (Gallo et al. 2012). The same authors also demonstrated regional vascular dysfunction and altered arterial stiffness in female rats born to placental insufficient mothers (Mazzuca et al. 2010).

Maternal Stress

Stress during pregnancy is known to cause IUGR due to alterations in the HPA axis via modulation of cortisol. Thus, animal models of maternal stress have mainly utilised dexamethasone (a synthetic corticosteroid). In both rats (Benediktsson et al. 1993) and sheep (Dodic et al. 2002) maternal dexamethasone exposure resulted in hypertensive, IUGR offspring. Exposure of rat fetuses to glucocorticoids also increased hepatic PEPCK and glucocorticoid receptor (GR) expression which caused glucose intolerance in adulthood (Nyrienda et al. 1998). Increases in GR expression have also been observed in visceral adipose tissue from rat offspring exposed to dexamethasone *in utero*, which could contribute to insulin resistance (Cleasby et al. 2003). In a non-human primate model of prenatal dexamethasone exposure, offspring had impaired glucose tolerance, and hyperinsulinaemia, reduced pancreatic β-cell number and were hypertensive (de Vries et al. 2007).

Maternal Iron Insufficiency

Global estimates suggest that 41.8% of all pregnant women and 30% of women of reproductive age are anaemic and this is particularly common in the developing world (Scholl 2011). In the rat, models of iron deficiency result in IUGR offspring (Lewis et al. 2001a, b) which were hypertensive at 3 months of age (Lewis et al. 2001a, b) and the elevated blood pressure was still observed at 16 months of age (Lewis et al. 2002). The increased blood pressure may, in part, be due to a deficit in nephron number (Lisle et al. 2003). Rat offspring from iron restricted mothers also have alterations in placental vascularisation (Lewis et al. 2001a, b). The developing fetus also had a more severe and less reversible phenotype, if exposed to maternal iron restriction during the first half of pregnancy (Gambling et al. 2004). Maternal iron restriction also resulted in the alteration of hepatic lipid metabolising enzymes in the fetuses (Zhang et al. 2005).

Mechanisms of Ageing in Developmental Programming

Oxidative Stress in the Aetiology of Developmental Programming

Generation of reactive oxygen species (ROS) or reactive nitrogen species (RNS) have long been associated with the development of an ageing phenotype. Originally, in the 1950's Harman proposed the free radical theory of ageing (Harman 1956), in which it was proposed that organisms age as a consequence of the accumulation of free radical oxidative damage. However one must take into account more recent evidence which shows that ROS accumulation is not solely responsible for causing damage to cellular macromolecules and thus facilitating the ageing process, but can also be involved in signaling functions, thereby activating protective and adaptive functions. However, many studies demonstrate accumulation of ROS and oxidative damage during the ageing process and these are prevalent in a host of age-associated diseases including CVD, T2D, Alzheimer's disease, cancers, liver and kidney dysfunction (Valko et al. 2006). There are three major sources of ROS generation in cells; i) excess stimulation of NAD(P)H oxidase, ii) xanthine oxidase overproduction and iii) alterations in mitochondrial electron transport chain (ETC) complex activity. It is well established that the generation of ROS is a common shared mechanism in suboptimal maternal and early life environments, and thereby may influence progression of age-associated pathologies. In our rat model of maternal protein restriction followed by accelerated postnatal growth ('recuperated'), we observed increased ROS generation in these offspring. This included increased 4-hydroxynonenal levels (a marker of lipid peroxidation), impairments in the activities and expression of the ETC complexes and alterations in xanthine oxidase and NADPH oxidase expression. This was observed in a variety of tissues, including the aorta (Tarry-Adkins et al. 2008), heart (Tarry-Adkins et al. 2013) liver (Tarry-Adkins et al. 2016a), pancreatic islets (Tarry-Adkins et al. 2010), skeletal muscle (Tarry-Adkins et al. 2016b) and ovaries (Aiken et al. 2013). Placental insufficiency mediated by bilateral uterine ligation also has a robust oxidative stress phenotype, particularly centred on mitochondrial dysfunction. Rat offspring born to mothers that underwent bilateral uterine ligation had impaired oxidative phosphorylation in liver (Peterside et al. 2003) and skeletal muscle (Selak et al. 2003) which was associated with alterations of mitochondrial gene expression and function in the skeletal muscle (Lane et al. 1998). In a rabbit model of placental insufficiency mediated by bilateral uterine ligation, offspring had increased protein carbonylation levels and HIF-1α, eNOS, p-eNOS, and iNOS induction (Figueroa et al. 2017). Models of maternal hypoxia have also demonstrated a defined pro-oxidative stress phenotype. Rat offspring born to hypoxic mothers had increased 3-nitrotyrosine staining (a marker of protein tyrosination) (Giussani et al. 2012) and increased lipid peroxidation (Richter et al. 2012). In models of maternal obesity, offspring from obese pregnancies had decreased activities of complexes II and III of the mitochondrial ETC in skeletal muscle which was associated with alterations in insulin signaling molecules (Shelley et al. 2009) and in the liver, reduced mtDNA copy number and alterations in mitochondrial genes in the aorta was also observed in offspring from obese rat mothers (Taylor et al. 2005).

Oxidative Stress and Telomeres

Mammals, telomeres are made of a variable number of tandem repeats of DNA $(TTAGGG)_n$. Due to the 'end-replication problem', telomeres from somatic cells shorten with every cellular division until they reach a critical length and activate cellular senescence and apoptosis pathways (Sharpless and DePhino 2004). This has led some investigators to believe that telomere length measurement is a robust proxy of cellular ageing (Blackburn et al. 2015, 2006). Moreover, the association between short telomeres and increased mortality risk has been shown in many studies (Needham et al. 2015, Bakayska et al. 2007, Deelen et al. 2014). Excessive ROS generation is known to damage cell macromolecules, especially telomeric DNA as ROS preferentially damages the guanine-rich regions of the telomere (Richter and von Zglincki 2007; von Zglincki 2002). Resultantly, increased oxidative stress can cause accelerated telomere attrition which can result in accelerated cellular ageing. Both epidemiological studies and animal models of developmental programming have investigated telomere length in the context of cellular ageing: Recently, shorter telomeres have been observed in low birth weight babies from a Chinese cohort (Lee et al. 2017) and in cohorts of pregnant women, those infants born to mothers with severe maternal psychological stress during pregnancy had significantly shorter leukocyte telomere length as newborns (Entringer et al. 2013; Marchetto et al. 2016) and in young adulthood (Entringer et al. 2011 compared to offspring from low-stress pregnancies. In our model of maternal protein restriction followed by rapid postnatal growth, in which we also observe a reduction in longevity, we demonstrated accelerated telomere shortening in the aorta (Tarry-Adkins et al. 2008), heart (Tarry-Adkins et al. 2013), pancreatic islets (Tarry-Adkins et al. 2009), and ovaries (Aiken et al. 2013) of these 'recuperated' offspring, which was associated with alterations in antioxidant defense capacity. Taken together, this may suggest that a suboptimal early life environment can lead to accelerated cellular ageing, which may be associated with increased disease pathology and reduced longevity.

Transgenerational Evidence and Epigenetic Programming

Recently, evidence has emerged that an individual may be 'programmed' for increased risk of disease in later life, without a direct exposure to a suboptimal maternal environment and that 'memory' of the suboptimal maternal milieu can be passed down from previous generations. This developing programming phenomenon is known as 'transgenerational transmission', and has been widely reported, in both epidemiological studies and animal models, (reviewed in Eriksson 2016; Lee 2015; Bianco-Miotto et al. 2017). Transmission can occur via the maternal line or paternal line where epigenetic modifications can be inherited via the somatic cell lineage or down the germline (Aiken and Ozanne 2014). Epigenetic modification to DNA can include DNA methylation, acetylation, phosphorylation, and sumolyation and causes changes in gene expression without altering the DNA sequence. Mechanisms of transgenerational programming are not, however, limited to epigenetic regulation. In fact, it is becoming clear that 'programmed' phenotypes can be transmitted through the maternal line de-novo, to beyond the F2 generation, due to the fetus developing in a suboptimal developed uterine tract (Aiken and Ozanne 2014).

Intervention Strategies

Given that a common mechanism of age-associated diseases as a result of a suboptimal early-life environment is the generation of oxidative stress, many studies have investigated the efficacy of antioxidant intervention in models of developmental programming. Most studies have focused on administration of antioxidants during pregnancy. An *in utero* dosage of ascorbate reversed placental oxidative stress, reversed IUGR (Richter et al. 2012) and ameliorated cardiac dysfunction in a rat model of maternal hypoxia (Kane et al. 2013). *In utero* melatonin administration has also increased placental expression of antioxidant enzymes and reversed IUGR in undernourished rat pregnancies (Richter et al. 2009). These studies however were at doses far higher than considered safe for use in pregnant women. Also, in many cases, evidence is not present for suboptimal *in utero* exposure until at the time of, or just after delivery. Therefore it is also important to address the potential beneficial effects of targeted postnatal antioxidant supplementation at a clinically relevant and safe dose. Over the last few years, our laboratory have investigated the effect of postnatal intervention of the endogenous antioxidant coenzyme Q_{10} (CoQ_{10}), in maternally protein restricted rats that underwent accelerated postnatal catch-up growth (recuperated). We have demonstrated a prevention of accelerated cellular ageing in many tissues from these 'programmed' rats including the heart (Tarry-Adkins et al. 2013), aorta (Tarry-Adkins et al. 2014), liver (Tarry-Adkins et al. 2016a, b) and adipose tissue (Tarry-Adkins et al. 2015). Recently, postnatal intervention of resveratrol also improved metabolic and cardiac dysfunction in a rat model of maternal hypoxia and postnatal high-fed feeding (Shah et al. 2016, Shah et al. 2017).

Conclusion

Over the last 25 years, the phenomenon of developmental programming in the aetiology of age-associated disease has been well established using a range of epidemiological evidence and animal models. Mechanisms underpinning this association are still being investigated, however over generation of ROS and epigenetic modification which have consequences for alterations in tissue structure and function are thought to play a role, and ultimately increase risk of age-associated disease.

References

Aiken CE, Ozanne SE (2014) Transgenerational developmental programming. Hum Reprod Update 20:63–75

Aiken CE, Tarry-Adkins JL, Ozanne SE (2013) Suboptimal nutrition in utero causes DNA damage and accelerated aging of the female reproductive tract. FASEB J 27:3959–3965

Alfaradhi MZ, Fernandez-Twinn DS, Martin-Gronert MS et al (2014) Oxidative stress and altered lipid homeostasis in the programming of offspring fatty liver by maternal obesity. Am J Physiol Regul Intergr Comp Physiol 307:R26–R34

Alfaradhi MZ, Kusinski LC, Fernandez-Twinn DS et al (2016) Maternal obesity in pregnancy developmentally programs adipose tissue inflammation in young, lean male mice. Endocrinology 157(11):4246–4256

Bakayska SL, Mucci LA, Slagbloom PE et al (2007) Telomere length predicts survival independent of genetic influences. Aging Cell 6:769–774

Barker DJ, Hales CN, Fall CH et al (1993) Type 2 (non-insulin dependent) diabetes mellitus, hypertension and hyperlipidaemia (syndrome X): relation to reduced fetal growth. Diabetologia 36:62–67

Belsky DW, Caspi A, Houts R et al (2015) Quantification of biological aging in young adults. PNAS 112(30):E4104–E4110

Benediktsson R, Lindsay R, Noble J et al (1993) Glucocorticoid exposure in utero: a new model for adult hypertension. Lancet 341:339–341

Bianco-Miotto T, Craig JM, Gasser YP (2017) Epigenetics and DOHaD: from basics to birth and beyond. J Dev Orig Health Dis 8:513–519

Bieswal F, Ahn MT, Reusens B et al (2006) The importance of catch-up growth after early malnutrition for the programming of obesity in the male rat. Obesity 14:1330–1334

Blackburn EH, Greider CW, Szostak JW (2006) Telomeres and telomerase: the path from maize, tetrahymena and yeast to human cancer and aging. Nat Med 12:1133–1138

Blackburn EH, Epel ES, Lin J (2015) Human telomere biology: a contributory and interactive factor in aging, disease risks and protection. Science 350(6265):1193–1198

Blackmore HL, Niu Y, Fernandez-Twinn DS et al (2014) Maternal diet-induced obesity programs cardiovascular dysfunction in adult male mouse offspring independent of current body weight. Endocrinology 155:3970–3980

Bol VV, Delattre AI, Reusens B et al (2009) Forced catch-up growth after fetal protein restriction alters the adipose tissue gene expression program leading to obesity in adult mice. Am J Physiol Regul Integr Comp Physiol 297:R291–R299

Boney CM, Verma A, Tucker R et al (2005) Metabolic syndrome in childhood: assocations with birthweight, maternal obesity, and gestational diabetes mellitus. Paediatrics 115:e290–e296

Bourque SL, Gragasin FS, Quon AL et al (2013) Prenatal hypoxia causes long-term alterations in vascular endothelin-1 function in aged male, but not female, offspring. Hypertension 62:753–758

Calabrese V, Cornelius C, Cuzzocrea S et al (2011) Hormesis, cellular stress response and vitagenes as critical determinants in aging and longevity. Mol Asp Med 32:279–304

Campisano SE, Echarte SM, Podaza E (2017) Protein malnutrition during fetal programming induces fatty liver in adult male offspring rats. J Physiol Biochem 73:275–285

Choi J, Li C, MacDonald TJ et al (2011) Emergence of insulin resistance in juvenile baboon offspring of mothers exposed to moderate maternal nutrient restriction. Am J Physiol Regul Integr Comp Physiol 301:R757–R762

Cleasby ME, Kelly PAT, Walker BR et al (2003) Programming of rat muscle and fat metabolism by in utero overexposure to glucocorticoids. Endocrinology 144:999–1007

Corstius HB, Zimanyi MA, Maka N et al (2005) Effect of intrauterine growth restriction on the number of cardiomyocytes in rat hearts. Pediatr Res 57:796–800

Crowther NJ, Cameron N, Trusler J et al (1998) Association with poor glucose tolerance and rapid postnatal growth in seven-year old children. Diabetologia 41:1163–1167

Crowther NJ, Trusler J, Cameron N et al (2000) Relation between weight gain and beta-cell secretory activity and non-esterified fatty acid production in 7 year old African children: results from the Birth to Ten study. Diabetologia 43:978–985

Dabelea D, Pettitt DJ, Hanson RL et al (1999) Birth weight, type 2 diabetes, and insulin resistance in Pima Indian children and adults. Diabetes Care 22:944–950

Dahri S, Snoeck A, Reusens-Billen B et al (1991) Islet function in offspring of mothers on low-protein diet during lactation. Diabetes 40:115–120

de Almeida Faria J, Duque-Guimaraes D, Carpenter AAM et al (2017) A post-weaning obesogenic diet exacerbates the detrimental effects of maternal obesity on offspring insulin signaling in adipose tissue. Sci Rep 7:44949

Deelen J, Beekman M, Codd V et al (2014) Leukocyte telomere length associates with prospective mortality independent of immune-related parameters and known genetic markers. Int J Epidemiol 43:878–886

Delahaye F, Breton C, Risold PY et al (2008) Maternal perinatal undernutrition drastically reduces postnatal leptin surge and affects the development of arcuate nucleus proopiomelanocortin neurons in neonatal male rat pups. Endocrinology 149:470–475

de Rooj SR, Painter RC, Phillips DI et al (2006) Impaired insulin secretion after prenatal exposure to the Dutch famine. Diabetes Care 29:1897–1901

de Vries A, Holmes MC, Heijnis A et al (2007) Prenatal dexamethasone exposure induces changes in nonhuman primate offspring cardiometabolic and hypothalamic-pituitary-adrenal axis function. J Clin Invest 117:1058–1067

Dodic M, Moritz K, Koukoulas I et al (2002) Programming effects of short prenatal exposure to cortisol. FASEB J 16:1017–1026

Dodson RB, Miller TA, Powers K et al (2017) Intrauterine growth restriction influences vascular remodelling and stiffening in the weanling rat more than sex or diet. Am J Physiol Heart Circ Physiol 312:H250–H264

Ekamper P, van Poppel F, Stein AD et al (2014) Independent and additive association of prenatal famine exposure and intermediary life conditions with adult mortality between 18–63 years. Soc Sci Med 119:232–239

Entringer S, Epel ES, Kumsta R et al (2011) Stress exposure in intrauterine life is associated with shorter telomere length in young adulthood. Proc Natl Acad Sci U S A 108:E513–E518

Entringer S, Epel ES, Lin J et al (2013) Maternal psychological stress during pregnancy is associated with newborn leukocyte telomere length. Am J Obestet Gynecol 208:e1–e7

Eriksson JG (2006) Early growth, and coronary heart disease and type 2 diabetes: experiences from the Helsinki Birth Cohort studies. Int J Obes 30(Suppl. 4):S18–S22

Eriksson JG (2016) Developmental origins of adult health and disease – from a small body size at birth to epigenetics. Ann Med 48:456–467

Eriksson J, Forsen T, Tuomilehto J et al (1999) Catch-up growth in childhood and death from coronary heart disease: longitudinal study. BMJ 318:427–443

Eriksson J, Forsen T, Tuomilehto J (2000) Fetal and childhood growth and hypertension in adult life. Hypertension 36:790–794

Estourgie-van Burk GF, Bartels M, Hoekstra RA et al (2009) A twin study of cognitive costs of low birth weight and catch-up growth. J Pediatr 154:29–32

Fall CH, Osmond C, Barker DJ et al (1995) Fetal and growth and cardiovascular risk factors for women. BMJ 310:428–432

Fall CHD, Stein CE, Kumaran K et al (1998) Size at birth, maternal weight and type 2 diabetes in South India. Diabet Med 15:220–227

Fernandez-Twinn DS, Wayman A, Ekizoglou S et al (2005) Maternal protein restriction leads to hyperinsulinaemia and reduced insulin signaling protein expression in 21-mo old offspring. Am J Physiol Regul Integr Comp Physiol 288:R368–R373

Figueroa H, Alvarado C, Cifuentes J et al (2017) Oxidative damage and nitric oxide synthase induction by surgical uteroplacental circulation in the rabbit. Prenat Diagn 37(5):453–459

Gallo LA, Denton KM, Moritz KM et al (2012) Long-term alteration in maternal blood pressure and renal function after pregnancy in normal and growth-restricted rats. Hypertension 60:206–213

Gambling L, Andersen HS, Czopek A et al (2004) Effect of timing of iron supplementation on maternal and neonatal growth and iron status of iron-deficit pregnant rats. J Physiol 561:195–203

Garafono A, Czernichow P, Breant B (1997) In utero undernutrition impairs beta cell development. Diabetologia 40:1231–1234

Garafono A, Czernichow P, Breant B (1998) Beta-cell mass and proliferation following late fetal and early postnatal malnutrition in the rat. Diabetologia 41:1114–1120

Garafono A, Czernichow P, Breant B (1999) Effect of ageing on beta cell mass and function in rats malnourished during the perinatal period. Diabetologia 42:711–718

Germani D, Puglianiello A, Cianfarani S (2008) Uteroplacental insufficiency down regulates insulin receptor and affects expression of key enzymes of long-chain fatty (LCFA) metabolism in skeletal muscle at birth. Cardiovasc Diabetol 7:7–14

Giussani DA, Camm EJ, Niu Y et al (2012) Developmental programming of cardiovascular dysfunction by prenatal hypoxia and oxidative stress. PLoS One 7:e31017

Grunnet L, Vielwerth S, Vaag A et al (2007) Birth weight is nongenetically associated with glucose intolerance in elderly twins, independent of adiposity. J Intern Med 262:96–103

Hales CN, Barker DJP, Clark PMS, Cox LJ, Fall C, Osmond C, Winter PD (1991) Fetal and infant growth and impaired glucose tolerance at age 64. BMJ 303:1019–1022

Halvorsen CP, Andolf E, Hu J et al (2006) Discordant twin growth in utero and differences in blood pressure and endothelial dysfunction at 8 years of age. J Intern Med 259:155–163

Harman D (1956) Aging: A theory based on free radical and radiation chemistry. J Gerontol 11:298–300

Hauton D, Al-Shammari A, Gaffney EA et al (2015) Maternal hypoxia decreases capillary supply and increases metabolic inefficiency leading to divergence in myocardial oxygen supply and demand. PLoS One 10:e0127424

Hemachandra AH, Howards PP, Furth SL et al (2007) Birth weight, postnatal growth and risk for high blood pressure at 7 years of age: results from the Collaborative Perinatal Project. Pediatrics 119:e1264–e1270

Hendrix N, Berghella V (2008) Non-placental causes of intrauterine growth restriction. Sem Perinatal 32:161–165

Heslehurst N, Rankin J, Wilkinson JR et al (2010) A nationally representative study of maternal obesity in England, UK: trends in incidence and demographic inequalities in 619 323 births, 1989–2007. Int J Obes 34:420–428

Hokke S, Puelles VG, Armitage JA et al (2016) Maternal fat feeding augments offspring nephron endowment in mice. PLoS One 11:e0161578

Hoppe CC, Evans RG, Bertram JF et al (2007) Effects of dietary protein restriction on nephron number in the mouse. Am J Physiol Regul Integr Comp Physiol 292:R1768–R1774

Huber HF, Ford SM, Bartlett TQ et al (2015) Increased aggressive and affiliative display behavior in intrauterine growth restricted baboons. J Med Primatol 44:143–147

Jakoubek V, Bibova J, Herget J et al (2008) Chronic hypoxia increases fetoplacental vascular resistance and vasoconstrictor reactivity in the rat. Am J Physiol Heart Circ Physiol 294:H1638–H1644

Kafka P, Vajnerova O, Hampi V (2016) Chronic hypoxia increases fetoplacental vascular resistance in rat placental perfused with blood. Bratisl Lek Listy 117:583–586

Kane AD, Herrara EA, Camm EJ (2013) Vitamin C prevents intrauterine programming of in-vivo cardiovascular dysfunction in the rat. Circ J 77:2604–2611

Kim DW, Young SL, Grattan DR et al (2014) Obesity during pregnancy disrupts placental morphology, cell proliferation, and inflammation in a sex-specific manner across gestation in the mouse. Biol Reprod 90:1–11

Kirk SL, Samuelsson AM, Argenton M et al (2009) Maternal obesity induced by diet in rats permanently influences central processes regulating food intake in offspring. PLoS One 4:e5870

Kuo AH, Li C, Huber HF et al (2017) Cardiac remodelling in a baboon model of intrauterine growth restriction mimics accelerated aging. J Physiol 595:1093–1110

Lane RH, Chandorkar AK, Flozak AS et al (1998) Intrauterine growth-retardation alters mitochondrial gene expression and function in fetal and juvenile rat skeletal muscle. Pediatr Res 43:563–570

Lane RH, Kelley DE, Gruetzmacher EM et al (2001) Uteroplacental insufficiency alters hepatic fatty acid-metabolizing enzymes in juvenile and adult rats. Am J Phys Regul Integr Comp Phys 280:R183–R190

Langley-Evans SC, Philips GJ, Jackson AA (1994) In utero exposure to maternal low protein diets induces hypertension in weanling rats, independently of maternal blood pressure changes. Clin Nutr 13:319–324

Langley-Evans SC, Welham SJ, Jackson AA (1999) Fetal exposure to a maternal low protein diet impairs nephrogenesis and promotes hypertension in the rat. Life Sci 64:965–974

Law CM, Shiell AW, Newsome CA et al (2002) Fetal, infant and childhood growth and adult blood pressure: a longitudinal study from birth to 22 years of age. Circulation 105:1088–1092

Lee HS (2015) Impact of maternal diet on the epigenome during in utero life and the developmental programming of diseases in childhood and adulthood. Nutrients 7:9492–9507

Lee SP, Hande P, Yeo GS et al (2017) Correlation of cord blood telomere length with birth weight. BMC Res Notes 10:469

Lesage J, Blondeau B, Grino M et al (2001) Maternal undernutrition during late gestation induces fetal overexposure to glucocorticoids and intrauterine growth retardation, and disturbs the hypothalamo-pituitary adrenal axis in the newborn rat. Endocrinology 142:1692–1702

Lewis RM, Petry CJ, Ozanne SE et al (2001a) Effects of maternal iron restriction in the rat on blood pressure, glucose tolerance, and serum lipids in 3-month old offspring. Metabolism 50:562–567

Lewis RM, Doherty CB, James LA et al (2001b) Effects of maternal iron restriction on placental vascularization in the rat. Placenta 22:534–539

Lewis RM, Forhead AJ, Petry CJ et al (2002) Long-term programming of blood pressure by maternal iron restriction in the rat. Br J Nutr 88(3):283–290

Lim K, Zimanyi M, Black MJ (2006) Effect of maternal protein restriction in rats on cardiac fibrosis and capillarization in adulthood. Pediatr Res 60:83–87

Lisle SJ, Lewis RM, Petry CJ et al (2003) Effect of maternal iron restriction during pregnancy on renal morphology in the adult rat offspring. Br J Nutr 90(1):33–39

Lithell HO, McKeigue PM, Berglund L et al (1996) Relation of size at birth to non-insulin dependent diabetes and insulin concentrations in men aged 50–60 years. BMJ 312:406–410

Marchetto NM, Glynn RA, Ferry ML et al (2016) Prenatal stress and newborn telomere length. Am J Obstet Gynecol 215:e1–e8

Mazzuca MQ, Wlodek ME, Dragomir NM et al (2010) Uteroplacental insufficiency programs regional vascular dysfunction and alters arterial stiffness in female offspring. J Physiol 588:1997–2010

McGillick EV, Orgeig S, Allison BJ et al (2017) Maternal chronic hypoxia increases expression of genes regulating lung liquid movement and surfactant maturation in male foetuses in late gestation. J Physiol 595:4329–4350

Mi J, Law C, Zhang K-L et al (2000) Effects of infant birth weight and maternal body mass index in pregnancy on components of the insulin resistance syndrome in China. Ann Intern Med 132:253–260

Monrad RN, Grunnet LG, Rasmussen EL et al (2009) Age-dependent nongenetic influences of birth weight and adult body fat in insulin sensitivity in twins. J Clin Endocrinol Metab 94:2394–2399

Moritz KM, Mazzuca MQ, Siebell AL et al (2009) Uteroplacental insufficiency causes a nephron deficit, modest renal insufficiency but no hypertension with ageing in female rats. J Physiol 587:2635–2646

Needham BL, Rehkopf D, Adler N et al (2015) Leukocyte telomere length and mortality in the National Health and Nutritional Examination Survey, 1999–2002. Epidemiology 26(4):528–535

Nyrienda MJ, Lindsay RS, Kenyon CJ et al (1998) Glucocorticoid exposure in late gestation permanently programs rat hepatic phosphoenolpyruvate carboxylase and glucocorticoid receptor expression and causes glucose intolerance in adult offspring. J Clin Invest 101:2174–2181

Ozanne SE, Hales CN (2004) Lifespan: catch up growth and obesity in male mice. Nature 427:411–412

Ozanne SE, Wang CL, Coleman N et al (1996) Altered muscle sensitivity in the male offspring of protein-malnourished rats. Am J Phys 271:E1128–E1134, 1996

Ozanne SE, Dorling MW, Wang CL et al (2001) Impaired PI-3 kinase activation in adipocytes from early growth-restricted male rats. Am J Physiol Endocrinol Metab 280:E543–E539

Ozanne SE, Jensen CB, Tingey KJ et al (2005) Low birthweight is associated with specific changes in muscle insulin-signaling protein expression. Diabetologia 48:547–552

Peterside IE, Selak MA, Simmons RA (2003) Impaired oxidative phosphorylation in hepatic mitochondria in growth-retarded rats. Am J Physiol Endocrinol Metab 285:E1258–E1266

Petrik J, Reusens B, Arany E et al (1999) A low protein diet alters the balance of islet cell replication and apoptosis in the fetal and neonatal rat and associated with a reduced pancreatic expression of insulin-like growth factor II. Endocrinology 140:4861–4873

Petry CJ, Dorling MW, Pawlak DB (2002) Diabetes in old male offspring of rat dams fed a reduced protein diet. Int J Exp Diabetes Res 2(2):139–143

Pettitt DJ, Jovanovic L (2001) Birth weight as a predictor of type 2 diabetes mellitus: The U-shaped curve. Curr Diab Rep 1:78–81

Poulsen P, Vaag AA, Kyvik KO et al (1996) Low birth weight is associated with NIDDM in discordant monozygotic and dizygotic twin pairs. Diabetologia 40(4):439–446

Poulsen P, Kyvik KO, Vaag A et al (1999) Heritability of type II (non-insulin dependent) diabetes mellitus and abnormal glucose tolerance – a population based twin study. Diabetologia 42:139–145

Poulsen P, Levin K, Beck-Nielsen H et al (2002) Age-dependent impact of zygosity and birth weight on insulin secretion and insulin action on twins. Diabetologia 45:1645–1659

Ravelli AC, van der Meulen HJP, Michels RPJ et al (1998) Glucose intolerance in adults after prenatal exposure to famine. Lancet 351:173–177

Reynolds RM, Allan KM, Raja EA et al (2013) Maternal obesity during pregnancy and premature mortality from cardiovascular event in adult offspring: follow up of 1 323 275 person years. BMJ 347:f4539

Richter T, von Zglincki T (2007) A continuous correlation between oxidative stress and telomere length in fibroblasts. Exp Gerontol 42(11):1039–1042

Richter HG, Hansell JA, Raut S et al (2009) Melatonin improves placental efficiency and birth weight and increases placental expression of antioxidant enzymes in undernourished pregnancy. J Pineal Res 46:357–364

Richter HG, Camm EJ, Modi BN et al (2012) Ascorbate prevents placental oxidative stress and enhances birth weight in hypoxic pregnancy rats. J Physiol 590:1377–1387

Roseboom T, de Rooj S, Painter R (2006) The Dutch famine and its long-term consequences for adult health. Early Hum Dev 82:485–491

Samuelsson AM, Matthews PA, Argenton M et al (2008) Diet-induced obesity in female mice leads to offspring hyperphagia, adiposity, hypertension, and insulin resistance. Hypertension 51:383–392

Scholl TO (2011) Maternal iron status: Relation to fetal growth, length of gestation and the neonate's iron endowment. Nutr Rev 69(suppl 1):S23–S29

Selak MA, Storey BT, Peterside I et al (2003) Impaired oxidative phosphorylation in skeletal muscle of intrauterine growth-retarded rats. Am J Physiol Endocrinol Metab 285:E130–E137

Shah A, Reyes LM, Morton JS et al (2016) Effect of resveratrol on metabolic and cardiovascular function in male and female adult offspring exposed to prenatal hypoxia and a high-fat diet. J Physiol 594:1465–1482

Shah A, Quon A, Morton JS et al (2017) Postnatal resveratrol supplementation improves cardiovascular function in male and female intrauterine growth restricted offspring. Physiol Rep 5(2):e13109

Sharpless N, DePhino RA (2004) Telomeres, stem cells, and cancer. J Clin Invest 113:160–168

Shelley P, Martin-Gronert MS, Rowlerson A (2009) Altered skeletal muscle insulin signaling and mitochondrial complex II-III linked activity in adult offspring of obese mice. Am J Physiol Regul Integr Comp Physiol 297:R675–R681

Simmons RA, Templeton LG, Gertz SJ (2001) Intrauterine growth retardation leads to the development of type 2 diabetes in the rat. Diabetes 50:2279–2286

Singhal A, Cole TJ, Lucas A (2001) Early nutrition in preterm infants and later blood pressure: two cohorts after randomised trials. Lancet 357:413–419

Singhal A, Farooqi IS, O'Rahilly S et al (2002) Early nutrition and leptin concentrations in later life. Am J Clin Nutr 75:993–999

Singhal A, Cole TJ, Fewtrell M et al (2004) Breastmilk feeding and lipoprotein profile in adolescents born preterm: follow up of a prospective randomised trial. Lancet 363:1571–1578

Singhal A, Kennedy K, Lanigan J et al (2010) Nutrition in infancy and long-term risk of obesity: evidence from 2 randomized control trials. Am J Clin Nutr 92:1133–1144

Snoeck A, Remacle C, Reusens B et al (1990) Effect of a low protein diet during pregnancy on the fetal rat endocrine pancreas. Biol Neonat 57:107–118

Tam WH, Ma RCW, Ozaki R et al (2017) In utero exposure to maternal hyperglycemia increases childhood cardiometabolic risk in the offspring. Diabetes Care 40:679–686

Tang J, Zhu Z, Xia S et al (2015) Chronic hypoxia in pregnancy affected vascular tone of renal interlobar arteries in the offspring. Sci Rep 5:9723

Tarry-Adkins JL, Martin-Gronert MS, Chen JH et al (2008) Maternal diet influences DNA damage, aortic telomere length, oxidative stress, and antioxidant defense capacity in rats. FASEB J 22:2037–2044

Tarry-Adkins JL, Chen JH, Smith NS et al (2009) Poor maternal nutrition followed by accelerated postnatal growth leads to telomere shortening and increased markers of cell senescence in rat islets. FASEB J 23:1521–1528

Tarry-Adkins JL, Chen JH, Jones RH et al (2010) Poor maternal nutrition leads to alterations in oxidative stress, antioxidant defense capacity, and markers of fibrosis in rat islets: potential underlying mechanisms for development of the diabetic phenotype in later life. FASEB J 24:2762–2771

Tarry-Adkins JL, Blackmore HL, Martin-Gronert MS et al (2013) Coenzyme Q10 prevents accelerated cardiac aging in a rat model of poor maternal nutrition and accelerated postnatal growth. Mol Metab 2:480–490

Tarry-Adkins JL, Fernandez-Twinn DS, Chen JH et al (2014) Nutritional programming of coenzyme Q10: potential for prevention and intervention? FASEB J 28:5398–5405

Tarry-Adkins JL, Fernandez-Twinn DS, Madsen R et al (2015) Coenzyme Q10 prevents insulin signaling dysregulation and inflammation prior to development of insulin resistance in male offspring of a rat model of poor maternal nutrition and accelerated postnatal growth. Endocrinology 156:3528–3537

Tarry-Adkins JL, Fernandez-Twinn DS, Hargreaves IP (2016a) Coenzyme Q10 prevents hepatic fibrosis, inflammation and oxidative stress in a male model of poor maternal nutrition and accelerated postnatal growth. Am J Clin Nutr 103:579–588

Tarry-Adkins JL, Fernandez-Twinn DS, Chen JH (2016b) Poor maternal nutrition and accelerated postnatal growth induces an accelerated aging phenotype and oxidative stress in skeletal muscle of male rats. Dis Model Mech 9:1221–1229

Taylor PD, McConnell JM, Khan IY et al (2005) Impaired glucose homeostasis and mitochondrial abnormalities in offspring of rat fed a fat-rich diet during pregnancy. Am J Physiol Regul Integr Comp Physiol 288:R134–R139

United Nations, Department of Economic and Social Affairs, Population Division (2015) World Population Ageing (ST/ESA/SER.A/390). www.un.org/en/development/desa/population/.../ pdf/ageing/WPA2015_Report.pdf. Accessed 20 Nov 2017

Valko M, Leibfritz D, Moncol J et al (2006) Free radicals and antioxidants in normal physiological functions and human disease. Int J Biochem Cell Biol 39:44–84

Vickers MH, Reddy S, Ikenasio A et al (2001) Dysregulation of the adipoinsular axis – a mechanism for the pathogenesis of hyperleptinemia and adipogenic diabetes induced by fetal programming. J Endocrinol 170:323–332

von Zglincki T (2002) Oxidative stress shortens telomeres. Trends Biochem Sci 27(7):339–344

Walton SL, Bielefeldt-Ohmann H, Singh RR et al (2017) Prenatal hypoxia leads to hypertension, renal renin-angiotensin system activation and exacerbates salt- induced pathology in a sex dependent manner. Sci Rep 7:8241

Welham SJ, Wade A, Woolfe AS (2002) Protein restriction in pregnancy is associated with increased apoptosis of mesenchymal cells at the start of rat metanephrogenesis. Kid Int 61:1231–1242

Welham SJ, Riley PR, Wade A et al (2005) Maternal diet programs embryonic kidney gene expression. Physiol Genomics 22:48–56

Wlodek ME, Westcott K, Siebel AL et al (2008) Growth restriction before or after birth reduces nephron number and increases blood pressure in male rats. Kid Int 74:187–195

Xiao D, Kou H, Zhang L et al (2017) Prenatal food restriction with postnatal high-fat diet alters glucose metabolic function in adult rat offspring. Arch Med Res 48:35–45

Zambrano E, Sosa-Larios T, Calzada L et al (2016) Decreased basal insulin secretion from pancreatic islets of pups in a rat model of maternal obesity. J Endocrinol 231:49–57

Zhang J, Lewis RM, Wang C et al (2005) Maternal dietary iron restriction modulates hepatic lipid metabolism in the fetuses. Am J Physiol Regul Intergr Comp Physiol 288:R104–R111

Chapter 2
The Immune System and Its Dysregulation with Aging

Ludmila Müller, Svetlana Di Benedetto, and Graham Pawelec

Abstract Aging leads to numerous changes that affect all physiological systems of the body including the immune system, causing greater susceptibility to infectious disease and contributing to the cardiovascular, metabolic, autoimmune, and neuro-degenerative diseases of aging. The immune system is itself also influenced by age-associated changes occurring in such physiological systems as the endocrine, nervous, digestive, cardio-vascular and muscle-skeletal systems. This chapter describes the multidimensional effects of aging on the most important components of the immune system. It considers the age-related changes in immune cells and molecules of innate and adaptive immunity and consequent impairments in their ability to communicate with each other and with their aged environment. The contribution of age-related dysregulation of hematopoiesis, required for continuous replenishment of immune cells throughout life, is discussed in this context, as is the developmentally-programmed phenomenon of thymic involution that limits the output of naïve T cells and markedly contributes to differences between younger and older people in the distribution of peripheral blood T-cell types. How all these changes may contribute to low-grade inflammation, sometimes dubbed "inflammaging", is considered. Due to findings implicating elevated inflammatory immuno-mediators in age-associated chronic autoimmune and neurodegenerative processes, evidence for their possible contribution to neuroinflammation is reviewed.

L. Müller (✉)
Max Planck Institute for Human Development, Berlin, Germany
e-mail: lmueller@mpib-berlin.mpg.de

S. Di Benedetto
Max Planck Institute for Human Development, Berlin, Germany

Center for Medical Research, University of Tübingen, Tübingen, Germany
e-mail: dibenedetto@mpib-berlin.mpg.de

G. Pawelec
Center for Medical Research, University of Tübingen, Tübingen, Germany

Health Sciences North Research Institute, Sudbury, ON, Canada
e-mail: graham.pawelec@uni-tuebingen.de

© Springer Nature Singapore Pte Ltd. 2019
J. R. Harris, V. I. Korolchuk (eds.), *Biochemistry and Cell Biology of Ageing: Part II Clinical Science*, Subcellular Biochemistry 91,
https://doi.org/10.1007/978-981-13-3681-2_2

Keywords Immune system · Immunosenescence · Thymic involution · Impaired hematopoiesis · Cytomegalovirus · T-cell diversity · Inflammaging · Neuroinflammation

Introduction

Aging is a highly complex process associated with numerous changes at the organismal, tissue, cellular and molecular levels, characterized by dysregulation of different physiological systems (such as the immune, central and peripheral nervous, endocrine, metabolic and other systems) from their optimal homeostatic state (Xu and Larbi 2017; Müller et al. 2013a; Müller and Pawelec 2015; Masters et al. 2017). Particularly, chronic oxidative stress affects cells of these central regulatory systems leading to modified interactions between them. This influences their functionality, disturbs homeostasis and may thereby affect longevity (De la Fuente et al. 2011).

Deleterious alterations in the immune system that accompany human aging are commonly known as immunosenescence, although many differences between young and old individuals are in fact only assumed to be detrimental without unequivocal proof. Thus, it is clear that aging per se does not always lead to inevitable decline in immune functions, but rather causes their modification. Paramount amongst these age-associated changes are the replacement of naïve T and B cells with memory cells, which is of course the basis of adaptive immune function. While some aspects of immunity deteriorate with age, some remain unchanged, while others tend to become overactive. In this sense, we should rather refer to dysregulation than deterioration that is likely to be driving force for immunosenescence. However, progressive increases in the incidence of infectious, metabolic, and autoimmune diseases, in cancer and neurological disorders occurring with aging appear likely to be at least partly caused by immunosenescence (Müller and Pawelec 2014; Menza et al. 2010; Michaud et al. 2013).

In this context, immunosenescence should be considered as a multistage process consisting of numerous developmentally regulated and interrelated changes, instead of conceiving it as an unidirectional general decay of the immune system (Accardi and Caruso 2018; Del Giudice et al. 2018). Thus far, it has not been clear which changes are truly age-related and which are adaptive and compensatory, arising as a consequence of immune exposures and other confounders (Nikolich-Zugich 2018; Pawelec 2017a, b; Fulop et al. 2017). Such events as hematopoietic stem cell dysfunction and thymus involution with decreased T-cell generation contribute to the age-related remodeling of the immune system leading to the decreased levels of naïve T-lymphocytes accompanied by increased numbers of memory cells but also manifesting as accumulations of dysfunctional potentially senescent immune cells. Additionally, intrinsic defects in development, maturation, migration, and homeostasis of peripheral immune cells contribute to this process (Gruver et al. 2007).

Along with such developmental/age-related changes and partly related to these events, low-level chronic inflammation commonly present in older adults is believed to be implicated in age-related mortality. There have been many attempts to explain the cause of chronic inflammation in aging, considering such reasons as mitochondrial damage, redox stress, age-related alterations in endocrine function, epigenetic changes and other phenomena, linked or not to immunosenescence. It appears that no single theory or mechanism has thus far proven itself able to explain all aspects of aging and immune aging. Therefore, it is more probable that multiple interrelated processes contribute to aging, but in some way, nearly all of them might contribute to or be related to inflammation (Jenny 2012; Müller and Pawelec 2015).

Maintaining well-balanced inflammatory and anti-inflammatory homeostasis seems to be decisive to maintain functional longevity required for optimal aging. If this equilibrium is disturbed for any reason, pathological pathways may become prevalent. Fülop and colleagues hypothesize that "canalized dysregulations occur as a result of adaptations to the aging process and thus reflect an optimized response to an imperfect situation. In contrast, non-canalized dysregulations reflect a true loss of homeostatic control and may themselves be the imperfect situation causing the canalized responses" (Fulop et al. 2017).

It is widely accepted that age-associated inflammatory conditions are closely interrelated with changes in immune system function. Additionally, they are under the influence of neuroendocrine hormones, including dehydroepiandrosterone, glucocorticoids, and the catecholamines, epinephrine, and norepinephrine. In the course of aging, alterations in endocrine function can potentially modulate their interrelationships and lead to deterioration of the regulatory feedback pathways. Chronic stress, including psychological and social stressors, is also known to negatively affect both neuroendocrine and immune functions and may contribute to increased age-associated diseases and mortality in old people (Heffner 2011; Hawkley and Cacioppo 2004; Müller and Pawelec 2014). Thus, the aging immune system influences and is influenced by these other physiological systems of the body.

Components of the Immune System

After barrier functions, our immune system is the most important defender of our body's integrity against external invaders or pathogens as well as altered or modified internal factors. At the same time, according to one school of thought, adaptive immunity has the important task "to facilitate the maintenance of a beneficial microbiota" (Pawelec 2012; Müller and Pawelec 2014). For this reason our immune system possess various sophisticated mechanisms of both the innate and adaptive branches, consisting of different cell populations, which are capable of responding to these challenges and defending the integrity of the body as well as its beneficial microbiota (Fig. 2.1). So called "first line of defense" is represented by the skin and mucosa of the gastro-intestinal, respiratory and urogenital tracts, serving not only as a mechanical and physiological barrier (Doran et al. 2013), but also containing fatty acids, anti-microbial molecules, proteases, digestive enzymes, lysozymes,

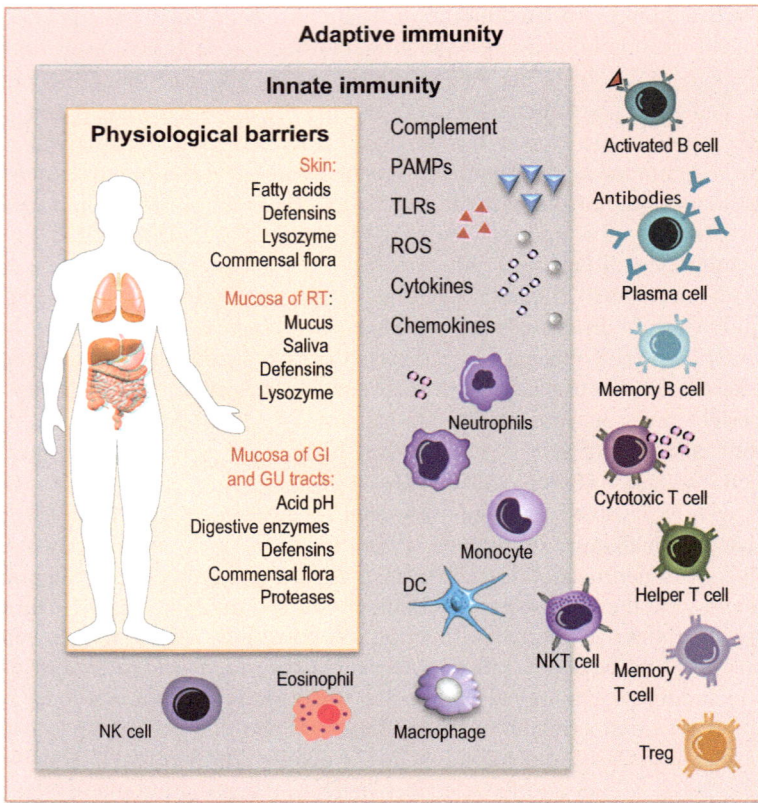

Fig. 2.1 Overview of the human immune system
RT respiratory tract, *GI* gastro-intestinal, *GU* urogenital, *PAMPs* pathogen-associated molecular patterns, *TLRs* Toll-like receptors, *ROS* reactive oxygen species, *DC* dendritic cells, *NK cells* natural killer cells, *NKT cells* natural killer T cells, *Treg* regulatory T cells

defensins, and is colonized by commensal flora to provide sophisticated defense (Müller et al. 2013b). But when barrier function is compromised, immune function is clearly paramount.

Thus, when a pathogen breaks these physiological barriers and enters the body, the innate immune system (Fig. 2.2) affords a rapid response involving immediate migration of innate cells with phagocytic activity to the lesion, and providing local release of toxic mediators. By means of specialized receptors, which are present on the surface of innate immune cells (Fig. 2.3, top), they recognise highly conserved molecules on microorganisms, so-called "PAMPs" (pathogen-associated molecular patterns), which are shared by many different pathogens but are not present on mammalian cells (Medzhitov and Janeway 2000). These interactions in turn activate different innate defence mechanisms, which include phagocytosis, release of inflammatory proteins, activation of the complement system, production of acute phase proteins, secretion

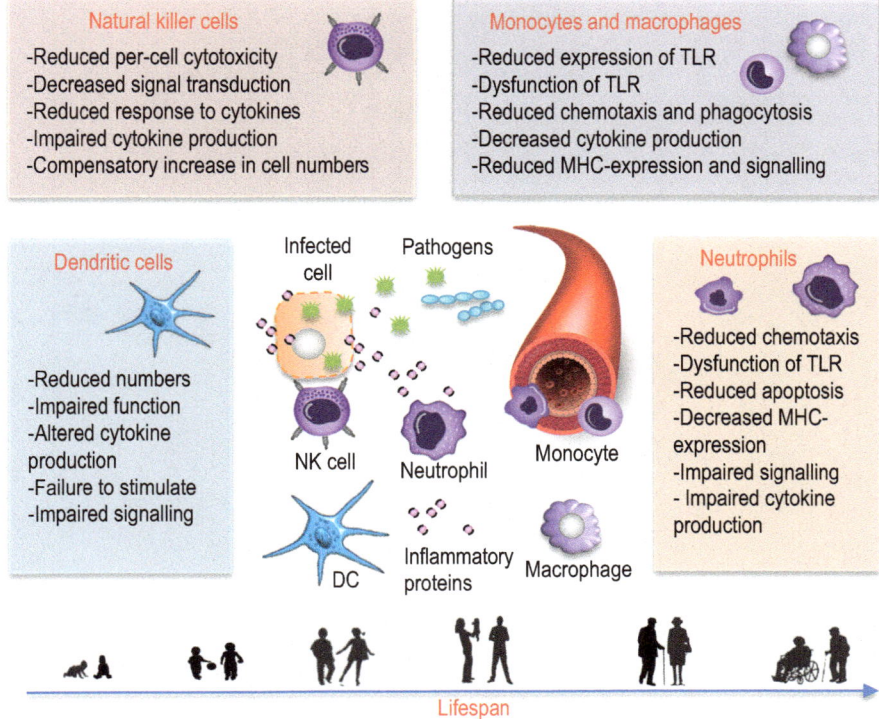

Fig. 2.2 Innate immune system and age-related changes
TLR Toll-like receptor, *MHC* major histocompatibility complex, *NK* cells natural killer cells, *DC* dendritic cells

chemokines and cytokines (Müller and Pawelec 2014). The ongoing responses of the innate immune system "call into play" adaptive immune responses and both arms act together to eliminate pathogens and develop immune memory.

Cells of the innate immune system are believed to provide much more rapid but less specific immune response than do the cells of the adaptive immune system. However, from a more recent point of view, the separation of these two arms of immunity, previously considered as completely distinct, has become muddied. It was shown that cells of the innate immune system may possess a kind of memory function under certain conditions and reciprocally, immune cells of the adaptive arm may express receptors specific for innate cells and act like these cells functionally. These receptors appear especially at late stages of differentiation, are normally elevated later in life and accompanied by other age-associated changes (Lanier and Sun 2009). Chronological aging was also shown to be associated with accumulations of immune cells combining characteristics of both the innate and adaptive arms of the immune system. This phenomenon might be partly explained by compensatory mechanisms for age-related functional defects of conventional immune cells (Pereira and Akbar 2016).

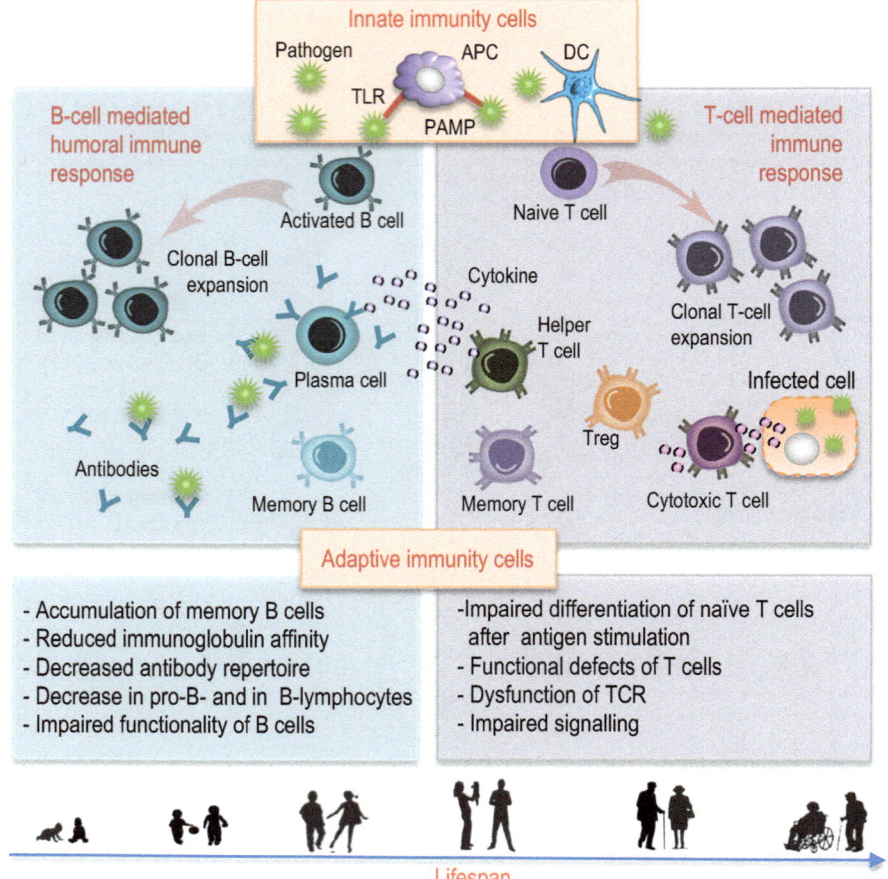

Fig. 2.3 Adaptive immune system age-related changes
APC antigen presenting cells, *PAMPs* pathogen-associated molecular patterns, *TLRs* Toll-like receptors, *DC* dendritic cells, *Treg* regulatory T cells, *TCR* T-cell receptor

Effects of Aging on the Innate Immune System

As mentioned above, "innate immune defences are not non-specific, but they respond to pathogens in a generic way" (Litman et al. 2010). Further important functions of innate immunity are activation of the complement cascade and generation and release of chemical humoral factors, rapid recruitment to sites of infection such cells as eosinophils, neutrophils, macrophages, natural killer cells (NK), and dendritic cells (DCs) (Figs. 2.1 and 2.2).

Neutrophils short-lived polymorphonuclear granulocytic cells representing the most abundant circulating leukocytes, and act as a very effective first line of defence against pathogenic bacteria, yeast and fungi (Fig. 2.2). They are immediately recruited to the site of infection and act rapidly through specialised mechanisms such as phagocytosis and killing target organisms. Neutrophils mediate high oxidative burst activity, triggered after recognition of PAMPs on the pathogen. These innate cells are characterised by a relatively high turnover and have a very short lifespan. Despite these features, various age-related functional impairments have been reported in phagocytic mechanisms, in chemotaxis and generation of toxic free radicals, as well as in their susceptibility to apoptosis (Shaw et al. 2010). These alterations presumably reflect aging processes at the level of the hematopoietic stem cell. The level of expression and functional integrity of receptors recognising PAMPs, such as Toll-like receptors, and also the expression of major histocompatibility complex class II (MHC-II) molecules, together with production and release of chemokines and cytokines, are also lower in neutrophils from old individuals (Fulop et al. 2004). Thus, disturbed function and impairments in these initial defence mechanisms against infectious agents are likely to have vital importance in elderly populations (Müller and Pawelec 2014).

Monocytes and macrophages also have important roles including multiple functions in terms of phagocytosis and antigen presentation (Fig. 2.2). Monocytes act very efficiently in controlling invading bacteria – the feature that is most important for the aged population characterized by high prevalence of infectious diseases. Macrophages are involved in the initiation of inflammatory responses, in the direct destruction of pathogens and elimination of malignant cells, as well as in interactions with and activation of the adaptive immune response by means of antigen presentation (Oishi and Manabe 2016). They are then able to eliminate their target antigens directly or indirectly – through production and secretion of immune molecules and immune factors, such as interleukin (IL)-1, Tumor Necrosis factor (TNF), and Interferon (IFN)-γ, which in turn, are capable of activating and recruiting additional immune cells (Derhovanessian et al. 2008; Keller 1993).

The proportion of circulating monocytes in peripheral blood of aged individuals does not seem to be significantly different to the young, but there are fewer macrophages in the bone marrow of 80–100 year old people. This decrease can be explained by accompanying processes of an increased apoptosis as well as reduced cellularity observed with advancing age. Also expression and functionality of MHC-II antigens on macrophages are decreased with aging, with apparently epigenetic changes being responsible for this impairment. The phagocytic and apoptotic features of macrophages are also impaired with age, accompanied by decreased release of reactive oxygen species, such as NO_2 and H_2O_2 and less secretion of TNF and IL-1 (Fernandez-Morera et al. 2010; Gonzalo 2010; Müller and Pawelec 2014).

Dendritic cells (Fig. 2.2) are important players in the immune response and are said to bridge the innate and adaptive immune systems. They are capable of recognising PAMPs and take up and process pathogens to present MHC-bound antigens to T lymphocytes. They are often designated "professional" antigen-presenting cells (APCs). In fact, they represent the cell fraction "without which adaptive immune responses cannot take place. Moreover, the manner in which the DCs present antigen and co-stimulate T cells influences the type of T-cell response initiated, and thus the quality and intensity of adaptive immune responses" as we reviewed previously (Müller et al. 2013a). Thus, it is understandable that even small age-related alterations in the function of dendritic cells could greatly affect T-cell functionality.

DCs do appear to be somewhat impaired by aging in terms of their migratory capacity and distribution, in their ability to process antigen, in their expression of costimulatory signals, and their ability to secrete cytokines (Fig. 2.2). Despite the fact that antigen presentation by DCs appears to be relatively unimpaired in the elderly, slightly decreased numbers of DC found in peripheral blood and follicles could probably explain reduced immune supportive capacity of this cell fraction in immune defence against pathogens and tumours (Gonzalo 2010; Della Bella et al. 2007; Derhovanessian et al. 2008). Results from animal studies demonstrate that the circulating levels of DCs are markedly reduced. The same may be true in elderly humans (Adema 2009; Müller and Pawelec 2014).

Natural killer (NK) cells are a class of cytotoxic lymphocytes. They are responsible for and are involved in early defence, with particular importance in recognizing virus-infected cells. They recognize and kill their target cell in an MHC-unrestricted manner, and are inhibited by the presence of self-MHC molecules (Fig. 2.2). This particular outstanding immune competence of NK cells makes them most important players in cancer immune surveillance during aging. Results of studies on elderly individuals have shown that although numbers of NK cells are often elevated with age, they tend to have an impaired cytotoxic function on a per-cell basis, and are more likely to have a mature phenotype compared to NK cells from the young. NK cells from the elderly are also characterized by decreased production and release of chemokines and cytokines, such as MIP1α, RANTES, and IL-8. In this regard, the elevated numbers of NK cells commonly found in aged people may be required to achieve a certain baseline level of functionality as a compensatory mechanism for the decreased function of each cell. Production and secretion of such cytokines as IL-2, IFN-γ, TNF and IL-12 were also shown to be decreased with aging – which may also contribute (among other factors) for immune impairments related to advanced age (Dewan et al. 2012; Müller and Pawelec 2014).

Adaptive Immune System

As stated by Bonilla and Oettgen, "A defining feature of adaptive immunity is the somatic development by genetic rearrangements of the inherited germline of a pool of clonally diverse lymphocytes with a large repertoire of different antigen-recognition receptors" (Bonilla and Oettgen 2010). Accordingly to this definition, the hallmarks of adaptive immunity are: (i) clonal expansion – in order to generate sufficient amounts of cells to confront the pathogen; (ii) differentiation to effector cells – in order to destroy and eliminate the pathogen; and finally (iii) maintenance of a small fraction of antigen-specific memory cells – in order to respond faster and more strongly to pathogen re-exposure (Müller et al. 2013b). In other words, adaptive immunity (Figs. 2.1 and 2.3) involves a highly regulated multidirectional interplay between innate cells such as DCs (or other antigen-presenting cells), and T- and B-lymphocytes. These specific interactions activate, induce, and promote pathogen-specific immunologic effector pathways and the generation of immunological memory mediated by subpopulations of long-lived memory T- and B cells. When the latter encounter and recognize the same pathogen later, they respond more rapidly with a stronger protective effect (Cooper 2010; Müller and Pawelec 2014).

During the course of aging, after long-term residence in the aging host, naïve T cells begin to suffer impaired differentiation into effector cells after antigen stimulation, as well as further functional impairments, such as reduced production of cytokines, and a more restricted T-cell receptor (TCR) repertoire (Ferrando-Martinez et al. 2011; Weiskopf et al. 2009). Epigenetic age-related changes, such as methylation of cytokine gene promoters can also lead to impaired immune function (Calvanese et al. 2009; Gonzalo 2010). The altered DNA methylation profiles were found to be present over the course of aging for both immune-specific genes and for the genes involved in common pathways normally related to cell homeostasis (Fernandez-Morera et al. 2010; Shanley et al. 2009).

The functional potency of T lymphocytes from aged individuals was found to be reduced relative to cells of the same phenotype from the young (Liu et al. 2002). Naïve T cells freshly exported from the thymus, are activated by APCs through binding of their TCRs to the appropriate MHC-peptide complex on the APC surface. The decisive event at the beginning of T-cell activation appears to be the reorganisation of the plasma membrane into a "lipid raft" or "immune synapse". These assemble at the place of contact between T cell and APC. However, this step of lipid raft clustering is impaired in T lymphocytes from older people. One possible explanation for this may be the higher levels of cholesterol commonly found in aged cells. The rise in the cholesterol content impedes the motility of their rafts and in this way reduces recruitment of signalling molecules to the locality that are required for successful activation (Larbi et al. 2011; Müller and Pawelec 2014; Fulop et al. 2017).

For appropriate activation and differentiation of T cells, a necessary requirement is the integrity of the genes coding for the main key factors. This in turn may be dependent on different environmental factors, such as exposure to hormones or chemicals, infection with different pathogens etc., which influence this process. Such factors can change the epigenetic profile of T lymphocytes, affecting their gene expression and modifying these processes. The consequence of such immune remodelling that usually accompanies aging is a significantly increased morbidity and mortality commonly observed in the elderly population (Dewan et al. 2012; Larbi et al. 2008; Solana et al. 2006; Fernandez-Morera et al. 2010).

Relatively less is known about age-related alterations occurring in B cells and their subpopulations than in T cells. Some studies report that memory B-cell populations accumulate with age, in terms of proportions and numbers of CD27$^+$ cells. Indeed, humoral immunity does appear to be different in older people both in quality and quantity. Such alterations as a reduced numbers of pro-B lymphocytes in the bone marrow, but also their reduced ability to differentiate into pre-B cells may be responsible for these impairments. Both antibody diversity and quality are different in older adults because of compromised somatic hypermutation in germinal centers, but the relative contribution of intrinsic changes or microenvironmental changes is unclear (Müller et al. 2013b).

It has been consistently reported that there are deceased numbers of functional immunoglobulin-secreting B cells in aged people, correlating with decreased titres of antigen-specific antibodies. Not only the B-cell subsets themselves but also the repertoire of immunoglobulins produced by them are altered in both specificity and isotype. The consequence of this may well be the reduced duration of humoral response with reduced B-cell capacity for providing specific primary and secondary responses seen in aged individuals (Colonna-Romano et al. 2008; Ademokun et al. 2010). The entire process appears to be generally related to the specific interactions inside the germinal center where antigen-activated B cells communicate with CD4+ "helper" T lymphocytes. Age-associated deficiencies in T cells can have modulatory effects on these cognate interactions and reduce the quality of antibodies in terms of their avidities, and their titer (Dewan et al. 2012). Age-related alterations in epigenetic status is likely to be playing an important role here as well, and might be responsible for altered B-cell function. Such epigenetic changes may induce important modifications in cell differentiation pathways as well as in rearrangement of B-cell receptor and immunoglobulin genes and their affinity maturation (Shanley et al. 2009; Colonna-Romano et al. 2008; Müller and Pawelec 2014).

Age-Related Dysregulation of Hematopoiesis

Figure 2.4 represents the development of lymphocytes that takes place in the functionally specialised environments of the bone marrow and thymus, in a simplified manner (Fig. 2.4). The original lymphoid B- and T-cell precursors emerge from a pluripotent hematopoietic stem cell (HSC) in the bone marrow (Fig. 2.4, left). The haematopoietic stem cell compartment serves as the origin and source for the continuous replenishment of blood cells including the cells of the immune system, providing and delivering all lymphoid and myeloid cellular components. Aging alters the HSC compartment (Geiger et al. 2013) such that the output of immune cells is skewed away from lymphocytes towards myeloid cells. The stromal matrix of the aging bone marrow that normally gives rise to and regulates HSC production, undergoes structural alterations because of the diminished stromal cell numbers and decrease in IL-7 production (Pangrazzi et al. 2017). In the course of aging the bone marrow haematopoietic compartment becomes invaded by fatty adipose tissue for reasons that remain unclear (Compston 2002; Gruver et al. 2007; Müller and Pawelec 2014).

Alterations within the HSC compartment occurring with aging may be partly related to intrinsic changes to the HSCs themselves, as well as changes in the microenvironment. Age-related accumulation of DNA damage, epigenetic changes, and telomere attrition, accompanied by elevations in intracellular ROS are characteristic features of aged HSCs (Dykstra and de Haan 2008; Warren and Rossi 2009). Accidents of genomic instability may lead to increasing malignant transformation of HSCs, especially in the myeloid lineage, and thus to the development of myelo-

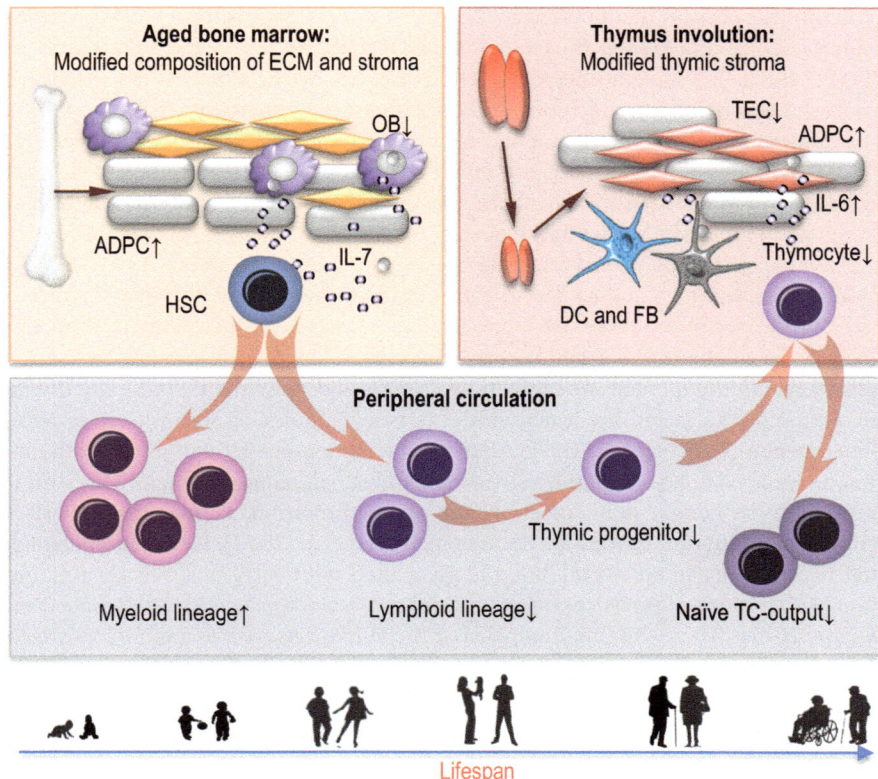

Fig. 2.4 Age-related dysregulation of hematopoiesis and thymopoiesis (Modified from Müller and Pawelec (2015))
ECM extra-cellular matrix, *ADPC* adipocytes, *OB* osteoblasts, *HSC* hematopoietic stem cell; *IL* interleukin, *TEC* thymic epithelial cells, *DC* dendritic cells, *FB* fibroblast, *TC* T cell

proliferative disorders. The consequent age-related changes in the lymphoid and myeloid lineage are likely to account for a major part of problem with immune function in the elderly and their increased susceptibility to myeloproliferative disease (Beerman et al. 2010; Dykstra and de Haan 2008; Müller and Pawelec 2014).

Thymic Involution

It is clear that a major event contributing to age-associated immune system changes is the physiological process of thymic involution (Fig. 2.4, right). After puberty, a progressive reduction in the size of the thymus size takes place, areas of active thymopoiesis decrease and the organ shrinks (Aspinall et al. 2010). This evolutionarily programmed process is supported by increased levels thymo-suppressive cytokines, such as leukaemia inhibitory factor, oncostatin M, and IL-6 (Sempowski et al. 2000), with a simultaneous decrease of IL-7 secretion. These alterations negatively affect the numbers of thymic epithelial cells and lead to decreases in thymopoiesis. The contraction of thymic output and consequently reduced capacity to replace the circulating naïve T lymphocytes after their activation and differentiation is believed to be an important element contributing to the development of immunosenescence (Aspinall et al. 2010; Müller and Pawelec 2014).

Thymus involution is characterised not only by a diminished efficiency of T-cell development but also and by the impaired migratory capacity of naïve T lymphocytes to the periphery (Fig. 2.4, bottom). The marked decline of thymic function leads to shrinkage of the T-cell repertoire and is a decisive contributor to the age-related rise in the incidence of infectious disease – despite the fact that the thymic involution per se, beginning early in life, is a developmental not in itself an aging phenomenon (Lynch et al. 2009). Additional to this process, alterations in thymic function may also be affected by environmental surroundings, prenatally as well as postnatally, by genetic predisposition, and sexual dimorphism – men display different patterns of thymic involution than women (Gui et al. 2012). It was demonstrated that thymectomy in early childhood is associated with early onset of several age-related alterations reminiscent of immunosenescence (Appay et al. 2010; Sauce and Appay 2011). These findings emphasize the importance of potentially powerful naïve T-cell output in early life for the generation of sufficient enough reserve naïve T cells for later life.

Impact of CMV Infection on Immunosenescence

An important driver of remodeling of the peripheral T-cell compartment over the lifespan is chronic immune activation by different pathogens and particularly persistent cytomegalovirus (CMV). CMV infection is usually asymptomatic unless the host is immunosuppressed and establishes a lifelong latent infection (Dowd et al. 2009; Britt 2008). CMV resides in latent form in cells broadly distributed throughout the body, including HSC, monocytes, macrophages, dendritic cells and endothelial cells (Nikolich-Zugich and van Lier 2017). Lifelong coexistence with CMV as significant implications for the immune system, leading to functional alterations and eventually driving immunosenescence (Müller et al. 2017). It is likely that latent CMV reactivates sporadically, but whether this occurs more often in the elderly than the young has not been unequivocally demonstrated (Stowe et al. 2007). In the predominantly subclinical reactivation process, it is believed that immunogenic viral transcripts are produced, which then cause expansion of virus-specific cytotoxic memory lymphocytes. The whole persistence process requires a complex equilibrium between virus immune evasion and host immune recognition over the lifetime (Jackson et al. 2017; Nikolich-Zugich et al. 2017).

As result of such persistent CMV infection an expansion of the total CD8+ T-cell pool occurs that is associated with accumulation of late-stage differentiated effector memory T cells in the peripheral circulation (Müller et al. 2017). In people infected with CMV and especially in the elderly, a large proportion of circulating CD8+ T cells is specific for CMV. This phenomenon of successive increases of CMV-specific CD8+ T-cells over the lifespan is known as "memory inflation" (Karrer et al. 2003; Kim et al. 2015). Therefore, the absolute number of CMV-specific T lymphocytes assessed as functional may be even higher in the old than in the young individuals, due to memory inflation, emphasizing the over-riding importance of immunosurveillance against this virus (Ouyang et al. 2003).

In a recent systematic review, Weltevrede and colleagues considered current evidence concerning the relationship between CMV and immunosenescence (Weltevrede et al. 2016). They concluded that in the majority of studies, CMV-seropositivity seems to enhance the accumulation of T-cell phenotypes commonly associated in the literature with immunosenescence (i.e. elevated levels of Effector Memory (EM) and TEMRA (Effector Memory T cells re-expressing CD45RA) cells in both CD4+ and the CD8+ T-cell subsets) in CMV-positive relative to CMV-negative elderly. No clear evidence was found for a lower level of naïve T cells in CMV-seropositive-vs-seronegative individuals.

Due to the observed high levels of both CD4+ and CD8+ memory T lymphocytes in older people, it was suggested that CMV might contribute to and promote immunosenescence in a clinically-relevant sense (i.e. may be associated with mortality (Ouyang et al. 2003; Pawelec et al. 2009, 2012; Pawelec and Derhovanessian 2011; Müller et al. 2017)). Many reports suggest that CMV infection increases mortality in the elderly and is related to frailty and to impaired survival (Savva et al. 2013; Wikby et al. 2005; Spyridopoulos et al. 2015). Functional decline in elderly individuals has been found to be associated with altered immune characteristics and the intensity of the immune response to CMV (Moro-Garcia et al. 2012; Müller et al. 2017). However, these data remain controversial.

Aging and Loss of T-Cell Diversity

As we have seen above, age-related changes in peripheral T-cell dynamics are associated with altered hematopoiesis, with thymic involution and with lifelong immune stimulation by multitudinous antigens, but especially CMV. Such alterations contribute to lower proportions of naïve T vcells due to lower HSC output and minimal remaining thymic function, and possibly filling of the "immunological space" by memory cells, leading to reduced diversity of the T-cell repertoire (Lynch et al. 2009; Naylor et al. 2005; Holder et al. 2016). A shift occurs towards accumulations of T-cell populations with memory and effector phenotypes, as well as to accumulation of putatively senescent immune cells. Higher levels of CD28− T cells in elderly Swedes, clustered together with certain other parameters, such as an altered CD4/CD8 ratio, fewer B-cells, and CMV-seropositivity, has led to the definition of a so-called "immune risk phenotype" associated with higher 2, 4 and 6-year mortality in longitudinal studies of people 85 years old at baseline (Pawelec et al. 2009). However, these risk parameters may be different in different populations.

A broad TCR repertoire is commonly considered to be essential for maintaining adequate immune competence in the face of new pathogens. In this regard, the age-related restraints on diversity are assumed to have negative implications. It has been reported that TCR repertoire contraction, demonstrated at least for $CD4^+$ T lympho-cytes, may occur quite suddenly: TCR diversity was found to be well-preserved up to an age of around 60–65 years, but was thereafter shown to be abruptly reduced (Naylor et al. 2005). This contraction may lead or at least contribute to the lowered responses to new pathogens and to poor vaccination response in this age group (Müller et al. 2017), but surprisingly few data are available on this issue.

The consistently lower absolute numbers and percentages of peripheral blood naïve $CD8^+$ T cells are clearly one of the universal biomarkers of human immune aging in non-Western populations as well (Fagnoni et al. 2000; Alam et al. 2013; Pawelec 2017a). While this phenomenon is consistently found in studies comparing younger and older adults, the accumulation of memory $CD8^+$ T cells is not univer-sally reported (Pawelec 2017a). This may be due to the fact that accumulation of memory T cells is apparently associated with persistent CMV infection, but not infection with other common herpesviruses, and that *de novo* infection with CMV may occur at any time of life.

Inflammaging and Its Contribution to Neuroinflammation

Aging is characterized by accumulation of senescent cells in many tissues. These cells secrete inflammatory cytokines, chemokines and other inflammatory media-tors modulating their microenvironment (Grabowska et al. 2017). Aging-related accumulation of such functionally exhausted memory T lymphocytes, commonly secreting the pro-inflammatory cytokines IFN-γ and TNF, together with mediators and factors of the innate immune system, is considered to be one source contribut-ing to the low-grade inflammation (inflammaging) often observed in old people (Franceschi et al. 2007).

Chronic activation of the adaptive immune system together with significant func-tional alterations in monocytes is believed to contribute to the process of aging. All these changes may also have implications for development and maintenance of a chronic state of low-grade inflammation and contribute to various age-related diseases (Hearps et al. 2012). Thus, inflammatory monocytes together with acti-vated macrophages may contribute to inflammaging (Franceschi et al. 2000, 2007).

It was postulated that inflammaging is a decisive source of and is associated with different age-related diseases, playing an important role in their pathology (Franceschi et al. 2000, 2007). Thus, "inflammaging seems to be a universal phe-nomenon accompanying human aging, and is associated with frailty, morbidity and mortality in elderly individuals" (Jenny 2012; Soysal et al. 2016; Vallejo 2007).

Nevertheless, it is often difficult to determine whether inflammation per se is the cause for these pathological conditions or rather the consequence (Rymkiewicz et al. 2012). Although nearly all aged individuals exhibit low-level chronic inflammation to some extent, not all suffer from age-related diseases. Thus, factors in addition to the chronic inflammatory state must be required for the onset of age-associated diseases (Bektas et al. 2017; Coder et al. 2017; Franceschi et al. 2017b; Nikolich-Zugich 2018).

Even in overtly healthy individuals, chronic low-grade inflammation has repeatedly been identified during aging, as reflected by increased levels of circulating pro-inflammatory cytokines such as IL-6, IL-1β, TNF, and IFN-γ (Franceschi et al. 2017a). These cytokines influence all physiological systems and affect their functional status, particularly neurological function, thus affecting behavioral and cognitive parameters (Alboni and Maggi 2015; Di Benedetto et al. 2017).

Peripheral immunosenescence, accompanied by inflammaging contributes at the systemic level to age-related alterations in the proportions and functions of cells in the circulation (Fig. 2.5). It is assumed that cytokines produced from such aged and functionally altered cells in the periphery can access the brain via several routes and affect neurological function in this way. "Chronic exposure to inflammatory mediators can compromise the blood-brain barrier and permit entry of immune cells and pro-inflammatory cytokines into the brain. These influence the phenotype and function of microglia due to low-grade brain inflammation. Other cells such as astrocytes and neurons, and including peripheral immune cells such as T cells, monocytes, and macrophages, thus all participate in neuroinflammation" (Di Benedetto et al. 2017; Liang et al. 2017).

Age-associated increased levels of certain cytokines and their modulators are found in the CNS of aged rodents (Barrientos et al. 2012; Ye and Johnson 1999; Scheinert et al. 2015). Aging microglia show characteristics of an increased inflammatory state referred to as a "primed profile" and defined by Norden et al. as (i) "increased baseline expression of inflammatory markers and mediators;" (ii) a "decreased threshold to be activated and to switch to a pro-inflammatory state"; and (iii) "exaggerated inflammatory response following immune activation" (Norden et al. 2015). In other words, immunosenescence and inflammaging in the periphery may contribute to neuroinflammation "by modulating glial cells towards a more active pro-inflammatory state". These altered conditions may lead to loss of neuroprotective function (normally provided by microglia) and hence to neuronal dysfunction contributing to elevated brain tissue damage (Giunta et al. 2008; von Bernhardi et al. 2010; Smith et al. 2018). Thus, systemic inflammation contributes to the risk of developing age-related cognitive impairment, neurodegenerative changes and neurological disorders (Pizza et al. 2011; Harrison 2016; Goldeck et al. 2016; Di Benedetto et al. 2017).

Currently, there is no definitive answer concerning the main underlying mechanisms involved in the induction of the neuroinflammation and its role in neurodegenerative processes occurring with aging. Nevertheless, neuroinflammation seems to be the basic contributor "that links together many factors associated with cognitive aging" (Ownby 2010). Greater cognitive impairment following immune

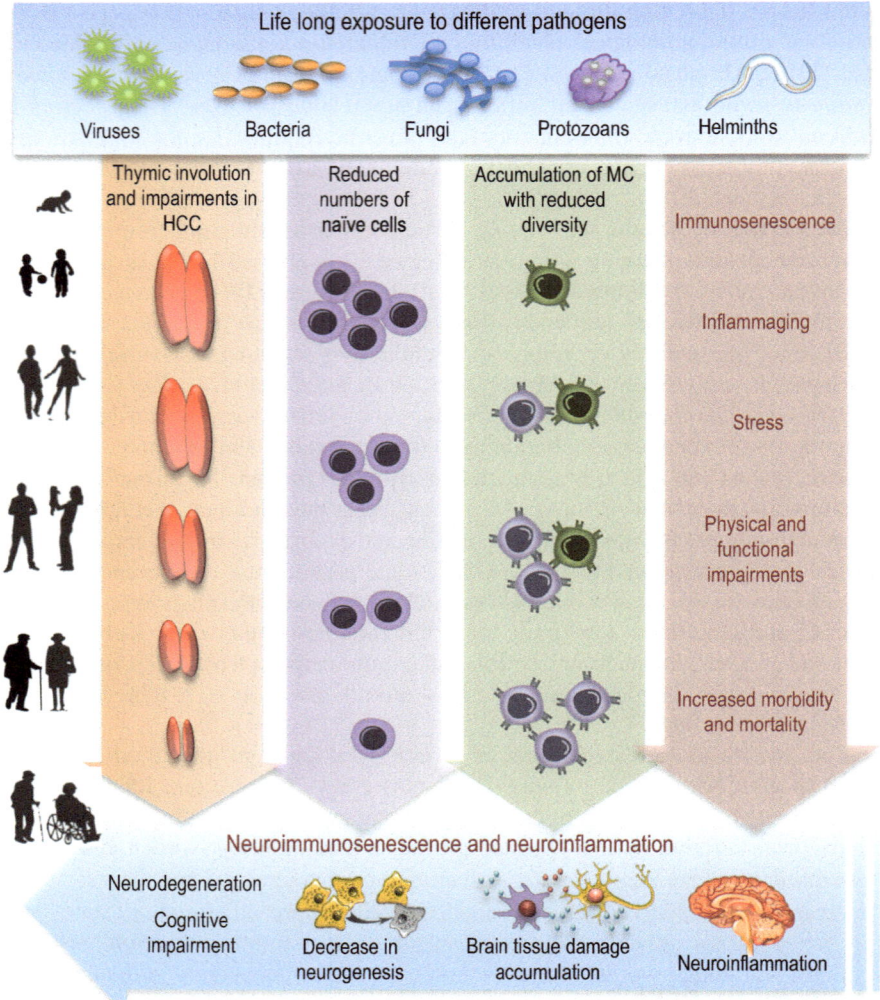

Fig. 2.5 Immunosenescence and inflammaging and their contribution to neuroinflammation (Modified from Di Benedetto et al. (2017))

"Immunosenescence affects both adaptive and innate immune systems. The most relevant changes to adaptive immunity are decreased peripheral naïve T cells and the concomitant accumulation of late-stage differentiated memory T cells with reduced antigen receptor repertoire diversity. This phenomenon results from age-related impairments in the hematopoietic stem cell compartment and thymic involution. Lifelong exposure to different pathogens is the major driver of the phenotypic changes in the distribution of T-cell subsets over the life course. Ageing is characterized by a chronic, low-grade inflammation (inflammaging). Peripheral immunosenescence and inflammaging may promote neuroinflammation by modulating glial cells towards a more active proinflammatory state, leading to a loss of neuroprotective function, to neuronal dysfunction accumulation of brain tissue damage and neurodegeneration" (Di Benedetto et al. 2017)

HCC hematopoietic cell compartment

challenge is often seen in the elderly compared to the young, accompanied by elevated and more prolonged release of pro-inflammatory cytokines in the otherwise healthy aged brain (Barrientos et al. 2015). It is broadly accepted that aging together with stress affects the neuroendocrine system, activating the hypothalamic-pituitary-adrenal axis to secrete corticotropin-releasing hormone from the paraventricular nucleus of the hypothalamus and further promoting the release of adrenocorticotropin by the anterior pituitary gland. As a consequence of this, the adrenal gland begins to produce glucocorticoids and release them into the circulation (Barrientos et al. 2012). The main glucocorticoid, cortisol, exerts modulatory effects on the immune system in a bidirectional way: by modulating the production and release of chemokines, cytokines, and adhesion molecules, as well as by altering maturation, differentiation, and migration of immune cells (Barrientos et al. 2015; Hansel et al. 2010). Elevated levels of cortisol can impair hippocampal neurogenesis either directly or indirectly, through modulation of cytokine release and through regulated expression of receptors on immune and brain cells (Di Benedetto et al. 2017). As a consequence of such modulation, an inflammatory milieu is produced that contains resident and peripheral immune cells, which all are engaged in complex interactions between secreted inflammatory mediators and cell surface receptors such as TLRs. TLRs are commonly expressed on cells that primarily participate in the inflammatory response, including microglia, astrocytes, and macrophages (Doty et al. 2015). Activated microglia and astrocytes alter their functional and morphological characteristics and start to release elevated levels of such pro-inflammatory cytokines as IL-6, TNF, and IL-1β. It was suggested that brain microglia undergo a process of senescence, similar to the immune cells in the peripheral circulation. Recent studies report detection of senescent and hyperactive microglia in the aged and diseased brain (Deleidi et al. 2015). It is assumed that also the aging brain is in turn capable of regulating the immune system in terms of supporting recruitment of immune cells from the periphery. This may further contribute to immunosenescence and neuroinflammation (Gemechu and Bentivoglio 2012; Di Benedetto et al. 2017).

Concluding Remarks

Like all somatic tissues, the immune system exhibits age-related changes, sharing certain characteristics in all mammals so far studied. The root-cause of immune aging is at least two-fold; (1) "intrinsic" aging manifesting as dysregulated immune cell generation from hematopoietic stem cells and altered peripheral selection processes related to differentiation of precursor cells, especially T cells due to thymic involution; and (2) "extrinsic" aging resulting from the effects of lifelong exposures of immune cells to antigenic and other challenges from the internal and external environment. Improved understanding of the mechanisms involved in these two disparate aspects of immune aging will be required to develop rational approaches to interventions aimed at restoring appropriate immune function in the elderly.

References

Accardi G, Caruso C (2018) Immune-inflammatory responses in the elderly: an update. Immun Ageing 15:11. https://doi.org/10.1186/s12979-018-0117-8

Adema GJ (2009) Dendritic cells from bench to bedside and back. Immunol Lett 122(2):128–130. https://doi.org/10.1016/j.imlet.2008.11.017

Ademokun A, Wu YC, Dunn-Walters D (2010) The ageing B cell population: composition and function. Biogerontology 11(2):125–137. https://doi.org/10.1007/s10522-009-9256-9

Alam I, Goldeck D, Larbi A, Pawelec G (2013) Aging affects the proportions of T and B cells in a group of elderly men in a developing country – a pilot study from Pakistan. Age (Dordr) 35(5):1521–1530. https://doi.org/10.1007/s11357-012-9455-1

Alboni S, Maggi L (2015) Editorial: cytokines as players of neuronal plasticity and sensitivity to environment in healthy and pathological brain. Front Cell Neurosci 9:508. https://doi.org/10.3389/fncel.2015.00508

Appay V, Sauce D, Prelog M (2010) The role of the thymus in immunosenescence: lessons from the study of thymectomized individuals. Aging 2(2):78–81

Aspinall R, Pitts D, Lapenna A, Mitchell W (2010) Immunity in the elderly: the role of the thymus. J Comp Pathol 142(Suppl 1):S111–S115. https://doi.org/10.1016/j.jcpa.2009.10.022

Barrientos RM, Frank MG, Watkins LR, Maier SF (2012) Aging-related changes in neuroimmune-endocrine function: implications for hippocampal-dependent cognition. Horm Behav 62(3):219–227. https://doi.org/10.1016/j.yhbeh.2012.02.010

Barrientos RM, Kitt MM, Watkins LR, Maier SF (2015) Neuroinflammation in the normal aging hippocampus. Neuroscience 309:84–99. https://doi.org/10.1016/j.neuroscience.2015.03.007

Beerman I, Bhattacharya D, Zandi S, Sigvardsson M, Weissman IL, Bryder D, Rossi DJ (2010) Functionally distinct hematopoietic stem cells modulate hematopoietic lineage potential during aging by a mechanism of clonal expansion. Proc Natl Acad Sci U S A 107(12):5465–5470. https://doi.org/10.1073/pnas.1000834107

Bektas A, Schurman SH, Sen R, Ferrucci L (2017) Human T cell immunosenescence and inflammation in aging. J Leukoc Biol 102(4):977–988. https://doi.org/10.1189/jlb.3RI0716-335R

Bonilla FA, Oettgen HC (2010) Adaptive immunity. J Allergy Clin Immunol 125(2 Suppl 2):S33–S40. https://doi.org/10.1016/j.jaci.2009.09.017

Britt W (2008) Manifestations of human cytomegalovirus infection: proposed mechanisms of acute and chronic disease. Curr Top Microbiol Immunol 325:417–470

Calvanese V, Lara E, Kahn A, Fraga MF (2009) The role of epigenetics in aging and age-related diseases. Ageing Res Rev 8(4):268–276. https://doi.org/10.1016/j.arr.2009.03.004

Coder B, Wang W, Wang L, Wu Z, Zhuge Q, Su DM (2017) Friend or foe: the dichotomous impact of T cells on neuro-de/re-generation during aging. Oncotarget 8(4):7116–7137. https://doi.org/10.18632/oncotarget.12572

Colonna-Romano G, Bulati M, Aquino A, Vitello S, Lio D, Candore G, Caruso C (2008) B cell immunosenescence in the elderly and in centenarians. Rejuvenation Res 11(2):433–439. https://doi.org/10.1089/rej.2008.0664

Compston JE (2002) Bone marrow and bone: a functional unit. J Endocrinol 173(3):387–394; doi: JOE04756 [pii]

Cooper MD (2010) 99th Dahlem conference on infection, inflammation and chronic inflammatory disorders: evolution of adaptive immunity in vertebrates. Clin Exp Immunol 160(1):58–61. https://doi.org/10.1111/j.1365-2249.2010.04126.x

De la Fuente M, Cruces J, Hernandez O, Ortega E (2011) Strategies to improve the functions and redox state of the immune system in aged subjects. Curr Pharm Des 17(36):3966–3993

Del Giudice G, Goronzy JJ, Grubeck-Loebenstein B, Lambert PH, Mrkvan T, Stoddard JJ, Doherty TM (2018) Fighting against a protean enemy: immunosenescence, vaccines, and healthy aging. NPJ Aging Mech Dis 4:1. https://doi.org/10.1038/s41514-017-0020-0

Deleidi M, Jaggle M, Rubino G (2015) Immune aging, dysmetabolism, and inflammation in neurological diseases. Front Neurosci 9:172. https://doi.org/10.3389/fnins.2015.00172

Della Bella S, Bierti L, Presicce P, Arienti R, Valenti M, Saresella M, Vergani C, Villa ML (2007) Peripheral blood dendritic cells and monocytes are differently regulated in the elderly. Clin Immunol 122(2):220–228. https://doi.org/10.1016/j.clim.2006.09.012

Derhovanessian E, Solana R, Larbi A, Pawelec G (2008) Immunity, ageing and cancer. Immun Ageing 5:11. https://doi.org/10.1186/1742-4933-5-11

Dewan SK, Zheng SB, Xia SJ, Bill K (2012) Senescent remodeling of the immune system and its contribution to the predisposition of the elderly to infections. Chin Med J 125(18):3325–3331

Di Benedetto S, Muller L, Wenger E, Duzel S, Pawelec G (2017) Contribution of neuroinflammation and immunity to brain aging and the mitigating effects of physical and cognitive interventions. Neurosci Biobehav Rev 75:114–128. https://doi.org/10.1016/j.neubiorev.2017.01.044

Doran KS, Banerjee A, Disson O, Lecuit M (2013) Concepts and mechanisms: crossing host barriers. Cold Spring Harb Perspect Med 3(7):a010090. https://doi.org/10.1101/cshperspect.a010090

Doty KR, Guillot-Sestier MV, Town T (2015) The role of the immune system in neurodegenerative disorders: adaptive or maladaptive? Brain Res 1617:155–173. https://doi.org/10.1016/j.brainres.2014.09.008

Dowd JB, Aiello AE, Alley DE (2009) Socioeconomic disparities in the seroprevalence of cytomegalovirus infection in the US population: NHANES III. Epidemiol Infect 137(1):58–65. https://doi.org/10.1017/S0950268808000551

Dykstra B, de Haan G (2008) Hematopoietic stem cell aging and self-renewal. Cell Tissue Res 331(1):91–101. https://doi.org/10.1007/s00441-007-0529-9

Fagnoni FF, Vescovini R, Passeri G, Bologna G, Pedrazzoni M, Lavagetto G, Casti A, Franceschi C, Passeri M, Sansoni P (2000) Shortage of circulating naive CD8(+) T cells provides new insights on immunodeficiency in aging. Blood 95(9):2860–2868

Fernandez-Morera JL, Calvanese V, Rodriguez-Rodero S, Menendez-Torre E, Fraga MF (2010) Epigenetic regulation of the immune system in health and disease. Tissue Antigens 76(6):431–439. https://doi.org/10.1111/j.1399-0039.2010.01587.x

Ferrando-Martinez S, Ruiz-Mateos E, Hernandez A, Gutierrez E, Rodriguez-Mendez Mdel M, Ordonez A, Leal M (2011) Age-related deregulation of naive T cell homeostasis in elderly humans. Age 33(2):197–207. https://doi.org/10.1007/s11357-010-9170-8

Franceschi C, Bonafe M, Valensin S, Olivieri F, De Luca M, Ottaviani E, De Benedictis G (2000) Inflamm-aging. An evolutionary perspective on immunosenescence. Ann N Y Acad Sci 908:244–254

Franceschi C, Capri M, Monti D, Giunta S, Olivieri F, Sevini F, Panourgia MP, Invidia L, Celani L, Scurti M, Cevenini E, Castellani GC, Salvioli S (2007) Inflammaging and anti-inflammaging: a systemic perspective on aging and longevity emerged from studies in humans. Mech Ageing Dev 128(1):92–105. https://doi.org/10.1016/j.mad.2006.11.016

Franceschi C, Garagnani P, Vitale G, Capri M, Salvioli S (2017a) Inflammaging and 'Garb-aging'. Trends Endocrinol Metab 28(3):199–212. https://doi.org/10.1016/j.tem.2016.09.005

Franceschi C, Salvioli S, Garagnani P, de Eguileor M, Monti D, Capri M (2017b) Immunobiography and the heterogeneity of immune responses in the elderly: a focus on Inflammaging and trained immunity. Front Immunol 8:982. https://doi.org/10.3389/fimmu.2017.00982

Fulop T, Larbi A, Douziech N, Fortin C, Guerard KP, Lesur O, Khalil A, Dupuis G (2004) Signal transduction and functional changes in neutrophils with aging. Aging Cell 3(4):217–226. https://doi.org/10.1111/j.1474-9728.2004.00110.x

Fulop T, Larbi A, Dupuis G, Le Page A, Frost EH, Cohen AA, Witkowski JM, Franceschi C (2017) Immunosenescence and Inflamm-aging as two sides of the same coin: friends or foes? Front Immunol 8:1960. https://doi.org/10.3389/fimmu.2017.01960

Geiger H, de Haan G, Florian MC (2013) The ageing haematopoietic stem cell compartment. Nat Rev Immunol 13(5):376–389. https://doi.org/10.1038/nri3433

Gemechu JM, Bentivoglio M (2012) T cell recruitment in the brain during normal aging. Front Cell Neurosci 6:38. https://doi.org/10.3389/fncel.2012.00038

Giunta B, Fernandez F, Nikolic WV, Obregon D, Rrapo E, Town T, Tan J (2008) Inflammaging as a prodrome to Alzheimer's disease. J Neuroinflammation 5:51. https://doi.org/10.1186/1742-2094-5-51

Goldeck D, Witkowski JM, Fulop T, Pawelec G (2016) Peripheral immune signatures in Alzheimer disease. Curr Alzheimer Res 13(7):739–749

Gonzalo S (2010) Epigenetic alterations in aging. J Appl Physiol 109(2):586–597. https://doi.org/10.1152/japplphysiol.00238.2010

Grabowska W, Sikora E, Bielak-Zmijewska A (2017) Sirtuins, a promising target in slowing down the ageing process. Biogerontology 18(4):447–476. https://doi.org/10.1007/s10522-017-9685-9

Gruver AL, Hudson LL, Sempowski GD (2007) Immunosenescence of ageing. J Pathol 211(2):144–156. https://doi.org/10.1002/path.2104

Gui J, Mustachio LM, Su DM, Craig RW (2012) Thymus size and age-related thymic involution: early programming, sexual dimorphism, progenitors and stroma. Aging Dis 3(3):280–290

Hansel A, Hong S, Camara RJ, von Kanel R (2010) Inflammation as a psychophysiological bio-marker in chronic psychosocial stress. Neurosci Biobehav Rev 35(1):115–121. https://doi.org/10.1016/j.neubiorev.2009.12.012

Harrison NA (2016) Brain structures implicated in inflammation-associated depression. Curr Top Behav Neurosci 31:221–248. https://doi.org/10.1007/7854_2016_30

Hawkley LC, Cacioppo JT (2004) Stress and the aging immune system. Brain Behav Immun 18(2):114–119

Hearps AC, Martin GE, Angelovich TA, Cheng WJ, Maisa A, Landay AL, Jaworowski A, Crowe SM (2012) Aging is associated with chronic innate immune activation and dys-regulation of monocyte phenotype and function. Aging Cell 11(5):867–875. https://doi.org/10.1111/j.1474-9726.2012.00851.x

Heffner KL (2011) Neuroendocrine effects of stress on immunity in the elderly: implications for inflammatory disease. Immunol Allergy Clin N Am 31(1):95–108. https://doi.org/10.1016/j.iac.2010.09.005

Holder A, Mella S, Palmer DB, Aspinall R, Catchpole B (2016) An age-associated decline in thymic output differs in dog breeds according to their longevity. PLoS One 11(11):e0165968. https://doi.org/10.1371/journal.pone.0165968

Jackson SE, Redeker A, Arens R, van Baarle D, van den Berg SPH, Benedict CA, Cicin-Sain L, Hill AB, Wills MR (2017) CMV immune evasion and manipulation of the immune system with aging. GeroScience 39(3):273–291. https://doi.org/10.1007/s11357-017-9986-6

Jenny NS (2012) Inflammation in aging: cause, effect, or both? Discov Med 13(73):451–460

Karrer U, Sierro S, Wagner M, Oxenius A, Hengel H, Koszinowski UH, Phillips RE, Klenerman P (2003) Memory inflation: continuous accumulation of antiviral CD8+ T cells over time. J Immunol 170(4):2022–2029

Keller R (1993) The macrophage response to infectious agents: mechanisms of macrophage acti-vation and tumour cell killing. Res Immunol 144(4):271–273; discussion 294–278

Kim J, Kim AR, Shin EC (2015) Cytomegalovirus infection and memory T cell inflation. Immune Netw 15(4):186–190. https://doi.org/10.4110/in.2015.15.4.186

Lanier LL, Sun JC (2009) Do the terms innate and adaptive immunity create conceptual barriers? Nat Rev Immunol 9(5):302–303. https://doi.org/10.1038/nri2547

Larbi A, Franceschi C, Mazzatti D, Solana R, Wikby A, Pawelec G (2008) Aging of the immune system as a prognostic factor for human longevity. Physiology 23:64–74. https://doi.org/10.1152/physiol.00040.2007

Larbi A, Pawelec G, Wong SC, Goldeck D, Tai JJ, Fulop T (2011) Impact of age on T cell sig-naling: a general defect or specific alterations? Ageing Res Rev 10(3):370–378. https://doi.org/10.1016/j.arr.2010.09.008

Liang Z, Zhao Y, Ruan L, Zhu L, Jin K, Zhuge Q, Su DM, Zhao Y (2017) Impact of aging immune system on neurodegeneration and potential immunotherapies. Prog Neurobiol 157:2–28. https://doi.org/10.1016/j.pneurobio.2017.07.006

Litman GW, Rast JP, Fugmann SD (2010) The origins of vertebrate adaptive immunity. Nat Rev Immunol 10(8):543–553. https://doi.org/10.1038/nri2807

Liu K, Catalfamo M, Li Y, Henkart PA, Weng NP (2002) IL-15 mimics T cell receptor crosslinking in the induction of cellular proliferation, gene expression, and cytotoxicity in CD8+ memory T cells. Proc Natl Acad Sci U S A 99(9):6192–6197. https://doi.org/10.1073/pnas.092675799

Lynch HE, Goldberg GL, Chidgey A, Van den Brink MR, Boyd R, Sempowski GD (2009) Thymic involution and immune reconstitution. Trends Immunol 30(7):366–373. https://doi.org/10.1016/j.it.2009.04.003

Masters AR, Haynes L, Su DM, Palmer DB (2017) Immune senescence: significance of the stromal microenvironment. Clin Exp Immunol 187(1):6–15. https://doi.org/10.1111/cei.12851

Medzhitov R, Janeway C Jr (2000) Innate immunity. N Engl J Med 343(5):338–344. https://doi.org/10.1056/NEJM200008033430506

Menza M, Dobkin RD, Marin H, Mark MH, Gara M, Bienfait K, Dicke A, Kusnekov A (2010) The role of inflammatory cytokines in cognition and other non-motor symptoms of Parkinson's disease. Psychosomatics 51(6):474–479. https://doi.org/10.1176/appi.psy.51.6.474

Michaud M, Balardy L, Moulis G, Gaudin C, Peyrot C, Vellas B, Cesari M, Nourhashemi F (2013) Proinflammatory cytokines, aging, and age-related diseases. J Am Med Dir Assoc 14(12):877–882. https://doi.org/10.1016/j.jamda.2013.05.009

Moro-Garcia MA, Alonso-Arias R, Lopez-Vazquez A, Suarez-Garcia FM, Solano-Jaurrieta JJ, Baltar J, Lopez-Larrea C (2012) Relationship between functional ability in older people, immune system status, and intensity of response to CMV. Age (Dordr) 34(2):479–495. https://doi.org/10.1007/s11357-011-9240-6

Müller L, Pawelec G (2014) Aging and immunity – impact of behavioral intervention. Brain Behav Immun 39:8–22. https://doi.org/10.1016/j.bbi.2013.11.015

Müller L, Pawelec G (2015) As we age: does slippage of quality control in the immune system lead to collateral damage? Ageing Res Rev 23(Pt A):116–123. https://doi.org/10.1016/j.arr.2015.01.005

Müller L, Fülop T, Pawelec G (2013a) Immunosenescence in vertebrates and invertebrates. Immun Ageing 10(1):12. https://doi.org/10.1186/1742-4933-10-12

Müller L, Pawelec G, Derhovanessian E (2013b) The immune system during aging. In: Calder P, Yaqoob P (eds) Diet, immunity and inflammation. Woodhead Publishing, Oxford, pp 631–651

Müller L, Hamprecht K, Pawelec G (2017) The role of CMV in "immunosenescence". In: Bueno V, Lord JM, Jackson TA (eds) The ageing immune system and health. Springer, Cham, pp 53–68. https://doi.org/10.1007/978-3-319-43365-3_4

Naylor K, Li G, Vallejo AN, Lee WW, Koetz K, Bryl E, Witkowski J, Fulbright J, Weyand CM, Goronzy JJ (2005) The influence of age on T cell generation and TCR diversity. J Immunol 174(11):7446–7452

Nikolich-Zugich J (2018) The twilight of immunity: emerging concepts in aging of the immune system. Nat Immunol 19(1):10–19. https://doi.org/10.1038/s41590-017-0006-x

Nikolich-Zugich J, van Lier RAW (2017) Cytomegalovirus (CMV) research in immune senescence comes of age: overview of the 6th International Workshop on CMV and Immunosenescence. GeroScience 39(3):245–249. https://doi.org/10.1007/s11357-017-9984-8

Nikolich-Zugich J, Goodrum F, Knox K, Smithey MJ (2017) Known unknowns: how might the persistent herpesvirome shape immunity and aging? Curr Opin Immunol 48:23–30. https://doi.org/10.1016/j.coi.2017.07.011

Norden DM, Muccigrosso MM, Godbout JP (2015) Microglial priming and enhanced reactivity to secondary insult in aging, and traumatic CNS injury, and neurodegenerative disease. Neuropharmacology 96(Pt A):29–41. https://doi.org/10.1016/j.neuropharm.2014.10.028

Oishi Y, Manabe I (2016) Macrophages in age-related chronic inflammatory diseases. NPJ Aging Mech Dis 2:16018. https://doi.org/10.1038/npjamd.2016.18

Ouyang Q, Wagner WM, Voehringer D, Wikby A, Klatt T, Walter S, Muller CA, Pircher H, Pawelec G (2003) Age-associated accumulation of CMV-specific CD8+ T cells expressing the inhibitory killer cell lectin-like receptor G1 (KLRG1). Exp Gerontol 38(8):911–920

Ownby RL (2010) Neuroinflammation and cognitive aging. Curr Psychiatry Rep 12(1):39–45. https://doi.org/10.1007/s11920-009-0082-1

Pangrazzi L, Meryk A, Naismith E, Koziel R, Lair J, Krismer M, Trieb K, Grubeck-Loebenstein B (2017) "Inflamm-aging" influences immune cell survival factors in human bone marrow. Eur J Immunol 47(3):481–492. https://doi.org/10.1002/eji.201646570

Pawelec G (2012) Hallmarks of human "immunosenescence": adaptation or dysregulation? Immun Ageing 9(1):15. https://doi.org/10.1186/1742-4933-9-15

Pawelec G (2017a) Age and immunity: what is "immunosenescence"? Exp Gerontol 105:4–9. https://doi.org/10.1016/j.exger.2017.10.024

Pawelec G (2017b) Does the human immune system ever really become "senescent"? F1000Research 6:1323. https://doi.org/10.12688/f1000research.11297.1

Pawelec G, Derhovanessian E (2011) Role of CMV in immune senescence. Virus Res 157(2):175–179. https://doi.org/10.1016/j.virusres.2010.09.010

Pawelec G, Derhovanessian E, Larbi A, Strindhall J, Wikby A (2009) Cytomegalovirus and human immunosenescence. Rev Med Virol 19(1):47–56. https://doi.org/10.1002/rmv.598

Pawelec G, McElhaney JE, Aiello AE, Derhovanessian E (2012) The impact of CMV infection on survival in older humans. Curr Opin Immunol 24(4):507–511. https://doi.org/10.1016/j.coi.2012.04.002

Pereira BI, Akbar AN (2016) Convergence of innate and adaptive immunity during human aging. Front Immunol 7:445. https://doi.org/10.3389/fimmu.2016.00445

Pizza V, Agresta A, D'Acunto CW, Festa M, Capasso A (2011) Neuroinflammation and ageing: current theories and an overview of the data. Rev Recent Clin Trials 6(3):189–203

Rymkiewicz PD, Heng YX, Vasudev A, Larbi A (2012) The immune system in the aging human. Immunol Res 53(1–3):235–250. https://doi.org/10.1007/s12026-012-8289-3

Sauce D, Appay V (2011) Altered thymic activity in early life: how does it affect the immune system in young adults? Curr Opin Immunol 23(4):543–548. https://doi.org/10.1016/j.coi.2011.05.001

Savva GM, Pachnio A, Kaul B, Morgan K, Huppert FA, Brayne C, Moss PA, Medical Research Council Cognitive F, Ageing S (2013) Cytomegalovirus infection is associated with increased mortality in the older population. Aging Cell 12(3):381–387. https://doi.org/10.1111/acel.12059

Scheinert RB, Asokan A, Rani A, Kumar A, Foster TC, Ormerod BK (2015) Some hormone, cytokine and chemokine levels that change across lifespan vary by cognitive status in male Fischer 344 rats. Brain Behav Immun 49:216–232. https://doi.org/10.1016/j.bbi.2015.06.005

Sempowski GD, Hale LP, Sundy JS, Massey JM, Koup RA, Douek DC, Patel DD, Haynes BF (2000) Leukemia inhibitory factor, oncostatin M, IL-6, and stem cell factor mRNA expression in human thymus increases with age and is associated with thymic atrophy. J Immunol 164(4):2180–2187

Shanley DP, Aw D, Manley NR, Palmer DB (2009) An evolutionary perspective on the mechanisms of immunosenescence. Trends Immunol 30(7):374–381. https://doi.org/10.1016/j.it.2009.05.001

Shaw AC, Joshi S, Greenwood H, Panda A, Lord JM (2010) Aging of the innate immune system. Curr Opin Immunol 22(4):507–513. https://doi.org/10.1016/j.coi.2010.05.003

Smith LK, White CW 3rd, Villeda SA (2018) The systemic environment: at the interface of aging and adult neurogenesis. Cell Tissue Res 371(1):105–113. https://doi.org/10.1007/s00441-017-2715-8

Solana R, Pawelec G, Tarazona R (2006) Aging and innate immunity. Immunity 24(5):491–494. https://doi.org/10.1016/j.immuni.2006.05.003

Soysal P, Stubbs B, Lucato P, Luchini C, Solmi M, Peluso R, Sergi G, Isik AT, Manzato E, Maggi S, Maggio M, Prina AM, Cosco TD, Wu YT, Veronese N (2016) Inflammation and frailty in the elderly: a systematic review and meta-analysis. Ageing Res Rev 31:1–8. https://doi.org/10.1016/j.arr.2016.08.006

Spyridopoulos I, Martin-Ruiz C, Hilkens C, Yadegarfar ME, Isaacs J, Jagger C, Kirkwood T, von Zglinicki T (2015) CMV seropositivity and T-cell senescence predict increased cardiovascular mortality in octogenarians: results from the Newcastle 85+ study. Aging Cell. https://doi. org/10.1111/acel.12430

Stowe RP, Kozlova EV, Yetman DL, Walling DM, Goodwin JS, Glaser R (2007) Chronic herpesvirus reactivation occurs in aging. Exp Gerontol 42(6):563–570. https://doi.org/10.1016/j. exger.2007.01.005

Vallejo AN (2007) Immune remodeling: lessons from repertoire alterations during chronological aging and in immune-mediated disease. Trends Mol Med 13(3):94–102. https://doi. org/10.1016/j.molmed.2007.01.005

von Bernhardi R, Tichauer JE, Eugenin J (2010) Aging-dependent changes of microglial cells and their relevance for neurodegenerative disorders. J Neurochem 112(5):1099–1114. https://doi. org/10.1111/j.1471-4159.2009.06537.x

Warren LA, Rossi DJ (2009) Stem cells and aging in the hematopoietic system. Mech Ageing Dev 130(1–2):46–53. doi: S0047-6374(08)00091-2 [pii]. https://doi.org/10.71016/j. mad.2008.03.010

Weiskopf D, Weinberger B, Grubeck-Loebenstein B (2009) The aging of the immune system. Transpl Int 22(11):1041–1050. https://doi.org/10.1111/j.1432-2277.2009.00927.x

Weltevrede M, Eilers R, de Melker HE, van Baarle D (2016) Cytomegalovirus persistence and T-cell immunosenescence in people aged fifty and older: a systematic review. Exp Gerontol 77:87–95. https://doi.org/10.1016/j.exger.2016.02.005

Wikby A, Ferguson F, Forsey R, Thompson J, Strindhall J, Lofgren S, Nilsson BO, Ernerudh J, Pawelec G, Johansson B (2005) An immune risk phenotype, cognitive impairment, and survival in very late life: impact of allostatic load in Swedish octogenarian and nonagenarian humans. J Gerontol A Biol Sci Med Sci 60(5):556–565

Xu W, Larbi A (2017) Immunity and inflammation: from Jekyll to Hyde. Exp Gerontol 107:98–101. https://doi.org/10.1016/j.exger.2017.11.018

Ye SM, Johnson RW (1999) Increased interleukin-6 expression by microglia from brain of aged mice. J Neuroimmunol 93(1–2):139–148

Chapter 3
Pulmonary Diseases and Ageing

Peter J. Barnes

Abstract Chronic obstructive pulmonary disease (COPD) and idiopathic pulmonary fibrosis are regarded as a diseases of accelerated lung ageing and show all of the hallmarks of ageing, including telomere shortening, cellular senescence, activation of PI3 kinase-mTOR signaling, impaired autophagy, mitochondrial dysfunction, stem cell exhaustion, epigenetic changes, abnormal microRNA profiles, immunosenescence and a low grade chronic inflammation due to senescence-associated secretory phenotype (SASP). Many of these ageing mechanisms are driven by exogenous and endogenous oxidative stress. There is also a reduction in anti-ageing molecules, such as sirtuins and Klotho, which further accelerate the ageing process. Understanding these molecular mechanisms has identified several novel therapeutic targets and several drugs and dietary interventions are now in development to treat chronic lung disease.

Keywords Autophagy · Cellular senescence · Immunosenescence · Mitochondria · Oxidative stress · Sirtuins

Introduction

Accelerated ageing plays a key role in several pulmonary diseases, particularly chronic obstructive pulmonary disease (COPD) and idiopathic pulmonary fibrosis (IPF), but also may play a role in asthma, bronchiectasis, pulmonary hypertension and lung infections. Although cystic fibrosis (CF) is a disease of childhood, with advances in management patients may survive into middle age so that ageing processes may interact with the disease. In this chapter most focus is on COPD as this is a prevalent disease and most research on lung ageing has been related to this disease.

P. J. Barnes (✉)
Airway Disease Section, National Heart and Lung Institute, Imperial College, London, UK
e-mail: p.j.barnes@imperial.ac.uk

© Springer Nature Singapore Pte Ltd. 2019
J. R. Harris, V. I. Korolchuk (eds.), *Biochemistry and Cell Biology of Ageing:*
Part II Clinical Science, Subcellular Biochemistry 91,
https://doi.org/10.1007/978-981-13-3681-2_3

Lung Ageing

In normal individuals lung functions, such as forced expiratory volume in 1 s (FEV_1) and forced vital capacity (FVC) and increased functional residual capacity (resting lung volume) decline slowly with age after a peak of around 25 years in men (Janssens et al. 1999). There is a change in the flow volume loop that suggests that the loss of lung volumes is largely due to narrowing of peripheral airways. This loss of lung function over time may be associated with loss of lung elasticity and enlargement of alveolar spaces (which has been termed senile emphysema), resulting in a reduction in gas transfer, although this is not associated with alveolar wall destruction as in COPD. None of these changes in lung function with age are sufficient to cause symptoms or impaired oxygenation and do not lead to respiratory symptoms. The aged lung appears to be more susceptible to damage by environmental stresses, such as cigarette smoking and more susceptiblity to infection.

COPD

COPD is a global epidemic, with rising prevalence as populations live longer because of reduced mortality from infectious and cardiovascular diseases. COPD is the third commonest cause of death in developed countries and the fifth ranked cause of disability, but is increasing to a greater extent in developing countries (Lozano et al. 2012). COPD is present in approximately 10% of people over 45 years and in developed countries it as common in women as in men, reflecting the prevalence of smoking in the population (Barnes et al. 2015). The major risk factor for COPD world-wide is cigarette smoking, but in developing countries exposure to biomass smoke and household air pollution, especially in rural areas is also common (Sood et al. 2018). COPD is also linked to low socioeconomic status, which is likely to be explained by a combination of factors, such as smoking, air pollution, poor nutrition and damp housing (Burney et al. 2014).

COPD is characterized by largely irreversible and progressive airway obstruction as a result of fibrosis and obstruction of small airways (chronic obstructive bronchiolitis) and destruction of the lung parenchyma (emphysema) (Hogg and Timens 2009). This results physiologically in air trapping, hyperinflated lungs and dynamic hyperinflation that leads to shortness of breath on exertion, the major symptom of COPD patients. This results in accelerated decline in lung function over time, although only half of the patients diagnosed with COPD show rapid decline, the remainder have a normal decline starting from low peak lung volumes as a result of poor lung growth during childhood (Lange et al. 2015). About 20% of smokers develop COPD but the reasons for this susceptibility have not been fully determined. Although several genes have been implicated in susceptibility, but do not account for most of this susceptibility, other factors such as epigenetic modifications may be important for inherited and environmental influences.

Table 3.1 Features of accelerated ageing found in COPD and IPF patients

Telomere shortening
Cellular senescence
Activation of PI3K-mTOR signaling
Decreased autophagy
Defective DNA repair
Mitochondrial dysfunction
Stem cell exhaustion
Immunosenescence
Abnormal microRNA pattern
Epigenetic changes
Decreased anti-ageing molecules

The underlying mechanisms of COPD are poorly understood but it is associated with chronic inflammation that is largely corticosteroid-resistant (Barnes 2016). This inflammation is similar to that found in normal smokers, but appears to be amplified. A significant mechanism of amplification involves a reduction in the nuclear enzyme histone deacetylase-2 (HDAC2), which plays an important role in switching off activated inflammatory genes (Barnes 2009). Many cells and mediators are involved in COPD (Barnes 2014) and increased oxidative stress from inhaled smoke, or endogenously from activated lung inflammatory cells, is a key driving mechanism and results in a reduction in HDAC2 (Kirkham and Barnes 2013).

The global increase in COPD is likely to be related to the ageing population, as this disease predominantly affects the elderly, with peak prevalence around 65 years. Airway obstruction in COPD is slowly progressive and represents an acceleration of the normal decline in lung function with age, so this has suggested that COPD, and especially emphysema, involves acceleration of the normal lung ageing process (Ito and Barnes 2009; Mercado et al. 2015). There is increasing evidence that in COPD there are all of the hallmarks of accelerated ageing, compared to smokers with normal lung function and non-smokers, and this is discussed below (Table 3.1). Age is the most important risk factor for several chronic diseases, including COPD, and drives morbidity and mortality (Kennedy et al. 2014).

Idiopathic Pulmonary Ibrosis

IPF involves slowly progressive fibrosis of the lung parenchyma that is not associated with any causative mechanisms (Sgalla et al. 2018). It is usually diagnosed in people over the age of 70 and has a poor prognosis. The fibrotic process is thought to be due to sustained alveolar microinjury and aberrant repair in genetically predisposed individuals. There is increasing evidence that IPF is due to accelerated ageing and accumulation of senescent cells (Schafer et al. 2017). Removal of senescent cells abrogates the development of fibrosis in animal models of IPF.

Some patients with IPF also have emphysema and this overlap syndrome called combined pulmonary fibrosis and emphysema (CPFE) is increasingly recognized (Cottin and Cordier 2012). It is related to cigarette smoking, which is a defined risk factor for IPF as well as COPD. CPFE is characterized by accelerated ageing and cellular senescence (Chilosi et al. 2013).

Hallmarks of Ageing in Lung Disease

Chronic lung diseases are characterized by increased oxidative stress, which may be a key mechanisms in driving accelerated ageing and cellular senescence in COPD and IPF (Kirkham and Barnes 2013; Fois et al. 2018). Excessive reactive oxygen species (ROS) induce the accumulation of molecular damage, which shortens lifespan and an optimal level of ROS is needed for healthy ageing (Hekimi et al. 2011). Oxidative stress that is greater than that associated with normal ageing, may induce age-related degeneration that favors chronic inflammation.

All of the classical hallmarks of ageing are seen in COPD and IPF, including telomere shortening, genomic instability, epigenetic alterations, loss of proteostasis, mitochondrial dysfunction, deregulated nutrient-sensing and stem cell exhaustion (Table 3.1) (Mercado et al. 2015; Meiners et al. 2015; Sgalla et al. 2018). There is an accumulation of senescent cells, which characteristically have irreversible cell cycle arrest, but are metabolically active and display what is termed a "senescence-associated secretory phenotype" (SASP) or "inflammaging", with secretion of pro-inflammatory cytokines, chemokines and matrix metalloproteinases (MMP) (Salama et al. 2014). These proinflammatory mediators may induce further senescence in the cell itself (autocrine) and surrounding cells (paracrine), thus amplifying and spreading cellular senescence (Fig. 3.1). Senescent cells accumulate with age in tissues and the SASP proinflammatory state may be an important driving mechanism in age-related lung diseases.

Telomere Shortening in Lung Disease

Telomere shortening is greater in COPD and IPF (Birch et al. 2017). Telomere shortening is found in circulating leukocytes from patients with COPD to a greater extent than in smokers with normal lung function and appears to be related to a greater risk of lung cancer (Lee et al. 2012; Houben et al. 2009; Savale et al. 2009; Rutten et al. 2016). Short telomeres increase the risks of developing emphysema amongst smokers (Houben et al. 2009). In a large observational study short telomeres in circulating leukocytes have been linked to reduced lung function, although the independent effect of COPD, after correction for age and smoking, is small. Shorter telomeres have also been described in alveolar epithelial and endothelial cells from patients with emphysema (Tsuji et al. 2006), although this may also be

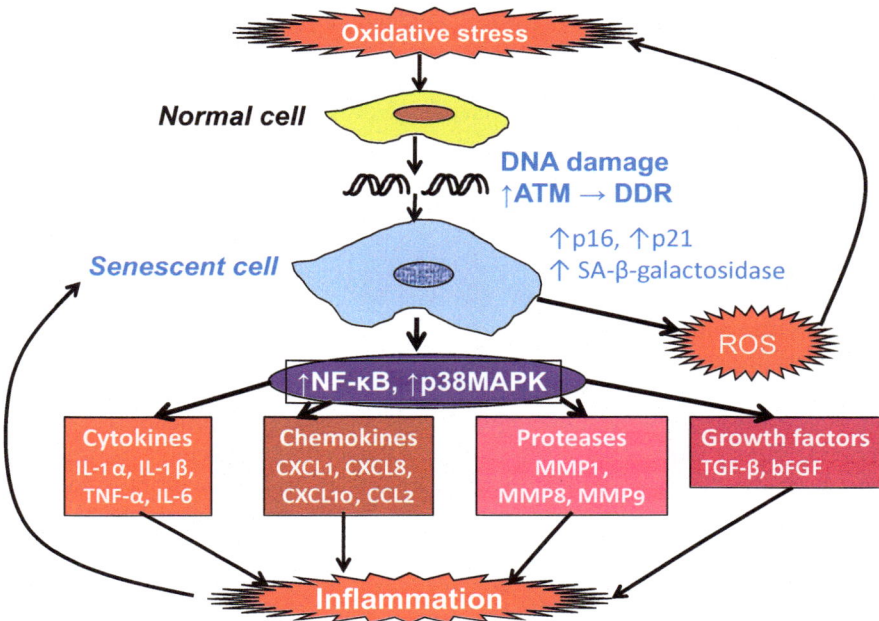

Fig. 3.1 *Senescence-associated secretory phenotype (SASP)*. Senescent cells that express p16 and p21 are metabolically active, with activation of nuclear factor-kappa B (NF-κB) and mitogen- activated protein kinase p38, resulting in the increased expression of multiple inflammatory proteins, including cytokines, chemokines, proteases and growth factors, all of which are increased in chronic lung diseases. This inflammatory response has paracrine effects inducing senescence in adjacent cells and senescent cells also produce reactive oxygen species that further drive the senescence process

found in smokers with normal lung function (Tomita et al. 2010). Telomere-associated DNA damage is found in small airway epithelial cells from COPD patients compared to age-matched controls (Birch et al. 2015). A family with a genetic defect in telomerase has been linked to early onset emphysema (Alder et al. 2011). Approximately 1% of patients (mainly female) with severe emphysema have mutations of telomere reverse transcriptase (TERT) resulting in short telomeres (Stanley et al. 2015). Telomerase null mice with short telomeres showed increased susceptibility to develop emphysema and increased cellular senescence after chronic cigarette exposure (Alder et al. 2011). Knock-out of telomerase in mice results in replicative senescence of alveolar cells and a low grade lung inflammation, with increased interleukin(IL)-1, IL-6, KC (murine IL-8) and CCL2, indicating how telomere shortening can result in COPD-like lung disease (Chen et al. 2015). Telomerase knockout mice also show increased emphysema after cigarette smoke exposure (Birch et al. 2015). Genetic determinants of telomere length may increase susceptibility to COPD but also to risk of developing comorbidities, such as cardiovascular and metabolic diseases.

The mechanisms leading to telomere shortening in association with chronic diseases such as COPD are not yet understood. Increased oxidative stress impairs telomerase activity and telomerase DNA is particularly susceptible to DNA damage (Passos et al. 2007). Telomere shortening in turn results in activation of p21, resulting in cellular senescence and the release of proinflammatory mediators, such as IL-6 and CXCL8 (IL-8). Cultured pulmonary endothelial cells from COPD patients have reduced telomerase activity, which is associated with shorter telomeres with increased cellular senescence and a SASP response compared to age-matched non-smoking control subjects (Amsellem et al. 2011).

Telomere shortening is also an important feature in IPF. Telomerase defiance mice develop pulmonary fibrosis (Povedano et al. 2015). IPF patients have shorter telomeres in circulating leukocytes and in alveolar type 2 cells (Alder et al. 2008) and there is an increase in telomere DNA damage foci (Schafer et al. 2017). Mutation in telomerase genes is commonly found in patients with IPF, particularly if there is a family history (Armanios 2012). Shortened telomeres and telomere damage have also been described in airway epithelial cells from patients with bronchiectasis (Birch et al. 2016). Shorter telomeres have also been described in circulating leukocytes of patients with life-long chronic persistent asthma, possibly as a result of chronic inflammation and oxidative stress (Belsky et al. 2014).

Cellular Senescence

Diseases of accelerated ageing are characterized by the accumulation of senescent cells in tissues, which cells enter a state of irreversible cell cycle arrest (Fig. 3.2) (Munoz-Espin and Serrano 2014; van Deursen 2014). Mammalian cells have a limited number of divisions and cells enter cellular senescence and subsequently death by apoptosis once DNA damage can no longer be repaired effectively. The accumulation of senescent cells results in age-related loss of function and removal of senescent cells prolongs lifespan of mice and delays the development of organ failure and development of cancers (Baker et al. 2011; Baker et al. 2016a). Cellular senescence results from the activation of the tumor suppression pathways p53 and p21$^{CIP1/WAF1}$ and the p16^{INK4a}/retinoblastoma protein pathways, which are activated by the DNA damage response (DDR) in response to critical telomere shortening and telomeric and non-telomeric DNA damage. Deletion of cells that express p16^{INK4a} increases the lifespan of normal mice (by up to 30%) and delays several age-related diseases and the onset of cancer (Baker et al. 2011, 2016a). In COPD and IPF there is an accumulation of senescent cells (Tsuji et al. 2006; Rutten et al. 2016; Schafer et al. 2017).

Cellular senescence may be enhanced by extraneous stressful stimuli, such as oxidative stress and ultraviolet radiation in the case of skin. Cigarette smoke exposure in mice increases the number of p16 expressing cells in the lung, indicating that smoking can accelerate lung ageing (Sorrentino et al. 2014). Unlike apoptotic cells, senescent cells remain metabolically active and therefore may influence other cells through the SASP response (Salama et al. 2014; Correia-Melo et al. 2014)., The

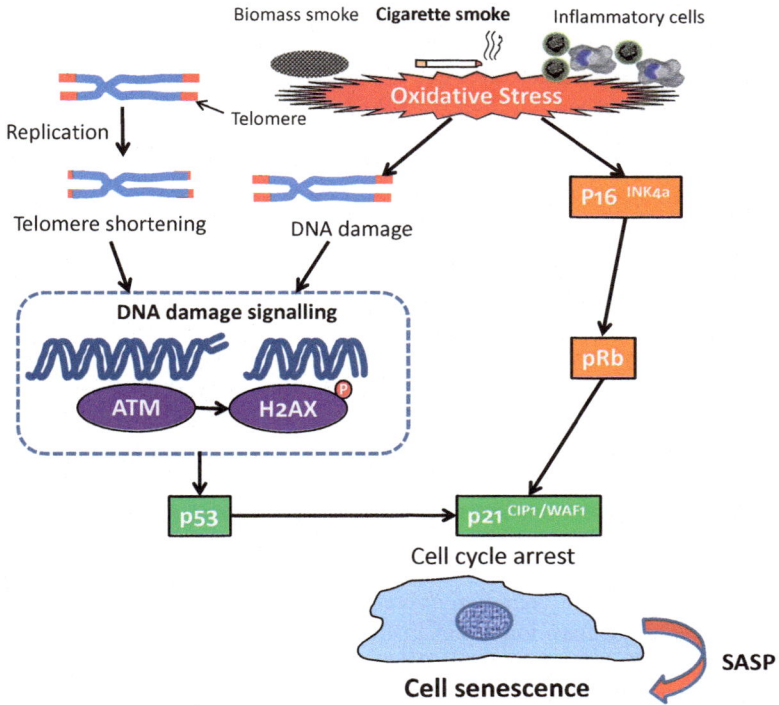

Fig. 3.2 *Cellular senescence pathways in lung disease.* Repeated cell division leads to progressive telomere shortening (replicative senescence) or DNA (particularly telomeres) is damaged by reactive oxygen species (ROS), activating DNA damage pathways, including the DNA repair kinase ataxia-telangectasia mutated kinase (ATM) which phosphorylates histone 2AX (H2AX), leading to activation of p53 and the activation of p21$^{CIP1/WAF1}$ which results in cell cycle arrest and senescence. Oxidative stress and other stresses acvtivate p16^{INK4a}, which is a cyclin inhibitor and phosphorylates retinoblastoma protein (pRb) which also activates the cyclin dependent kinase inhibitor p21. Senescent cells secrete cytokines and chemokines – the senescence-activated secretory phenotype (SASP), which amplifies and spreads cellular senescence

SASP response is triggered by p21, which activates p38 mitogen-activated protein (MAP) kinase and Janus-activated kinases (JAK). This results in the activation of the proinflammatory transcription factor nuclear factor-κB (NF-κB), resulting in secretion of cytokines, such as IL-1β, IL-6, TNF-α, growth factors, such as transforming growth factor (TGF)-β and chemokines, such as CXCL1, CXCL8 and CCL2, and elastolytic matrix metalloproteinases (MMP), such as MMP-2 and MMP-9, all of which are increased in diseases of accelerated ageing including COPD. Plasminogen activator inhibitor-1 (PAI-1) is another characteristic SASP protein and is increased in the sputum of COPD patients (To et al. 2013). This makes it likely that the chronic inflammation seen in COPD is a result of cellular senescence. CXCL8 acts via the chemokine receptor CXCR2, which induces cellular senescence and DNA damage, whereas blocking CXCR2 reduces both replicative and stress-induced senescence (Acosta et al. 2008). The SASP response itself mediates cellular senescence thus amplifying and spreading senescence. Activation of the p16^{INK4a} pathway also activates NADPH oxidases, resulting in increased oxidative stress, which further activates NF-κB (Takahashi et al. 2006). JAK inhibitors inhibit the SASP response and thereby reduce frailty in ageing mice (Xu et al. 2015). SASP is also inhibited by rapamycin through inhibition of MK2, which is downstream of p38 MAPK (Herranz et al. 2015). The SASP also includes activation of the NRLP3 inflammasome and secretion of IL-1β (Acosta et al. 2013).

There is an accumulating of senescent cells in murine lungs with age, which is associated with emphysematous changes and deposition of collagen (Calhoun et al. 2016). Elimination of senescent cells (positive for p19^{ARF}) appears to abrogate the decline in lung elasticity in ageing mice (Hashimoto et al. 2016). Cellular senescence is found in emphysematous lungs with enhanced expression of p21$^{CIP1/WAF1}$, p16^{INK4a} and senescence-associated β-galactosidase (SA-βGal) activity (Chilosi et al. 2013; Amsellem et al. 2011). Lung macrophages from COPD patients also express the senescence marker p21$^{CIP1/WAF1}$ (Tomita et al. 2002). Furthermore, in COPD there is an increase expression of SASP components, including IL-1, IL-6, TNF-α, CXCL8, CCL2, TBG-β, MMP2 and MMP9 (Barnes 2004) (Fig. 3.1). Similarly lung fibroblasts and alveolar epithelial cells from IPF patients also show cellular senescence, with increased expression of SA-βGal, p16, p21, p53 and the SASP response, that may important in the fibrotic process (Alvarez et al. 2017; Schafer et al. 2017; Minagawa et al. 2011).

Increased number of senescent airway epithelial cells that express p16 are also seen in cystic fibrosis (Fischer et al. 2013) and in bronchiectasis (Birch et al. 2016).

Senolytic Therapy

Removal of senolytic cells expressing p16^{INK4a} using a caspase-activated system extends the lifespan of mice and delays the development of organ failure (Baker et al. 2016a). Using this approach pulmonary fibrosis induced by bleomycin and its associated inflammatory response is prevented (Schafer et al. 2017). Senolytics are drugs that mimic this effect by inducing apoptosis in senescent cells and have little or no effect on proliferating cells (Kirkland et al. 2017). Several senolytic drugs have been identified by screening and include the naturally occurring polyphonic quercitin and the cytotoxic agent navitoclax (ABT263) which inhibits the Bcl2 family of anti-apoptotic proteins (Zhu et al. 2016). A combination of quercitin and navitoclax is effective in protecting against bleomycin-induced pulmonary fibrosis in mice (Schafer et al. 2017) and navitoclax also reduces the senescence associated with lung irradiation by selectively eliminating senescent type 2 pneumocytes (Pan et al. 2017). A senolytic combination of quercitin and dasitinib also reduced the numbers of senescent type 2 pneumocytes in bleomycin-induced fibrosis in mice, with reduced p16 and SASP proteins (Lehmann et al. 2017). Another approach is to target FOXO4, which is important in maintaining senescent cell viability through binding to p53. A cell penetrant D-retro-inverso peptide FOXO4-DRI interferes with this interaction so that cells enter apoptosis and *in vivo* it counteracts chemotherapy-induced senescence in normal and a strain of fast ageing mice (Baar et al. 2017). Several biotechnology and pharmaceutical companies are now searching for safe and effective senolytic therapies.

PI3K-mTOR Signaling

Signaling though the phosphoinositide-3-kinase (PI3K)-AKT-mammalian target of rapamycin (mTOR) pathway plays a key role in inducing cellular senescence and ageing and inhibiting this pathway extends the lifespan of many species, from yeast to mammals (Johnson et al. 2013). Rapamycin Inhibits mTOR and increases longevity in mice (Harrison et al. 2009). This may partly be due to a reduction in SASP, which reduces the paracrine spread of senescence (Laberge et al. 2015; Herranz et al. 2015). mTOR comprises two complexes, mTORC1 and mTORC2, with mTORC1 playing the most well defined role on growth and inhibited by rapamycin through its binding to the FK506-binding protein FKB12. mTORC1 is activated by growth factors, stress (including oxidative stress) and caloric nutrients, whereas mTORC2 is activated by growth factors. Inhibition of mTORC1 by rapamycin may results in up-regulation of mTORC2, which may reduce its efficacy so that dual inhibitors of mTORC1 and 2 might be more effective in chronic therapy. mTORC1 activation by oxidative and other cellular stresses and by nutrients results in increased protein synthesis through ribosomal S6 kinases with secretion of growth factors, such as TGF-β and vascular-endothelial growth factor (VEGF).

The mTOR pathway has multiple effects, including inhibition of FOXO transcription factors that are linked to longevity. There is evidence for PI3K activation in the lungs and cells of COPD patients as shown by increased expression of the downstream kinase phosphorylated AKT, which in turn activates mTOR (To et al. 2010) (Fig. 3.3). The mTORC1 complex is activated in COPD peripheral lung and peripheral blood mononuclear cells (PBMC), as shown by increased phosphorylated S6 kinase and is effectively inhibited by rapamycin (Mitani et al. 2016).

Not surprisingly, there are important endogenous inhibitory mechanisms to limit the activation of this pathway. PI3K is inhibited endogenously by the membrane tyrosine phosphatases phosphatase and tensin homologue (PTEN) and SH2-containing inositol-5'-phosphatase-1 (SHIP-1). Both have oxidation-susceptible cysteine residues in the active site that is readily modified by oxidative stress, and thus excessive oxidative stress reduces their catalytic activity (Worby and Dixon 2014). PTEN polymorphisms have been shown to be a genetic risk factor for COPD (Hosgood et al. 2009). PTEN activity and expression are markedly reduced in COPD lungs and this is likely to be due to oxidative stress and leads to activation of PI3K (Yanagisawa et al. 2017a).

mTOR is inhibited endogenously by 5'-adenosine monophosphate activating kinase (AMPK) which is activated by low ATP levels in the cell. AMPK is activated

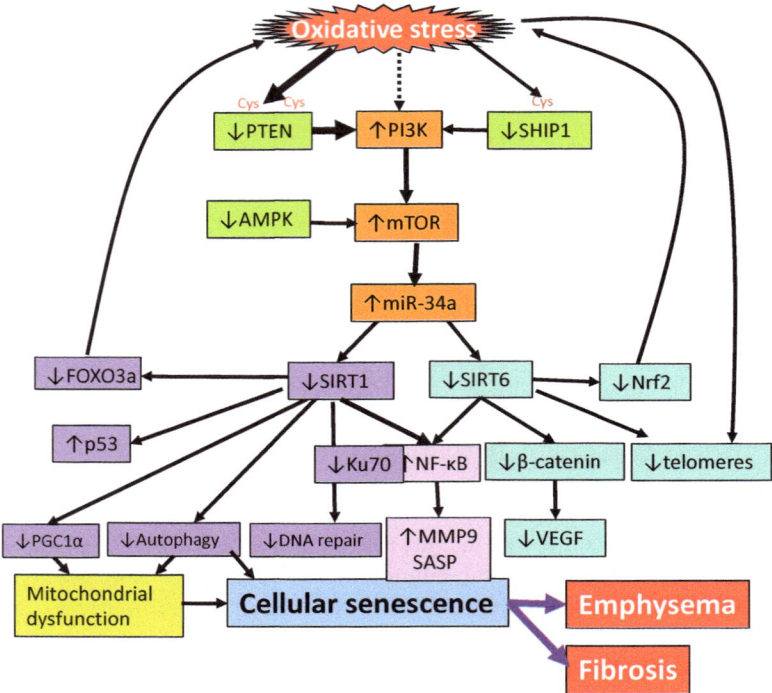

Fig. 3.3 *Reduced sirtuins and senescence*. Sirtuin-1 is reduced by reactive oxygen species (ROS) through the activation of phosphoinositide-3-kinase (PI3K), mammalian target of rapamycin (mTOR) and microRNA-34a (MiR-34a). This accelerates ageing through increased acetylation of several proteins, including Ku70 that is important in double-stranded DNA repair, peroxisome proliferator gamma coactivator-1 (PGC-1), resulting in mitochondrial dysfunction, nuclear factor-kappa B (NF-κB), which orchestrates the senescence-associated secretory phenotype (SASP) and matrix metalloproteinase-9 (MMP-9), forkhead transcription factor-3a (FOXO3a) which reduces antioxidants and further increases oxidative stress. Sirtuin-6 is also reduced by oxidative stress through the same pathway, including miR-34a and this leads to activation of NF-κB, reduced β-catenin leading to reduced vascular-endothelial growth factor (VEGF), reduced telomere length and reduced Nrf2, further increasing oxidative stress. The coordinated reduction of these sirtuins leads to cellular senescence, accelerated aging and emphysema

by the biguanide metformin, which therefore inhibits mTOR signaling and has been shown to extend the lifespan of several species, including mammals (Anisimov et al. 2008). AMPK is also activated by caloric restriction, also extends lifespan (Colman et al. 2014). This suggests that activation of the mTOR pathway may play an important role in COPD and inhibition of this pathway offers a future therapeutic opportunity (Lamming et al. 2013).

Mitochondrial Dysfunction

Mitochondrial dysfunction is an important feature of ageing and mitochondria are an important intracellular source of ROS (Correia-Melo and Passos 2015). There is a gradual accumulation of mutations in mitochondrial DNA during ageing, with a reduced resistance to damage by oxidative stress (Zheng et al. 2012). Mitochondrial function is regulated by the transcription factors peroxisome proliferator-activated receptor-gamma co-activator PGC-1α and PGC-1β, which are inhibited by telomere shortening and by mTOR activation so that with senescence there is a reduction in mitochondrial membrane potential, fragmentation and reduced mitochondria numbers. In turn mitochondria derived ROS may maintain DNA damage, thus driving further senescence through feedback loops. Damaged mitochondria are removed by autophagy (mitophagy) which is impaired in senescent cells (Garcia-Prat et al. 2016). Mitochondria play an important role in senescence and in driving the SASP response in lung disease (Birch et al. 2017).

COPD is associated with increased mitochondrial ROS production, decreased intracellular anti-oxidants and reduced numbers of mitochondria (Kirkham and Barnes 2013; Sureshbabu and Bhandari 2013). COPD cells show increased mitochondrial ROS generation (Wiegman et al. 2015). Mitochondrial dysfunction is also found in skeletal muscle of COPD patients with muscle weakness, with increased mitochondrial ROS production and decreased numbers (Meyer et al. 2013). Prohibitins (PHB), which are localized to the inner membrane of mitochondria, play a role in mitochondrial biogenesis and maintaining normal function (Artal-Sanz and Tavernarakis 2009). However, PHB1 expression is reduced in epithelial cells of smokers and to a greater extent in COPD patients, suggesting a mechanism for mitochondrial dysfunction in COPD patients (Soulitzis et al. 2012). Mitochondrial fragmentation with increased expression of mitochondrial fission/fusion markers, oxidative phosphorylation (OXPHOS) proteins (Complex II, III and V) are found in epithelial cells from COPD patients to a greater extent than cells from normal smokers, along with SASP expression with increased IL-1β, IL-6, CXCL8 secretion (Hoffmann et al. 2013). The mitochondrial stress markers Parkin and PTEN-induced protein kinase-1 (PINK1) are also increased in COPD patients (Hoffmann et al. 2013; Mizumura et al. 2014). Knockdown of the PINK1 protects mice against mitochondrial oxidative stress induced by cigarette smoke (Mizumura et al. 2014). Damaged mitochondria are normally efficiently removed by mitophagy through highly regulated pathways, including PTEN, which may be reduced by oxidative stress and PINK1. Failure of mitophagy leads to generation of intracellular ROS and eventually cell death. PGC-1α, which is a key regulator of mitochondrial biogenesis and generation of mitochondrial ROS is increased in epithelial cells of mild COPD patients but reduced with increasing COPD severity (Li et al. 2010).

Mitochondrial dysfunction is also evident in IPF and type 2 pneumocytes show the accumulation of abnormal and dysfunctional mitochondria associated with reduced PINK1 and increased expression of profibrotic mediators (Bueno et al. 2015). PINK1 gene knockout in mice is linked to increased susceptibility to lung fibrosis.

Defective Autophagy

Autophagy is a highly regulated process that removes degraded proteins, damaged organelles (including mitochondria, as discussed above) and foreign organisms (such as bacteria), in order to maintain normal cellular function. Defective autophagy is a key characteristic of ageing cells and age-related diseases, including COPD (Mizumura et al. 2012; Mizushima et al. 2008). Autophagy removes damaged proteins, organelles and pathogens via lysosomal degradation. However, ageing cells accumulate damaged and misfolded proteins through a decline in autophagy, leading eventually to cellular senescence. Although autophagy plays a protective role in response to exogenous stress, prolonged and excessive autophagy has been associated with cell death when cytosol and organelles are destroyed irreparably. Inhibition of autophagy increases susceptibility to oxidative damage and apoptosis, whereas activation of autophagy leads to inhibition of apoptosis (Murrow and Debnath 2013). Cigarette smoke and oxidative stress activate autophagy, indicating that autophagy may be a protective response designed to remove damaged proteins and organelles. Alveolar macrophages from smokers show defective autophagy and this could contribute to accumulation of aggregates, abnormal mitochondrial function and defective clearance of bacteria (Monick et al. 2010). Patients and mice with emphysema show increased autophagy markers in lung tissue, with increased activation of autophagic proteins, such as light chain-3 (LC3-II), and autophagy may be contributory to apoptosis and alveolar cell destruction (Chen et al. 2010). Other studies have demonstrated increased autophagic vacuoles (autophagosomes) in COPD. However, it is possible that completion of autophagy may not occur (defective autophagic flux). Macrophages from smokers show defective autophagic flux, resulting in accumulation of the substrate of autophagy, p62, and misfolded proteins due to dysfunctional lysosomal digestion of the autophagosomal load caused by a reduction in the lysosomal protein LAMP2 (Monick et al. 2010). Inhibition of autophagy after cigarette smoke exposure leads to accumulation of p62 and ubiquitinated proteins in airway epithelial cells, resulting in increased cellular senescence and SASP with secretion of CXCL8, thereby mimicking the changes seen in COPD cells (Fujii et al. 2012). Loss of autophagy might account for reduced mitophagy and also contribute to defective phagocytosis of bacteria in COPD (Donnelly and Barnes 2012). Autophagy may be impaired through the activation of PI3K-mTOR signaling, resulting in inhibition of unc-51 like autophagy-activating kinase-1 (ULK1) complex that normally activates autophagy, thereby linking defective autophagy to the accelerated ageing mechanisms (Dunlop and Tee 2014). Defective autophagy and mitochondrial function play a key role in IPF and the generation of profibrotic mediators, such as TGF-β (Mora et al. 2017).

Stem Cell Exhaustion

The clearance of senescent cells means that progenitor cells are needed to maintain cell numbers, but this may become less efficient or may exhaust the regenerative capacity of stem cells. Stem cell exhaustion and depletion are characteristic feature of ageing (Lopez-Otin et al. 2013).

Senescence of mesenchymal stem cells (fibroblast, endothelial cells) may be a mechanism of emphysema and the failure to repair lung injury. Damaged alveolar cells may be replaced by progenitor cells, particularly alveolar type 2 (AT2) cells (Kubo 2012). However, once the alveolar architecture is destroyed by elastases, such as MMP9, progenitors cannot rebuild the appropriate functional lung structure. In stem cells, ROS force cells out of quiescence and into a more proliferative state by activating PI3K/AKT signaling and further promoting the production ROS, thus repressing FOXO-mediated stress response and autophagy. AT2 cells are progenitors for type I alveolar cells and show evidence of senescence in patients with emphysema (Tsuji et al. 2006).

Circulating endothelial progenitor cells (EPC) and specifically blood outgrowth endothelial cells are made in the bone marrow and are important in maintaining endothelial integrity (Fadini et al. 2012). EPC from smokers and COPD patients show cellular senescence and increased DNA double-strand breaks compared to cells from non-smokers and this is correlated with reduced expression of sirtuin-1 (Paschalaki et al. 2013). These senescent stem cells are poorly effective in repairing endothelial damage and provide a link between COPD and vascular ageing that leads to ischaemic heart disease.

In diseases of accelerated ageing, stem cells, including AT2 cells and EPCs, show features of cellular senescence, oxidative stress-induced DNA damage. This leads to loss of regenerative capacity and eventually to organ failure, including lungs (Sousounis et al. 2014). Understanding the molecular mechanisms of stem cell ageing is critical to elucidating accelerated ageing in the lungs (Oh et al. 2014). Senescent stem cells characteristically show increased generation of mitochondrial ROS, mitochondrial damage and dysfunction, activation of the PI3K-mTOR signaling, DNA damage with double-stranded DNA breaks and a defect in the DNA damage-sensing kinase ATM (ataxia telangiectasia mutated), defective proteostasis and deficiency of anti-ageing molecules, such as sirtuin-1 and FOXO transcription factors. EPCs from COPD patients show activation of PI3K, reduced ATM and reduced sirtuin-1, all of which may be reversible by resveratrol, a sirtuin activator (Paschalaki et al. 2013). Accumulation of various types of DNA damage in stem cells with age may be a major mechanism leading to progressive failure of stem cells to repair damage of organs and thus to slowly progressive organ failure (Behrens et al. 2014). Epigenetic changes in stem cells, including DNA methylation and histone acetylation, also play an important role in stem cell senescence.

MicroRNA and Lung Ageing

MicroRNA (miRNA) are small non-coding single-stranded RNAs of 18-22 nucleotides that regulate the post-transcriptional expression of genes by inhibiting their translation or inducing degradation of mRNAs by binding to complimentary sequences in the 3'-UTR of mRNAs. There is increasing evidence that miRNAs play an important role in in the ageing process and may regulate several key proteins involved in cellular senescence, such as p16 (miR-24), p53 (miR-885-5p), SASP (miR-146) and reduced sirtuin-1 expression (miR-34a) (Gorospe and Abdelmohsen 2011; Munk et al. 2017). Genome-wide assessment of miRNA expression in monocytes of elderly compared to young individuals show that miRNA tend to decrease with age and regulate PI3K-mTOR signaling and DNA repair (Noren Hooten et al. 2010). MiRNAs that have been implicated in cellular senescence are also increased in diseases of accelerated ageing, including COPD (Dimmeler and Nicotera 2013).

MiR-34a, which is known to down-regulate sirtuin-1, shows increased expression in peripheral lungs and epithelial cells of COPD patients and is correlated with increased markers of cellular senescence in lung cells (Baker et al. 2016b). MiR-34a also regulates sirtuin-6, which is reduced in COPD along with sirtuin-1 (Nakamaru et al. 2009). MiR-34a is increased by oxidative stress through activation of PI3K-mTOR signaling pathways and this leads to a parallel reduction in sirtuin-1 and sirtuin-6, whereas other sirtuins are unchanged. Importantly, an antagomir of miR-34a restores sirtuin-1 and -6 in small airway epithelial cells from COPD patients, reduces markers of cellular senescence (p16, p21, p53), reduces the SASP response (IL-1β, IL-6, TNF-α, CCL2, MMP9) and increases the growth of senescent epithelial cells (Baker et al. 2016b). This suggests that blocking specific miRNA may result in rejuvenation of these cells. MiR-34a is also increased in macrophages of COPD patients and this may be associated with the impaired phagocytosis and uptake of apoptotic cells (efferocytosis) that is seen in COPD patients (McCubbrey et al. 2016). MiR-126, which plays a key role in maintain vascular integrity and endothelial cell function, is reduced in endothelial progenitor cells and airway epithelial cells and from COPD patients and plays a key role in regulating the DNA damage response pathway that is linked to cellular senescence (Paschalaki et al. 2018).

MiRNAs are exported from cells in extracellular vesicles, which include microvesicles and exosomes and then may be taken up by other cells (Fujita et al. 2015; Turturici et al. 2014). This might provide a means of spreading senescence from cell to cell within the lung, but also via the circulation to other organs, resulting in accelerated ageing in other systems or multimorbidity (Barnes 2015; Kadota et al. 2018). Thus, targeting miRNAs may be a future therapeutic strategy for treating and preventing cellular senescence and delivery of miRNA via extracellular vesicles and nanoparticles might be a novel way to deliver there therapies (Jiang and Gao 2017).

Epigenetic Mechanisms

There is little evidence that genes have important effects on the ageing process, but increasing evidence that epigenetic mechanisms may be important, with long-term changes in gene expression due to environmental exposures. These epigenetic changes include DNA methylation and modifications in histone proteins (for example, by acetylation, methylation, phosphorylation), which result in increased or decreased gene transcription (Song and Johnson 2018). Chromatin structure changes with age and becomes more "open", which means it is more transcriptionally active (Feser and Tyler 2011). There is a decline in histone proteins with ageing and consistent changes in histone modifications and associated modifying enzymes, with increased histone acetylation and either increased or decreased histone methylation. Changes in histone methylation with age are likely to be important in accelerated ageing and changes in trimethylation of lysine 4 on histone-3 (H3K4me3) occur during ageing and regulate cellular senescence (Zhang et al. 2014). At present there is little information about histone methylation status in diseases of accelerated ageing, such as COPD. Histone methylation plays an important role in regulating the various proteins involved in autophagy, DNA repair and SASP. The role of histone methyltransferases and demethylases in the regulation of cellular senescence is under intense investigation now that selective enzymes inhibitors are in development. Coactivator-associated arginine methyltransferase-1 (CARM1) deficiency causes cellular senescence in type 2 pneumocytes and accelerates the development of emphysema in mice, indicating a key role for histone methylation in protecting against lung ageing (Sarker et al. 2015). Changes in DNA methylation have been described in COPD cells and may play a role in cellular senescence (Clifford et al. 2018). For example, DNA methylation of the FOXA2 and SPDEF genes in airway epithelial cells of COPD patients has been linked to mucus hypersecretion (Song et al. 2017).

Epigenetic changes associated with ageing are of particular relevance in stem cell populations and lead to dysfunction and depletion (Armstrong et al. 2014). Mesenchymal stem cells from elderly people appear to have less potential for differentiating into different cells types and this is linked to changes in DNA methylation (Horvath 2013).

Immunosenescence

The immune system becomes less effective during ageing, with impaired immune responses to new antigens, a greater susceptibility to infection and an increased tendency to develop autoimmunity. Immunosenescence affects both innate and adaptive immunity resulting in loss of function and has been implicated in chronic inflammatory diseases. Immunosenescence is important in several pulmonary diseases, including COPD, severe asthma and IPF (Murray and Chotirmall 2015). Ageing of innate immunity leads to defective function in innate cells, with impaired cell migration and signaling through pattern recognition receptors, such as Toll-like receptors (TLR) (Hazeldine and Lord 2015). Neutrophils show decreased phagocytosis, chemotaxis and apoptosis and have reduced directional function that is corrected by PI3K inhibition (Sapey et al. 2014). This defect is exaggerated in patients with COPD (Sapey et al. 2011). Macrophages show reduced phagocytosis with age and this is increased to a greater extent in COPD macrophages (Donnelly and Barnes 2012). Dendritic cells produce less interferon and reduced TLR signaling, which is associated with decreased T-cell-mediated immunity, resulting in reduced ability to protect against pathogens in the elderly and an increased risk of carcinogenesis (Mahbub et al. 2011). However in patients with severe COPD there is greater expression of TLR4, which is associated with an amplified inflammatory response to colonizing bacteria (Di Stefano et al. 2017).

Defects in adaptive immunity are also commonly seen in COPD and may be related to ageing. There is a loss of naïve T cells and B cells as the thymus involutes with age, as well as telomere shortening, with consequent reduced responses to new antigens. A characteristic of T cells from elderly individuals is a decrease in $CD4^+/CD8^+$ ratios and a loss of the co-stimulatory molecule CD28, with an increase in both $CD4^+CD28^{null}$ and $CD8^+CD28^{null}$ cells, which have reduced immune and vaccine responses (Arnold et al. 2011). In COPD CD8+CD28^{null} cells have reduced HDAC2 expression and are corticosteroid-resistant (Hodge et al. 2015).

Autoimmunity is also a manifestation of immunosenescence with increased production of autoantibodies that may lead to further tissue damage. In COPD there are increased autoantibodies directed against endothelial cells and against carbonylated proteins formed by exposure to oxidative stress in severe disease (Feghali-Bostwick et al. 2008; Karayama et al. 2010; Kirkham et al. 2011). This increase in autoimmunity has been associated with an imbalance between Th17 and regulatory T cells (T_{reg}). An increased ratio of Th17 to T_{reg} cells is found in sputum In COPD patients (Maneechotesuwan et al. 2013). T_{reg} cells may even transform into Th17 cells under disease conditions with loss of the key T_{reg} transcription factor Foxp3 (Komatsu et al. 2014).

There is evidence for activation of the PI3K signaling pathway. In ageing neutrophils the impaired chemotaxis response is associated with increased PI3K and restored by a PI3K inhibitor (Sapey et al. 2014). A rapamycin analog, everolimus, reduces immunosenescence in normal elderly individuals, with an increased antigenic response to influenza vaccination and a reduction in T cells expressing programmed death-1 (PD-1) receptors that are increased in senescent $CD4^+$ and $CD8^+$ cells (Mannick et al. 2014).

Reduced Anti-ageing Molecules

Many endogenous anti-ageing molecules that counteract the mechanisms of senescence and a reduction in their expression may accelerate the ageing process. Defective anti-ageing molecules have been suggested as a mechanism for accelerated lung ageing in COPD and other pulmonary diseases (Ito and Barnes 2009; Faner et al. 2012; Mercado et al. 2015).

Sirtuins

Silent information regulators or sirtuins are well recognized as anti-ageing molecules that regulate lifespan. Sirtuins are highly conserved NAD^+-dependent enzymes that play a role in resistance to stress, genomic stability and energy metabolism (Finkel et al. 2009; Watroba et al. 2017). Of the 7 sirtuins found in mammals most attention has centered on sirtuin-1 and sirtuin-6, as both are associated with prolongation of lifespan in many species, including mammals (Guarente 2011). Sirtuin-1 deacetylates many key regulatory proteins and transcription factors that are known to be involved in DNA repair, inflammation, antioxidant gene expression and cellular senescence, including the PI3K-AKT-mTOR pathway and autophagy (Fig. 3.3). Sirtuin-1 deacetylates the transcription factor FOXO3a, which enhances antioxidant responses (particularly superoxide dismutases) and inhibits p53-induced cellular senescence, inhibits NF-κB leading to suppression of inflammation and activates PGC-1α, which maintains mitochondrial function. Sirtuin-1 is markedly reduced in peripheral lung and circulating PBMC of patients with COPD (Nakamaru et al. 2009; Rajendrasozhan et al. 2008; Baker et al. 2016b). Sirtuin-1 is reduced by oxidative stress via reduction in PTEN and activation of the PI3K-mTOR pathway and in turn sirtuin-1 inhibits mTOR signaling. Sirtuin-1 also activates autophagy by inhibiting mTOR (Lee et al. 2008). The polyphenol resveratrol, found in the skin of red fruits and in red wine, activates sirtuin-1, but is poorly bioavailable, so this has led to the development of more potent sirtuin-1 activating compounds (STACs), which are now in clinical development for the treatment of age-related diseases (Hubbard and Sinclair 2014; Bonkowski and Sinclair 2016). Sirtuin-6 is an ADP-ribosylase as well as protein deacetylase and plays a key role in regulating DNA repair, telomere maintenance, inflammation and metabolic homeostasis and, like sirtuin-1, is linked to extension of lifespan (Kugel and Mostoslavsky 2014). Sirtuin-6 expression is reduced in the lungs of patients with COPD (Nakamaru et al. 2009; Takasaka et al. 2014). Reduced sirtuin-6 is found in airway epithelial cells of COPD patients and is reduced by cigarette smoke exposure, resulting in cellular senescence and impaired autophagy (Takasaka et al. 2014). As discussed above, oxidative stress activates PI3K-mTOR signaling, which increases the expression of miR-34a that directly inhibits sirtuin-1 and sirtuin-6 mRNA and protein expression without any effects on the other sirtuins (Baker et al. 2016b). Circulating sirtuin-1 concentrations are also reduced in patients with COPD (Yanagisawa et al. 2017b).

The role of sirtuins in other lung diseases is less well defined. Reduced sirtuin-1 is found in PBMC from elderly patients with severe asthma and appears to increase IL-4 secretion though increased acetylation of the transcription factor GATA-3 (Colley et al. 2016). Sirtuin-1 knockout in mice enhances lung fibrosis, and sirtuin-1 inhibits TGF-β fibroblast activation. Paradoxically sirtuin-1 is increased in the lungs of IPF patients, perhaps as a compensatory mechanisms (Zeng et al. 2017).

Klotho

Animal models of accelerated ageing have identified other key molecules involved in senescence, Klotho is defective in a mouse model of premature ageing, which has a substantially decreased lifespan, with features of accelerated ageing, including emphysema (Dermaku-Sopjani et al. 2013). Klotho is a transmembrane protein that is a co-receptor of fibroblast growth factor (FGF23) and regulates insulin/IGF signaling, phosphate homeostasis, cell survival and proliferation. It is protective against oxidative stress and decreased expression of Klotho is found in epithelial cells of COPD patients and is reduced by oxidative stress, resulting in increased release of inflammatory cytokines from these cells (Gao et al. 2015). Circulating concentrations of Klotho are also reduced in patients with COPD (Kureya et al. 2016).

SMP-30

Senescence marker protein-30 (SMP30) is an anti-ageing molecule that was identified in a mouse model of ageing, which regulates calcium homeostasis and is also sensitive to oxidative stress (Feng et al. 2004). Reduced expression of SMP30 is found in aged tissues, including lung. In SMP30 knockout mice there is a increased susceptibility to development of emphysema after exposure to cigarette smoke (Sato et al. 2006).

Implications for Therapy

Considerable progress has been made in understanding the molecular mechanisms involved in accelerated ageing in chronic lung diseases. Although it is unrealistic (but not impossible) to expect reversal of the normal ageing process, it may be possible to reduce the mechanisms that accelerate senescence in these diseases. Several novel therapeutic targets that have been identified leading to the development of geroprotectors (Moskalev et al. 2017). Existing drugs (such as metformin and rapamycin) may be repurposed and novel drugs developed by screening and rational drug design. The potential for these drugs is that they may treat multimorbidities, which are diseases of ageing that occur together in the elderly, as there appear to be common pathways (Barnes 2015). Future therapies may also involve lifestyle interventions, such as diet and increased physical activity (de Cabo et al. 2014).

Pharmacological Therapies

Better understanding of cellular senescence mechanisms has identified several potential therapies for inhibiting accelerated ageing in lung disease (Fig. 3.4) (Ito et al. 2012). As discussed above, the PI3K- mTOR pathway plays a key role in cellular senescence and inhibition of autophagy, and inhibitors of this pathway may extend lifespan in several species, including mammals. Rapamycin induces autophagy, increases sirtuin-1 and extends the lifespan of all organisms, including mammals (Bjedov and Partridge 2011; Lamming et al. 2013). Rapamycin is very effective in inhibiting activated mTOR in COPD cells *in vitro* (Mitani et al. 2016). However, rapamycin and its analogues (rapalogs) have several adverse effects, including mouth ulceration, anaemia, pneumonitis and delayed wound healing, making it unsuitable for long-term use. A pilot study in elderly humans demonstrated its potential in lower doses in reversing immunosenescence and this was not associated with significant adverse effects (Mannick et al. 2014). Metformin is widely used to treat type 2 diabetes and indirectly activates AMPK, resulting in inhibition of mTOR and extension of lifespan in mice, possibly though increasing Nrf2-induced antioxidant gene expression (Anisimov et al. 2008; Martin-Montalvo et al. 2013). Metformin is relatively well-tolerated, it might be a suitable therapy for treating accelerated ageing in COPD and other pulmonary diseases. It has been shown to be well-tolerated in COPD patients (Hitchings et al. 2016). PI3K signaling may also be inhibited by activators of the endogenous inhibitor SHIP-1 (Stenton et al. 2013) and these drugs have already entered into clinical trials for treatment of allergic disease (Leaker et al. 2014).

Naturally occurring molecules may also be effective and may be obtained via dietary supplementation (neutraceuticals). Quercitin, a polyphenol found in apples, is also an activator of AMPK (Mitani et al. 2017) as well as a senolytic (Schafer et al. 2017). Resveratrol, found in the skins of red fruit and in red wine and increases lifespan in several species, including mice, through the activation of sirtuin-1. In vitro resveratrol inhibits the SASP response in COPD epithelial cells (Culpitt et al. 2003) and is effective against neutrophilic inflammation induced by lipopolysaccharide in mice (Birrell et al. 2005). However, resveratrol and related dietary polyphenols have poor oral bioavailability, rapid metabolism and low potency, which has led to the development of novel potent synthetic analogs known as STACs, which work on an allosteric mechanism to stimulate sirtuin-1 activity (Hubbard and

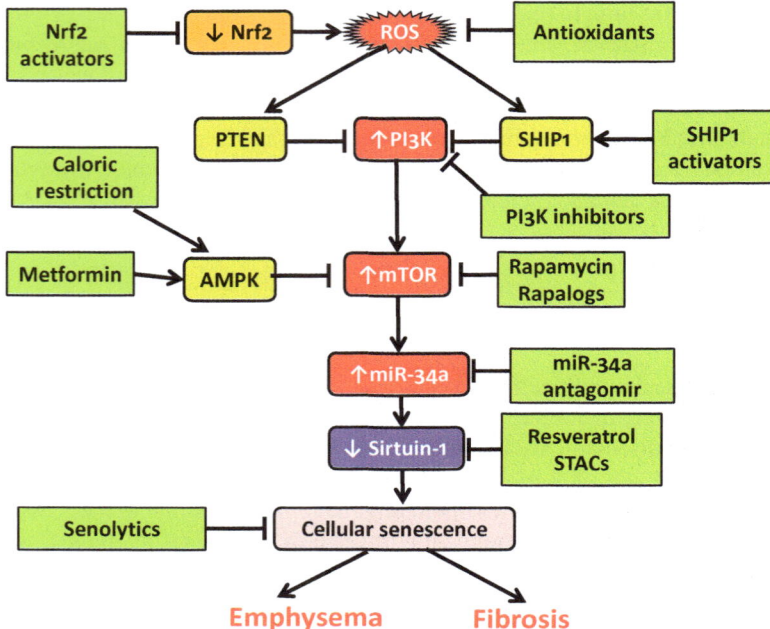

Fig. 3.4 *Targeting cellular senescence in lung disease.* The PI3K-mTOR pathway driven by reactive oxygen species (ROS), which activate phosphoinositide-3-kinase (PI3K) and mammalian target of rapamycin (mTOR), which activate microRNA-34a (miR-34a) to inactivate and reduce sirtuin-1 activity, which accelerates the ageing process. There are several endogenous and exogenous inhibitors of this pathway. Increased ROS may be the result of reduced activity of the antioxidant regulating transcription factor Nrf2. There are several ways to modulate this pathway (green boxes). Nrf2 may be activated by small molecule activators, ROS may also be counteracted by antioxidants. PI3K is inhibited by the phosphatases PTEN (phosphatase and tensin homolog) and SHIP-1 (SH2-containing inositol-5'-phosphatase-1). SHIP-1 activators are now in development and there are several PI3K inhibitors. mTOR is inhibited by rapamycin and its less toxic analogs and endogenously by AMP-kinase (AMPK), which may be activated by caloric restriction and by metformin. Increased miR-34a may be targeted specifically by antagomirs. Sirtuin-1 may be activated by resveratrol and novel sirtuin-activating compounds (STACs). Finally senescent cells may be removed by senolytic drugs

Sinclair 2014; Bonkowski and Sinclair 2016). In mice exposed to cigarette smoke the STAC SRT-2171 prevents the increase in MMP-9 that is associated with neutrophilic inflammation and improves lung function (Nakamaru et al. 2009). In addition, resveratrol reduces senescence in EPCs from COPD patients through an increase in sirtuin-1, indicating that exhausted stem cells may be an important target of STACs (Paschalaki et al. 2013). A resveratrol derivative isorhapontigenin, which is found in Chinese herbal remedies used to treat inflammatory diseases, is more potent than resveratrol at inhibiting the SASP response in COPD cells and has a better oral bioavailability (Yeo et al. 2017).

Oxidative stress is an important driving mechanism leading to accelerated ageing in lung disease, suggesting that antioxidants should be effective therapies. However, existing antioxidants, such as N-acetyl cysteine (NAC), are poorly effective as they are thiol derivatives that become inactivated by oxidative stress. Novel antioxidants include NADPH oxidase (NOX) inhibitors, superoxide dismutase mimetics and activators of Nrf2 (Saso and Firuzi 2014). Since there is strong evidence for mitochondria-derived oxidative stress in age-related lung diseases an intracellular antioxidant may be more effective. The mitochondrial antioxidant SkQ1 reverses ageing-related biomarkers in rats, whereas NAC is ineffective (Kolosova et al. 2012). Nrf2 activators are of particular interest as Nrf2 may be defective in several disease models of accelerated ageing, including COPD. Sulforaphane, which occurs naturally in broccoli, is an Nrf2 activator but is non-specific and toxic in high concentrations, leading to a search for new drugs. Bardoxelone methyl is more potent but also has shown toxicity in clinical studies.

Lifestyle Interventions

Caloric restriction (CR) has been shown to prolong lifespan in species from yeast to mammals, including primates (Colman et al. 2014). This leads to inhibition of PI3K-AKT-mTOR signaling though activation of AMPK and reduces the release of insulin and insulin-like growth factor(IGF)-1 and increases sirtuin-1 (Fontana et al. 2010). Several intermittent fasting regimes that give sufficient CR to activate anti-ageing pathways are now being explored. Periodic fasting has been found to be effective in several animal models of age-related diseases but has not been explored in age-related lung diseases. The Mediterranean diet, is rich in fruit, vegetables, red wine and olive oil, contains flavones, polyphenols and stilbenes that may activate sirtuin-1 and increases healthy life, with reduced incidence of neurodegenerative, cardiovascular and metabolic diseases and cancer (Perez-Lopez et al. 2009). For example, a Mediterranean diet may improve the function of EPCs from elderly patients and thus improve endothelial function in cardiovascular disease (Marin et al. 2013). Several compounds that occur in foodstuffs, as outlined above, may also reduce senescence.

Physical inactivity is a risk factor for the development of diseases of ageing, such as COPD, and is an important determinant of mortality (Gulsvik et al. 2012). Aerobic exercise training provides significant clinical benefits in several age-related biomarkers, including lipid profiles, blood pressure, glucose tolerance, bone density, depression, loss of skeletal muscle (sarcopenia) and quality of life (Fleg 2012). Exercise is the major component of pulmonary rehabilitation in COPD, which results in improved lung function, reduced exacerbations and improved quality of life, although it is not certain whether it is able to reduce mechanisms of accelerated ageing (Casaburi and ZuWallack 2009).

References

Acosta JC, O'Loghlen A, Banito A, Guijarro MV, Augert A, Raguz S, Fumagalli M, Da Costa M, Brown C, Popov N, Takatsu Y, Melamed J, d'Adda di Fagagna F, Bernard D, Hernando E, Gil J (2008) Chemokine signaling via the CXCR2 receptor reinforces senescence. Cell 133:1006–1018

Acosta JC, Banito A, Wuestefeld T, Georgilis A, Janich P, Morton JP, Athineos D, Kang TW, Lasitschka F, Andrulis M, Pascual G, Morris KJ, Khan S, Jin H, Dharmalingam G, Snijders AP, Carroll T, Capper D, Pritchard C, Inman GJ, Longerich T, Sansom OJ, Benitah SA, Zender L, Gil J (2013) A complex secretory program orchestrated by the inflammasome controls paracrine senescence. Nat Cell Biol 15:978–990

Alder JK, Chen JJ, Lancaster L, Danoff S, Su SC, Cogan JD, Vulto I, Xie M, Qi X, Tuder RM, Phillips JA 3rd, Lansdorp PM, Loyd JE, Armanios MY (2008) Short telomeres are a risk factor for idiopathic pulmonary fibrosis. Proc Natl Acad Sci U S A 105:13051–13056

Alder JK, Guo N, Kembou F, Parry EM, Anderson CJ, Gorgy AI, Walsh MF, Sussan T, Biswal S, Mitzner W, Tuder RM, Armanios M (2011) Telomere length is a determinant of emphysema susceptibility. Am J Respir Crit Care Med 184:904–912

Alvarez D, Cardenes N, Sellares J, Bueno M, Corey C, Hanumanthu VS, Peng Y, D'Cunha H, Sembrat J, Nouraie M, Shanker S, Caufield C, Shiva S, Armanios M, Mora AL, Rojas M (2017) IPF lung fibroblasts have a senescent phenotype. Am J Phys Lung Cell Mol Phys 313:L1164–l1173

Amsellem V, Gary-Bobo G, Marcos E, Maitre B, Chaar V, Validire P, Stern JB, Noureddine H, Sapin E, Rideau D, Hue S, Le Corvoisier P, Le Gouvello S, Dubois-Rande JL, Boczkowski J, Adnot S (2011) Telomere dysfunction causes sustained inflammation in chronic obstructive pulmonary disease. Am J Respir Crit Care Med 184:1358–1366

Anisimov VN, Berstein LM, Egormin PA, Piskunova TS, Popovich IG, Zabezhinski MA, Tyndyk ML, Yurova MV, Kovalenko IG, Poroshina TE, Semenchenko AV (2008) Metformin slows down aging and extends life span of female SHR mice. Cell Cycle 7:2769–2773

Armanios M (2012) Telomerase and idiopathic pulmonary fibrosis. Mutat Res 730(1-2):52–58

Armstrong L, Al-Aama J, Stojkovic M, Lako M (2014) Concise review: the epigenetic contribution to stem cell ageing: can we rejuvenate our older cells? Stem Cells (Dayton, Ohio) 32(9):2291–2298. https://doi.org/10.1002/stem.1720

Arnold CR, Wolf J, Brunner S, Herndler-Brandstetter D, Grubeck-Loebenstein B (2011) Gain and loss of T cell subsets in old age–age-related reshaping of the T cell repertoire. J Clin Immunol 31(2):137–146

Artal-Sanz M, Tavernarakis N (2009) Prohibitin and mitochondrial biology. Trends Endocrinol Metab 20:394–401

Baar MP, Brandt RMC, Putavet DA, Klein JDD, Derks KWJ, Bourgeois BRM, Stryeck S, Rijksen Y, van Willigenburg H, Feijtel DA, van der Pluijm I, Essers J, van Cappellen WA, van IWF, Houtsmuller AB, Pothof J, de Bruin RWF, Madl T, Hoeijmakers JHJ, Campisi J, de Keizer PLJ (2017) Targeted apoptosis of senescent cells restores tissue homeostasis in response to chemotoxicity and aging. Cell 169:132–147.e116

Baker DJ, Wijshake T, Tchkonia T, LeBrasseur NK, Childs BG, van de Sluis B, Kirkland JL, van Deursen JM (2011) Clearance of p16Ink4a-positive senescent cells delays ageing-associated disorders. Nature 479:232–236

Baker DJ, Childs BG, Durik M, Wijers ME, Sieben CJ, Zhong J, Saltness RA, Jeganathan KB, Verzosa GC, Pezeshki A, Khazaie K, Miller JD, van Deursen JM (2016a) Naturally occurring p16(Ink4a)-positive cells shorten healthy lifespan. Nature 530:184–189

Baker J, Vuppusetty C, Colley T, Papaioannou A, Fenwick P, Donnelly L, Ito K, Barnes PJ (2016b) Oxidative stress dependent microRNA-34a activation via PI3Kα reduces the expression of sirtuin-1 and sirtuin-6 in epithelial cells. Sci Rep 6:35871

Barnes PJ (2004) Mediators of chronic obstructive pulmonary disease. Pharm Rev 56:515–548

Barnes PJ (2009) Role of HDAC2 in the pathophysiology of COPD. Annu Rev Physiol 71:451–464

Barnes PJ (2014) Cellular and molecular mechanisms of chronic obstructive pulmonary disease. Clin Chest Med 35:71–86

Barnes PJ (2015) Mechanisms of development of multimorbidity in the elderly. Eur Respir J 45:790–806

Barnes PJ (2016) Inflammatory mechanisms in COPD. J Allergy Clin Immunol 138(1):16–27

Barnes PJ, Burney PGJ, Silverman EK, Celli BR, Vestbo J, Wedzicha JA, Wouters EFM (2015) Chronic obstructive pulmonary disease. Nat Rev Primers 1:1–21

Behrens A, van Deursen JM, Rudolph KL, Schumacher B (2014) Impact of genomic damage and ageing on stem cell function. Nat Cell Biol 16:201–207

Belsky DW, Shalev I, Sears MR, Hancox RJ, Lee Harrington H, Houts R, Moffitt TE, Sugden K, Williams B, Poulton R, Caspi A (2014) Is chronic asthma associated with shorter leukocyte telomere length at midlife? Am J Respir Crit Care Med 190:384–391

Birch J, Anderson RK, Correia-Melo C, Jurk D, Hewitt G, Marques FM, Green NJ, Moisey E, Birrell MA, Belvisi MG, Black F, Taylor JJ, Fisher AJ, De Soyza A, Passos JF (2015) DNA damage response at telomeres contributes to lung aging and chronic obstructive pulmonary disease. Am J Phys Lung Cell Mol Phys 309:L1124–L1137

Birch J, Victorelli S, Rahmatika D, Anderson RK, Jiwa K, Moisey E, Ward C, Fisher AJ, De Soyza A, Passos JF (2016) Telomere dysfunction and senescence-associated pathways in bronchiectasis. Am J Respir Crit Care Med 193:929–932

Birch J, Barnes PJ, Passos JF (2017) Mitochondria, telomeres and cell senescence: implications for lung ageing and disease. Pharmacol Ther 183:34–49. https://doi.org/10.1016/j.pharmthera.2017.10.005

Birrell MA, McCluskie K, Wong S, Donnelly LE, Barnes PJ, Belvisi MG (2005) Resveratrol, an extract of red wine, inhibits lipopolysaccharide induced airway neutrophilia and inflammatory mediators through an NF-κB-independent mechanism. FASEB J 19:840–841

Bjedov I, Partridge L (2011) A longer and healthier life with TOR down-regulation: genetics and drugs. Biochem Soc Trans 39:460–465

Bonkowski MS, Sinclair DA (2016) Slowing ageing by design: the rise of NAD(+) and sirtuin-activating compounds. Nat Rev Mol Cell Biol 17:679–690

Bueno M, Lai YC, Romero Y, Brands J, St Croix CM, Kamga C, Corey C, Herazo-Maya JD, Sembrat J, Lee JS, Duncan SR, Rojas M, Shiva S, Chu CT, Mora AL (2015) PINK1 deficiency impairs mitochondrial homeostasis and promotes lung fibrosis. J Clin Invest 125:521–538

Burney P, Jithoo A, Kato B, Janson C, Mannino D, Nizankowska-Mogilnicka E, Studnicka M, Tan W, Bateman E, Kocabas A, Vollmer WM, Gislason T, Marks G, Koul PA, Harrabi I, Gnatiuc L, Buist S (2014) Chronic obstructive pulmonary disease mortality and prevalence: the associations with smoking and poverty–a BOLD analysis. Thorax 69:465–473

Calhoun C, Shivshankar P, Saker M, Sloane LB, Livi CB, Sharp ZD, Orihuela CJ, Adnot S, White ES, Richardson A, Le Saux CJ (2016) Senescent cells contribute to the physiological remodeling of aged lungs. J Gerontol 71:153–160

Casaburi R, ZuWallack R (2009) Pulmonary rehabilitation for management of chronic obstructive pulmonary disease. N Engl J Med 360:1329–1335

Chen ZH, Lam HC, Jin Y, Kim HP, Cao J, Lee SJ, Ifedigbo E, Parameswaran H, Ryter SW, Choi AM (2010) Autophagy protein microtubule-associated protein 1 light chain-3B (LC3B) activates extrinsic apoptosis during cigarette smoke-induced emphysema. Proc Natl Acad Sci U S A 107:18880–18885

Chen R, Zhang K, Chen H, Zhao X, Wang J, Li L, Cong Y, Ju Z, Xu D, Williams BR, Jia J, Liu JP (2015) Telomerase deficiency causes alveolar stem cell senescence-associated low-grade inflammation in lungs. J Biol Chem 290:30813–30829

Chilosi M, Carloni A, Rossi A, Poletti V (2013) Premature lung aging and cellular senescence in the pathogenesis of idiopathic pulmonary fibrosis and COPD/emphysema. J Lab Clin Med 116:156–173

Clifford RL, Fishbane N, Patel J, MacIsaac JL, McEwen LM, Fisher AJ, Brandsma CA, Nair P, Kobor MS, Hackett TL, Knox AJ (2018) Altered DNA methylation is associated with aberrant gene expression in parenchymal but not airway fibroblasts isolated from individuals with COPD. Clin Epigenetics 10:32

Colley T, Mercado N, Kunori Y, Brightling C, Bhavsar PK, Barnes PJ, Ito K (2016) Defective sirtuin-1 increases IL-4 expression through acetylation of GATA-3 in patients with severe asthma. J Allergy Clin Immunol 137:1595–1597

Colman RJ, Beasley TM, Kemnitz JW, Johnson SC, Weindruch R, Anderson RM (2014) Caloric restriction reduces age-related and all-cause mortality in rhesus monkeys. Nat Commun 5:3557. https://doi.org/10.1038/ncomms4557

Correia-Melo C, Passos JF (2015) Mitochondria: are they causal players in cellular senescence? Biochim Biophys Acta 1847:1373–1379

Correia-Melo C, Hewitt G, Passos JF (2014) Telomeres, oxidative stress and inflammatory factors: partners in cellular senescence? Longev Healthsp 3:1

Cottin V, Cordier JF (2012) Combined pulmonary fibrosis and emphysema in connective tissue disease. Curr Opin Pulm Med 18:418–427

Culpitt SV, Rogers DF, Fenwick PS, Shah P, de Matos C, Russell RE, Barnes PJ, Donnelly LE (2003) Inhibition by red wine extract, resveratrol, of cytokine release by alveolar macrophages in COPD. Thorax 58:942–946

de Cabo R, Carmona-Gutierrez D, Bernier M, Hall MN, Madeo F (2014) The search for antiaging interventions: from elixirs to fasting regimens. Cell 157:1515–1526

Dermaku-Sopjani M, Kolgeci S, Abazi S, Sopjani M (2013) Significance of the anti-aging protein Klotho. Mol Membr Biol 30:369–385

Di Stefano A, Ricciardolo FLM, Caramori G, Adcock IM, Chung KF, Barnes PJ, Brun P, Leonardi A, Ando F, Vallese D, Gnemmi I, Righi L, Cappello F, Balbi B (2017) Bronchial inflammation and bacterial load in stable COPD is associated with TLR4 overexpression. Eur Respir J 49(5). https://doi.org/10.1183/13993003.02006-2016

Dimmeler S, Nicotera P (2013) MicroRNAs in age-related diseases. EMBO Mol Med 5:180–190

Donnelly LE, Barnes PJ (2012) Defective phagocytosis in airways disease. Chest 141:1055–1062

Dunlop EA, Tee AR (2014) mTOR and autophagy: a dynamic relationship governed by nutrients and energy. Sem Cell Devel Biol 36C:121–129

Fadini GP, Losordo D, Dimmeler S (2012) Critical reevaluation of endothelial progenitor cell phenotypes for therapeutic and diagnostic use. Circ Res 110(4):624–637. https://doi.org/10.1161/circresaha.111.243386

Faner R, Rojas M, Macnee W, Agusti A (2012) Abnormal lung aging in chronic obstructive pulmonary disease and idiopathic pulmonary fibrosis. Am J Respir Crit Care Med 186:306–313

Feghali-Bostwick CA, Gadgil AS, Otterbein LE, Pilewski JM, Stoner MW, Csizmadia E, Zhang Y, Sciurba FC, Duncan SR (2008) Autoantibodies in patients with chronic obstructive pulmonary disease. Am J Respir Crit Care Med 177:156–163

Feng D, Kondo Y, Ishigami A, Kuramoto M, Machida T, Maruyama N (2004) Senescence marker protein-30 as a novel antiaging molecule. Ann N Y Acad Sci 1019:360–364

Feser J, Tyler J (2011) Chromatin structure as a mediator of aging. FEBS Lett 585(13):2041–2048. https://doi.org/10.1016/j.febslet.2010.11.016

Finkel T, Deng CX, Mostoslavsky R (2009) Recent progress in the biology and physiology of sirtuins. Nature 460:587–591

Fischer BM, Wong JK, Degan S, Kummarapurugu AB, Zheng S, Haridass P, Voynow JA (2013) Increased expression of senescence markers in cystic fibrosis airways. Am J Phys Lung Cell Mol Phys 304:L394–L400

Fleg JL (2012) Aerobic exercise in the elderly: a key to successful aging. Discov Med 13(70):223–228

Fois AG, Paliogiannis P, Sotgia S, Mangoni AA, Zinellu E, Pirina P, Carru C, Zinellu A (2018) Evaluation of oxidative stress biomarkers in idiopathic pulmonary fibrosis and therapeutic applications: a systematic review. Respir Res 19:51

Fontana L, Partridge L, Longo VD (2010) Extending healthy life span–from yeast to humans. Science 328:321–326

Fujii S, Hara H, Araya J, Takasaka N, Kojima J, Ito S, Minagawa S, Yumino Y, Ishikawa T, Numata T, Kawaishi M, Hirano J, Odaka M, Morikawa T, Nishimura S, Nakayama K, Kuwano K (2012) Insufficient autophagy promotes bronchial epithelial cell senescence in chronic obstructive pulmonary disease. Oncoimmunology 1(5):630–641

Fujita Y, Kosaka N, Araya J, Kuwano K, Ochiya T (2015) Extracellular vesicles in lung microenvironment and pathogenesis. Trends Mol Med 21:533–542

Gao W, Yuan C, Zhang J, Li L, Yu L, Wiegman CH, Barnes PJ, Adcock IM, Huang M, Yao X (2015) Klotho expression is reduced in COPD airway epithelial cells: effects on inflammation and oxidant injury. Clin Sci 129:1011–1023

Garcia-Prat L, Martinez-Vicente M, Perdiguero E, Ortet L, Rodriguez-Ubreva J, Rebollo E, Ruiz-Bonilla V, Gutarra S, Ballestar E, Serrano AL, Sandri M, Munoz-Canoves P (2016) Autophagy maintains stemness by preventing senescence. Nature 529:37–42

Gorospe M, Abdelmohsen K (2011) MicroRegulators come of age in senescence. Trends Genet 27:23

Guarente L (2011) Sirtuins, aging, and metabolism. Cold Spring Harbor Symp 76:81–90

Gulsvik AK, Thelle DS, Samuelsen SO, Myrstad M, Mowe M, Wyller TB (2012) Ageing, physical activity and mortality–a 42-year follow-up study. Int J Epidemiol 41:521–530

Harrison DE, Strong R, Sharp ZD, Nelson JF, Astle CM, Flurkey K, Nadon NL, Wilkinson JE, Frenkel K, Carter CS, Pahor M, Javors MA, Fernandez E, Miller RA (2009) Rapamycin fed late in life extends lifespan in genetically heterogeneous mice. Nature 460:392–395

Hashimoto M, Asai A, Kawagishi H, Mikawa R, Iwashita Y, Kanayama K, Sugimoto K, Sato T, Maruyama M, Sugimoto M (2016) Elimination of p19(ARF)-expressing cells enhances pulmonary function in mice. JCI Insight 1:e87732

Hazeldine J, Lord JM (2015) Innate immunesenescence: underlying mechanisms and clinical relevance. Biogerontology 16:187–201

Hekimi S, Lapointe J, Wen Y (2011) Taking a "good" look at free radicals in the aging process. Trends Cell Biol 21:569–576

Herranz N, Gallage S, Mellone M, Wuestefeld T, Klotz S, Hanley CJ, Raguz S, Acosta JC, Innes AJ, Banito A, Georgilis A, Montoya A, Wolter K, Dharmalingam G, Faull P, Carroll T, Martinez-Barbera JP, Cutillas P, Reisinger F, Heikenwalder M, Miller RA, Withers D, Zender L, Thomas GJ, Gil J (2015) mTOR regulates MAPKAPK2 translation to control the senescence-associated secretory phenotype. Nat Cell Biol 17:1205–1217

Hitchings AW, Lai D, Jones PW, Baker EH (2016) Metformin in severe exacerbations of chronic obstructive pulmonary disease: a randomised controlled trial. Thorax 71:587–593

Hodge G, Jersmann H, Tran HB, Roscioli E, Holmes M, Reynolds PN, Hodge S (2015) Lymphocyte senescence in COPD is associated with decreased histone deacetylase 2 expression by pro-inflammatory lymphocytes. Respir Res 16:130

Hoffmann RF, Zarrintan S, Brandenburg SM, Kol A, de Bruin HG, Jafari S, Dijk F, Kalicharan D, Kelders M, Gosker HR, Ten Hacken NH, van der Want JJ, van Oosterhout AJ, Heijink IH (2013) Prolonged cigarette smoke exposure alters mitochondrial structure and function in airway epithelial cells. Respir Res 14:97

Hogg JC, Timens W (2009) The pathology of chronic obstructive pulmonary disease. Annu Rev Pathol 4:435–459

Horvath S (2013) DNA methylation age of human tissues and cell types. Genome Biol 14(10):R115

Hosgood HD III, Menashe I, He X, Chanock S, Lan Q (2009) PTEN identified as important risk factor of chronic obstructive pulmonary disease. Respir Med 103:1866–1870

Houben JM, Mercken EM, Ketelslegers HB, Bast A, Wouters EF, Hageman GJ, Schols AM (2009) Telomere shortening in chronic obstructive pulmonary disease. Respir Med 103:230–236

Hubbard BP, Sinclair DA (2014) Small molecule SIRT1 activators for the treatment of aging and age-related diseases. Trends Pharmacol Sci 35:146–154

Ito K, Barnes PJ (2009) COPD as a disease of accelerated lung aging. Chest 135:173–180

Ito K, Colley T, Mercado N (2012) Geroprotectors as a novel therapeutic strategy for COPD, an accelerating aging disease. Int J Chron Obstruct Pulmon Dis 7:641–652

Janssens JP, Pache JC, Nicod LP (1999) Physiological changes in respiratory function associated with ageing. Eur Respir J 13:197–205

Jiang XC, Gao JQ (2017) Exosomes as novel bio-carriers for gene and drug delivery. Int J Pharm 521:167–175

Johnson SC, Rabinovitch PS, Kaeberlein M (2013) mTOR is a key modulator of ageing and age-related disease. Nature 493:338–345

Kadota T, Fujita Y, Yoshioka Y, Araya J, Kuwano K, Ochiya T (2018) Emerging role of extracellular vesicles as a senescence-associated secretory phenotype: insights into the pathophysiology of lung diseases. Mol Asp Med 60:92–103

Karayama M, Inui N, Suda T, Nakamura Y, Nakamura H, Chida K (2010) Antiendothelial cell antibodies in patients with COPD. Chest 138(6):1303–1308

Kennedy BK, Berger SL, Brunet A, Campisi J, Cuervo AM, Epel ES, Franceschi C, Lithgow GJ, Morimoto RI, Pessin JE, Rando TA, Richardson A, Schadt EE, Wyss-Coray T, Sierra F (2014) Geroscience: linking aging to chronic disease. Cell 159:709–713

Kirkham PA, Barnes PJ (2013) Oxidative stress in COPD. Chest 144:266–273

Kirkham PA, Caramori G, Casolari P, Papi AA, Edwards M, Shamji B, Triantaphyllopoulos K, Hussain F, Pinart M, Khan Y, Heinemann L, Stevens L, Yeadon M, Barnes PJ, Chung KF, Adcock IM (2011) Oxidative stress-induced antibodies to carbonyl-modified protein correlate with severity of chronic obstructive pulmonary disease. Am J Respir Crit Care Med 184(7):796–802

Kirkland JL, Tchkonia T, Zhu Y, Niedernhofer LJ, Robbins PD (2017) The clinical potential of senolytic drugs. J Am Geriatr Soc 65:2297–2301

Kolosova NG, Stefanova NA, Muraleva NA, Skulachev VP (2012) The mitochondria-targeted antioxidant SkQ1 but not N-acetylcysteine reverses aging-related biomarkers in rats. Aging 4:686–694

Komatsu N, Okamoto K, Sawa S, Nakashima T, Oh-hora M, Kodama T, Tanaka S, Bluestone JA, Takayanagi H (2014) Pathogenic conversion of Foxp3+ T cells into TH17 cells in autoimmune arthritis. Nat Med 20:62–68

Kubo H (2012) Concise review: clinical prospects for treating chronic obstructive pulmonary disease with regenerative approaches. Stem Cells Transl Med 1(8):627–631

Kugel S, Mostoslavsky R (2014) Chromatin and beyond: the multitasking roles for SIRT6. Trends Biochem Sci 39:72–81

Kureya Y, Kanazawa H, Ijiri N, Tochino Y, Watanabe T, Asai K, Hirata K (2016) Down-regulation of soluble alpha-klotho is associated with reduction in serum irisin levels in chronic obstructive pulmonary disease. Lung 194:345–351

Laberge RM, Sun Y, Orjalo AV, Patil CK, Freund A, Zhou L, Curran SC, Davalos AR, Wilson-Edell KA, Liu S, Limbad C, Demaria M, Li P, Hubbard GB, Ikeno Y, Javors M, Desprez PY, Benz CC, Kapahi P, Nelson PS, Campisi J (2015) MTOR regulates the pro-tumorigenic senescence-associated secretory phenotype by promoting IL1A translation. Nat Cell Biol 17:1049–1061

Lamming DW, Ye L, Sabatini DM, Baur JA (2013) Rapalogs and mTOR inhibitors as anti-aging therapeutics. J Clin Invest 123:980–989

Lange P, Celli B, Agusti A, Boje Jensen G, Divo M, Faner R, Guerra S, Marott JL, Martinez FD, Martinez-Camblor P, Meek P, Owen CA, Petersen H, Pinto-Plata V, Schnohr P, Sood A, Soriano JB, Tesfaigzi Y, Vestbo J (2015) Lung-function trajectories leading to chronic obstructive pulmonary disease. N Engl J Med 373:111–122

Leaker BR, Barnes PJ, O'Connor BJ, Ali FY, Tam P, Neville J, Mackenzie LF, MacRury T (2014) The effects of the novel SHIP1 activator AQX-1125 on allergen-induced responses in mild-to-moderate asthma. Clin Exp Allergy 44:1146–1153

Lee IH, Cao L, Mostoslavsky R, Lombard DB, Liu J, Bruns NE, Tsokos M, Alt FW, Finkel T (2008) A role for the NAD-dependent deacetylase Sirt1 in the regulation of autophagy. Proc Natl Acad Sci U S A 105:3374–3379

Lee J, Sandford AJ, Connett JE, Yan J, Mui T, Li Y, Daley D, Anthonisen NR, Brooks-Wilson A, Man SF, Sin DD (2012) The relationship between telomere length and mortality in chronic obstructive pulmonary disease (COPD). PLoS One 7:e35567

Lehmann M, Korfei M, Mutze K, Klee S, Skronska-Wasek W, Alsafadi HN, Ota C, Costa R, Schiller HB, Lindner M, Wagner DE, Gunther A, Konigshoff M (2017) Senolytic drugs target alveolar epithelial cell function and attenuate experimental lung fibrosis ex vivo. Eur Respir J 50(2):1602367

Li J, Dai A, Hu R, Zhu L, Tan S (2010) Positive correlation between PPARγ/PGC-1α and γ-GCS in lungs of rats and patients with chronic obstructive pulmonary disease. Acta Biochim Biophys Sin 42:603–614

Lopez-Otin C, Blasco MA, Partridge L, Serrano M, Kroemer G (2013) The hallmarks of aging. Cell 153(6):1194–1217. https://doi.org/10.1016/j.cell.2013.05.039

Lozano R, Naghavi M, Foreman K (2012) Global and regional mortality from 235 causes of death for 20 age groups in 1990 and 2010: a systematic analysis for the Global Burden of Disease Study 2010. Lancet 380:2095–2128

Mahbub S, Brubaker AL, Kovacs EJ (2011) Aging of the innate immune system: an update. Curr Immunol Rev 7:104–115

Maneechotesuwan K, Kasetsinsombat K, Wongkajornsilp A, Barnes PJ (2013) Decreased indoleamine 2,3-dioxygenase activity and IL-10/IL-17A ratio in patients with COPD. Thorax 68:330–337

Mannick JB, Del Giudice G, Lattanzi M, Valiante NM, Praestgaard J, Huang B, Lonetto MA, Maecker HT, Kovarik J, Carson S, Glass DJ, Klickstein LB (2014) mTOR inhibition improves immune function in the elderly. Sci Transl Med 6:268

Marin C, Yubero-Serrano EM, Lopez-Miranda J, Perez-Jimenez F (2013) Endothelial aging associated with oxidative stress can be modulated by a healthy mediterranean diet. Int J Mol Sci 14:8869–8889

Martin-Montalvo A, Mercken EM, Mitchell SJ, Palacios HH, Mote PL, Scheibye-Knudsen M, Gomes AP, Ward TM, Minor RK, Blouin MJ, Schwab M, Pollak M, Zhang Y, Yu Y, Becker KG, Bohr VA, Ingram DK, Sinclair DA, Wolf NS, Spindler SR, Bernier M, de Cabo R (2013) Metformin improves healthspan and lifespan in mice. Nat Commun 4:2192

McCubbrey AL, Nelson JD, Stolberg VR, Blakely PK, McCloskey L, Janssen WJ, Freeman CM, Curtis JL (2016) MicroRNA-34a negatively regulates efferocytosis by tissue macrophages in part via SIRT1. J Immunol 196:1366–1375

Meiners S, Eickelberg O, Konigshoff M (2015) Hallmarks of the ageing lung. Eur Respir J 45:807–827

Mercado N, Ito K, Barnes PJ (2015) Accelerated ageing in chronic obstructive pulmonary disease: new concepts. Thorax 70:482–489

Meyer A, Zoll J, Charles AL, Charloux A, de Blay F, Diemunsch P, Sibilia J, Piquard F, Geny B (2013) Skeletal muscle mitochondrial dysfunction during chronic obstructive pulmonary disease: central actor and therapeutic target. Exp Physiol 98:1063–1078

Minagawa S, Araya J, Numata T, Nojiri S, Hara H, Yumino Y, Kawaishi M, Odaka M, Morikawa T, Nishimura SL, Nakayama K, Kuwano K (2011) Accelerated epithelial cell senescence in IPF and the inhibitory role of SIRT6 in TGF-beta-induced senescence of human bronchial epithelial cells. Am J Phys Lung Cell Mol Phys 300:L391–L401

Mitani A, Ito K, Vuppusetty C, Barnes PJ, Mercado N (2016) Restoration of corticosteroid sensitivity in chronic obstructive pulmonary disease by inhibition of mammalian target of rapamycin. Am J Respir Crit Care Med 193:143–153

Mitani A, Azam A, Vuppusetty C, Ito K, Mercado N, Barnes PJ (2017) Quercetin restores corticosteroid sensitivity in cells from patients with chronic obstructive pulmonary disease. Exp Lung Res 43:417–425

Mizumura K, Cloonan SM, Haspel JA, Choi AM (2012) The emerging importance of autophagy in pulmonary diseases. Chest 142:1289–1299

Mizumura K, Cloonan SM, Nakahira K, Bhashyam AR, Cervo M, Kitada T, Glass K, Owen CA, Mahmood A, Washko GR, Hashimoto S, Ryter SW, Choi AM (2014) Mitophagy-dependent necroptosis contributes to the pathogenesis of COPD. J Clin Invest 124:3987–4003

Mizushima N, Levine B, Cuervo AM, Klionsky DJ (2008) Autophagy fights disease through cellular self-digestion. Nature 451(7182):1069–1075

Monick MM, Powers LS, Walters K, Lovan N, Zhang M, Gerke A, Hansdottir S, Hunninghake GW (2010) Identification of an autophagy defect in smokers' alveolar macrophages. J Immunol 185:5425–5435

Mora AL, Bueno M, Rojas M (2017) Mitochondria in the spotlight of aging and idiopathic pulmonary fibrosis. J Clin Invest 127:405–414

Moskalev A, Chernyagina E, Kudryavtseva A, Shaposhnikov M (2017) Geroprotectors: a unified concept and screening approaches. Aging Dis 8:354–363

Munk R, Panda AC, Grammatikakis I, Gorospe M, Abdelmohsen K (2017) Senescence-associated MicroRNAs. Int Rev Cell Mol Biol 334:177–205

Munoz-Espin D, Serrano M (2014) Cellular senescence: from physiology to pathology. Nat Rev Mol Cell Biol 15:482–496

Murray MA, Chotirmall SH (2015) The Impact of immunosenescence on pulmonary disease. Mediat Inflamm 2015:692546

Murrow L, Debnath J (2013) Autophagy as a stress-response and quality-control mechanism: implications for cell injury and human disease. Annu Rev Pathol 8:105–137

Nakamaru Y, Vuppusetty C, Wada H, Milne JC, Ito M, Rossios C, Elliot M, Hogg J, Kharitonov S, Goto H, Bemis JE, Elliott P, Barnes PJ, Ito K (2009) A protein deacetylase SIRT1 is a negative regulator of metalloproteinase-9. FASEB J 23:2810–2819

Noren Hooten N, Abdelmohsen K, Gorospe M, Ejiogu N, Zonderman AB, Evans MK (2010) microRNA expression patterns reveal differential expression of target genes with age. PLoS One 5:e10724

Oh J, Lee YD, Wagers AJ (2014) Stem cell aging: mechanisms, regulators and therapeutic opportunities. Nat Med 20:870–880

Pan J, Li D, Xu Y, Zhang J, Wang Y, Chen M, Lin S, Huang L, Chung EJ, Citrin DE, Wang Y, Hauer-Jensen M, Zhou D, Meng A (2017) Inhibition of Bcl-2/xl With ABT-263 selectively kills senescent type II pneumocytes and reverses persistent pulmonary fibrosis induced by ionizing radiation in mice. Int J Radiat Oncol Biol Phys 9:353–361

Paschalaki KE, Starke RD, Hu Y, Mercado N, Margariti A, Gorgoulis VG, Randi AM, Barnes PJ (2013) Dysfunction of endothelial progenitor cells from smokers and COPD patients due to increased DNA damage and senescence. Stem Cells 31:2813–2826

Paschalaki K, Zampetaki A, Baker J, Birrell M, Starke RD, Belvisi M, Donnelly LE, Mayr M, Randi AM, Barnes PJ (2018) Downregulation of microRNA-126 augments DNA damage response in cigarette smokers and COPD patients. Am J Respir Crit Care Med 197:665–668

Passos JF, Saretzki G, von Zglinicki T (2007) DNA damage in telomeres and mitochondria during cellular senescence: is there a connection? Nucleic Acids Res 35:7505–7513

Perez-Lopez FR, Chedraui P, Haya J, Cuadros JL (2009) Effects of the Mediterranean diet on longevity and age-related morbid conditions. Maturitas 64:67–79

Povedano JM, Martinez P, Flores JM, Mulero F, Blasco MA (2015) Mice with pulmonary fibrosis driven by telomere dysfunction. Cell Rep 12:286–299

Rajendrasozhan S, Yang SR, Kinnula VL, Rahman I (2008) SIRT1, an antiinflammatory and anti-aging protein, is decreased in lungs of patients with chronic obstructive pulmonary disease. Am J Respir Crit Care Med 177(8):861–870

Rutten EP, Gopal P, Wouters EF, Franssen FM, Hageman GJ, Vanfleteren LE, Spruit MA, Reynaert NL (2016) Various mechanistic pathways representing the aging process are altered in COPD. Chest 149:53–61

Salama R, Sadaie M, Hoare M, Narita M (2014) Cellular senescence and its effector programs. Genes Dev 28:99–114

Sapey E, Stockley JA, Greenwood H, Ahmad A, Bayley D, Lord JM, Insall RH, Stockley RA (2011) Behavioral and structural differences in migrating peripheral neutrophils from patients with chronic obstructive pulmonary disease. Am J Respir Crit Care Med 183:1176–1186

Sapey E, Greenwood H, Walton G, Mann E, Love A, Aaronson N, Insall RH, Stockley RA, Lord JM (2014) Phosphoinositide 3-kinase inhibition restores neutrophil accuracy in the elderly: toward targeted treatments for immunosenescence. Blood 123:239–248

Sarker RS, John-Schuster G, Bohla A, Mutze K, Burgstaller G, Bedford MT, Konigshoff M, Eickelberg O, Yildirim AO (2015) Coactivator-associated arginine methyltransferase-1 function in alveolar epithelial senescence and elastase-induced emphysema susceptibility. Am J Respir Cell Mol Biol 53:769–781

Saso L, Firuzi O (2014) Pharmacological applications of antioxidants: lights and shadows. Curr Drug Targets 15:1177–1199

Sato T, Seyama K, Sato Y, Mori H, Souma S, Akiyoshi T, Kodama Y, Mori T, Goto S, Takahashi K, Fukuchi Y, Maruyama N, Ishigami A (2006) Senescence marker protein-30 protects mice lungs from oxidative stress, aging and smoking. Am J Respir Crit Care Med 174:530–537

Savale L, Chaouat A, Bastuji-Garin S, Marcos E, Boyer L, Maitre B, Sarni M, Housset B, Weitzenblum E, Matrat M, Le Corvoisier P, Rideau D, Boczkowski J, Dubois-Rande JL, Chouaid C, Adnot S (2009) Shortened telomeres in circulating leukocytes of patients with chronic obstructive pulmonary disease. Am J Respir Crit Care Med 179:566–571

Schafer MJ, White TA, Iijima K, Haak AJ, Ligresti G, Atkinson EJ, Oberg AL, Birch J, Salmonowicz H, Zhu Y, Mazula DL, Brooks RW, Fuhrmann-Stroissnigg H, Pirtskhalava T, Prakash YS, Tchkonia T, Robbins PD, Aubry MC, Passos JF, Kirkland JL, Tschumperlin DJ, Kita H, LeBrasseur NK (2017) Cellular senescence mediates fibrotic pulmonary disease. Nat Commun 8:14532

Sgalla G, Iovene B, Calvello M, Ori M, Varone F, Richeldi L (2018) Idiopathic pulmonary fibrosis: pathogenesis and management. Respir Res 19:32

Song S, Johnson FB (2018) Epigenetic mechanisms impacting aging: a focus on histone levels and telomeres. Genes 9(4):201

Song J, Heijink IH, Kistemaker LEM, Reinders-Luinge M, Kooistra W, Noordhoek JA, Gosens R, Brandsma CA, Timens W, Hiemstra PS, Rots MG, Hylkema MN (2017) Aberrant DNA methylation and expression of SPDEF and FOXA2 in airway epithelium of patients with COPD. Clin Epigenetics 9:42

Sood A, Assad NA, Barnes PJ, Churg A, Gordon SB, Harrod KS, Irshad H, Kurmi OP, Martin WJ 2nd, Meek P, Mortimer K, Noonan CW, Perez-Padilla R, Smith KR, Tesfaigzi Y, Ward T, Balmes J (2018) ERS/ATS workshop report on respiratory health effects of household air pollution. Eur Respir J 51:1700698

Sorrentino JA, Krishnamurthy J, Tilley S, Alb JG Jr, Burd CE, Sharpless NE (2014) p16INK4a reporter mice reveal age-promoting effects of environmental toxicants. J Clin Invest 124:169–173

Soulitzis N, Neofytou E, Psarrou M, Anagnostis A, Tavernarakis N, Siafakas N, Tzortzaki EG (2012) Downregulation of lung mitochondrial prohibitin in COPD. Respir Med 106:954–961

Sousounis K, Baddour JA, Tsonis PA (2014) Aging and regeneration in vertebrates. Curr Top Dev Biol 108:217–246

Stanley SE, Chen JJ, Podlevsky JD, Alder JK, Hansel NN, Mathias RA, Qi X, Rafaels NM, Wise RA, Silverman EK, Barnes KC, Armanios M (2015) Telomerase mutations in smokers with severe emphysema. J Clin Invest 125:563–570

Stenton GR, Mackenzie LF, Tam P, Cross JL, Harwig C, Raymond J, Toews J, Wu J, Ogden N, MacRury T, Szabo C (2013) Characterization of AQX-1125, a small-molecule SHIP1 activator: Part 1. Effects on inflammatory cell activation and chemotaxis in vitro and pharmacokinetic characterization in vivo. Br J Pharmacol 168:1506–1518

Sureshbabu A, Bhandari V (2013) Targeting mitochondrial dysfunction in lung diseases: emphasis on mitophagy. Front Physiol 4:384

Takahashi A, Ohtani N, Yamakoshi K, Iida S, Tahara H, Nakayama K, Nakayama KI, Ide T, Saya H, Hara E (2006) Mitogenic signalling and the p16INK4a-Rb pathway cooperate to enforce irreversible cellular senescence. Nat Cell Biol 8:1291–1297

Takasaka N, Araya J, Hara H, Ito S, Kobayashi K, Kurita Y, Wakui H, Yoshii Y, Yumino Y, Fujii S, Minagawa S, Tsurushige C, Kojima J, Numata T, Shimizu K, Kawaishi M, Kaneko Y, Kamiya N, Hirano J, Odaka M, Morikawa T, Nishimura SL, Nakayama K, Kuwano K (2014) Autophagy induction by SIRT6 through attenuation of insulin-like growth factor signaling is involved in the regulation of human bronchial epithelial cell senescence. J Immunol 192:958–968

To Y, Ito K, Kizawa Y, Failla M, Ito M, Kusama T, Elliot M, Hogg JC, Adcock IM, Barnes PJ (2010) Targeting phosphoinositide-3-kinase-δ with theophylline reverses corticosteroid insensitivity in COPD. Am J Respir Crit Care Med 182:897–904

To M, Takagi D, Akashi K, Kano I, Haruki K, Barnes PJ, Ito K (2013) Sputum PAI-1 elevation by oxidative stress-dependent NF-kappaB activation in chronic obstructive pulmonary disease. Chest 144:515–521

Tomita K, Caramori G, Lim S, Ito K, Hanazawa T, Oates T, Chiselita I, Jazrawi E, Chung KF, Barnes PJ, Adcock IM (2002) Increased p21CIP1/WAF1 and B cell lymphoma leukemia-xL expression and reduced apoptosis in alveolar macrophages from smokers. Am J Respir Crit Care Med 166:724–731

Tomita K, Caramori G, Ito K, Lim S, Sano H, Tohda Y, Adcock IM, Barnes PJ (2010) Telomere shortening in alveolar macrophages of smokers and COPD patients. Open Path J 4:23–29

Tsuji T, Aoshiba K, Nagai A (2006) Alveolar cell senescence in patients with pulmonary emphysema. Am J Respir Crit Care Med 174:886–893

Turturici G, Tinnirello R, Sconzo G, Geraci F (2014) Extracellular membrane vesicles as a mechanism of cell-to-cell communication: advantages and disadvantages. Am J Phys Cell Physiol 306:C621–C633

van Deursen JM (2014) The role of senescent cells in ageing. Nature 509:439–446

Watroba M, Dudek I, Skoda M, Stangret A, Rzodkiewicz P, Szukiewicz D (2017) Sirtuins, epigenetics and longevity. Ageing Res Rev 40:11–19

Wiegman CH, Michaeloudes C, Haji G, Narang P, Clarke CJ, Russell KE, Bao W, Pavlidis S, Barnes PJ, Kanerva J, Bittner A, Rao N, Murphy MP, Kirkham PA, Chung KF, Adcock IM (2015) Oxidative stress-induced mitochondrial dysfunction drives inflammation and airway smooth muscle remodeling in patients with chronic obstructive pulmonary disease. J Allergy Clin Immunol 136:769–780

Worby CA, Dixon JE (2014) PTEN. Annu Rev Biochem 83:641–669

Xu M, Tchkonia T, Ding H, Ogrodnik M, Lubbers ER, Pirtskhalava T, White TA, Johnson KO, Stout MB, Mezera V, Giorgadze N, Jensen MD, LeBrasseur NK, Kirkland JL (2015) JAK inhibition alleviates the cellular senescence-associated secretory phenotype and frailty in old age. Proc Natl Acad Sci U S A 112:E6301–E6310

Yanagisawa S, Baker JR, Vuppusetty C, Fenwick P, Donnelly LE, Ito K, Barnes PJ (2017a) Decreased phosphatase PTEN amplifies PI3K signaling and enhances pro-inflammatory cytokine release in COPD. Am J Phys Lung Cell Mol Phys 313:L230–L239

Yanagisawa S, Papaioannou AI, Papaporfyriou A, Baker J, Vuppusetty C, Loukides S, Barnes PJ, Ito K (2017b) Decreased serum sirtuin-1 in chronic obstructive pulmonary disease. Chest 152(2):343–352

Yeo SC, Fenwick PS, Barnes PJ, Lin HS, Donnelly LE (2017) Isorhapontigenin, a bioavailable dietary polyphenol, suppresses airway epithelial cell inflammation through a corticosteroid-independent mechanism. Br J Pharmacol 174:2043–2059

Zeng Z, Cheng S, Chen H, Li Q, Hu Y, Wang Q, Zhu X, Wang J (2017) Activation and overexpression of Sirt1 attenuates lung fibrosis via P300. Biochem Biophys Res Commun 486:1021–1026

Zhang W, Hu D, Ji W, Yang L, Yang J, Yuan J, Xuan A, Zou F, Zhuang Z (2014) Histone modifications contribute to cellular replicative and hydrogen peroxide-induced premature senescence in human embryonic lung fibroblasts. Free Radic Res 48:550–559

Zheng S, Wang C, Qian G, Wu G, Guo R, Li Q, Chen Y, Li J, Li H, He B, Chen H, Ji F (2012) Role of mtDNA haplogroups in COPD susceptibility in a southwestern Han Chinese population. Free Radic Biol Med 53:473–481

Zhu Y, Tchkonia T, Fuhrmann-Stroissnigg H, Dai HM, Ling YY, Stout MB, Pirtskhalava T, Giorgadze N, Johnson KO, Giles CB, Wren JD, Niedernhofer LJ, Robbins PD, Kirkland JL (2016) Identification of a novel senolytic agent, navitoclax, targeting the Bcl-2 family of anti-apoptotic factors. Aging Cell 15:428–435

Chapter 4
Neurodegenerative Diseases and Ageing

Lauren Walker, Kirsty E. McAleese, Daniel Erskine, and Johannes Attems

Abstract This chapter describes the main neuropathological features of the most common age associated neurodegenerative diseases including Alzheimer's disease, Lewy body diseases, vascular dementia and the various types of frontotemporal lobar degeneration. In addition, the more recent concepts of primary age-related tauopathy and ageing-related tau astrogliopathy as well as chronic traumatic encephalopathy are briefly described. One section is dedicated to cerebral multi-morbidity as it is becoming increasingly clear that the old brain is characterised by the presence of multiple pathologies (to varying extent) rather than by one single, disease specific pathology alone. The main aim of this chapter is to inform the reader about the neuropathological basics of age associated neurodegenerative diseases as we feel this is crucial to meaningfully interpret the vast literature that is published in the broad field of dementia research.

Keywords Neuropathology · Neurodegeneration · Cerebrovascular disease · Tau · Amyloid-β · α-synuclein · Alzheimer's disease · Lewy body disease · Primary ageing related tauopathy · Cerebral amyloid angiopathy

Alzheimer's Disease

Clinical Aspects

Alzheimer's disease (AD) is the leading cause of dementia caused by neurodegenerative disease, accounting for between 50% and 80% of cases (Blennow et al. 2006; Mayeux and Stern 2012). Clinically, AD is characterised by chronic and irreversible decline in cognition of insidious onset which renders the patient unable to perform activities of daily living, as outlined by a collaborative report by the

L. Walker · K. E. McAleese · D. Erskine · J. Attems (✉)
Institute of Neuroscience, Newcastle University, Newcastle upon Tyne, UK
e-mail: Lauren.Walker1@newcastle.ac.uk; Kirsty.McAleese@newcastle.ac.uk; Daniel.Erskine@newcastle.ac.uk; Johannes.Attems@ncl.ac.uk

© Springer Nature Singapore Pte Ltd. 2019   75
J. R. Harris, V. I. Korolchuk (eds.), *Biochemistry and Cell Biology of Ageing: Part II Clinical Science*, Subcellular Biochemistry 91,
https://doi.org/10.1007/978-981-13-3681-2_4

National Institute of Neurological and Communicative Disorders and Stroke (NINCDS) and Alzheimer's Disease and Related Disorders Association (ADRDA) (McKhann et al. 1984). More recently the National Institute on Ageing (NIA) and Alzheimer's Association (AA) proposed that in addition to the core features of an amnestic presentation and impairments in visuospatial abilities and language functions, consideration should be given to magnetic resonance imageing (MRI), positron emission tomography (PET) imaging imageing and cerebral spinal fluid (CSF) bio-markers (Jack et al. 2011; McKhann et al. 2011).

Macroscopic Observations

Macroscopic changes of AD brains include the characteristic marked atrophy of the medial temporal lobe (inclusive of entorhinal cortex, hippocampus and temporal lobe) and atrophy of the frontal lobe, while parietal and occipital lobes are less severely affected. Loss of both grey and white matter leads to the enlargement of the ventricles in particular the inferior horn of the lateral ventricles (Halliday et al. 2003), and to a general reduction in brain weight.

Hallmark Pathologies

The hallmark pathological lesions associated with AD mainly constitute intracellular accumulations of hyperphosphorylated tau protein forming neurofibrillary tangles (NFTs) and neuropil threads (NTs) (Fig. 4.1) and extracellular amyloid-β (Aβ), which may appear as fleecy Aβ and Aβ plaques (Fig. 4.2a–c). Neuritic plaques are a combination of Aβ and tau as they have an Aβ core surrounded by dystrophic neurites positive for hyperphosphprylated tau (Fig. 4.2d). Each type of protein accumulation has a different topographical pattern of distribution (*see* Stageing Criteria,

Fig. 4.1 Photomicrographs demonstrating neurofibrillary tangles (arrow) and neuropil threads (arrowhead) positive for hyperphosphorylated tau in the CA2 subregion of the hippocampus in *post-mortem* tissue from a patient with Alzheimer's disease. Scale bar represents 50 μm

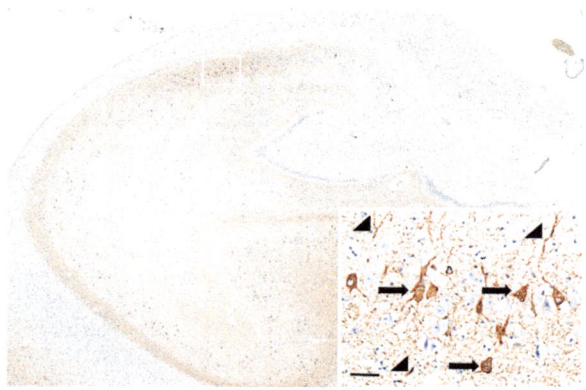

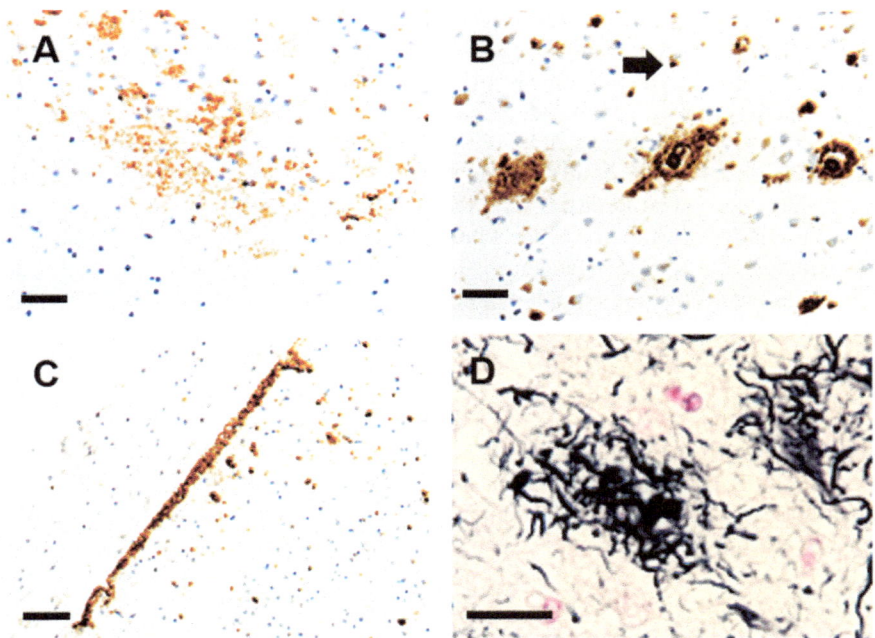

Fig. 4.2 Photomicrographs of different Aβ depositions: fleecy amyloid (**a**), cored and focal plaques with intracellular amyloid precursor protein (arrow) (**b**), subpial deposits (**c**), and a neuritic plaque stained with Gallyas silver stain with an Aβ core surrounded by dystrophic neurites positive for hyperphosphorylated tau. Scale bars: 50 μm in **a** and **b**; 100 μm in **c**, and 20 μm in **d**

below). In its physiological, state the microtubule associated protein (MAP) tau can been found in abundance in the axons of neurons, where it regulates the interaction between microtubule motor proteins and microtubules, and controls the movement of axonal organelles *i.e.* mitochondria and vesicles, maintaining the functionality and viability of the neurons (Hong et al. 1998). Human tau protein is encoded on chromosome 17q21, which consists of 16 exons, and alternate splicing of exons 2, 3 and 10 producing six isoforms ranging from 352 to 441 amino acids in length (Binder et al. 1985; Ingram and Spillantini 2002). Each isoform differs by the number of tubulin binding repeats they possess (either 3 or 4, known as 3R or 4R respectively) and the presence or absence of either one or two 29-amino acid long inserts at the N-terminal portion of the protein (Binder et al. 1985). Although the functionality of the six isoforms is largely similar they are differentially expressed during development. In the adult human brain 3R and 4R isoforms are expressed in a 1:1 ratio and deviations from this are characteristic of neurodegenerative tauopathies (*see* below) (Hong et al. 1998). To allow effective axonal transport, tau is in a constant state of flux, on and off the microtubules between phosphorylated and dephosphorylated states. Dysregulation of the phosphorylation/dephosphorylation system is observed in AD brains (Grundke-Iqbal et al. 1986). Whether it is triggered by an increased rate of phosphorylation and/or a decreased rate in dephosphorylation is yet to be fully determined, it results in a three- to four fold increase of hyperphosphorylated tau compared to normally aged brains (Ksiezak-Reding et al. 1992; Kopke et al. 1993). Tau phosphorylation (which detaches tau from the microtubule) is mediated by numerous kinases, the key candidate protein kinases associated with AD being glycogen synthase-3β (GSK-3β), cyclin-dependant kinase 5 (cdk5), casein kinase 1 (CK1), cyclic AMP dependant protein kinase (PKA), and calcium/calmodulin-dependent kinase II (CaMK-II). There are currently over 80 potential phosphorylation sites on the longest tau isoform, with over 40 being implicated in the pathogenesis of AD (Hanger et al. 2009). Current therapeutic strategies are aimed at the specific and complete inhibition of individual tau kinases and as it is thought the relationship between phosphorylation site and putative kinase is not mutually exclusive, this can be an additional challenge to designing effective treatments (Hanger et al. 2009).

Down-regulation of phosphatases (PPs; involved in attaching tau to the microtubule) have also been implicated in the pathogenesis of AD. PP-2A is the most active enzyme in de-phosphorylating tau to its physiological state, followed by PP5 (Liu et al. 2005a). Experiments investigating the expression and activity of both phosphates in AD revealed activity levels and mRNA expression were decreased by 30% in PP2-A and activity levels but not mRNA expression of PP5 were decreased by 20% compared to aged-matched controls (Gong et al. 1995; Liu et al. 2005b).

Another hallmark pathological lesion involved in the pathogenesis of AD are accumulations of Aβ which form plaques in the parenchyma, which can take various forms such as diffuse depositions, focal and core plaques, and subpial deposits (Fig. 4.2). Aβ is a 4 kDa peptide, which has isoforms terminating at carboxy terminus 40 ($A\beta_{40}$) and 42 ($A\beta_{42}$). The physiological role of Aβ in non-disease states is unclear, however evidence suggests it may be involved in learning and memory as Aβ 1-42 facilitated induction and maintenance of long term potentiation in hippocampal slices (Morley et al. 2010). Depositions of Aβ in the brain are considered to be a result of a gradual and chronic imbalance in the production and clearance of Aβ from the brain (Selkoe 2001a). Aβ is derived by proteolytic cleavage of the transmembrane protein, amyloid-β precursor protein (Selkoe 2001b). The Aβ peptide sequence is located at the junction between the integral membrane domain and the extracellular domain of APP (Kang et al. 1987). There are two pathways in which APP can produce amyloid fragments the amyloidgenic and non-amyloidogenic pathways. In AD, APP is cleaved by two proteases. Firstly β secretase cleaves APP at its extracellular domain leaving fragment C99 which then undergoes further cleavage by γ secretase liberating the Aβ peptide. The alternative proteolytic pathway of APP, which is more dominant in healthy controls, involves processing of APP by α secretases. The cleavage site for α secretase lies within the Aβ sequence and therefore processing down this pathway precludes Aβ formation. Aβ can be degraded by multiple enzymes i.e. neprilysin and insulin degrading enzyme. In addition to degradation Aβ can be cleared from the brain parenchyma through interstitial fluid and perivascular spaces (Weller et al. 2009).

Stageing Criteria

The progression of neuropathological lesions associated with AD follows a step-wise progression: -

1. Braak neurofibrillary stageing (Braak and Braak 1991; Braak et al. 2006) describes the progression of NFTs/NTs. Stages I and II describe the presence of pathology in the trans-entorhinal and entorhinal regions, stages III and IV highlights progression to the hippocampus, whilst stages V and VI demonstrate progression to the neocortex.
2. Thal phases (Thal et al. 2002) describe the deposition of Aβ plaques; the neocortex is affected in phase 1, the hippocampus and cingulate become involved in phase 2, the striatum in phase 3, Aβ plaques can be seen in the brainstem nuclei in phase 4, and finally the cerebellum is affected in phase 5.
3. The Consortium to Establish a Registry for Alzheimer's disease (CERAD) (Mirra et al. 1991) is based on the semi-quantitative assessment (sparse, moderate and frequent) of neuritic plaques in the middle frontal gyrus, superior/middle temporal gyrus, and inferior parietal lobe.
4. NIA-AA criteria have been proposed (Montine et al. 2012), to determine the degree of neuropathologic change (not, low, intermediate, or high) regardless of clinical history, by taking into consideration each of the above pathologies.

It is important to note that the presence of both hyperphosphorylated tau pathology and Aβ plaques is required for a neuropathological diagnosis of AD, however limited amounts of both pathologies can be frequently observed in the brains of individuals without any signs of cognitive impairment (i.e. Braak stages below V, Thal phases usually below 4). Braak stages V and VI are highly associated with clinical dementia (Bancher et al. 1996), with the influence of Aβ1-42 only becoming clinically relevant (in combination with tau pathology) at Thal phases 4 and 5 (Thal et al. 2002). However it is becoming apparent that Aβ can be post-translationally modified, and N terminal truncations, such as Aβ pyroglutamylated at position 3 (pE(3)-Aβ), may play a role in the pathogenesis of AD as it has a greater propensity to aggregate into plaques, and demonstrates an increased neuronal toxicity compared to full length Aβ (Russo et al. 2002; Wirths et al. 2009). Additionally pE(3)-Aβ has been shown to associate with hyperphosphorylated tau and disease severity in AD, and may represent an interesting target for diagnostics and therapeutic intervention (Mandler et al. 2014).

Cerebral Amyloid Angiopathy

Cerebral amyloid angiopathy (CAA) describes the pathological changes that occur in the cerebral blood vessels resulting from the deposition of Aβ (Revesz et al. 2003). CAA exists in two forms, the first (CAA type 1) affecting capillaries, with or without the involvement of leptomeningeal and cortical arteries, arterioles, veins and venules. The second (CAA type 2) exhibits Aβ pathology in the leptomeningeal and cortical vessels, but not in capillaries (Thal et al. 2010) (*see also below*, Cerebrovascular Disease). The accumulation of β-amyloid in the vessel walls is thought to occur by two mechanisms. Incorporation of Aβ into vessel walls may be the result of the failure of the perivascular drainage system to clear Aβ (via the basement membrane), whilst capillary CAA is thought to be the failure of transendothelial clearance of Aβ via *ApoE* ε4 complexes (Weller et al. 2000). CAA is common in the ageing brain but is also thought to have an association with AD as it has been shown to increase in line with AD neuropathologic change (Attems et al. 2005).

Lewy Body Disease

The Lewy body diseases (LBD) comprise Parkinson's disease (PD), Parkinson's disease dementia (PDD) and dementia with Lewy bodies (DLB) (McKeith et al. 2017). The LBDs are neuropathologically characterised by the aggregation of the synaptic protein α-synuclein within neuronal somata as Lewy bodies, and within cellular processes as Lewy neurites (Spillantini et al. 1997). The protein α-synuclein is also implicated in multiple systems atrophy (MSA), though accumulations are typically observed in oligodendrocytes as Papp-Lantos bodies in MSA (Jellinger and Lantos 2010).

Hallmark Pathologies

The protein α-synuclein is the product of the *SNCA* gene positioned on the long arm of chromosome 4 (Shibasaki et al. 1995), and belongs to the family of synucleins, which also contains β- and γ- synuclein, all of which are exclusively found amongst vertebrates (Surguchov 2015). The synucleins are characterised by substantial sequence homology in their N-terminal regions, characterised by KTKEGV repeats with alpha-helical propensity in the presence of lipid membranes, though their C-termini are specific for each family member (Surguchov 2013). A notable aspect of synucleins is that they have long been considered to be natively unstructured proteins without secondary structure. However, this long-perpetuated view has been challenged in the past decade with evidence demonstrating α-synuclein may exist as a helically folded tetramer resistant to aggregation (Bartels et al. 2011), with protein aggregation resulting from a shift from structured tetramers to unstructured monomers (Dettmer et al. 2015). However, these findings were immediately contested and comprehensive studies have suggested α-synuclein is predominantly monomeric and disordered (Coelho-Cerqueira et al. 2013; Fauvet et al. 2012). Nevertheless, the suggestion that α-synuclein may exist as a functional oligomer under certain conditions cannot be entirely discounted considering its substantial conformational plasticity.

α-Synuclein is thought to be a predominantly synaptic protein that interacts with the SNARE complex, promoting its assembly by binding to synaptobrevin-2 (Burre et al. 2010). However, α-synuclein is not a component of all synapses (Braak et al. 2000), and, curiously, is expressed at low levels in brain regions that do not typically develop Lewy body pathology (Erskine et al. 2018), perhaps suggesting α-synuclein expression levels under normal conditions may influence pathological propensity. Elucidation of its physiological role under normal conditions has remained elusive, not least because of the dynamic structure of α-synuclein, where it cycles from disorganisation to α-helices in the presence of lipid membranes (Burre et al. 2013). Nevertheless, there is some evidence to suggest a role in organising pools of synaptic vesicles in the presynaptic terminal and acting as a molecular chaperone (for review *see* Burre 2015). The central role of α-synuclein in Lewy body diseases originated from the almost simultaneous findings of it being the major protein component of Lewy bodies (Spillantini et al. 1997) and mutations in the α-synuclein gene causing familial PD (Polymeropoulos et al. 1997). Since then, it has been established that polymorphisms in the α-synuclein gene are associated with elevated risk of developing PD (Han et al. 2015). For reasons that are not yet clear, in pathological conditions α-synuclein misfolds into beta-pleated sheet conformations, and aggregates into high molecular weight oligomers. Oligomers subsequently adopt a fibrillar structure and ultimately develop into large intracellular aggregates, termed Lewy bodies, which may be visualised on *post-mortem* neurohistology.

Two types of Lewy bodies are typically encountered in the brain in LBD:

1. Classical Lewy bodies are spherical cytoplasmic inclusions characterised by a pale-staining core surrounding a dark staining halo on α-synuclein immunohistochemistry, and are typically found in the pigmented neurons of the substantia nigra, locus coeruleus and dorsal motor nucleus of the vagal nerve. Notably, classical Lewy bodies can often by identified using the basic histological stain haematoxylin and eosin, whilst neurites and cortical Lewy bodies typically require α-synuclein antibodies to be reliably detected (Figs. 4.3a, b).
2. Cortical Lewy bodies lack a halo and are typically observed in cortical neurons. Cortical Lewy bodies typically require antibodies against α-synuclein to be identified in *post-mortem* tissue (Fig. 4.3c).

The hallmark pathology in the Lewy body diseases is the presence of α-synuclein aggregates within neuronal somata and processes, visible on *post-mortem* histology. Lewy bodies are the largest and most distinct feature, though neuritic pathology in neuronal processes and small dots are also typically observed. The antibody used to label α-synuclein has a substantial bearing on the degree of pathology visible, with antibodies against phosphorylated α-synuclein and conformation-specific antibodies against aggregated α-synuclein demonstrating more widespread pathology, particularly in the form of small thread-like structures and dots (Fig. 4.3d) (Kovacs et al. 2012). Less consistently, astroglial pathology may also be observed in cases

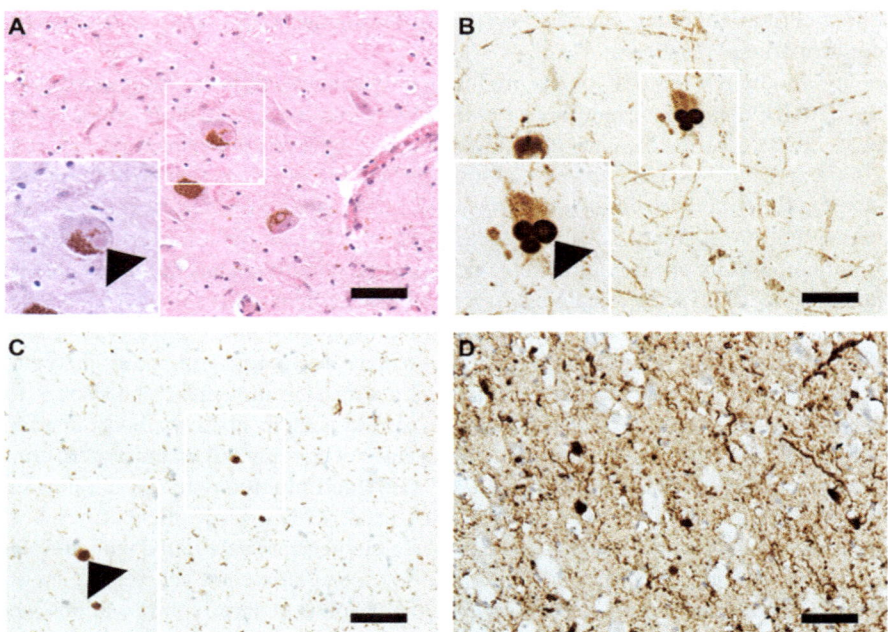

Fig. 4.3 Images of Lewy body pathology in a case with dementia with Lewy bodies. Haematoxylin and eosin staining demonstrates a Lewy body within a pigmented substantia nigra neuron, characterised by a spherical eosinophilic structure (**a**); immunohistochemistry for α-synuclein with the KM51 antibody recognises a classical Lewy body (arrow head) in the substantia nigra, notable for its pale core and deep-staining halo (**b**); cortical Lewy bodies in the temporal cortex identified by the KM51 antibody, like classical Lewy bodies they have a spherical structure but lack a core and halo (**c**); cortical Lewy bodies in the temporal cortex in the same case as (**c**) but stained with the 5G4 antibody that recognises aggregated α-synuclein, revealing more widespread pathology, particularly in the form of thread- and dot- like profiles (**d**). Scale bars = 50 μm, inset images are 3x the original image

with LBD, particularly in limbic and cortical regions highly vulnerable to Lewy body pathology, such as the amygdala, cingulate gyrus and anterior temporal cortex.

Clinical Features

Parkinson's Disease

PD is one of the most prevalent neurodegenerative disorders, coming second only to AD (Ascherio and Schwarzschild 2016). The prevalence of PD is estimated at 0.3% but rises to 1% in individuals over the age of 60 years in European populations (Nussbaum and Ellis 2003; Pringsheim et al. 2014). The characteristic clinical features of PD are resting tremor (a tremor in the limbs when they are not engaged in action), rigidity, bradykinesia (slow movement) and postural instability (Postuma et al. 2015). However, PD can present with a variety of additional clinical features and further clinical sub-types have been described (Lawton et al. 2015).

Parkinson's Disease Dementia

Up to 80% of PD patients will develop dementia over the course of their disease (Walker et al. 2015). PDD is characterised by dementia in the context of established PD, with motor dysfunction preceding the occurrence of cognitive symptoms by a minimum of 1 year (McKeith et al. 2017). The cognitive profile of PDD typically involves impairments in the domains of attention, executive function, visuospatial function and free recall (Emre et al. 2007). However, behavioural abnormalities, such as apathy, depression, anxiety, hallucinations, delusions and excessive daytime sleepiness may also occur, but their absence is not prohibitive to a diagnosis of PDD.

Dementia with Lewy Bodies

DLB is the second most common form of neurodegenerative dementia after AD (Heidebrink 2002) though misdiagnosis is common, particularly in primary care settings (Vann Jones and O'Brien 2014). The characteristic clinical features of DLB are fluctuating cognition (periods of inattention interspersed with periods of lucidity), complex visual hallucinations, parkinsonian motor features and REM sleep behaviour disorder (failure to enter paralysis during REM sleep and subsequent physical acting of dreams) (McKeith et al. 2017). Additional features, such as severe autonomic dysfunction, hallucinations outside of the visual modality and depression may also occur, but lack the specificity to be included in formal clinical diagnostic criteria. In contrast to PDD, DLB is characterised by cognitive features preceding, or occurring contemporaneous to, the onset of motor features (McKeith et al. 2017).

Macroscopic Observations

The most remarkable macroscopic observation of the brain in PD and PDD is the marked pallor of the substantia nigra, the result of the widespread loss of pigmented dopaminergic neurons (Fig. 4.4). The brain in DLB is more variable, with the substantia nigra varying from normal to pale, and the locus coeruleus also typically depigmented. The degree of nigral pallor in DLB patients is predictive of clinical parkinsonism *intra vitam* (McKeith et al. 2017). Cortical atrophy is typically not observed in PD/PDD though DLB is less consistent, with cases varying from diffuse atrophy reminiscent of AD to that of normal appearance.

Fig. 4.4 Macroscopic
image of the substantia
nigra showing a normally
pigmented substantia nigra
(arrowhead; **a**) and a
depigmented substantia
nigra (arrowhead; **b**) in a
case of Lewy body disease.
Scale bar represents 1 cm.
Photographs courtesy of
Newcastle Brain Tissue
Resource

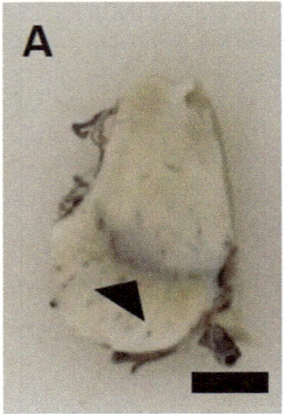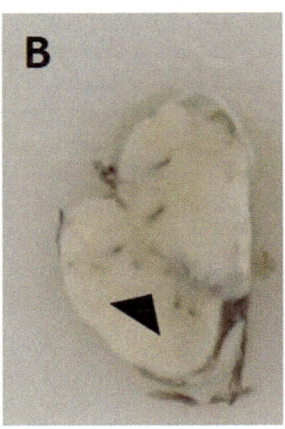

Potential Mechanisms of Pathogenesis

Considerable converging evidence indicates that misfolded and aggregated forms of
α-synuclein may spread through the brain in a manner reminiscent of that previously
described for prion protein (Masuda-Suzukake et al. 2014; Woerman et al. 2015;
Kovacs et al. 2014). Under this hypothesis, aggregated forms of α-synuclein are
transported in a retrograde manner, acting as a template for the misfolding and aggregation of endogenous α-synuclein within recipient cells. However, the suggestion of
simple spreading on the basis of anatomical connectivity is difficult to reconcile with
the observation that regions with significant connectivity to early predilection sites
do not develop substantial Lewy body pathology (for review *see* Surmeier et al.
2017). Furthermore, even within severely affected areas, not all sub-types of neurons
manifest Lewy bodies, even if they are connected to populations affected by Lewy
bodies. Therefore, it is apparent that cell- or region-autonomous factors may govern
the ability of aggregated α-synuclein to spread in neuronal networks.

Stageing Criteria

Two major stageing schemes exist to categorise the extent to which Lewy body
pathology has spread through the brain:

1. Braak stages (Braak et al. 2003) for α-synuclein pathology postulate that the
   dorsal motor nucleus of the vagal nerve is the initial site of Lewy body deposition (stage 1), prior to spreading to the locus coeruleus in the pons (stage 2) and
   midbrain substantia nigra (stage 3), entorhinal cortex and hippocampus (stage 4)
   before finally affecting the neocortex (stages 5 and 6). Under this scheme, PD
   would reach stage 3 whilst DLB would reach at least stage 4, though *see* below.

2. The updated Newcastle-McKeith criteria (McKeith et al. 2017) do not suggest a hierarchical progression of pathology but rather distinguish between brainstem-predominant, limbic and neocortical stages of Lewy body pathology. Thus, a case may have severe neocortical involvement but only minimal affection of the brainstem. The updated criteria also recognise olfactory bulb only and amygdala-predominant sub-types, the latter of which is observed in 24% of otherwise pure AD cases (McKeith et al. 2017). This stageing scheme also takes into account the degree of concomitant AD-type pathology in assigning the probability that the individual had DLB, and the neuronal loss of the substantia nigra in assigning the probability that the individual experienced parkinsonian features.

The Braak stageing scheme, which assumes a hierarchical progression from the brainstem, is inconsistent with both clinical and neuropathological findings in DLB. Not all individuals with DLB experience parkinsonism, with prevalence ranging from 60% to 92% and, when present, parkinsonism develops on average 2 years after dementia (McKeith et al. 2005). Furthermore, not all DLB cases evidence nigral Lewy bodies and neurites. Thus, DLB cases do not fit into any Braak stage, indicating that a different topographical spreading pattern may exist in DLB compared to PD.

Vascular Dementia

Sporadic vascular cognitive impairment (O'Brien et al. 2003) can be defined as acquired cognitive impairment as a result of global or focal effects of cerebrovascular lesion (CVL) that arise from cerebrovascular disease (CVD). Sporadic vascular cognitive impairment itself comprises of a continuum of different clinical manifestations, *i.e.* 'mild' classification comprising of patients who's vascular cognitive impairment (VCI) falls short of dementia, and a 'major' classification that describes vascular dementia (VaD). VaD is a broad term that describes four main subtypes, i.e., subcortical ischaemic vascular dementia (SIVaD), multiinfarct dementia (MID) or cortical dementia, post-stroke dementia and mixed dementia, of which further classification with the respective neurodegenerative pathology, e.g. VaD-AD, Va-DLB, of which the pathogenesis and underlying CVD and resulting CVL differ. Due to the high variability of VaD, no generally accepted morphologic scheme for stageing cerebrovascular lesions and no validated neuropathological criteria for VaD have been established to date (*see* Jellinger 2007, 2008). However, the recent Vascular Cognitive Impairment Neuropathology Guidelines (VCING) (Skrobot et al. 2016) from the UK, which is based upon previously published neuropathological criteria for various CVD/CVL, is useful in determining low, intermediate and high likelihood of CVD contributing to cognitive impairment.

Cerebrovascular Pathology

The three-major CVD, or vessel disorders, in the ageing brain are atherosclerosis (AS), small vessel disease (SVD) and cerebral amyloid angiopathy (CAA). All three CVD can lead to brain tissue destruction, *i.e.* CVL, leading to neurological symptoms. The most common CVL in the ageing brain are infarction, haemorrhage and white matter lesions (WML), with the size and location of the CVL dependent upon the vessel affected. Here we will briefly describe the main CVD, associated CVL and dementias in the ageing human brain: for a more comprehensive review please *see* Jellinger (2007) and Grinberg and Thal (2010).

Cerebrovascular Disease

Atherosclerosis (AS) is a degenerative vessel disorder affecting the cranial and extracranial large to medium sized cerebral arteries, most commonly affecting the arteries of the Circle of Willis. AS leads to the formation of atherosclerotic plaques as a result of high levels of blood-derived lipids: accumulation of blood-derived lipids in macrophages, *i.e.* foam cells, the *tunica intima* and proliferation of fibroblasts and matrix components leads to structural disorganization and thickening of the *intima*. Continued expansion of the lipid core leads to occlusion of the artery and further disruption of the *tunic media* resulting in stiffening of the vessel. In the later stages a neuritic core and calcification of the plaque can occur, which are very prone to rupture and result in subsequent thrombosis, and therefore vessel occlusion, or thromboembolisms that may occlude smaller arteries (Jellinger 2007; Grinberg and Thal 2010; Stary 2000). SVD primarily affects the perforating cerebral arteries and arterioles that penetrate into the basal ganglia and deep white matter (WM). SVD itself encompasses three alterations of the vessels walls that affect vessels of varying sizes: (*i*) SVD-AS has a similar pathogenesis to large vessel AS, however, these atherosclerotic plaques do not calcify and affects the leptomeningeal and intracerebral arteries (200–800 μm in diameter) (Fig. 4.5a). (*ii*) Lipohyalinosis affects smaller arteries and arterioles (40–300 μm in diameter) and its hallmark characteristic is asymmetric fibrosis/hyalinosis (Fig. 4.5b). Foam cell infiltration, fibroid necrosis, *i.e.* breakdown of the vessel wall matrix proteins, and blood-derived plasma protein leakage as a result of blood-brain barrier breakdown (BBB) may also be present (Fig. 4.5c). (*iii*) Arteriolosclerosis is present in the arterioles (40–150 μm) with the hallmark characteristic of concentric hyaline thickening (Fig. 4.5d). SVD alterations lead to stiffening of the vessel walls, hence, vessels cannot adjust to blood pressure changes, and loss of autoregulation meaning the vessel wall is unable to respond and adjust to required changes in blood perfusion: overall resulting in a chronic hypoperfusion and ischaemia. SVD initially manifests as lipohyalinosis in the putamen and globus pallidus of the basal ganglia, and as SVD-AS in leptomeningeal arteries. Lipohyalinosis and arteriolosclerosis then progress to

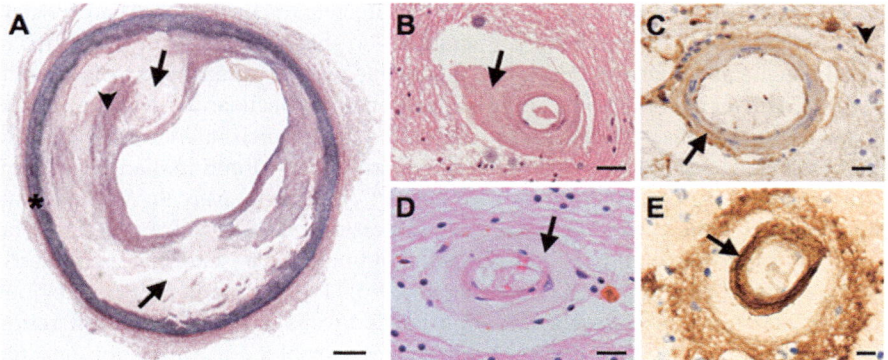

Fig. 4.5 (**a**) Internal carotid artery indicating displaced tunica intima due to a severe atherosclerotic plaque (arrow) and proliferation of myofibroblast/fibroblast (arrowhead). Note the preserved smooth cell and elastin layer of the tunic media (asterix). (**b**) Lipohyalinosis of a white matter arteriole showing severe asymmetrical fibrosis/hyalinosis (arrow). (**c**) Blood-brain barrier dysfunction indicated by the plasma protein fibrinogen leakage into the vessel walls of a white matter artery (arrow) and white matter tissue (arrowhead). (**d**) Arteriolosclerosis of a white matter arteriole showing severe concentric hyalinosis (arrow). (**e**) Cortical cerebral amyloid angiopathy; Aβ deposition in the vessel wall of a cortical artery (arrow). Histochemistry: **a**, Van Gieson; **b** and **c**, H&E. Immunohistichemistry: C fibrinogen antibody; **e**, 4G8 antibody. Scale bars; **a**, 10 mm; **b**, **c**, **d** and **e**, 20 μm

affect the white matter, and only in severe end stage are the cerebrellum and brain stem affected. Overall, cortical vessels remain relatively free of SVD pathology (Thal et al. 2003). Finally, CAA is characterised by the deposition of Aβ (predominately Aβ-40) in the vessel walls of leptomeningeal and cortical arteries, arterioles, capillaries and, rarely, veins (Vinters 1987) (*see also* Cerebral Amyloid Angiopathy, above) (Fig. 4.5e). Aβ deposition in the vessel walls leads to loss of smooth muscle cells and disruption of vessel architecture resulting in disturbances to cerebral blood flow and vessel fragility. Topographically, CAA usually presents in the neocortex, with more frequent and severe deposition seen in the occipital region, followed by the allocortex and cerebellum, and in severe cases, the basal ganglia, thalamus and white matter (Attems 2005).

Cerebrovascular Lesions

Insufficient cerebral blood flow, below a critical level required to maintain normal cellular conditions, can result in cerebral ischaemic infarction, *i.e.* necrosis of cerebral tissues. The size of the infarction is dependent upon the vessel affected: large infarcts (>15mm^3) are frequently the result of atherosclerotic plaque rupture that lead to the occlusion of large/medium arteries lumen due to thrombosis or thromboembolisms (Brun and Englund 1986). This type of infarction is within a focal and defined territory of the specific artery/arteries affected. Lacunar infarcts, are cavitating infarcts (5–15 mm^3), are largely confined to the subcortical grey matter, *i.e.* basal ganglia and thalamus, and white matter and, hence, primarily associated with SVD (Challa et al. 1990). The size and location of the lacunar infarct corresponds to the territory of the perforated artery affected. Finally, microinfarcts are ischaemic infarcts that can only been seen under the microscope (<5 mm in diameter). They can be present in both the cortex, where they are associated upstream AS embolism or CAA, and white matter, where they are associated with SVD (Grinberg and Thal 2010). Haemorrhagic infarct describes blood influx into the ischaemic infarct territory as a result of reperfusion (Hacke et al. 1996). This occurs in infarcted regions in which the remaining vessels have fragile vessel walls as a result of the infarction or due to SVD or CAA. Cerebral haemorrhages, or blood extravasation into the brain parenchyma, over 10 mm in diameter can result from all types of vessel disorders. Haemorrhages located in the lobes of the brain are most commonly associated with CAA, whereas those located in the subcortical, ventricular region, deep white matter and brain stem are strongly associated with SVD. Small haemorrhages and microbleeds (<10 mm in diameter) may histologically appear as extravasations of erythrocytes or as haemosiderin-laden macrophages in the perivascular space. In the cortex, small haemorrhages and microbleeds are associated with CAA (Greenberg et al. 2009), while those located in the white matter, subcortical grey matter and brain stem are associated with SVD (Jeong et al. 2002). Finally, WML encompass white matter rarefaction, *i.e.* demyelination and axonal loss with additional oedema, mild astrocytosis, and macrophage infiltration (Grinberg and Thal 2010). Morphologically there are two types of WML: periventricular (PV) WML that surround the lateral ventricles, and the deep WML that affect the deep lobar WM. Deep WML are associated with SVD-related chronic hypoperfusion and BBB alterations (Fazekas et al. 1993; Hachinski et al. 2006; Schmidt et al. 2011), unlike PV-WML, of which the pathogenesis is still unclear.

Pathological lassifications of Vascular Dementia

Classification of vascular dementia (VaD) can be divided into four main subtypes that are dependent upon the CVL(s) present. SIVaD is the result of SVD as the main CVD and subcortical WML and/or lacunar infarcts are the primary CVL present. MID/cortical relates to the involvement, and likely contribution, of multiple large/small cortical infarcts of which the total sum of damaged cerebral tissue results in a significant decrease in functional brain capacity, surpassing the threshold for cognitive impairment. Mixed dementia occurs when pathologically there is the presence of both CVD and associated CVL and a neuropathological disease, *e.g.* AD; Braak NFT stage V/VI, CERAD score of 3, and Aβ phase 5 according to the NIA–AA guidelines (Braak et al. 2006; Hyman et al. 2012; Mirra et al. 1991; Thal et al. 2002) (see above, Stageing Criteria), and both pathologies could account for the dementia. Where possible, this should be further classified to the respective neurodegenerative pathology, *e.g.* VaD-AD. Finally, post-stroke dementia occurs when the patient exhibits permanent cognitive decline within 6 months of stroke and can be the result of multiple CVD and associated CVL including singular/multiple cortical and subcortical infarcts, as well as mixed dementia.

Frontotemporal Lobar Dementia (Including Tauopathies)

Neurodegenerative disorders that are characterised by the deposition of the abnormal tau protein tau are termed 'tauopathies'. Frontotemporal lobar degeneration (FTLD) is classified in this category and refers to a group of disorders caused by progressive neuronal loss and abnormal protein accumulation mainly affecting the frontal and temporal lobes, and account for up to 20% of presenile dementia cases (Neary et al. 1998; Armstrong et al. 2005). In addition to tau pathology, additional lesions include inclusions of TDP-43, a protein encoded by the fused in sarcoma gene (FUS) (Cairns et al. 2007), ubiquitin, neurofilament inclusions, and basophilic inclusion bodies (Bigio et al. 2003).

Traditionally, Pick's disease, corticobasal degeneration (CBD), progressive supranuclear palsy (PSP) and agyrophilic grain (AGD) are also frequently listed in the tauopathies group and are characterised by different morphological configuration of tau pathology. More recently primary age-related tauopathy (PART) (Crary et al. 2014), and chronic traumatic encephalopathy (CTE) (McKee et al. 2013) have been recognised as part of the tauopathy spectrum.

FTLD with TDP-43 Pathology

Transactive response DNA-binding protein 43 (TDP-43) is a ubiquitously expressed, highly conserved RNA- and DNA-binding nuclear protein (Ou et al. 1995). Under pathological conditions TDP-43 can be sequestered into the cytoplasm and cleaved into C-terminal fragments that are abnormally hyperphosphorylated and subsequently aggregate forming intracellular inclusions. These inclusions are observed as neuronal cytoplasmic inclusions (NCIs), neuronal intranuclear inclusions (NIIs), and/or dystrophic neurites (DNs) and are collectively referred to as TDP-43 pathology. TDP-43 pathology is a hallmark pathology associated with the FTLD-TDP (Bigio 2011), of which there are four subtypes classified by the topological appearance and distribution of hallmark lesions (subtypes A, B, C, and D) (Mackenzie et al. 2011; Mackenzie and Neumann 2017). TDP-43 positive inclusions are also present in the spinal motor neurones in motor neurone disease (MND) (Neumann et al. 2006) where approximately 20% of patients develop FTLD during their disease course (Morita et al. 2006).

FTLD-FUS

The RNA-binding fused in sarcoma (FUS) protein is composed of 526 amino acids, encoded by 15 exons on chromosome 16, which is ubiquitously expressed, and belongs to the FET/TET family of DNA/RNA binding proteins (Bertolotti et al. 1996). Although it's exact function is poorly understood, FUS is thought to be implicated in numerous cellular processes including cell proliferation, DNA repair, and RNA and microRNA processing (Lagier-Tourenne et al. 2010; Janknecht 2005). FUS is continuously shuttled between the nucleus and cytoplasm and therefore shows expression in both cellular compartments (Zinszner et al. 1997). Pathological lesions in FUS proteinopathies are characterised by FUS immunoreactive inclusions in both neuronal and glial cells although the distributions can differ between 3 distinct pathological entities (*see* review (Mackenzie et al. 2010)).

Pick's Disease

Pick's disease is a rare cause of FTLD accounting with non-fluent progressive aphasia or behavioural abnormalities presenting as the most common clinical features (Dickson et al. 2011) Macroscopically it is characterised by severe circumscribed lobar atrophy also involving the limbic regions (cingulate, entorhinal cortex and amygdala), however the brainstem and cerebellum are relatively spared (for review *see* Dickson 1998). Microscopically the predominating hallmark lesions are spherical cytoplasmic inclusions termed Pick bodies, composed of 3R tau (Dickson 1998; Buee and Delacourte 1999).

Cortico-Basal Degeneration

The classical clinical presentation associated with cortico-basal degeneration (CBD) is asymmetrical rigidity and apraxia, often with dystonia (Litvan et al. 2000), however atypical presentations are often common such as frontal type dementia (Dickson et al. 2011). The clinical heterogeneity observed in CBD is reflective of the high variability in the findings observed at neuropathological assessment. However the classical presentation shows asymmetrical atrophy of the cortical gyri in particular in the superior frontal and parietal regions and loss of neuromelanin pigment in the substantia nigra of the midbrain. The hallmark pathological lesions seen at *post-mortem* examination are the accumulation of 4R tau in neurones and astrocytes in the cortex, basal ganglia, thalamus and brainstem (Dickson et al. 2002). Astrocytic lesions are termed astrocytic plaques due to the plaque like arrangement of cell processes, whilst swollen cortical neurones positive for 4R tau are known as ballooned neurones (Dickson et al. 2002). Oligodendroglial depositions known as coiled bodies are also present, usually in the white matter (for review *see* Dickson et al. 2011).

Progressive Supranuclear Palsy

Affecting men and women equally, progressive supranuclear palsy (PSP) presents with an atypical parkinsonism with axial rigidity, with most patients developing progressive vertical gaze palsy (Steele et al. 1964). Brain regions most commonly affected include the basal ganglia, subthalamic nuclei, and the substantia nigra. Similar to CBD, PSP is characterised neuropathologically as a 4R tauopathy, however the immunohistochemistry for 4R tau labels tufted astrocytes which are often found in the motor cortex and corpus striatum (Dickson et al. 2007). 4R tau positive globose NFTs and NTs are seen in the grey and white matter of cortical and subcortical regions, whilst coiled bodies can be seen in the white matter. Although presenting as distinct entities there is considerable clinical, genetic, and neuropathological overlap between CBD and PSP with a recent GWAS study identifying shared novel genetic risk factors beyond the MAPT locus (Yokoyama et al. 2017).

Argyrophilic Grain Disease

Clinico-pathological studies have failed to show a robust clinical or imaging correlate for argyrophilic grain disease (AGD) other than a slowly progressing mild cognitive impairment, and currently diagnosis is only possible at *post-mortem* examination (Saito and Murayama 2007). Associated pathology of AGD takes the form of Argyrophilic grains which are spindle or comma shaped structures found within neuronal processes, mainly in the dendrites and dendritic spines (Tolnay et al. 1998).

Primary Age-Related Tauopathy

Primary age related tauopathy (PART) is a term used to describe the pathologic continuum of tau pathology that encompasses cognitively normal aged individuals that harbour focally distributed NFTs, through to patients with dementia that have been referred to as 'tangle predominant senile dementia' (Crary et al. 2014). As NFTs are practically universal in the ageing brain, the term PART was introduced to identify a group of cases with only a few or no Aβ deposits (0–2 Thal Aβ phase (Thal et al. 2002)). Recent neuropathologic criteria have been proposed to further subdivide PART cases into "definite" (Braak stage ≤IV, Thal Aβ phase 0) and "possible" (Braak stage ≤IV, Thal Aβ phase 1–2) (Crary et al. 2014).

Ageing-Related Tau Astrogliopathy (ARTAG)

Ageing-related tau astrogliopathy (ARTAG) is characterised by deposits of abnormally hyperphosphorylated tau protein seen in astrocytes in the brains of people with and without AD related pathology (Arima et al. 1998; Ferrer et al. 2014; Kovacs et al. 2016). The predilection sites for ARTAG pathology are the subpial layer of the cortex, subependymal zone, perivascular regions, the basal forebrain, and white matter (Kovacs et al. 2017; Liu et al. 2016). The clinical relevance of white and grey matter ARTAG is poorly understood, however some studies suggest associations with cognitive decline in a subset of neurodegenerative disorders (Kovacs et al. 2011; Munoz et al. 2007), whilst others have failed to demonstrate such associations (Lace et al. 2012; Wharton et al. 2016). There are two main morphological conformations of ARTAG in the ageing brain; thorn-shaped astrocytes (TSA) where tau immunoreactivity is localised to astrocytic perikarya with extension to the proximal part of the astrocytic processes, and granular or fuzzy immunoreactivity in the processes of astrocytes (GFA) with fine granular depositions of branching processes. For a comprehensive overview *see* Kovacs et al. (2016).

Chronic Traumatic Encephalopathy (CTE)

Historically referred to as 'dementia pugilistica', chronic traumatic encephalopathy (CTE) is a more recently coined term to describe the neurodegenerative consequences as a result of repetitive subconcussive or concussive head impacts (Ling et al. 2017). Clinical manifestations of CTE include irritability, impulsivity, aggression, depression and long-term memory deterioration (McKee et al. 2009). Multiple neuropathologic lesions have been observed at *post-mortem* examination of individuals with a history of repetitive head impacts including Aβ plaques and TDP-43 positive inclusions, however it has been suggested that neocortical NFT formation is a the predominating feature in these individuals (Ling et al. 2017). Similar to AD, hyperphosphorylated tau deposits seen in CTE are composed of both 3R and 4R isoforms (Schmidt et al. 2001), and predilection sites for abnormal hyperphosphorylated tau deposits include the frontal and temporal cortices, particularly at the depths of the cerebral sulci and around small cerebral vessels, limbic regions, diencephalon, and brainstem nuclei. As well as neuronal tau aggregates, ARTAG (*see* above) is a common feature of CTE with a significant overlap identified in the brain regions affected (McKee et al. 2016; Kovacs et al. 2016). Recently, a progressive staging criteria (I-IV) has been proposed to assess the neuropathological distribution of pathology in CTE cases (McKee et al. 2013).

Cerebral Multimorbidity and Mixed Dementia

The traditional method used to ascertain the diagnosis of a neurodegenerative disease is based on a consensus clinico-pathological diagnosis; where the combination of signature neuropathological lesions identified at *post-mortem* examination and clinical symptoms observed during life lead to a clinico-pathological diagnosis (*e.g.* fulfilling the core clinical criteria (McKhann et al. 2011) and high AD neuropathologic change for AD (Montine et al. 2012). The neuropathological diagnosis of neurodegenerative disorders is based on the most prevalent pathology, however it is common to observe pathologies associated with multiple neurodegenerative diseases in the brain at *post-mortem* examination, which brings into question the idea of neurodegenerative disorders being distinct disease entities and making the design of specific therapeutics challenging (Attems and Jellinger 2013).

Overlap Between AD and LBD

Historically AD and DLB were distinguished by the presence of Aβ plaques and NFTs versus LBs. However, a large proportion of AD cases (50%) exhibit concomitant Lewy body pathology in addition to plaques and tangles, but do not fulfil neuropathological criteria for a diagnosis of AD (Kovacs et al. 2008) (*see* Stageing Criteria, above), and similarly LBD cases can exhibit concomitant AD related pathology (Fig. 4.6). The clinical relevance of this subpopulation of patients revealed the presence of multiple pathologies resulted in a faster decline in cognition and an accelerated mortality compared to those with pathology associated with a single neurodegenerative disease (Olichney et al. 1998; Serby et al. 2003; Kraybill et al. 2005). Furthermore a recent clinico-pathological study investigated AD related and other concomitant pathologies in a cohort of autopsy confirmed synucleinopathies (inclusive of DLB, PD, and PDD). An increase in AD (in particular neurofibrillary

Fig. 4.6 Multiple pathologies are frequent in the brains of patients with dementia. Fluorescent images demonstrating *post-mortem* tissue from the temporal lobe of a patient with dementia with Lewy bodies exhibits concomitant Alzheimer's disease related pathology. (**a**) AT8 antibody highlights a hyperphosphorylated tau positive neurofibrillary tangle (arrow). (**b**) Lewy body (arrowhead) is labeled with an antibody for α-synuclein phosphorylated at Serine 129. (**c**) merged images demonstrating both pathologies with neuronal nuclei conterstained with DAPI. Scale bar represents 20 μm

tau pathology) and LB pathology was associated with a shorter interval between onset of motor symptoms and onset and onset of dementia, and a shorter disease duration (Irwin et al. 2017). Given the effects of multiple pathologies on clinical phenotype, research to elucidate putative interactions between pathological protein aggregates is ongoing; α-synuclein has been demonstrated to initiate the polymerization of tau *in vitro* (Giasson et al. 2003) whilst tau inclusions were observed in approximately 50% of Ala53Thr transgenic mice expressing human α-synuclein leading to severe motor dysfunction (Lee et al. 2004).

TDP-43 in AD

TDP-43 pathology is a hallmark pathology associated with the FTLD-TDP subtype (Bigio 2011) and motor neuron disease (Neumann et al. 2006) (see section "FTLD with TDP-43 pathology"); however, it is also frequently present in AD, present in 29–74% of AD cases (Amador-Ortiz et al. 2007; Josephs et al. 2008, 2014a, b, 2015; Arai et al. 2009; Davidson et al. 2011; Wilson et al. 2013; Uchino et al. 2015; James et al. 2016). The presence of TDP-43 pathology in AD has been shown to be associated with more severe cognitive impairment (Josephs et al. 2014b; McAleese et al. 2017) episodic and working memory (Wilson et al. 2013) and significantly more advanced hippocampal atrophy (Josephs et al. 2008, 2017). The topographical distribution of TDP-43 pathology in AD differs from that seen in FTLD-TDP, manifesting in the amygdala (stage I), followed by the entorhinal cortex and/or subiculum (stage II), dentate gyrus and/or occipitotemporal cortex (OTC) (stage III), inferior temporal cortex (ITC) (stage IV), and finally the mid-frontal cortex and/or striatum (stage V) (Josephs et al. 2014a, 2016). Here, the term TDP-43 pathology refers only to TDP-43 deposition in AD and not to the hallmark pathology of FTLD-TDP and motor neuron disease. In addition, TDP-43 pathology has also been identified in DLB (Pollanen et al. 1993), though different studies have reported highly divergent prevalence rates, ranging from 0% (Nakashima-Yasuda et al. 2007) to 56% (Arai et al. 2009). The neuroanatomical distribution of TDP-43 pathology in DLB has been shown to be similar to that seen in AD, predominantly affecting the amygdala and hippocampal structures (Higashi et al. 2007).

Stratification for Clinical Trials

Neurodegenerative diseases by nature are heterogeneous which makes identifying potential genetic risk factors or targets for therapeutic intervention challenging. In the biggest DLB GWAS study to date (which included 1743 patients with dementia with Lewy bodies, 1324 with a pathologically confirmed diagnosis), the most significant genome-wide association was seen at the *APOE* locus (Guerreiro et al.

2018), and whilst it has been reported to affect the levels of Aβ and LB pathology in the brains of DLB patients (Tsuang et al. 2013) the ε4 allele association in DLB has been shown to be driven largely by concomitant AD related pathology (Peuralinna et al. 2015). Larger cohorts and stratification by level of concomitant pathology will ultimately allow identification of disease specific targets for drug discovery studies (Guerreiro et al. 2018).

Disease biomarkers such as cerebrospinal fluid markers (Blennow et al. 2006; Irwin et al. 2018), blood based biomarkers (for review *see* Olsson et al. 2016), and imaging correlates (Weiner et al. 2013) may not be a realistic option for routine diagnostics, however they may provide invaluable information when stratifying patients for clinical trials leading to identification of more accurate and disease specific targets for therapeutic intervention.

Future Considerations

Clinico-pathological correlative studies to date have provided significant contributions to our understanding of how pathological protein aggregates correspond to the clinical manifestations of dementia. Current assessment utilises semi-quantitative diagnostic staging criteria based on grading pathology as mild, moderate or severe (Alafuzoff et al. 2008; McKeith et al. 2017), which may mask subtle differences in pathological burden. However, the recent advances in quantitative automated image analysis technologies have enabled the assessment of large scale cohorts and offers accurate and reproducible methods that can be implemented by researchers with varying degrees of neuropathology experience (Neltner et al. 2012; Attems et al. 2014). A study demonstrating the importance of quantification techniques in AD identified three distinct clinico-pathological subtypes of AD. In the large-scale study consisting of 889 *post-mortem* cases, 11% of cases were found to have hippocampal sparing AD whilst 14% demonstrated limbic predominant AD in addition to those with typical AD (Murray et al. 2011). Moreover, these distinct subgroups of AD could be predicted *intra vitam* (Whitwell et al. 2012).

In addition to producing large-scale studies capable of detecting such discreet clinico-pathological subtypes of neurodegenerative disease, due to the complex nature of network connectivity in the brain and to gain knowledge of the global impact of pathological protein aggregates, brain regions should be studied simultaneously rather than in isolation. Tissue Microarray (TMA) is a technique most commonly employed in tumour studies, which allows a large number of samples from individual cases to be relocated into a single block suitable for high throughput analysis (Kononen et al. 1998; Bubendorf et al. 2001). This application has been developed to assess 15 anatomically distinct brain regions from a single *post-mortem* brain (Walker et al. 2017). Furthermore as serial sections can be taken from the same tissue block and stained for multiple markers it provides the ideal platform to investigate co-morbidities in age related neurodegenerative diseases.

Acknowledgments We thank the Newcastle Brain Tissue Resource for providing the photographic images, which is funded in part by a grant from the UK Medical Research Council (G0400074) and by Brains for Dementia Research, a joint venture between Alzheimer's Society and Alzheimer's Research UK. LW and KMc are funded by the Alzheimer's Society, and DE is funded by Alzheimer's Research UK.

References

Alafuzoff I, Arzberger T, Al-Sarraj S, Bodi I, Bogdanovic N, Braak H, Bugiani O, Del-Tredici K, Ferrer I, Gelpi E, Giaccone G, Graeber MB, Ince P, Kamphorst W, King A, Korkolopoulou P, Kovacs GG, Larionov S, Meyronet D, Monoranu C, Parchi P, Patsouris E, Roggendorf W, Seilhean D, Tagliavini F, Stadelmann C, Streichenberger N, Thal DR, Wharton SB, Kretzschmar H (2008) Stageing of neurofibrillary pathology in Alzheimer's disease: a study of the BrainNet Europe Consortium. Brain Pathol (Zurich, Switzerland) 18(4):484–496. https://doi.org/10.1111/j.1750-3639.2008.00147.x

Amador-Ortiz C, Lin WL, Ahmed Z, Personett D, Davies P, Duara R, Graff-Radford NR, Hutton ML, Dickson DW (2007) TDP-43 immunoreactivity in hippocampal sclerosis and Alzheimer's disease. Ann Neurol 61(5):435–445. https://doi.org/10.1002/ana.21154

Arai T, Mackenzie IR, Hasegawa M, Nonoka T, Niizato K, Tsuchiya K, Iritani S, Onaya M, Akiyama H (2009) Phosphorylated TDP-43 in Alzheimer's disease and dementia with Lewy bodies. Acta Neuropathol 117(2):125–136. https://doi.org/10.1007/s00401-008-0480-1

Arima K, Izumiyama Y, Nakamura M, Nakayama H, Kimura M, Ando S, Ikeda K, Takahashi K (1998) Argyrophilic tau-positive twisted and non-twisted tubules in astrocytic processes in brains of Alzheimer-type dementia: an electron microscopical study. Acta Neuropathol 95(1):28–39

Armstrong RA, Lantos PL, Cairns NJ (2005) Overlap between neurodegenerative disorders. Neuropathology 25(2):111–124

Ascherio A, Schwarzschild MA (2016) The epidemiology of Parkinson's disease: risk factors and prevention. Lancet Neurol 15(12):1257–1272. https://doi.org/10.1016/s1474-4422(16)30230-7

Attems J (2005) Sporadic cerebral amyloid angiopathy: pathology, clinical implications, and possible pathomechanisms. Acta Neuropathol 110(4):345–359. https://doi.org/10.1007/s00401-005-1074-9

Attems J, Jellinger K (2013) Neuropathological correlates of cerebral multimorbidity. Curr Alzheimer Res 10(6):569–577

Attems J, Jellinger KA, Lintner F (2005) Alzheimer's disease pathology influences severity and topographical distribution of cerebral amyloid angiopathy. Acta Neuropathol 110(3):222–231. https://doi.org/10.1007/s00401-005-1064-y

Attems J, Neltner JH, Nelson PT (2014) Quantitative neuropathological assessment to investigate cerebral multi-morbidity. Alzheimers Res Ther 6(9):85. https://doi.org/10.1186/s13195-014-0085-y

Bancher C, Jellinger K, Lassmann H, Fischer P, Leblhuber F (1996) Correlations between mental state and quantitative neuropathology in the Vienna Longitudinal Study on Dementia. Eur Arch Psychiatry Clin Neurosci 246(3):137–146

Bartels T, Choi JG, Selkoe DJ (2011) Alpha-Synuclein occurs physiologically as a helically folded tetramer that resists aggregation. Nature 477(7362):107–110. https://doi.org/10.1038/nature10324

Bertolotti A, Lutz Y, Heard DJ, Chambon P, Tora L (1996) hTAF(II)68, a novel RNA/ssDNA-binding protein with homology to the pro-oncoproteins TLS/FUS and EWS is associated with both TFIID and RNA polymerase II. EMBO J 15(18):5022–5031

Bigio EH (2011) TDP-43 variants of frontotemporal lobar degeneration. J Mol Neurosci 45(3):390–401. https://doi.org/10.1007/s12031-011-9545-z

Bigio EH, Lipton AM, White CL 3rd, Dickson DW, Hirano A (2003) Frontotemporal and motor neurone degeneration with neurofilament inclusion bodies: additional evidence for overlap between FTD and ALS. Neuropathol Appl Neurobiol 29(3):239–253

Binder LI, Frankfurter A, Rebhun LI (1985) The distribution of tau in the mammalian central nervous system. J Cell Biol 101(4):1371–1378

Blennow K, de Leon MJ, Zetterberg H (2006) Alzheimer's disease. Lancet 368(9533):387–403. https://doi.org/10.1016/s0140-6736(06)69113-7

Braak H, Braak E (1991) Neuropathological stageing of Alzheimer-related changes. Acta Neuropathol 82(4):239–259

Braak H, Del Tredici K, Gai WP, Braak E (2000) Alpha-synuclein is not a requisite component of synaptic boutons in the adult human central nervous system. J Chem Neuroanat 20(3–4):245–252

Braak H, Del Tredici K, Rub U, de Vos RA, Jansen Steur EN, Braak E (2003) Stageing of brain pathology related to sporadic Parkinson's disease. Neurobiol Ageing 24(2):197–211

Braak H, Alafuzoff I, Arzberger T, Kretzschmar H, Del Tredici K (2006) Stageing of Alzheimer disease-associated neurofibrillary pathology using paraffin sections and immunocytochemistry. Acta Neuropathol 112(4):389–404. https://doi.org/10.1007/s00401-006-0127-z

Brun A, Englund E (1986) A white matter disorder in dementia of the Alzheimer type: a pathoanatomical study. Ann Neurol 19(3):253–262. https://doi.org/10.1002/ana.410190306

Bubendorf L, Nocito A, Moch H, Sauter G (2001) Tissue microarray (TMA) technology: miniaturized pathology archives for high-throughput in situ studies. J Pathol 195(1):72–79. https://doi.org/10.1002/path.893

Buee L, Delacourte A (1999) Comparative biochemistry of tau in progressive supranuclear palsy, corticobasal degeneration, FTDP-17 and Pick's disease. Brain Path (Zurich, Switzerland) 9(4):681–693

Burre J (2015) The synaptic function of Alpha-synuclein. J Parkinson's Dis 5(4):699–713. https://doi.org/10.3233/jpd-150642

Burre J, Sharma M, Tsetsenis T, Buchman V, Etherton MR, Sudhof TC (2010) Alpha-synuclein promotes SNARE-complex assembly in vivo and in vitro. Science (New York, NY) 329(5999):1663–1667. https://doi.org/10.1126/science.1195227

Burre J, Vivona S, Diao J, Sharma M, Brunger AT, Sudhof TC (2013) Properties of native brain alpha-synuclein. Nature 498(7453):E4–E6.; discussion E6-7. https://doi.org/10.1038/nature12125

Cairns NJ, Neumann M, Bigio EH, Holm IE, Troost D, Hatanpaa KJ, Foong C, White CL 3rd, Schneider JA, Kretzschmar HA, Carter D, Taylor-Reinwald L, Paulsmeyer K, Strider J, Gitcho M, Goate AM, Morris JC, Mishra M, Kwong LK, A Stieber YX, Forman MS, Trojanowski JQ, Lee VM, Mackenzie IR (2007) TDP-43 in familial and sporadic frontotemporal lobar degeneration with ubiquitin inclusions. Am J Pathol 171(1):227–240. https://doi.org/10.2353/ajpath.2007.070182

Challa VR, Bell MA, Moody DM (1990) A combined hematoxylin-eosin, alkaline phosphatase and high-resolution microradiographic study of lacunes. Clin Neuropathol 9(4):196–204

Coelho-Cerqueira E, Carmo-Goncalves P, Pinheiro AS, Cortines J, Follmer C (2013) Alpha-Synuclein as an intrinsically disordered monomer–fact or artefact? FEBS J 280(19):4915–4927. https://doi.org/10.1111/febs.12471

Crary JF, Trojanowski JQ, Schneider JA, Abisambra JF, Abner EL, Alafuzoff I, Arnold SE, Attems J, Beach TG, Bigio EH, Cairns NJ, Dickson DW, Gearing M, Grinberg LT, Hof PR, Hyman BT, Jellinger K, Jicha GA, Kovacs GG, Knopman DS, Kofler J, Kukull WA, Mackenzie IR, Masliah E, McKee A, Montine TJ, Murray ME, Neltner JH, Santa-Maria I, Seeley WW, Serrano-Pozo A, Shelanski ML, Stein T, Takao M, Thal DR, Toledo JB, Troncoso JC, Vonsattel JP, White CL 3rd, Wisniewski T, Woltjer RL, Yamada M, Nelson PT (2014) Primary age-related tauopathy

(PART): a common pathology associated with human ageing. Acta Neuropathol 128(6):755–766. https://doi.org/10.1007/s00401-014-1349-0

Davidson YS, Raby S, Foulds PG, Robinson A, Thompson JC, Sikkink S, Yusuf I, Amin H, Duplessis D, Troakes C, Al-Sarraj S, Sloan C, Esiri MM, Prasher VP, Allsop D, Neary D, Pickering-Brown SM, Snowden JS, Mann DM (2011) TDP-43 pathological changes in early onset familial and sporadic Alzheimer's disease, late onset Alzheimer's disease and Down's syndrome: association with age, hippocampal sclerosis and clinical phenotype. Acta Neuropathol 122(6):703–713. https://doi.org/10.1007/s00401-011-0879-y

Dettmer U, Newman AJ, Soldner F, Luth ES, Kim NC, von Saucken VE, Sanderson JB, Jaenisch R, Bartels T, Selkoe D (2015) Parkinson-causing alpha-synuclein missense mutations shift native tetramers to monomers as a mechanism for disease initiation. Nat Commun 6:7314. https://doi.org/10.1038/ncomms8314

Dickson DW (1998) Pick's disease: a modern approach. Brain Path (Zurich, Switzerland) 8(2):339–354

Dickson DW, Bergeron C, Chin SS, Duyckaerts C, Horoupian D, Ikeda K, Jellinger K, Lantos PL, Lippa CF, Mirra SS, Tabaton M, Vonsattel JP, Wakabayashi K, Litvan I (2002) Office of rare diseases neuropathologic criteria for corticobasal degeneration. J Neuropathol Exp Neurol 61(11):935–946

Dickson DW, Rademakers R, Hutton ML (2007) Progressive supranuclear palsy: pathology and genetics. Brain Pat (Zurich, Switzerland) 17(1):74–82. https://doi.org/10.1111/j.1750-3639.2007.00054.x

Dickson DW, Kouri N, Murray ME, Josephs KA (2011) Neuropathology of frontotemporal lobar degeneration-tau (FTLD-tau). J Mol Neurosci 45(3):384–389. https://doi.org/10.1007/s12031-011-9589-0

Emre M, Aarsland D, Brown R, Burn DJ, Duyckaerts C, Mizuno Y, Broe GA, Cummings J, Dickson DW, Gauthier S, Goldman J, Goetz C, Korczyn A, Lees A, Levy R, Litvan I, McKeith I, Olanow W, Poewe W, Quinn N, Sampaio C, Tolosa E, Dubois B (2007) Clinical diagnostic criteria for dementia associated with Parkinson's disease. Mov Disord 22(12):1689–1707.; quiz 1837. https://doi.org/10.1002/mds.21507

Erskine D, Patterson L, Alexandris A, Hanson PS, McKeith IG, Attems J, Morris CM (2018) Regional levels of physiological alpha-synuclein are directly associated with Lewy body pathology. Acta Neuropathol 135(1):153–154. https://doi.org/10.1007/s00401-017-1787-6

Fauvet B, Mbefo MK, Fares MB, Desobry C, Michael S, Ardah MT, Tsika E, Coune P, Prudent M, Lion N, Eliezer D, Moore DJ, Schneider B, Aebischer P, El-Agnaf OM, Masliah E, Lashuel HA (2012) Alpha-Synuclein in central nervous system and from erythrocytes, mammalian cells, and Escherichia coli exists predominantly as disordered monomer. J Biol Chem 287(19):15345–15364. https://doi.org/10.1074/jbc.M111.318949

Fazekas F, Kleinert R, Offenbacher H, Schmidt R, Kleinert G, Payer F, Radner H, Lechner H (1993) Pathologic correlates of incidental MRI white matter signal hyperintensities. Neurology 43(9):1683–1689

Ferrer I, Lopez-Gonzalez I, Carmona M, Arregui L, Dalfo E, Torrejon-Escribano B, Diehl R, Kovacs GG (2014) Glial and neuronal tau pathology in tauopathies: characterization of disease-specific phenotypes and tau pathology progression. J Neuropathol Exp Neurol 73(1):81–97. https://doi.org/10.1097/nen.0000000000000030

Giasson BI, Forman MS, Higuchi M, Golbe LI, Graves CL, Kotzbauer PT, Trojanowski JQ, Lee VM (2003) Initiation and synergistic fibrillization of tau and alpha-synuclein. Science (New York, NY) 300(5619):636–640. https://doi.org/10.1126/science.1082324

Gong CX, Shaikh S, Wang JZ, Zaidi T, Grundke-Iqbal I, Iqbal K (1995) Phosphatase activity toward abnormally phosphorylated tau: decrease in Alzheimer disease brain. J Neurochem 65(2):732–738

Greenberg SM, Vernooij MW, Cordonnier C, Viswanathan A, Al-Shahi Salman R, Warach S, Launer LJ, Van Buchem MA, Breteler MM (2009) Cerebral microbleeds: a guide to detection and interpretation. Lancet Neurol 8(2):165–174. https://doi.org/10.1016/s1474-4422(09)70013-4

Grinberg LT, Thal DR (2010) Vascular pathology in the aged human brain. Acta Neuropathol 119(3):277–290. https://doi.org/10.1007/s00401-010-0652-7

Grundke-Iqbal I, Iqbal K, Tung YC, Quinlan M, Wisniewski HM, Binder LI (1986) Abnormal phosphorylation of the microtubule-associated protein tau (tau) in Alzheimer cytoskeletal pathology. Proc Natl Acad Sci U S A 83(13):4913–4917

Guerreiro R, Ross OA, Kun-Rodrigues C, Hernandez DG, Orme T, Eicher JD, Shepherd CE, Parkkinen L, Darwent L, Heckman MG, Scholz SW, Troncoso JC, Pletnikova O, Ansorge O, Clarimon J, Lleo A, Morenas-Rodriguez E, Clark L, Honig LS, Marder K, Lemstra A, Rogaeva E, St George-Hyslop P, Londos E, Zetterberg H, Barber I, Braae A, Brown K, Morgan K, Troakes C, Al-Sarraj S, Lashley T, Holton J, Compta Y, Van Deerlin V, Serrano GE, Beach TG, Lesage S, Galasko D, Masliah E, Santana I, Pastor P, Diez-Fairen M, Aguilar M, Tienari PJ, Myllykangas L, Oinas M, Revesz T, Lees A, Boeve BF, Petersen RC, Ferman TJ, Escott-Price V, Graff-Radford N, Cairns NJ, Morris JC, Pickering-Brown S, Mann D, Halliday GM, Hardy J, Trojanowski JQ, Dickson DW, Singleton A, Stone DJ, Bras J (2018) Investigating the genetic architecture of dementia with Lewy bodies: a two-stage genome-wide association study. Lancet Neurol 17(1):64–74. https://doi.org/10.1016/s1474-4422(17)30400-3

Hachinski V, Iadecola C, Petersen RC, Breteler MM, Nyenhuis DL, Black SE, Powers WJ, Decarli C, Merino JG, Kalaria RN, Vinters HV, Holtzman DM, Rosenberg GA, Wallin A, Dichgans M, Marler JR, Leblanc GG (2006) National Institute of Neurological Disorders and Stroke-Canadian Stroke Network vascular cognitive impairment harmonization standards. Stroke 37(9):2220–2241. https://doi.org/10.1161/01.str.0000237236.88823.47

Hacke W, Schwab S, Horn M, Spranger M, De Georgia M, von Kummer R (1996) 'Malignant' middle cerebral artery territory infarction: clinical course and prognostic signs. Arch Neurol 53(4):309–315

Halliday GM, Double KL, MacDonald V, Kril JJ (2003) Identifying severely atrophic cortical subregions in Alzheimer's disease. Neurobiol Ageing 24(6):797–806

Han W, Liu Y, Mi Y, Zhao J, Liu D, Tian Q (2015) Alpha-synuclein (SNCA) polymorphisms and susceptibility to Parkinson's disease: a meta-analysis. Am J Med Genet Part B Neuropsychiatr Genet 168b(2):123–134. https://doi.org/10.1002/ajmg.b.32288

Hanger DP, Anderton BH, Noble W (2009) Tau phosphorylation: the therapeutic challenge for neurodegenerative disease. Trends Mol Med 15(3):112–119. https://doi.org/10.1016/j.molmed.2009.01.003

Heidebrink JL (2002) Is dementia with Lewy bodies the second most common cause of dementia? J Geriatr Psychiatry Neurol 15(4):182–187. https://doi.org/10.1177/089198870201500402

Higashi S, Iseki E, Yamamoto R, Minegishi M, Hino H, Fujisawa K, Togo T, Katsuse O, Uchikado H, Furukawa Y, Kosaka K, Arai H (2007) Concurrence of TDP-43, tau and alpha-synuclein pathology in brains of Alzheimer's disease and dementia with Lewy bodies. Brain Res 1184:284–294. https://doi.org/10.1016/j.brainres.2007.09.048

Hong M, Zhukareva V, Vogelsberg-Ragaglia V, Wszolek Z, Reed L, Miller BI, Geschwind DH, Bird TD, McKeel D, Goate A, Morris JC, Wilhelmsen KC, Schellenberg GD, Trojanowski JQ, Lee VM (1998) Mutation-specific functional impairments in distinct tau isoforms of hereditary FTDP-17. Science (New York, NY) 282(5395):1914–1917

Hyman BT, Phelps CH, Beach TG, Bigio EH, Cairns NJ, Carrillo MC, Dickson DW, Duyckaerts C, Frosch MP, Masliah E, Mirra SS, Nelson PT, Schneider JA, Thal DR, Thies B, Trojanowski JQ, Vinters HV, Montine TJ (2012) National Institute on Ageing-Alzheimer's Association guidelines for the neuropathologic assessment of Alzheimer's disease. Alzheimers Dement 8(1):1–13. https://doi.org/10.1016/j.jalz.2011.10.007

Ingram EM, Spillantini MG (2002) Tau gene mutations: dissecting the pathogenesis of FTDP-17. Trends Mol Med 8(12):555–562

Irwin DJ, Grossman M, Weintraub D, Hurtig HI, Duda JE, Xie SX, Lee EB, Van Deerlin VM, Lopez OL, Kofler JK, Nelson PT, Jicha GA, Woltjer R, Quinn JF, Kaye J, Leverenz JB, Tsuang D, Longfellow K, Yearout D, Kukull W, Keene CD, Montine TJ, Zabetian CP, Trojanowski JQ (2017) Neuropathological and genetic correlates of survival and dementia onset in synucleinopathies: a retrospective analysis. Lancet Neurol 16(1):55–65. https://doi.org/10.1016/s1474-4422(16)30291-5

Irwin DJ, Xie SX, Coughlin D, Nevler N, Akhtar RS, McMillan CT, Lee EB, Wolk DA, Weintraub D, Chen-Plotkin A, Duda JE, Spindler M, Siderowf A, Hurtig HI, Shaw LM, Grossman M, Trojanowski JQ (2018) CSF tau and beta-amyloid predict cerebral synucleinopathy in autopsied Lewy body disorders. Neurology 90(12):e1038–e1046. https://doi.org/10.1212/wnl.0000000000005166

Jack CR Jr, Albert MS, Knopman DS, McKhann GM, Sperling RA, Carrillo MC, Thies B, Phelps CH (2011) Introduction to the recommendations from the National Institute on Ageing-Alzheimer's Association workgroups on diagnostic guidelines for Alzheimer's disease. Alzheimers Dement 7(3):257–262. https://doi.org/10.1016/j.jalz.2011.03.004

James BD, Wilson RS, Boyle PA, Trojanowski JQ, Bennett DA, Schneider JA (2016) TDP-43 stage, mixed pathologies, and clinical Alzheimer's-type dementia. Brain J Neurol 139(11):2983–2993. https://doi.org/10.1093/brain/aww224

Janknecht R (2005) EWS-ETS oncoproteins: the linchpins of Ewing tumors. Gene 363:1–14. https://doi.org/10.1016/j.gene.2005.08.007

Jellinger KA (2007) The enigma of vascular cognitive disorder and vascular dementia. Acta Neuropathol 113(4):349–388. https://doi.org/10.1007/s00401-006-0185-2

Jellinger KA (2008) The pathology of "vascular dementia": a critical update. J Alzheimers Dis 14(1):107–123

Jellinger KA, Lantos PL (2010) Papp-Lantos inclusions and the pathogenesis of multiple system atrophy: an update. Acta Neuropathol 119(6):657–667. https://doi.org/10.1007/s00401-010-0672-3

Jeong JH, Yoon SJ, Kang SJ, Choi KG, Na DL (2002) Hypertensive pontine microhemorrhage. Stroke 33(4):925–929

Josephs KA, Whitwell JL, Knopman DS, Hu WT, Stroh DA, Baker M, Rademakers R, Boeve BF, Parisi JE, Smith GE, Ivnik RJ, Petersen RC, Jack CR Jr, Dickson DW (2008) Abnormal TDP-43 immunoreactivity in AD modifies clinicopathologic and radiologic phenotype. Neurology 70(19 Pt 2):1850–1857. https://doi.org/10.1212/01.wnl.0000304041.09418.b1

Josephs KA, Murray ME, Whitwell JL, Parisi JE, Petrucelli L, Jack CR, Petersen RC, Dickson DW (2014a) Stageing TDP-43 pathology in Alzheimer's disease. Acta Neuropathol 127(3):441–450. https://doi.org/10.1007/s00401-013-1211-9

Josephs KA, Whitwell JL, Weigand SD, Murray ME, Tosakulwong N, Liesinger AM, Petrucelli L, Senjem ML, Knopman DS, Boeve BF, Ivnik RJ, Smith GE, Jack CR Jr, Parisi JE, Petersen RC, Dickson DW (2014b) TDP-43 is a key player in the clinical features associated with Alzheimer's disease. Acta Neuropathol 127(6):811–824. https://doi.org/10.1007/s00401-014-1269-z

Josephs KA, Whitwell JL, Tosakulwong N, Weigand SD, Murray ME, Liesinger AM, Petrucelli L, Senjem ML, Ivnik RJ, Parisi JE, Petersen RC, Dickson DW (2015) TAR DNA-binding protein 43 and pathological subtype of Alzheimer's disease impact clinical features. Ann Neurol 78(5):697–709. https://doi.org/10.1002/ana.24493

Josephs KA, Murray ME, Whitwell JL, Tosakulwong N, Weigand SD, Petrucelli L, Liesinger AM, Petersen RC, Parisi JE, Dickson DW (2016) Updated TDP-43 in Alzheimer's disease stageing scheme. Acta Neuropathol 131(4):571–585. https://doi.org/10.1007/s00401-016-1537-1

Josephs KA, Dickson DW, Tosakulwong N, Weigand SD, Murray ME, Petrucelli L, Liesinger AM, Senjem ML, Spychalla AJ, Knopman DS, Parisi JE, Petersen RC, Jack CR Jr, Whitwell JL (2017) Rates of hippocampal atrophy and presence of post-mortem TDP-43 in patients with Alzheimer's disease: a longitudinal retrospective study. Lancet Neurol. https://doi.org/10.1016/s1474-4422(17)30284-3

Kang J, Lemaire HG, Unterbeck A, Salbaum JM, Masters CL, Grzeschik KH, Multhaup G, Beyreuther K, Muller-Hill B (1987) The precursor of Alzheimer's disease amyloid A4 protein resembles a cell-surface receptor. Nature 325(6106):733–736. https://doi.org/10.1038/325733a0

Kononen J, Bubendorf L, Kallioniemi A, Barlund M, Schraml P, Leighton S, Torhorst J, Mihatsch MJ, Sauter G, Kallioniemi OP (1998) Tissue microarrays for high-throughput molecular profiling of tumor specimens. Nat Med 4(7):844–847

Kopke E, Tung YC, Shaikh S, Alonso AC, Iqbal K, Grundke-Iqbal I (1993) Microtubule-associated protein tau. Abnormal phosphorylation of a non-paired helical filament pool in Alzheimer disease. J Biol Chem 268(32):24374–24384

Kovacs GG, Alafuzoff I, Al-Sarraj S, Arzberger T, Bogdanovic N, Capellari S, Ferrer I, Gelpi E, Kovari V, Kretzschmar H, Nagy Z, Parchi P, Seilhean D, Soininen H, Troakes C, Budka H (2008) Mixed brain pathologies in dementia: the BrainNet Europe consortium experience. Dement Geriatr Cogn Disord 26(4):343–350. https://doi.org/10.1159/000161560

Kovacs GG, Molnar K, Laszlo L, Strobel T, Botond G, Honigschnabl S, Reiner-Concin A, Palkovits M, Fischer P, Budka H (2011) A peculiar constellation of tau pathology defines a subset of dementia in the elderly. Acta Neuropathol 122(2):205–222. https://doi.org/10.1007/s00401-011-0819-x

Kovacs GG, Wagner U, Dumont B, Pikkarainen M, Osman AA, Streichenberger N, Leisser I, Verchere J, Baron T, Alafuzoff I, Budka H, Perret-Liaudet A, Lachmann I (2012) An antibody with high reactivity for disease-associated alpha-synuclein reveals extensive brain pathology. Acta Neuropathol 124(1):37–50. https://doi.org/10.1007/s00401-012-0964-x

Kovacs GG, Breydo L, Green R, Kis V, Puska G, Lorincz P, Perju-Dumbrava L, Giera R, Pirker W, Lutz M, Lachmann I, Budka H, Uversky VN, Molnar K, Laszlo L (2014) Intracellular processing of disease-associated alpha-synuclein in the human brain suggests prion-like cell-to-cell spread. Neurobiol Dis 69:76–92. https://doi.org/10.1016/j.nbd.2014.05.020

Kovacs GG, Ferrer I, Grinberg LT, Alafuzoff I, Attems J, Budka H, Cairns NJ, Crary JF, Duyckaerts C, Ghetti B, Halliday GM, Ironside JW, Love S, Mackenzie IR, Munoz DG, Murray ME, Nelson PT, Takahashi H, Trojanowski JQ, Ansorge O, Arzberger T, Baborie A, Beach TG, Bieniek KF, Bigio EH, Bodi I, Dugger BN, Feany M, Gelpi E, Gentleman SM, Giaccone G, Hatanpaa KJ, Heale R, Hof PR, Hofer M, Hortobagyi T, Jellinger K, Jicha GA, Ince P, Kofler J, Kovari E, Kril JJ, Mann DM, Matej R, McKee AC, McLean C, Milenkovic I, Montine TJ, Murayama S, Lee EB, Rahimi J, Rodriguez RD, Rozemuller A, Schneider JA, Schultz C, Seeley W, Seilhean D, Smith C, Tagliavini F, Takao M, Thal DR, Toledo JB, Tolnay M, Troncoso JC, Vinters HV, Weis S, Wharton SB, White CL 3rd, Wisniewski T, Woulfe JM, Yamada M, Dickson DW (2016) Ageing-related tau astrogliopathy (ARTAG): harmonized evaluation strategy. Acta Neuropathol 131(1):87–102. https://doi.org/10.1007/s00401-015-1509-x

Kovacs GG, Robinson JL, Xie SX, Lee EB, Grossman M, Wolk DA, Irwin DJ, Weintraub D, Kim CF, Schuck T, Yousef A, Wagner ST, Suh E, Van Deerlin VM, Lee VM, Trojanowski JQ (2017) Evaluating the patterns of ageing-related tau Astrogliopathy unravels novel insights into brain ageing and neurodegenerative diseases. J Neuropathol Exp Neurol 76(4):270–288. https://doi.org/10.1093/jnen/nlx007

Kraybill ML, Larson EB, Tsuang DW, Teri L, McCormick WC, Bowen JD, Kukull WA, Leverenz JB, Cherrier MM (2005) Cognitive differences in dementia patients with autopsy-verified AD, Lewy body pathology, or both. Neurology 64(12):2069–2073. https://doi.org/10.1212/01.wnl.0000165987.89198.65

Ksiezak-Reding H, Liu WK, Yen SH (1992) Phosphate analysis and dephosphorylation of modified tau associated with paired helical filaments. Brain Res 597(2):209–219

Lace G, Ince PG, Brayne C, Savva GM, Matthews FE, de Silva R, Simpson JE, Wharton SB (2012) Mesial temporal astrocyte tau pathology in the MRC-CFAS ageing brain cohort. Dement Geriatr Cogn Disord 34(1):15–24. https://doi.org/10.1159/000341581

Lagier-Tourenne C, Polymenidou M, Cleveland DW (2010) TDP-43 and FUS/TLS: emerging roles in RNA processing and neurodegeneration. Hum Mol Genet 19(R1):R46–R64. https://doi.org/10.1093/hmg/ddq137

Lawton M, Baig F, Rolinski M, Ruffman C, Nithi K, May MT, Y Ben-Shlomo MTH (2015) Parkinson's disease subtypes in the Oxford Parkinson Disease Centre (OPDC) discovery cohort. J Parkinson's Dis 5(2):269–279. https://doi.org/10.3233/jpd-140523

Lee VM, Giasson BI, Trojanowski JQ (2004) More than just two peas in a pod: common amyloido-genic properties of tau and alpha-synuclein in neurodegenerative diseases. Trends Neurosci 27(3):129–134. https://doi.org/10.1016/j.tins.2004.01.007

Ling H, Neal JW, Revesz T (2017) Evolving concepts of chronic traumatic encephalopathy as a neuropathological entity. Neuropathol Appl Neurobiol 43(6):467–476. https://doi.org/10.1111/nan.12425

Litvan I, Grimes DA, Lang AE (2000) Phenotypes and prognosis: clinicopathologic studies of corticobasal degeneration. Adv Neurol 82:183–196

Liu F, Grundke-Iqbal I, Iqbal K, Gong CX (2005a) Contributions of protein phosphatases PP1, PP2A, PP2B and PP5 to the regulation of tau phosphorylation. Eur J Neurosci 22(8):1942–1950. https://doi.org/10.1111/j.1460-9568.2005.04391.x

Liu F, Iqbal K, Grundke-Iqbal I, Rossie S, Gong CX (2005b) Dephosphorylation of tau by protein phosphatase 5: impairment in Alzheimer's disease. J Biol Chem 280(3):1790–1796. https://doi.org/10.1074/jbc.M410775200

Liu AK, Goldfinger MH, Questari HE, Pearce RK, Gentleman SM (2016) ARTAG in the basal forebrain: widening the constellation of astrocytic tau pathology. Acta Neuropathol Commun 4(1):59. https://doi.org/10.1186/s40478-016-0330-7

Mackenzie IR, Neumann M (2017) Reappraisal of TDP-43 pathology in FTLD-U subtypes. Acta Neuropathol 134(1):79–96. https://doi.org/10.1007/s00401-017-1716-8

Mackenzie IR, Rademakers R, Neumann M (2010) TDP-43 and FUS in amyotrophic lateral scle-rosis and frontotemporal dementia. Lancet Neurol 9(10):995–1007. https://doi.org/10.1016/s1474-4422(10)70195-2

Mackenzie IR, Neumann M, Baborie A, Sampathu DM, Du Plessis D, Jaros E, Perry RH, Trojanowski JQ, Mann DM, Lee VM (2011) A harmonized classification system for FTLD-TDP pathology. Acta Neuropathol 122(1):111–113. https://doi.org/10.1007/s00401-011-0845-8

Mandler M, Walker L, Santic R, Hanson P, Upadhaya AR, Colloby SJ, Morris CM, Thal DR, Thomas AJ, Schneeberger A, Attems J (2014) Pyroglutamylated amyloid-beta is associ-ated with hyperphosphorylated tau and severity of Alzheimer's disease. Acta Neuropathol 128(1):67–79. https://doi.org/10.1007/s00401-014-1296-9

Masuda-Suzukake M, Nonaka T, Hosokawa M, Kubo M, Shimozawa A, Akiyama H, Hasegawa M (2014) Pathological alpha-synuclein propagates through neural networks. Acta Neuropathol Commun 2(1):88. https://doi.org/10.1186/preaccept-1296467154135944

Mayeux R, Stern Y (2012) Epidemiology of Alzheimer disease. Cold Spring Harb Perspect Med 2(8). https://doi.org/10.1101/cshperspect.a006239

McAleese KE, Walker L, Erskine D, Thomas AJ, McKeith IG, Attems J (2017) TDP-43 pathol-ogy in Alzheimer's disease, dementia with Lewy bodies and ageing. Brain Path (Zurich, Switzerland) 27(4):472–479. https://doi.org/10.1111/bpa.12424

McKee AC, Cantu RC, Nowinski CJ, Hedley-Whyte ET, Gavett BE, Budson AE, Santini VE, Lee HS, Kubilus CA, Stern RA (2009) Chronic traumatic encephalopathy in athletes: progressive tauopathy after repetitive head injury. J Neuropathol Exp Neurol 68(7):709–735. https://doi.org/10.1097/NEN.0b013e3181a9d503

McKee AC, Stern RA, Nowinski CJ, Stein TD, Alvarez VE, Daneshvar DH, Lee HS, Wojtowicz SM, Hall G, Baugh CM, Riley DO, Kubilus CA, Cormier KA, Jacobs MA, Martin BR, Abraham CR, Ikezu T, Reichard RR, Wolozin BL, Budson AE, Goldstein LE, Kowall NW, Cantu RC (2013) The spectrum of disease in chronic traumatic encephalopathy. Brain J Neurol 136(Pt 1):43–64. https://doi.org/10.1093/brain/aws307

McKee AC, Cairns NJ, Dickson DW, Folkerth RD, Keene CD, Litvan I, Perl DP, Stein TD, Vonsattel JP, Stewart W, Tripodis Y, Crary JF, Bieniek KF, Dams-O'Connor K, Alvarez VE, Gordon WA (2016) The first NINDS/NIBIB consensus meeting to define neuropathological criteria for the diagnosis of chronic traumatic encephalopathy. Acta Neuropathol 131(1):75–86. https://doi.org/10.1007/s00401-015-1515-z

McKeith IG, Dickson DW, Lowe J, Emre M, O'Brien JT, Feldman H, Cummings J, Duda JE, Lippa C, Perry EK, Aarsland D, Arai H, Ballard CG, Boeve B, Burn DJ, Costa D, Del Ser T, Dubois B,

Galasko D, Gauthier S, Goetz CG, Gomez-Tortosa E, Halliday G, Hansen LA, Hardy J, Iwatsubo T, Kalaria RN, Kaufer D, Kenny RA, Korczyn A, Kosaka K, Lee VM, Lees A, Litvan I, Londos E, Lopez OL, Minoshima S, Mizuno Y, Molina JA, Mukaetova-Ladinska EB, Pasquier F, Perry RH, Schulz JB, Trojanowski JQ, Yamada M, DLB Consortium on DLB (2005) Diagnosis and management of dementia with Lewy bodies: third report of the DLB Consortium. Neurology 65(12):1863–1872. https://doi.org/10.1212/01.wnl.0000187889.17253.b1

McKeith IG, Boeve BF, Dickson DW, Halliday G, Taylor JP, Weintraub D, Aarsland D, Galvin J, Attems J, Ballard CG, Bayston A, Beach TG, Blanc F, Bohnen N, Bonanni L, Bras J, Brundin P, Burn D, Chen-Plotkin A, Duda JE, El-Agnaf O, Feldman H, Ferman TJ, Ffytche D, Fujishiro H, Galasko D, Goldman JG, Gomperts SN, Graff-Radford NR, Honig LS, Iranzo A, Kantarci K, Kaufer D, Kukull W, Lee VMY, Leverenz JB, Lewis S, Lippa C, Lunde A, Masellis M, Masliah E, McLean P, Mollenhauer B, Montine TJ, Moreno E, Mori E, Murray M, O'Brien JT, Orimo S, Postuma RB, Ramaswamy S, Ross OA, Salmon DP, Singleton A, Taylor A, Thomas A, Tiraboschi P, Toledo JB, Trojanowski JQ, Tsuang D, Walker Z, Yamada M, Kosaka K (2017) Diagnosis and management of dementia with Lewy bodies: fourth consensus report of the DLB Consortium. Neurology 89(1):88–100. https://doi.org/10.1212/wnl.0000000000004058

McKhann G, Drachman D, Folstein M, Katzman R, Price D, Stadlan EM (1984) Clinical diagnosis of Alzheimer's disease: report of the NINCDS-ADRDA Work Group under the auspices of Department of Health and Human Services Task Force on Alzheimer's Disease. Neurology 34(7):939–944

McKhann GM, Knopman DS, Chertkow H, Hyman BT, Jack CR Jr, Kawas CH, Klunk WE, Koroshetz WJ, Manly JJ, Mayeux R, Mohs RC, Morris JC, Rossor MN, Scheltens P, Carrillo MC, Thies B, Weintraub S, Phelps CH (2011) The diagnosis of dementia due to Alzheimer's disease: recommendations from the National Institute on Ageing-Alzheimer's Association workgroups on diagnostic guidelines for Alzheimer's disease. Alzheimers Dement 7(3):263–269. https://doi.org/10.1016/j.jalz.2011.03.005

Mirra SS, Heyman A, McKeel D, Sumi SM, Crain BJ, Brownlee LM, Vogel FS, Hughes JP, van Belle G, Berg L (1991) The Consortium to Establish a Registry for Alzheimer's Disease (CERAD). Part II. Standardization of the neuropathologic assessment of Alzheimer's disease. Neurology 41(4):479–486

Montine T, Phelps C, Beach T, Bigio E, Cairns N, Dickson D, Duyckaerts C, Frosch M, Masliah E, Mirra S, Nelson P, Schneider J, Thal D, Trojanowski J, Vinters H, Hyman B (2012) National Institute on Ageing–Alzheimer's Association guidelines for the neuropathologic assessment of Alzheimer's disease: a practical approach. Acta Neuropathol 123:1–11

Morita M, Al-Chalabi A, Andersen PM, Hosler B, Sapp P, Englund E, Mitchell JE, Habgood JJ, de Belleroche J, Xi J, Jongjaroenprasert W, Horvitz HR, Gunnarsson LG, Brown RH Jr (2006) A locus on chromosome 9p confers susceptibility to ALS and frontotemporal dementia. Neurology 66(6):839–844. https://doi.org/10.1212/01.wnl.0000200048.53766.b4

Morley JE, Farr SA, Banks WA, Johnson SN, Ka Yamada LX (2010) A physiological role for amyloid-beta protein: enhancement of learning and memory. J Alzheimers Dis 19(2):441–449. https://doi.org/10.3233/jad-2009-1230

Munoz DG, Woulfe J, Kertesz A (2007) Argyrophilic thorny astrocyte clusters in association with Alzheimer's disease pathology in possible primary progressive aphasia. Acta Neuropathol 114(4):347–357. https://doi.org/10.1007/s00401-007-0266-x

Murray ME, Graff-Radford NR, Ross OA, Petersen RC, Duara R, Dickson DW (2011) Neuropathologically defined subtypes of Alzheimer's disease with distinct clinical characteristics: a retrospective study. Lancet Neurol 10(9):785–796. https://doi.org/10.1016/s1474-4422(11)70156-9

Nakashima-Yasuda H, Uryu K, Robinson J, Xie SX, Hurtig H, Duda JE, Arnold SE, Siderowf A, Grossman M, Leverenz JB, Woltjer R, Lopez OL, Hamilton R, Tsuang DW, Galasko D, Masliah E, Kaye J, Clark CM, Montine TJ, Lee VM, Trojanowski JQ (2007) Co-morbidity of TDP-43 proteinopathy in Lewy body related diseases. Acta Neuropathol 114(3):221–229. https://doi.org/10.1007/s00401-007-0261-2

Neary D, Snowden JS, Gustafson L, Passant U, Stuss D, Black S, Freedman M, Kertesz A, Robert PH, Albert M, Boone K, Miller BL, Cummings J, Benson DF (1998) Frontotemporal lobar degeneration: a consensus on clinical diagnostic criteria. Neurology 51(6):1546–1554

Neltner JH, Abner EL, Schmitt FA, Denison SK, Anderson S, Patel E, Nelson PT (2012) Digital pathology and image analysis for robust high-throughput quantitative assessment of Alzheimer disease neuropathologic changes. J Neuropathol Exp Neurol 71(12):1075–1085. https://doi.org/10.1097/NEN.0b013e3182768de4

Neumann M, Sampathu DM, Kwong LK, Truax AC, Micsenyi MC, Chou TT, Bruce J, Schuck T, Grossman M, Clark CM, McCluskey LF, Miller BL, Masliah E, Mackenzie IR, Feldman H, Feiden W, Kretzschmar HA, Trojanowski JQ, Lee VM (2006) Ubiquitinated TDP-43 in frontotemporal lobar degeneration and amyotrophic lateral sclerosis. Science (New York, NY) 314(5796):130–133. https://doi.org/10.1126/science.1134108

Nussbaum RL, Ellis CE (2003) Alzheimer's disease and Parkinson's disease. N Engl J Med 348(14):1356–1364. https://doi.org/10.1056/NEJM2003ra020003

O'Brien JT, Erkinjuntti T, Reisberg B, Roman G, Sawada T, Pantoni L, Bowler JV, Ballard C, Decarli C, Gorelick PB, Rockwood K, Burns A, Gauthier S, Dekosky ST (2003) Vascular cognitive impairment. Lancet Neurol 2(2):89–98

Olichney JM, Galasko D, Salmon DP, Hofstetter CR, Hansen LA, Katzman R, Thal LJ (1998) Cognitive decline is faster in Lewy body variant than in Alzheimer's disease. Neurology 51(2):351–357

Olsson B, Lautner R, Andreasson U, Ohrfelt A, Portelius E, Bjerke M, Holtta M, Rosen C, Olsson C, G Strobel EW, Dakin K, Petzold M, Blennow K, Zetterberg H (2016) CSF and blood biomarkers for the diagnosis of Alzheimer's disease: a systematic review and meta-analysis. Lancet Neurol 15(7):673–684. https://doi.org/10.1016/s1474-4422(16)00070-3

Ou SH, Wu F, Harrich D, Garcia-Martinez LF, Gaynor RB (1995) Cloning and characterization of a novel cellular protein, TDP-43, that binds to human immunodeficiency virus type 1 TAR DNA sequence motifs. J Virol 69(6):3584–3596

Peuralinna T, Myllykangas L, Oinas M, Nalls MA, Keage HA, Isoviita VM, Valori M, Polvikoski T, Paetau A, Sulkava R, Ince PG, Zaccai J, Brayne C, Traynor BJ, Hardy J, Singleton AB, Tienari PJ (2015) Genome-wide association study of neocortical Lewy-related pathology. Ann Clin Transl Neurol 2(9):920–931. https://doi.org/10.1002/acn3.231

Pollanen MS, Dickson DW, Bergeron C (1993) Pathology and biology of the Lewy body. J Neuropathol Exp Neurol 52(3):183–191

Polymeropoulos MH, Lavedan C, Leroy E, Ide SE, Dehejia A, Dutra A, Pike B, Root H, Rubenstein J, Boyer R, Stenroos ES, Chandrasekharappa S, Athanassiadou A, Papapetropoulos T, Johnson WG, Lazzarini AM, Duvoisin RC, Di Iorio G, Golbe LI, Nussbaum RL (1997) Mutation in the alpha-synuclein gene identified in families with Parkinson's disease. Science (New York, NY) 276(5321):2045–2047

Postuma RB, Berg D, Stern M, Poewe W, Olanow CW, Oertel W, Obeso J, Marek K, Litvan I, Lang AE, Halliday G, Goetz CG, Gasser T, Dubois B, Chan P, Bloem BR, Adler CH, Deuschl G (2015) MDS clinical diagnostic criteria for Parkinson's disease. Mov Disord 30(12):1591–1601. https://doi.org/10.1002/mds.26424

Pringsheim T, Jette N, Frolkis A, Steeves TD (2014) The prevalence of Parkinson's disease: a systematic review and meta-analysis. Mov Disord 29(13):1583–1590. https://doi.org/10.1002/mds.25945

Revesz T, Ghiso J, Lashley T, Plant G, Rostagno A, Frangione B, Holton JL (2003) Cerebral amyloid angiopathies: a pathologic, biochemical, and genetic view. J Neuropathol Exp Neurol 62(9):885–898

Russo C, Violani E, Salis S, Venezia V, Dolcini V, Damonte G, Benatti U, D'Arrigo C, Patrone E, Carlo P, Schettini G (2002) Pyroglutamate-modified amyloid beta-peptides–AbetaN3(pE)–strongly affect cultured neuron and astrocyte survival. J Neurochem 82(6):1480–1489

Saito Y, Murayama S (2007) Neuropathology of mild cognitive impairment. Neuropathology 27(6):578–584. https://doi.org/10.1111/j.1440-1789.2007.00806.x

Schmidt ML, Zhukareva V, Newell KL, Lee VM, Trojanowski JQ (2001) Tau isoform profile and phosphorylation state in dementia pugilistica recapitulate Alzheimer's disease. Acta Neuropathol 101(5):518–524

Schmidt R, Schmidt H, Haybaeck J, Loitfelder M, Weis S, Cavalieri M, Seiler S, Enzinger C, Ropele S, Erkinjuntti T, Pantoni L, Scheltens P, Fazekas F, Jellinger K (2011) Heterogeneity in age-related white matter changes. Acta Neuropathol 122(2):171–185. https://doi.org/10.1007/s00401-011-0851-x

Selkoe DJ (2001a) Alzheimer's disease results from the cerebral accumulation and cytotoxicity of amyloid beta-protein. J Alzheimers Dis 3(1):75–80

Selkoe DJ (2001b) Alzheimer's disease: genes, proteins, and therapy. Physiol Rev 81(2):741–766

Serby M, Brickman AM, Haroutunian V, Purohit DP, Marin D, Lantz M, Mohs RC, Davis KL (2003) Cognitive burden and excess Lewy-body pathology in the Lewy-body variant of Alzheimer disease. Am J Geriatr Psychiatry 11(3):371–374

Shibasaki Y, Baillie DA, St Clair D, Brookes AJ (1995) High-resolution mapping of SNCA encoding alpha-synuclein, the non-A beta component of Alzheimer's disease amyloid precursor, to human chromosome 4q21.3-->q22 by fluorescence in situ hybridization. Cytogenet Cell Genet 71(1):54–55. https://doi.org/10.1159/000134061

Skrobot OA, Attems J, Esiri M, Hortobagyi T, Ironside JW, Kalaria RN, King A, Lammie GA, Mann D, Neal J, Ben-Shlomo Y, Kehoe PG, Love S (2016) Vascular cognitive impairment neuropathology guidelines (VCING): the contribution of cerebrovascular pathology to cognitive impairment. Brain J Neurol 139(11):2957–2969. https://doi.org/10.1093/brain/aww214

Spillantini MG, Schmidt ML, Lee VM, Trojanowski JQ, Jakes R, Goedert M (1997) Alpha-synuclein in Lewy bodies. Nature 388(6645):839–840. https://doi.org/10.1038/42166

Stary HC (2000) Natural history and histological classification of atherosclerotic lesions: an update. Arterioscler Thromb Vasc Biol 20(5):1177–1178

Steele JC, Richardson JC, Olszewski J (1964) Progressive supranuclear palsy. A heterogeneous degeneration involving the brain stem, basal ganglia and cerebellum with vertical gaze and pseudobulbar palsy, nuchal dystonia and dementia. Arch Neurol 10:333–359

Surguchov A (2013) Synucleins: are they two-edged swords? J Neurosci Res 91(2):161–166. https://doi.org/10.1002/jnr.23149

Surguchov A (2015) Intracellular dynamics of synucleins: "here, there and everywhere". Int Rev Cell Mol Biol 320:103–169. https://doi.org/10.1016/bs.ircmb.2015.07.007

Surmeier DJ, Obeso JA, Halliday GM (2017) Selective neuronal vulnerability in Parkinson disease. Nat Rev Neurosci 18(2):101–113. https://doi.org/10.1038/nrn.2016.178

Thal DR, Rub U, Orantes M, Braak H (2002) Phases of A beta-deposition in the human brain and its relevance for the development of AD. Neurology 58(12):1791–1800

Thal DR, Ghebremedhin E, Orantes M, Wiestler OD (2003) Vascular pathology in Alzheimer disease: correlation of cerebral amyloid angiopathy and arteriosclerosis/lipohyalinosis with cognitive decline. J Neuropathol Exp Neurol 62(12):1287–1301

Thal DR, Papassotiropoulos A, Saido TC, Griffin WS, Mrak RE, Kolsch H, Del Tredici K, Attems J, Ghebremedhin E (2010) Capillary cerebral amyloid angiopathy identifies a distinct APOE epsilon4-associated subtype of sporadic Alzheimer's disease. Acta Neuropathol 120(2):169–183. https://doi.org/10.1007/s00401-010-0707-9

Tolnay M, Mistl C, Ipsen S, Probst A (1998) Argyrophilic grains of Braak: occurrence in dendrites of neurons containing hyperphosphorylated tau protein. Neuropathol Appl Neurobiol 24(1):53–59

Tsuang D, Leverenz JB, Lopez OL, Hamilton RL, Bennett DA, Schneider JA, Buchman AS, Larson EB, Crane PK, Kaye JA, Kramer P, Woltjer R, Trojanowski JQ, Weintraub D, Chen-Plotkin AS, Irwin DJ, Rick J, Schellenberg GD, Watson GS, Kukull W, Nelson PT, Jicha GA, Neltner JH, Galasko D, Masliah E, Quinn JF, Chung KA, Yearout D, Mata IF, Wan JY, Edwards KL, Montine TJ, Zabetian CP (2013) APOE epsilon4 increases risk for dementia in pure synucleinopathies. JAMA Neurol 70(2):223–228. https://doi.org/10.1001/jamaneurol.2013.600

Uchino A, Takao M, Hatsuta H, Sumikura H, Nakano Y, Nogami A, Saito Y, Arai T, Nishiyama K, Murayama S (2015) Incidence and extent of TDP-43 accumulation in ageing human brain. Acta Neuropathol Commun 3:35. https://doi.org/10.1186/s40478-015-0215-1

Vann Jones SA, O'Brien JT (2014) The prevalence and incidence of dementia with Lewy bodies: a systematic review of population and clinical studies. Psychol Med 44(4):673–683. https://doi.org/10.1017/s0033291713000494

Vinters HV (1987) Cerebral amyloid angiopathy. Crit Rev Stroke 18(2):311–324

Walker Z, Possin KL, Boeve BF, Aarsland D (2015) Lewy body dementias. Lancet 386(10004):1683–1697. https://doi.org/10.1016/s0140-6736(15)00462-6

Walker L, McAleese KE, Johnson M, Khundakar AA, Erskine D, Thomas AJ, McKeith IG, Attems J (2017) Quantitative neuropathology: an update on automated methodologies and implications for large scale cohorts. J Neural Transm (Vienna) 124(6):671–683. https://doi.org/10.1007/s00702-017-1702-2

Weiner MW, Veitch DP, Aisen PS, Beckett LA, Cairns NJ, Green RC, Harvey D, Jack CR, Jagust W, Liu E, Morris JC, Petersen RC, Saykin AJ, Schmidt ME, Shaw L, Shen L, Siuciak JA, Soares H, Toga AW, Trojanowski JQ (2013) The Alzheimer's Disease Neuroimageing Initiative: a review of papers published since its inception. Alzheimers Demen 9(5):e111–e194. https://doi.org/10.1016/j.jalz.2013.05.1769

Weller RO, Massey A, Kuo YM, Roher AE (2000) Cerebral amyloid angiopathy: accumulation of A beta in interstitial fluid drainage pathways in Alzheimer's disease. Ann N Y Acad Sci 903:110–117

Weller RO, Djuanda E, Yow HY, Carare RO (2009) Lymphatic drainage of the brain and the pathophysiology of neurological disease. Acta Neuropathol 117(1):1–14. https://doi.org/10.1007/s00401-008-0457-0

Wharton SB, Minett T, Drew D, Forster G, Matthews F, Brayne C, Ince PG (2016) Epidemiological pathology of Tau in the ageing brain: application of staging for neuropil threads (BrainNet Europe protocol) to the MRC cognitive function and ageing brain study. Acta Neuropathol Commun 4:11. https://doi.org/10.1186/s40478-016-0275-x

Whitwell JL, Dickson DW, Murray ME, Weigand SD, Tosakulwong N, Senjem ML, Knopman DS, Boeve BF, Parisi JE, Petersen RC, Jack CR Jr, Josephs KA (2012) Neuroimageing correlates of pathologically defined subtypes of Alzheimer's disease: a case-control study. Lancet Neurol 11(10):868–877. https://doi.org/10.1016/s1474-4422(12)70200-4

Wilson RS, Yu L, Trojanowski JQ, Chen EY, Boyle PA, Bennett DA, Schneider JA (2013) TDP-43 pathology, cognitive decline, and dementia in old age. JAMA Neurol 70(11):1418–1424. https://doi.org/10.1001/jamaneurol.2013.3961

Wirths O, Breyhan H, Cynis H, Schilling S, Demuth HU, Bayer TA (2009) Intraneuronal pyroglutamate-Abeta 3-42 triggers neurodegeneration and lethal neurological deficits in a transgenic mouse model. Acta Neuropathol 118(4):487–496. https://doi.org/10.1007/s00401-009-0557-5

Woerman AL, Stohr J, Aoyagi A, Rampersaud R, Krejciova Z, Watts JC, Ohyama T, Patel S, Widjaja K, Oehler A, Sanders DW, Diamond MI, Seeley WW, Middleton LT, Gentleman SM, Mordes DA, Sudhof TC, Giles K, Prusiner SB (2015) Propagation of prions causing synucleinopathies in cultured cells. Proc Natl Acad Sci U S A 112(35):E4949–E4958. https://doi.org/10.1073/pnas.1513426112

Yokoyama JS, Karch CM, Fan CC, Bonham LW, Kouri N, Ross OA, Rademakers R, Kim J, Wang Y, Hoglinger GU, Muller U, Ferrari R, Hardy J, Momeni P, Sugrue LP, Hess CP, James Barkovich A, Boxer AL, Seeley WW, Rabinovici GD, Rosen HJ, Miller BL, Schmansky NJ, Fischl B, Hyman BT, Dickson DW, Schellenberg GD, Andreassen OA, Dale AM, Desikan RS (2017) Shared genetic risk between corticobasal degeneration, progressive supranuclear palsy, and frontotemporal dementia. Acta Neuropathol 133(5):825–837. https://doi.org/10.1007/s00401-017-1693-y

Zinszner H, Sok J, Immanuel D, Yin Y, Ron D (1997) TLS (FUS) binds RNA in vivo and engages in nucleo-cytoplasmic shuttling. J Cell Sci 110(Pt 15):1741–1750

Chapter 5
Ageing and Cognition

Sydney M. A. Juan and Paul A. Adlard

Abstract With an increasingly ageing population that is expected to double by 2050 in the U.S., it is paramount that we further understand the neurological changes that occur during ageing. This is relevant not only in the context of "pathological" ageing, where the development of many neurodegenerative disorders is typically a feature of only the older population (and indeed, age is the primary risk factor for many conditions such as Alzheimer's disease), but also for what is considered to be "normal" or "healthy" ageing. Specifically, a significant proportion of the older population are affected by "age-related cognitive decline" (ARCD), which is both independent of dementia and has an incidence 70% higher than dementia alone. However, whilst it is reported that there are pathogenic and phenotypic overlaps between healthy and pathological ageing, it is clear that there is a need to identify the pathways and understand the mechanisms that contribute to this loss of cognitive function with normal ageing, particularly in light of the increasing life expectancy of the global population. Importantly, there is an increasing body of evidence implicating zinc homeostasis as a key player in learning and memory and also potentially ARCD. Further research will ultimately contribute to the development of targeted therapeutics that will promote successful brain ageing. In this chapter we will explore the notion of ARCD, with a perspective on potential key neurochemical pathways that can be targeted for future intervention.

Keywords Ageing · Cognition · Zinc · Therapeutics

S. M. A. Juan · P. A. Adlard (✉)
The Florey Institute of Neuroscience and Mental Health and The University of Melbourne, Parkville 3052, VIC, Australia
e-mail: sydney.juan@florey.edu.au; paul.adlard@florey.edu.au

© Springer Nature Singapore Pte Ltd. 2019
J. R. Harris, V. I. Korolchuk (eds.), *Biochemistry and Cell Biology of Ageing: Part II Clinical Science*, Subcellular Biochemistry 91,
https://doi.org/10.1007/978-981-13-3681-2_5

Introduction

By the year 2100, the population is estimated to increase by nearly 3.6 billion people worldwide (United Nations 2017). This growth is largely due to previous advances in modern medicine, improved hygiene and the implementation of antibiotics and pesticides which have resulted in an increase in life expectancy, especially in developed countries (Lunenfeld 2008). That being said, developing countries are experiencing the highest growth in the ageing population. For instance, the elderly population in Singapore and Malaysia is expected to increase by 372% and 277% respectively by the year 2030 (Kinsella and Velkoff 2002). In contrast, France and the United Kingdom are predicted to see a 56% and 55% increase in the ageing population by the same year (Kinsella and Velkoff 2002). The elderly population is, therefore, growing at a faster rate than ever before. These demographic differences can be explained by the increase in birth and mortality rates. Indeed, birth rates remain much higher in developing countries (3.3 births per woman in Malaysia in 2000) as compared to developed countries (1.7 births per woman in United Kingdom in 2000) (Kinsella and Velkoff 2002). Whereas developed countries have seen a large decline in mortality rates (39.3% of individuals in Japan aged 65 years or more will be older than 80 by 2030), so have developing countries (27.3% of individuals in Argentina aged 65 years or more will be older than 80 by 2030) (Kinsella and Velkoff 2002). Furthermore, it is estimated that 40% of the population aged 60 years or more is affected by ARCD (VanGuilder and Freeman 2011). Age-related cognitive decline significantly hampers quality of life and these individuals therefore require specialized care. Unfortunately, developing countries with low economic power and less resources will experience more difficulty in taking care of the elderly as compared to developed countries (Lunenfeld 2008). For these reasons, a better understanding of the neurochemical basis underlying the ageing process will aid us in determining the specific biochemical pathways that contribute to cognitive decline, and thus will be critical for the development of targeted therapeutics in the pursuit of promoting healthy ageing.

What is Ageing and What Causes It?

Ageing, which is a normal and complex biological process that remains poorly understood, is characterized by a steady decline in various physiological functions that result in both physical and cognitive impairment (Deary et al. 2009). Physiologically, a number of biological changes occur to impact upon normal daily function. For instance, body composition is altered with age, such that lean muscle mass declines while fat mass increases (Bigby 2004). Because the amount of collagen decreases with age, flexibility of ligaments, tendons, muscles and joints decline and this affects muscle function over time. This reduction in muscle tone is directly linked to reduced lung capacity and therefore reduced oxygen consumption

(Bigby 2004). Ageing also causes thinning of arterial walls and increases fat deposition within arteries. These cardiovascular problems, along with reduced lung capacity, render physical activity a difficult task. It is therefore no surprise that the elderly are susceptible to a range of medical complications over time. Moreover, the skin undergoes changes in pigmentation, water content and elasticity via a decrease in elastin and collagen content which presents as wrinkles on the face and body (Bigby 2004). Hair loss, which is perhaps one of the most well-known physical consequence of ageing, is dependent upon genetics and involves thinning and de-pigmentation of hair follicles perceived as greying of the hair. In addition, bone density decreases with age and this is threefold greater in women than men (Bigby 2004). Finally, vision and hearing loss are debilitating consequences of ageing to many individuals as well as decreased sensitivity to touch, smell and taste over time (Bigby 2004). It is therefore suggested that ageing is, for the most part, caused by the deterioration of these core biological functions (Deary et al. 2009).

There are, however, several factors that are likely to impact upon the rate and magnitude of the ageing process which varies greatly between individuals (Deary et al. 2009), such as changes in diet and lifestyle, inflammation, genetics and the presence of confounding illness. Individuals with co-morbidities such as cardiovascular disease, for example, display increased ageing rates compared to healthy individuals (VanGuilder and Freeman 2011). Interestingly, diets high in antioxidants have been shown to improve cognitive function during ageing (Robertson et al. 2013). Furthermore, age-related chronic illness and stress may lead to inflammation initiating the release of pro-inflammatory cytokines such as tumor necrosis factor alpha (TNF-α) and interleukin-6 (IL-6), which only exacerbate the inflammatory response (Robertson et al. 2013). Finally, testosterone, which decreases with age, is suggested to have neuroprotective properties through the enhancement of hippocampal synaptic plasticity as well as limiting the aggregation of proteins involved in neurodegenerative diseases, namely amyloid beta (Maggio et al. 2012). Hence, it is suggested that amelioration of these factors that can be relatively easily controlled, such as diet for example, might promote "healthy" ageing and this is discussed in further detail in a subsequent section.

López-Otín et al. (2013) have proposed nine probable hallmarks contributing to the ageing process and these include genomic instability, mitochondrial dysfunction, telomere attrition, cellular senescence, stem cell exhaustion and altered intercellular communication, epigenetics, proteostasis and nutrient sensing. Indeed, as individuals age, their genes become increasingly susceptible to damage from endogenous and exogenous sources. Errors in DNA replication, point mutations, reactive oxygen species (ROS) generation and alterations in DNA methylation all contribute to accelerated ageing (Moskalev et al. 2013). Evidence for this is supported by studies in transgenic mice that overexpress the BubR1 gene (a gene involved in chromosomal separation during mitosis), such that these mice exhibit an increased life expectancy and reduced chances of developing cancer (Baker et al. 2013). As such, ageing is favored by DNA damage and suppressed by DNA repair (Kirkwood 2005). DNA damage as a result of ageing occurs in a fairly random fashion, whereas the telomeres at the end of chromosomes are specifically vulnerable to damage with increasing age (López-Otín et al. 2013). That is, telomeres bind to shelterin, a protein complex that prevents repair processes from restoring damaged DNA thereby causing persistent damage at telomeres with increasing age (Palm and de Lange 2008). Experimental studies in mice have demonstrated that telomere attrition causes a decrease in lifespan (Armanios et al. 2009), and conversely that lifespan can be prolonged by telomere activation (López-Otín et al. 2013). As we age, ATP generation becomes less efficient causing mitochondrial dysfunction and increased ROS generation which ultimately leads to widespread cellular damage. Furthermore, adequate proteostasis is essential for the normal functioning of the proteome where misfolded proteins are repaired (Powers et al. 2009). During ageing, proteins are misfolded and/or aggregated and this can lead to a variety of neurodegenerative pathologies (Bishop et al. 2010). In addition, nutrient sensing, in which glucose is sensed by insulin and insulin-like growth factor 1 (IGF-1) is dysregulated in ageing such that genetic alterations of these factors or even dietary restriction has been shown to increase lifespan (Fontana et al. 2010). Finally, ageing has been linked with decreased stem cell function and altered intercellular communication ultimately leading to global cellular senescence (López-Otín et al. 2013). As such, the following sections will more specifically interrogate the influence of ageing on the brain.

The Impact of Ageing on the Brain

Ageing is the biggest risk factor for the development of neurodegenerative diseases including Alzheimer's disease (AD) and other disorders such as Parkinson's disease and Huntington's disease. Whilst such diseases will precipitate degenerative changes in the brain, their evolution and progression may be impacted by normal age-related changes in various pathways. These concepts on the intersection between ageing and neurodegenerative disease have been reviewed extensively in the past (Celsis 2000; Harada et al. 2013; Hulette et al. 1998) and are not the focus of this current chapter. Rather, we will explore the biochemical and anatomical changes that occur across the lifespan that may contribute to ARCD.

During normal ageing, various neuroanatomical regions are affected but these structural changes are not identical in all brain regions. In early adulthood, grey matter volume but more specifically the prefrontal cortex, decreases in size (Harada et al. 2013). In fact, the whole brain volume, as estimated in longitudinal studies, decreases in size by 0.2–0.5% every year (Fjell and Walhovd 2010). That being said, white matter changes during ageing are more prominent than that of grey matter changes (Harada et al. 2013). While there is a steady increase in structural grey matter alterations from childhood to adulthood, the white matter undergoes significant alterations in adulthood where this plateaus and resumes in old age (Madden et al. 2008). For example, diffusion tensor magnetic resonance images (DT-MRI) comparing white matter integrity between young and old individuals show greater anisotropic diffusion of water molecules and therefore greater white matter density in young individuals as compared to older individuals (Head et al. 2004). Depending on where these structural alterations occur in the white matter tracts, these will have varying functional outcomes. The largest volumetric changes recorded include the putamen, accumbens, thalamus and frontal and temporal cortices (Fjell and Walhovd 2010). Interestingly, the ventricular system has been shown to increase in size during ageing. Clinical data from Rogalski et al. (2012) demonstrates reductions in parahippocampal white matter volumes in healthy aged participants. They suggest that this plays a role in age-related memory impairment as the parahippocampal tract is involved in the relay of sensory information between the hippocampus and the entorhinal cortex. Indeed, axon myelination greatly reduces with increasing age therefore impeding upon axonal signal transduction (Fjell and Walhovd 2010). Similarly, Madden et al. (2008) suggest that decreased integrity of a portion of the corpus callosum during ageing leads to reduced perceptual speed and memory retrieval. Furthermore, functionality of posterior regions of the brain also decline with age, particularly the ventral visual cortex that exhibits less neural selectivity to inputs from the visual system (Park et al. 2004).

It is clear, therefore, that there are widespread changes throughout various regions of the brain across age. The hippocampus, however, is an area of particular interest in ageing and cognition as it governs spatial learning and memory; processes that are intimately linked to cognitive decline (VanGuilder and Freeman 2011). Hippocampal atrophy is one of the defining anatomical alterations that occur as a result of ageing, and is likely due to a reduction in synaptic density and neuronal size (Harada et al. 2013). The hippocampus is made up of several structures namely the *cornu ammonis* areas CA1-4 which together form what is known as the "hippocampus proper" and along with the dentate gyrus and the subiculum are collectively known as the "hippocampal formation". During normal ageing, neuronal loss has been demonstrated in the dentate gyrus and the subiculum but not in the hippocampus proper (Jagust 2013). Some have suggested that the hippocampus may serve in the discrimination of "healthy" and "pathological" ageing. Indeed, studies have shown that neuronal loss in the dentate gyrus and subiculum of the hippocampus are due to age-related changes only, whereas neuronal loss in CA1 and the entorhinal cortex are implicated in neurodegenerative diseases and particularly in Alzheimer's disease (Jagust 2013). In contrast to this view, others have suggested that older adults display decreased volumes of the dentate gyrus and the CA3 region of the hippocampus (Shing et al. 2011) while others have found smaller volumes of CA1 in memory impaired individuals (Yassa et al. 2010). Therefore, while there is a current disagreement in the field as to which hippocampal regions are affected or spared during ARCD, this may be due to the fact that few studies have investigated the link between volumetric changes in hippocampal sub-regions and subsequent changes in functional endpoints – a notion discussed in greater detail in the following section. Nonetheless, these morphometric changes in the ageing brain may only be partly attributed to neuronal loss, rather, it seems that reductions in spine density and decreased neuron sizes and number of synapses more likely contribute to the volumetric alterations observed during normal ageing (Fjell and Walhovd 2010). During ageing, neurons undergo drastic morphological changes via a decrease in synaptic branching and the number and length of dendritic spines (Harada et al. 2013) and this has been demonstrated in humans, rats and non-human primates (Morisson and Baxter 2012). Interestingly, increased neuronal activity in the prefrontal cortex has been reported during normal ageing and it is suggested that this functions to compensate for the overall dampened activity in other neuroanatomical regions (Reuter-Lorenz 2002). This compensatory mechanism is thought to be evidence of the brain's adaptation to ageing (Ballesteros et al. 2009). Indeed, during memory tasks, the activation of the brain's memory networks becomes bilateral perhaps due to the decreased activation in other areas specifically involved in memory (Peters 2006). Furthermore, functional reorganization of dendritic synapses via dendritic sprouting has been suggested to compensate for cell death during ageing (Peters 2006).

Other pathways contributing to the ageing process include reduced vascular density, elevations in inflammation and oxidative stress, compromised neurotransmitter performance such as GABA (Rozycka and Liguz-Lecznar 2017) and the dysregulation of a number of key hippocampal proteins namely the glutamate receptors AMPA (alpha-amino-3-hydroxy-5-methyl-4-isoxazolepropionic acid) and NMDA (N-methyl-D-aspartate) and other pre and post-synaptic proteins involved in learning and memory processes (VanGuilder and Freeman 2011). A number of these processes, which will be discussed in further detail in the next section, play key roles in the development of ARCD.

The Impact of Ageing on Cognition

Cognition involves complex information processing, planning and reasoning (Buckner and Fridland 2017). During ageing, specific cognitive domains decline such as reasoning, memory and processing speed (Ballesteros et al. 2009). While some of these domains decline throughout the lifespan, some decline only during late life whilst other cognitive domains are relatively preserved (Hedden and Gabrieli 2004). For example, a consequence of ARCD that is widely agreed upon in the field is a decrease in episodic memory which occurs at around 70 years of age (Celsis 2000; Salthouse 2011). In an MRI study consisting of 50 elderly healthy participants, Kramer et al. (2007) reported that decreased grey matter and hippocampal volumes were associated with impaired episodic memory. Furthermore, processing speed and working memory have also been reported to decrease linearly across the lifespan (Hedden and Gabrieli 2004). For example, Meijer et al. (2009) showed that middle-aged adults (50–60 years old) displayed longer reaction times to a word processing task than did younger adults (25–35 years old), demonstrating that this cognitive domain is already susceptible to decline in mid-life. The decline in this cognitive domain can pose an important threat to society as the elderly are at higher risk of car accidents due to their reduced processing speed, executive function and visual processing (Harada et al. 2013). In contrast, short-term memory and tasks involving world knowledge show little decline until late life (Hedden and

Gabrieli 2004). In addition, emotional processing is a cognitive function that has been reported to be well preserved into old age (Hedden and Gabrieli 2004). For instance, Mather (2006) reports that older individuals remember positive memories more in comparison to younger adults who recall negative information more, suggesting that age-related changes in processing emotional information dictates which emotional memories are remembered. Moreover, there have been inconsistent reports regarding verbal ability during normal ageing. Longitudinal data, for example, have suggested that verbal ability declines with age particularly after 60 years of age (Hedden and Gabrieli 2004). In contrast, Ballesteros et al. (2009) suggest that verbal ability and general knowledge do not deteriorate with age but instead show improvements, perhaps owing to the compensatory process discussed above. Indeed, Meunier et al. (2014) reported that age-related reductions in grey matter density in language areas lead to functional reorganization of the network via an increase in functional connectivity which results in the recruitment of new areas closely located to core language regions. Overall, these data highlight that ARCD has the potential to negatively impact upon the daily life of those affected and over time can greatly limit the breadth of activities one can perform.

It is important to note that the rate and degree of cognitive decline varies widely across individuals and there are various reasons to explain this. In the same way that there is heterogeneity in the onset and progression of disease, there is heterogeneity in the lifestyle of each individual which can ultimately influence the degree and susceptibility to ARCD development. For example, individuals may have inherent differences in plasticity based on differing life experiences or even due to differing levels of physical activity throughout their lives (Hedden and Gabrieli 2004). Based on their genetic background and epigenetic influences throughout the lifespan, certain individuals are more susceptible to cognitive decline. Education level has also been implicated, with highly educated individuals being less susceptible to cognitive impairment such as memory loss and language ability (Celsis 2000).

Cognition is intimately linked to synaptic function and neuronal activity (VanGuilder and Freeman 2011). During ageing, neuronal synapses at the hippocampus become dysregulated which may be due to morphological alterations of the synapse itself, impaired neurotransmitter signaling or changes in protein and/or gene expression. Decreased synaptic density has been demonstrated in the dentate gyrus and the CA3 region of the hippocampus but not in the CA1 region, as evidenced in aged rats exhibiting a loss of axospinous synapses in these regions (Hara and Morrison 2014). Similarly, in a study of aged female monkeys, there was a reduction in the number of presynaptic boutons and thus, fewer connections made in multiple synapses of the dentate gyrus and this reduction in synaptic strength was correlated to ARCD (Hara et al. 2011). Interestingly, the loss of excitatory synapses is correlated with cognitive disability in aged monkeys, even though the amount of both excitatory and inhibitory synaptic loss during ageing is similar (Petralia et al. 2014). Morisson and Baxter (2012) report that during ageing, the hippocampus loses the large perforated mushroom synapses that modulate learning and memory while the prefrontal cortex loses thin spines, which they suggest causes a decrease in cognitive performance. Interestingly, old rats in which protein kinase C is activated (a protein believed to modulate synaptic growth), demonstrate an increase of mushroom-like spines which improves spatial memory (Petralia et al. 2014). Moreover, synaptic release of neurotransmitters modulates memory, executive function, hormone release and motor function amongst other things and is therefore essential for normal functioning of the central nervous system (Azpurua and Eaton 2015). During ageing, both excitatory and inhibitory neurotransmitters play a role in the destabilization of the synaptic machinery which significantly contributes to cognitive decline. Indeed, changes in the activity of excitatory glutamatergic synapses during ageing, which are heavily implicated in learning and memory, are evidenced via a number of anatomical and chemical mechanisms. Firstly, key hippocampal proteins are decreased in aged mice namely the glutamate receptors AMPA and NMDA, the presynaptic protein SNAP25 and postsynaptic protein PSD95 (Adlard et al. 2010; Canas et al. 2009). In addition, several markers of synaptic plasticity such as CaMKII (calmodulin-dependent kinase II) and synaptophysin have also been shown to decrease in aged mice (Adlard et al. 2014; VanGuilder and Freeman 2011). Smith et al. (2000) reported a decrease in CA3 hippocampal synaptic terminals in aged rats that correlated with the observed deficits in spatial learning, from which they concluded that synaptic loss within the hippocampus is implicated in ARCD. On the other hand, inhibitory neurons function to control overall excitability of neural circuits, thereby maintaining homeostatic balance in the brain (Rozycka and Liguz-Lecznar 2017). During ageing, this control is impaired and therefore contributes to cognitive disability in old age. As such, neuroplasticity therefore functions to overcome these alterations in order to promote homeostatic balance of excitatory and inhibitory activity within the synapse (Rozycka and Liguz-Lecznar 2017). Indeed, impaired GABA signaling has been well characterized in the ageing synapse. In the hippocampus of aged rats, Stanley et al. (2012) observed a decrease in inhibitory dendritic input and reduced GABA release in the entorhinal cortex. Additionally, it has been demonstrated via patch-clamp electrophysiology that the ageing hippocampus exhibits a decrease in inhibitory postsynaptic potential frequency and in the frequency and amplitude of currents mediated by GABA (McQuail et al. 2015). Together, these data suggest that the disruption in synaptic integrity and neurotransmission greatly contributes to ARCD (Petralia et al. 2014). Furthermore, these synaptic alterations further contribute to cognitive decline by prohibiting the induction of long-term potentiation (LTP; one of the primary cellular mechanisms underlying learning and memory) and encouraging long-term depression (LTD; which opposes LTP) (Barnes et al. 2000). For instance, Barnes et al. (2000) found that there was an increased threshold for LTP induction in cognitively impaired old rats compared to middle-aged and young controls, suggesting that reduced NMDA signaling at the synapse causes a decrease in neuroplasticity in aged mice. Over time, this decrease in synaptic health and plasticity results in impaired learning and memory abilities (Celsis 2000). These cognitive deficits are substantiated via a decreased ability of aged mice to perform on the Morris Water maze, a well-established spatial memory task (Adlard et al. 2014). On the other hand, we can genetically modify mice to increase specific proteins that are normally decreased with age in order to enhance learning and memory via increases in LTP. An example of this are mice where the voltage-dependent Ca^{2+} channel β subunit is knocked out, causing them to exhibit an increase in NMDA receptor expression in the hippocampus and therefore an increase in LTP (Lee and Silva 2009). This was evidenced by increased performance on a range of learning and memory tasks such as novel-object recognition and contextual fear conditioning (Lee and Silva 2009). Another example of this are transgenic mice where increasing CREB levels, a transcription factor regulated by cAMP-PKA/phosphatase signaling, leads to increased memory (Lee and Silva 2009). Indeed, increasing CREB levels is suggested to increase the expression of proteins required for memory and neuroplasticity (Lee and Silva 2009).

Finally, the dysregulation of neuronal synapses observed during ageing are also influenced by age-related changes in gene expression, otherwise known as epigenetics. Ageing individuals undergo changes in histone methylation, acetylation and DNA demethylation which have been reported to modulate the observed age-related changes in synapse loss and structural changes (Azpurua and Eaton 2015). The expression of certain gene categories changes with increasing age. For example, the gene categories involved in the stress response, inflammatory response, metal ion homeostasis, myelin-related proteins and glial genes are all increased during ageing in humans (Bishop et al. 2010). In contrast, those gene categories involved in neural plasticity, synaptic function and inhibitory interneuron function decrease in ageing humans (Bishop et al. 2010). Guan et al. (2009) overexpressed histone deacetylase 2 (HDAC2) in wild-type mice and demonstrated a decrease in the number of synapses, dendritic spine density but also synaptic plasticity and memory formation in the CA1 pyramidal neurons of the hippocampus. Moreover, memory formation has been shown to transiently increase acetylation of histone H3, and the administration of HDAC inhibitors in mice was demonstrated to increase LTP induction (Levenson et al. 2004). Similarly, Peleg et al. (2010) showed that aged mice exhibit changes in acetylation of histone H4 lysine 12 (H4K12) during learning which prevents the expression of specific hippocampal genes that are required for memory consolidation. They further demonstrate that expression of these memory-consolidating genes can be repaired via reestablishment of H4K12 acetylation and that this improves cognitive ability.

How Can We Intervene to Prevent Cognitive Failure with Age?

It is clear, then, that there are a variety of pathways (genetic, environmental, behavioural etc) that can lead to cognitive decline, many of which are no doubt additive in nature. As such, there is a large number of different pharmacological and behavioural approaches that have been used in an attempt to enhance cognitive function with ageing. For instance, increased physical activity, social involvement and taking part in intellectually demanding activities such as computer use, reading and music have shown to increase cognitive performance (Harada et al. 2013). Indeed, physical activity induces proteins that improve cognitive function, learning and memory (Chen et al. 2007) with, for example, profound effects shown on critical proteins such as brain-derived neurotrophic factor (BDNF) in both health and disease (Adlard and Cotman 2004, Adlard et al. 2005a, b). Indeed, BDNF levels were shown to significantly increase following 4 weeks of exercise in rats who also showed improved performance on the Morris water maze, a well characterized learning task (Adlard et al. 2004). Alternatively, cognitive training has been shown to increase individuals' performance in day to day activities. The advantage to this is that cognitive training is a method accessible to most of the population as it can be delivered within a clinical setting or at home on video (Harada et al. 2013). These behavioural methods of either preserving, or enhancing, cognition with age may also be augmented through the use of targeted therapeutics. In this regard, there have been innumerable compounds tested for their ability to enhance cognition (Froestl et al. 2012, 2013a, b), largely via actions on discrete proteins or signaling cascades, or on other auxiliary pathways that may either directly or indirectly influence cognition (such as through the use of growth factors to enhance neuronal health and function with the hope of a downstream impact on gross function).

With the number of possibilities open to testing, the use of techniques such as gene profiling in the hippocampus during aging, for example, has enabled the more targeted study of specific pathways that impact upon cognitive function (VanGuilder and Freeman 2011). These advancements in the field have allowed for the identification of specific molecules and pathways that are dysfunctional or altered during ageing and which may represent tractable therapeutic targets.

Finally, whilst efforts over many decades have gone into targeting specific hippocampal proteins that are dysregulated in ageing, recent evidence suggests that there may be a crucial link between zinc levels in the brain and cognitive function. Specifically, the decline in cognitive performance observed during ageing has been linked to the dysregulation of zinc homeostasis (Hancock et al. 2014; Mocchegiani et al. 2006).

Is Zinc Dyshomeostasis a Common Pathway for Age-Related Cognitive Decline?

Zinc is essential for the normal functioning of the brain and there is an increasing body of evidence to suggest that zinc plays an important role in ARCD. The levels of zinc in the brain are tightly controlled as it is essential for various cellular and molecular processes, enzymatic activity and transcription factors (Levenson and Tassabehji 2007). In fact, the concentration of zinc in the brain surpasses that of the body by ten-fold (Portbury and Adlard 2017). While the majority of zinc is bound to proteins or amino acids, around 10% of zinc remains as chelatable, or free, zinc (Levenson and Tassabehji 2007). Under normal conditions, chelatable zinc is located within vesicles of presynaptic glutamatergic neurons and therefore plays a crucial role in modulating neuroplasticity and cognition (Besser et al. 2009). Approximately 300 μM of zinc is released with glutamate into the synapse upon depolarization (Assaf and Chung 1984; Frederickson et al. 2005). Zinc acts via a range of transporters, ion channels (such as voltage-gated Ca^{2+} channels), synaptic receptors namely the AMPA and NMDA receptors and post-synaptic factors such as TrkB for example (Besser et al. 2009), in order to mediate synaptic communication and neuroplasticity (Paoletti et al. 2009).

Under healthy conditions, zinc levels are maintained by a variety of zinc transporters and other transporter proteins that balance its export/import into discrete cellular compartments (in fact, when these transporters are decreased or dysfunctional, significant pathologies can result) (Kambe et al. 2015). Of particular relevance here is the zinc transporter-3 protein (ZnT3), which was shown to be responsible for loading zinc into pre-synaptic vesicles in the hippocampus (Cole et al. 1999, Linkous et al. 2008, Palmiter et al. 1996). For example, Saito et al. (2000) demonstrated that reduced ZnT3 expression causes a decrease in synaptic zinc levels in the hippocampal mossy fiber pathway. This decreased synaptic zinc leads to excitotoxicity due to excessive glutamate release ultimately impairing learning and memory. Furthermore, studies in mice exhibiting accelerated ageing present with brain atrophy and impaired learning and memory functions probably due to reduced zinc levels at the synapse (Saito et al. 2000). Indeed, studies have shown that the onset of age-related cognitive decline is influenced by ZnT3 mutations (Rocha et al. 2014). Impaired memory has also been observed in ZnT3 knock out mice and there is a reduction in ZnT3 protein levels in the cortex of aged mice and humans (in both health and disease) (Adlard et al. 2010).

Furthermore, metallothioneins (MTs) which are molecules that bind zinc and other metals, have been shown to increase with age. Because plasma levels of zinc decrease with age, this suggests that MTs retain zinc intracellularly thereby causing an overall decrease in extracellular zinc (Mocchegiani et al. 2006). However, although both vesicular and synaptic zinc levels have been demonstrated to reduce with increasing age, the overall amount of zinc is unchanged in ageing humans and rodents (Takahashi et al. 2001).

It is becoming increasingly apparent that zinc dyshomeostasis may play an important role in ARCD. As such, several pharmacological strategies that target zinc have been assessed for their therapeutic efficacy in both cognitive ageing and a range of neurodegenerative diseases over the last decade. An example of this is a recent study that showed that PBT2 (Prana Biotechnology), an 8-hydroxy quinolone that functions as a cellular chaperone for zinc, works by redistributing zinc within the brain and increasing the amount of total zinc within the hippocampus of aged mice (Adlard et al. 2014). This lead to an improvement in cognitive performance, neurogenesis and increased synaptic plasticity that was underscored by changes in a suite of relevant synaptic proteins (Adlard et al. 2014). A related compound, clioquinol, was similarly shown to restore cognition in ZnT3 KO mice (Adlard et al. 2015). These studies corroborate the idea that redistribution of zinc within different compartments of the brain, and therefore re-establishment of zinc ion homeostasis may prove to be a valuable avenue for preventing age-related cognitive decline.

Zinc dyshomeostasis has also been shown to have a number of implications for other age-related and non-ageing diseases. For example, zinc dyshomeostasis has been identified as a key player in the onset and progression of AD. In fact, PBT2 treatment of transgenic mice exhibiting an AD-like phenotype has previously been shown to improve cognitive function and to reduce disease pathology (Adlard et al. 2008). Further, zinc has also been implicated in the cognitive failure associated with non-ageing disorders such as schizophrenia, depression, attention-deficit hyperactivity-disorder (ADHD) and malnutrition in children. Research has shown that malnutrition in young infants may lead to zinc deficiency and that dietary zinc supplements may have favorable effects on cognitive performance (Black 2003). Zinc deficiency has also been linked to schizophrenia, ADHD and depression. In depression, when low plasma zinc levels are restored, the depressive symptoms are reduced (Levenson and Tassabehji 2007). Similarly, Scarr et al. (2016) increased the expression of zinc transporter SLC39A12 (which was shown to be higher in schizophrenia patients in comparison to controls) in CHO cells and found an increased uptake of zinc in these cells, suggesting that zinc homeostasis may be altered in schizophrenia due to dysregulation of the SLC39A12 zinc transporter. Finally, ADHD is caused by the dysfunction of the dopamine transporter which has been shown to have a high affinity binding site for zinc (Lepping and Huber 2010). As zinc binds to the dopamine receptor, this causes inhibition of dopamine re-uptake. There has been cumulative evidence to suggest that zinc deficiency is involved in ADHD such that various groups have reported that children with ADHD have decreased plasma zinc levels and that this zinc deficiency is positively correlated with more severe ADHD symptoms (Arnold et al. 2005). Taken together, these data highlight the impact that zinc dyshomeostasis can have on cognitive decline and demonstrate its potential as a therapeutic target in the prevention of age-related cognitive decline.

Summary and Conclusion

Currently, there is no treatment available for age-related cognitive decline. This is due to the fact that our current knowledge on normal ageing is still relatively limited, perhaps because there is more urgency in understanding the pathology of neurodegenerative diseases. However, future research into the biological implications of ageing will help reveal what aberrant processes contribute to cognitive decline in both normal and pathological ageing. As outlined in this chapter, one avenue under investigation is zinc, which may represent a tractable therapeutic target for a number of disorders where a failure in zinc ion homeostasis is implicated either more generally in the pathogenesis of disease or specifically in the cognitive deficits associated with the condition. While the research is ongoing, immediate efforts should work to maximize and preserve cognitive function in order to improve the quality of life of those affected.

References

Adlard P, Cotman C (2004) Voluntary exercise protects against stress-induced decreases in brain-derived neurotrophic factor protein expression. Neuroscience 124:985–992

Adlard PA, Perreau VM, Engesser-Cesar C, Cotman CW (2004) The timecourse of induction of brain-derived neurotrophic factor mRNA and protein in the rat hippocampus following voluntary exercise. Neurosci Lett 363:43–48

Adlard PA, Perreau VM, Cotman CW (2005a) The exercise-induced expression of BDNF within the hippocampus varies across life-span. Neurobiol Aging 26:511–520

Adlard PA, Perreau VM, Pop V, Cotman CW (2005b) Voluntary exercise decreases amyloid load in a transgenic model of Alzheimer's disease. J Neurosci 25:4217–4221

Adlard PA, Cherny RA, Finkelstein DI, Gautier E, Robb E et al (2008) Rapid restoration of cognition in Alzheimer's transgenic mice with 8-hydroxy quinoline analogs is associated with decreased interstitial Aβ. Neuron 59:43–55

Adlard PA, Parncutt JM, Finkelstein DI, Bush AI (2010) Cognitive loss in zinc transporter-3 knock-out mice: a phenocopy for the synaptic and memory deficits of Alzheimer's disease? J Neurosci 30:1631–1636

Adlard PA, Sedjahtera A, Gunawan L, Bray L, Hare D et al (2014) A novel approach to rapidly prevent age-related cognitive decline. Aging Cell 13:351–359

Adlard PA, Parncutt J, Lal V, James S, Hare D et al (2015) Metal chaperones prevent zinc-mediated cognitive decline. Neurobiol Dis 81:196–202

Armanios M, Alder JK, Parry EM, Karim B, Strong MA, Greider CW (2009) Short telomeres are sufficient to cause the degenerative defects associated with aging. Am J Hum Genet 85:823–832

Arnold LE, Bozzolo H, Hollway J, Cook A, DiSilvestro RA et al (2005) Serum zinc correlates with parent-and teacher-rated inattention in children with attention-deficit/hyperactivity disorder. J Child Adolesc Psychopharmacol 15:628–636

Assaf S, Chung S-H (1984) Release of endogenous Zn2+ from brain tissue during activity. Nature 308:734

Azpurua J, Eaton BA (2015) Neuronal epigenetics and the aging synapse. Front Cell Neurosci 9:208

Baker DJ, Dawlaty MM, Wijshake T, Jeganathan KB, Malureanu L et al (2013) Increased expression of BubR1 protects against aneuploidy and cancer and extends healthy lifespan. Nat Cell Biol 15:96

Ballesteros S, Nilsson L-G, Lemaire P (2009) Ageing, cognition, and neuroscience: an introduction. Eur J Cogn Psychol 21:161–175

Barnes C, Rao G, Houston F (2000) LTP induction threshold change in old rats at the perforant path–granule cell synapse. Neurobiol Aging 21:613–620

Besser L, Chorin E, Sekler I, Silverman WF, Atkin S et al (2009) Synaptically released zinc triggers metabotropic signaling via a zinc-sensing receptor in the hippocampus. J Neurosci 29:2890–2901

Bigby C (2004) Ageing with a lifelong disability: a guide to practice, program, and policy issues for human services professionals. Jessica Kingsley, London

Bishop NA, Lu T, Yankner BA (2010) Neural mechanisms of ageing and cognitive decline. Nature 464:529

Black MM (2003) The evidence linking zinc deficiency with children's cognitive and motor functioning, 2. J Nutr 133:1473S–1476S

Buckner C, Fridland E (2017) What is cognition? angsty monism, permissive pluralism (s), and the future of cognitive science. Springer, Heidelberg

Canas PM, Duarte JM, Rodrigues RJ, Köfalvi A, Cunha RA (2009) Modification upon aging of the density of presynaptic modulation systems in the hippocampus. Neurobiol Aging 30:1877–1884

Celsis P (2000) Age-related cognitive decline, mild cognitive impairment or preclinical Alzheimer's disease? Ann Med 32:6–14

Chen WQ, Viidik A, Skalicky M, Höger H, Lubec G (2007) Hippocampal signaling cascades are modulated in voluntary and treadmill exercise rats. Electrophoresis 28:4392–4400

Cole TB, Wenzel HJ, Kafer KE, Schwartzkroin PA, Palmiter RD (1999) Elimination of zinc from synaptic vesicles in the intact mouse brain by disruption of the ZnT3 gene. Proc Natl Acad Sci 96:1716–1721

Deary IJ, Corley J, Gow AJ, Harris SE, Houlihan LM et al (2009) Age-associated cognitive decline. Br Med Bull 92:135–152

Fjell AM, Walhovd KB (2010) Structural brain changes in aging: courses, causes and cognitive consequences. Rev Neurosci 21:187–222

Fontana L, Partridge L, Longo VD (2010) Extending healthy life span—from yeast to humans. Science 328:321–326

Frederickson CJ, Koh J-Y, Bush AI (2005) The neurobiology of zinc in health and disease. Nat Rev Neurosci 6:449

Froestl W, Muhs A, Pfeifer A (2012) Cognitive enhancers (nootropics). Part 1: drugs interacting with receptors. J Alzheimers Dis 32:793–887

Froestl W, Muhs A, Pfeifer A (2013a) Cognitive enhancers (nootropics). Part 2: Drugs interacting with enzymes. J Alzheimers Dis 33:547–658

Froestl W, Pfeifer A, Muhs A (2013b) Cognitive enhancers (nootropics). Part 3: drugs interacting with targets other than receptors or enzymes. disease-modifying drugs. J Alzheimers Dis 34:1–114

Guan J-S, Haggarty SJ, Giacometti E, Dannenberg J-H, Joseph N et al (2009) HDAC2 negatively regulates memory formation and synaptic plasticity. Nature 459:55

Hancock SM, Finkelstein DI, Adlard PA (2014) Glia and zinc in ageing and Alzheimer's disease: a mechanism for cognitive decline? Front Aging Neurosci 6:137

Hara Y, Morrison JH (2014) Synaptic correlates of aging and cognitive decline. In: The synapse: structure and function. Elsevier Inc, Amsterdam, pp 301–342

Hara Y, Park CS, Janssen WG, Punsoni M, Rapp PR, Morrison JH (2011) Synaptic characteristics of dentate gyrus axonal. boutons and their relationships with aging, menopause, and memory in female rhesus monkeys. J Neurosci 31:7737–7744

Harada CN, Love MCN, Triebel KL (2013) Normal cognitive aging. Clin Geriatr Med 29:737–752

Head D, Buckner RL, Shimony JS, Williams LE, Akbudak E et al (2004) Differential vulnerability of anterior white matter in nondemented aging with minimal acceleration in dementia of the Alzheimer type: evidence from diffusion tensor imaging. Cereb Cortex 14:410–423

Hedden T, Gabrieli JD (2004) Insights into the ageing mind: a view from cognitive neuroscience. Nat Rev Neurosci 5:87

Hulette CM, Welsh-Bohmer KA, Murray MG, Saunders AM, Mash DC, McIntyre LM (1998) Neuropathological and neuropsychological changes in "normal" aging: evidence for preclinical Alzheimer disease in cognitively normal individuals. J Neuropathol Exp Neurol 57:1168–1174

Jagust W (2013) Vulnerable neural. systems and the borderland of brain aging and neurodegeneration. Neuron 77:219–234

Kambe T, Tsuji T, Hashimoto A, Itsumura N (2015) The physiological, biochemical, and molecular roles of zinc transporters in zinc homeostasis and metabolism. Physiol Rev 95:749–784

Kinsella K, Velkoff VA (2002) The demographics of aging. Aging Clin Exp Res 14:159–169

Kirkwood TB (2005) Understanding the odd science of aging. Cell 120:437–447

Kramer JH, Mungas D, Reed BR, Wetzel ME, Burnett MM et al (2007) Longitudinal MRI and cognitive change in healthy elderly. Neuropsychology 21:412

Lee Y-S, Silva AJ (2009) The molecular and cellular biology of enhanced cognition. Nat Rev Neurosci 10:126

Lepping P, Huber M (2010) Role of zinc in the pathogenesis of attention-deficit hyperactivity disorder. CNS Drugs 24:721–728

Levenson C, Tassabehji N (2007) Role and regulation of copper and zinc transport proteins in the central nervous system. In: Handbook of neurochemistry and molecular neurobiology. Springer, New York, pp 257–284

Levenson JM, O'Riordan KJ, Brown KD, Trinh MA, Molfese DL, Sweatt JD (2004) Regulation of histone acetylation during memory formation in the hippocampus. J Biol Chem 279:40545–40559

Linkous DH, Flinn JM, Koh JY, Lanzirotti A, Bertsch PM et al (2008) Evidence that the ZNT3 protein controls the total amount of elemental zinc in synaptic vesicles. J Histochem Cytochem 56:3–6

López-Otín C, Blasco MA, Partridge L, Serrano M, Kroemer G (2013) The hallmarks of aging. Cell 153:1194–1217

Lunenfeld B (2008) An Aging World – demographics and challenges. Gynecol Endocrinol 24:1–3

Madden DJ, Spaniol J, Costello MC, Bucur B, White LE et al (2008) Cerebral white matter integrity mediates adult age differences in cognitive performance. J Cogn Neurosci 21:289–302

Maggio M, Dall'Aglio E, Lauretani F, Cattabiani C, Ceresini G et al (2012) The hormonal pathway to cognitive impairment in older men. J Nutr Health Aging 16:40–54

Mather M (2006) Why memories may become more positive as people age. Blackwell Publishing, Malden

McQuail JA, Frazier CJ, Bizon JL (2015) Molecular aspects of age-related cognitive decline: the role of GABA signaling. Trends Mol Med 21:450–460

Meijer WA, de Groot RH, van Gerven PW, van Boxtel MP, Jolles J (2009) Level of processing and reaction time in young and middle-aged adults and the effect of education. Eur J Cogn Psychol 21:216–234

Meunier D, Stamatakis EA, Tyler LK (2014) Age-related functional reorganization, structural changes, and preserved cognition. Neurobiol Aging 35:42–54

Mocchegiani E, Costarelli L, Giacconi R, Cipriano C, Muti E et al (2006) Zinc homeostasis in aging: two elusive faces of the same "metal". Rejuvenation Res 9:351–354

Morisson JH, Baxter MG (2012) The ageing cortical synapse: hallmarks and implications for cognitive decline. Nat Rev Neurosci 13:240

Moskalev AA, Shaposhnikov MV, Plyusnina EN, Zhavoronkov A, Budovsky A et al (2013) The role of DNA damage and repair in aging through the prism of Koch-like criteria. Ageing Res Rev 12:661–684

Palm W, de Lange T (2008) How shelterin protects mammalian telomeres. Annu Rev Genet 42:301–334

Palmiter RD, Cole TB, Quaife CJ, Findley SD (1996) ZnT-3, a putative transporter of zinc into synaptic vesicles. Proc Natl Acad Sci 93:14934–14939

Paoletti P, Vergnano A, Barbour B, Casado M (2009) Zinc at glutamatergic synapses. Neuroscience 158:126–136

Park DC, Polk TA, Park R, Minear M, Savage A, Smith MR (2004) Aging reduces neural specialization in ventral visual cortex. Proc Natl Acad Sci U S A 101:13091–13095

Peleg S, Sananbenesi F, Zovoilis A, Burkhardt S, Bahari-Javan S et al (2010) Altered histone acetylation is associated with age-dependent memory impairment in mice. Science 328:753–756

Peters R (2006) Ageing and the brain. Postgrad Med J 82:84–88

Petralia RS, Mattson MP, Yao PJ (2014) Communication breakdown: the impact of ageing on synapse structure. Ageing Res Rev 14:31–42

Plassman BL, Langa KM, Fisher GG, Heeringa SG, Weir DR et al (2007) Prevalence of dementia in the United States: the aging, demographics, and memory study. Neuroepidemiology 29:125–132

Plassman BL, Langa KM, Fisher GG, Heeringa SG, Weir DR et al (2008) Prevalence of cognitive impairment without dementia in the United States. Ann Intern Med 148:427–434

Portbury SD, Adlard PA (2017) Zinc signal in brain diseases. Int J Mol Sci 18:2506

Powers ET, Morimoto RI, Dillin A, Kelly JW, Balch WE (2009) Biological and chemical approaches to diseases of proteostasis deficiency. Annu Rev Biochem 78:959–991

Reuter-Lorenz PA (2002) New visions of the aging mind and brain. Trends Cogn Sci 6:394–400

Robertson DA, Savva GM, Kenny RA (2013) Frailty and cognitive impairment—a review of the evidence and causal mechanisms. Ageing Res Rev 12:840–851

Rocha TJ, Blehm CJ, Bamberg DP, Fonseca TLR, Tisser LA et al (2014) The effects of interactions between selenium and zinc serum concentration and SEP15 and SLC30A3 gene polymorphisms on memory scores in a population of mature and elderly adults. Genes Nutr 9:377

Rogalski E, Stebbins G, Barnes C, Murphy C, Stoub T et al (2012) Age-related changes in parahippocampal white matter integrity: a diffusion tensor imaging study. Neuropsychologia 50:1759–1765

Rozycka A, Liguz-Lecznar M (2017) The space where aging acts: focus on the GABAergic synapse. Aging Cell 16(4):634–643

Saito T, Takahashi K, Nakagawa N, Hosokawa T, Kurasaki M et al (2000) Deficiencies of hippocampal Zn and ZnT3 accelerate brain aging of Rat. Biochem Biophys Res Commun 279:505–511

Salthouse TA (2011) Neuroanatomical substrates of age-related cognitive decline. Psychol Bull 137:753

Scarr E, Udawela M, Greenough MA, Neo J, Seo MS et al (2016) Increased cortical expression of the zinc transporter SLC39A12 suggests a breakdown in zinc cellular homeostasis as part of the pathophysiology of schizophrenia. NPJ Schizophrenia 2:16002

Shing YL, Rodrigue KM, Kennedy KM, Fandakova Y, Bodammer N et al (2011) Hippocampal subfield volumes: age, vascular risk, and correlation with associative memory. Front Aging Neurosci 3:2

Smith TD, Adams MM, Gallagher M, Morrison JH, Rapp PR (2000) Circuit-specific alterations in hippocampal synaptophysin immunoreactivity predict spatial learning impairment in aged rats. J Neurosci 20:6587–6593

Stanley EM, Fadel JR, Mott DD (2012) Interneuron loss reduces dendritic inhibition and GABA release in hippocampus of aged rats. Neurobiol Aging 33:431. e1–31. e13

Takahashi S, Takahashi I, Sato H, Kubota Y, Yoshida S, Muramatsu Y (2001) Age-related changes in the concentrations of major and trace elements in the brain of rats and mice. Biol Trace Elem Res 80:145–158

United Nations, Department of Economic and Social Affairs, Population Division (2017) World population prospects: the 2017 revision, key findings and advance tables. Working Paper No. ESA/P/WP/248

VanGuilder H, Freeman W (2011) The hippocampal neuroproteome with aging and cognitive decline: past progress and future directions. Front Aging Neurosci 3:8

Vincent GK, Velkoff VA (2010) The next four decades: the older population in the United States: 2010 to 2050. US Department of Commerce, Economics and Statistics Administration, US Census Bureau

Yassa MA, Stark SM, Bakker A, Albert MS, Gallagher M, Stark CE (2010) High-resolution structural and functional MRI of hippocampal CA3 and dentate gyrus in patients with amnestic mild cognitive impairment. NeuroImage 51:1242–1252

Chapter 6
Ageing and Osteoarthritis

Pradeep Kumar Sacitharan

Abstract The increase in global lifespan has in turn increased the prevalence of osteoarthritis which is now the most common type of arthritis. Cartilage tissue located on articular joints erodes during osteoarthritis which causes pain and may lead to a crippling loss of function in patients. The pathophysiology of osteoarthritis has been understudied and currently no disease modifying treatments exist. The only current end-point treatment remains joint replacement surgery. The primary risk factor for osteoarthritis is age. Clinical and basic research is now focused on understanding the ageing process of cartilage and its role in osteoarthritis. This chapter will outline the physiology of cartilage tissue, the clinical presentation and treatment options for the disease and the cellular ageing processes which are involved in the pathophysiology of the disease.

Keywords Osteoarthritis · Aging · Cartilage · Chondrocytes

Introduction

Osteoarthritis (OA) is the most common form of arthritis worldwide (Woolf and Pfleger 2003; Glyn-Jones et al. 2015). The number of OA sufferers will increase as life expectancy of the global population rises (Woolf and Pfleger 2003; Glyn-Jones et al. 2015). OA is a highly heterogeneous disease that affects all synovial joints, including the hand, knee, hip and spine. The disease is mainly characterised by the progressive degradation of the articular cartilage along with secondary episodic synovitis and bone remodelling (Vincent and Watt 2014). Ageing is the most important risk factor for OA. Several cellular mechanisms have been proposed by which ageing impacts on the progression of joint degeneration in OA. This chapter at first will outline the physiology of the tissues and cellular components involved in OA, then the clinical aspects of the disease and will finally discuss the cellular mechanisms that are dysregulated in ageing cartilage and which may contribute to OA pathogenesis.

P. K. Sacitharan (✉)
Institute of Ageing and Chronic Disease, University of Liverpool, Liverpool, UK

© Springer Nature Singapore Pte Ltd. 2019
J. R. Harris, V. I. Korolchuk (eds.), *Biochemistry and Cell Biology of Ageing:
Part II Clinical Science*, Subcellular Biochemistry 91,
https://doi.org/10.1007/978-981-13-3681-2_6

Structure of Articular Cartilage

Articular cartilage is located on the surfaces of joints which are involved in mechanical movement and is avascular and aneural (Pearle et al. 2005). Its key role is to provide a smooth surface which results in frictionless articulation to absorb and distribute loads (Buckwalter and Mankin 1998). Articular cartilage is mainly composed of water, collagen, proteoglycans and chondrocytes (the only resident cells) (Pearle et al. 2005). It also contains other minor collagens (type VI, IX and XI) and several non-collagenous proteins and glycoproteins (Heinegård and Saxne 2011). These components make up an extensive hydrated extracellular matrix (ECM). Both cell content and matrix synthesis of cartilage decrease with age and the organization of articular cartilage reflects its functional role (Buckwalter and Mankin 1998).

Histomorphological Organization of Articular Cartilage

Cartilage can be divided into four zones (Fig. 6.1). The superficial zone, which is closest to the articular surface and comprises of collagen fibres, which are densely packed around flattened ellipsoid shaped chondrocytes (Stockwell 1978). The morphology of the superficial zone provides high tensile strength of cartilage to accommodate the forces generated during joint loading and articulation (Buckwalter and

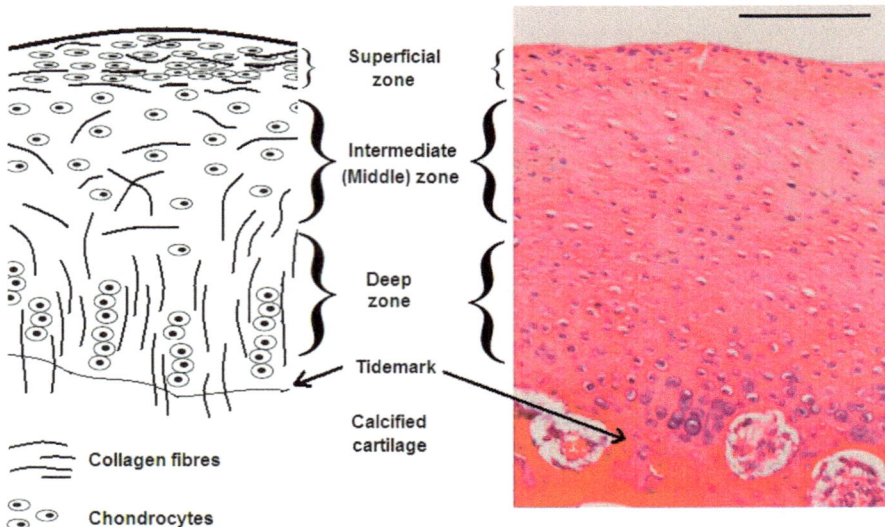

Fig. 6.1 Histomorphological organization of articular cartilage. Normal articular cartilage is composed of three zones and the tidemark. The formation of chondrocytes and alignment of collagen fibres differ in each zone. The tidemark divides normal articular cartilage from deeper calcified cartilage. (Adapted from Matsiko et al. 2013)

Mankin 1998). The collagen fibres in the middle zone of articular cartilage are randomly arranged with chondrocytes spherical in shape and at times found in pairs (Pearle et al. 2005). The deep zone of articular cartilage comprises of columns of ellipsoid shaped chondrocytes distributed between radially oriented collagen fibres that extend into the calcified zone (Stockwell 1978). The calcified zone of articular cartilage contains chondrocytes expressing a hypertrophic phenotype which synthesise calcified matrix providing an interface between uncalcified cartilage and the underlying subchondral bone (Stockwell 1991).

Chondrocytes

Chondrocytes are responsible for maintaining a balanced cartilage turnover and respond to changes induced by joint loading, cytokines, growth factors and the presence of fragmented matrix molecules in the ECM of cartilage (Stockwell 1978; Wieland et al. 2005). During embryogenesis, chondrocytes are responsible for the longitudinal bone growth within the epiphyseal growth plate (Hall and Miyake 1992; Archer and Francis-West 2003). Chondrocytes differentiate from colony-forming unit-fibroblast and later mesenchymal stem cells and initially express collagen type I prior to differentiation into foetal chondrocytes (Goldring 2012). Foetal chondrocytes switch to the expression of chondrocyte specific type II collagen (COL2) (Ryan and Sandell 1990; Sandell et al. 1991) in addition to type IX and XI collagens, while levels of type I collagen expression are significantly inhibited (Luo et al. 1995). During postnatal bone growth, foetal chondrocytes in growth plates enlarge (chondrocyte hypertrophy) which results in terminal differentiation (Drissi et al. 2005). Hypertrophic chondrocytes increase in volume and express high levels of collagen X (COL10) and calcium deposition within the matrix (Alvarez et al. 2001). Differentiated chondrocytes (mature) do not exhibit proliferation potential (Buckwalter and Mankin 1998; Drissi et al. 2005). As cartilage is avascular in nature, chondrocytes exist in hypoxic conditions and rely on nutrients obtained from the articular surface by diffusion as well as signals obtained by the extracellular ionic environment and ECM interactions (Stockwell 1991). The transcription factor SRY (sex determining region Y)-box 9 (SOX-9) is a key regulator of chondrocyte differentiation in both mesenchymal stem cells and foetal chondrocytes (Bell et al. 1997; DeLise et al. 2000). SOX-9 expression level remains elevated in differentiated chondrocytes (DeLise et al. 2000). The expression of chondrocyte-specific genes including COL2 is tightly regulated by SOX-9 transcription factor (De Crombrugghe et al. 2000). Mutations in SOX-9 lead to skeletal malformation syndromes such as Campomelic Dysplasia (Sock et al. 2003). Substantial loss of chondrocytes by necrosis or apoptosis is observed in the superficial and middle zones of cartilage throughout osteoarthritis progression (Lotz and Caramés 2011). Chondrocyte apoptosis may be the result of various inflammatory mediators, decreased cellular metabolism and decreased protection from mechanical force (Heinegård and Saxne 2011; Lotz and Caramés 2011). The loss of chondrocytes reduces cartilage volume and effective ECM turnover which in turn increases cartilage loss leading to OA (Stockwell 1991; Buckwalter and Mankin 1998).

Collagens

Collagens account for around 10% of the wet and over 50% of the dry weight of articular cartilage (Sophia Fox et al. 2009). Each molecule of collagen comprises of three α polypeptide chains coiled into a rigid helical structure either in homotrimers or heterotrimers states (Shoulders and Raines 2010). A repeating tripeptide motif, G-X-Y, within the polypeptide chains allows collagens to form into a triple helix formation (Shoulders and Raines 2010). There are four types of collagens in articular cartilage: fibrillar, fibrillar-associated collagens, fibrillar-associated collagens with interrupted triple helices and non-fibrillar collagens (Thomas et al. 1994).

Types II and XI collagen are fibril forming collagens (Thomas et al. 1994). COL2 accounts for 90–95% of the collagen in articular cartilage and forms the primary component of cross-banded fibrils (Buckwalter and Mankin 1998). COL2 is composed of three α1 (II)-chains which are arranged to form a triple helix (Thomas et al. 1994). Covalent cross-links between individual collagen molecules lead to formation of fibrils, with many fibrils forming fibres which provide the cartilage with tensile strength (Buckwalter and Mankin 1998). COL2 is synthesized as a procollagen molecule with N- and C-propeptides and both propeptides are cleaved before the collagen is embedded into the matrix (Thomas et al. 1994). In contrast, Type XI collagen retains its N-propeptide domain and is formed as a heterotrimer of three different α-chains (α1, α2, α3), which then form the core of COL2 fibrils (Thomas et al. 1994).

Type IX collagen is heterotrimeric molecule consisting of three different α-chains (α1(IX), α2(IX), and α3(IX)) with domains interrupted with short-non helical regions (Thomas et al. 1994). Type IX collagen alongside type XI collagen is thought to bind covalently to the surface of COL2 fibrils in antiparallel directions, which stabilizes the collagen network (Thomas et al. 1994). COL10 is a homotrimeric collagen with a large C-terminal, a short N-terminal domain. It is normally secreted by hypertrophic chondrocytes, plays a role in endochondral ossification and matrix calcification and in articular cartilage it is expressed in the calcified zone (Pearle et al. 2005). Type VI is a monomer which is a heterotrimer consisting of three distinct α chains (α1, α2 and α3). Type VI is found in the pericelluar matrix (surrounds individual chondrocytes) and binds to other components of cartilage such as biglycan, hyaluron, decorin and fibronectin (Heinegård and Saxne 2011).

Proteoglycans

Proteoglycans found in the ECM of articular cartilage enable the tissue to cope with dynamic swelling pressures to resist compressive loads (Maroudas and Bullough 1968). Proteoglycans consist of a core protein which is covalently attached to glycosaminoglycan (GAG) side chains (Ferguson et al. 2009). The core protein can also be attached to chondroitin sulphate, dermatan sulphate, keratan sulphate or heparan sulphate both alone or in combination (Heinegård and Saxne 2011). These side chains can be sulphated which leads to a high negative charge that is responsible for the hydrophilic properties of the ECM (Heinegard et al. 1987).

Aggrecan (gene name ACAN) comprises about 90% of the total ECM proteoglycan weight (Sophia Fox et al. 2009). The aggrecan core protein has three globular domains G1, G2 and G3 (Knudson and Knudson 2001). A long GAG chain covalently bound to keratan sulphate separates the G1 and G2 domains (Knudson and Knudson 2001). Whereas, the G1 domain of aggrecan interacts with GAG hyaluronan; this interaction is stabilised by a link protein (Knudson and Knudson 2001). Both interactions of aggrecan to hyaluronan and the link protein forms the proteoglycan aggregates (Knudson and Knudson 2001). Hyaluronan also has the ability to bind to the cluster differentiation (CD) 44, integrin receptor located on the chondrocyte (Knudson et al. 1996). Versican is the other large proteoglycan in articular cartilage which also contains chondroitin sulfate (CS) but is much less abundant than aggrecan (Olin et al. 2001).

Decorin and biglycan proteoglycans are similar in structure and contain an N-terminal domain that is substituted with one or two dermatan sulphate/CS side chains respectively (Knudson and Knudson 2001). Decorin is found in the interterritorial matrix and binds to COL2 fibrils (Sophia Fox et al. 2009). In comparison, Biglycan is found in the pericellular matrix and binds to type VI collagen (Knudson and Knudson 2001). Lumican and fibromodulin proteoglycans are also found in articular cartilage and are primarily substituted with keratan sulphate chains (Chakravarti 2002). Fibromodulin attaches to COL2 fibres and with decorin and biglycan may play a role in regulating collagen fibril organisation and cartilage stiffness (Knudson and Knudson 2001). Other proteoglycans in articular cartilage include epiphycan, basement membrane proteoglycan perlecan, cell surface proteoglycans such as the syndecans and the phosphatidylinositol linked heparan sulfate proteoglycan and glypican (Knudson and Knudson 2001).

Other Components of Articular Cartilage

The ECM of articular cartilage also contains proteins which are neither collagens nor proteoglycans, but are critical for ECM function and maintenance (Fig. 6.2) (Heinegård and Saxne 2011). Fibronectin is a high molecular weight found in the pericellular and territorial matrix of articular cartilage (Rees et al. 1987). Fibronectin is upregulated in OA tissue but is expressed at low levels in normal, healthy cartilage (Rees et al. 1987). Cleaved fibronectin fragments can activate toll like receptor (TLR) 4 signalling, which can increase inflammatory cytokine expression, upregulate matrix metalloproteinases (MMPs) and induce proteoglycan loss and enhance COL2 cleavage in cultured cartilage explants (Okamura et al. 2001). Cartilage oligomeric matrix protein (COMP), also termed thrombospondin-5, is another abundant macromolecule in cartilage (Oldberg et al. 1992). It is expressed at high levels in proliferating chondrocytes located in the growth plate and in the

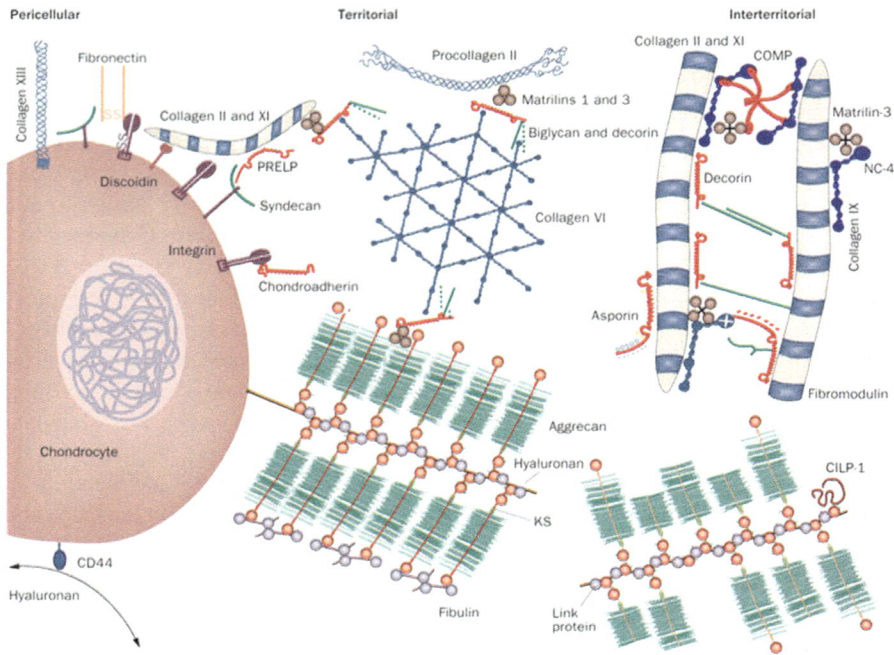

Fig. 6.2 Molecular organisation of normal cartilage. The ECM around chondrocytes in normal articular cartilage is organised into zones; pericellular matrix, territorial matrix and the interterritorial matrix. Each zone consists of different collagens and proteins supporting the integrity of articular cartilage. Abbreviations: Cartilage Intermediate Layer Protein 1 (CILP-1); Proline-Arginine-Rich End Leucine-rich repeat Protein (PRELP). (Image used with permission from Springer Nature. Originally image found in Heinegård and Saxne (2011))

pericellular matrix of mature cartilage (Carlsén et al. 1998). COMP is able to bind to collagens I and II in a zinc-dependent manner and it is involved in collagen fibril organization (Rosenberg et al. 1998). Tenascin-C is another large glycoprotein composed of six subunits and is expressed in normal articular cartilage, upregulated in OA and following stimulation with Interleukin 1-beta (IL-1β), and mechanical loading (Järvinen et al. 1999; Pfander et al. 2004). Another structural glycoprotein present in cartilage is fibrillin-1 (Ferguson et al. 2009). Fibrillin-1 is localized in the pericellular matrix and constitutes the structural backbone of microfibrils which in turn provide a scaffold for the deposition of elastin (Keene et al. 1997). The formation of elastin fibers gives elasticity and resilience to cartilage (Rock et al. 2004). However, unlike collagen fibrils, fibrillin-1 microfibrils do not have well-defined patterns of organization (Keene et al. 1997).

Osteoarthritis (OA)

OA is the most common form of arthritis worldwide (Woolf and Pfleger 2003; Glyn-Jones et al. 2015). OA mostly affects the joints of the hand, knee and hip, and is characterized by the progressive degradation of articular cartilage with secondary synovitis (Vincent and Watt 2014). Increased age is a primary risk factor for the development of OA, along with obesity, genetic predisposition and joint injury (Fig. 6.3) (Bijlsma et al. 2011). Among older adults, OA is one of the most common causes of chronic disability (Loeser 2012a). Functional impairment and pain can lead to severe distress and depression in these patients (Bijlsma et al. 2011). The increase in the lifespan of the global population is set to increase the prevalence of the disease (Woolf and Pfleger 2003; Glyn-Jones et al. 2015).

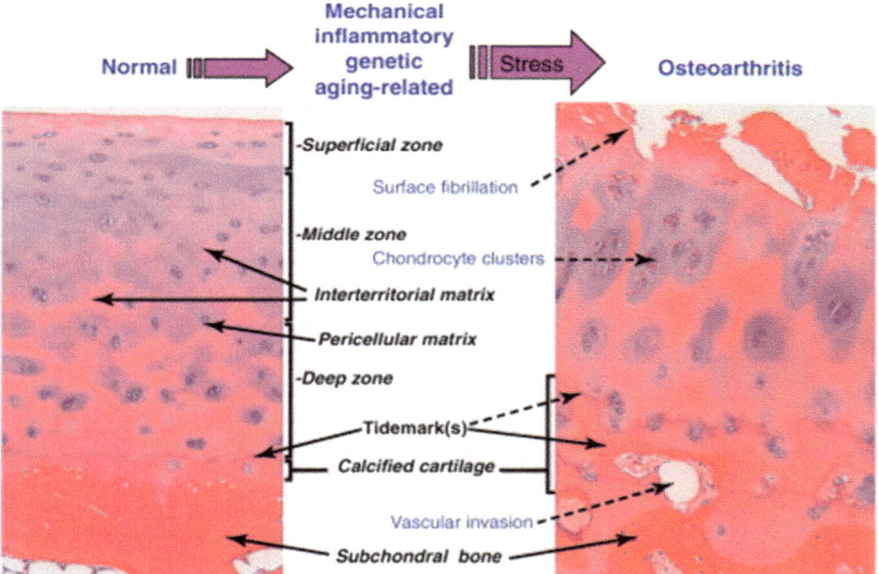

Fig. 6.3 Degradation and changes of cartilage in OA. Normal cartilage has a smooth articular layer and a uniform distribution of chondrocytes. In comparison, OA cartilage displays fissuring of the articular surface, extensive fibrillations and increased chondrocyte clustering in the superficial layer. (Image used with permission from Elsevier. Originally image found in Goldring and Marcu (2012))

Pathophysiology of OA

The degradation of articular cartilage is caused by principle matrix degrading enzymes such as members of the 'a disintegrin and metalloproteinase with thrombospondin motif' (ADAMTS) family and the MMPs; which degrade aggrecan (ADAMTS) and collagen (MMPs) respectively (Nagase et al. 2006). OA chondrocytes secrete higher levels of cytokines such as IL-1β and Tumour Necrosis Factor-alpha (TNFα) that are capable of inducing and activating these catabolic enzymes (Goldring and Otero 2011). The significance of joint inflammation in contributing to tissue breakdown is unclear. Infiltration of mononuclear cells into the synovial membrane is observed in human OA (Scanzello and Goldring 2012) and there is much evidence to support activation of the innate immune system in disease (Fig. 6.4) (Orlowsky and Kraus 2015). Inflammation, when present, will exacerbate tissue breakdown and contribute to painful episodes of disease (Glyn-Jones et al. 2015).

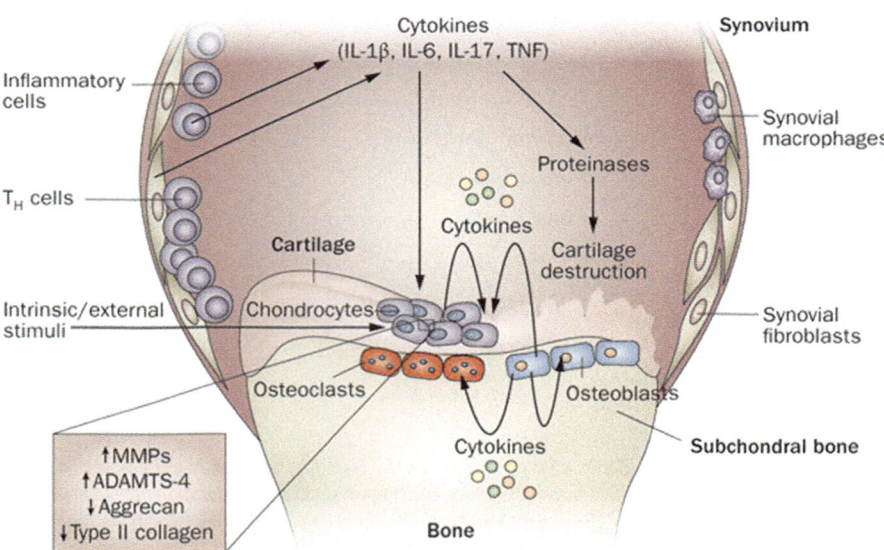

Fig. 6.4 Pathophysiological mechanisms underlying OA. The cellular and molecular mechanisms undying OA are complex and involve multiple components. In general, cells in OA joints release increased concentrations of pro-inflammatory cytokines resulting in the secretion of proteases from chondrocytes which degrade cartilage. (Image used with permission from Springer Nature. Originally figure found in Kapoor et al. (2011))

Cytokines

Cytokines regulate cellular activities and cell-cell interactions (Murphy et al. 2008). Cytokines are secreted and can act in an autocrine, paracrine or in some instances endocrine action (Murphy et al. 2008). They can be secreted in a cascade and have overlapping functions which can result in catabolic or anabolic events or may inhibit either of these processes (Murphy et al. 2008).

Proinflammatory Cytokines

The role of catabolic or proinflammatory cytokines such as IL-1 and TNFα in the pathophysiology of rheumatoid arthritis (RA) is well established (McInnes and Schett 2007). Blockade of these cytokines, especially TNF-α has resulted in effective treatments of RA (Feldmann and Maini 2001). However, the roles of proinflammatory cytokines in OA are still being addressed. IL-1 was originally termed 'Catabolin' and initially purified from pig synovium (Saklatvala 1981). Subsequently, IL-1 has been demonstrated to be upregulated in OA cartilage and both IL-1 and TNF-α are detected in OA synovial fluid (Smith et al. 1997). Immunohistochemical staining of IL-1β showed its expression in the superficial layers of OA cartilage (Melchiorri et al. 1998). Studies in animal models demonstrated intra-articular IL-1 injections induce an OA-like disease (Pettipher et al. 1986) and interleukin-1 receptor antagonist decreased cartilage loss in animal models of OA (Pelletier et al. 1997). Experiments inhibiting the IL-1 receptor in humans have still not been carried out. However, IL-1 antagonists have showed low efficacy in patients (Calich et al. 2010). Treatment of chondrocytes with TNFα also results in the release of matrix degrading enzymes and suppression of matrix synthesis (Saklatvala 1986; Nagase et al. 2006). TNFα is expressed in the middle and deep zones of cartilage explants from OA knee joints (Moos et al. 1999). Both TNFα and TNFα-converting enzyme messenger ribonucleic acid (mRNA) are upregulated in OA (Amin 1999). Only a few experimental trials have investigated the efficacy of TNFα in OA and showed that blockade of the cytokine had modest decreases in disease (Calich et al. 2010). IL-1β inhibition in clinical trials have shown significant benefit in knee and hand OA patients (Chevalier et al. 2013). Blockade of IL-1 receptor improved pain and global disability during a in hand and knee OA patients but patients did display upper respiratory tract infections (Chevalier et al. 2013). These promising studies have not yet been followed up with larger controlled trials.

Both IL-1 and TNFα directly upregulate the expression of Cyclooxygenase-2 (COX-2) and inducible nitric oxide synthase (iNOS) which in turn induce chondrocytes to synthesise prostaglandin E2 (PGE2) and nitric oxide (NO) (Goldring and Berenbaum 2015). COX-2 and iNOS are elevated in OA cartilage compared to normal human cartilage (Amin and Abramson 1998). Nitric oxide (NO) is a cytotoxic free radical and mediates a number of processes associated with OA including endoplasmic reticulum (ER) stress and apoptosis (Amin and Abramson 1998). In addition, PGE2 has been shown to oppose the effects of IL-1 on cartilage matrix synthesis by stimulating COL2 expression which could be a counteractive mechanism to IL-1 induced chondrocyte dedifferentiation (Melchiorri et al. 1998).

Transforming Growth Factor Beta (TGF-β)

TGF-β is a large family of growth factors, which plays a critical role in embryonic development cell proliferation, differentiation, apoptosis and cell migration (Massagué 2012). There are three isotypes of TGF-β: TGF-β -1, -2 and -3, which share a conserved homology of around 90% and are differentially expressed in different tissues and cell populations (Massagué 2012). High TGF-β levels have been detected in the synovial fluid of OA patients and the correlation between the genetic variants of TGF-β signalling pathway components and OA is also reported in patients (Blom et al. 2004; Shen et al. 2014). Both TGF-β isoforms and TGF-β receptors are broadly expressed in cartilage, bone and synovial tissues and play different roles in these tissues. Hence, the role of TGF-β in OA pathogenesis is unclear (Shen et al. 2014). Transgenic mice that overexpress the dominant-negative type II TGF-β receptor in skeletal tissue exhibit progressive skeletal degeneration (Serra et al. 1997; Hiramatsu et al. 2011). Global and cartilage-specific deletion of Mothers against decapentaplegic homolog 3 (SMAD3), a key downstream signalling molecule of TGF-β, causes increased cartilage tissue, articular cartilage fibrillation, clefting, subchondral bone sclerosis and osteophyte formation (Yang et al. 2001; Wu et al. 2009). In addition, deletion of TGF-β Receptor type II in articular chondrocytes also leads to a progressive OA-like phenotype in mice (Shen et al. 2013). Moreover, overexpression of TGF-β1 in osteoblastic cells in mice leads to spontaneous OA and subchondral bone sclerosis whereas inhibition of TGF-β activity in subchondral bone, stabilizes the subchondral bone microarchitecture and decreases OA pathology in mice (Zhen et al. 2013). In addition, inducible knockout of the TGF-β type II receptor in Nestin-positive MSCs leads to fewer changes in the subchondral bone and less articular cartilage degeneration (Zhen et al. 2013). Blaney Davidson et al. (2007) also showed that the cells in the outer layer of osteophytes strongly express TGF-β1 to activate the TGF-β signalling pathway in the experimental OA murine models; which further illustrates the diverse role of TGF-β in OA.

Insulin-Like Growth Factor 1 (IGF1)

IGF1 was identified by Salmon and Daughaday (1957) and originally named as "sulphation factor" because of its ability to stimulate 35-sulphate incorporation into rat cartilage. Exogenous IGF1, when added to monolayers of bovine chondrocytes or cartilage explants, increases proteoglycan synthesis (Sah et al. 1994). Furthermore, a combination of IGF1 and TGF-β has been shown to regulate proliferation and differentiation of periosteal mesenchymal cells during chondrogenesis (Fukumoto et al. 2003). However, addition of IGF1 alone did not have the same effect on chondrogenesis (Fukumoto et al. 2003). This result validates the experiments conducted by Tsukazaki et al. (1994) who showed the ability of TGF-β1 to increase the number of IGF1 receptors in chondrocytes without changing their affinity. IGF1 deficient rats developed articular cartilage lesions (Ekenstedt et al. 2006) and in murine and equine models addition of IGF1 to chondrocyte grafts enhanced chondrogenesis in cartilage defects (Fortier et al. 2002; Goodrich et al. 2007). These studies further support the need for IGF1 to maintain articular cartilage integrity. Several studies have demonstrated that the ability of chondrocytes to respond to IGF1 decreases with age and in OA (Loeser et al. 2000; Morales 2008).

Matrix Metalloproteinases (MMPs)

MMPs are a family of proteinases in humans with a major role in regulating ECM turnover (Nagase et al. 2006). All of the MMPs share three domains: the pro-peptide, the catalytic domain which is linked to a third similar domain the haemopexin-like C-terminal domain (Nagase et al. 2006). All the MMPs contain a zinc atom in the catalytic domain and are synthesised as inactive proenzymes that require activation (Nagase et al. 2006). The MMPs can be classified into four main groups: collagenases, stromelysins, gelatinases and membrane type MMPs (MT-MMPs) (Verma and Hansch 2007). Gelatinases (MMP-2 and MMP-9) degrade partially denatured collagen and have been implicated in angiogenesis and neurogenesis (Singh et al. 2015). Stromelysins (MMP-3, MMP-10, MMP-11, and MMP-7) are small proteases that degrade segments of the extracellular matrix (Nagase et al. 2006). Stromelysins have a broad range of substrates including proteoglycans, laminin and fibronectin, and they are believed to play central role in the activation of other MMPs through cleavage of their pro-peptide domains (Nagase et al. 2006). MT-MMPs (MMP-14 to 17 and 24) activate a few proteases and are involved in the synthesis of components at the cell surface (Page-McCaw et al. 2007). Collagenases are the main type of MMPs involved in cartilage biology and OA (Heinegård and Saxne 2011). These collagenases cleave fibrillar collagens at a single point in collagen strands (Nagase et al. 2006). MMP-13 preferentially cleaves COL2 over collagen type I and III whereas MMP-1 preferentially cleaves type III and MMP-8 type I collagen (Nagase et al. 2006). Overexpression of MMP-13 in the mouse leads to an OA phenotype (Neuhold et al. 2001) and MMP-13 knockout mice are protected from surgically induced osteoarthritis (Little et al. 2009). Zymogens (pro-MMPs) are expressed in normal cartilage tissue but MMP-2, -9, -13, -14, -16 and -28 are overexpressed in osteoarthritic cartilage and synovial tissue. Expression of MMP-1, -3 and -10 are down regulated in disease (Hofmann et al. 1992).

A Disintegrin and Metalloproteinase with Thrombospondin Motifs (ADAMTS)

The ADAMTS are a family of (ADAMTS1-ADAMTS19) metalloproteinases. (Stanton et al. 2011). ADAMTS have at least one thrombospondin type 1 sequence motif familiar with ECM proteins and similarly to MMPs they contain a zinc-binding sequence within their catalytic domain (Nagase and Kashiwagi 2003). ADAMTS can also cleave matrix proteins such as decorin, fibromodulin and COMP (Heinegård and Saxne 2011). Aggrecanase activity was first observed in bovine articular cartilage treated with IL-1 (Westling et al. 2002). Subsequently, ADAMTS4 (Tortorella et al. 1999) and ADAMTS5 (Thirunavukkarasu et al. 2007) were identified in cartilage. It has since been established that ADAMTS5 is the main enzyme active in mouse cartilage (Glasson et al. 2004), as deletion of ADAMTS5, but not ADAMTS4, protects against cartilage degradation and inflammatory arthritis. ADAMTS16, -2, -14 and -12 are up regulated in diseased cartilage whereas ADAMTS1, -9, -5 and -15 are down regulated (Milner et al. 2006).

Tissue Inhibitor of Metalloproteinases (TIMPs)

The regulation of the activity of MMPs and ADAMTS is controlled by the TIMPs (Nagase et al. 2006). TIMPs (1-4) are endogenous inhibitors of MMPs and ADAMTSs and are thought to regulate proteolytic activity (Nagase et al. 2006). TIMP-3 has been identified as a potent inhibitor of ADAMTS4 and ADAMTS5 in vitro (Kashiwagi et al. 2001). TIMP-3 can also inhibit aggrecanase-mediated aggrecan degradation from cartilage explants stimulated with IL-1 or retinoic acid (Gendron et al. 2003).

Mechanical Forces

Some researchers consider all OA to be secondary to abnormal mechanical loading or the failure of the joint to cope with regular mechanical stresses (Cicuttini and Wluka 2014). Joint structures including cartilage, menisci, ligament and subchondral bone are subject to daily mechanical stress and load (Brandt et al. 2009). Occupational associated OA such as 'miners knee' (McMillan and Nichols 2005) or 'foundry workers elbow' (Mintz and Fraga 1973) and the predisposition of OA after knee injury (Nelson et al. 2006; Stefan et al. 2007) contributed to the idea that OA was due to 'wear and tear' (Jensen and Eenberg 1996). Mechano-protective mechanism such as due to muscle hypertrophy declines in elderly populations and is very likely to contribute to loss of joint protection during the normal gait cycle with age (Brook et al. 2015). This is amplified by loss of gait reflexes whereby deceleration of movement upon heel strike would normally occur in a younger individual but is lost in the elderly (Brandt et al. 2009). In addition, proteases and aggrecanases have been shown to be induced rapidly upon surgical joint destabilization in a highly mechano-sensitive manner; gene regulation and disease is abrogated if the joint is immobilized following surgery. This suggests that mechanical factors can also initiate pathogenic pathways (Burleigh et al. 2012). These findings accord well with epidemiological evidence that OA may principally be driven by mechanical joint overload and injury (Brandt et al. 2009).

Clinical Aspects of OA

OA affects an estimated 10% of men and 18% of women over 60 years of age and the socioeconomic burden is large, costing between 1·0% and 2·5% of gross domestic product (Woolf and Pfleger 2003; Hiligsmann et al. 2014). OA can present as mono- or polyarticular with clinical presentation usually occurring late in disease; associated with pain, stiffness, decreased mobility, deformity and at times with depression (Vincent and Watt 2014). Symptomatic joints display chronic pain with radiographic joint space narrowing indicating cartilage loss, bone cysts and osteophyte formation and subchondral sclerosis (Fig. 6.5) (Bijlsma et al. 2011). The diagnosis of OA can be based on clinical criteria and symptom-based case definitions such as those recommended by the American Rheumatism Association (Hochberg et al. 2012), or on radiological criteria, such as the Kellgren-Lawrence grading of plain joint radiography (Kellgren and Lawrence 1957). This overall joint scoring system grades OA in five levels from 0 to 4, defining OA by the presence of a definite osteophyte (Grade ≥2), and more severe grades by the appearance of joint

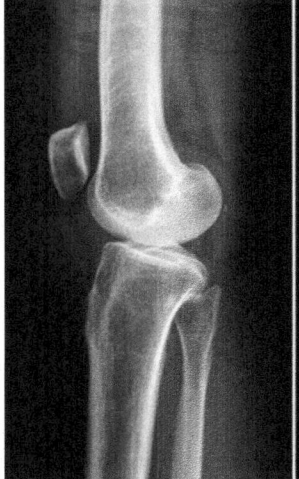 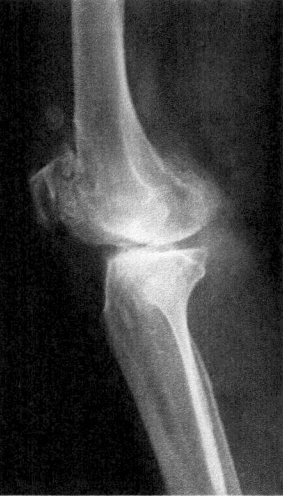

Fig. 6.5 Radiographic imaging of OA. Normal knee (left image) and osteoarthritis Knee (right image) (lateral view). OA images display joint space narrowing, morphological changes of the joint and surrounding bone structures and importantly a thinner cartilage layer. Image courtesy of stockdevil at FreeDigitalPhotos.net

space narrowing, sclerosis, cysts, and deformity (Kellgren and Lawrence 1957; Neogi and Zhang 2013). More sensitive imaging methods such as magnetic resonance imaging can visualize multiple structures in a joint however are very expensive for general clinical use (Glyn-Jones et al. 2015).

Current Treatments for OA

Depending on the severity of OA, patients are first advised to lose weight and modify their lifestyles (Fig. 6.6) (Glyn-Jones et al. 2015). Weight loss in overweight and obese patients reduces the risk of symptomatic osteoarthritis and improves symptoms once evidence of disease is found (Gudbergsen et al. 2013). In addition, exercises to improve muscle strength and aerobic capacity decrease OA symptoms and have benefits in cardiovascular health and all-cause mortality (Uthman et al. 2013; Glyn-Jones et al. 2015). Current pharmacological treatments for OA aim to alleviate pain and target joint inflammation (Hunter 2011). Use of supplements including glucosamine and chondroitin has been widely recommended and shown to be safe; however, the efficacy of these supplements remains controversial (Wieland et al. 2005). Non-steroidal anti-inflammatory drugs such as aspirin and ibuprofen are prescribed to reduce inflammation and pain (Cheng and Visco 2012). Another form of treatment is the local administration of corticosteroids via intra-articular injections

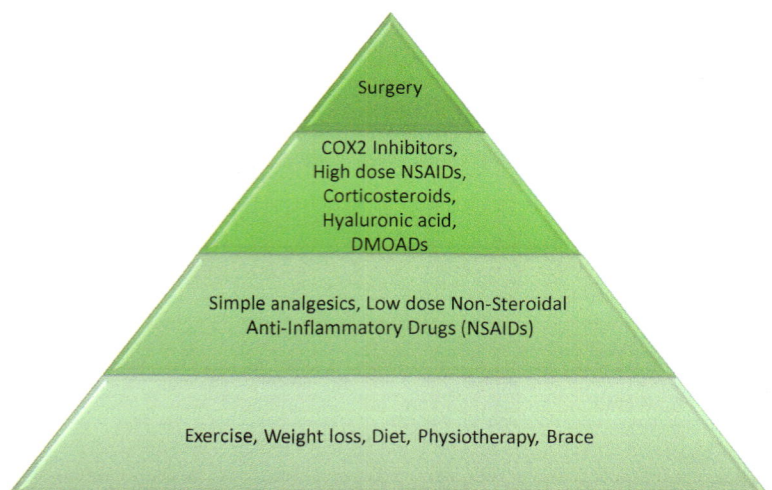

Fig. 6.6 Treatment regime for OA. The treatment for OA is customized for each individual. Patients presenting early, and mild symptoms usually are recommended changes in life-style choices. As the disease worsens, patients are monitored more carefully and treated with stronger therapeutics usually to manage the pain. The end-point treatment for OA is joint replacement surgery

to reduce the production of pro-inflammatory mediators and the infiltration of mononuclear cells into the synovium (Buckingham 2006). The current strategy of pharmaceutical companies is to develop disease-modifying OA drugs (DMOADs) which include inhibitors of cyclooxygenases, MMPs, cytokines and NO (Wieland et al. 2005). However, with poor results in clinical trials and low efficacy of DMOADs, the current endpoint treatment for OA still remains joint replacement surgery (Hunter 2011).

Risk Factors of OA

OA can be caused by a complex interplay between systemic and local factors due to multiple risk factors (Fig. 6.7) (Neogi and Zhang 2013). For example, an inherited predisposition to develop OA may only initiate in old age or if an insult to the joint has occurred (Neogi and Zhang 2013). The significance of risk factors may vary for different joints, for different stages of the disease, or even for the development as opposed to the progression of disease (Reynard and Loughlin 2012). These risk factors include genetics, diet and obesity, gender, mechanical injury and age (Glyn-Jones et al. 2015).

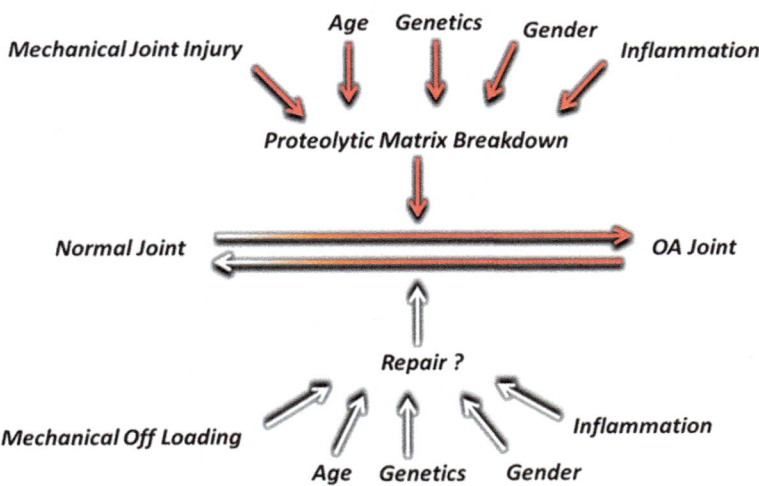

Fig. 6.7 Factors involved in OA pathogenesis. The heterogeneous nature of OA means the disease has multiple risk factors. The occurrence of OA increases with age, making the ageing process the primary risk factor for the disease. However, it is well accepted varying stages of OA in different individuals might be caused by the combination of multiple risk factors. The goal of treating each or multiple risk factors to initiate a repair process in cartilage during OA still remains elusive

Genetics

Many studies have investigated the association between genetics and OA (Peach et al. 2005). Twin and familial studies have identified hereditary factors as a component of primary OA development (Spector et al. 1996). Ten to thirty percent of knee OA is estimated to be hereditary and 50% of hip OA and 70% of hand OA is estimated to be hereditary (Hunter et al. 2004; Peach et al. 2005). There is also evidence that suggests the joint which might develop OA in a person has hereditary link too (Valdes and Spector 2009).

Genome wide association and linkage studies have identified candidate genes and susceptibility gene loci for OA (Loughlin 2005). These studies identified regions with OA susceptibility on chromosomes 2p, 4q, 7p, 8q, 9p, 9q, 10p and 16p (Valdes and Spector 2014). Moreover, genes on chromosomes 2, 4, and 16 are likely to be the strongest candidates for hereditary OA onset and progression (Loughlin 2001). Some of these gene candidates include IL-1 cluster genes at chromosome 2p11.2-q13 associated with increased susceptibility for knee and hip OA; IL4 receptor chain at 16p12.1 and frizzled related protein 3 gene at 2q32.1 associated with increased susceptibility to hip OA; missense mutation in Matrilin 3 gene at 2p24.1 associated with hand OA and Metalloproteinase gene ADAM12 at 10q26.2 associated with knee OA alone (reviewed in Reynard and Loughlin 2012). In addition, severe osteoarthritis of the hand has been associated with variants within the ALDH1A2 gene at 15q22 (Styrkarsdottir et al. 2014).

Out of approximately 80 published genes only one gene has shown consistent and robust evidence of genetic association: growth and differentiation factor 5 gene (GDF5); Reynard and Loughlin 2012). A single nucleotide polymorphism rs143383, located in the 3' untranslated region of the GDF5 gene is associated with OA (Miyamoto et al. 2007). GDF5 is an extracellular signalling molecule and a member of the TGF-β superfamily and participates in the development, maintenance and repair of synovial joint tissues (Reynard et al. 2011). Many studies have found that the genetic deficit in GDF5 is not restricted to cartilage but is found in all of the joint tissues so far examined in OA patients (Reynard and Loughlin 2012). Mouse studies have confirmed that a reduction in GDF5 mRNA and protein can result in an OA-like phenotype (Daans et al. 2011).

Dietary Factors

Dietary factors such as vitamin supplementation are the subject of considerable interest and debate in OA. The Framingham Study demonstrated subjects in the lowest (<27 ng/ml) and middle (27.0–33.0 ng/ml) tertile of serum 25- hydroxyvitamin D had a threefold increased risk for progressive knee OA compared with those in the highest tertile. However, no such effect was observed for the risk of developing hip OA (McAlindon et al. 1996a). In contrast, in the Study of Osteoporotic Fractures, women in the middle (23–29 ng/mL) and lowest (8–22 ng/mL) tertiles of serum 25-vitamin D were 3 times as likely to develop incident hip OA, compared to those in the highest tertile (30–72 ng/mL) (Lane et al. 1999). Low vitamin C dietary intake was associated with an increased risk of progression, but not incidence, of knee OA among the participants in the Framingham Study (McAlindon et al. 1996b). Neogi et al. (2008) showed high serum vitamin K was associated with a low prevalence of radiographic hand OA and the presence of large osteophytes but the relation of serum vitamin K levels to knee radiographic OA was less clear. However, a placebo-controlled trial of vitamin K supplementation did not confirm a protective effect of vitamin K on the severity of radiographic hand OA (Neogi and Zhang 2013). In areas of China and Eastern Asia the prevalence of Kashin-Beck Disease, an early onset of osteoarthropathy is high (Fang et al. 2003; Moreno-Reyes et al. 2003). In these areas, the levels of selenium (a non-metal element with anti-oxidative properties) are low in soil (Moreno-Reyes et al. 2003). Interestingly, supplementation of selenium decreased the incidence of this disease (Moreno-Reyes et al. 2003). However, others have reported that high selenium intake is associated with increased risk of both hip and knee OA (Neogi et al. 2008). Studies on dietary factors such as the ones outlined above and other studies in the past focusing on nutritional supplementation such as glucosamine or chondroitin have been conflicting and produced poor clinical efficacy (McAlindon et al. 1996b; Hunter 2011; Neogi and Zhang 2013).

Obesity

Being obese or overweight are potent risk factors for OA, especially OA of the knee (King et al. 2013). The increase load caused by obesity and being overweight might be a leading cause of knee or hip OA by placing undue stress on joint and supporting structures (Neogi and Zhang 2013). The Framingham Study showed that women who lose approximately 5 kg had a 50% reduction in the risk of development of symptomatic knee OA and a strong association with a reduced risk of radiographic knee OA (Felson et al. 1992). Losing weight has also been shown to decrease pain and disability in established knee OA patients (Messier et al. 2004). Interestingly, a clinical trial showed patients who combined weight loss with exercise were effective in decreasing pain and improving function (Messier et al. 2004). However, exercise and/or weight loss alone did not improve pain and functional outcomes (Messier et al. 2004). Meta-analysis of the study demonstrated the effects of weight loss on pain were not consistent but weight reduction by about 5% was associated with an improvement of physical function (Christensen et al. 2007). The association of being overweight and hip OA is inconsistent and weaker than knee OA (Neogi et al. 2008); however, obesity has been shown to increase the risk of bilateral radiographic as well as symptomatic hip OA (Tepper and Hochberg 1993). High body mass index, especially at a young age has been also associated with an increased risk of total hip replacement therapy (Karlson et al. 2003).

Gender

Women are more likely to have OA than men (Srikanth et al. 2005). There is an increase in the prevalence of OA in women around the time of menopause (Wluka et al. 2000; Srikanth et al. 2005). This raises the question of whether hormonal factors might play a role in the development of OA. However, there has been conflicting observational data to support the positive effect of estrogen, either endogenous or exogenous, in OA (Neogi and Zhang 2013). No significant difference was found in the prevalence of knee pain or its associated disability between those taking estrogen plus progestin therapy or those taking placebo (Nevitt et al. 2001). In contrast, Women's Health Initiative showed women on estrogen replacement therapy were 15% less likely to require total knee or hip arthroplasty than those not taking such therapy (Cirillo et al. 2006).

Mechanical Injury

Numerous studies have shown that knee injury is a major risk factor for OA. Injury to the structures of a joint, particularly a trans-articular fracture, meniscal tear requiring meniscectomy, or anterior cruciate ligament injury increases the prevalence of OA (Neogi and Zhang 2013). In the Framingham Study the prevalence of meniscal damage was much higher among subjects with radiographic knee OA (82%) than those without OA (25%) (Englund et al. 2008). A systemic review of studies involving patients with anterior cruciate ligament injury, either isolated or combined with medial collateral ligament, reported the prevalence of OA was up to 48% in subjects with anterior cruciate ligament and additional meniscal injury (Øiestad et al. 2009). In addition, Stefan et al. (2007) reported that 50% of those with a diagnosed anterior cruciate ligament or meniscus tear develop osteoarthritis with associated pain and functional impairment 10–20 years after the original diagnosis.

Age

Age is the strongest risk factor for OA in all joints (Shane Anderson and Loeser 2010). The increased prevalence and incidence of OA with age may be a consequence of cumulative exposure to various risk factors and biological changes that occur with ageing (Lotz and Loeser 2012). The Framingham Osteoarthritis Study showed the prevalence of radiographic OA increased with each decade of life from 33% among those aged 60–70 to 43.7% among those over 80 years of age, whereas, the prevalence of symptomatic knee OA in all subjects was 9.5% and increased with age in women but not men (Felson et al. 1987). In addition, a population-based cohort of 3000 study participants conducted in the United States of America showed the incidence of radiographic knee OA rose from 26.2% in the 55–64 year range to nearly 50% in the 75+ group (Jordan et al. 2009). Likewise, the prevalence of symptomatic knee OA increased from 16.3% to 32.8% between these groups (Jordan et al. 2009). The Chingford Women's study, which is a community-based cohort living in the United Kingdom, showed that at baseline, when the median age was 53 years, 13.7% of the patients had radiographic knee osteoarthritis; by a median age of 68 years the incidence of radiographic knee osteoarthritis increased to 47.8% (Leyland et al. 2012).

Hip OA is less common in the ageing population than knee OA but is still quite prevalent (Shane Anderson and Loeser 2010). The prevalence of primary radiographic hip OA increased from 0.7% in the 40–44 age group to 14% in the 85+ age group in a systematic review which examined the prevalence of primary hip OA (Dagenais et al. 2009). An American cohort study demonstrated a higher prevalence of symptomatic hip OA in their population with age: of 5.9% in the 45–54 age groups increasing to 17% in the 75+ age group (Jordan et al. 2009). Moreover, a study of a cohort of patients with osteoporotic fractures reported an increase in the prevalence and incidence of hip OA in women over age 65 (Arden et al. 2009).

The hand is probably the joint most commonly affected by OA in the ageing population (Loeser 2012b). However, hand OA is not as disabling as OA of the knee or hip (Shane Anderson and Loeser 2010). The Zoetermeer survey found that radiographic involvement of the distal interphalangeal joint affected more than half of men over the age of 65 and more than half of women over the age of 55 (van Saase et al. 1989). The Framingham study found 13% of men and 26% of women over the age of 70 have symptomatic hand OA involving at least one joint (Zhang et al. 2002). The Fallon Community Health Plan reported that yearly incidence rates for hand OA were 0.35% and 0.21% for men and women respectively over the age of 60 and the incidence rates younger than 60 years were much lower (Oliveria et al. 1995).

Cellular Aging Pathways in OA

One significant advance from the late 90s been the characterization of the cellular metabolic pathways that are affected by ageing (Fig. 6.8). Largely, these pathways have been defined in other model systems and in recent years studied in OA.

Autophagy

Autophagy is a cellular homeostasis mechanism for the removal of dysfunctional organelles and proteins (Boya et al. 2013). Defective autophagy is involved in the pathogenesis of age related diseases and recent observations indicate that this process is compromised in aging cartilage (Rubinsztein et al. 2011; Lotz and Caramés

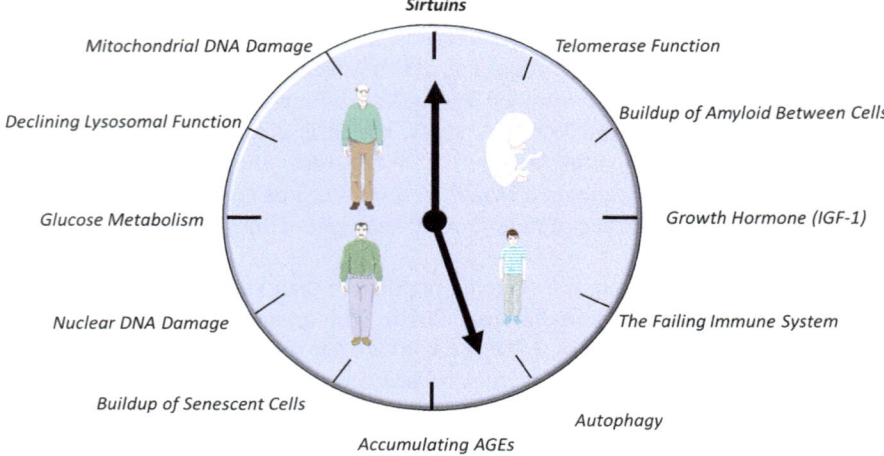

Fig. 6.8 Cellular ageing mechanisms. The ageing process at the cellular level is complicated and involves multiple signalling pathways. These pathways interact together to impact the cellular response to external and internal stimulus during ageing, from early embryonic stages to old age

2011). Caramés et al. (2010) initially examined key autophagy proteins in human and murine cartilage. These markers were Unc-51-like kinase 1 (ULK1), an inducer of autophagy, Beclin1, a regulator of autophagy, and microtubule-associated protein 1 light chain 3 (LC3), which executes autophagy (Lotz and Caramés 2011). ULK1, Beclin1, and LC3 protein expression was reduced in human OA chondrocytes and cartilage (Caramés et al. 2010). In mouse knee joints, protein expression of ULK1, Beclin1, and LC3 decreased together with glycosaminoglycans (GAG) loss at ages 9 months and 12 months and in the DMM OA disease model, 8 weeks after surgery (Caramés et al. 2010). However, Poly(ADP-ribose) polymerase (PARP) p85 (a marker for apoptosis) expression was increased at these time points (Caramés et al. 2010). A further study by Caramés and colleagues (2015) showed 28-month-old mice had significant reduction in the total number of autophagic vesicles per cell and in the total area of vesicles per cell in articular cartilage compared to young 6-month-old mice. With increasing age, the expression of Autophagy Related 5 (ATG5) and LC3 decreased in these mice, and this was followed by a reduction in cartilage cellularity, OA disease score and an increase in PARP p85 (Caramés et al. 2015). Interestingly, Bouderlique et al. (2015) generated a cartilage-specific ATG5 conditional KO (ATG5cKO) mice which developed serve osteoarthritis when aged and also displayed increased cell death, cleaved caspase 3 and cleaved caspase 9. Surprisingly, no difference in the development of post-traumatic osteoarthritis was observed between Atg5cKO and control mice post DMM (Bouderlique et al. 2015).

Studies have also examined the role of mammalian target of rapamycin (mTOR) signalling which when activated inhibits ULK1 which in turn inhibits autophagy. Zhang et al. (2014) reported upregulation of mTOR expression correlates with increased chondrocyte apoptosis and reduced expression of key autophagy genes in human OA tissue. In addition, inducible cartilage-specific mTOR KO mice displayed increased autophagy signalling and significant protection from DMM induced OA associated with a significant reduction in the apoptosis and synovial fibrosis (Zhang et al. 2014). In another study, Carames et al. (2012) utilised rapamycin to pharmacological inhibit mTOR in mice. The severity of cartilage degradation was significantly reduced in the rapamycin-treated group compared with the control group and this was associated with a significant decrease in synovitis 10 weeks post DMM (Carames et al. 2012). Rapamycin treatment also maintained cartilage cellularity and decreased ADAMTS5 and IL-1β expression in articular cartilage (Carames et al. 2012). These results suggest that rapamycin, at least in part by autophagy activation, reduces the severity of experimental osteoarthritis. However, further studies are required to investigate if other key regulators of autophagy exist in aging cartilage apart from the classical mTOR/ULK1 signalling cascade.

Sirtuins

The well-conserved Sirtuin gene family have been strongly associated with longevity since Sir2 (Silent information regulator) was shown to extend lifespan in budding yeast, worms, flies and mice (Giblin et al. 2014). In mammals, Sirtuins are a 7 member family (SirT1-7) of NAD-dependent deacetylases, with SirT4 and SirT6 also demonstrating ADP-ribosyltransferase activity (Fig. 6.9) (Verdin 2014). SirT1 is found in the cytosol and nucleus, whereas SirT2 is predominately located in the cytosol. SirT3, SirT4 and SirT5 are found in the mitochondria, SirT6 in the nucleus and SirT7 in the nucleolus (Nakagawa and Guarente 2011). The Sirtuins are involved in a wide range of physiological systems and cellular functions including cell metabolism, apoptosis, growth and development, inflammatory and stress responses (Sacitharan et al. 2012; Morris 2013). The diverse roles of the Sirtuin family have resulted in them being studied extensively over the past decade in the fields of age-related pathologies (Giblin et al. 2014).

Until now no research investigating the roles of SirT 2,3,4,5 or 7 in OA has been conducted. However, some research has been conducted on SirT6. Mice with a SirT6 haploinsufficiency fed on high fat diet had increased OA scores at 6 months

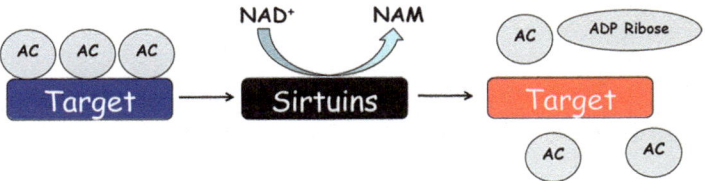

Fig. 6.9 Mechanism of action of sirtuins. Sirtuins cleave NAD to produce nicotinamide which results in the acetyl group of the substrate being transferred to the ADP-ribose moiety of NAD. This reaction yields an O-acetyl-ADP ribose and a deacetylated substrate

of age compared to wild type controls fed on a high fat diet (Ailixiding et al. 2015). No change in OA score was reported between mice with SirT6 haploinsufficiency and control mice fed on a normal diet (Ailixiding et al. 2015). Furthermore Wu et al. (2015) reported SirT6 protein expression levels were significantly decreased in the articular chondrocytes of OA patients compared to normal human chondrocytes. In addition, intraarticular injection of Lenti-Sirt6 in mice reduced OA disease score 8 week post DMM compared to mice receiving Lenti-negative control (Wu et al. 2015). More evidence using SirT6 cartilage-specific KO mice or specific pharmacological agents are required to confirm the protective role of SirT6 in OA.

The most researched into Sirtuin until now in the OA field has been SirT1. Initially, Dvir-Ginzberg et al. (2008) showed that overexpression of SirT1 in chondrocytes isolated from articular cartilage of human joints resulted in an increase in cartilage specific gene expression of collagen 2(α1) (COL2A1) through the deacetylation of SOX-9 (Dvir-Ginzberg et al. 2008). Similarly, decreased expression of crucial matrix components including aggrecan, COL2A1 and collagen 9(α1) was observed in SirT1 siRNA treated human knee chondrocytes (Fujita et al. 2011). In contrast, decreasing SirT1 resulted in the upregulation of collagen 10(α1) and ADAMTS5 gene expressions (Fujita et al. 2011). Furthermore, chondrocytes from human cartilage explants when subjected to stresses (nutritional, catabolic and mechanical shear stress) associated with aging and OA demonstrated a decrease in SirT1 expression (Takayama et al. 2009).

Nine-month-old mice (with an allele carrying a point mutation that encodes SirT1 with no enzymatic activity) had increased levels of apoptotic chondrocytes and increased OA disease scores when compared to age-matched wild type control mice (Gabay et al. 2013). In addition, Gabay et al. (2012) demonstrated SirT1 constitutive whole body knockout mice at 3 weeks of age (constitutive SirT1 KO mice can only survive up to approximately 3 weeks) have increased OA disease scores and exhibited low levels of COL2A1, ACAN, GAG release and high protein levels of MMP-13. Matsuzaki et al. (2013) generated cartilage specific SirT1 conditional knockout mice using COL2A1ER^{T2} Cre (Sirt1-CKO) which showed accelerated OA progression at 2 and 4 (but not surprisingly 8) weeks post DMM surgery compared control mice. In addition, OA disease score was significantly higher in 1 year old Sirt1-CKO mice than in control mice (Matsuzaki et al. 2013). Together, these studies to date suggest that SirT1 may have a protective role in cartilage against the development of OA. Hence, increasing SirT1 expression or targeting its unknown mechanism of action in aging chondrocytes could prove to be beneficial in treating OA.

Oxidative Stress

Reactive oxygen species (ROS) are produced as a result of cellular metabolism and environmental factors (Nathan and Cunningham-Bussel 2013). ROS can damage nucleic acids and proteins, thus altering their functions (Gorrini et al. 2013). Hence, cells produce antioxidants to counteract oxidative stress caused by ROS (Gorrini et al. 2013; Nathan and Cunningham-Bussel 2013). Initial studies by Baker and Feigan (1988) demonstrated hydrogen peroxide (H_2O_2) suppressed proteoglycan synthesis in human cartilage explant culture. Further suppression of proteoglycan synthesis by H_2O_2 was reported when catalase and the glutathione peroxidase/reductase systems were inhibited (Baker and Feigan 1988). Oxidative stress with aging has also been shown to make human chondrocytes more susceptible to apoptosis through the dysregulation of the glutathione antioxidant system (Carlo and Loeser 2003). In addition, reduced levels of catalase were observed in aged rat chondrocytes (15–18 months) than chondrocytes from young adult rats (6 months old) (Jallali et al. 2005). Furthermore, reductions in mitochondrial superoxide dismutase, manganese-superoxide dismutase and glutathione peroxidase were all observed in isolated chondrocytes from OA tissue samples compared to chondrocytes from normal cartilage explants (Aigner et al. 2006; Ruiz-Romero et al. 2006, 2009). Loeser et al. (2002) observed Nitrotyrosine; a maker of protein oxidation was also increased in aging normal human and monkey cartilage and OA human cartilage. ROS are also produced by chondrocytes in response to stimulation by cytokines and growth factors, including IL-1, TNFα, and TGF-β1 (Lo and Cruz 1995; Jallali et al. 2007; Loeser 2012a). All these studies to date prove the importance of reducing oxidative stress in chondrocytes. However, it still remains to be seen if targeting any of these anti-oxidant enzymes could be a therapeutic option in OA.

Endoplasmic Reticulul (ER) Stress and Unfolded Protein Response (UPR)

The ER is arranged in a dynamic tubular network and protein maturation at the ER is vital for the correct folding of proteins (Hetz 2012). Several feedback mechanisms are in place to cope with impaired protein folding which causes ER stress (Hetz 2012). These coping mechanisms together are known as the Unfolded Protein Response (UPR) (Ron and Walter 2007). The UPR signalling network is complex and requires several key components (Ron and Walter 2007). In summary, UPR stress sensors, Inositol-requiring Protein 1α (IRE1α), Protein Kinase RNA-like Endoplasmic Keticulum Kinase (PERK) and Activating Transcription Factor 6 (ATF6), signal information about the folding status of the ER to the cytosol and nucleus to restore protein-folding capacity (Ron and Walter 2007; Hetz 2012). IRE1α signalling activates X Box-binding Protein 1 (XBP1) transcription factor (Ron and Walter 2007; Hetz 2012). PERK signalling activates Translation Initiator Factor 2α (eIF2α) which in turn activates ATF4 and CCAAT/-enhancer-binding Protein Homologous Protein (CHOP) transcription factors (Ron and Walter 2007; Hetz 2012). ATF6 signalling releases its cytosolic domain fragment (ATF6f) which is also a UPR transcription factor (Ron and Walter 2007; Hetz 2012). Thereafter, all the transcription factors mentioned above activate UPR target genes involved in pathways such as autophagy, apoptosis, lipid synthesis and NF-kB signalling which counterbalance the ER stress (Ron and Walter 2007; Hetz 2012).

In vitro studies have shown increased UPR in OA articular chondrocytes predominately through the PERK and IRE1 pathways (Liu-Bryan and Terkeltaub 2015). In addition, Husa et al. (2013) showed biomechanical injury, IL-1β and NO increase ER stress and the UPR in cultured bovine chondrocytes. Interestingly, transfection of CHOP in 'gain of function' experiments sensitized normal human chondrocytes to IL-1β induced NO and MMP3 release without inducing these responses by itself (Husa et al. 2013). It must be noted excessive CHOP signalling induces ER stress as well; by increasing protein synthesis (Ron and Walter 2007; Hetz 2012). Hence CHOP signalling must be finely regulated in chondrocytes. Surgically induced OA in mice revealed that pan tissue KO of CHOP partially protected against increased chondrocyte apoptosis and cartilage degradation (Uehara et al. 2014). Interestingly, there was no difference in ER stress between CHOP KO mice and control mice (Uehara et al. 2014).

IRE1α signalling of XBP1 in UPR has been demonstrated to be important in chondrocyte differentiation. Liu et al. (2012) overexpressed XBP1 in chondrocyte cell lines and revealed this to accelerate chondrocyte hypertrophy by increasing expression of type II collagen, type X collagen and Runt-related transcription factor 2 (RUNX2). Conversely, knockdown of XBP1 via siRNA abolished hypertrophic chondrocyte differentiation (Liu et al. 2012). In addition, Takada et al. (2011) highlighted XBP1 expression induced ATF6 signalling which enhanced apoptosis in osteoarthritic cartilage. This highlights the high degree of crosstalk between the components of UPR signalling in chondrocytes. However, additional in vivo studies are required in the field using cartilage specific UPR protein KO mice to dissect the important regulators of the UPR in OA models and aging mice.

Adenosine Monophosphate-Activated Protein Kinase (AMPK)

AMPK is a master regulator of cellular energy and adjusts to changes in energy demand (Salminen and Kaarniranta 2012). Terkeltaub et al. (2011) first described normal human knee articular chondrocytes expressed AMPKα1, α2, β1, β2, and γ1 subunits with constitutive and robust activity. However, AMPK activity was observed to decrease in OA articular chondrocytes and cartilage and in normal chondrocytes treated with IL-1β and TNFα (Terkeltaub et al. 2011). Knockdown of AMPKα resulted in enhanced catabolic responses to IL-1β and TNFα in chondrocytes (Terkeltaub et al. 2011). Interestingly, using AMPK activators (AICAR and A-769662) suppressed pro-catabolic responses to IL-1β and TNFα from chondrocytes (Terkeltaub et al. 2011). A further study by the same group revealed AMPK activity decreased in mouse knee OA post DMM and aged knee cartilage (6–24 months) and in bovine chondrocytes after biomechanical injury (Petursson et al. 2013). This study further identified a upstream kinase, the liver protein kinase

B1 (LKB1) as the promoter of AMPK activity in chondrocytes (Petursson et al. 2013). Knockdown of LKB1 attenuated chondrocyte AMPK activity and increased NO, MMP3 and MMP13 release in response to IL-1β and TNFα (Petursson et al. 2013). LKB1 similar to AMPK was also observed to decrease in mouse knee OA post DMM and aged knee cartilage (6–24 months) (Petursson et al. 2013). In addition, pre-treatment of bovine chondrocytes with AMPK activators inhibited the catabolic response of NO after biomechanical injury (Petursson et al. 2013). Dysregulation of LKB1 might have a major role in the suppression of AMPK activity in chondrocytes. Targeting this pathway with AMPK activators (AICAR and A-769662) might be a future therapeutic target for OA. Further studies using AMPK KO mice can aid in validating the role of this master regulator in cartilage aging and OA disease pathogenesis.

Forkhead Box (FOXO) Transcription Factors

FOXO transcription factors are involved in the regulation of the cell cycle, apoptosis, metabolism and even autophagy (Eijkelenboom and Burgering 2013). Multiple and diverse upstream pathways regulate FOXO activity through post-translational modifications and nuclear–cytoplasmic shuttling (Eijkelenboom and Burgering 2013). Normal human cartilage (from humans aged between 23 and 90) expresses FOXO1 and FOXO3 but not FOXO4 protein subtypes (Akasaki et al. 2014b). Akasaki et al. (2014a, b) went on to report during aging expression of FOXO1 and FOXO3 was markedly reduced in the superficial zone of human cartilage regions exposed to maximal weight bearing. Similar patterns of FOXO expression was observed in aging (4–24 months) and OA (post DMM) mouse models (Akasaki et al. 2014a, b). In addition, FOXO1 protein expression was suppressed when human chondrocytes were cultured with IL-1β and TNFα while TGFβ increased FOXO1 and FOXO3 protein expression (Akasaki et al. 2014a, b). A further study by Akasaki et al. (2014a, b) reported oxidative stress caused by the oxidant tert-butyl-hydroperoxide in chondrocytes reduced expression of FOXO transcription factors which increased cell death. Interestingly, this increase in cell death was accompanied by reduced levels of antioxidant proteins (glutathione peroxidase 1 and catalase) and autophagy-related proteins (LC3 and Beclin1) (Akasaki et al. 2014a, b). These studies point to a tissue specific signature of FOXO expression and its partial role in regulating oxidative stress resistance and autophagy. However, in vivo studies using FOXO cartilage specific KO mice or FOXO activators and inhibitors are still lacking in the field. This will help validate if FOXO transcription factors can be targeted in the future OA therapy.

Peroxisome Proliferator-Activated Receptors (PPAR) Transcription Factors

Another family of transcription factors known as Peroxisome Proliferator-Activated receptors (PPARs) have been shown in recent studies to be involved in OA. PPARs are ligand-activated transcription factors that are involved in regulating glucose and lipid homeostasis, inflammation, proliferation and differentiation (Peters et al. 2012). Three PPAR isoforms, PPARα, PPARβ/δ (also known as PPARβ or PPARδ) and PPARγ, are found in all mammals (Peters et al. 2012). It has been known for some time that PPARγ inhibits IL-1β and induced proteoglycan degradation (Francois et al. 2006). However, recent advances in transgenic technology have allowed in-depth dissection of the roles of PPARs in OA pathogenesis. Vasheghani et al. (2013) reported that (constitutive) cartilage-specific disruption of PPARγ results in spontaneous OA in mice (14 months of age). The same group went on to generate a conditional cartilage specific PPARγ KO mouse line (PPARγ-cKO) (Vasheghani et al. 2015). Postnatal deletion of PPARγ in chondrocytes upon admin-istration of doxycycline did not lead to spontaneous OA in mice but the mice became more susceptible to experimental destabilization of the medial meniscus (DMM) OA compared to control mice (Vasheghani et al. 2015). PPARγ-cKO mice displayed increased expression of MMP13, ADAMTS5, and higher number of apoptotic chondrocytes in knee joints post DMM (Vasheghani et al. 2015). This same group previously reported mTOR KO protected mice from experimental OA by increasing autophagy (Zhang et al. 2014). Therefore, the group tested the hypothesis that the worse OA outcome in PPARγ-cKO mice was due to enhanced mTOR signalling. PPARγ-cKO had increased expression of mTOR and a decrease in autophagy mark-ers in naive and post DMM joint (Vasheghani et al. 2015). However, double KO mice of mTOR and PPARγ were protected from DMM-induced OA with a pheno-type similar to that of the previously published mTOR KO (Zhang et al. 2014; Vasheghani et al. 2015). These studies demonstrated PPARγ is dependent on mTOR in certain conditions and explains the protective role of PPARγ in OA. Interestingly, Pioglitazone (PPARγ agonist) has been shown to reduce cartilage lesions in an experimental dog model of OA (Boileau et al. 2007). Studies to-date make PPARγ an attractive future drug target for OA.

Studies have also been conducted to elucidate the role of PPARα and PPARβ/δ in OA. Clockaerts and collegues (2011) reported PPARα agonist Wy-14643 inhib-ited the mRNA expression of MMP1, MMP3 and MMP13 in human OA cartilage explants after IL-1β treatment but did not have an effect on COL2A1 or aggrecan mRNA expression. This study suggests a possible partial protective effect of PPARα in OA. A recent in vivo study by Ratneswaran et al. (2015) looked into the role of PPARβ/δ in OA. PPARβ/δ activation by GW501516 increased expression of several proteases (MMP2, MMP3, ADAMTS2) in murine chondrocytes and increased aggrecan degradation and GAG release in knee joint explants. Constitutive cartilage specific PPARβ/δ KO displayed no developmental phenotype and showed marked protection in the DMM model of OA (Ratneswaran et al. 2015). This study suggests

PPARβ/δ to have a protective role in OA. The role of each PPAR isomer seems to be very specific in OA and highly context specific. Hence careful targeting of these transcription factors might be required in OA. Furthermore, the redundancy between the PPAR isoforms in cartilage biology is still not known.

Conclusions

Advances in basic science have improved our understanding of how age contributes to the pathogenesis of OA. We now better appreciate the complex signalling networks that control metabolic processes and regulators in articular chondrocytes. The dysregulation of these factors can refine inflammatory responses and cartilage degradation leading to progressive OA over time which may lead to more preventative and therapeutic strategies to treat OA.

Acknowledgements Work by this author was supported by a U.S.-U.K. All Disciplines Fulbright Research Scholar Award.

References

Aigner T, Fundel K, Saas J et al (2006) Large-scale gene expression profiling reveals major pathogenetic pathways of cartilage degeneration in osteoarthritis. Arthritis Rheum 54:3533–3544. https://doi.org/10.1002/art.22174

Ailixiding M, Aibibula Z, Iwata M et al (2015) Pivotal role of Sirt6 in the crosstalk among ageing, metabolic syndrome and osteoarthritis. Biochem Biophys Res Commun 466:319–326. https://doi.org/10.1016/j.bbrc.2015.09.019

Akasaki Y, Alvarez-Garcia O, Saito M, et al (2014a) FOXO transcription factors support oxidative stress resistance in human chondrocytes. Arthritis Rheumatol (Hoboken, NJ). https://doi.org/10.1002/art.38868

Akasaki Y, Hasegawa A, Saito M et al (2014b) Dysregulated FOXO transcription factors in articular cartilage in aging and osteoarthritis. Osteoarthr Cartil 22:162–170. https://doi.org/10.1016/j.joca.2013.11.004

Alvarez J, Balbin M, Fernandez M, Lopez JM (2001) Collagen metabolism is markedly altered in the hypertrophic cartilage of growth plates from rats with growth impairment secondary to chronic renal failure. J Bone Miner Res 16:511–524

Amin AR (1999) Regulation of tumor necrosis factor-alpha and tumor necrosis factor converting enzyme in human osteoarthritis. Osteoarthr Cartil 7:392–394

Amin AR, Abramson SB (1998) The role of nitric oxide in articular cartilage breakdown in osteoarthritis. Curr Opin Rheumatol 10:263–268. https://doi.org/10.1097/00002281-199805000-00018

Archer CW, Francis-West P (2003) The chondrocyte. Int J Biochem Cell Biol 35:401–404

Arden NK, Lane NE, Parimi N et al (2009) Defining incident radiographic hip osteoarthritis for epidemiologic studies in women. Arthritis Rheum 60:1052–1059. https://doi.org/10.1002/art.24382

Baker MS, Feigan JLD (1988) Chondrocyte antioxidant defences: the roles of catalase and glutathione peroxidase in protection against H_2O_2 dependent inhibition of proteoglycan biosynthesis. J Rheumatol 15:670–677

Bell DM, Leung KK, Wheatley SC et al (1997) SOX9 directly regulates the type-II collagen gene. Nat Genet 16:174–178. https://doi.org/10.1038/ng0697-174

Bijlsma JWJ, Berenbaum F, Lafeber FPJG (2011) Osteoarthritis: an update with relevance for clinical practice. Lancet 377:2115–2126. https://doi.org/10.1016/S0140-6736(11)60243-2

Blaney Davidson EN, Vitters EL, van Beuningen HM et al (2007) Resemblance of osteophytes in experimental osteoarthritis to transforming growth factor beta-induced osteophytes: limited role of bone morphogenetic protein in early osteoarthritic osteophyte formation. Arthritis Rheum 56:4065–4073. https://doi.org/10.1002/art.23034

Blom AB, van Lent PLEM, Holthuysen AEM et al (2004) Synovial lining macrophages mediate osteophyte formation during experimental osteoarthritis. Osteoarthr Cartil 12:627–635. https://doi.org/10.1016/j.joca.2004.03.003

Boileau C, Martel-pelletier J, Fahmi H, Boily M (2007) The peroxisome proliferator – activated receptor gamma agonist pioglitazone reduces the development of cartilage lesions in an experimental dog model of osteoarthritis: in vivo protective effects mediated through the inhibition of key signaling and catabolic pathways. Arthritis Rheum 56:2288–2298. https://doi.org/10.1002/art.22726

Bouderlique T, Vuppalapati KK, Newton PT, et al (2015) Targeted deletion of Atg5 in chondrocytes promotes age-related osteoarthritis. Ann Rheum Dis annrheumdis-2015-207742. https://doi.org/10.1136/annrheumdis-2015-207742

Boya P, Reggiori F, Codogno P (2013) Emerging regulation and functions of autophagy. Nat Cell Biol 15:1017–1017. https://doi.org/10.1038/ncb2815

Brandt KD, Dieppe P, Radin E (2009) Etiopathogenesis of osteoarthritis. Med Clin North Am 93:1–24

Brook MS, Wilkinson DJ, Phillips BE et al (2015) Skeletal muscle homeostasis and plasticity in youth and ageing: impact of nutrition and exercise. Acta Physiol n/a-n/a. https://doi.org/10.1111/apha.12532

Buckingham JC (2006) Glucocorticoids: exemplars of multi-tasking. Br J Pharmacol 147(Suppl):S258–S268. https://doi.org/10.1038/sj.bjp.0706456

Buckwalter JA, Mankin HJ (1998) Articular cartilage: tissue design and chondrocyte-matrix interactions. Instr Course Lect 47:477–486

Burleigh A, Chanalaris A, Gardiner MD et al (2012) Joint immobilization prevents murine osteoarthritis and reveals the highly mechanosensitive nature of protease expression in vivo. Arthritis Rheum 64:2278–2288. https://doi.org/10.1002/art.34420

Calich ALG, Domiciano DS, Fuller R (2010) Osteoarthritis: can anti-cytokine therapy play a role in treatment? Clin Rheumatol 29:451–455. https://doi.org/10.1007/s10067-009-1352-3

Caramés B, Taniguchi N, Otsuki S et al (2010) Autophagy is a protective mechanism in normal cartilage, and its aging-related loss is linked with cell death and osteoarthritis. Arthritis Rheum 62:791–801. https://doi.org/10.1002/art.27305

Carames B, Hasegawa A, Taniguchi N et al (2012) Autophagy activation by rapamycin reduces severity of experimental osteoarthritis. Ann Rheum Dis 71:575–581. https://doi.org/10.1136/annrheumdis-2011-200557

Caramés B, Olmer M, Kiosses WB, Lotz MK (2015) The relationship of autophagy defects to cartilage damage during joint aging in a mouse model. Arthritis Rheumatol 67:1568–1576. https://doi.org/10.1002/art.39073

Carlo MD, Loeser RF (2003) Increased oxidative stress with aging reduces chondrocyte survival: correlation with intracellular glutathione levels. Arthritis Rheum 48:3419–3430. https://doi.org/10.1002/art.11338

Carlsén S, Hansson AS, Olsson H et al (1998) Cartilage oligomeric matrix protein (COMP)-induced arthritis in rats. Clin Exp Immunol 114:477–484. https://doi.org/10.1046/j.1365-2249.1998.00739.x

Chakravarti S (2002) Functions of lumican and fibromodulin: lessons from knockout mice. Glycoconj J 19:287–293

Cheng DS, Visco CJ (2012) Pharmaceutical therapy for osteoarthritis. PM R 4:S82–S88. https://doi.org/10.1016/j.pmrj.2012.02.009

Chevalier X, Eymard F, Richette P (2013) Biologic agents in osteoarthritis: hopes and disappointments. Nat Rev Rheumatol 9:400–410. https://doi.org/10.1038/nrrheum.2013.44

Christensen R, Bartels EM, Astrup A, Bliddal H (2007) Effect of weight reduction in obese patients diagnosed with knee osteoarthritis: a systematic review and meta-analysis. Ann Rheum Dis 66:433–439. https://doi.org/10.1136/ard.2006.065904

Cicuttini FM, Wluka AE (2014) Osteoarthritis: is OA a mechanical or systemic disease? Nat Rev Rheumatol 10:1–2. https://doi.org/10.1038/nrrheum.2014.114

Cirillo DJ, Wallace RB, Wu L, Yood RA (2006) Effect of hormone therapy on risk of hip and knee joint replacement in the women's health initiative. Arthritis Rheum 54:3194–3204. https://doi.org/10.1002/art.22138

Clockaerts S, Bastiaansen-jenniskens YM, Feijt C et al (2011) Peroxisome proliferator activated receptor alpha activation decreases inflammatory and destructive responses in osteoarthritic cartilage. Osteoarthr Cartil 19:895–902. https://doi.org/10.1016/j.joca.2011.03.010

Daans M, Luyten FP, Lories RJU (2011) GDF5 deficiency in mice is associated with instability-driven joint damage, gait and subchondral bone changes. Ann Rheum Dis 70:208–213. https://doi.org/10.1136/ard.2010.134619

Dagenais S, Garbedian S, Wai EK (2009) Systematic review of the prevalence of radiographic primary hip osteoarthritis. Clin Orthop Relat Res 467(3):623–637

De Crombrugghe B, Lefebvre V, Behringer RR et al (2000) Transcriptional mechanisms of chondrocyte differentiation. Matrix Biol 19:389–394

DeLise AM, Fischer L, Tuan RS (2000) Cellular interactions and signaling in cartilage development. Osteoarthr Cartil 8:309–334. https://doi.org/10.1053/joca.1999.0306

Drissi H, Zuscik M, Rosier R, O'Keefe R (2005) Transcriptional regulation of chondrocyte maturation: potential involvement of transcription factors in OA pathogenesis. Mol Aspects Med 26:169–179

Dvir-Ginzberg M, Gagarina V, Lee E-J, Hall DJ (2008) Regulation of cartilage-specific gene expression in human chondrocytes by SirT1 and nicotinamide phosphoribosyltransferase. J Biol Chem 283:36300–36310. https://doi.org/10.1074/jbc.M803196200

Eijkelenboom A, Burgering BMT (2013) FOXOs: signalling integrators for homeostasis maintenance. Nat Rev Mol Cell Biol 14:83–97. https://doi.org/10.1038/nrm3507

Ekenstedt KJ, Sonntag WE, Loeser RF et al (2006) Effects of chronic growth hormone and insulin-like growth factor 1 deficiency on osteoarthritis severity in rat knee joints. Arthritis Rheum 54:3850–3858. https://doi.org/10.1002/art.22254

Englund M, Guermazi A, Gale D et al (2008) Incidental meniscal findings on knee MRI in middle-aged and elderly persons. N Engl J Med 359:1108–1115. https://doi.org/10.1056/NEJMoa0800777

Fang W, Wu P, Hu R, Huang Z (2003) Environmental Se-Mo-B deficiency and its possible effects on crops and Keshan-beck disease (KBD) in the Chousang area, Yao County, Shaanxi Province. China Environ Geochem Health 25:267–280. https://doi.org/10.1023/A:1023271403310

Feldmann M, Maini RN (2001) Anti-TNF alpha therapy of rheumatoid arthritis: what have we learned? Annu Rev Immunol 19:163–196. doi: 19/1/163[pii]\r10.1146/annurev.immunol.19.1.163

Felson DT, Naimark A, Anderson J et al (1987) The prevalence of knee osteoarthritis in the elderly. The Framingham Osteoarthritis Study. Arthritis Rheum 30:914–918. https://doi.org/10.1002/art.1780300811

Felson DT, Zhang Y, Anthony JM et al (1992) Weight loss reduces the risk for symptomatic knee osteoarthritis in women: the Framingham Study. Ann Intern Med 116:535–539. https://doi.org/10.1059/0003-4819-116-7-535

Ferguson MA, Kinoshita T, Hart GW (2009) Essentials of glycobiology, 2nd edn. Cold Spring Harbor Laboratory Press, New York

Fortier LA, Mohammed HO, Lust G, Nixon a J (2002) Cell-based repair of articular cartilage. J Bone Jt Surg Br 84:276–288

Francois M, Richette P, Tsagris L et al (2006) Activation of the peroxisome proliferator-activated receptor alpha pathway potentiates interleukin-1 receptor antagonist production in cytokine-treated chondrocytes. Arthritis Rheum 54:1233–1245. https://doi.org/10.1002/art.21728

Fujita N, Matsushita T, Ishida K et al (2011) Potential involvement of SIRT1 in the pathogenesis of osteoarthritis through the modulation of chondrocyte gene expressions. J Orthop Res 29:511–515. https://doi.org/10.1002/jor.21284

Fukumoto T, Sperling JW, Sanyal A et al (2003) Combined effects of insulin-like growth factor-1 and transforming growth factor-beta1 on periosteal mesenchymal cells during chondrogenesis in vitro. Osteoarthr Cartil 11:55–64. https://doi.org/10.1053/joca.2002.0869

Gabay O, Oppenhiemer H, Meir H, et al (2012) NIH Public Access. 71:613–616. https://doi.org/10.1136/ard.2011.200504.

Gabay O, Sanchez C, Dvir-Ginzberg M et al (2013) Sirtuin 1 enzymatic activity is required for cartilage homeostasis in vivo in a mouse model. Arthritis Rheum 65:159–166. https://doi.org/10.1002/art.37750

Gendron C, Kashiwagi M, Hughes C et al (2003) TIMP-3 inhibits aggrecanase-mediated glycosaminoglycan release from cartilage explants stimulated by catabolic factors. FEBS Lett 555:431–436. https://doi.org/10.1016/S0014-5793(03)01295-X

Giblin W, Skinner ME, Lombard DB (2014) Sirtuins: guardians of mammalian healthspan. Trends Genet 30:271–286. https://doi.org/10.1016/j.tig.2014.04.007

Glasson SS, Askew R, Sheppard B et al (2004) Characterization of and osteoarthritis susceptibility in ADAMTS-4-knockout mice. Arthritis Rheum 50:2547–2558. https://doi.org/10.1002/art.20558

Glyn-Jones S, Palmer AJR, Agricola R et al (2015) Osteoarthritis. Lancet 386:376–387. https://doi.org/10.1016/S0140-6736(14)60802-3

Goldring MB (2012) Chondrogenesis, chondrocyte differentiation, and articular cartilage metabolism in health and osteoarthritis. Ther Adv Musculoskelet Dis 4:269–285. https://doi.org/10.1177/1759720X12448454

Goldring MB, Berenbaum F (2015) Emerging targets in osteoarthritis therapy. Curr Opin Pharmacol 22:51–63. https://doi.org/10.1016/j.coph.2015.03.004

Goldring MB, Marcu KB (2012) Epigenomic and microRNA-mediated regulation in cartilage development, homeostasis, and osteoarthritis. Trends Mol Med 18:109–118. https://doi.org/10.1016/j.molmed.2011.11.005

Goldring MB, Otero M (2011) Inflammation in osteoarthritis. Curr Opin Rheumatol 23:471–478. https://doi.org/10.1097/BOR.0b013e328349c2b1

Goodrich LR, Hidaka C, Robbins PD et al (2007) Genetic modification of chondrocytes with insulin-like growth factor-1 enhances cartilage healing in an equine model. J Bone Jt Surg Br 89–B:672–685. https://doi.org/10.1302/0301-620X.89B5.18343

Gorrini C, Harris IS, Mak TW (2013) Modulation of oxidative stress as an anticancer strategy. Nat Rev Drug Discov 12:931–947. https://doi.org/10.1038/nrd4002

Gudbergsen H, Lohmander LS, Jones G et al (2013) Correlations between radiographic assessments and MRI features of knee osteoarthritis – a cross-sectional study. Osteoarthr Cartil 21:535–543. https://doi.org/10.1016/j.joca.2012.12.010

Hall BK, Miyake T (1992) The membranous skeleton: the role of cell condensations in vertebrate skeletogenesis. Anat Embryol (Berl) 186:107–124

Heinegård D, Saxne T (2011) The role of the cartilage matrix in osteoarthritis. Nat Rev Rheumatol 7:50–56. https://doi.org/10.1038/nrrheum.2010.198

Heinegard D, Inerot S, Olsson SE, Saxne T (1987) Cartilage proteoglycans in degenerative joint disease. J Rheumatol 14:110–112

Hetz C (2012) The unfolded protein response: controlling cell fate decisions under ER stress and beyond. Nat Publ Gr 13:89–102. https://doi.org/10.1038/nrm3270

Hiligsmann M, Cooper C, Guillemin F et al (2014) A reference case for economic evaluations in osteoarthritis: an expert consensus article from the European Society for Clinical and Economic Aspects of Osteoporosis and Osteoarthritis (ESCEO). Semin Arthritis Rheum 44:271–282

Hiramatsu K, Iwai T, Yoshikawa H, Tsumaki N (2011) Expression of dominant negative TGF-β receptors inhibits cartilage formation in conditional transgenic mice. J Bone Miner Metab 29:493–500. https://doi.org/10.1007/s00774-010-0248-2

Hochberg MC, Altman RD, April KT et al (2012) American College of Rheumatology 2012 recommendations for the use of nonpharmacologic and pharmacologic therapies in osteoarthritis of the hand, hip, and knee. Arthritis Care Res 64:465–474. https://doi.org/10.1002/acr.21596

Hofmann C, Gropp R, von der Mark K (1992) Expression of anchorin CII, a collagen-binding protein of the annexin family, in the developing chick embryo. Dev Biol 151:391–400. https://doi.org/10.1016/0012-1606(92)90179-K

Hunter DJ (2011) Pharmacologic therapy for osteoarthritis – the era of disease modification. Nat Rev Rheumatol 7:13–22. https://doi.org/10.1038/nrrheum.2010.178

Hunter DJ, Demissie S, Cupples LA et al (2004) A genome scan for joint-specific hand osteoarthritis susceptibility: the Framingham Study. Arthritis Rheum 50:2489–2496. https://doi.org/10.1002/art.20445

Husa M, Petursson F, Lotz M et al (2013) C/EBP homologous protein drives pro-catabolic responses in chondrocytes. Arthritis Res Ther 15:R218. https://doi.org/10.1186/ar4415

Jallali N, Ridha H, Thrasivoulou C et al (2005) Vulnerability to ROS-induced cell death in ageing articular cartilage: the role of antioxidant enzyme activity. Osteoarthr Cartil 13:614–622. https://doi.org/10.1016/j.joca.2005.02.011

Jallali N, Ridha H, Thrasivoulou C et al (2007) Modulation of intracellular reactive oxygen species level in chondrocytes by IGF-1, FGF, and TGF-beta1. Connect Tissue Res 48:149–158. https://doi.org/10.1080/03008200701331516

Järvinen TA, Jozsa L, Kannus P et al (1999) Mechanical loading regulates tenascin-C expression in the osteotendinous junction. J Cell Sci 112(Pt 18):3157–3166

Jensen LK, Eenberg W (1996) Occupation as a risk factor for knee disorders. Scand J Work Environ Heal 22:165–175

Jordan JM, Helmick CG, Renner JB et al (2009) Prevalence of hip symptoms and radiographic and symptomatic hip osteoarthritis in African Americans and Caucasians: the Johnston County osteoarthritis project. J Rheumatol 36:809–815. https://doi.org/10.3899/jrheum.080677

Kapoor M, Martel-Pelletier J, Lajeunesse D et al (2011) Role of proinflammatory cytokines in the pathophysiology of osteoarthritis. Nat Rev Rheumatol 7:33–42. https://doi.org/10.1038/nrrheum.2010.196

Karlson EW, Mandl LA, Aweh GN et al (2003) Total hip replacement due to osteoarthritis: the importance of age, obesity, and other modifiable risk factors. Am J Med 114:93–98. https://doi.org/10.1016/S0002-9343(02)01447-X

Kashiwagi M, Tortorella M, Nagase H, Brew K (2001) TIMP-3 is a potent inhibitor of aggrecanase 1 (ADAM-TS4) and aggrecanase 2 (ADAM-TS5). J Biol Chem 276:12501–12504. https://doi.org/10.1074/jbc.C000848200

Keene DR, Jordan CD, Reinhardt DP et al (1997) Fibrillin-1 in human cartilage: developmental expression and formation of special banded fibers. J Histochem Cytochem 45:1069–1082. https://doi.org/10.1177/002215549704500805

Kellgren J, Lawrence J (1957) Radiological assessment of osteoarthritis. Ann Rheum Dis 16:494

King LK, March L, Anandacoomarasamy A (2013) Obesity & osteoarthritis. Indian J Med Res 138:185–193

Knudson CB, Knudson W (2001) Cartilage proteoglycans. Semin Cell Dev Biol 12:69–78. https://doi.org/10.1006/scdb.2000.0243

Knudson W, Aguiar DJ, Hua Q, Knudson CB (1996) CD44-anchored hyaluronan-rich pericellular matrices: an ultrastructural and biochemical analysis. Exp Cell Res 228:216–228. https://doi.org/10.1006/excr.1996.0320

Lane NE, Gore LR, Cummings SR et al (1999) Serum vitamin D levels and incident changes of radiographic hip osteoarthritis: a longitudinal study. Study of Osteoporotic Fractures Research Group. Arthritis Rheum 42:854–860. https://doi.org/10.1002/1529-0131(199905)42:5<854::AID-ANR3>3.0.CO;2-I

Leyland KM, Hart DJ, Javaid MK et al (2012) The natural history of radiographic knee osteoarthritis: a fourteen-year population-based cohort study. Arthritis Rheum 64:2243–2251. https://doi.org/10.1002/art.34415

Little CB, Barai A, Burkhardt D et al (2009) Matrix metalloproteinase 13-deficient mice are resistant to osteoarthritic cartilage erosion but not chondrocyte hypertrophy or osteophyte development. Arthritis Rheum 60:3723–3733. https://doi.org/10.1002/art.25002

Liu Y, Zhou J, Zhao W et al (2012) XBP1S associates with RUNX2 and regulates chondrocyte hypertrophy. J Biol Chem 287:34500–34513. https://doi.org/10.1074/jbc.M112.385922

Liu-Bryan R, Terkeltaub R (2015) Emerging regulators of the inflammatory process in osteoarthritis. Nat Rev Rheumatol 11:35–44. https://doi.org/10.1038/nrrheum.2014.162

Lo YY, Cruz TF (1995) Involvement of reactive oxygen species in cytokine and growth factor induction of c-fos expression in chondrocytes. J Biol Chem 270:11727–11730

Loeser RF (2012a) NIH Public Access. 23:492–496. https://doi.org/10.1097/BOR.0b013e3283494005.Aging

Loeser RF (2012b) Aging processes and the development of osteoarthritis. Curr Opin Rheumatol 25(1). https://doi.org/10.1097/BOR.0b013e32835a9428

Loeser RF, Shanker G, Carlson CS et al (2000) Reduction in the chondrocyte response to insulin-like growth factor 1 in aging and osteoarthritis: studies in a non-human primate model of naturally occurring disease. Arthritis Rheum 43:2110–2120. https://doi.org/10.1002/1529-0131(200009)43:9<2110::AID-ANR23>3.0.CO;2-U

Loeser RF, Carlson CS, Del Carlo M, Cole A (2002) Detection of nitrotyrosine in aging and osteoarthritic cartilage: correlation of oxidative damage with the presence of interleukin-1beta and with chondrocyte resistance to insulin-like growth factor 1. Arthritis Rheum 46:2349–2357. https://doi.org/10.1002/art.10496

Lotz MK, Caramés B (2011) Autophagy and cartilage homeostasis mechanisms in joint health, aging and OA. Nat Rev Rheumatol 7:579–587. https://doi.org/10.1038/nrrheum.2011.109

Lotz M, Loeser RF (2012) Effects of aging on articular cartilage homeostasis. Bone 51:241–248. https://doi.org/10.1016/j.bone.2012.03.023

Loughlin J (2001) Genetic epidemiology of primary osteoarthritis. Curr Opin Rheumatol 13:111–116. https://doi.org/10.1097/00002281-200103000-00004

Loughlin J (2005) The genetic epidemiology of human primary osteoarthritis: current status. Expert Rev Mol Med 7:1–12. https://doi.org/10.1017/S1462399405009257

Luo G, D'Souza R, Hogue D, Karsenty G (1995) The matrix Gla protein gene is a marker of the chondrogenesis cell lineage during mouse development. J Bone Miner Res 10:325–334. https://doi.org/10.1002/jbmr.5650100221

Maroudas A, Bullough P (1968) Permeability of articular cartilage. Nature 219:1260–1261. https://doi.org/10.1038/2191260a0

Massagué J (2012) TGFβ signalling in context. Nat Rev Mol Cell Biol 13:616–630. https://doi.org/10.1038/nrm3434

Matsiko A, Levingstone TJ, O'Brien FJ (2013) Advanced strategies for articular cartilage defect repair. Materials (Basel) 6:637–668. https://doi.org/10.3390/ma6020637

Matsuzaki T, Matsushita T, Takayama K et al (2013) Disruption of Sirt1 in chondrocytes causes accelerated progression of osteoarthritis under mechanical stress and during ageing in mice. Ann Rheum Dis. https://doi.org/10.1136/annrheumdis-2012-202620

McAlindon TE, Felson DT, Zhang Y et al (1996a) Relation of dietary intake and serum levels of vitamin D to progression of osteoarthritis of the knee among participants in the Framingham Study. Ann Intern Med 125:353–359. https://doi.org/10.7326/0003-4819-125-5-199609010-00001

McAlindon TE, Jacques P, Zhang Y et al (1996b) Do antioxidant micronutrients protect against the development and progression of knee osteoarthritis? Arthritis Rheum 39:648–656. https://doi.org/10.1002/art.1780390417

McInnes IB, Schett G (2007) Cytokines in the pathogenesis of rheumatoid arthritis. Nat Rev Immunol 7:429–442. https://doi.org/10.1038/nri2094

McMillan G, Nichols L (2005) Osteoarthritis and meniscus disorders of the knee as occupational diseases of miners. Occup Environ Med 62:567–575. https://doi.org/10.1136/oem.2004.017137

Melchiorri C, Meliconi R, Frizziero L et al (1998) Enhanced and coordinated in vivo expression of inflammatory cytokines and nitric oxide synthase by chondrocytes from patients with osteoarthritis. Arthritis Rheum 41:2165–2174. https://doi.org/10.1002/1529-0131(199812)41:12<2165::AID-ART11>3.0.CO;2-O

Messier SP, Loeser RF, Miller GD et al (2004) Exercise and dietary weight loss in overweight and obese older adults with knee osteoarthritis: the arthritis, diet, and activity promotion trial. Arthritis Rheum 50:1501–1510. https://doi.org/10.1002/art.20256

Milner JM, Rowan AD, Cawston TE, D a Y (2006) Metalloproteinase and inhibitor expression profiling of resorbing cartilage reveals pro-collagenase activation as a critical step for collagenolysis. Arthritis Res Ther 8:R142. https://doi.org/10.1186/ar2034

Mintz G, Fraga A (1973) Severe osteoarthritis of the elbow in foundry workers. Arch Environ Heal An Int J 27:78–80. https://doi.org/10.1080/00039896.1973.10666322

Miyamoto Y, Mabuchi A, Shi DQ et al (2007) A functional polymorphism in the 5' UTR of GDF5 is associated with susceptibility to osteoarthritis. Nat Genet 39:529–533. https://doi.org/10.1038/Ng2005

Moos V, Fickert S, Müller B et al (1999) Immunohistological analysis of cytokine expression in human osteoarthritic and healthy cartilage. J Rheumatol 26:870–879

Morales TI (2008) The quantitative and functional relation between insulin-like growth factor-I (IGF) and IGF-binding proteins during human osteoarthritis. J Orthop Res 26:465–474. https://doi.org/10.1002/jor.20549

Moreno-Reyes R, Mathieu F, Boelaert M et al (2003) Selenium and iodine supplementation of rural Tibetan children affected by Kashin-Beck osteoarthropathy. Am J Clin Nutr 78:137–144

Morris BJ (2013) Seven sirtuins for seven deadly diseases of aging. Free Radic Biol Med 56:133–171. https://doi.org/10.1016/j.freeradbiomed.2012.10.525

Murphy K, Travers P, Walport M et al (2008) Janeway's immunobiology. New York: Garland Science Janeway's immunobiology, 7th edn. Shock 29:770

Nagase H, Kashiwagi M (2003) Aggrecanases and cartilage matrix degradation. Arthritis Res Ther 5:94–103. https://doi.org/10.1186/ar630

Nagase H, Visse R, Murphy G (2006) Structure and function of matrix metalloproteinases and TIMPs. Cardiovasc Res 69:562–573. https://doi.org/10.1016/j.cardiores.2005.12.002

Nakagawa T, Guarente L (2011) Sirtuins at a glance. J Cell Sci 124:833–838. https://doi.org/10.1242/jcs.081067

Nathan C, Cunningham-Bussel A (2013) Beyond oxidative stress: an immunologist's guide to reactive oxygen species. Nat Rev Immunol 13:349–361. https://doi.org/10.1038/nri3423

Nelson F, Billinghurst RC, Pidoux RT et al (2006) Early post-traumatic osteoarthritis-like changes in human articular cartilage following rupture of the anterior cruciate ligament. Osteoarthr Cartil 14:114–119. https://doi.org/10.1016/j.joca.2005.08.005

Neogi T, Zhang Y (2013) Epidemiology of osteoarthritis. Rheum Dis Clin North Am 39:1–19

Neogi T, Felson DT, Sarno R, Booth SL (2008) Vitamin K in hand osteoarthritis: results from a randomised clinical trial. Ann Rheum Dis 67:1570–1573. https://doi.org/10.1136/ard.2008.094771

Neuhold LA, Killar L, Zhao W et al (2001) Postnatal expression in hyaline cartilage of constitutively active human collagenase-3 (MMP-13) induces osteoarthritis in mice. J Clin Invest 107:35–44. https://doi.org/10.1172/JCI10564

Nevitt MC, Felson DT, Williams EN, Grady D (2001) The effect of estrogen plus progestin on knee symptoms and related disability in postmenopausal women: The Heart and Estrogen/

Progestin Replacement Study, a randomized, double-blind, placebo-controlled trial. Arthritis Rheum 44(4):811–818

Øiestad BE, Engebretsen L, Storheim K, Risberg MA (2009) Knee osteoarthritis after anterior cruciate ligament injury: a systematic review. Am J Sports Med 37:1434–1443. https://doi.org/10.1177/0363546509338827

Okamura Y, Watari M, Jerud ES et al (2001) The extra domain A of fibronectin activates Toll-like receptor 4. J Biol Chem 276:10229–10233. https://doi.org/10.1074/jbc.M100099200

Oldberg A, Antonsson P, Lindblom K, Heinegård D (1992) COMP (cartilage oligomeric matrix protein) is structurally related to the thrombospondins. J Biol Chem 267:22346–22350

Olin AI, Mörgelin M, Sasaki T et al (2001) The proteoglycans aggrecan and versican form networks with fibulin-2 through their lectin domain binding. J Biol Chem 276:1253–1261. https://doi.org/10.1074/jbc.M006783200

Oliveria SA, Felson DT, Reed JI et al (1995) Incidence of symptomatic hand, hip, and knee osteoarthritis among patients in a health maintenance organization. Arthritis Rheum 38:1134–1141. https://doi.org/10.1002/art.1780380817

Orlowsky EW, Kraus VB (2015) The role of innate immunity in osteoarthritis: When our first line of defense goes on the offensive. J Rheumatol 42(3):363–371. https://doi.org/10.3899/jrheum.140382

Page-McCaw A, Ewald AJ, Werb Z (2007) Matrix metalloproteinases and the regulation of tissue remodelling. Nat Rev Mol Cell Biol 8:221–233. https://doi.org/10.1038/nrm2125

Peach CA, Carr AJ, Loughlin J (2005) Recent advances in the genetic investigation of osteoarthritis. Trends Mol Med 11:186–191

Pearle AD, Warren RF, S a R (2005) Basic science of articular cartilage and osteoarthritis. Clin Sports Med 24:1–12. https://doi.org/10.1016/j.csm.2004.08.007

Pelletier JP, Caron JP, Evans C et al (1997) In vivo suppression of early experimental osteoarthritis by interleukin-1 receptor antagonist using gene therapy. Arthritis Rheum 40:1012–1019. https://doi.org/10.1002/1529-0131(199706)40:6<1012::AID-ART3>3.0.CO;2-#

Peters JM, Shah YM, Gonzalez FJ (2012) The role of peroxisome proliferator-activated receptors in carcinogenesis and chemoprevention. Nat Publ Gr 12:181–195. https://doi.org/10.1038/nrc3214

Pettipher ER, Higgs G a, Henderson B (1986) Interleukin 1 induces leukocyte infiltration and cartilage proteoglycan degradation in the synovial joint. Proc Natl Acad Sci U S A 83:8749–8753. https://doi.org/10.1073/pnas.83.22.8749

Petursson F, Husa M, June R et al (2013) Linked decreases in liver kinase B1 and AMP-activated protein kinase activity modulate matrix catabolic responses to biomechanical injury in chondrocytes. Arthritis Res Ther 15:R77. https://doi.org/10.1186/ar4254

Pfander D, Heinz N, Rothe P et al (2004) Tenascin and aggrecan expression by articular chondrocytes is influenced by interleukin 1beta: a possible explanation for the changes in matrix synthesis during osteoarthritis. Ann Rheum Dis 63:240–244. https://doi.org/10.1136/ard.2002.003749

Ratneswaran A, LeBlanc EAA, Walser E et al (2015) Peroxisome proliferator-activated receptor δ promotes the progression of posttraumatic osteoarthritis in a mouse model. Arthritis Rheumatol (Hoboken, NJ) 67:454–464. https://doi.org/10.1002/art.38915

Rees J a, Ali SY, Brown R a (1987) Ultrastructural localisation of fibronectin in human osteoarthritic articular cartilage. Ann Rheum Dis 46:816–822. https://doi.org/10.1136/ard.46.11.816

Reynard LN, Loughlin J (2012) Genetics and epigenetics of osteoarthritis. Maturitas 71:200–204. https://doi.org/10.1016/j.maturitas.2011.12.001

Reynard LN, Bui C, Canty-laird EG et al (2011) Expression of the osteoarthritis-associated gene GDF5 is modulated epigenetically by DNA methylation. Hum Mol Genet 20:3450–3460. https://doi.org/10.1093/hmg/ddr253

Rock MJ, Cain SA, Freeman LJ et al (2004) Molecular basis of elastic fiber formation: critical interactions and a tropoelastin-fibrillin-1 cross-link. J Biol Chem 279:23748–23758. https://doi.org/10.1074/jbc.M400212200

Ron D, Walter P (2007) Signal integration in the endoplasmic reticulum unfolded protein response. Nat Rev Mol Cell Biol 8:519–529. doi: nrm2199 [pii]\n10.1038/nrm2199

Rosenberg K, Olsson H, Mörgelin M, Heinegård D (1998) Cartilage oligomeric matrix protein shows high affinity zinc-dependent interaction with triple helical collagen. J Biol Chem 273:20397–20403. https://doi.org/10.1074/jbc.273.32.20397

Rubinsztein DC, Mariño G, Kroemer G (2011) Autophagy and aging. Cell 146:682–695. https://doi.org/10.1016/j.cell.2011.07.030

Ruiz-Romero C, López-Armada MJ, Blanco FJ (2006) Mitochondrial proteomic characterization of human normal articular chondrocytes. Osteoarthr Cartil 14:507–518. https://doi.org/10.1016/j.joca.2005.12.004

Ruiz-Romero C, Calamia V, Mateos J et al (2009) Mitochondrial dysregulation of osteoarthritic human articular chondrocytes analyzed by proteomics: a decrease in mitochondrial superoxide dismutase points to a redox imbalance. Mol Cell Proteomics 8:172–189. https://doi.org/10.1074/mcp.M800292-MCP200

Ryan MC, Sandell LJ (1990) Differential expression of a cysteine-rich domain in the amino-terminal propeptide of type II (cartilage) procollagen by alternative splicing of mRNA. J Biol Chem 265:10334–10339

Sacitharan PK, Snelling SJB, Edwards JR (2012) Aging mechanisms in arthritic disease. Discov Med 14:345–352

Sah R, Chen AC, Grodzinsky AJ, Trippel S (1994) Differential effect of bFGF and IGF-I on matrix metabolism in calf and adult bovine cartilage explants. Arch Biochem Biophys 308:137–147

Saklatvala J (1981) Characterization of catabolin, the major product of pig synovial tissue that induces resorption of cartilage proteoglycan in vitro. Biochem J 199:705–714

Saklatvala J (1986) Tumour necrosis factor alpha stimulates resorption and inhibits synthesis of proteoglycan in cartilage. Nature 322:547–549. https://doi.org/10.1038/322547a0

Salminen A, Kaarniranta K (2012) AMP-activated protein kinase (AMPK) controls the aging process via an integrated signaling network. Ageing Res Rev 11:230–241. https://doi.org/10.1016/j.arr.2011.12.005

Salmon WD, Daughaday WH (1957) A hormonally controlled serum factor which stimulates sulfate incorporation by cartilage in vitro. J Lab Clin Med 49:825–836. https://doi.org/10.5555/uri:pii:0022214357900914

Sandell LJ, Morris N, Robbins JR, Goldring MB (1991) Alternatively spliced type II procollagen mRNAs define distinct populations of cells during vertebral development: differential expression of the amino-propeptide. J Cell Biol 114:1307–1319. https://doi.org/10.1083/jcb.114.6.1307

Scanzello CR, Goldring SR (2012) The role of synovitis in osteoarthritis pathogenesis. Bone 51:249–257. https://doi.org/10.1016/j.bone.2012.02.012

Serra R, Johnson M, Filvaroff EH et al (1997) Expression of a truncated, kinase-defective TGF-β type II receptor in mouse skeletal tissue promotes terminal chondrocyte differentiation and osteoarthritis. J Cell Biol 139:541–552. https://doi.org/10.1083/jcb.139.2.541

Shane Anderson A, Loeser RF (2010) Why is osteoarthritis an age-related disease? Best Pract Res Clin Rheumatol 24:15–26

Shen J, Li J, Wang B et al (2013) Deletion of the transforming growth factor β receptor type II gene in articular chondrocytes leads to a progressive osteoarthritis-like phenotype in mice. Arthritis Rheum 65:3107–3119. https://doi.org/10.1002/art.38122

Shen J, Li S, Chen D (2014) TGF-β signaling and the development of osteoarthritis. Bone Res 2:14002. https://doi.org/10.1038/boneres.2014.2

Shoulders MD, Raines RT (2010) Collagen structure and stability. Annu Rev Biochem 78:929–958. https://doi.org/10.1146/annurev.biochem.77.032207.120833.COLLAGEN

Singh D, Srivastava SK, Chaudhuri TK, Upadhyay G (2015) Multifaceted role of matrix metalloproteinases (MMPs). Front Mol Biosci 2:19. https://doi.org/10.3389/fmolb.2015.00019

Smith MD, Triantafillou S, Parker a, et al (1997) Synovial membrane inflammation and cytokine production in patients with early osteoarthritis. J Rheumatol 24:365–371

Sock E, Pagon RA, Keymolen K et al (2003) Loss of DNA-dependent dimerization of the transcription factor SOX9 as a cause for campomelic dysplasia. Hum Mol Genet 12:1439–1447. https://doi.org/10.1093/hmg/ddg158

Sophia Fox AJ, Bedi A, Rodeo SA (2009) The basic science of articular cartilage: structure, composition, and function. Sports Health 1:461–468. https://doi.org/10.1177/1941738109350438

Spector TD, Cicuttini F, Baker J et al (1996) Genetic influences on osteoarthritis in women: a twin study. BMJ 312:940–943. https://doi.org/10.1136/bmj.312.7036.940

Srikanth VK, Fryer JL, Zhai G et al (2005) A meta-analysis of sex differences prevalence, incidence and severity of osteoarthritis. Osteoarthr Cartil 13:769–781. https://doi.org/10.1016/j.joca.2005.04.014

Stanton H, Melrose J, Little CB, Fosang AJ (2011) Proteoglycan degradation by the ADAMTS family of proteinases. Biochim Biophys Acta Mol Basis Dis 1812:1616–1629

Stefan L, Martin P, Stefan Lohmander OL et al (2007) The long-term consequence of ACL and meniscus injuries. Am J Sport Med 35:1756–1769. https://doi.org/10.1177/0363546507307396

Stockwell RA (1978) Chondrocytes. J Clin Pathol (Royal Coll Pathol) 12:7–13

Stockwell R a (1991) Cartilage failure in osteoarthritis: relevance of normal structure and function. A review. Clin Anat 4:161–191. https://doi.org/10.1002/ca.980040303

Styrkarsdottir U, Thorleifsson G, Helgadottir HT et al (2014) Severe osteoarthritis of the hand associates with common variants within the ALDH1A2 gene and with rare variants at 1p31. Nat Genet 46:498–502. https://doi.org/10.1038/ng.2957

Takada K, Hirose J, Senba K et al (2011) Enhanced apoptotic and reduced protective response in chondrocytes following endoplasmic reticulum stress in osteoarthritic cartilage. Int J Exp Pathol 92:232–242. https://doi.org/10.1111/j.1365-2613.2010.00758.x

Takayama K, Ishida K, Matsushita T et al (2009) SIRT1 regulation of apoptosis of human chondrocytes. Arthritis Rheum 60:2731–2740. https://doi.org/10.1002/art.24864

Tepper S, Hochberg MC (1993) Factors associated with hip osteoarthritis: data from the First National Health and Nutrition Examination Survey (NHANES-I). Am J Epidemiol 137:1081–1088

Terkeltaub R, Yang B, Lotz M, Liu-Bryan R (2011) Chondrocyte AMP-activated protein kinase activity suppresses matrix degradation responses to proinflammatory cytokines interleukin-1β and tumor necrosis factor α. Arthritis Rheum 63:1928–1937. https://doi.org/10.1002/art.30333

Thirunavukkarasu K, Pei Y, Wei T (2007) Characterization of the human ADAMTS-5 (aggrecanase-2) gene promoter. Mol Biol Rep 34:225–231. https://doi.org/10.1007/s11033-006-9037-3

Thomas JT, Ayad S, Grant ME (1994) Cartilage collagens: strategies for the study of their organisation and expression in the extracellular matrix. Ann Rheum Dis 53:488–496. https://doi.org/10.1136/ard.53.8.488

Tortorella MD, Burn TC, Pratta MA et al (1999) Purification and cloning of aggrecanase-1: a member of the ADAMTS family of proteins. Science 284:1664–1666. https://doi.org/10.1126/science.284.5420.1664

Tsukazaki T, Matsumoto T, Enomoto H, Usa T, Ohtsuru A, Namba H, Iwasaki KYS (1994) Growth hormone directly and indirectly stimulates articular chondrocyte cell growth. Osteoarthr Cartil 2:259–267

Uehara Y, Hirose J, Yamabe S et al (2014) Endoplasmic reticulum stress-induced apoptosis contributes to articular cartilage degeneration via C/EBP homologous protein. Osteoarthr Cartil 22:1007–1017. https://doi.org/10.1016/j.joca.2014.04.025

Uthman OA, van der Windt DA, Jordan JL et al (2013) Exercise for lower limb osteoarthritis: systematic review incorporating trial sequential analysis and network meta-analysis. BMJ 347:f5555. https://doi.org/10.1136/bmj.f5555

Valdes AM, Spector TD (2009) The contribution of genes to osteoarthritis. Med Clin North Am 93:45–66

Valdes AM, Spector TD (2014) Genetics of osteoarthritis. Rheumatology: sixth edition. Elsevier Academic Press Inc, Massachusetts, pp 1477–1482 van Saase JL, van Romunde LK, Cats a, et al (1989) Epidemiology of osteoarthritis: zoetermeer survey. Comparison of radiological osteoarthritis in a Dutch population with that in 10 other populations. Ann Rheum Dis 48:271–280. doi: https://doi.org/10.1136/ard.48.4.271

Vasheghani F, Monemdjou R, Fahmi H et al (2013) SHORT COMMUNICATION Adult cartilage-specific peroxisome proliferator-activated receptor gamma knockout mice exhibit the spontaneous osteoarthritis phenotype. Am J Pathol 182:1099–1106. https://doi.org/10.1016/j.ajpath.2012.12.012

Vasheghani F, Zhang Y, Li Y et al (2015) PPAR γ deficiency results in severe, accelerated osteoarthritis associated with aberrant mTOR signalling in the articular cartilage. Ann Rheum Dis:569–578. https://doi.org/10.1136/annrheumdis-2014-205743

Verdin E (2014) The many faces of sirtuins: coupling of NAD metabolism, sirtuins and lifespan. Nat Med 20:25–27. https://doi.org/10.1038/nm.3447

Verma RP, Hansch C (2007) Matrix metalloproteinases (MMPs): chemical-biological functions and (Q)SARs. Bioorganic Med Chem 15:2223–2268

Vincent TL, Watt FE (2014) Osteoarthritis. Medicine (Baltimore) 42:213–219. https://doi.org/10.1016/j.mpmed.2014.01.010

Westling J, Fosang AJ, Last K et al (2002) ADAMTS4 cleaves at the aggrecanase site (Glu373-Ala374) and secondarily at the matrix metalloproteinase site (Asn341-Phe342) in the aggrecan interglobular domain. J Biol Chem 277:16059–16066. https://doi.org/10.1074/jbc.M108607200

Wieland HA, Michaelis M, Kirschbaum BJ, Rudolphi KA (2005) Osteoarthritis – an untreatable disease? Nat Rev Drug Discov 4:331–344. https://doi.org/10.1038/nrd1693

Wluka AE, Cicuttini FM, Spector TD (2000) Menopause, oestrogens and arthritis. Maturitas 35:183–199

Woolf AD, Pfleger B (2003) Burden of major musculoskeletal conditions. Bull World Health Organ 81:646–656. doi: S0042-96862003000900007 [pii]

Wu Q, Huang JH, Sampson ER et al (2009) Smurf2 induces degradation of GSK-3beta and upregulates beta-catenin in chondrocytes: a potential mechanism for Smurf2-induced degeneration of articular cartilage. Exp Cell Res 315:2386–2398. https://doi.org/10.1016/j.yexcr.2009.05.019

Wu Y, Chen L, Wang Y et al (2015) Overexpression of Sirtuin 6 suppresses cellular senescence and NF-κB mediated inflammatory responses in osteoarthritis development. Sci Rep 5:17602. https://doi.org/10.1038/srep17602

Yang X, Chen L, Xu X et al (2001) TGF-beta/Smad3 signals repress chondrocyte hypertrophic differentiation and are required for maintaining articular cartilage. J Cell Biol 153:35–46. https://doi.org/10.1083/jcb.153.1.35

Zhang Y, Niu J, Kelly-Hayes M et al (2002) Prevalence of symptomatic hand osteoarthritis and its impact on functional status among the elderly: the Framingham Study. Am J Epidemiol 156:1021–1027. https://doi.org/10.1093/aje/kwf141

Zhang Y, Vasheghani F, Li Y-H et al (2014) Cartilage-specific deletion of mTOR upregulates autophagy and protects mice from osteoarthritis. Ann Rheum Dis 1(9). https://doi.org/10.1136/annrheumdis-2013-204599

Zhen G, Wen C, Jia X et al (2013) Inhibition of TGF-β signaling in mesenchymal stem cells of subchondral bone attenuates osteoarthritis. Nat Med 19:704–712. https://doi.org/10.1038/nm.3143

Chapter 7
Down Syndrome, Ageing and Epigenetics

Noémie Gensous, Claudio Franceschi, Stefano Salvioli, Paolo Garagnani, and Maria Giulia Bacalini

Abstract During the past decades, life expectancy of subjects with Down syndrome (DS) has greatly improved, but age-specific mortality rates are still important and DS subjects are characterized by an acceleration of the ageing process, which

N. Gensous
DIMES- Department of Experimental, Diagnostic and Specialty Medicine, Alma Mater Studiorum, Bologna, Italy

C. Franceschi
DIMES- Department of Experimental, Diagnostic and Specialty Medicine, Alma Mater Studiorum, Bologna, Italy

CIG, Interdepartmental Center 'L. Galvani', Alma Mater Studiorum, Bologna, Italy

IRCCS Istituto delle Scienze Neurologiche di Bologna, Bologna, Italy
e-mail: claudio.franceschi@unibo.it

S. Salvioli
DIMES- Department of Experimental, Diagnostic and Specialty Medicine, Alma Mater Studiorum, Bologna, Italy

CIG, Interdepartmental Center 'L. Galvani', Alma Mater Studiorum, Bologna, Italy
e-mail: stefano.salvioli@unibo.it

P. Garagnani (✉)
DIMES- Department of Experimental, Diagnostic and Specialty Medicine, Alma Mater Studiorum, Bologna, Italy

CIG, Interdepartmental Center 'L. Galvani', Alma Mater Studiorum, Bologna, Italy

Department of Laboratory Medicine, Karolinska Institutet, Stockholm, Sweden

CNR IAC "Mauro Picone", Roma, Italy

Applied Biomedical Research Center, S. Orsola-Malpighi Polyclinic, Bologna, Italy

Institute of Molecular Genetics (IGM)-CNR, Unit of Bologna, Bologna, Italy

Laboratory of Musculoskeletal Cell Biology, Rizzoli Orthopaedic Institute, Bologna, Italy
e-mail: paolo.garagnani2@unibo.it

M. G. Bacalini
IRCCS Istituto delle Scienze Neurologiche di Bologna, Bologna, Italy
e-mail: mariagiuli.bacalini2@unibo.it

© Springer Nature Singapore Pte Ltd. 2019
J. R. Harris, V. I. Korolchuk (eds.), *Biochemistry and Cell Biology of Ageing: Part II Clinical Science*, Subcellular Biochemistry 91,
https://doi.org/10.1007/978-981-13-3681-2_7

affects particularly the immune and central nervous systems. In this chapter, we will first review the characteristics of the ageing phenomenon in brain and in immune system in DS and we will then discuss the biological hallmarks of ageing in this specific population. Finally, we will also consider in detail the knowledge on epigenetics in DS, particularly DNA methylation.

Keywords Down syndrome · Ageing · Epigenetics · Epigenetic clock

Introduction

Down syndrome (DS) or trisomy 21 (OMIM #190685) is a complex genetic condition, caused by a chromosomal disorder, corresponding to a total or partial trisomy of the chromosome 21 (HSA21). It occurs in approximately 1 out of every 600–700 live births and it is the most common known genetic cause associated with moderate to severe intellectual disability. In the past decades, thanks to improvements in medical care for children and adults, life expectancy of DS people has rapidly increased (Bittles and Glasson 2004; Glasson et al. 2016; Leonard et al. 2000; Yang et al. 2002), but age-specific mortality rates are still important compared to other populations (Strauss and Eyman 1996). Subjects with DS appear to age differently to individuals without DS (Zigman 2013), and the increase in life expectancy does not follow the one observed in the general population or other groups with intellectual disability (Coppus 2013). DS is characterized by premature ageing, frequently described as a segmental progeroid syndrome (Martin 1982; Patterson and Cabelof 2012). DS subjects commonly display in middle adulthood health-related problems that do not normally arise in the general population before 70 years of age and that resemble those of the geriatric population: dermatological changes (skin wrinkling, hair loss and/or greying), presbycusis with prevalence of hearing loss of more than 90% after 50 years old (Picciotti et al. 2017), osteoporosis (Baptista et al. 2005; Carfì et al. 2014; McKelvey et al. 2013), early menopause onset, but most prominently, immune impairment and cognitive decline (Zigman 2013). Accelerated ageing in DS is atypical and segmental: it is associated with many, but not all, of the classical ageing signs (Esbensen 2010) and affects particularly the immune and central nervous systems. DS can represent a model to study events that occur with age, to identify molecular markers and potential therapeutic targets. As life expectancy in DS individuals increase, new health issues emerge with a great need of specialized therapeutic tools. In this chapter, we will review evidence on the premature ageing of the immune and central nervous systems in DS, as well as the presence of biological hallmarks of ageing, and finally, the epigenetics processes associated with this disease.

Ageing of the Central Nervous and Immune Systems in Down Syndrome

Brain Ageing

In addition to their congenital cognitive impairment, individuals with DS experience age-related cognitive decline and subsequent dementia (essentially Alzheimer's disease, AD) more frequently and at an earlier age than people without DS. This represents a great health problem (Glasson et al. 2002) and has major implications from a care perspective. In the general population aged 60 and over, in Western Europe, prevalence rates of dementia are around 7% ("WHO|Dementia" 2012). These rates are higher in subjects with intellectual disability and specially in the population with DS (Coppus et al. 2006; Franceschi et al. 1990; Holland et al. 1998, 2000; Lai and Williams 1989; Margallo-Lana et al. 2007; McCarron et al. 2014, 2017; Oliver et al. 1998; Sekijima et al. 1998; Stancliffe et al. 2012; Tyrrell et al. 2001; Visser et al. 1997). In DS people aged 45 years and older, prevalence rates of dementia are reported between 15 and 45%, and approximately 50–80% of individuals with DS will develop dementia before they reach the age of 60–70 years old. The median age of dementia onset is below 60 years old, but with an important inter-individual variability in the age of the first clinical symptoms (Ballard et al. 2016; Lott and Dierssen 2010; Zigman and Lott 2007). Important variations are present in reported prevalence rates: discrepancies can be due to studies designs (longitudinal or cross-sectional studies), patient heterogeneity or choices in the instruments used to measure cognitive decline (Strydom et al. 2007).

Individuals with DS experience important cognitive deterioration, which follows frequently a course similar to the one seen in AD. Dementia is frequently preceded by changes in language skills and in executive functions (Ghezzo et al. 2014; Holland et al. 2000; Iacono et al. 2010; Kittler et al. 2006). Non-demented adult DS subjects over 40 years old have lower neuropsychological functions and adaptive skills as compared to younger ones, with a particular impact on language and short memory skills, frontal lobe functions, visuo-spatial abilities and adaptive behaviour (Ghezzo et al. 2014). Changes in cognitive abilities were age-associated, as performance at tests (semantic fluency, token test, phonemic fluency, Tower of London, Frontal Assessment Battery, etc.) were found to be inversely correlated with age (Ghezzo et al. 2014). Diagnosis of dementia in DS can be a major challenge, related to the presence of pre-existing congenital cognitive impairment, communication issues or difficulties in the choice of standardized tests in line with a limited capacity of individuals with DS to be assessed with traditional cognitive measures. In the youngest individuals, diagnosis can be delayed due to the initial presentation with atypical symptoms, like changes in behavior, in personality or psychological symptoms (Dekker et al. 2015). Finally, diagnosis is complicated by the presence of comorbidities affecting the neurological phenotype, such as epilepsy and depression, whose prevalence increase with age. Epilepsy has a particularly important prevalence in DS subjects (McCarron et al. 2014, 2017), recently estimated at 77.9% in a prospective longitudinal study (McCarron et al. 2017) and among subjects with dementia, a substantial proportion are also diagnosed with depression (McCarron et al. 2017; Shooshtari et al. 2011).

Two neuropathological lesions are considered as hallmarks of AD: senile plaques and neurofibrillary tangles. While neurofibrillary tangles are aggregates of the abnormally hyper-phosphorylated protein tau (τ) within the cytoplasm of neurons, plaques are extracellular deposits in the cerebral cortex, containing the beta-amyloid (Aβ) peptide. Aβ is produced by sequential cleavage of the amyloid precursor protein (APP). Several peptides of different lengths are generated from APP by beta- and gamma-secretases and Aβ40 and Aβ42 are believed to be the most pathogenic. These neuropathological changes typical of AD start to develop in the childhood of DS subjects, decades earlier compared to aged control brains (Lemere et al. 1996; Leverenz and Raskind 1998; Lott and Dierssen 2010; Teller et al. 1996; Zigman and Lott 2007). Virtually all adults with DS over 40 years of age have sufficient senile plaques and neurofibrillary tangles for a neuropathologically based diagnosis of AD. The increase in Aβ charge with age, assessed in a longitudinal study, has been found to be related to the cognitive decline in the DS population (Hartley et al. 2017).

From a neuroimaging perspective, as compared to AD patients in the general population, people with DS have reduced brain volumes, specially in the frontal and temporal lobes, and important cortical thickness (Lott 2012; Mullins et al. 2013; Pinter et al. 2001). Neuroimaging changes are observed before the onset of dementia (Krasuski et al. 2002; Sabbagh et al. 2015; Teipel et al. 2004) and are mostly associated with age (Beacher et al. 2010; Koran et al. 2014; Romano et al. 2016). Non-demented individuals with DS show an 'accelerated ageing' of some brain regions from a morphological point of view: they have significantly greater age-related reductions in volume of frontal, temporal and parietal lobes, and significantly greater age-related changes in ventricle volumes (Beacher et al. 2010; Koran et al. 2014). These neuroimaging changes are associated with the development of fibrillary β-amyloidosis (Annus et al. 2017) and correlated to cognitive and memory impairments (Koran et al. 2014; Krasuski et al. 2002). To directly measure by magnetic resonance imaging how brain structure changes with ageing, structural neuroimaging data have been used in a general population to generate indexes that predict ageing of brain. This model of 'brain-predicted age' was applied recently to DS subjects (Cole et al. 2017): brain-predicted age difference (equal to 'brain-predicted age' – 'chronological age') in DS participants was significantly greater than in controls, suggesting a premature structural brain ageing in DS subjects. The variability of the score was associated with the presence and the magnitude of Aβ deposition measured by positron emission tomography scans, and also with levels of cognitive performance (Cole et al. 2017).

Search for genetic risk factors for dementia in people with DS has been important during the past decades. As mentioned previously, Aβ, the major contributor to AD pathology, is produced from the precursor protein APP, which gene is located on the proximal part of the long arm of HSA21 and is thus present in triplicate in DS. This triplication yields to higher levels of APP and its splicing products, and was considered as the key contributor to strong increase in risk for AD in DS (Mao et al. 2003). However, APP is not the only gene present in triplicate in DS subjects and others appear to be also important determinants in the development of dementia by their impact either on Aβ pathology or on neuroinflammation, such as beta amyloid converting enzyme 2 (*BACE2*) or astrocyte-derived neurotrophic factor S100 beta (*S100β*) (Leclerc et al. 2010). Of particular interest is the gene for dual-specificity tyrosine phosphorylated and regulated kinase 1a (*DYRK1A*), which is involved in the appearance of AD-like pathology. It has emerged as a key determinant in neuronal loss, neurofibrillary degeneration and cognitive impairment, and is considered as a potential therapeutic target (De la Torre et al. 2014; Duchon and Herault 2016). It is worthy of note that in neurons derived from Induced Pluripotent Stem Cells (iPSC) of patients with DS, increased expression levels of genes such as *APP* and *BACE2* have been reported, as compared to healthy controls, and have been associated with abnormal metabolism of Aβ *in vitro* (Dashinimaev et al. 2017).

Additional genetic risk factors, not located on HSA21, have been investigated in dementia in DS subjects. One of the most consistent genetic risk factor found for AD dementia in the general population is the gene for Apolipoprotein E (*APOE*). ApoE is polymorphic with three different isoforms (ε2, ε3 and ε4) and is considered as a possible chaperone for Aβ deposition. *APOEε4* allele has been associated with dementia in the general population (Liu et al. 2013). Also in DS subjects, the presence of one or two *APOEε4* allele predisposes to a greater risk of AD, to an early-onset of symptoms (before 45 years old) and to a more rapid progression to death, as compared to individuals without DS or DS adults without any *APOEε4* allele (Deb et al. 2000; Lai et al. 1999; Prasher et al. 2008). The *APOEε4* allele is also considered as associated with earlier mortality in the DS population, independently of the risk of dementia (Rohn et al. 2014). Some authors (Prasher et al. 2008) recommend early screening for APOE genotype in individuals with DS, for early identification of individuals at increased risk of cognitive decline.

In the general population, multiple genome-wide association studies (GWAS) have identified at least 20 genes that are significantly associated with AD, but these genomic variants were not necessarily found in the specific DS population (Patel et al. 2014). In adults with DS, several studies have examined relations between single-nucleotide polymorphisms (SNPs) and dementia, using a candidate gene approach, identifying associations between AD in adults with DS and genetic variants in sortilin-related receptor 1 (*SORL1*) (Lee et al. 2007), estrogen-receptor 2 (*ESR2*) (Zhao et al. 2011) or *APP* and *CST3* (Lee et al. 2017). SNPs in genes *CAHLM1*, *IDE*, *SOD1* or *SORCS1* were found to be associated with plasma levels of Aβ peptides (Aβ40, Aβ42, Aβ42/Aβ40 ratio) (Schupf et al. 2015). Finally, some genetic polymorphisms were found to be associated with the age at the onset of dementia in DS: *APP* locus (Margallo-Lana et al. 2004), *PICALM* and *APOE* loci (Jones et al. 2013a), *HSD17B1* locus (Lee et al. 2012) or *BACE2* (Mok et al. 2014).

In addition to neuropathology related to Aβ, several other types of pathologic processes, which could contribute to neurodegeneration and to the development of dementia, have been reported in DS subjects. Individuals with DS show higher levels of oxidative stress at all ages as compared to controls, notably in brains (Aivazidis et al. 2017; Cenini et al. 2012; Garlet et al. 2013; Jovanovic et al. 1998; Odetti et al. 1998; Reynolds and Cutts 1993). Cortical neurons from fetal DS exhibit an *in vitro* increase in intracellular reactive oxygen species (ROS) and elevated levels of lipid peroxidation, as compared to age-matched normal brains, leading to neuronal apoptosis (Busciglio and Yankner 1995). A critical process associated with oxidative damage in DS individuals with age is mitochondrial dysfunction (Arbuzova et al. 2002; Busciglio et al. 2002; Helguera et al. 2013; Valenti et al. 2011). In DS neurons, altered mitochondrial morphology and function have been reported, regarding in particular impairment of mitochondrial transport (Helguera et al. 2013). Mitochondrial dysfunction leads to an alteration in the metabolism of APP, resulting into the intracellular deposition of the insoluble form of Aβ (Busciglio et al. 2002). Finally, neuroinflammation, considered as a key contributor to neurodegenerative disorders, may lead to an increase in the vulnerability of DS neurons in the presence of senile plaques, neurofibrillary tangles and oxidative damage (Wilcock 2012; Wilcock et al. 2015).

Ageing of the Immune System

In the spectrum of the symptoms presented by DS subjects, several have led to the hypothesis that the syndrome is associated with impairment in the immune system: DS subjects have a higher susceptibility to bacterial infections and to develop hematological malignancies (Bruwier and Chantrain 2012; Goldacre et al. 2004) or organ-specific autoimmune disorders (hypothyroidism, celiac disease or insulin-dependent diabetes mellitus) (Guaraldi et al. 2017; Karlsson et al. 1998; Storm 1990). Immune abnormalities in DS have been described for more than 30 years and attention has been mainly focused on the immunologic impairment of the T cell compartment. DS subjects have reductions in the number of circulating CD4+ T cells as compared to age-matched controls (Barrena et al. 1993; Cocchi et al. 2007; Cossarizza et al. 1990; Joshi et al. 2011; Schoch et al. 2017; Trotta et al. 2011), with alterations in maturation and differentiation. Regarding CD4+ T lymphocytes subsets, an imbalance of subpopulations in the peripheral blood is noted (Barrena et al. 1993; Burgio et al. 1978; Guazzarotti et al. 2009), with reduction in naïve lymphocytes (Guazzarotti et al. 2009) or increased proportions of peripheral regulatory T cells with a defective inhibitory activity compared to age-matched controls (Pellegrini et al. 2012; Roat et al. 2008; Schoch et al. 2017). Recently, Schoch and colleagues reported higher percentages of Th1 and Th17 lymphocytes in children and adolescents with DS, in comparison to healthy controls, with similar percentages of Th2 cells (Schoch et al. 2017). The vast expansion of naïve helper (CD4+ CD45RA+) and cytotoxic (CD8+ CD45RA+

CD27+) T lymphocytes, normally observed in healthy children, is lacking in DS subjects in the first years of life (Kusters et al. 2010). From a functional point of view, an impairment is also supported, with weak proliferative responses against specific mitogens (Franceschi et al. 1981; Karttunen et al. 1984; Lockitch et al. 1987; Park et al. 2000), as well as changes in cytokine production (Cetiner et al. 2010; Park et al. 2000; Roat et al. 2008). However, in their report recently published, Schoch et al. (2017) observed a normal functionality of lymphocytes in DS subjects and considered that they are able to mount effector T-cell responses with normal functional characteristics. Observed alterations in the peripheral T-cell compartment have led to numerous joint reports on thymic abnormalities in DS subjects (Fabris et al. 1984; Larocca et al. 1988; Murphy et al. 1990; Murphy and Epstein 1990, 1992; Musiani et al. 1990; Papadopoulos et al. 2003), with reduced thymic size, impaired intrathymic expansion of immature T cells and inefficient intrathymic maturation. Reduced thymic expression of large set of genes may be associated with the observed abnormalities (Lima et al. 2011). For example, *AIRE* (autoimmune regulator, located on HSA21) expression was found significantly reduced in DS thymic medullary epithelial cells, as compared to age-matched individuals.

Besides the T-cell compartment, the humoral immune system seems also disturbed. DS subjects have a lower number of total circulating B cells compared to healthy controls (Cossarizza et al. 1991; Schoch et al. 2017; Verstegen et al. 2010), with decreased transitional and naïve B lymphocytes (Verstegen et al. 2010) and also severe defects in memory B cells (Carsetti et al. 2015; Joshi et al. 2011; Valentini et al. 2015; Verstegen et al. 2014), whereas germinal centers and plasma cells in tonsils appeared normal (Verstegen et al. 2014). Humoral immune responses after vaccination are altered in DS subjects (Kusters et al. 2011; Valentini et al. 2015). These impairments could be due to a true B cell defect, but also to a disturbance in T-lymphocyte help or a combination of both.

Some reports also pointed out modifications in the innate immune system. Bloemers et al. observed lower absolute total granulocytes counts in peripheral blood, lower absolute numbers of myeloid dendritic cells, but higher absolute numbers of pro-inflammatory CD14^{dim}CD16$^+$monocytes (Bloemers et al. 2010). Regarding natural killer (NK) cells, literature results are contradictory. First reports identified an age-related expansion of CD57+ cells in comparison with age- and sex-matched healthy controls, associated with a low NK activity (Cossarizza et al. 1990, 1991), but more recent reports demonstrated that absolute numbers of NK cells were low (de Hingh et al. 2005) or normal (Bloemers et al. 2010) and discrepancies were attributed to differences of surface markers used. However, in the ultimate study published by Schoch and colleagues, they reported a significantly higher percentage of NK cells among lymphocytes of individuals with DS compared to controls (Schoch et al. 2017).

The picture is extremely complex, and it is still debated whether the immune system in DS is intrinsically deficient from the very beginning or another victim of a generalized process of precocious ageing. For some authors, observed alterations in immune system are related to a primary immunodeficiency, intrinsically present in DS subjects (de Hingh et al. 2005; Kusters et al. 2009, 2010). The findings of decreased naïve T and B cells are related to a deficient production from birth onwards, with absence of any vast expansion of lymphocytes during first years of life. Other authors favor the opinion of early senescence of the immune system (Barrena et al. 1993; Cossarizza et al. 1990; Cuadrado and Barrena 1996; Roat et al. 2008). Considering that important incidences of the conditions previously mentioned (higher susceptibility to bacterial infections, hematological malignancies, autoimmune disorders) are normally seen in elderly individuals, they supposed that DS is associated with a premature ageing in the immune system and that most of the phenotypic abnormalities observed are associated with an abnormal ageing of the thymus-dependent system and are reminiscent of immunosenescence seen in the elderly. Age-related alterations were thus reported: for example, progressive decrease in the values of CD4+ lymphocytes was observed in the first years of life in DS subjects (Cocchi et al. 2007) or levels of lymphocytes expressing the signal-joint T cell receptor rearrangement excision circles (sj-TREC+), that were lower in DS children as compared to controls, were strongly correlated with age (Roat et al. 2008).

Biological Hallmarks of Ageing in Down Syndrome

Ageing is a really complex and multifactorial process. Different conceptualizations have proposed sets of biological pillars of ageing, *i.e.* cellular and molecular processes that promote ageing in a highly interconnected way (Kennedy et al. 2014; López-Otín et al. 2013). DS individuals exhibit not only clinical accelerated ageing, but also an earlier appearance of these age-associated markers.

DNA Damage Accumulation

Genetic damages, which can be of endogenous or exogenous origin, accumulate throughout life and participate to the ageing process. Several lines of evidence suggest that DS is associated with a DNA repair defect, like many other progeroid syndromes (Patterson and Cabelof 2012). It has been observed that cells from DS donors (skin fibroblasts or peripheral blood lymphocytes) have higher basal endogenous levels of DNA damage, as compared to control cells (Franceschi et al. 1992; Maluf and Erdtmann 2001; Morawiec et al. 2008; Tiano et al. 2005; Zana et al. 2006). Moreover, DS cells were found to be more sensitive to DNA damaging agents as compared with the controls, with defective DNA repair processes (Morawiec et al. 2008; Necchi et al. 2015; Raji and Rao 1998; Zana et al. 2006). Loss of base excision repair phenomena, associated with a reduction in β-polymerase activity, were found in DS subjects (Cabelof et al. 2009; Raji and Rao 1998).

Telomere Shortening

Telomere attrition is considered as a contributor to cell senescence and ageing (Blackburn et al. 2015) and has been investigated for many years as a potential biomarker of ageing, with telomere length diminishing progressively with age (Jylhävä et al. 2017; Sanders and Newman 2013). Regarding DS, several teams have investigated telomere length in this syndrome. In the first report on telomere length quantification, a significantly higher rate of telomere loss was observed in blood lymphocytes from DS subjects (age 0–45 years) in comparison to controls (Vaziri et al. 1993). The same profile of acceleration of telomere loss, assessed by mean telomere restriction fragment length, was observed in another study published later where accelerated telomere shortening was associated with stem cell deficiency and was described as already present in fetal life (Holmes et al. 2006). Sukenik-Halevy et al. also investigated telomere length in pre-natal samples (Sukenik-Halevy et al. 2011). They studied telomere length and the *hTERC* gene copy number, which encodes the telomerase RNA subunit, in amniocytes of trisomy 21 conceptions and normal pregnancies. They observed a telomere shortening, associated with an increase in the copy number of *hTERC* in amniocytes of trisomy 21 fetuses compared to the control group with normal karyotype. Fetal DS fibroblast primary cell lines were also found to have more pronounced telomere attrition than their control counterparts (Gimeno et al. 2014).

Regarding DS newborns, Wenger et al. reported results of a small study on ten subjects: telomere length in newborns with DS was significantly shorter than in newborns with normal karyotype (Wenger et al. 2014). These results were in conflict with those published the same year by Nakamura et al.: telomere lengths were assessed by Q-FISH in a larger number of newborns (n = 31), with chromosomal abnormalities (trisomy 21 or trisomy 18) and controls with diploid karyotypes. These authors did not find marked differences of whole telomere length between the groups (Nakamura et al. 2014). Finally, it was recently observed that DS babies have longer telomeres than controls, with impact of maternal age on the length of telomeres (Bhaumik et al. 2017). These results were in accordance with those of Gruszecka et al. who showed that blood leukocytes from juvenile DS patients (mean age = 4.5; range: 2–21 years) have longer telomeres than age matched controls (Gruszecka et al. 2015).

Discrepancies in some published results could be attributed to different causes: differences in the age of the samples studied, differences in the cell type analyzed or differences in the technique for the measurement of telomere length. If the question of the presence of shorter telomeres at birth as an inherent trait is still debated, an accelerated rate loss during lifespan seems to be present in DS subjects. This acceleration of telomere erosion could serve as a clinical biomarker, informative on cognitive status. Jenkins et al. evaluated in several publications the relationship between telomere length in older DS subjects (>40 years old) and cognitive decline (Jenkins et al. 2006, 2008, 2010, 2012, 2016, 2017). They reported, in cross-sectional studies, that telomere length was associated with the presence of mild cognitive impairment and dementia (Jenkins et al. 2006, 2008, 2010, 2012). T lymphocytes in DS subjects with dementia (age 45–60 years old) have shorter telomeres than those of DS subjects without dementia (Jenkins et al. 2006). In their latest publications in 2016 and 2017, these authors observed that longitudinal changes in telomere length were associated to transition to dementia (Jenkins et al. 2016, 2017). Telomeres became shorter with the declining of the clinical status (transition from 'clinically normal ageing' to mild-cognitive impairment or dementia) (Jenkins et al. 2016).

Loss of Proteostasis

Proteostasis, or protein homeostasis, referring to the processes associated with biogenesis, folding, trafficking and degradation of proteins in cells, is disturbed during ageing (Morimoto and Cuervo 2014). Cells (skin fibroblasts or frontal cortex) derived from DS individuals exhibit extensively remodeled proteostasis networks, as compared to controls (Di Domenico et al. 2013; Liu et al. 2017). Alterations in different key elements of these networks compared to euploid controls, such as chaperone systems, unfolded protein responses or proteasomal degradation have been observed (Aivazidis et al. 2017). When DS fibroblasts were exposed to a moderate heat stress, they were unable to cope with this increased proteomic stress, leading to diminished cell viability compared to controls (Aivazidis et al. 2017). This disturbance of the proteostasis network can participate to the accumulation of misfolded proteins, notably in the brain (Di Domenico et al. 2013).

Oxidative Stress and Mitochondrial Dysfunction

As previously mentioned, individuals with DS are characterized by higher levels of oxidative stress as compared to the general population, from a local and a systemic perspective (Aivazidis et al. 2017; Cenini et al. 2012; Garlet et al. 2013; Jovanovic et al. 1998; Odetti et al. 1998; Reynolds and Cutts 1993). They have increased basal levels of endogenous oxidative stress, which seems to occur early in life. An important number of genes coding for proteins relevant to oxidative damage are located on HSA21 and over-expressed in DS, among which superoxide dismutase 1 (*SOD1*) located in the Down Syndrome Critical Region 1 (DSCR1) seems to be the most relevant (Murakami et al. 2011). Increased SOD1 activity in DS subjects leads to a higher production of hydrogen peroxide (H_2O_2), not adequately compensated (Garlet et al. 2013). Iron metabolism dysregulation (Barone et al. 2017) and aberrant mammalian target of rapamycin (mTOR) pathway signaling (Di Domenico et al. 2017) are considered as potential contributors to the increase in oxidative stress observed in DS subjects.

The increased production of reactive oxygen species (ROS) in DS individuals is accompanied by mitochondrial dysfunction (Arbuzova et al. 2002; Busciglio et al.

2002; Helguera et al. 2013; Valenti et al. 2011). Mitochondrial DNA (mtDNA) mutations, along with defective repair of mtDNA damage, are found at a high rate in DS subjects (Arbuzova 1998; Coskun et al. 2010; Druzhyna et al. 1998).

Cellular Senescence

Cellular senescence is a complex phenomenon, that is considered to be a contributor to the ageing process (López-Otín et al. 2013). Regarding DS individuals, literature data are not concordant. It was previously found that fibroblasts from DS subjects have a diminished rate of cellular proliferation than those from normal donors, but that there was no difference in terms of cumulative number of population doublings until replicative senescence or for the beta-galactosidase staining between the two populations (Kimura et al. 2005). More recently, Adorno et al. reported a strong proliferation defect, associated with a premature senescent phenotype according to beta-galactosidase staining in fibroblasts derived from DS subjects (Adorno et al. 2013).

Stem Cells Ageing

Evidence is now compelling on the presence of a decline in stem cell function during ageing, specially for hematopoietic stem cells (HSC) or neural stem cells (NSC) (Liu and Rando 2011). In DS individuals, some of the anomalies seem to be recapitulated earlier in life. A first report was published in 2002, regarding neuronal progenitors (Bahn et al. 2002). NSC derived from the central nervous system of DS or control post-mortem fetal tissues were investigated and authors reported abnormalities in neuronal proliferation and/or survival capacity in DS stem cells. Assessing gene expression, they identified a dysregulation of the network of genes regulated by the transcription factor REST (neuron-restrictive silencer factor). Regarding HSC, peripheral blood from fetuses and bone marrow samples from children with DS exempt of malignant hematological diseases were analyzed for their content in stem cells (assessed by the percentage of CD34+ cells) and compared to samples from non-trisomic controls (Holmes et al. 2006) and a marked stem cell deficiency was seen in both trisomic fetuses and children. To gain insights into the molecular mechanisms associated with the stem cell exhaustion, gene expression has been investigated and compared to those of non-trisomic controls (Cairney et al. 2009). HSC obtained from the iliac crest of DS children of 1–5 years old and NSC obtained from fetal DS cortex were studied and 430 genes were found differentially expressed. Analysis of these genes revealed an enrichment in pathways previously associated with cellular ageing. In particular, a down-regulation of DNA repair genes and an up-regulation of genes involved in apoptosis, inflammation and angiogenesis were observed (Cairney et al. 2009). Authors identified a dyregulation of the Notch/Wnt pathway as a potential hub that may drive stem cell ageing. Finally, changes in DS stem cells were similar to those observed in stem cells of older people (individuals of 60–80 years old). The same trends were observed in HSC, suggesting the existence of shared molecular events between the two types of stem cells (Cairney et al. 2009). Adorno et al. took the advantage of the study of two murine models of DS and identified the *USP16* gene, which is triplicated in DS, as an important contributor to the lack of self-renewal in DS stem cells (Adorno et al. 2013). An over-expression of USP16 in normal human fibroblasts decreased their proliferation, whereas its down-regulation in DS fibroblasts, which show proliferation defects and premature senescent phenotype in culture, promoted their proliferation. Finally, over-expression of the enzyme in human neural progenitor cells reduced their *in vitro* expansion potential and the formation of neurospheres (Adorno et al. 2013).

Inflammation

Ageing is characterized by a peculiar chronic inflammatory status, called 'inflammageing' (Franceschi et al. 2000). As mentioned earlier, DS subjects exhibit markers of early immunosenescence and marks of chronic inflammation can be observed: PBMC of DS subjects are characterized by spontaneous higher production of pro-inflammatory cytokines, such as tumor necrosis factor (TNFα) and interferon (IFNγ), as compared to control cells in *in vitro* culture (Trotta et al. 2011). Serum levels of TNFα and IFNγ, associated with decreased levels of interleukin (IL)-10, were found to be higher in DS children as compared to healthy controls (Nateghi Rostami et al. 2012) and a similar pattern of pro-inflammatory cytokines was observed in adults with DS (Carta et al. 2002). However, all data do not tend to the same conclusion and some studies reported on the contrary an increase in anti-inflammatory cytokines: for example, besides the increase in TNFα and IFNγ, Trotta et al. observed also an increase in the production of IL-10 (Trotta et al. 2011). Levels of IL-4 and IL-10 were found significantly increased in children with DS, whereas IL-6 and TNF-α levels were decreased (Cetiner et al. 2010). Recently, Zhang et al. performed a meta-analysis on 19 studies, demonstrating that patients with DS, despite great heterogeneity, have significantly increased circulating TNFα, IFNγ and IL-1β, without differences in concentrations of IL-4, IL-6, IL-8 and IL-10 (Zhang et al. 2017).

Biological Clocks in Down Syndrome: Evidence for a Biological Accelerated Ageing

Twenty years ago, Nakamura and Tanaka estimated the biological age of 11 DS individuals according to a global index developed on 436 healthy subjects involving 14 clinical and biological variables (body mass index, systolic blood pressure, diastolic blood pressure, total protein, ratio of albumin to globulin, concentrations of glutamate oxaloacetate transaminase, glutamic pyruvic transaminase, total cholesterol, triglyceride and blood urea nitrogen, white and red blood cell counts, hemoglobin concentration and hematocrit) (Nakamura and Tanaka 1998). They observed an increase in the biological age of the subjects with DS as compared to healthy subjects. It was also noted that important fluctuations of biological ages from year to year in DS subjects, a pattern that was not observed in controls, and could be in part due, according to the authors, to early senescence of the immune system (white blood cell counts were included in the index) (Nakamura and Tanaka 1998). More recently, new biomarkers of ageing, referred as biological ages or clocks, have been developed and applied to DS populations.

Epigenetic Clocks

During the last few years, DNA methylation (DNAm)-based biomarkers of ageing, called epigenetic clocks, have gained particular relevance in the field of ageing research and at present, three different models have been proposed (Hannum et al. 2013; Horvath 2013; Weidner et al. 2014). Horvath's epigenetic clock, developed in 2013 (Horvath 2013), based on the DNAm levels of 353 specific CpG sites, was applied to four datasets (including 89 DS individuals in total) in which genome-wide DNA methylation levels were assessed in peripheral blood leukocytes, various brain regions, whole blood or buccal epithelium, using Illumina Infinium 27K or 450K platforms (Horvath et al. 2015). DS subjects exhibited a highly significant age acceleration effect (defined as a residual resulting from a linear model that regressed the calculated epigenetic age, also called DNAm age, on chronological age) in three independent data sets involving blood and brain tissues. Age acceleration in brain was estimated at 11.5 years, whereas it was at 4.25 years in blood, for an average of 6.6 years. Results were similar even after correction for blood cell type abundance measures. No significant age acceleration effect was observed in the dataset on buccal epithelium DNA. Cole et al. in their report on 'brain-predicted age' using neuroimaging methods in DS subjects, noticed that they observed, taking the same approach as Horvath et al., an effect of similar magnitude (7.4 vs 6.6 years of added ageing) (Cole et al. 2017).

   In 2016, Obeid et al. investigated DNA methylation by pyrosequencing in three specific regions, named *ASPA*, *ITGA2B* and *PDE4C*, in 31 young subjects with DS and in controls (Obeid et al. 2016). These regions include the CpG sites that have been reported to predict ageing in adults according to Weidner's clock (Weidner et al. 2014). They observed hypomethylation of *ASPA* and *ITGA2B* in subjects with DS, associated with a strong negative association with age. DNA methylation of *PDE4C* did not differ between the two groups but showed a positive correlation with age. According to the prediction of age using DNAm data of the three loci in combination with plasma concentrations of $A\beta$ (1-42) measured by high sensitive ELISA assay, subjects with DS in their second decade were on average 3.1 (95% CI

1.5–4.6) years older than their predicted age based on regression extrapolated from their counterparts without DS.

N-glycome Signature

In recent years, plasma N-glycans have also emerged as biomarkers associated with ageing (Dall'Olio et al. 2013; Yu et al. 2016). Plasma N-glycome changes in 76 DS individuals of different ages have been investigated, as well as in controls (mothers and siblings), using two different methods (MALDI-TOF-MS and DSA-FACE) (Borelli et al. 2015). Glycomic changes associated with DS were identified, resulting in a specific plasma N-glycomic signature with 24 plasma N-glycans differentially expressed. GlycoAge, an index developed to monitor galactosylation changes that mark ageing, was evaluated in this cohort of subjects and increased values were found in DS as compared to their siblings, in particular at young age. Thus, from a glycomic point of view, DS individuals are older than their age-matched controls. Moreover, GlycoAge Test values were negatively correlated with the score of Performances IQ.

Epigenetics in Down Syndrome

In addition to the previously mentioned hallmarks of ageing, epigenetic mechanisms have emerged as playing a major role in ageing and age-related diseases (Kennedy et al. 2014; López-Otín et al. 2013). DNA methylation, histone modifications or non-coding RNAs are also involved in neurodevelopment, learning and memory, as well as neurodegenerative diseases (Day and Sweatt 2011; Della Ragione et al. 2014; Gräff and Tsai 2013; Saab and Mansuy 2014; Sanchez-Mut et al. 2016), and some importance has been attributed to epigenetics in the pathogenesis of DS over the past decade. Epigenetic influences have been investigated, particularly to explore their impact on the variability of the phenotypes observed in the disease, on the accelerated ageing process or on the occurrence of acute megakaryoblastic leukaemia (Malinge et al. 2013).

DNA Methylation

Different DNA Methylation Patterns in Subjects with Down Syndrome

One of the first reports on the presence of differential DNA methylation in DS individuals was published in 2001 (Pogribna et al. 2001): DNA extracted from lymphocytes of DS children was found globally hypermethylated, as compared to the DNA of their siblings, according to radiolabel incorporation. In agreement, in 2006, Chango et al. used a method based on a combination of methylation-sensitive arbitrarily primed polymerase chain reaction (MS-AP- PCR) and quantification of DNA fragments, to investigate relative levels of DNA methylation in peripheral blood lymphocytes: they identified six fragments that were hypermethylated in DS subjects as compared to controls (Chango et al. 2006). Some years later, using alternative techniques to assess DNA methylation levels, studies started to dissect in greater details the epigenetic characteristics of DS. DNA methylation patterns of samples from DS subjects were investigated at the genome-wide level, using Illumina Infinium platforms 27K and 450K (Eckmann-Scholz et al. 2012; Jones et al. 2013b; Kerkel et al. 2010) or reduced representation bisulfite sequencing (RRBS) (Jin et al. 2013; Sailani et al. 2015). Different tissues were investigated: peripheral blood lymphocytes from adults with DS (Kerkel et al. 2010), buccal epithelium (Jones et al. 2013a), skin fibroblasts (Sailani et al. 2015) or placenta samples (Eckmann-Scholz et al. 2012; Jin et al. 2013). Results were globally concordant, with shared common findings: marked DNA methylation alterations in DS cells with predominantly hypermethylation as compared to controls and genome-wide perturbance of DNA methylation without enrichment on HSA21.

In 2010, performing microarray-based genome-wide DNA methylation profiling of white blood cells and T-lymphocytes from adults with DS and normal controls, Kerkel et al. observed consistent hypo or hypermethylation in 118 genes in DS subjects, with corresponding differential expression for some of them (*TMEM131*, *TCF7*, *NPDC1*) and without specific enrichment for genes on HSA21 (Kerkel et al. 2010). Many of the differentially methylated genes are involved in lymphocyte development and function. The alterations in methylation were generally stable in a given individual and were independent of the differential cell counts. In samples derived from buccal epithelial cells of adults, 9982 probes were found differentially methylated between DS and control samples, with 3300 of them having an absolute difference between means of methylation of more than 10% (Jones et al. 2013b). Here also no enrichment for HSA21 was observed. Within the differentially methylated CpG sites, authors identified a number of genes known to be involved in the pathology of the disease and several probes overlapped with the ones identified in other published works (Jin et al. 2013; Kerkel et al. 2010). Interestingly, cognitive function was assessed with Dalton Brief Praxis Test and correlations with epigenetic data were investigated: five CpG sites were found to be correlated with cognitive impairment, including two probes located in the *TSC2* gene, a component of the mTOR pathway that has previously been associated with AD pathology (Jones et al. 2013b). Recently, Sailani et al. investigated DNA methylation profiles of skin fibroblasts of monozygotic twins (MZ) discordant for DS (Sailani et al. 2015): they observed 35 differentially methylated gene promoter regions, that were also concordant with data obtained by comparing unrelated controls and DS individuals and that are mostly related to embryonic organ morphogenesis and development. Here again no enrichment for HSA21 was found and global higher levels of methylation were observed in the twin affected by DS, compared to his unaffected sibling. Interestingly, the differentially methylated regions observed were stably maintained in iPSCs generated from fibroblasts obtained from the twin pair discordant for DS.

Apart from samples derived from children or adults with DS, Eckmann-Scholz et al. and Jin et al. investigated DNA methylation patterns in placenta samples and observed a general hypermethylation across all chromosomes in placentas with trisomy 21, compared to normal ones. The first team used the Illumina Infinium 27K BeadChips on chorionic villi samples and identified 464 loci corresponding to 404 genes differentially methylated in samples with trisomy 21 as compared to samples with a normal karyotype (Eckmann-Scholz et al. 2012). 387 Genes, significantly enriched in developmental processes, were found hypermethylated in the trisomic samples and three of them are located on HSA21 (*COL6A2*, *H2BFS*, *RUNX1*). Jin et al. used an improved version of RRBS to quantify DNA methylation levels in 17 placenta villi samples (11 DS and 6 control samples) and performed also RNA-Seq analysis in five normal and four DS placenta villi samples (Jin et al. 2013). A global hypermethylation was observed in all genomic regions in the samples with trisomy 21, with a predominance in promoter regions. Out of the 589 sites found hypermethylated, significant down-regulation of gene expression was observed in 207 genes. Interestingly, three genes (*TCF7*, *FAM62C* and *CPT1B*), that were found differentially methylated between DS and controls in the study published by Kerkel et al. on adult samples (Kerkel et al. 2010), were also found similarly differentially methylated in the placental samples in this study, suggesting a possible conservation of the methylation patterns in different tissues and across the life course.

Three recent reports analyzed DNA methylation patterns in developing fetal cortex (El Hajj et al. 2016; Lu et al. 2016; Mendioroz et al. 2015). They all observed hypermethylation in the brains with trisomy 21 as compared to controls. In the report by Lu et al. four sites overlapped with the ones found by Kerkel et al. in adult peripheral blood lymphocytes (Kerkel et al. 2010) and 88 probes were also shared with the adult buccal epithelium (Jones et al. 2013b). Mendioroz et al. evaluated DNA methylation in cerebrum of fetuses as well as cerebral and cerebellar cortex of adults: some genes that were differentially methylated in fetal brains were also in adult brain cells, suggesting again an early onset of the epigenetic changes (Mendioroz et al. 2015). Differentially methylated sites found in samples with trisomy 21 were enriched in CpG in or near specific transcription factor binding sites and some of them were also found differentially expressed. In their study, Lu et al. identified an alteration in the signaling pathway of several genes involved in ubiquitination and suggested this pathway as a possible key player in the development of DS neuropathology (Lu et al. 2016).

Using Illumina Infinium 450K platform, our team analyzed whole blood samples from a family-based model of DS, chosen in order to minimize confounding genetic and environmental factors, with 29 trios composed by the DS person, his mother and his non-affected sibling (Bacalini et al. 2015b). Data were processed through a pipeline specifically tailored to identify differentially methylated regions (DMRs) (Bacalini et al. 2015a). Our analysis confirmed a prevalent hypermethylation, as well as the majority of DMRs previously identified by Jones et al. (2013b), and our gene ontology analysis identified an enrichment in genes involved in morphogenetic and developmental processes (*HOXA* family, *RUNX1*, *EBF4*, *NCAM1*) as well as regulation of chromatin structure (*PRMD8, KDM2B, TET1*). However, unlike other published studies, we observed a enrichment in HSA21 for the DMRs identified in DS subjects as compared to their controls. We selected a short list of 68 DMRs whose DNA methylation status was remarkably different between DS subjects and healthy siblings (methylation difference greater than 0.15), which constituted an epigenetic signature of DS.

Five-hydroxymethylcytosine (5hmC) is also an epigenetic modification that occurs on cytosine-bases and that is produced by the activity of TET enzymes. 5hmC acts as an intermediate during the DNA demethylation but has been also considered as acting as a stable epigenetic marker. During ageing, 5hmC content was found to be negatively correlated with age in blood or PBMC (Buscarlet et al. 2016; Valentini et al. 2016) and its importance in brain development and ageing has been recently investigated (Kraus et al. 2015). Alterations in DNA hydroxymethylation in DS were evaluated by Ciccarone et al.: the content of 5hmC was measured by dot-blot assay on DNA extracted from PBMC and levels were found lower in PBMC from DS subjects as compared to controls (Ciccarone et al. 2017).

Finally, it is important to note that DNA methylation patterns are used in prenatal diagnosis of DS. Studies have been investigating differentially methylated regions that could differentiate fetuses with trisomy 21 and fetuses with normal karyotypes, in order to develop non-invasive pre-natal diagnosis techniques analyzing free fetal-specific DNA methylation in maternal blood (Hatt et al. 2015, 2016; Jin et al. 2013; Lee et al. 2016; Lim et al. 2011; Papageorgiou et al. 2009, 2011; Sifakis et al. 2012; Yin et al. 2014).

Regulation of DNA Methylation in Down Syndrome

Several proteins are involved in the regulation of DNA methylation patterns, which result from a balanced state between methylation and demethylation processes. In mammals, DNA methylation is related to the addition of a methyl group to cytosine bases, generally located in CpG dinucleotides, by the DNA methyltransferase (DNMT) family of enzymes (DNMT1, DNMT3A and DNMT3B), whereas demethylation is mostly catalyzed by TET enzymes (TET1, TET2 and TET3). In order to understand potential mechanisms driving the epigenetic changes seen in DS, levels of these enzymes were investigated. Expression levels of *DNMT1* and *DNMT3B* were found similar between DS subjects and controls, whereas *DNMT3A* was found down-regulated in PBMC from DS subjects compared to controls (Ciccarone et al. 2017). On the contrary, *DNMT3L*, found on HSA21, encodes a protein that has no catalytic function on its own, but assists DNMT3A and DNMT3B in establishing *de novo* DNA methylation marks. *DNMT3L* was found over-expressed in DS neural progenitors (Lu et al. 2016). Regarding TET enzymes, they were all previously found down-regulated in DS (Jin et al. 2013) and reduced levels of *TET1* and *TET2* expression in DS subjects as compared to controls were confirmed recently (Ciccarone et al. 2017). This down-regulation may lead to hypermethylation through decreased DNA demethylation.

DNA methylation reactions need a universal methyl donor, represented by S-adenosylmethionine (SAM), which is generated in a metabolic network called one-carbon metabolism. DS subjects are characterized by perturbations in this metabolism, related to the over-expression of cystathionine beta-synthase (*CBS*) gene located on HSA21. CBS is a central enzyme in this specific metabolic pathway, catalyzing the conversion of homocysteine into cystathionine. In 2001, plasma levels of homocysteine, methionine, S-adenosylhomocysteine (SAH) and SAM were found decreased in children with DS, whereas plasma levels of cystathionine and cysteine were found increased (Pogribna et al. 2001), consistent with an increase in CBS activity (Chadefaux et al. 1985). Discordant results were published later: cystathionine, cysteine, SAH and SAM were found at higher levels in DS compared to the controls, whereas levels of methionine did not differ significantly (Obeid et al. 2012; Obermann-Borst et al. 2011). There are also discordant results regarding homocysteine levels, that were found higher (Song et al. 2015) or lower (Meguid et al. 2010) in DS subjects as compared to controls.

Non-coding RNAs

Another epigenetic mechanism implicated in ageing, neurodevelopmental disorders and neurodegenerative diseases is represented by non-coding RNAs (ncRNAs) (Jung and Suh 2014; Tan et al. 2013). In recent years, different research groups have studied the potential contribution of micro-RNAs (miRNAs) to the regulation of DS transcriptome and *in fine*, to DS phenotypes and phenotypic variability. Various miRNAs are encoded on HSA21 and therefore likely over-expressed in DS, and some of them have been implicated in the development of some DS-related pathologies. Principal HSA21-encoded miRNAs that have been implicated in DS and found over-expressed in the disease or adequate models are miRNA-125b-2, miRNA-155, miR-99a, let-7c and miRNA-802 (Siew et al. 2013). Using real-time quantitative PCR to study the expression of miRNA-155 in fibroblasts from a MZ twins discordant for DS, Sethupathy et al. observed that miR-155 was over-expressed in the fibroblasts from the twin with DS (Sethupathy et al. 2007). miR-155 is also over-expressed in DS brain, spleen and liver (Li et al. 2012), as well as in iPSCs generated from human DS amniotic fluid cells (Lu et al. 2013). miR-155 have been associated to brain pathology, as altered expression has been found in AD: miRNA-125b and miRNA-155 are significantly up-regulated in sporadic AD and have emerged as key contributors to the sporadic AD process (Zhao et al. 2015). Up-regulation of miR-125b and miR-155 have been associated to the pathogenic mechanism of complement factor H deficiency that drives inflammatory neurodegeneration in AD and in age-related macular degeneration (Lukiw et al. 2012). Finally, miR 125b-2 has been also identified as a potential onco-miR associated with megakaryoblastic leukemia in DS individuals (Klusmann et al. 2010).

Two studies have investigated miRNA expressions in DS placentas, without significant match between the two datasets (Lim et al. 2015; Svobodová et al. 2016). In 2015, Lim et al. identified 34 differentially expressed miRNAs in DS placenta samples compared to normal ones (16 up-regulated and 18 down-regulated miRNAs), distributed on various chromosomes, without inclusion of any HSA21-derived miRNA (Lim et al. 2015). In 2016, Svobodová et al. identified 7 miRNAs over-expressed in DS placentas as compared to euploïd samples and three of them were located on HSA21 (let-7c, miR-125b and miR-99a) (Svobodová et al. 2016). Analyzing of genome-wide expression of miRNAs in cord blood mononuclear cells from fetuses with DS, Xu et al. observed that most of the mRNA targets of differentially expressed miRNAs were associated with immune modulation (Xu et al. 2013).

Resembling miRNAs, long noncoding RNAs (lncRNAs) have been also demonstrated to have various regulatory roles in gene expression and to contribute to neurological diseases (Qureshi et al. 2010). A large number of lncRNAs were identified differentially expressed in iPSCs generated from DS subjects as compared to normal iPSCs. Most of the differentially expressed lncRNAs were closely associated with mitochondrial functions and they could thus be associated to the dysfunction of mitochondria observed in DS (Qiu et al. 2017).

Conclusion

For many decades DS has been described as a progeroid syndrome, as subjects affected are characterized by the appearance early in life of many typical age-related conditions, involving particularly neurological, immune, endocrine, musculoskeletal and sensorial systems. As DS subjects now live longer than before, we observe more and more age-related diseases in this specific population. Notwithstanding the recent advances in the characterization of the biological basis underpinning DS age-related phenotype, several aspects are still to be elucidated. One of the unsolved questions is whether DS atypical ageing is precocious or accelerated. Some data from literature are in favor of the presence of a phenomenon of precocious ageing in DS, present from fetal life and birth and which could be considered as an intrinsic characteristic of the syndrome. On the contrary other reports, using the latest generation of biological markers of ageing, consider DS as an accelerated-ageing condition. The combination of both phenomena is a possibility that should be further explored, also in the framework of the research for early anti-ageing interventions in this population.

Acknowledgements This work was supported by the European Union's H2020 Project (grant number 634821, PROPAG-AGEING); by JPco-fuND (ADAGE). This project has received funding from the European Union's Horizon 2020 research and innovation programme under the Marie Skłodowska-Curie grant agreement No 675003 (http://www.birmingham.ac.uk/panini).

References

Adorno M, Sikandar S, Mitra SS, Kuo A, Nicolis Di Robilant B, Haro-Acosta V, Ouadah Y, Quarta M, Rodriguez J, Qian D, Reddy VM, Cheshier S, Garner CC, Clarke MF (2013) Usp16 contributes to somatic stem-cell defects in Down's syndrome. Nature 501:380–384. https://doi.org/10.1038/nature12530

Aivazidis S, Coughlan CM, Rauniyar AK, Jiang H, Liggett LA, Maclean KN, Roede JR (2017) The burden of trisomy 21 disrupts the proteostasis network in Down syndrome. PloS One 12:e0176307. https://doi.org/10.1371/journal.pone.0176307

Annus T, Wilson LR, Acosta-Cabronero J, Cardenas-Blanco A, Hong YT, Fryer TD, Coles JP, Menon DK, Zaman SH, Holland AJ, Nestor PJ (2017) The Down syndrome brain in the presence and absence of fibrillar β-amyloidosis. Neurobiol Ageing 53:11–19. https://doi.org/10.1016/j.neurobiolageing.2017.01.009

Arbuzova S (1998) Why it is necessary to study the role of mitochondrial genome in trisomy 21 pathogenesis. Syndr Res Pract 5:126–130. https://doi.org/10.3104/reports.88

Arbuzova S, Hutchin T, Cuckle H (2002) Mitochondrial dysfunction and Down's syndrome. BioEssays News Rev Mol Cell Dev Biol 24:681–684. https://doi.org/10.1002/bies.10138

Bacalini MG, Boattini A, Gentilini D, Giampieri E, Pirazzini C, Giuliani C, Fontanesi E, Remondini D, Capri M, Del Rio A, Luiselli D, Vitale G, Mari D, Castellani G, Di Blasio AM, Salvioli S, Franceschi C, Garagnani P (2015a) A meta-analysis on age-associated changes in blood DNA methylation: results from an original analysis pipeline for Infinium 450k data. Ageing 7:97–109. https://doi.org/10.18632/ageing.100718

Bacalini MG, Gentilini D, Boattini A, Giampieri E, Pirazzini C, Giuliani C, Fontanesi E, Scurti M, Remondini D, Capri M, Cocchi G, Ghezzo A, Del Rio A, Luiselli D, Vitale G, Mari D, Castellani G, Fraga M, Di Blasio AM, Salvioli S, Franceschi C, Garagnani P (2015b) Identification of a DNA methylation signature in blood cells from persons with Down syndrome. Ageing 7:82–96. https://doi.org/10.18632/ageing.100715

Bahn S, Mimmack M, Ryan M, Caldwell MA, Jauniaux E, Starkey M, Svendsen CN, Emson P (2002) Neuronal target genes of the neuron-restrictive silencer factor in neurospheres derived from fetuses with Down's syndrome: a gene expression study. Lancet Lond Engl 359:310–315. https://doi.org/10.1016/S0140-6736(02)07497-4

Ballard C, Mobley W, Hardy J, Williams G, Corbett A (2016) Dementia in Down's syndrome. Lancet Neurol 15:622–636. https://doi.org/10.1016/S1474-4422(16)00063-6

Baptista F, Varela A, Sardinha LB (2005) Bone mineral mass in males and females with and without Down syndrome. Osteoporos Int 16:380–388. https://doi.org/10.1007/s00198-004-1687-1

Barone E, Arena A, Head E, Butterfield DA, Perluigi M (2017) Disturbance of redox homeostasis in Down syndrome: role of iron dysmetabolism. Free Radic Biol Med 114:84–93. https://doi.org/10.1016/j.freeradbiomed.2017.07.009

Barrena MJ, Echaniz P, Garcia-Serrano C, Cuadrado E (1993) Imbalance of the CD4+ subpopulations expressing CD45RA and CD29 antigens in the peripheral blood of adults and children with Down syndrome. Scand J Immunol 38:323–326

Beacher F, Daly E, Simmons A, Prasher V, Morris R, Robinson C, Lovestone S, Murphy K, Murphy DGM (2010) Brain anatomy and ageing in non-demented adults with Down's syndrome: an in vivo MRI study. Psychol Med 40:611–619. https://doi.org/10.1017/S0033291709990985

Bhaumik P, Bhattacharya M, Ghosh P, Ghosh S, Kumar Dey S (2017) Telomere length analysis in Down syndrome birth. Mech Ageing Dev 164:20–26. https://doi.org/10.1016/j.mad.2017.03.006

Bittles AH, Glasson EJ (2004) Clinical, social, and ethical implications of changing life expectancy in Down syndrome. Dev Med Child Neurol 46:282–286

Blackburn EH, Epel ES, Lin J (2015) Human telomere biology: a contributory and interactive factor in ageing, disease risks, and protection. Science 350:1193–1198. https://doi.org/10.1126/science.aab3389

Bloemers BLP, van Bleek GM, Kimpen JLL, Bont L (2010) Distinct abnormalities in the innate immune system of children with Down syndrome. J Pediatr 156:804–809., 809.e1–809.e5. https://doi.org/10.1016/j.jpeds.2009.12.006

Borelli V, Vanhooren V, Lonardi E, Reiding KR, Capri M, Libert C, Garagnani P, Salvioli S, Franceschi C, Wuhrer M (2015) Plasma N-glycome signature of Down syndrome. J Proteome Res 14:4232–4245. https://doi.org/10.1021/acs.jproteome.5b00356

Bruwier A, Chantrain CF (2012) Hematological disorders and leukemia in children with Down syndrome. Eur J Pediatr 171:1301–1307. https://doi.org/10.1007/s00431-011-1624-1

Burgio GR, Lanzavecchia A, Maccario R, Vitiello A, Plebani A, Ugazio AG (1978) Immunodeficiency in Down's syndrome: T-lymphocyte subset imbalance in trisomic children. Clin Exp Immunol 33:298–301

Buscarlet M, Tessier A, Provost S, Mollica L, Busque L (2016) Human blood cell levels of 5-hydroxymethylcytosine (5hmC) decline with age, partly related to acquired mutations in TET2. Exp Hematol 44:1072–1084. https://doi.org/10.1016/j.exphem.2016.07.009

Busciglio J, Yankner BA (1995) Apoptosis and increased generation of reactive oxygen species in Down's syndrome neurons in vitro. Nature 378:776–779. https://doi.org/10.1038/378776a0

Busciglio J, Pelsman A, Wong C, Pigino G, Yuan M, Mori H, Yankner BA (2002) Altered metabolism of the amyloid beta precursor protein is associated with mitochondrial dysfunction in Down's syndrome. Neuron 33:677–688

Cabelof DC, Patel HV, Chen Q, van Remmen H, Matherly LH, Ge Y, Taub JW (2009) Mutational spectrum at GATA1 provides insights into mutagenesis and leukemogenesis in Down syndrome. Blood 114:2753–2763. https://doi.org/10.1182/blood-2008-11-190330

Cairney CJ, Sanguinetti G, Ranghini E, Chantry AD, Nostro MC, Bhattacharyya A, Svendsen CN, Keith WN, Bellantuono I (2009) A systems biology approach to Down syndrome: identification of Notch/Wnt dysregulation in a model of stem cells ageing. Biochim Biophys Acta 1792:353–363. https://doi.org/10.1016/j.bbadis.2009.01.015

Carfì A, Antocicco M, Brandi V, Cipriani C, Fiore F, Mascia D, Settanni S, Vetrano DL, Bernabei R, Onder G (2014) Characteristics of adults with down syndrome: prevalence of age-related conditions. Front Med 1:51. https://doi.org/10.3389/fmed.2014.00051

Carsetti R, Valentini D, Marcellini V, Scarsella M, Marasco E, Giustini F, Bartuli A, Villani A, Ugazio AG (2015) Reduced numbers of switched memory B cells with high terminal differentiation potential in Down syndrome. Eur J Immunol 45:903–914. https://doi.org/10.1002/eji.201445049

Carta MG, Serra P, Ghiani A, Manca E, Hardoy MC, Del Giacco GS, Diaz G, Carpiniello B, Manconi PE (2002) Chemokines and pro-inflammatory cytokines in Down's syndrome: an early marker for Alzheimer-type dementia? Psychother Psychosom 71:233–236. https://doi.org/10.1159/000063649

Cenini G, Dowling ALS, Beckett TL, Barone E, Mancuso C, Murphy MP, Levine H, Lott IT, Schmitt FA, Butterfield DA, Head E (2012) Association between frontal cortex oxidative damage and beta-amyloid as a function of age in Down syndrome. Biochim Biophys Acta 1822:130–138. https://doi.org/10.1016/j.bbadis.2011.10.001

Cetiner S, Demirhan O, Inal TC, Tastemir D, Sertdemir Y (2010) Analysis of peripheral blood T-cell subsets, natural killer cells and serum levels of cytokines in children with Down syndrome. Int J Immunogenet 37:233–237. https://doi.org/10.1111/j.1744-313X.2010.00914.x

Chadefaux B, Rethoré MO, Raoul O, Ceballos I, Poissonnier M, Gilgenkranz S, Allard D (1985) Cystathionine beta synthase: gene dosage effect in trisomy 21. Biochem Biophys Res Commun 128:40–44

Chango A, Abdennebi-Najar L, Tessier F, Ferré S, Do S, Guéant J-L, Nicolas JP, Willequet F (2006) Quantitative methylation-sensitive arbitrarily primed PCR method to determine differential genomic DNA methylation in Down syndrome. Biochem Biophys Res Commun 349:492–496. https://doi.org/10.1016/j.bbrc.2006.08.038

Ciccarone F, Valentini E, Malavolta M, Zampieri M, Bacalini MG, Calabrese R, Guastafierro T, Reale A, Franceschi C, Capri M, Breusing N, Grune T, Moreno-Villanueva M, Bürkle A, Caiafa

P (2017) DNA hydroxymethylation levels are altered in blood cells from Down syndrome persons enrolled in the MARK-AGE project. J Gerontol A Biol Sci Med Sci 73:737–744. https://doi.org/10.1093/gerona/glx198

Cocchi G, Mastrocola M, Capelli M, Bastelli A, Vitali F, Corvaglia L (2007) Immunological patterns in young children with Down syndrome: is there a temporal trend? Acta Paediatr Oslo Nor 1992(96):1479–1482. https://doi.org/10.1111/j.1651-2227.2007.00459.x

Cole JH, Annus T, Wilson LR, Remtulla R, Hong YT, Fryer TD, Acosta-Cabronero J, Cardenas-Blanco A, Smith R, Menon DK, Zaman SH, Nestor PJ, Holland AJ (2017) Brain-predicted age in Down syndrome is associated with beta amyloid deposition and cognitive decline. Neurobiol Ageing 56:41–49. https://doi.org/10.1016/j.neurobiolageing.2017.04.006

Coppus AMW (2013) People with intellectual disability: what do we know about adulthood and life expectancy? Dev Disabil Res Rev 18:6–16. https://doi.org/10.1002/ddrr.1123

Coppus A, Evenhuis H, Verberne G-J, Visser F, van Gool P, Eikelenboom P, van Duijin C (2006) Dementia and mortality in persons with Down's syndrome. J Intellect Disabil Res JIDR 50:768–777. https://doi.org/10.1111/j.1365-2788.2006.00842.x

Coskun PE, Wyrembak J, Derbereva O, Melkonian G, Doran E, Lott IT, Head E, Cotman CW, Wallace DC (2010) Systemic mitochondrial dysfunction and the etiology of Alzheimer's disease and down syndrome dementia. J Alzheimers Dis JAD 20(Suppl 2):S293–S310. https://doi.org/10.3233/JAD-2010-100351

Cossarizza A, Monti D, Montagnani G, Ortolani C, Masi M, Zannotti M, Franceschi C (1990) Precocious ageing of the immune system in Down syndrome: alteration of B lymphocytes, T-lymphocyte subsets, and cells with natural killer markers. Am J Med Genet Suppl 7:213–218

Cossarizza A, Ortolani C, Forti E, Montagnani G, Paganelli R, Zannotti M, Marini M, Monti D, Franceschi C (1991) Age-related expansion of functionally inefficient cells with markers of natural killer activity in Down's syndrome. Blood 77:1263–1270

Cuadrado E, Barrena MJ (1996) Immune dysfunction in Down's syndrome: primary immune deficiency or early senescence of the immune system? Clin Immunol Immunopathol 78:209–214

Dall'Olio F, Vanhooren V, Chen CC, Slagboom PE, Wuhrer M, Franceschi C (2013) N-glycomic biomarkers of biological ageing and longevity: a link with inflammageing. Ageing Res Rev 12:685–698. https://doi.org/10.1016/j.arr.2012.02.002

Dashinimaev EB, Artyuhov AS, Bolshakov AP, Vorotelyak EA, Vasiliev AV (2017) Neurons derived from induced pluripotent stem cells of patients with Down syndrome reproduce early stages of Alzheimer's disease type pathology in vitro. J Alzheimers Dis JAD 56:835–847. https://doi.org/10.3233/JAD-160945

Day JJ, Sweatt JD (2011) Cognitive neuroepigenetics: a role for epigenetic mechanisms in learning and memory. Neurobiol Learn Mem 96:2–12. https://doi.org/10.1016/j.nlm.2010.12.008

de Hingh YCM, van der Vossen PW, Gemen EFA, Mulder AB, Hop WCJ, Brus F, de Vries E (2005) Intrinsic abnormalities of lymphocyte counts in children with down syndrome. J Pediatr 147:744–747. https://doi.org/10.1016/j.jpeds.2005.07.022

De la Torre R, De Sola S, Pons M, Duchon A, de Lagran MM, Farré M, Fitó M, Benejam B, Langohr K, Rodriguez J, Pujadas M, Bizot JC, Cuenca A, Janel N, Catuara S, Covas MI, Blehaut H, Herault Y, Delabar JM, Dierssen M (2014) Epigallocatechin-3-gallate, a DYRK1A inhibitor, rescues cognitive deficits in Down syndrome mouse models and in humans. Mol Nutr Food Res 58:278–288. https://doi.org/10.1002/mnfr.201300325

Deb S, Braganza J, Norton N, Williams H, Kehoe PG, Williams J, Owen MJ (2000) APOE epsilon 4 influences the manifestation of Alzheimer's disease in adults with Down's syndrome. Br J Psychiatry J Ment Sci 176:468–472

Dekker AD, Strydom A, Coppus AMW, Nizetic D, Vermeiren Y, Naudé PJW, Van Dam D, Potier M-C, Fortea J, De Deyn PP (2015) Behavioural and psychological symptoms of dementia in Down syndrome: early indicators of clinical Alzheimer's disease? Cortex J Devoted Study Nerv Syst Behav 73:36–61. https://doi.org/10.1016/j.cortex.2015.07.032

Della Ragione F, Gagliardi M, D'Esposito M, Matarazzo MR (2014) Non-coding RNAs in chromatin disease involving neurological defects. Front Cell Neurosci 8:54. https://doi.org/10.3389/fncel.2014.00054

Di Domenico F, Coccia R, Cocciolo A, Murphy MP, Cenini G, Head E, Butterfield DA, Giorgi A, Schinina ME, Mancuso C, Cini C, Perluigi M (2013) Impairment of proteostasis network in Down syndrome prior to the development of Alzheimer's disease neuropathology: redox proteomics analysis of human brain. Biochim Biophys Acta 1832:1249–1259. https://doi.org/10.1016/j.bbadis.2013.04.013

Di Domenico F, Tramutola A, Foppoli C, Head E, Perluigi M, Butterfield DA (2017) mTOR in Down syndrome: role in Aß and tau neuropathology and transition to Alzheimer disease-like dementia. Free Radic Biol Med 114:94–101. https://doi.org/10.1016/j.freeradbiomed.2017.08.009

Druzhyna N, Nair RG, LeDoux SP, Wilson GL (1998) Defective repair of oxidative damage in mitochondrial DNA in Down's syndrome. Mutat Res 409:81–89

Duchon A, Herault Y (2016) DYRK1A, a dosage-sensitive gene involved in neurodevelopmental disorders, is a target for drug development in Down syndrome. Front Behav Neurosci 10:104. https://doi.org/10.3389/fnbeh.2016.00104

Eckmann-Scholz C, Bens S, Kolarova J, Schneppenheim S, Caliebe A, Heidemann S, von Kaisenberg C, Kautza M, Jonat W, Siebert R, Ammerpohl O (2012) DNA-methylation profiling of fetal tissues reveals marked epigenetic differences between chorionic and amniotic samples. PloS One 7:e39014. https://doi.org/10.1371/journal.pone.0039014

El Hajj N, Dittrich M, Böck J, Kraus TFJ, Nanda I, Müller T, Seidmann L, Tralau T, Galetzka D, Schneider E, Haaf T (2016) Epigenetic dysregulation in the developing Down syndrome cortex. Epigenetics 11:563–578. https://doi.org/10.1080/15592294.2016.1192736

Esbensen AJ (2010) Health conditions associated with ageing and end of life of adults with Down syndrome. Int Rev Res Ment Retard 39:107–126. https://doi.org/10.1016/S0074-7750(10)39004-5

Fabris N, Mocchegiani E, Amadio L, Zannotti M, Licastro F, Franceschi C (1984) Thymic hormone deficiency in normal ageing and Down's syndrome: is there a primary failure of the thymus? Lancet Lond Engl 1:983–986

Franceschi C, Licastro F, Chiricolo M, Bonetti F, Zannotti M, Fabris N, Mocchegiani E, Fantini MP, Paolucci P, Masi M (1981) Deficiency of autologous mixed lymphocyte reactions and serum thymic factor level in Down's syndrome. J Immunol Baltim Md 1950(126):2161–2164

Franceschi M, Comola M, Piattoni F, Gualandri W, Canal N (1990) Prevalence of dementia in adult patients with trisomy 21. Am J Med Genet Suppl 7:306–308

Franceschi C, Monti D, Scarfí MR, Zeni O, Temperani P, Emilia G, Sansoni P, Lioi MB, Troiano L, Agnesini C (1992) Genomic instability and ageing. Studies in centenarians (successful ageing) and in patients with Down's syndrome (accelerated ageing). Ann N Y Acad Sci 663:4–16

Franceschi C, Bonafè M, Valensin S, Olivieri F, De Luca M, Ottaviani E, De Benedictis G (2000) Inflamm-ageing. An evolutionary perspective on immunosenescence. Ann N Y Acad Sci 908:244–254

Garlet TR, Parisotto EB, de Medeiros G d S, Pereira LCR, Moreira EADM, Dalmarco EM, Dalmarco JB, Wilhelm Filho D (2013) Systemic oxidative stress in children and teenagers with Down syndrome. Life Sci 93:558–563. https://doi.org/10.1016/j.lfs.2013.08.017

Ghezzo A, Salvioli S, Solimando MC, Palmieri A, Chiostergi C, Scurti M, Lomartire L, Bedetti F, Cocchi G, Follo D, Pipitone E, Rovatti P, Zamberletti J, Gomiero T, Castellani G, Franceschi C (2014) Age-related changes of adaptive and neuropsychological features in persons with Down syndrome. PloS One 9:e113111. https://doi.org/10.1371/journal.pone.0113111

Gimeno A, García-Giménez JL, Audí L, Toran N, Andaluz P, Dasí F, Viña J, Pallardó FV (2014) Decreased cell proliferation and higher oxidative stress in fibroblasts from Down syndrome fetuses. Preliminary study. Biochim Biophys Acta 1842:116–125. https://doi.org/10.1016/j.bbadis.2013.10.014

Glasson EJ, Sullivan SG, Hussain R, Petterson BA, Montgomery PD, Bittles AH (2002) The changing survival profile of people with Down's syndrome: implications for genetic counselling. Clin Genet 62:390–393

Glasson EJ, Jacques A, Wong K, Bourke J, Leonard H (2016) Improved survival in Down syndrome over the last 60 years and the impact of perinatal factors in recent decades. J Pediatr 169:214–220.e1. https://doi.org/10.1016/j.jpeds.2015.10.083

Goldacre MJ, Wotton CJ, Seagroatt V, Yeates D (2004) Cancers and immune related diseases associated with Down's syndrome: a record linkage study. Arch Dis Child 89:1014–1017. https://doi.org/10.1136/adc.2003.046219

Gräff J, Tsai L-H (2013) Histone acetylation: molecular mnemonics on the chromatin. Nat Rev Neurosci 14:97–111. https://doi.org/10.1038/nrn3427

Gruszecka A, Kopczyński P, Cudziło D, Lipińska N, Romaniuk A, Barczak W, Rozwadowska N, Totoń E, Rubiś B (2015) Telomere shortening in Down syndrome patients – when does it start? DNA Cell Biol 34:412–417. https://doi.org/10.1089/dna.2014.2746

Guaraldi F, Rossetto Giaccherino R, Lanfranco F, Motta G, Gori D, Arvat E, Ghigo E, Giordano R (2017) Endocrine autoimmunity in Down's syndrome. Front Horm Res 48:133–146. https://doi.org/10.1159/000452912

Guazzarotti L, Trabattoni D, Castelletti E, Boldrighini B, Piacentini L, Duca P, Beretta S, Pacei M, Caprio C, Vigan Ago A, di Natale B, Zuccotti GV, Clerici M (2009) T lymphocyte maturation is impaired in healthy young individuals carrying trisomy 21 (Down syndrome). Am J Intellect Dev Disabil 114:100–109. https://doi.org/10.1352/2009.114.100-109

Hannum G, Guinney J, Zhao L, Zhang L, Hughes G, Sadda S, Klotzle B, Bibikova M, Fan J-B, Gao Y, Deconde R, Chen M, Rajapakse I, Friend S, Ideker T, Zhang K (2013) Genome-wide methylation profiles reveal quantitative views of human ageing rates. Mol Cell 49:359–367. https://doi.org/10.1016/j.molcel.2012.10.016

Hartley SL, Handen BL, Devenny D, Mihaila I, Hardison R, Lao PJ, Klunk WE, Bulova P, Johnson SC, Christian BT (2017) Cognitive decline and brain amyloid-β accumulation across 3 years in adults with Down syndrome. Neurobiol Ageing 58:68–76. https://doi.org/10.1016/j.neurobiolageing.2017.05.019

Hatt L, Aagaard MM, Graakjaer J, Bach C, Sommer S, Agerholm IE, Kølvraa S, Bojesen A (2015) Microarray-based analysis of methylation status of CpGs in placental DNA and maternal blood DNA – potential new epigenetic biomarkers for cell free fetal DNA-based diagnosis. PloS One 10:e0128918. https://doi.org/10.1371/journal.pone.0128918

Hatt L, Aagaard MM, Bach C, Graakjaer J, Sommer S, Agerholm IE, Kølvraa S, Bojesen A (2016) Microarray-based analysis of methylation of 1st trimester trisomic placentas from Down syndrome, Edwards syndrome and Patau syndrome. PloS One 11:e0160319. https://doi.org/10.1371/journal.pone.0160319

Helguera P, Seiglie J, Rodriguez J, Hanna M, Helguera G, Busciglio J (2013) Adaptive down-regulation of mitochondrial function in down syndrome. Cell Metab 17:132–140. https://doi.org/10.1016/j.cmet.2012.12.005

Holland AJ, Hon J, Huppert FA, Stevens F, Watson P (1998) Population-based study of the prevalence and presentation of dementia in adults with Down's syndrome. Br J Psychiatry J Ment Sci 172:493–498

Holland AJ, Hon J, Huppert FA, Stevens F (2000) Incidence and course of dementia in people with Down's syndrome: findings from a population-based study. J Intellect Disabil Res JIDR 44. (Pt 2:138–146

Holmes DK, Bates N, Murray M, Ladusans EJ, Morabito A, Bolton-Maggs PHB, Johnston TA, Walkenshaw S, Wynn RF, Bellantuono I (2006) Hematopoietic progenitor cell deficiency in fetuses and children affected by Down's syndrome. Exp Hematol 34:1611–1615. https://doi.org/10.1016/j.exphem.2006.10.013

Horvath S (2013) DNA methylation age of human tissues and cell types. Genome Biol 14:R115. https://doi.org/10.1186/gb-2013-14-10-r115

Horvath S, Garagnani P, Bacalini MG, Pirazzini C, Salvioli S, Gentilini D, Di Blasio AM, Giuliani C, Tung S, Vinters HV, Franceschi C (2015) Accelerated epigenetic ageing in Down syndrome. Ageing Cell 14:491–495. https://doi.org/10.1111/acel.12325

Iacono T, Torr J, Wong HY (2010) Relationships amongst age, language and related skills in adults with Down syndrome. Res Dev Disabil 31:568–576. https://doi.org/10.1016/j.ridd.2009.12.009

Jenkins EC, Velinov MT, Ye L, Gu H, Li S, Jenkins EC, Brooks SS, Pang D, Devenny DA, Zigman WB, Schupf N, Silverman WP (2006) Telomere shortening in T lymphocytes of older individuals with Down syndrome and dementia. Neurobiol Ageing 27:941–945. https://doi.org/10.1016/j.neurobiolageing.2005.05.021

Jenkins EC, Ye L, Gu H, Ni SA, Duncan CJ, Velinov M, Pang D, Krinsky-McHale SJ, Zigman WB, Schupf N, Silverman WP (2008) Increased "absence" of telomeres may indicate Alzheimer's disease/dementia status in older individuals with Down syndrome. Neurosci Lett 440:340–343. https://doi.org/10.1016/j.neulet.2008.05.098

Jenkins EC, Ye L, Gu H, Ni SA, Velinov M, Pang D, Krinsky-McHale SJ, Zigman WB, Schupf N, Silverman WP (2010) Shorter telomeres may indicate dementia status in older individuals with Down syndrome. Neurobiol Ageing 31:765–771. https://doi.org/10.1016/j.neurobiolageing.2008.06.001

Jenkins EC, Ye L, Velinov M, Krinsky-McHale SJ, Zigman WB, Schupf N, Silverman WP (2012) Mild cognitive impairment identified in older individuals with Down syndrome by reduced telomere signal numbers and shorter telomeres measured in microns. Am J Med Genet Part B Neuropsychiatr Genet Off Publ Int Soc Psychiatr Genet 159B:598–604. https://doi.org/10.1002/ajmg.b.32066

Jenkins EC, Ye L, Krinsky-McHale SJ, Zigman WB, Schupf N, Silverman WP (2016) Telomere longitudinal shortening as a biomarker for dementia status of adults with Down syndrome. Am J Med Genet Part B Neuropsychiatr Genet Off Publ Int Soc Psychiatr Genet 171B:169–174. https://doi.org/10.1002/ajmg.b.32389

Jenkins EC, Marchi EJ, Velinov MT, Ye L, Krinsky-McHale SJ, Zigman WB, Schupf N, Silverman WP (2017) Longitudinal telomere shortening and early Alzheimer's disease progression in adults with down syndrome. Am J Med Genet Part B Neuropsychiatr Genet Off Publ Int Soc Psychiatr Genet 174:772–778. https://doi.org/10.1002/ajmg.b.32575

Jin S, Lee YK, Lim YC, Zheng Z, Lin XM, Ng DPY, Holbrook JD, Law HY, Kwek KYC, Yeo GSH, Ding C (2013) Global DNA hypermethylation in down syndrome placenta. PLoS Genet 9:e1003515. https://doi.org/10.1371/journal.pgen.1003515

Jones EL, Mok K, Hanney M, Harold D, Sims R, Williams J, Ballard C (2013a) Evidence that PICALM affects age at onset of Alzheimer's dementia in Down syndrome. Neurobiol Ageing 34(2441):e1–e5. https://doi.org/10.1016/j.neurobiolageing.2013.03.018

Jones MJ, Farré P, McEwen LM, Macisaac JL, Watt K, Neumann SM, Emberly E, Cynader MS, Virji-Babul N, Kobor MS (2013b) Distinct DNA methylation patterns of cognitive impairment and trisomy 21 in Down syndrome. BMC Med Genomics 6:58. https://doi.org/10.1186/1755-8794-6-58

Joshi AY, Abraham RS, Snyder MR, Boyce TG (2011) Immune evaluation and vaccine responses in Down syndrome: evidence of immunodeficiency? Vaccine 29:5040–5046. https://doi.org/10.1016/j.vaccine.2011.04.060

Jovanovic SV, Clements D, MacLeod K (1998) Biomarkers of oxidative stress are significantly elevated in Down syndrome. Free Radic Biol Med 25:1044–1048

Jung HJ, Suh Y (2014) Circulating miRNAs in ageing and ageing-related diseases. J Genet Genomics Yi Chuan Xue Bao 41:465–472. https://doi.org/10.1016/j.jgg.2014.07.003

Jylhävä J, Pedersen NL, Hägg S (2017) Biological age predictors. EBioMedicine 21:29–36. https://doi.org/10.1016/j.ebiom.2017.03.046

Karlsson B, Gustafsson J, Hedov G, Ivarsson SA, Annerén G (1998) Thyroid dysfunction in Down's syndrome: relation to age and thyroid autoimmunity. Arch Dis Child 79:242–245

Karttunen R, Nurmi T, Ilonen J, Surcel HM (1984) Cell-mediated immunodeficiency in Down's syndrome: normal IL-2 production but inverted ratio of T cell subsets. Clin Exp Immunol 55:257–263

Kennedy BK, Berger SL, Brunet A, Campisi J, Cuervo AM, Epel ES, Franceschi C, Lithgow GJ, Morimoto RI, Pessin JE, Rando TA, Richardson A, Schadt EE, Wyss-Coray T, Sierra F (2014) Geroscience: linking ageing to chronic disease. Cell 159:709–713. https://doi.org/10.1016/j.cell.2014.10.039

Kerkel K, Schupf N, Hatta K, Pang D, Salas M, Kratz A, Minden M, Murty V, Zigman WB, Mayeux RP, Jenkins EC, Torkamani A, Schork NJ, Silverman W, Croy BA, Tycko B (2010)

Altered DNA methylation in leukocytes with trisomy 21. PLoS Genet 6:e1001212. https://doi.org/10.1371/journal.pgen.1001212

Kimura M, Cao X, Skurnick J, Cody M, Soteropoulos P, Aviv A (2005) Proliferation dynamics in cultured skin fibroblasts from Down syndrome subjects. Free Radic Biol Med 39:374–380. https://doi.org/10.1016/j.freeradbiomed.2005.03.023

Kittler P, Krinsky-McHale SJ, Devenny DA (2006) Verbal intrusions precede memory decline in adults with Down syndrome. J Intellect Disabil Res JIDR 50:1–10. https://doi.org/10.1111/j.1365-2788.2005.00715.x

Klusmann J-H, Li Z, Böhmer K, Maroz A, Koch ML, Emmrich S, Godinho FJ, Orkin SH, Reinhardt D (2010) miR-125b-2 is a potential oncomiR on human chromosome 21 in megakaryoblastic leukemia. Genes Dev 24:478–490. https://doi.org/10.1101/gad.1856210

Koran MEI, Hohman TJ, Edwards CM, Vega JN, Pryweller JR, Slosky LE, Crockett G, Villa de Rey L, Meda SA, Dankner N, Avery SN, Blackford JU, Dykens EM, Thornton-Wells TA (2014) Differences in age-related effects on brain volume in Down syndrome as compared to Williams syndrome and typical development. J Neurodev Disord 6:8. https://doi.org/10.1186/1866-1955-6-8

Krasuski JS, Alexander GE, Horwitz B, Rapoport SI, Schapiro MB (2002) Relation of medial temporal lobe volumes to age and memory function in nondemented adults with Down's syndrome: implications for the prodromal phase of Alzheimer's disease. Am J Psychiatry 159:74–81. https://doi.org/10.1176/appi.ajp.159.1.74

Kraus TFJ, Guibourt V, Kretzschmar HA (2015) 5-Hydroxymethylcytosine, the "Sixth Base", during brain development and ageing. J Neural Transm Vienna Austria 1996 122:1035–1043. https://doi.org/10.1007/s00702-014-1346-4

Kusters MAA, Verstegen RHJ, Gemen EFA, de Vries E (2009) Intrinsic defect of the immune system in children with Down syndrome: a review. Clin Exp Immunol 156:189–193. https://doi.org/10.1111/j.1365-2249.2009.03890.x

Kusters MAA, Gemen EFA, Verstegen RHJ, Wever PC, de Vries E (2010) Both normal memory counts and decreased naive cells favor intrinsic defect over early senescence of Down syndrome T lymphocytes. Pediatr Res 67:557–562. https://doi.org/10.1203/PDR.0b013e3181d4eca3

Kusters MA, Jol-Van Der Zijde ECM, Gijsbers RHJM, de Vries E (2011) Decreased response after conjugated meningococcal serogroup C vaccination in children with Down syndrome. Pediatr Infect Dis J 30:818–819. https://doi.org/10.1097/INF.0b013e31822233f9

Lai F, Williams RS (1989) A prospective study of Alzheimer disease in Down syndrome. Arch Neurol 46:849–853

Lai F, Kammann E, Rebeck GW, Anderson A, Chen Y, Nixon RA (1999) APOE genotype and gender effects on Alzheimer disease in 100 adults with Down syndrome. Neurology 53:331–336

Larocca LM, Piantelli M, Valitutti S, Castellino F, Maggiano N, Musiani P (1988) Alterations in thymocyte subpopulations in Down's syndrome (trisomy 21). Clin Immunol Immunopathol 49:175–186

Leclerc E, Sturchler E, Vetter SW (2010) The S100B/RAGE axis in Alzheimer's disease. Cardiovasc Psychiatry Neurol 2010:539581. https://doi.org/10.1155/2010/539581

Lee JH, Chulikavit M, Pang D, Zigman WB, Silverman W, Schupf N (2007) Association between genetic variants in sortilin-related receptor 1 (SORL1) and Alzheimer's disease in adults with Down syndrome. Neurosci Lett 425:105–109. https://doi.org/10.1016/j.neulet.2007.08.042

Lee JH, Gurney S, Pang D, Temkin A, Park N, Janicki SC, Zigman WB, Silverman W, Tycko B, Schupf N (2012) Polymorphisms in HSD17B1: early onset and increased risk of Alzheimer's disease in women with Down syndrome. Curr Gerontol Geriatr Res 2012:361218. https://doi.org/10.1155/2012/361218

Lee DE, Lim JH, Kim MH, Park SY, Ryu HM (2016) Novel epigenetic markers on chromosome 21 for noninvasive prenatal testing of fetal trisomy 21. J Mol Diagn JMD 18:378–387. https://doi.org/10.1016/j.jmoldx.2015.12.002

Lee JH, Lee AJ, Dang L-H, Pang D, Kisselev S, Krinsky-McHale SJ, Zigman WB, Luchsinger JA, Silverman W, Tycko B, Clark LN, Schupf N (2017) Candidate gene analysis for Alzheimer's disease in adults with Down syndrome. Neurobiol Ageing 56:150–158. https://doi.org/10.1016/j.neurobiolageing.2017.04.018

Lemere CA, Blusztajn JK, Yamaguchi H, Wisniewski T, Saido TC, Selkoe DJ (1996) Sequence of deposition of heterogeneous amyloid beta-peptides and APO E in Down syndrome: implications for initial events in amyloid plaque formation. Neurobiol Dis 3:16–32. https://doi.org/10.1006/nbdi.1996.0003

Leonard S, Bower C, Petterson B, Leonard H (2000) Survival of infants born with Down's syndrome: 1980–96. Paediatr Perinat Epidemiol 14:163–171

Leverenz JB, Raskind MA (1998) Early amyloid deposition in the medial temporal lobe of young Down syndrome patients: a regional quantitative analysis. Exp Neurol 150:296–304. https://doi.org/10.1006/exnr.1997.6777

Li YY, Alexandrov PN, Pogue AI, Zhao Y, Bhattacharjee S, Lukiw WJ (2012) miRNA-155 upregulation and complement factor H deficits in Down's syndrome. Neuroreport 23:168–173. https://doi.org/10.1097/WNR.0b013e32834f4eb4

Lim JH, Kim SY, Park SY, Lee SY, Kim MJ, Han YJ, Lee SW, Chung JH, Kim MY, Yang JH, Ryu HM (2011) Non-invasive epigenetic detection of fetal trisomy 21 in first trimester maternal plasma. PloS One 6:e27709. https://doi.org/10.1371/journal.pone.0027709

Lim JH, Kim DJ, Lee DE, Han JY, Chung JH, Ahn HK, Lee SW, Lim DH, Lee YS, Park SY, Ryu HM (2015) Genome-wide microRNA expression profiling in placentas of fetuses with Down syndrome. Placenta 36:322–328. https://doi.org/10.1016/j.placenta.2014.12.020

Lima FA, Moreira-Filho CA, Ramos PL, Brentani H, Lima L d A, Arrais M, Bento-de-Souza LC, Bento-de-Souza L, Duarte MI, Coutinho A, Carneiro-Sampaio M (2011) Decreased AIRE expression and global thymic hypofunction in Down syndrome. J Immunol Baltim Md 1950(187):3422–3430. https://doi.org/10.4049/jimmunol.1003053

Liu L, Rando TA (2011) Manifestations and mechanisms of stem cell ageing. J Cell Biol 193:257–266. https://doi.org/10.1083/jcb.201010131

Liu C-C, Liu C-C, Kanekiyo T, Xu H, Bu G (2013) Apolipoprotein E and Alzheimer disease: risk, mechanisms and therapy. Nat Rev Neurol 9:106–118. https://doi.org/10.1038/nrneurol.2012.263

Liu Y, Borel C, Li L, Müller T, Williams EG, Germain P-L, Buljan M, Sajic T, Boersema PJ, Shao W, Faini M, Testa G, Beyer A, Antonarakis SE, Aebersold R (2017) Systematic proteome and proteostasis profiling in human trisomy 21 fibroblast cells. Nat Commun 8:1212. https://doi.org/10.1038/s41467-017-01422-6

Lockitch G, Singh VK, Puterman ML, Godolphin WJ, Sheps S, Tingle AJ, Wong F, Quigley G (1987) Age-related changes in humoral and cell-mediated immunity in Down syndrome children living at home. Pediatr Res 22:536–540. https://doi.org/10.1203/00006450-198711000-00013

López-Otín C, Blasco MA, Partridge L, Serrano M, Kroemer G (2013) The hallmarks of ageing. Cell 153:1194–1217. https://doi.org/10.1016/j.cell.2013.05.039

Lott IT (2012) Neurological phenotypes for Down syndrome across the life span. Prog Brain Res 197:101–121. https://doi.org/10.1016/B978-0-444-54299-1.00006-6

Lott IT, Dierssen M (2010) Cognitive deficits and associated neurological complications in individuals with Down's syndrome. Lancet Neurol 9:623–633. https://doi.org/10.1016/S1474-4422(10)70112-5

Lu H-E, Yang Y-C, Chen S-M, Su H-L, Huang P-C, Tsai M-S, Wang T-H, Tseng C-P, Hwang S-M (2013) Modeling neurogenesis impairment in Down syndrome with induced pluripotent stem cells from Trisomy 21 amniotic fluid cells. Exp Cell Res 319:498–505. https://doi.org/10.1016/j.yexcr.2012.09.017

Lu J, Mccarter M, Lian G, Esposito G, Capoccia E, Delli-Bovi LC, Hecht J, Sheen V (2016) Global hypermethylation in fetal cortex of Down syndrome due to DNMT3L overexpression. Hum Mol Genet 25:1714–1727. https://doi.org/10.1093/hmg/ddw043

Lukiw WJ, Surjyadipta B, Dua P, Alexandrov PN (2012) Common micro RNAs (miRNAs) target complement factor H (CFH) regulation in Alzheimer's disease (AD) and in age-related macular degeneration (AMD). Int J Biochem Mol Biol 3:105–116

Malinge S, Chlon T, Doré LC, Ketterling RP, Tallman MS, Paietta E, Gamis AS, Taub JW, Chou ST, Weiss MJ, Crispino JD, Figueroa ME (2013) Development of acute megakaryoblastic leukemia in Down syndrome is associated with sequential epigenetic changes. Blood 122:e33–e43. https://doi.org/10.1182/blood-2013-05-503011

Maluf SW, Erdtmann B (2001) Genomic instability in Down syndrome and Fanconi anemia assessed by micronucleus analysis and single-cell gel electrophoresis. Cancer Genet Cytogenet 124:71–75

Mao R, Zielke CL, Zielke HR, Pevsner J (2003) Global up-regulation of chromosome 21 gene expression in the developing Down syndrome brain. Genomics 81:457–467

Margallo-Lana M, Morris CM, Gibson AM, Tan AL, Kay DWK, Tyrer SP, Moore BP, Ballard CG (2004) Influence of the amyloid precursor protein locus on dementia in Down syndrome. Neurology 62:1996–1998

Margallo-Lana ML, Moore PB, Kay DWK, Perry RH, Reid BE, Berney TP, Tyrer SP (2007) Fifteen-year follow-up of 92 hospitalized adults with Down's syndrome: incidence of cognitive decline, its relationship to age and neuropathology. J Intellect Disabil Res JIDR 51:463–477. https://doi.org/10.1111/j.1365-2788.2006.00902.x

Martin GM (1982) Syndromes of accelerated ageing. Natl Cancer Inst Monogr 60:241–247

McCarron M, McCallion P, Reilly E, Mulryan N (2014) A prospective 14-year longitudinal follow-up of dementia in persons with Down syndrome. J Intellect Disabil Res JIDR 58:61–70. https://doi.org/10.1111/jir.12074

McCarron M, McCallion P, Reilly E, Dunne P, Carroll R, Mulryan N (2017) A prospective 20-year longitudinal follow-up of dementia in persons with Down syndrome. J Intellect Disabil Res JIDR 61:843–852. https://doi.org/10.1111/jir.12390

McKelvey KD, Fowler TW, Akel NS, Kelsay JA, Gaddy D, Wenger GR, Suva LJ (2013) Low bone turnover and low bone density in a cohort of adults with Down syndrome. Osteoporos Int J Establ Result Coop Eur Found Osteoporos Natl Osteoporos Found USA 24:1333–1338. https://doi.org/10.1007/s00198-012-2109-4

Meguid NA, Dardir AA, El-Sayed EM, Ahmed HH, Hashish AF, Ezzat A (2010) Homocysteine and oxidative stress in Egyptian children with Down syndrome. Clin Biochem 43:963–967. https://doi.org/10.1016/j.clinbiochem.2010.04.058

Mendioroz M, Do C, Jiang X, Liu C, Darbary HK, Lang CF, Lin J, Thomas A, Abu-Amero S, Stanier P, Temkin A, Yale A, Liu M-M, Li Y, Salas M, Kerkel K, Capone G, Silverman W, Yu YE, Moore G, Wegiel J, Tycko B (2015) Trans effects of chromosome aneuploidies on DNA methylation patterns in human Down syndrome and mouse models. Genome Biol 16:263. https://doi.org/10.1186/s13059-015-0827-6

Mok KY, Jones EL, Hanney M, Harold D, Sims R, Williams J, Ballard C, Hardy J (2014) Polymorphisms in BACE2 may affect the age of onset Alzheimer's dementia in Down syndrome. Neurobiol Ageing 35(1513):e1–e5. https://doi.org/10.1016/j.neurobiolageing.2013.12.022

Morawiec Z, Janik K, Kowalski M, Stetkiewicz T, Szaflik J, Morawiec-Bajda A, Sobczuk A, Blasiak J (2008) DNA damage and repair in children with Down's syndrome. Mutat Res 637:118–123. https://doi.org/10.1016/j.mrfmmm.2007.07.010

Morimoto RI, Cuervo AM (2014) Proteostasis and the ageing proteome in health and disease. J Gerontol A Biol Sci Med Sci 69(Suppl 1):S33–S38. https://doi.org/10.1093/gerona/glu049

Mullins D, Daly E, Simmons A, Beacher F, Foy CM, Lovestone S, Hallahan B, Murphy KC, Murphy DG (2013) Dementia in Down's syndrome: an MRI comparison with Alzheimer's disease in the general population. J Neurodev Disord 5:19. https://doi.org/10.1186/1866-1955-5-19

Murakami K, Murata N, Noda Y, Tahara S, Kaneko T, Kinoshita N, Hatsuta H, Murayama S, Barnham KJ, Irie K, Shirasawa T, Shimizu T (2011) SOD1 (copper/zinc superoxide dismutase) deficiency drives amyloid β protein oligomerization and memory loss in mouse model of Alzheimer disease. J Biol Chem 286:44557–44568. https://doi.org/10.1074/jbc.M111.279208

Murphy M, Epstein LB (1990) Down syndrome (trisomy 21) thymuses have a decreased proportion of cells expressing high levels of TCR alpha, beta and CD3. A possible mechanism for diminished T cell function in Down syndrome. Clin Immunol Immunopathol 55:453–467

Murphy M, Epstein LB (1992) Down syndrome (DS) peripheral blood contains phenotypically mature CD3+TCR alpha, beta+ cells but abnormal proportions of TCR alpha, beta+, TCR gamma, delta+, and CD4+ CD45RA+ cells: evidence for an inefficient release of mature T cells by the DS thymus. Clin Immunol Immunopathol 62:245–251

Murphy M, Lempert MJ, Epstein LB (1990) Decreased level of T cell receptor expression by Down syndrome (trisomy 21) thymocytes. Am J Med Genet Suppl 7:234–237

Musiani P, Valitutti S, Castellino F, Larocca LM, Maggiano N, Piantelli M (1990) Intrathymic deficient expansion of T cell precursors in Down syndrome. Am J Med Genet Suppl 7:219–224

Nakamura E, Tanaka S (1998) Biological ages of adult men and women with Down's syndrome and its changes with ageing. Mech Ageing Dev 105:89–103

Nakamura K-I, Ishikawa N, Izumiyama N, Aida J, Kuroiwa M, Hiraishi N, Fujiwara M, Nakao A, Kawakami T, Poon SSS, Matsuura M, Sawabe M, Arai T, Takubo K (2014) Telomere lengths at birth in trisomies 18 and 21 measured by Q-FISH. Gene 533:199–207. https://doi.org/10.1016/j.gene.2013.09.086

Nateghi Rostami M, Douraghi M, Miramin Mohammadi A, Nikmanesh B (2012) Altered serum pro-inflammatory cytokines in children with Down's syndrome. Eur Cytokine Netw 23:64–67. https://doi.org/10.1684/ecn.2012.0307

Necchi D, Pinto A, Tillhon M, Dutto I, Serafini MM, Lanni C, Govoni S, Racchi M, Prosperi E (2015) Defective DNA repair and increased chromatin binding of DNA repair factors in Down syndrome fibroblasts. Mutat Res 780:15–23. https://doi.org/10.1016/j.mrfmmm.2015.07.009

Obeid R, Hartmuth K, Herrmann W, Gortner L, Rohrer TR, Geisel J, Reed MC, Nijhout HF (2012) Blood biomarkers of methylation in Down syndrome and metabolic simulations using a mathematical model. Mol Nutr Food Res 56:1582–1589. https://doi.org/10.1002/mnfr.201200162

Obeid R, Hübner U, Bodis M, Geisel J (2016) Plasma amyloid beta 1-42 and DNA methylation pattern predict accelerated ageing in young subjects with Down syndrome. Neuromolecular Med 18:593–601. https://doi.org/10.1007/s12017-016-8413-y

Obermann-Borst SA, van Driel LMJW, Helbing WA, de Jonge R, Wildhagen MF, Steegers EAP, Steegers-Theunissen RPM (2011) Congenital heart defects and biomarkers of methylation in children: a case-control study. Eur J Clin Invest 41:143–150. https://doi.org/10.1111/j.1365-2362.2010.02388.x

Odetti P, Angelini G, Dapino D, Zaccheo D, Garibaldi S, Dagna-Bricarelli F, Piombo G, Perry G, Smith M, Traverso N, Tabaton M (1998) Early glycoxidation damage in brains from Down's syndrome. Biochem Biophys Res Commun 243:849–851. https://doi.org/10.1006/bbrc.1998.8186

Oliver C, Crayton L, Holland A, Hall S, Bradbury J (1998) A four year prospective study of age-related cognitive change in adults with Down's syndrome. Psychol Med 28:1365–1377

Papadopoulos N, Simopoulos C, Venizelos J, Kotini A, Skaphida P, Tamiolakis D (2003) Fetal thymic medulla functional alterations in Down's syndrome. Minerva Med 94:181–185

Papageorgiou EA, Fiegler H, Rakyan V, Beck S, Hulten M, Lamnissou K, Carter NP, Patsalis PC (2009) Sites of differential DNA methylation between placenta and peripheral blood: molecular markers for noninvasive prenatal diagnosis of aneuploidies. Am J Pathol 174:1609–1618. https://doi.org/10.2353/ajpath.2009.081038

Papageorgiou EA, Karagrigoriou A, Tsaliki E, Velissariou V, Carter NP, Patsalis PC (2011) Fetal-specific DNA methylation ratio permits noninvasive prenatal diagnosis of trisomy 21. Nat Med 17:510–513. https://doi.org/10.1038/nm.2312

Park E, Alberti J, Mehta P, Dalton A, Sersen E, Schuller-Levis G (2000) Partial impairment of immune functions in peripheral blood leukocytes from aged men with Down's syndrome. Clin Immunol Orlando Fla 95:62–69. https://doi.org/10.1006/clim.2000.4834

Patel A, Rees SD, Kelly MA, Bain SC, Barnett AH, Prasher A, Arshad H, Prasher VP (2014) Genetic variants conferring susceptibility to Alzheimer's disease in the general population; do they also predispose to dementia in Down's syndrome. BMC Res Notes 7:42. https://doi.org/10.1186/1756-0500-7-42

Patterson D, Cabelof DC (2012) Down syndrome as a model of DNA polymerase beta haploin-sufficiency and accelerated ageing. Mech Ageing Dev 133:133–137. https://doi.org/10.1016/j.mad.2011.10.001

Pellegrini FP, Marinoni M, Frangione V, Tedeschi A, Gandini V, Ciglia F, Mortara L, Accolla RS, Nespoli L (2012) Down syndrome, autoimmunity and T regulatory cells. Clin Exp Immunol 169:238–243. https://doi.org/10.1111/j.1365-2249.2012.04610.x

Picciotti PM, Carfì A, Anzivino R, Paludetti G, Conti G, Brandi V, Bernabei R, Onder G (2017) Audiologic assessment in adults with Down syndrome. Am J Intellect Dev Disabil 122:333–341. https://doi.org/10.1352/1944-7558-122.4.333

Pinter JD, Eliez S, Schmitt JE, Capone GT, Reiss AL (2001) Neuroanatomy of Down's syndrome: a high-resolution MRI study. Am J Psychiatry 158:1659–1665. https://doi.org/10.1176/appi.ajp.158.10.1659

Pogribna M, Melnyk S, Pogribny I, Chango A, Yi P, James SJ (2001) Homocysteine metabolism in children with Down syndrome: in vitro modulation. Am J Hum Genet 69:88–95. https://doi.org/10.1086/321262

Prasher VP, Sajith SG, Rees SD, Patel A, Tewari S, Schupf N, Zigman WB (2008) Significant effect of APOE epsilon 4 genotype on the risk of dementia in Alzheimer's disease and mortality in persons with Down syndrome. Int J Geriatr Psychiatry 23:1134–1140. https://doi.org/10.1002/gps.2039

Qiu J-J, Liu Y-N, Ren Z-R, Yan J-B (2017) Dysfunctions of mitochondria in close association with strong perturbation of long noncoding RNAs expression in down syndrome. Int J Biochem Cell Biol 92:115–120. https://doi.org/10.1016/j.biocel.2017.09.017

Qureshi IA, Mattick JS, Mehler MF (2010) Long non-coding RNAs in nervous system function and disease. Brain Res 1338:20–35. https://doi.org/10.1016/j.brainres.2010.03.110

Raji NS, Rao KS (1998) Trisomy 21 and accelerated ageing: DNA-repair parameters in peripheral lymphocytes of Down's syndrome patients. Mech Ageing Dev 100:85–101

Reynolds GP, Cutts AJ (1993) Free radical damage in Down's syndrome brain. Biochem Soc Trans 21:221S

Roat E, Prada N, Lugli E, Nasi M, Ferraresi R, Troiano L, Giovenzana C, Pinti M, Biagioni O, Mariotti M, Di Iorio A, Consolo U, Balli F, Cossarizza A (2008) Homeostatic cytokines and expansion of regulatory T cells accompany thymic impairment in children with Down syndrome. Rejuvenation Res 11:573–583. https://doi.org/10.1089/rej.2007.0648

Rohn TT, McCarty KL, Love JE, Head E (2014) Is apolipoprotein E4 an important risk factor for dementia in persons with Down syndrome? J Park Dis Alzheimers Dis 1(1):pii: 7

Romano A, Cornia R, Moraschi M, Bozzao A, Chiacchiararelli L, Coppola V, Iani C, Stella G, Albertini G, Pierallini A (2016) Age-related cortical thickness reduction in non-demented Down's syndrome subjects. J Neuroimaging Off J Am Soc Neuroimaging 26:95–102. https://doi.org/10.1111/jon.12259

Saab BJ, Mansuy IM (2014) Neuroepigenetics of memory formation and impairment: the role of microRNAs. Neuropharmacology 80:61–69. https://doi.org/10.1016/j.neuropharm.2014.01.026

Sabbagh MN, Chen K, Rogers J, Fleisher AS, Liebsack C, Bandy D, Belden C, Protas H, Thiyyagura P, Liu X, Roontiva A, Luo J, Jacobson S, Malek-Ahmadi M, Powell J, Reiman EM (2015) Florbetapir PET, FDG PET, and MRI in Down syndrome individuals with and without Alzheimer's dementia. Alzheimers Dement J Alzheimers Assoc 11:994–1004. https://doi.org/10.1016/j.jalz.2015.01.006

Sailani MR, Santoni FA, Letourneau A, Borel C, Makrythanasis P, Hibaoui Y, Popadin K, Bonilla X, Guipponi M, Gehrig C, Vannier A, Carre-Pigeon F, Feki A, Nizetic D, Antonarakis SE (2015) DNA-methylation patterns in trisomy 21 using cells from monozygotic twins. PloS One 10:e0135555. https://doi.org/10.1371/journal.pone.0135555

Sanchez-Mut JV, Heyn H, Vidal E, Moran S, Sayols S, Delgado-Morales R, Schultz MD, Ansoleaga B, Garcia-Esparcia P, Pons-Espinal M, de Lagran MM, Dopazo J, Rabano A, Avila J, Dierssen M, Lott I, Ferrer I, Ecker JR, Esteller M (2016) Human DNA methylomes of neurodegenerative diseases show common epigenomic patterns. Transl Psychiatry 6:e718. https://doi.org/10.1038/tp.2015.214

Sanders JL, Newman AB (2013) Telomere length in epidemiology: a biomarker of ageing, age-related disease, both, or neither? Epidemiol Rev 35:112–131. https://doi.org/10.1093/epirev/mxs008

Schoch J, Rohrer TR, Kaestner M, Abdul-Khaliq H, Gortner L, Sester U, Sester M, Schmidt T (2017) Quantitative, phenotypical, and functional characterization of cellular immunity in children and adolescents with down syndrome. J Infect Dis 215:1619–1628. https://doi.org/10.1093/infdis/jix168

Schupf N, Lee A, Park N, Dang L-H, Pang D, Yale A, Oh DK-T, Krinsky-McHale SJ, Jenkins EC, Luchsinger JA, Zigman WB, Silverman W, Tycko B, Kisselev S, Clark L, Lee JH (2015) Candidate genes for Alzheimer's disease are associated with individual differences in plasma levels of beta amyloid peptides in adults with Down syndrome. Neurobiol Ageing 36:2907.e1-10. https://doi.org/10.1016/j.neurobiolageing.2015.06.020

Sekijima Y, Ikeda S, Tokuda T, Satoh S, Hidaka H, Hidaka E, Ishikawa M, Yanagisawa N (1998) Prevalence of dementia of Alzheimer type and apolipoprotein E phenotypes in aged patients with Down's syndrome. Eur Neurol 39:234–237

Sethupathy P, Borel C, Gagnebin M, Grant GR, Deutsch S, Elton TS, Hatzigeorgiou AG, Antonarakis SE (2007) Human microRNA-155 on chromosome 21 differentially interacts with its polymorphic target in the AGTR1 3' untranslated region: a mechanism for functional single-nucleotide polymorphisms related to phenotypes. Am J Hum Genet 81:405–413. https://doi.org/10.1086/519979

Shooshtari S, Martens PJ, Burchill CA, Dik N, Naghipur S (2011) Prevalence of depression and dementia among adults with developmental disabilities in manitoba, Canada. Int J Fam Med 2011:319574. https://doi.org/10.1155/2011/319574

Siew W-H, Tan K-L, Babaei MA, Cheah P-S, Ling K-H (2013) MicroRNAs and intellectual disability (ID) in Down syndrome, X-linked ID, and Fragile X syndrome. Front Cell Neurosci 7:41. https://doi.org/10.3389/fncel.2013.00041

Sifakis S, Papantoniou N, Kappou D, Antsaklis A (2012) Noninvasive prenatal diagnosis of Down syndrome: current knowledge and novel insights. J Perinat Med 40:319–327. https://doi.org/10.1515/jpm-2011-0282

Song C, He J, Chen J, Liu Y, Xiong F, Wang Y, Li T (2015) Effect of the one-carbon unit cycle on overall DNA methylation in children with Down's syndrome. Mol Med Rep 12:8209–8214. https://doi.org/10.3892/mmr.2015.4439

Stancliffe RJ, Lakin KC, Larson SA, Engler J, Taub S, Fortune J, Bershadsky J (2012) Demographic characteristics, health conditions, and residential service use in adults with Down syndrome in 25 U.S. states. Intellect Dev Disabil 50:92–108. https://doi.org/10.1352/1934-9556-50.2.92

Storm W (1990) Prevalence and diagnostic significance of gliadin antibodies in children with Down syndrome. Eur J Pediatr 149:833–834

Strauss D, Eyman RK (1996) Mortality of people with mental retardation in California with and without Down syndrome, 1986–1991. Am J Ment Retard AJMR 100:643–653

Strydom A, Livingston G, King M, Hassiotis A (2007) Prevalence of dementia in intellectual disability using different diagnostic criteria. Br J Psychiatry J Ment Sci 191:150–157. https://doi.org/10.1192/bjp.bp.106.028845

Sukenik-Halevy R, Biron-Shental T, Sharony R, Fejgin MD, Amiel A (2011) Telomeres in trisomy 21 amniocytes. Cytogenet Genome Res 135:12–18. https://doi.org/10.1159/000329714

Svobodová I, Korabečná M, Calda P, Břešťák M, Pazourková E, Pospíšilová Š, Krkavcová M, Novotná M, Hořínek A (2016) Differentially expressed miRNAs in trisomy 21 placentas. Prenat Diagn 36:775–784. https://doi.org/10.1002/pd.4861

Tan L, Yu J-T, Hu N, Tan L (2013) Non-coding RNAs in Alzheimer's disease. Mol Neurobiol 47:382–393. https://doi.org/10.1007/s12035-012-8359-5

Teipel SJ, Alexander GE, Schapiro MB, Möller H-J, Rapoport SI, Hampel H (2004) Age-related cortical grey matter reductions in non-demented Down's syndrome adults determined by MRI with voxel-based morphometry. Brain J Neurol 127:811–824. https://doi.org/10.1093/brain/awh101

Teller JK, Russo C, DeBusk LM, Angelini G, Zaccheo D, Dagna-Bricarelli F, Scartezzini P, Bertolini S, Mann DM, Tabaton M, Gambetti P (1996) Presence of soluble amyloid beta-peptide precedes amyloid plaque formation in Down's syndrome. Nat Med 2:93–95

Tiano L, Littarru GP, Principi F, Orlandi M, Santoro L, Carnevali P, Gabrielli O (2005) Assessment of DNA damage in Down Syndrome patients by means of a new, optimised single cell gel electrophoresis technique. Biofactors Oxf Engl 25:187–195

Trotta MB, Serro Azul JB, Wajngarten M, Fonseca SG, Goldberg AC, Kalil JE (2011) Inflammatory and immunological parameters in adults with Down syndrome. Immun Ageing A 8:4. https://doi.org/10.1186/1742-4933-8-4

Tyrrell J, Cosgrave M, McCarron M, McPherson J, Calvert J, Kelly A, McLaughlin M, Gill M, Lawlor BA (2001) Dementia in people with Down's syndrome. Int J Geriatr Psychiatry 16:1168–1174

Valenti D, Manente GA, Moro L, Marra E, Vacca RA (2011) Deficit of complex I activity in human skin fibroblasts with chromosome 21 trisomy and overproduction of reactive oxygen species by mitochondria: involvement of the cAMP/PKA signalling pathway. Biochem J 435:679–688. https://doi.org/10.1042/BJ20101908

Valentini D, Marcellini V, Bianchi S, Villani A, Facchini M, Donatelli I, Castrucci MR, Marasco E, Farroni C, Carsetti R (2015) Generation of switched memory B cells in response to vaccination in Down syndrome children and their siblings. Vaccine 33:6689–6696. https://doi.org/10.1016/j.vaccine.2015.10.083

Valentini E, Zampieri M, Malavolta M, Bacalini MG, Calabrese R, Guastafierro T, Reale A, Franceschi C, Hervonen A, Koller B, Bernhardt J, Slagboom PE, Toussaint O, Sikora E, Gonos ES, Breusing N, Grune T, Jansen E, Dollé MET, Moreno-Villanueva M, Sindlinger T, Bürkle A, Ciccarone F, Caiafa P (2016) Analysis of the machinery and intermediates of the 5hmC-mediated DNA demethylation pathway in ageing on samples from the MARK-AGE Study. Ageing 8:1896–1922. https://doi.org/10.18632/ageing.101022

Vaziri H, Schächter F, Uchida I, Wei L, Zhu X, Effros R, Cohen D, Harley CB (1993) Loss of telomeric DNA during ageing of normal and trisomy 21 human lymphocytes. Am J Hum Genet 52:661–667

Verstegen RHJ, Kusters MAA, Gemen EFA, DE Vries E (2010) Down syndrome B-lymphocyte subpopulations, intrinsic defect or decreased T-lymphocyte help. Pediatr Res 67:563–569. https://doi.org/10.1203/PDR.0b013e3181d4ecc1

Verstegen RHJ, Driessen GJ, Bartol SJW, van Noesel CJM, Boon L, van der Burg M, van Dongen JJM, de Vries E, van Zelm MC (2014) Defective B-cell memory in patients with Down syndrome. J Allergy Clin Immunol 134:1346–1353.e9. https://doi.org/10.1016/j.jaci.2014.07.015

Visser FE, Aldenkamp AP, van Huffelen AC, Kuilman M, Overweg J, van Wijk J (1997) Prospective study of the prevalence of Alzheimer-type dementia in institutionalized individuals with Down syndrome. Am J Ment Retard AJMR 101:400–412

Weidner CI, Lin Q, Koch CM, Eisele L, Beier F, Ziegler P, Bauerschlag DO, Jöckel K-H, Erbel R, Mühleisen TW, Zenke M, Brümmendorf TH, Wagner W (2014) Ageing of blood can be tracked by DNA methylation changes at just three CpG sites. Genome Biol 15:R24. https://doi.org/10.1186/gb-2014-15-2-r24

Wenger SL, Hansroth J, Shackelford AL (2014) Decreased telomere length in metaphase and interphase cells from newborns with trisomy 21. Gene 542:87. https://doi.org/10.1016/j.gene.2014.03.019

WHO|Dementia: a public health priority [WWW Document], 2012 WHO. http://www.who.int/mental_health/publications/dementia_report_2012/en/. Accessed 15 Nov 2017

Wilcock DM (2012) Neuroinflammation in the ageing down syndrome brain; lessons from Alzheimer's disease. Curr Gerontol Geriatr Res 2012:170276. https://doi.org/10.1155/2012/170276

Wilcock DM, Hurban J, Helman AM, Sudduth TL, McCarty KL, Beckett TL, Ferrell JC, Murphy MP, Abner EL, Schmitt FA, Head E (2015) Down syndrome individuals with Alzheimer's disease have a distinct neuroinflammatory phenotype compared to sporadic Alzheimer's disease. Neurobiol Ageing 36:2468–2474. https://doi.org/10.1016/j.neurobiolageing.2015.05.016

Xu Y, Li W, Liu X, Ma H, Tu Z, Dai Y (2013) Analysis of microRNA expression profile by small RNA sequencing in Down syndrome fetuses. Int J Mol Med 32:1115–1125. https://doi.org/10.3892/ijmm.2013.1499

Yang Q, Rasmussen SA, Friedman JM (2002) Mortality associated with Down's syndrome in the USA from 1983 to 1997: a population-based study. Lancet Lond Engl 359:1019–1025

Yin Y-Z, She Q, Zhang J, Zhang P-Z, Zhang Y, Lin J-W, Ye Y-C (2014) Placental methylation markers in normal and trisomy 21 tissues. Prenat Diagn 34:63–70. https://doi.org/10.1002/pd.4256

Yu X, Wang Y, Kristic J, Dong J, Chu X, Ge S, Wang H, Fang H, Gao Q, Liu D, Zhao Z, Peng H, Pucic Bakovic M, Wu L, Song M, Rudan I, Campbell H, Lauc G, Wang W (2016) Profiling IgG N-glycans as potential biomarker of chronological and biological ages: a community-based study in a Han Chinese population. Medicine (Baltimore) 95:e4112. https://doi.org/10.1097/MD.0000000000004112

Zana M, Szécsényi A, Czibula A, Bjelik A, Juhász A, Rimanóczy A, Szabó K, Vetró A, Szucs P, Várkonyi A, Pákáski M, Boda K, Raskó I, Janka Z, Kálmán J (2006) Age-dependent oxidative stress-induced DNA damage in Down's lymphocytes. Biochem Biophys Res Commun 345:726–733. https://doi.org/10.1016/j.bbrc.2006.04.167

Zhang Y, Che M, Yuan J, Yu Y, Cao C, Qin X-Y, Cheng Y (2017) Aberrations in circulating inflammatory cytokine levels in patients with Down syndrome: a meta-analysis. Oncotarget 8:84489–84496. https://doi.org/10.18632/oncotarget.21060

Zhao Q, Lee JH, Pang D, Temkin A, Park N, Janicki SC, Zigman WB, Silverman W, Tycko B, Schupf N (2011) Estrogen receptor-Beta variants are associated with increased risk of Alzheimer's disease in women with down syndrome. Dement Geriatr Cogn Disord 32:241–249. https://doi.org/10.1159/000334522

Zhao Y, Pogue AI, Lukiw WJ (2015) MicroRNA (miRNA) signaling in the human CNS in sporadic Alzheimer's disease (AD)-novel and unique pathological features. Int J Mol Sci 16:30105–30116. https://doi.org/10.3390/ijms161226223

Zigman WB (2013) Atypical ageing in Down syndrome. Dev Disabil Res Rev 18:51–67. https://doi.org/10.1002/ddrr.1128

Zigman WB, Lott IT (2007) Alzheimer's disease in Down syndrome: neurobiology and risk. Ment Retard Dev Disabil Res Rev 13:237–246. https://doi.org/10.1002/mrdd.20163

Chapter 8
The Vestibular System and Ageing

Sonja Brosel and Michael Strupp

Abstract The world's population is ageing due to increased hygiene and improved medical care. Dizziness and imbalance frequently affect the elderly and is most common among individuals over the age of 60. In this age group approximately 30% of the population experience these debilitating symptoms at some point. They contribute to falls and frailty, which often result in hospitalization causing tremendous cost for the health care systems, and increased mortality. To make the matters worse balance disorders are often complex. Physicians face the difficulty of diagnosing the patient with the exact disorder especially since each disorder may manifest differently in each patient. In addition, several treatment options exist, however, with a low level of evidence. This chapter summarizes the underlying degenerative processes of the peripheral as well as the central vestibular system, diagnostic tools, the most common balance disorders in the elderly, and possible treatment options of these disorders.

Keywords Vestibular system · Aging · Vestibular disorders · Treatment · Diagnostic tools · Degeneration

S. Brosel (✉)
Department Biology II, Division of Neurobiology, Ludwig-Maximilians-University Munich, Munich, Germany
e-mail: stb2105@caa.columbia.edu

M. Strupp
Department of Neurology and German Center for Vertigo and Balance Disorders, Ludwig Maximilians University, Munich, Munich, Germany
e-mail: michael.strupp@med.uni-muenchen.de

© Springer Nature Singapore Pte Ltd. 2019
J. R. Harris, V. I. Korolchuk (eds.), *Biochemistry and Cell Biology of Ageing: Part II Clinical Science*, Subcellular Biochemistry 91,
https://doi.org/10.1007/978-981-13-3681-2_8

Vestibular System

In general terms the vestibular system is both a sensory system and a motor system. In mammals, the vestibular system is the most important system to coordinate spatial orientation and balance. The vestibular system performs these essential tasks by sending signals to the brain, for example to control eye movement and to the muscles to allow the body to maintain an upright position. In addition, it engages a number of reflex pathways that are responsible for generating compensatory movements and adjustments in body position.

The vestibular system is an integral part of the labyrinth of the inner ear that lays in the optic capsule in the petrous portion of the temporal bone, which also contains a part of the auditory system, the cochlea. Movement in general consists of two components, a linear and a rotational component. Linear or translational movement is sensed by the otolith organs, whereas rotational movement is detected by the semicircular canals. There are three semicircular canals necessary to recognize all possible movements of the body in a three-dimensional environment. The semicircular canals are filled with endolymph and are arranged orthogonally meaning that each canal is at a right angle to the other two canals and are called the horizontal or lateral, the anterior or superior and the posterior or inferior semicircular canal. The superior and the posterior canal are at 45° angles to the sagittal plane and the horizontal canal is 30° to the axial plane. Furthermore, each canal is maximally sensitive to rotations in the plane in which it is situated and is paired with a canal on the contralateral side so that stimuli that are excitatory to one canal are inhibitory to the other canal.

Each of the three semicircular canals has at its base a spherical enlargement called the ampulla. This is where the sensory epithelium is located, the so-called crista. The crista contains hair cells, which have stereocilia oriented in the same direction and extend out of the crista into a bulbous, wedge-shaped, gelatinous mass called the cupula. The cupula extends from the surface of the cristae to the roof and lateral walls of the membranous labyrinth, forming a fluid-tight partition through which endolymph cannot circulate. As a result, the cupula is distorted by movements of the endolymphatic fluid. During movement in the plane of one of the semicircular canals, the inertia of the endolymph produces a force across the cupula, resulting in the displacement of the hair bundles within the crista. However, during linear accelerations of the head equal forces on the two sides of the cupula are generated, so the hair cells stay in place. The distortion of the hair cells is then transduced to electrical signals by opening potassium channels (described later).

Mammals have two otolith organs: the utricle, which – with the head in an upright position – detects motion in the horizontal plane and the saccule responsible for the sensation of motion in the sagittal plane and gravity. Both organs are located between the semicircular channels and the cochlea. The larger utricle is situated in a ca. 90° angle to the saccule in the spherical recess on the medial wall where it is connected to the endolymphatic duct via the utriculosaccular duct. All semicircular channels are connected via five openings. The saccule is linked to the cochlear duct via the ductus reuniens and to the endolymphatic duct via the utriculasaccular duct. Both otoliths contain hair cells and associated supporting cells termed the macula. A gelatinous layer covers the hair cells, which is topped by the otolithic membrane that contains calcium carbonate crystals called otoconia. The otoconia make the otolithic membrane heavier than the surrounding structures and fluids causing the membrane to shift relative to the macula during head tilting and displacement of the hair cells. Each hair cell of the macula has 40–70 stereocilia and one true cilium called a kinocilium. In contrast to the kinocilia in the semicircular channels, the kinocilia in the otoliths are not oriented in the same direction. Here the kinocilia are organized relative to the striola: utricular hair cells point towards and saccular hair cells point away from the striola resulting in opposing morphological polarization. Thus, when the head is tilted hair cells on one side of the striola receive excitatory inputs where at the same time hair cells on the other side of the striola are inhibited. These sensory inputs are then transduced to the brain.

The afferent activity is caused by an influx of potassium ions and is transduced via the VIIIth cranial nerve to the vestibular nuclei in the brainstem as well as the rostral medulla. The cell bodies of the afferents are located in the vestibular or Scarpa's ganglion, which is at the distal end of the internal auditory meatus, whereas the axons pass through the internal auditory meatus and enter the brain stem at the junction between pons and medulla. The two smaller vestibular nuclei, the lateral and the superior, are located in the pons, whereas the medial and the inferior nuclei are usually close to each other in the rostral medulla. The vestibular nuclei project to many locations including the cerebellum, spinal cord, extraocular nuclei, parietal cortex, vagal nucleus, nucleus solitarius, and reticular formation. For example, signals sent to the cerebellum are relayed back as muscle movements of the head, eyes, and posture, whereas stimulation of spinal cord results in quick reflexes in both limbs and trunk. The extraocular nuclei allow for fixation of the eyes on a moving object. On the one hand the lateral and medial vestibular nuclei send information via the descending lateral and medial vestibulospinal tracts respectively along the entire length of the spinal cord to allow for upright walking. On the other hand the medial vestibular tract extends from the medial vestibular nucleus bilaterally through the mid-thoracic level of the spinal cord into the medial longitudinal fasciculus enabling coordinated head movements and the integration of head and eye movements. Ascending information to the extraocular muscles is sent through the medial longitudinal fasciculus by the superior and the medial vestibular nuclei.

In general the connections formed by the vestibular nuclei to other regions of the brain relay information regarding the movement and the position of the body in space.

Experience from the Vestibular System

Experience from the vestibular system is called equilibrioception. It is mainly used for the sense of balance and for spatial orientation. When the vestibular system is stimulated without any other inputs, one experiences a sense of self-motion. For example, a person in complete darkness and sitting in a chair will feel that he or she has turned to the left if the chair is turned to the left. A person in an elevator, with essentially constant visual input, will feel she is descending as the elevator starts to descend. There are a variety of direct and indirect vestibular stimuli, which can make people feel they are moving when they are not, not moving when they are, tilted when they are not, or not tilted when they are. Although the vestibular system is a very fast sense used to generate reflexes, including the righting reflex, to maintain perceptual and postural stability, compared to the other senses of vision, touch and audition, vestibular input is perceived with delay.

Importance for Balance

In every-day life one of the most important abilities of the body is to maintain its position with respect to gravity both at rest and in movement. Static as well as dynamic balance is highly depended on the visual system, the somatosensory system and the vestibular system. The importance of the latter system is most evident in situation in which visual and somatosensory inputs are limited such as in the dark. The vestibular system is comprised of a complex neuronal network and a number of peripheral end organs. Only with an intact vestibular system one can accurately process movement and self-orient the body in the environment. As such, the vestibular system acts mostly autonomously coordinating effectively postural and ocular reflexes to ensure a continuous balance and maintain visual acuity during movement. However, the motor response is highly depended on accurate sensory input. As mentioned above, the neuronal relay center are the vestibular nuclei that integrate and process the different sensory inputs, while the peripheral vestibular organs are entirely responsible for sensing on the one hand the orientation of the head with respect to gravity and on the other hand the direction and the amount of acceleration of the body within space.

Vestibular Disorders

When the vestibular system fails or falsely transmits sensory inputs, one realizes the importance of this system for the maintenance of one's equilibrium with respect to gravity. The result is the loss of balance that often is associated with multiple symptoms such as dizziness, vertigo, disequilibrium, nausea, pallor, falls and overall functional decline. As symptoms get more severe patients feel isolated due to limitations in their physical abilities and sometimes also mental abilities. This social isolation can lead to further diseases and as a result increase costs for the health care system.

Today much has been discovered about the complex function and the anatomy of the vestibular system and how it contributes to balance and equilibrium, but there is still much to be discovered. The complexity of the system makes it difficult to evaluate the entire system at once, but at the same time the access to individual components of the system to assess their functionality is difficult as well. Additionally, subtle changes of vestibular function or vestibular disorders in older patients are very heterogenous and the site of the lesions is not always consistent. Therefore, physicians many times are unable to identify and differentiate between the individual disorders.

Prevalence in the Elderly

Non-specific dizziness is the most frequently reported health issue among the elderly and results in approximately 8 million physician consultations per year (Marchetti and Whitney 2005). This corresponds to 30–35% of adults over 40 years of age, which is equivalent to 69 million individuals in the US (Agrawal et al. 2009; Neuhauser et al. 2008). The prevalence increases with age as shown in numerous studies, however there are larger discrepancies in the findings. At age 60 30% and at the age 85 50% of the Swedish population are affected (Jönsson et al. 2004). In contrast, Tinetti conducted a population-based study in the United States and reported that only 24% of people older than 72 years have dizziness (Tinetti et al. 2000). Yet another study by Cutson et al. reported an increase in dizziness from 22% for adults between 65 and 69 years of age to over 40% for adults between the ages of 80 and 84 years (Cutson 1994). Three times as high was the estimate by Hobeika et al. with 65% of individuals older than 60 years of age experiencing dizziness or loss of balance (Hobeika 1999). In the United Kingdom a population-based study showed that 30% of people older than 65 years have dizziness (Colledge et al. 1994). Finally, the National Center for Health Statistics (NCHS) reports the prevalence of balance impairment in the United States to be 75.3% over the age of 70 years (Dillon et al. 2010).

Dizziness and imbalance in the elderly are a growing public health concern since dizziness increases the risk of falling significantly. This results in injuries that often cause hospitalization and lead to mobility restrictions, an increase in the fear of falling again and consequently a loss of independence (Ekvall Hansson and Magnusson 2013; Graafmans et al. 1996; O'Loughlin et al. 1994; Rubenstein and Josephson 2006; Stel et al. 2003; Xie et al. 2017). Furthermore, vertigo and unsteadiness lead to a fear of falling, which is a strong predictor for those who will suffer one or more subsequent falls (Anson and Jeka 2016; Delbaere et al. 2004; Li et al. 2003).

Dizziness is also associated with a 1.7 times higher mortality (Corrales and Bhattacharyya 2016).

The great variance in the numbers of reported individuals in the population that are affected by balance impairment can be attributed to the fact that dizziness is one of the symptoms of central nervous system disorders such as stroke, trauma and neurodegeneration as well as system disorders such as cardiovascular diseases, inflammation and osteoarthritis as well as sensory system disorders of the visual and of course the vestibular system. In addition the side effect of the treatment with a number of medications can lead to balance problems (Shoair et al. 2011). In addition to the complexity in etiology of dizziness there are many different diagnostic tools and criteria being used leading to discrepancies as well. Last, only some of the studies control for risk factors, which include hypertension, diabetes and smoking (Agrawal et al. 2009).

Affected Organs

About 50 years ago Lars-Göuran Johnsson microdissected the temporal bone from 150 patients ranging from newborn to 97 years old and studied the gross anatomy. He found signs of degeneration in the vestibular system histologically (Johnsson 1971).

Age-related degeneration has been verified histologically in all parts of the vestibular system: otoconia of the macula organs, the sensory neuroepithelium, Scarpa's ganglion neurons, the vestibular nerve, the vestibular nuclei and Purkinje cells in the cerebellum (Anniko 1983; Baloh et al. 1993; Brosel et al. 2016; Campos et al. 1990; Furman and Redfern 2001; Gluth and Nelson 2017; Jang et al. 2006; Lim 1984; Maes et al. 2010; Matheson et al. 1999; Rauch et al. 2001; Walther and Westhofen 2007). This said histological studies of the entire vestibular system face two major obstacles namely that inner ear structures are safely 'buried' in the hardest bone of the human body where the vestibular and cochlear structures lie in different anatomical planes and the difficulty of characterizing the peripheral and the central vestibular system in parallel. Therefore, most studies only consider a single vestibular structure or a small part of the vestibular system. In addition, the correlation between clinical and histological findings is poor (Rosenhall and Rubin 1975) most probably due to effective vestibular compensation mechanisms, which are able to mask the vestibular degeneration process. There are also methodological limitations. First, the decalcification process takes many days implicating the possibility of alterations in the tissue such as distortions and shrinkage. Second, optimal orientation for serial sectioning of one structure restrains analysis of the other, which is of particular relevance for unbiased stereology (Tang et al. 2002). Third, the lengthy preparation processes interfere quite often with the purification of high quality DNA, RNA or proteins. Lastly, the quantification of changes in the neuroepithelium has been limited by the difficulty of distinguishing hair cells from supporting cells and type I from type II hair cells.

Otoconial Degeneration

Otoconia are minute biocrystals composed of glycoproteins, proteoglycans, and calcium carbonate, and are indispensable for sensory processing in the utricle and saccule. Otoconia abnormalities and degeneration can cause or facilitate crystal dislocation to the ampulla, leading to vertigo and imbalance in humans.

Degeneration or defects in otoconia can lead to crystal dislocation to the ampulla in the semi circular canals, leading to vertigo and imbalance in humans. The maculae of the saccule and the utricle are covered by a dense gelatinous layer consisting of filaments linking the attached otoconia (Lins et al. 2000; Lundberg et al. 2006; Thalmann et al. 2001). Especially in the saccule an age-related reduction in the number of otoconia and linking filaments has been reported (Igarashi et al. 1993; Ross et al. 1976). Jang et al. reported otoconia degeneration in the saccule as early as 50 years of age that increases in severity with advancing age (Jang et al. 2006). Another group has also reported a tendency for an increased rate of age-related saccular degeneration (Johnsson 1971; Johnsson and Hawkins 1967, 1972). The reasons behind the 'preferred' loss of otoconia in the saccule is poorly understood. However, some groups have proposed possible reasons. One reason might be the vertical orientation of the saccule within the temporal bone that might make it more susceptible to otoconial loss. Thalman et al. suggested that the loss of otoconia might be secondary to the lack of dark cells in the saccule (Thalmann et al. 2001). Lim et al. proposed yet has another hypothesis, namely the renewal of otoconia by a number of different processes. When any one of these processes or a combination thereof fails during the aging process otoconia degeneration results (Lim 1984).

The most common vestibular disorder 'benign paroxysmal positional vertigo (BPPV)' is caused by the accumulation of dislodged filaments and degenerated otoconia in the endolymphtic duct of the semicircular canals (Brandt and Steddin 1993; Brandt et al. 1994; Jang et al. 2006). One possible cause for otoconial degeneration may be dysregulation of the ionic components in the microenvironment surrounding the otoconia, such as calcium metabolism (Jeong et al. 2009; Lundberg et al. 2006). This theory is supported by the fact that 75% of women who suffer from BPPV also show signs of osteopenia or osteoporosis (Vibert et al. 2003). The underlying causes as suggested by the authors might be otoconial degeneration or decreased estrogen levels leading to a reduced capacity to resolve otolithic debris. Tabtabai et al. reported increased blood levels of the otoconia matrix protein otolin-1 as a result of demineralization of otoconia with increasing age. In a group of 79 individuals ages 22–95 years otolin-1 levels were significantly increased in patients 65 or older. The authors further suggest that otolin-1 blood levels can be used as a biomarker of otoconia degeneration, especially in patients with BPPV (Tabtabai et al. 2017).

As indicated by these studies, calcium metabolism and the presence of linking filaments are factors that play a role in age-related degeneration of otoconia. While this does not translate into therapeutic measures, the established therapy of osteoporosis and the supplementation of trace elements might also be beneficial in preventing otoconia degeneration.

Sensory Neuroepithelium

Over the last 50 years, numerous histopathological studies of the human vestibular neuroepithelium showed age-related degeneration of vestibular hair cells (Engström et al. 1977; Lopez et al. 2005; Rauch et al. 2001; Richter 1980; Rosenhall 1973). The majority of these investigations found an onset of significant hair cell decline between 65 and 70 years. As early as 1975 Rosenhall and Rubin microdissected the vestibular epithelium to demonstrate that the number of hair cells remains relatively constant in the first 70 years of life and then starts to gradually decline with increasing age (Rosenhall and Rubin 1975). More recently, the vestibular system of 67 individuals without signs of vestibular dysfunction was analyzed by histopathology. The individuals ranged from newborns to 100 years of age (Merchant et al. 2000). The researchers found that at birth the number of hair cells was as follows: 76–79 cells per 0.01 mm^2 in the cristae, 68 cells per 0.01 mm^2 in the utricle, and 61 cells per 0.01 mm^2 in the saccule (Merchant et al. 2000; Rauch et al. 2001), where the ratio of type I to type II hair cells was 2.4:1 in the cristae and 1.3:1 in the maculae. Merchant et al. also reported that hair cell loss starts as early as 20 years of age. The same study found that the mean hair cell count in individuals 70 years or older decreased by 21% in the maculae utriculi, 24% in the maculae sacculi and 40% in the cristae ampullares. The hair cell densities were greater at the periphery than the central regions of cristae. All in all, there was no significant difference in hair cell loss between the three semicircular canals, the two maculae, men and women, or right versus left ears. However, this study applied Nomarski microscopy (differential interference contrast) on serial sections, which may introduce a bias due to several assumptions, such as the spherical shape and uniform size of the hair cells, as well as a constant shrinkage and thickness of the specimen during the decalcification, fixation and cutting processes (Iwasaki and Yamasoba 2014). Supporting the data from Merchant et al. (2000) greater susceptibility to age-related degeneration has also been shown by Anniko (1983) in the cristae of the semicircular canals when compared with the utricle or the saccule in individuals 70–95 years of age (40% hair cell loss in the cristae, 20–25% in the utricle and saccule). In contrast, when using the physical fractionator method of unbiased stereology no significant age-related hair cell destruction was found in the utricle in 10 patients aged 42–96 years (Gopen et al. 2003). In a later study, the same group counted the number of hair cells in the semicircular canals and showed that hair cells decreased by 12% in adults in their 80s rising to 25% in their 90s (Lopez et al. 2005). Data from Rosenhall (1973) also supports that there are differences in the rate of hair cell degeneration between the saccule and the utricle with 24% loss in the saccule and 21% in the utricle. There seems to be a bias of type I hair cell loss in the cristae compared to the maculae, whereas type II hair cells degenerated at the same rate. This is supported by findings of Anniko (1983) who proposed that the preferential degeneration of type I hair cells may be due to the fact that type I hair cells are younger phylogenetically speaking and are therefore more specialized and differentiated. Type I hair cells have a distinct morphology, particularly in the cuticular plate and the apical part of the cell. These cells harbor rod-like inclusion bodies that proliferate. This process according to the author might present a weakness that is not present in type II hair cells (Anniko 1983). A faster degeneration rate of type I hair cells versus type II hair cells was also found in a study of 67 temporal bones. This observation was independent of the sensory epithelium under investigation (Rauch et al. 2001). Summarizing the findings there seems to be a more rapid loss of type I versus type II hair cells, which tends to be greater in the central epithelial regions of the sensory end organs than the periphery with a concomitant higher rate of hair cell loss for the cristae than the maculae. Even though there is much agreement that the number of vestibular sensory hair cells decreases significantly during the 'normal' aging process, on reviewing the evidence, the overall degree of hair cell degeneration appears to be low and poses the question whether there is a functional consequence. However, when considering only clinically affected individuals as in the aminoglycoside ototoxicity study of Tsuji et al. a correlation of clinical function and the degree of hair cell degeneration is clearly observed (Tsuji et al. 2000). Patients with absent caloric response had > 80% hair cell loss in the crista of the lateral semicircular canal. One also has to consider that hair cells can appear morphologically normal by light microscopy and still malfunction at the physiological level. Underlying age-related changes include disarrangement and loss of cilia, aggregations of lipofuscin pigments, multivesiculated bodies, disintegration of the cuticular plate, and rod-shaped inclusions from the cuticle area into the hair cell (Anniko 1983; Richter 1980; Rosenhall and Rubin 1975). Furthermore, Baloh et al. reported mitochondrial changes in a patient suffering from imbalance and oscillopsia in addition to accumulation of lipofuscin granules (Baloh et al. 1997). These changes included an increase in the number of mitochondria, which is generally accepted to reflect compensation mechanisms to mitochondrial dysfunction in metabolically active tissues, abnormalities in mitochondrial shape as well as partial disintegration of mitochondrial cristae, pointing to compromised organelle function. As a result neuronal function may be compromised long before cell death mechanisms are initiated and before hair cells are replaced by fibrotic scar tissue.

Scarpa's Ganglion Neurons

Scarpa's ganglion neurons are bipolar neurons of the vestibular nerve with their cell bodies located in the bilateral vestibular nerve ganglions. The axons of the vestibular afferents travel in the vestibular portion of the VIIIth cranial nerve and enter into the brain stem. Most of the axons project to one of the vestibular nuclei; some travel through the inferior cerebellar peduncle to the cerebellum, the location of coordinated motion to maintain balance. The ganglion can be divided in its superior and inferior part. The superior region on the one hand consists of the lateral and anterior ampullary nerves, as well as the utricular nerve. On the other the inferior part contains nerve fibers from the posterior canal and the saccule. These vestibular nerve fibers are myelinated. Bergström et al. reported in individuals 75–85 years of age a decrease in the number of myelinated vestibular nerve fibers in the periphery of Scarpa's ganglion (SG) neurons of about 37% (Bergström 1973a). Additionally, several studies have reported an age-dependent reduction in SG neuron count and density (Park et al. 2001; Richter 1980; Velázquez-Villaseñor et al. 2000). These studies disagree to some extent in the age of onset, the degree of observed degeneration of SG neurons as well as the correlation between age and rate of degeneration. While Velázquez-Villaseñor et al. found a linear decline of SG neurons from birth to senium (Velázquez-Villaseñor et al. 2000), data reported by Richter indicates a rather abrupt decline at about age 60 (Richter 1980). Yet another study proposed a non-linear pattern of SG neuron decline with the number of SG neurons being roughly constant at a young age and gradually declining at mid age. This group counted on average 28,952 cells in individuals 30 years of age or younger and 23,349 cells in older individuals (Park et al. 2001). Contradicting findings exist with respect to differences of SG neuron decline in the superior versus the inferior part of the ganglions. Richter found no difference in age-related decline between both parts of the ganglion (Richter 1980). In contrast, Velázquez-Villaseñor et al. reported that the superior part of the ganglion is significantly more affected than the inferior part (Velázquez-Villaseñor et al. 2000). Furthermore, this group found a gender bias towards women (Velázquez-Villaseñor et al. 2000). It is unclear whether SG neurons undergo cell death or if the degeneration of SG neurons is secondary to the loss of hair cells and therefore deafferentiation. One study by Richter supports the deafferentiation process. He observed hair cell degeneration prior to SG neuron decline (Richter 1980). Proximal to SG, the vestibular nerve continues to join the cochlear nerve. In 24 individuals 54–90 years of age, Moriyama et al. counted myelinated axons and measured the transverse area from the proximal vestibular nerve and found no significant age-related changes of total axon number, but the average transverse area of myelinated axons was decreased (Moriyama et al. 2007).

Vestibular Nuclei

Most afferent fibers from hair cells terminate in the vestibular nuclei in the brainstem. The nuclei are located in the pons and the medulla on the floor of the IVth ventricle confined medially by the pontine reticular formation, laterally by the restiform body, rostrally by the brachium conjunctivum, and ventrally by the nucleus and spinal tract of the trigeminal nerve. Most nuclei receive afferent and efferent fibers from the cerebellum, reticular formation, spinal cord, and contralateral vestibular nuclei, however, some nuclei receive only primary vestibular afferents. The vestibular nucleus complex (VNC) is composed of four major nuclei: the superior (SuVe), the medial (MVe), the lateral (LVe), and the spinal (SpVe) vestibular nuclei (Tascioglu 2005) and other small accessory neuronal groups. The nuclei are defined mainly by cytoarchitectonic features with the size and number of neurons varying within the different nuclei. In addition, the nuclei are subdivided by neurochemical differences implicating functional subdivisions within these nuclei. In addition, the vestibular system connects to sub-regions of the individual nuclei (Diaz et al. 1993, 1996; Suarez et al. 1989, 1997, 1993). Moreover, structures beyond the VNC receive primary afferent information from the vestibular nerve, for example the posterior cerebellum or the parasolitary nucleus (Barmack 2003). Age-related loss of neurons in the human VNC has been reported. The rate of neuronal loss was estimated to be approximately 5% per decade in the MVe in subjects between the age of 40 and 93 (Tang et al. 2002). Alverez et al. reported a near 40% loss of neuronal cells within the vestibular nuclei by 89 years of age that mainly impacted MVe (Alvarez et al. 1998). At age 50 Bergström found a significant reduction in vestibular nerve fibers that gradually increased to 40% at age 75–80 (Bergström 1973a, b). Lopez et al. analyzed the VNC of 15 normal human subjects 40–93 years old using computer-based microscopy to determine the number of neurons, nuclear volume, neuronal density, and nuclear length in the four major VN. The group observed a 3% decline in neuronal number per decade starting at age 40. The age-dependent reduction in nuclear volume and neuronal density was also significant, although to a lesser degree than the decline in number of neurons. The greatest loss of neurons was observed in the SVe, whereas the least amount of neuronal degeneration was seen in MVe (Lopez et al. 1997). Finally, age-related accumulation of lipofuscin in the cytoplasm of neurons was evident, especially in the LVe (Alvarez et al. 2000; Lopez et al. 1997).

The degeneration of vestibular and cochlear neurons is likely a sign of the physiological aging process. This is in contrast to other associated brainstem nuclei, such as the abducens or trochlear nuclei, which show no age-related neuronal cell loss (Vijayashankar and Brody 1977a, b). Therefore, neuronal loss in the vestibular system is presumably not the only cause of dizziness in the elderly, but rather a contributing factor to a decreasing potential for compensation. The degree of individual compensation depends on the overall physical status, the functional status of the remaining sensory systems, integrity of central brain mechanisms, and higher sensory functions such as memory, motor coordination, and cognitive ability—all functions that decline naturally with age (Giardino et al. 2002; Wiesmeier et al. 2015).

Types of Disorders

Underlying this phenomenon is the progressive multimodal impairment of balance, including the loss of vestibular and proprioceptive functions, and the impairment of central integration of these and other sensory inputs associated with aging, which may also be called presbystasis, presbyequilibrium, or multisensory dizziness (Ekvall Hansson and Magnusson 2013; Tuunainen et al. 2011, 2012).

Dizziness, vertigo and imbalance rank among the most common symptoms individuals 75 years or older are presented with and are a growing health concern since they increase the risk of falling in this age group. Although the causes of dizziness in the elderly are multifactorial, senescence of the vestibular periphery and the central vestibular system play a critical role (Iwasaki and Yamasoba 2014; Katsarkas 1994; Lawson et al. 1999). The most commonly diagnosed vestibular disorders include benign paroxysmal positional vertigo (BPPV), vestibular neuritis, Menière's disease, and labyrinthine infarction. Less common vestibular disorders include superior semicircular canal dehiscence, vestibular schwannoma, perilymph fistula/ superior canal dehiscence syndrome, ototoxicity, enlarged vestibular aqueduct, vestibular migraine, functional dizziness and mal de débarquement. Only disorders with an increasing prevalence with increasing age will be discussed in more detail.

Benign Paroxysmal Positional Vertigo (BPPV)

BPPV was first described by a colleague of Robert Bárány in 1921 and is today the most common peripheral vestibular disorder in the elderly (Baloh et al. 1987; von Brevern et al. 2007; Neuhauser et al. 2001). Approximately 2.4% of the population will at some point in their life be affected by BPPV, among individuals who live into their 80s the rate is as high as 10% with the onset typically being between 50 and 70 years. There is a bias towards females as on average twice as many women suffer from BPPV. Patients typically suffer from episodic vertigo provoked by changes in head position, concomitant nystagmus observed during the positioning maneuver as well as nausea and vomiting. In a study of elderly patients who visited an emergency room due to chronic dizziness 50% suffered from extremely weak, horizontal, direction changing apogeotropic nystagmus. This form of nystagmus is characteristic of horizontal canal BPPV (Johkura et al. 2008). The extremely high percentage of patients in this study that showed signs of BPPV implies that the actual prevalence of BPPV in the elderly population is much higher. The dizziness patients experience last anywhere from several seconds to minutes and are usually triggered by head movement. These head movements can be as simple as lifting one's head upward, sudden head movements or turning over in bed (for current diagnostic criteria see von Brevern (von Brevern et al. 2015). The underlying mechanism that causes BPPV is believed to be the aging otolitic membrane. Otoconia dislodge from the utricle and migrate into one of the semicircular canals with the PVe being the most commonly affected because of its anatomical position. The displaced otoconia result in abnormal endolymph movement in the affected semicircular canal leading to the sensation of dizziness (Parnes and McClure 1992). This condition is often also referred to as canalithiasis (Brandt and Steddin 1993). However, the infrequent possibility exists that the otoconial debris attaches to the cupula of the canal making the cupula heavier. This results in immediate and continuous excitation. This rare form of BPPV is also denoted cupulolithiasis. As mentioned above evidence exists that a high proportion of otoconia of the utricular macula degenerates in the elderly or shows signs of fractionation (Igarashi et al. 1993; Johnsson and Hawkins 1972).

Menière's Disease

Menière's disease is named after its discoverer Prosper Menière who documented the disorder in 1861 (Ekvall Hansson and Magnusson 2013; Harcourt et al. 2014; Tuunainen et al. 2011). The prevalence in the general population is between 0.3 and

1.9 in 1000 people. Among patients that suffer from dizziness 3% to 11% are diagnosed with Ménière's disease (Wladislavosky-Waserman et al. 1984). The common age of onset is between 40 and 60 years of age with females being more often affected than males (Lopez-Escamez et al. 2015). There is evidence that the prevalence of Ménière's disease is age-dependent. Ballester et al. reported that 15% of patients with Ménière's disease are older than 65. Forty percent of patients suffered from long-lasting Ménière's disease and 60% of patients received their first diagnosis (Ballester et al. 2002). Typical symptoms include vertigo, tinnitus, hearing loss and the feeling of fullness of the ear(s). In the beginning usually only one ear is affected; however over time both ears may be affected. Symptoms last from 20 min to a few hours with time between incidents being variable. Some patients will experience constant hearing loss and tinnitus after some time. However, vertigo often stops after 5–15 years and patients will experience a mild form of balance loss, reduced hearing ability in the affected ear and tinnitus.

Ménière's disease results from the built-up of endolymph in the semicircular canals. What triggers Meniere's disease is unknown. However, theories do exist. Some researchers believe that tightening of blood vessels, viral infections, allergies or autoimmune disorders play a role. Others suspect genetic alterations to be responsible since in 10% of Ménière's cases a familial history can be traced (Lopez-Escamez et al. 2015).

Vestibular Neuritis/Acute Unilateral Peripheral Vestibulopathy

The leading symptoms are acute onset of spinning vertigo, nausea, oscillopsia and a tendency to fall towards the affected side. The prevalence of vestibular neuritis is estimated at 3.5 in 1,000,000 and is first diagnosed most often in individuals between 40 and 60 years of age (Greco et al. 2014). Unfortunately, not many population-based studies exist. In an outpatient study 3–10% of patients with vertigo were diagnosed with vestibular neuritis (Guilemany et al. 2004). In Japan an epidemiological survey was conducted by otolaryngologists over a period of 3 years. The researchers found that no bias towards either sex exists, most affected individuals were between 40 and 50 years of age and in 60% of cases symptoms ceased within 3 months (Sekitani et al. 1993). In 60–70% of cases symptoms cease after 3–6 weeks due to vestibular compensation; in all other patients chronic dizziness persists (Godemann et al. 2005; Murofushi et al. 2006). It is believed that vestibular compensation mechanisms slowly deteriorate with age as a result of the functional decline of the peripheral vestibular organs, vision and proprioception. The etiology of vestibular neuritis is largely unknown. A wide spectrum of preceding conditions exists such as a viral or bacterial infection, head injury, extreme stress, allergies or a reaction towards a medication (Katsarkas 1994). 30% of patients report that they suffered from a common cold or the flu prior to being diagnosed with vestibular neuritis; others have a herpes simplex type 1 virus infection of the vestibular ganglion. However, this is most likely not the only virus that can infect the vestibular nerve and therefore be a cause of vestibular neuritis (Greco et al. 2014). Furthermore, acute localized ischemia of the vestibular nerve might be another cause. This can occur when the person experiences pressure changes during flying or scuba diving (Kennedy 1974; Martin-Saint-Laurent et al. 1990). But there are still a large number of patients that have not experienced any of these conditions prior to their diagnosis and further research is deemed necessary to clarify the cause in these cases.

Labyrinthine Infarction

The blood supply to the inner ear originates in the vertebrobasilar system and blood solely flows through the internal auditory artery (IAA). The IAA is a branch of the anterior inferior cerebellar artery (AICA), which in turn is a branch of the basilar artery (Lee et al. 2004). The IAA irrigates the cochlea and vestibular labyrinth, and occlusion of the IAA causes loss of auditory and vestibular function. This system is extremely vulnerable to ischemia since the IAA is an end artery with minimal collaterals from the otic capsule. IAA infarction mostly occurs due to thrombotic narrowing of the AICA itself, or in the basilar artery at the orifice of the AICA. Although a variety of conditions can cause labyrinthine infarction, the condition is most commonly associated with thromboembolic disease of the AICA or the basilar artery. Other causes for labyrinthine ischemia include thromboemboli of the posterior circulation, fat emboli, thromboangiitis obliterans, migrainous infarction, decompression illness, hyperlipidemia, macroglobulinemia, sickle cell disease, leukemia, polycythemia vera, and other causes of hypercoagulation or hyperviscosity. Patients often not only report vestibular dysfunction but also auditory dysfunction. Labyrinthine infarction manifests in the peripheral, central or both parts of the vestibular system with acute prolonged vertigo, imbalance, nausea, vomiting and tinnitus. In some cases vertigo is caused by small infarcts of the brainstem or cerebellum (Amarenco and Hauw 1990). These patients show no signs of other localizing neurologic symptoms. However, these infarcts only occur in roughly 1 in 10 patients. Most often they suffer also from nystagmus, and postural unsteadiness mimicking acute peripheral vestibular disorders (Kim and Lee 2010). Combining the data from different studies, the data suggests that 8–31% of patients suffer from acute audiovestibular disturbance with vertigo and hearing loss 1–10 days before symptoms or signs from a larger territory infarction occur (Kim and Lee 2017; Kim et al. 2014, 2009; Lee et al. 2009; Yi et al. 2005). This might present an intervention point in the future as there are no beneficial therapies currently known. Unfortunately there are no studies that correlate the prevalence of labyrinthine infarction with age. However, the risk of stroke increases in the elderly and patients that reported a history of stroke or other known vascular risk factors are at a 2–5% greater risk of developing labyrinthine infarction (Lee et al. 2011).

Tests to Identify Disorders and Their Shortcomings

The diagnosis of vestibular disorders in the elderly remains an important challenge even for most experienced physicians. There is not a single symptom that can predict with certainty the disorder that causes dizziness, and most of the time, elderly patients have more than one cause of dizziness (Lawson et al. 1999; Newman-Toker et al. 2007; Walther 2017). One of the most common testing methods is caloric testing. The responses from these tests depend on several factors that are potentially affected by age, such as ear canal volume, temporal bone thickness, and blood supply to the temporal bone (Enrietto et al. 1999). Several studies have found that caloric responses tend to increase in middle age with a peak between 50 and 70 years, and then decline modestly thereafter (Bruner and Norris 1971). Therefore, a systematic assessment with the numerous available tests is necessary.

Vestibular Ocular Reflex (VOR)

The impairment of each of the three semicircular canals can be examined by measuring the vestibular ocular reflex (VOR). The VOR allows us to stabilize an image in the center of the retina during head movement by producing eye movement in the opposite direction with respect to the head movement. This reflex to head movement is independent of the visual system and is also present in the dark. Many patients with an impaired VOR find it challenging to read. Electronystagmography records involuntary eye movements that result from an impaired vestibular system. This test consists of oculomotor evaluation, positional testing and caloric stimulation of the vestibular system. However, video nystagmography is slowly replacing electronystagmography because it is a more accurate way of tracing eye movements. Rotational test are most commonly used to assess age related deterioration of the VOR in patients (Paige 1994; Peterka et al. 1990). These sinusoidal harmonic acceleration tests detect bilateral vestibular lesions (Kaplan et al. 2001). During these tests the VOR gain and the VOR time constant are measured. Both parameters have been shown to decrease during aging. Baloh et al. exposed a cohort of normal adults (75 years and older) to sinusoidal rotation tests and found that both the VOR gain and the VOR time constant were decreased; this was most prominent during high velocity stimulation (Baloh et al. 1993). In a long term study, in which normal subjects (> 75 years) were tested 5 times per year, the VOR gain continuously got worse while the phase lead increased (Baloh et al. 2001). Yet another study found a decline in the response amplitude and less of a compensatory response phase with increasing age (normal subjects between 7 and 81 years old) (Peterka et al. 1990). The VOR is often also tested with the caloric reflex test, which allows for the detection of unilateral vestibular dysfunction. During this test cold or warm water is irrigated into the external auditory canal. The dissimilarity between the body temperature and the injected water temperature results in a convective current in the endolymph of the adjacent horizontal semicircular canal whereby hot and cold water produce currents in opposite directions causing a horizontal nystagmus in opposite directions. In healthy individuals warm water causes the endolymph to rise in the ipsilateral horizontal canal which in turn increases the firing rate of afferent vestibular neurons mimicking a head turn to the ipsilateral side. Both eyes will turn toward the contralateral ear, with horizontal nystagmus to the ipsilateral ear. However, if the irrigated water is cold the opposite occurs mimicking a head turn to the contralateral side. Horizontal nystagmus is observed towards the contralateral ear. In patients who have an impaired VOR nystagmus this is absent when the affected ear is being stimulated. If both phases are absent, this suggests the patient's brainstem reflexes are also damaged.

Yet another test to assess VOR is the head impulse test (HIT), or head thrust test. The examiner briskly rotates the patient's head while having the patient fixate on a target, usually the examiner's nose. During slow head movements the ocular smooth pursuit system normally maintains fixation. Rapid head movements exceed the capability of smooth pursuit and the VOR then must maintain fixation. When the VOR is impaired, ocular tracking fails when the head is turned towards the side of the vestibular lesion requiring a corrective saccade to be back at the target at the end of the movement. More recently, the accuracy of this test has been improved by recording the eye movements with high speed, low weight video goggles (vHIT) (McGarvie et al. 2015). This test allows the assessment of VOR at high frequency of each semicircular canal individually by calculating the duration ratio between the head impulse and gaze deviation. The vHIT test can diagnose vestibular weakness by measuring a reduction in gain and the appearance of overt and covert saccades. vHIT is more sensitive than HIT especially in patient with isolated covert saccades (Halmagyi et al. 2017).

Vestibular Evoked Myogenic Potentials (VEMPs)

Age related deterioration of the peripheral system can also be assessed by measuring ocular and cervical vestibular evoked myogenic potentials (VEMPs). VEMPs are short-latency muscle responses either recorded from the neck muscles (cervical VEMPs/cVEMPs) or the eye muscles (ocular VEMPs/oVEMPs) allowing for the assessment of utricular and saccular function independently. cVEMPs reflect the function of the saccule and the inferior vestibular nerve whereas oVEMPs reflects the function of the utricle and the superior vestibular nerve (Brantberg 2009; Colebatch and Halmagyi 1992; Colebatch et al. 2016; Iwasaki and Yamasoba 2014; Macambira et al. 2017). Furthermore, the non-vestibular proprioceptive and visual sensory components of balance and their central integration in overall equilibrium performance can be thoroughly assessed by dynamic computed posturography

(Soto-Varela et al. 2015). In patients ranging from 7 to 91 years of age an age-dependent decrease in cVEMP amplitude and a simultaneous increase in cVEMP latency was detected (Brantberg et al. 2007). oVEMPs show a similar age-dependent decrease in amplitude while latency increases (Iwasaki et al. 2008). These studies show a clear age dependent deterioration of both, saccule and utricle, as well as their central pathway.

In a study by Agrawal et al. both VOR and VEMP measurements were performed in the same healthy subjects who were more than 70 years old (Agrawal et al. 2012). The authors observed a simultaneous decline of the semicircular canal function and otolith organ function with age, although the magnitude of impairment was greater for the semicircular canals than the otolith organs. However, one has to keep in mind that the head thrust test that was used in this study might reflect oculomotor function as well as the semicircular canal function (Agrawal et al. 2012). In a study of 1521 patients, 227 (15%) were found to have abnormal oVEMPs and/or cVEMP responses with normal caloric responses. Most of these patients experienced multiple episodes of vertigo lasting from seconds to hours and were diagnosed with one of the following three disorders BBPV, Ménière's disease, or vestibular migraine. Eighty-one patients could not be diagnosed with a recognizable disease. These finding indicate that BPPV, Ménière's disease, and vestibular migraine are the most frequent diagnoses showing abnormal oVEMP and/or cVEMPs without canal paresis (Iwasaki and Yamasoba 2014).

Posturography

Age-related changes in postural stability have been examined using posturography, in which changes of the center of pressure are measured during quiet standing (Baloh et al. 1995; Fujimoto et al. 2009; Wolfson et al. 1992). These non-vestibular proprioceptive and visual sensory components of balance and their central integration in overall equilibrium performance can be thoroughly assessed by dynamic computed posturography. For this test a moving platform or a foam rubber surface have been developed. Foam posturography uses the technique of sway referencing to force reliance on the vestibular apparatus for the maintenance of upright stance (Alahmari et al. 2014; Weber and Cass 1993). The individual stands on a dense foam pad about 10 cm thick with arms folded across the chest and is observed for sway, first with eyes open and then with eyes closed. The compliance of the foam pad degrades somatosensory input and, with eyes closed, only vestibular signals are available for orientation. Excessive sway suggests a disorder of the vestibulospinal system. Foam posturography has high sensitivity (95%), specificity (90%) and shows almost identical results ($p < 0.005$) to laboratory-based platform posturography. An age-related decline in balance control has been observed during posturography testing. Teasdale et al. have demonstrated that alteration in any two of the three sensory inputs (visual, vestibular and somatosensory) had a significantly greater effect on older subjects than in younger subjects whereas alteration in one input did not have a significant effect due to age (Teasdale et al. 1991).

Romberg's Test

The Romberg test has been used since 1871 to diagnose a sensory deficit arising from the vestibular and/or somatosensory system. During the test the subject is asked to stand erect with feet together and eyes closed. In this situation with the visual input cancelled, postural control relies on vestibular input and proprioceptive input alone. The movement of the patient's body in relation to a perpendicular object is observed for 1 min. If swaying, irregular swaying or even toppling over occurs the test is positive (Lanska and Goetz 2000). A positive Romberg test suggests that the ataxia is sensory in nature. If Romberg's test is negative, it suggests that ataxia is depended on localized cerebellar dysfunction instead. Although a patient with an acute peripheral vestibular lesion is usually inclined to move towards the side of the problem, it has been shown that chronic vestibular damage when partially compensated does not produce deficits in the standard Romberg test (Black et al. 1982). More sensitive is the Romberg test on rubber foam since it mainly probes the vestibulospinal reflexes, as the foam minimizes the proprioceptive input from the feet (Shumway-Cook and Horak 1986). As a result, the test is very sensitive to identify patients with unilateral or bilateral vestibular loss (Petersen et al. 2013). Caution has to be implied because even in the absence of a vestibular deficit, the Romberg test on foam can be positive when patients suffer from midline cerebellar disorders.

Dix Hallpike Maneuver

The Dix-Hallpike maneuver is the gold standard for diagnosis of benign positional paroxysmal vertigo. The patient begins sitting up, and the head is oriented 45° toward the ear to be tested. The physician then lies the patient down quickly with their head past the end of the bed and extends the patient's neck 20° below the horizontal, maintaining the initial rotation of the head. The physician then watches the patient's eyes for torsional and up-beating nystagmus, which should start after a brief delay and persist for no more than 1 min. If nystagmus is observed BPPV is the cause of vertigo and this test can be readily transitioned into the Epley maneuver, in which the position of the otolith continues to be manipulated until it is out of the posterior canal, ending the sensation of vertigo with positional changes and curing the disease process. However, there is a high rate of recurrence (Bhattacharyya et al. 2017; Dix and Hallpike 1952).

Methods of Intervention

The spectrum of vestibular disorders is quite extensive and so are the possible treatment options which include physical therapy, pharmacology, psychotherapy or, rarely, surgery. The choice of intervention most often depends on the diagnosis and the severity of symptoms. Most peripheral vestibular lesions have a benign etiology and undergo spontaneous resolution due to compensatory mechanisms. Vestibular compensation results from active neuronal changes in the cerebellum and brainstem in response to sensory conflicts produced by vestibular pathology. However, if compensation does not occur spontaneously medical intervention is necessary. The following highlights the most common possibilities (Strupp and Brandt 2009).

Vestibular Rehabilitation Therapy

Vestibular rehabilitation therapy (VRT) is an exercise-based treatment program that is based on central mechanisms of neuroplasticity, known as adaptation, habituation and substitution, which promote vestibular compensation (Han et al. 2011; Ricci et al. 2010; Silva et al. 2016). Cawthorne and Cooksey developed the first exercises for VRT in 1946 for patients with labyrinth injury resulting from surgery or head injury (Cawthorne 1946; Cooksey 1946). They found that exercises designed to encourage head and eye movements hastened the patient's recovery. It soon became apparent that these exercises are useful to treat all forms of peripheral vestibular disorders and they are still used today in clinical practice.

The goals of VRT are to enhance gaze and postural stability to lead to improved vertigo and therefore, an improvement of quality of life. The key exercises for VRT are head-eye movements with various body postures and activities, and maintaining balance with a reduced support base with various orientations of the head and trunk, while performing various upper-extremity tasks, repeating the movements provoking vertigo, and exposing patients gradually to various sensory and motor environments (Balaban et al. 2012; Wiesmeier et al. 2017). These exercises can be beneficial for patients who suffer from vestibular dysfunction that is poorly compensated. Success of this therapy is neither influenced by age nor by the duration and intensity of symptoms. The key to success is the performance of the exercises several times a day. Even brief periods of exercising are sufficient to facilitate vestibular recovery.

To treat BPPV, in which otoconia are displaced from the utricle and move into the semicircular canals, canalith repositioning procedures (CRP) are implemented. Head movements shift the detached canaliths and stimulate sensitive nerve hairs to send false signals to the brain, causing dizziness and other symptoms. The goal of the CRP is to move the displaced otoconia to stop false stimulation of the hair cells within the canals that causes BPPV symptoms. This is achieved through a series of head position changes that cause the otoconia to move back to the utricle where they either readhere to the otolitic membrane or dissolve. There are three primary maneuvers: the Epley maneuver, the Semont-Liberatory maneuver and the Brandt-Daroff exercises. The choice of maneuver depends on results of the Dix-Hallpike test, which reveals the semicircular canal involved and whether or not the otoconia are inside the canal also termed canalithiasis or hung up on the cupula of the canal also referred to as cupulolithiasis. Brandt and Daroff introduced the first exercises for patients with cupulolithiasis in the 1980s. These exercises are referred as the

Brandt-Daroff exercises, which are performed multiple times over a period of 2 days after the bouts of positional vertigo (Brandt et al. 1994). In addition, Semont et al. proposed that the patient should be tilted 180° to the side opposite to the symptom inducing side (Semont et al. 1988). Epley yet employed another form of maneuver where the patient is moved into a head-hanging position (Epley 1994). Overall CRP is very effective, with an approximate cure rate of 95% and a very low recurrence rate (Strupp and Brandt 2009).

Psychotherapy

There are some studies that provide preliminary evidence that psychotherapy may be effective in patients suffering from dizziness. In the cognitive–behavioural therapeutic approach patients learn about dizziness and the vestibular system, discuss their experience with dizziness and their reactions, learn to self expose them to situations that trigger dizziness combined with relaxation techniques and exercises. The patients then complete questionnaires to assess their psychological state. The purpose of this therapeutic approach is to facilitate exposure to exercises and to increase awareness of strategies to cope with dizziness-associated fears. Johannson et al. treated 9 patients age 65–81 years old with vestibular rehabilitation in combination with a cognitive behavioral therapy for 7 weeks and compared the outcome to 10 patients who only did vestibular exercises. They found that 89% of patients who received the combined therapy improved significantly (Johansson et al. 2001). These findings are supported by another study in which 15 patients showed significant improvement after treatment with the combination therapy (Andersson et al. 2006). Holmberg et al. reported that 1 year after the treatment with cognitive behavioral therapy patients had no significant remaining positive treatment effects and therefore these authors questioned the effect of the therapeutic approach (Holmberg et al. 2007). In contrast, in a much larger study of 1159 patients, 72.1% still showed significant improvement in a follow-up assessment 2 years after the treatment (Obermann et al. 2015). Overall, the studies indicate beneficial outcomes to the cognitive behavioral treatment that should be kept in mind as a supportive therapy of vestibular rehabilitation.

Medication

Acute vestibular neuritis is most commonly treated with a vestibular suppressant, followed by vestibular rehabilitation exercises. Three vestibular suppressant classes exist including anticholinergics, antihistamines, and benzodiazepines. To just a list a few: meclizine, transdermal scopolamine, promethazine, amitriptyline, dimenhydinate, lorazepam, diazepam, and cis-platinum. In addition steroids and/or antiviral drugs are used when a viral infection is suspected during acute vestibular neuritis.

141 patients with acute vestibular neuritis were included in a double-blinded trial and either treated with placebo or methylprednisolone or valacyclovir or a combination of the two medications. Vestibular function was determined 3 days and 12 months after symptom onset. The greatest improvement in symptoms was experienced by patients who only received methylprednisolone (62.4%) closely followed by patients treated with the combination therapy (59.2%). The effect of valacyclovir alone was similar to the placebo effect (36% and 39.6% respectively). These results indicate that methylprednisolone has a significant effect in patients with vestibular neuritis (Strupp et al. 2004).

Meniere's disease patients often respond to a combination of a low-salt diet and diuretics (*e.g.* hydrochlorothiazide, triamterene), but also to intratympanic injections of gentamicin and corticosteroids. In Europe betahistine is often used to treat Meniere's disease. When UK otolaryngologists were asked about their choice of treatment for patients who suffer from Meniere's disease, 94% used betahistine, 63% diuretics, 71% salt restriction, 52% sac decompression, and approximately 50% inserted a grommet (Smith et al. 2005).

De Beer et al. treated 57 patients for Ménière's disease with intratympanic gentamicin. Patients received between one and ten intratympanic injections of gentamicin. In 49.1% of patients one injection was sufficient, the remaining 50.9% received multiple injections with a minimum time of 27 days in between. Complete or at least substantial decline in vertigo was experienced by 80.7% of patients 6 months after treatment. Although gentamicin is toxic for hair cells only in 15.8% did hearing worsen (De Beer et al. 2007). These results are supported by another study of 57 patients in which 53% only received one gentamicin injection and 32% two or three. At the follow up two to 4 years later 95% reported that their vertigo attacks were controlled (Lange et al. 2004). These studies support a beneficial effect of gentamicin in the treatment of Ménière's disease, which is due to the damage of hair cells. However, a complete ablation of function is unnecessary to control vertigo (Carey et al. 2002a, b).

In a small double blinded study with a two-year follow up of 22 patients dexamethasone resolved symptoms in 82% of patients over 52% with placebo (Garduño-Anaya et al. 2005). In another study the treatment with gentamicin was compared to dexamethasone. Of the 60 patients included in the study 32 were treated with one or two gentamicin injections and 28 patients were treated with three dexamethasone injections. Two years after the treatment, 81% of the gentamicin group reported complete and 12.5% substantial control of vertigo; whereas, in the dexamethasone group only 43% achieved complete and 18% substantial control of vertigo. The authors concluded that gentamicin injections are better in controlling vertigo attacks in Ménière's patients than dexamethasone (Casani et al. 2012).

Betahistine improves the microcirculation by acting on the precapillary sphincters of the stria vascularis (Dziadziola et al. 1999). In addition, a reduction of the production and an increase in the absorption of endolymph has been observed under the treatment with betahistine. Strupp et al. found that a high dosage of betahistine was significantly more effective than a low dosage treatment in 112 Ménière's patients when comparing the number of vertigo attacks 12 months after the treatment (Strupp et al. 2008). In a large scale, long term, multicenter, double blind, randomized, placebo controlled, dose defining trial (BEMED trial) 221 patients were either treated with low or high dosage betahistine or placebo for 9 months (equal number of patients in each group). The authors found no significant decline of attacks in either group over the nine-month treatment period. They were unable to confirm the different effect of low versus high dosage betahistine (Adrion et al. 2016). However, when Albu et al. compared the treatment of 62 Ménière's patients with dexamethasone injections to the treatment with dexamethasone in combination with betahistine, they were able to conclude that the combination of dexamethasone injections with betahistine is significantly more successful in the treatment of vertigo attacks of Ménière's disease. All patients received three dexamethasone injections and half also received a high dosage betahistine. 24 months after treatment 44% of patients who only received dexamethasone reported complete vertigo control, whereas 73.3% of the group with the combination therapy had achieved complete vertigo control (Albu et al. 2016).

Vestibular migraine headaches generally improve with dietary changes, a tricyclic antidepressant, and a betablocker (metoprolol or propranol) or calcium channel blocker (valproic acid). In a study of 81 patients of whom 31 received a tricyclic antidepressants in combination with a diet 77% reported a significant relief (Reploeg and Goebel 2002). Another open trial on 10 patients demonstrated that lamotrigine had a significant effect on the occurrence of headache and a more marked effect on vertigo (Bisdorff 2004). However, so far, only the standard treatment of migraine with aura can be recommended for vestibular migraine.

Dietary Adjustments

For patients suffering from Ménière's disease a restriction of salt and fluid intake and diuretics was proposed to be beneficial as early as 1934. This diet reduces the volume of fluid retained in the body and therefore lowers the chance of fluid build up in the inner ear, which is thought to be responsible for Ménière's disease. In a more recent study of 136 Ménière's patients who followed a low sodium and caffeine-free diet for at least 6 months a significant number showed improvement of their symptoms (Luxford et al. 2013). Contradicting this study are the findings of van-Deelen and Huizing that indicate no effect of salt restriction and diuretics in their double-blind study (van Deelen and Huizing 1986). Büki et al. observed in patients with BPPV low serum levels of vitamin D and found that supplementation with vitamin D abolished vertigo attacks in these patients. The authors recommend that patients with BPPV should have their vitamin D levels tested and in case of insufficient vitamin D they suggest vitamin D supplementation (Büki et al. 2013).

Surgery

Surgery is very rarely the method of choice to treat vertigo. Surgical treatments can be broken down into corrective and destructive types. For Ménière's patients shunt surgery is a corrective option. During the most common type of shunt surgery a small tube or plastic sheet is inserted into the endolymphatic sac (Belal and House 1979). Another procedure is termed endolymphatic sac enhancement, in which the lateral sinus is decompressed (Sajjadi et al. 1998). However, there is controversy about the long-term effectiveness of the procedure and unfortunately shunt surgery is as unfavorable as transtympanic gentamicin injections (see destructive procedures). Nevertheless, it is possible that procedures that either destroy the sac or remove surrounding bone through which lymphocytes migrate into the sac, might alter the immune function enough to cause a remission of Ménière's disease. The purpose of destructive treatments is to eliminate vertigo, but possibly sacrifice hearing at the same time. These procedures are warranted when all other therapies have failed to elevate symptoms. For Ménière's disease, destructive procedures are associated with better control of vertigo than shunt surgery, showing good control in over 90% of patients followed for five or more years. During these procedures the vestibular nerve is sectioned via the middle fossa, retrolabyrinthine, and retrosigmoid approaches, with similar efficacy. In addition, transtympanic gentamicin injections are available, which show a similar effectiveness to vestibular nerve sectioning but have a much lower risk. Another method is labyrinthectomy. This is preferably used in patients where hearing loss causes vertigo Labyrinthectomy can be performed in combination with vestibular nerve resection, which may provide better outcomes than labyrinthectomy alone (De La Cruz et al. 2007).

New surgical techniques currently in development include laser occlusion of the posterior semicircular canal for benign paroxysmal positional vertigo (Lin et al. 2010). In this study one patient with BPPV had no symptoms for 5 years after the surgery. Nomura used argon laser irradiation of the posterior and lateral semicircular canals in one patient, and of only the posterior canal in the other. Both patients have no signs of vertigo for over 6 years (Nomura 2002). Another method is selective microsurgical vestibular neurectomy (SMVN). SMVN has proven effective for intractable vertigo such as observed in Ménière's disease in a small study of nine patients, with the added benefit of hearing preservation (Bademci et al. 2004). Yet another procedure that preserves hearing is the placement of ventilation tubes in the affected ear of patients with Ménière's disease. Seven patients were treated and six experienced short-term relief over 24 months. Long-term the symptoms reoccurred (Sugawara et al. 2003). Singular neurectomy (Pournaras et al. 2008) and semicircular canal occlusion (Beyea et al. 2012; Shaia et al. 2006) are possible ways to treat BPPV. Both techniques are reported to provide similar symptomatic benefit, with low risk of hearing loss and balance impairment. However, singular neurectomy is the more challenging technique.

Conclusion on Current Treatment

The number of disorders that are associated with vertigo in the elderly is large and so are the underlying mechanisms. Many treatment regimens have been proposed over the years but few sufficiently powered placebo-controlled multicenter trails have been done, so the level of evidence is low. Therefore, physicians unfortunately still often rely on their experience rather than large data collections when deciding on the treatment of choice. Many more trails are necessary on the treatment of the various vestibular disorders to provide this data in the future.

References

Adrion C, Fischer CS, Wagner J, Gürkov R, Mansmann U, Strupp M (2016) Efficacy and safety of betahistine treatment in patients with Meniere's disease: primary results of a long term, multicentre, double blind, randomised, placebo controlled, dose defining trial (BEMED trial). BMJ *352*:h6816

Agrawal Y, Carey JP, Della Santina CC, Schubert MC, Minor LB (2009) Disorders of balance and vestibular function in us adults: data from the national health and nutrition examination survey, 2001-2004. Arch Intern Med 169:938–944

Agrawal Y, Zuniga MG, Davalos-Bichara M, Schubert MC, Walston JD, Hughes J, Carey JP (2012) Decline in semicircular canal and otolith function with age. Otol Neurotol: Off Publ Am Otol Soc Am Neurotol Soc Eur Acad Otol Neurotol 33:832–839

Alahmari KA, Marchetti GF, Sparto PJ, Furman JM, Whitney SL (2014) Estimating postural control with the balance rehabilitation unit: measurement consistency, accuracy, validity, and comparison with dynamic posturography. Arch Phys Med Rehabil 95:65–73

Albu S, Nagy A, Doros C, Marceanu L, Cozma S, Musat G, Trabalzini F (2016) Treatment of Meniere's disease with intratympanic dexamethazone plus high dosage of betahistine. Am J Otolaryngol 37:225–230

Alvarez JC, Díaz C, Suárez C, Fernández JA, González Del Rey C, Navarro A, Tolivia J (1998) Neuronal loss in human medial vestibular nucleus. Anat Rec 251:431–438

Alvarez JC, Díaz C, Suárez C, Fernández JA, González Del Rey C, Navarro A, Tolivia J (2000) Aging and the human vestibular nuclei: morphometric analysis. Mech Ageing Dev 114:149–172

Amarenco P, Hauw JJ (1990) Cerebellar infarction in the territory of the superior cerebellar artery: a clinicopathologic study of 33 cases. Neurology 40:1383–1390

Andersson G, Asmundson GJG, Denev J, Nilsson J, Larsen HC (2006) A controlled trial of cognitive-behavior therapy combined with vestibular rehabilitation in the treatment of dizziness. Behav Res Ther 44:1265–1273

Anniko M (1983) The aging vestibular hair cell. Am J Otolaryngol 4:151–160

Anson E, Jeka J (2016) Perspectives on aging vestibular function. Front Neurol 6:269

Bademci G, Batay F, Yorulmaz I, Küçük B, Cağlar S (2004) Selective microsurgical vestibular neurectomy: an option in the treatment of intractable vertigo and related microsurgical landmarks. Minim Invasive Neurosurg MIN 47:54–57

Balaban CD, Hoffer ME, Gottshall KR (2012) Top-down approach to vestibular compensation: translational lessons from vestibular rehabilitation. Brain Res 1482C:101–111

Ballester M, Liard P, Vibert D, Häusler R (2002) Menière's disease in the elderly. Otol Neurotol: Off Publ Am Otol Soc Am Neurotol Soc Eur Acad Otol Neurotol 23:73–78

Baloh RW, Honrubia V, Jacobson K (1987) Benign positional vertigo: clinical and oculographic features in 240 cases. Neurology 37:371–378

Baloh RW, Jacobson KM, Socotch TM (1993) The effect of aging on visual-vestibuloocular responses. Exp Brain Res 95:509–516

Baloh RW, Spain S, Socotch TM, Jacobson KM, Bell T (1995) Posturography and balance problems in older people. J Am Geriatr Soc 43:638–644

Baloh RW, Lopez I, Beykirch K, Ishiyama A, Honrubia V (1997) Clinical-pathologic correlation in a patient with selective loss of hair cells in the vestibular endorgans. Neurology 49:1377–1382

Baloh RW, Enrietto J, Jacobson KM, Lin A (2001) Age-related changes in vestibular function: a longitudinal study. Ann N Y Acad Sci 942:210–219

Barmack NH (2003) Central vestibular system: vestibular nuclei and posterior cerebellum. Brain Res Bull 60:511–541

Belal A, House WF (1979) Histopathology of endolymphatic subarachnoid shunt surgery for Meniere's disease. Am J Otolaryngol 1:37–44

Bergström B (1973a) Morphology of the vestibular nerve. II. The number of myelinated vestibular nerve fibers in man at various ages. Acta Otolaryngol (Stockh) 76:173–179

Bergström B (1973b) Morphology of the vestibular nerve. I. Anatomical studies of the vestibular nerve in man. Acta Otolaryngol (Stockh) 76:162–172

Beyea JA, Agrawal SK, Parnes LS (2012) Transmastoid semicircular canal occlusion: a safe and highly effective treatment for benign paroxysmal positional vertigo and superior canal dehiscence. Laryngoscope 122:1862–1866

Bhattacharyya N, Gubbels SP, Schwartz SR, Edlow JA, El-Kashlan H, Fife T, Holmberg JM, Mahoney K, Hollingsworth DB, Roberts R et al (2017) Clinical practice guideline: benign paroxysmal positional Vertigo (update). Otolaryngol--Head Neck Surg Off J Am Acad Otolaryngol-Head Neck Surg 156:S1–S47

Bisdorff AR (2004) Treatment of migraine related vertigo with lamotrigine an observational study. Bull Soc Sci Med Grand Duche Luxemb 24:103–108

Black FO, Wall C, Rockette HE, Kitch R (1982) Normal subject postural sway during the Romberg test. Am J Otolaryngol 3:309–318

Brandt T, Steddin S (1993) Current view of the mechanism of benign paroxysmal positioning vertigo: cupulolithiasis or canalolithiasis? J Vestib Res Equilib Orientat 3:373–382

Brandt T, Steddin S, Daroff RB (1994) Therapy for benign paroxysmal positioning vertigo, revisited. Neurology 44:796–800

Brantberg K (2009) Vestibular evoked myogenic potentials (VEMPs): usefulness in clinical neurotology. Semin Neurol 29:541–547

Brantberg K, Granath K, Schart N (2007) Age-related changes in vestibular evoked myogenic potentials. Audiol Neurootol 12:247–253

Brosel S, Laub C, Averdam A, Bender A, Elstner M (2016) Molecular aging of the mammalian vestibular system. Ageing Res Rev 26:72–80

Bruner A, Norris TW (1971) Age-related changes in caloric nystagmus. Acta Otolaryngol Suppl 282:1–24

Büki B, Ecker M, Jünger H, Lundberg YW (2013) Vitamin D deficiency and benign paroxysmal positioning vertigo. Med Hypotheses 80:201–204

Campos A, Cañizares FJ, Sánchez-Quevedo MC, Romero PJ (1990) Otoconial degeneration in the aged utricle and saccule. Adv Otorhinolaryngol 45:143–153

Carey JP, Minor LB, Peng GCY, Della Santina CC, Cremer PD, Haslwanter T (2002a) Changes in the three-dimensional angular vestibulo-ocular reflex following intratympanic gentamicin for Ménière's disease. J Assoc Res Otolaryngol JARO 3:430–443

Carey JP, Hirvonen T, Peng GCY, Della Santina CC, Cremer PD, Haslwanter T, Minor LB (2002b) Changes in the angular vestibulo-ocular reflex after a single dose of intratympanic gentamicin for Ménière's disease. Ann N Y Acad Sci 956:581–584

Casani AP, Piaggi P, Cerchiai N, Seccia V, Sellari Franceschini S, Dallan I (2012) Intratympanic treatment of intractable unilateral Ménière disease: gentamicin or dexamethasone? A randomized controlled trial. Otolaryngol Neck Surg 146:430–437

Cawthorne T (1946) Vestibular injuries. Proc R Soc Lond B Biol Sci 39:270–273

Colebatch JG, Halmagyi GM (1992) Vestibular evoked potentials in human neck muscles before and after unilateral vestibular deafferentation. Neurology 42:1635–1636

Colebatch JG, Rosengren SM, Welgampola MS (2016) Vestibular-evoked myogenic potentials. Handb Clin Neurol 137:133–155

Colledge NR, Wilson JA, Macintyre CC, MacLennan WJ (1994) The prevalence and characteristics of dizziness in an elderly community. Age Ageing 23:117–120

Cooksey FS (1946) Rehabilitation in vestibular injuries. Proc R Soc Lond B Biol Sci 39:273–278

Corrales CE, Bhattacharyya N (2016) Dizziness and death: an imbalance in mortality. Laryngoscope 126:2134–2136

Cutson TM (1994) Falls in the elderly. Am Fam Physician 49:149–156

De Beer L, Stokroos R, Kingma H (2007) Intratympanic gentamicin therapy for intractable Ménière's disease. Acta Otolaryngol (Stockh) 127:605–612

De La Cruz A, Borne Teufert K, Berliner KI (2007) Transmastoid labyrinthectomy versus translabyrinthine vestibular nerve section: does cutting the vestibular nerve make a difference in outcome? Otol Neurotol Off Publ Am Otol Soc Am Neurotol Soc Eur Acad Otol Neurotol 28:801–808

Delbaere K, Crombez G, Vanderstraeten G, Willems T, Cambier D (2004) Fear-related avoidance of activities, falls and physical frailty. A prospective community-based cohort study. Age Ageing 33:368–373

Diaz C, Suarez C, Navarro A, Gonzalez del Rey C, Tolivia J (1993) Rostrocaudal changes in neuronal cell size in human lateral vestibular nucleus. Neurosci Lett 157:4–6

Díaz C, Suárez C, Navarro A, González Del Rey C, Alvarez JC, Méndez E, Tolivia J (1996) Rostrocaudal and ventrodorsal change in neuronal cell size in human medial vestibular nucleus. Anat Rec 246:403–409

Dillon CF, Gu Q, Hoffman HJ, Ko C-W (2010) Vision, hearing, balance, and sensory impairment in Americans aged 70 years and over: United States, 1999–2006: (665372010–001) (American Psychological Association)

Dix MR, Hallpike CS (1952) The pathology, symptomatology and diagnosis of certain common disorders of the vestibular system. Ann Otol Rhinol Laryngol 61:987–1016

Dziadziola JK, Laurikainen EL, Rachel JD, Quirk WS (1999) Betahistine increases vestibular blood flow. Otolaryngol--Head Neck Surg Off J Am Acad Otolaryngol-Head Neck Surg 120:400–405

Ekvall Hansson E, Magnusson M (2013) Vestibular asymmetry predicts falls among elderly patients with multi- sensory dizziness. BMC Geriatr 13:77

Engström H, Ades HW, Engström B, Gilchrist D, Bourne G (1977) Structural changes in the vestibular epithelia in elderly monkeys and humans. Adv Otorhinolaryngol 22:93–110

Enrietto JA, Jacobson KM, Baloh RW (1999) Aging effects on auditory and vestibular responses: a longitudinal study. Am J Otolaryngol 20:371–378

Epley JM (1994) Canalith repositioning maneuver. Otolaryngol--Head Neck Surg Off J Am Acad Otolaryngol-Head Neck Surg 111:688–690

Fujimoto C, Murofushi T, Chihara Y, Ushio M, Sugasawa K, Yamaguchi T, Yamasoba T, Iwasaki S (2009) Assessment of diagnostic accuracy of foam posturography for peripheral vestibular disorders: analysis of parameters related to visual and somatosensory dependence. Clin Neurophysiol Off J Int Fed Clin Neurophysiol 120:1408–1414

Furman JM, Redfern MS (2001) Effect of aging on the otolith-ocular reflex. J Vestib Res 11:91–103

Garduño-Anaya MA, De Toledo HC, Hinojosa-González R, Pane-Pianese C, Ríos-Castañeda LC (2005) Dexamethasone inner ear perfusion by Intratympanic injection in unilateral Ménière's disease: a two-year prospective, placebo-controlled, double-blind, randomized trial. Otolaryngol Neck Surg 133:285–294

Giardino L, Zanni M, Fernandez M, Battaglia A, Pignataro O, Calzà L (2002) Plasticity of GABA(a) system during ageing: focus on vestibular compensation and possible pharmacological intervention. Brain Res 929:76–86

Gluth MB, Nelson EG (2017) Age-related change in vestibular ganglion cell populations in individuals with Presbycusis and Normal hearing. Otol Neurotol 38:540–546

Godemann F, Siefert K, Hantschke-Brüggemann M, Neu P, Seidl R, Ströhle A (2005) What accounts for vertigo one year after neuritis vestibularis – anxiety or a dysfunctional vestibular organ? J Psychiatr Res 39:529–534

Gopen Q, Lopez I, Ishiyama G, Baloh RW, Ishiyama A (2003) Unbiased Stereologic type I and type II hair cell counts in human utricular macula. Laryngoscope 113:1132–1138

Graafmans WC, Ooms ME, Hofstee HM, Bezemer PD, Bouter LM, Lips P (1996) Falls in the elderly: a prospective study of risk factors and risk profiles. Am J Epidemiol 143:1129–1136

Greco A, Macri GF, Gallo A, Fusconi M, De Virgilio A, Pagliuca G, Marinelli C, de Vincentiis M (2014) Is vestibular neuritis an immune related vestibular neuropathy inducing vertigo? J Immunol Res 2014:459048

Guilemany JM, Martínez P, Prades E, Sañudo I, De España R, Cuchi A (2004) Clinical and epidemiological study of vertigo at an outpatient clinic. Acta Otolaryngol (Stockh) 124:49–52

Halmagyi GM, Chen L, MacDougall HG, Weber KP, McGarvie LA, Curthoys IS (2017) The video head impulse test. Front Neurol 8:258

Han BI, Song HS, Kim JS (2011) Vestibular rehabilitation therapy: review of indications, mechanisms, and key exercises. J Clin Neurol Seoul Korea 7:184–196

Harcourt J, Barraclough K, Bronstein AM (2014) Meniere's disease. BMJ 349:g6544

Hobeika CP (1999) Equilibrium and balance in the elderly. Ear Nose Throat J 78:558–562. 565–566

Holmberg J, Karlberg M, Harlacher U, Magnusson M (2007) One-year follow-up of cognitive behavioral therapy for phobic postural vertigo. J Neurol 254:1189–1192

Igarashi M, Saito R, Mizukoshi K, Alford BR (1993) Otoconia in young and elderly persons: a temporal bone study. Acta Otolaryngol Suppl 504:26–29

Iwasaki S, Yamasoba T (2014) Dizziness and imbalance in the elderly: age-related decline in the vestibular system. Aging Dis 6:38–47

Iwasaki S, Smulders YE, Burgess AM, McGarvie LA, MacDougall HG, Halmagyi GM, Curthoys IS (2008) Ocular vestibular evoked myogenic potentials to bone conducted vibration of the midline forehead at Fz in healthy subjects. Clin Neurophysiol Off J Int Fed Clin Neurophysiol 119:2135–2147

Jang YS, Hwang CH, Shin JY, Bae WY, Kim LS (2006) Age-related changes on the morphology of the otoconia. Laryngoscope 116:996–1001

Jeong SH, Choi SH, Kim JY, Koo JW, Kim HJ, Kim JS (2009) Osteopenia and osteoporosis in idiopathic benign positional vertigo. Neurology 72:1069–1076

Johansson M, Akerlund D, Larsen HC, Andersson G (2001) Randomized controlled trial of vestibular rehabilitation combined with cognitive-behavioral therapy for dizziness in older people. Otolaryngol Neck Surg 125:151–156

Johkura K, Momoo T, Kuroiwa Y (2008) Positional nystagmus in patients with chronic dizziness. J Neurol Neurosurg Psychiatry 79:1324–1326

Johnsson L-G (1971) Degenerative changes and anomalies of the vestibular system in man. Laryngoscope 81:1682–1694

Johnsson LG, Hawkins JE (1967) Otolithic membranes of the saccule and utricle in man. Science 157:1454–1456

Johnsson LG, Hawkins JE (1972) Vascular changes in the human inner ear associated with aging. Ann Otol Rhinol Laryngol 81:364–376

Jönsson R, Sixt E, Landahl S, Rosenhall U (2004) Prevalence of dizziness and vertigo in an urban elderly population. J Vestib Res 14:47–52

Kaplan DM, Marais J, Ogawa T, Kraus M, Rutka JA, Bance ML (2001) Does high-frequency pseudo-random rotational chair testing increase the diagnostic yield of the eng caloric test in detecting bilateral vestibular loss in the dizzy patient? Laryngoscope 111:959–963

Katsarkas A (1994) Dizziness in aging: a retrospective study of 1194 cases. Otolaryngol--Head Neck Surg Off J Am Acad Otolaryngol-Head Neck Surg 110:296–301

Kennedy RS (1974) General history of vestibular disorders in diving. Undersea Biomed Res 1:73–81

Kim H-A, Lee H (2010) Isolated vestibular nucleus infarction mimicking acute peripheral vestibulopathy. Stroke 41:1558–1560

Kim H-A, Lee H (2017) Recent advances in understanding audiovestibular loss of a vascular cause. J Stroke 19:61–66

Kim JS, Cho K-H, Lee H (2009) Isolated labyrinthine infarction as a harbinger of anterior inferior cerebellar artery territory infarction with normal diffusion-weighted brain MRI. J Neurol Sci 278:82–84

Kim H-J, Lee S-H, Park JH, Choi J-Y, Kim J-S (2014) Isolated vestibular nuclear infarction: report of two cases and review of the literature. J Neurol 261:121–129

Lange G, Maurer J, Mann W (2004) Long-term results after interval therapy with Intratympanic gentamicin for Menière's disease. Laryngoscope 114:102–105

Lanska DJ, Goetz CG (2000) Romberg's sign: development, adoption, and adaptation in the 19th century. Neurology 55:1201–1206

Lawson J, Fitzgerald J, Birchall J, Aldren CP, Kenny RA (1999) Diagnosis of geriatric patients with severe dizziness. J Am Geriatr Soc 47:12–17

Lee H, Ahn B-H, Baloh RW (2004) Sudden deafness with vertigo as a sole manifestation of anterior inferior cerebellar artery infarction. J Neurol Sci 222:105–107

Lee H, Kim JS, Chung E-J, Yi H-A, Chung I-S, Lee S-R, Shin J-Y (2009) Infarction in the territory of anterior inferior cerebellar artery: spectrum of audiovestibular loss. Stroke 40:3745–3751

Lee C-C, Su Y-C, Ho H-C, Hung S-K, Lee M-S, Chou P, Huang Y-S (2011) Risk of stroke in patients hospitalized for isolated vertigo: a four-year follow-up study. Stroke 42:48–52

Li F, Fisher KJ, Harmer P, McAuley E, Wilson NL (2003) Fear of falling in elderly persons: association with falls, functional ability, and quality of life. J Gerontol B Psychol Sci Soc Sci 58:P283–P290

Lim DJ (1984) Otoconia in health and disease. A review. Ann Otol Rhinol Laryngol Suppl 112:17–24

Lin S-Z, Fan J-P, Sun A-H, Guan J, Liu H-B, Zhu Q-B (2010) Efficacy of laser occlusion of posterior semicircular canal for benign paroxysmal positional vertigo: case report. J Laryngol Otol 124:e5

Lins U, Farina M, Kurc M, Riordan G, Thalmann R, Thalmann I, Kachar B (2000) The Otoconia of the Guinea pig utricle: internal structure, surface exposure, and interactions with the filament matrix. J Struct Biol 131:67–78

Lopez I, Honrubia V, Baloh RW (1997) Aging and the human vestibular nucleus. J Vestib Res Equilib Orientat 7:77–85

Lopez I, Ishiyama G, Tang Y, Tokita J, Baloh RW, Ishiyama A (2005) Regional estimates of hair cells and supporting cells in the human crista ampullaris. J Neurosci Res 82:421–431

Lopez-Escamez JA, Carey J, Chung W-H, Goebel JA, Magnusson M, Mandalà M, Newman-Toker DE, Strupp M, Suzuki M, Trabalzini F et al (2015) Diagnostic criteria for Menière's disease. J Vestib Res Equilib Orientat 25:1–7

Lundberg YW, Zhao X, Yamoah EN (2006) Assembly of the otoconia complex to the macular sensory epithelium of the vestibule. Brain Res 1091:47–57

Luxford E, Berliner KI, Lee J, Luxford WM (2013) Dietary modification as adjunct treatment in Menière's disease: patient willingness and ability to comply. Otol Neurotol 34:1438–1443

Macambira YK d S, Carnaúba ATL, Fernandes LCBC, Bueno NB, Menezes P d L (2017) Aging and wave-component latency delays in oVEMP and cVEMP: a systematic review with meta-analysis. Braz J Otorhinolaryngol 83:475–487

Maes L, Dhooge I, D'Haenens W, Bockstael A, Keppler H, Philips B, Swinnen F, Vinck BM (2010) The effect of age on the sinusoidal harmonic acceleration test, pseudorandom rotation test, velocity step test, caloric test, and vestibular-evoked myogenic potential test. Ear Hear 31:84–94

Marchetti GF, Whitney SL (2005) Older adults and balance dysfunction. Neurol Clin 23:785–805. vii

Martin-Saint-Laurent A, Lavernhe J, Casano G, Simkoff A (1990) Clinical aspects of inflight incapacitations in commercial aviation. Aviat Space Environ Med 61:256–260

Matheson AJ, Darlington CL, Smith PF (1999) Further evidence for age-related deficits in human postural function. J Vestib Res Equilib Orientat 9:261–264

McGarvie LA, MacDougall HG, Halmagyi GM, Burgess AM, Weber KP, Curthoys IS (2015) The video head impulse test (vHIT) of semicircular canal function – age-dependent normative values of VOR gain in healthy subjects. Front Neurol 6:154

Merchant SN, Velázquez-Villaseñor L, Tsuji K, Glynn RJ, Wall C, Rauch SD (2000) Temporal bone studies of the human peripheral vestibular system. Normative vestibular hair cell data. Ann Otol Rhinol Laryngol Suppl 181:3–13

Moriyama H, Itoh M, Shimada K, Otsuka N (2007) Morphometric analysis of fibers of the human vestibular nerve: sex differences. Eur Arch Oto-Rhino-Laryngol Off J Eur Fed Oto-Rhino-Laryngol Soc EUFOS Affil Ger Soc Oto-Rhino-Laryngol – Head Neck Surg 264:471–475

Murofushi T, Iwasaki S, Ushio M (2006) Recovery of vestibular evoked myogenic potentials after a vertigo attack due to vestibular neuritis. Acta Otolaryngol (Stockh) 126:364–367

Neuhauser H, Leopold M, von Brevern M, Arnold G, Lempert T (2001) The interrelations of migraine, vertigo, and migrainous vertigo. Neurology 56:436–441

Neuhauser HK, Radtke A, von Brevern M, Lezius F, Feldmann M, Lempert T (2008) Burden of dizziness and Vertigo in the community. Arch Intern Med 168:2118–2124

Newman-Toker DE, Cannon LM, Stofferahn ME, Rothman RE, Hsieh Y-H, Zee DS (2007) Imprecision in patient reports of dizziness symptom quality: a cross-sectional study conducted in an acute care setting. Mayo Clin Proc 82:1329–1340

Nomura Y (2002) Argon laser irradiation of the semicircular canal in two patients with benign paroxysmal positional vertigo. J Laryngol Otol 116:723–725

O'Loughlin JL, Boivin JF, Robitaille Y, Suissa S (1994) Falls among the elderly: distinguishing indoor and outdoor risk factors in Canada. J Epidemiol Community Health 48:488–489

Obermann M, Bock E, Sabev N, Lehmann N, Weber R, Gerwig M, Frings M, Arweiler-Harbeck D, Lang S, Diener H-C (2015) Long-term outcome of vertigo and dizziness associated disorders following treatment in specialized tertiary care: the dizziness and Vertigo registry (DiVeR) study. J Neurol 262:2083–2091

Paige GD (1994) Senescence of human visual-vestibular interactions: smooth pursuit, optokinetic, and vestibular control of eye movements with aging. Exp Brain Res 98:355–372

Park JJ, Tang Y, Lopez I, Ishiyam A (2001) Unbiased estimation of human vestibular ganglion neurons. Ann N Y Acad Sci 942:475–478

Parnes LS, McClure JA (1992) Free-floating endolymph particles: a new operative finding during posterior semicircular canal occlusion. Laryngoscope 102:988–992

Peterka RJ, Black FO, Schoenhoff MB (1990) Age-related changes in human vestibulo-ocular reflexes: sinusoidal rotation and caloric tests. J Vestib Res Equilib Orientat 1:49–59

Petersen JA, Straumann D, Weber KP (2013) Clinical diagnosis of bilateral vestibular loss: three simple bedside tests. Ther Adv Neurol Disord 6:41–45

Pournaras I, Kos I, Guyot J-P (2008) Benign paroxysmal positional vertigo: a series of eight singular neurectomies. Acta Otolaryngol (Stockh) 128:5–8

Rauch SD, Velazquez-Villaseñor L, Dimitri PS, Merchant SN (2001) Decreasing hair cell counts in aging humans. Ann N Y Acad Sci 942:220–227

Reploeg MD, Goebel JA (2002) Migraine-associated dizziness: patient characteristics and management options. Otol Neurotol 23:364–371

Ricci NA, Aratani MC, Doná F, Macedo C, Caovilla HH, Ganança FF (2010) A systematic review about the effects of the vestibular rehabilitation in middle-age and older adults. Rev Bras Fisioter Sao Carlos Sao Paulo Braz 14:361–371

Richter E (1980) Quantitative study of human Scarpa's ganglion and vestibular sensory epithelia. Acta Otolaryngol (Stockh) 90:199–208

Rosenhall U (1973) Degenerative patterns in the aging human vestibular neuro-epithelia. Acta Otolaryngol (Stockh) 76:208–220

Rosenhall U, Rubin W (1975) Degenerative changes in the human vestibular sensory epithelia. Acta Otolaryngol 79(1-2):67–80

Ross MD, Peacor D, Johnsson LG, Allard LF (1976) Observations on normal and degenerating human otoconia. Ann Otol Rhinol Laryngol 85:310–326

Rubenstein LZ, Josephson KR (2006) Falls and their prevention in elderly people: what does the evidence show? Med Clin North Am 90:807–824

Sajjadi H, Paparella MM, Williams T (1998) Endolymphatic sac enhancement surgery in elderly patients with Ménière's disease. Ear Nose Throat J 77:975–982

Sekitani T, Imate Y, Noguchi T, Inokuma T (1993) Vestibular neuronitis: epidemiological survey by questionnaire in Japan. Acta Otolaryngol Suppl 503:9–12

Semont A, Freyss G, Vitte E (1988) Curing the BPPV with a Liberatory maneuver. Clin Test Vestib Syst 42:290–293

Shaia WT, Zappia JJ, Bojrab DI, Larouere ML, Sargent EW, Diaz RC (2006) Success of posterior semicircular canal occlusion and application of the dizziness handicap inventory. Otolaryngol Neck Surg 134:424–430

Shoair OA, Nyandege AN, Slattum PW (2011) Medication-related dizziness in the older adult. Otolaryngol Clin N Am 44:455–471. x

Shumway-Cook A, Horak FB (1986) Assessing the influence of sensory interaction of balance. Suggestion from the field. Phys Ther 66:1548–1550

Silva DCM e, Bastos VH, Sanchez M d O, Nunes MKG, Orsini M, Ribeiro P, Velasques B, Teixeira SS (2016) Effects of vestibular rehabilitation in the elderly: a systematic review. Aging Clin Exp Res 28:599–606

Smith WK, Sankar V, Pfleiderer AG (2005) A national survey amongst UK otolaryngologists regarding the treatment of Ménière's disease. J Laryngol Otol 119:102–105

Soto-Varela A, Faraldo-García A, Rossi-Izquierdo M, Lirola-Delgado A, Vaamonde-Sánchez-Andrade I, del-Río-Valeiras M, Gayoso-Diz P, Santos-Pérez S (2015) Can we predict the risk of falls in elderly patients with instability? Auris Nasus Larynx 42:8–14

Stel VS, Pluijm SMF, Deeg DJH, Smit JH, Bouter LM, Lips P (2003) A classification tree for predicting recurrent falling in community-dwelling older persons. J Am Geriatr Soc 51:1356–1364

Strupp M, Brandt T (2009) Current treatment of vestibular, ocular motor disorders and nystagmus. Ther Adv Neurol Disord 2:223–239

Strupp M, Zingler VC, Arbusow V, Niklas D, Maag KP, Dieterich M, Bense S, Theil D, Jahn K, Brandt T (2004) Methylprednisolone, valacyclovir, or the combination for vestibular neuritis. N Engl J Med 351:354–361

Strupp M, Hupert D, Frenzel C, Wagner J, Hahn A, Jahn K, Zingler V-C, Mansmann U, Brandt T (2008) Long-term prophylactic treatment of attacks of vertigo in Ménière's disease – comparison of a high with a low dosage of betahistine in an open trial. Acta Otolaryngol (Stockh) 128:520–524

Suarez C, Garcia C, Tolivia J (1989) Central vestibular projections of primary cervical fibers in the frog. Laryngoscope 99:1063–1071

Suárez C, González del Rey C, Tolivia J, Llorente JL, Díaz C, Navarro A, Gómez J (1993) Morphometric analysis of the vestibular complex in the rat. Laryngoscope 103:762–773

Suarez C, Diaz C, Tolivia J, Alvarez J c, Gonzalez Del Ray C, Navarro A (1997) Morphometric analysis of the human vestibular nuclei. Anat Rec 247:271–288

Sugawara K, Kitamura K, Ishida T, Sejima T (2003) Insertion of tympanic ventilation tubes as a treating modality for patients with Meniere's disease: a short- and long-term follow-up study in seven cases. Auris Nasus Larynx 30:25–28

Tabtabai R, Haynes L, Kuchel GA, Parham K (2017) Age-related increase in blood levels of Otolin-1 in humans. Otol Neurotol Off Publ Am Otol Soc Am Neurotol Soc Eur Acad Otol Neurotol 38:865–869

Tang Y, Lopez I, Baloh RW (2002) Age-related change of the neuronal number in the human medial vestibular nucleus: a stereological investigation. J Vestib Res Equilib Orientat 11:357

Tascioglu AB (2005) Brief review of vestibular system anatomy and its higher order projections. Neuroanatomy 4:24–27

Teasdale N, Stelmach GE, Breunig A (1991) Postural sway characteristics of the elderly under normal and altered visual and support surface conditions. J Gerontol 46:B238–B244

Thalmann R, Ignatova E, Kachar B, Ornitz DM, Thalmann I (2001) Development and maintenance of Otoconia. Ann N Y Acad Sci 942:162–178

Tinetti ME, Williams CS, Gill TM (2000) Dizziness among older adults: a possible geriatric syndrome. Ann Intern Med 132:337–344

Tsuji K, Velázquez-Villaseñor L, Rauch SD, Glynn RJ, Wall C, Merchant SN (2000) Temporal bone studies of the human peripheral vestibular system. Meniere's disease. Ann Otol Rhinol Laryngol Suppl 181:26–31

Tuunainen E, Poe D, Jäntti P, Varpa K, Rasku J, Toppila E, Pyykkö I (2011) Presbyequilibrium in the oldest old, a combination of vestibular, oculomotor and postural deficits. Aging Clin Exp Res 23:364–371

Tuunainen E, Jäntti P, Poe D, Rasku J, Toppila E, Pyykkö I (2012) Characterization of presbyequilibrium among institutionalized elderly persons. Auris Nasus Larynx 39:577–582

van Deelen GW, Huizing EH (1986) Use of a diuretic (Dyazide) in the treatment of Menière's disease. A double-blind cross-over placebo-controlled study. ORL J Oto-Rhino-Laryngol Its Relat Spec 48:287–292

Velázquez-Villaseñor L, Merchant SN, Tsuji K, Glynn RJ, Wall C, Rauch SD (2000) Temporal bone studies of the human peripheral vestibular system. Normative Scarpa's ganglion cell data. Ann Otol Rhinol Laryngol Suppl 181:14–19

Vibert D, Kompis M, Häusler R (2003) Benign paroxysmal positional vertigo in older women may be related to osteoporosis and osteopenia. Ann Otol Rhinol Laryngol 112:885–889

Vijayashankar N, Brody H (1977a) A study of aging in the human abducens nucleus. J Comp Neurol 173:433–437

Vijayashankar N, Brody H (1977b) Aging in the human brain stem. A study of the nucleus of the trochlear nerve. Acta Anat (Basel) 99:169–172

von Brevern M, Radtke A, Lezius F, Feldmann M, Ziese T, Lempert T, Neuhauser H (2007) Epidemiology of benign paroxysmal positional vertigo: a population based study. J Neurol Neurosurg Psychiatry 78:710–715

von Brevern M, Bertholon P, Brandt T, Fife T, Imai T, Nuti D, Newman-Toker D (2015) Benign paroxysmal positional vertigo: diagnostic criteria. J Vestib Res 25:105–117

Walther LE (2017) Current diagnostic procedures for diagnosing vertigo and dizziness. GMS Curr Top Otorhinolaryngol Head Neck Surg 16:Doc02

Walther LE, Westhofen M (2007) Presbyvertigo-aging of otoconia and vestibular sensory cells. J Vestib Res Equilib Orientat 17:89–92

Weber PC, Cass SP (1993) Clinical assessment of postural stability. Am J Otolaryngol 14:566–569

Wiesmeier IK, Dalin D, Maurer C (2015) Elderly use proprioception rather than visual and vestibular cues for postural motor control. Front Aging Neurosci 7:97

Wiesmeier IK, Dalin D, Wehrle A, Granacher U, Muehlbauer T, Dieterle J, Weiller C, Gollhofer A, Maurer C (2017) Balance training enhances vestibular function and reduces overactive proprioceptive feedback in elderly. Front Aging Neurosci 9:273

Wladislavosky-Waserman P, Facer GW, Mokri B, Kurland LT (1984) Meniere's disease: a 30-year epidemiologic and clinical study in Rochester, Mn, 1951-1980. Laryngoscope 94:1098–1102

Wolfson L, Whipple R, Derby CA, Amerman P, Murphy T, Tobin JN, Nashner L (1992) A dynamic posturography study of balance in healthy elderly. Neurology 42:2069–2075

Xie Y, Bigelow RT, Frankenthaler SF, Studenski SA, Moffat SD, Agrawal Y (2017) Vestibular loss in older adults is associated with impaired spatial navigation: data from the triangle completion task. Front Neurol 8:173

Yi H-A, Lee S-R, Lee H, Ahn B-H, Park B-R, Whitman GT (2005) Sudden deafness as a sign of stroke with normal diffusion-weighted brain MRI. Acta Otolaryngol (Stockh) 125:1119–1121

Chapter 9
Signal Transduction, Ageing and Disease

Lei Zhang, Matthew J. Yousefzadeh, Yousin Suh, Laura J. Niedernhofer, and Paul D. Robbins

Abstract Ageing is defined by the loss of functional reserve over time, leading to a decreased tissue homeostasis and increased age-related pathology. The accumulation of damage including DNA damage contributes to driving cell signaling pathways that, in turn, can drive different cell fates, including senescence and apoptosis, as well as mitochondrial dysfunction and inflammation. In addition, the accumulation of cell autonomous damage with time also drives ageing through non-cell autonomous pathways by modulation of signaling pathways. Interestingly, genetic and pharmacologic analysis of factors able to modulate lifespan and healthspan in model organisms and even humans have identified several key signaling pathways including IGF-1, NF-κB, FOXO3, mTOR, Nrf-2 and sirtuins. This review will discuss the roles of several of these key signaling pathways, in particular NF-κB and Nrf2, in modulating ageing and age-related diseases.

Keywords Signaling pathways · Age-related disease · Senescence · Apoptosis · Sirtuins · Inflammation · NF-κB Nrf2

L. Zhang · M. J. Yousefzadeh · L. J. Niedernhofer · P. D. Robbins (✉)
Institute on the Biology of Aging and Metabolism and Department of Biochemistry, Molecular Biology and Biophysics, University of Minnesota, Minneapolis, MN, USA
e-mail: leizhang@umn.edu; MYousefz@umn.edu; LNiedern@umn.edu; probbins@umn.edu

Y. Suh
Departments of Genetics and Medicine and the Institute for Ageing Research, Albert Einstein College of Medicine, Bronx, NY, USA
e-mail: yousin.suh@einstein.yu.edu

© Springer Nature Singapore Pte Ltd. 2019
J. R. Harris, V. I. Korolchuk (eds.), *Biochemistry and Cell Biology of Ageing: Part II Clinical Science*, Subcellular Biochemistry 91,
https://doi.org/10.1007/978-981-13-3681-2_9

Introduction

Mechanisms Underlying Ageing

Ageing is a complex process involving a number of different pathways with both genetic and environmental components (Kirkwood 2008; Passos et al. 2007; Vijg and Campisi 2008; Campisi and Vijg 2009; Hoeijmakers 2009). It is defined by the loss of functional reserve over time, leading to a decreased tissue homeostasis and increased age-related pathology. The biological processes driven by the accumulation of damage with time that underlie ageing phenotypes include chronic, low-grade, "sterile" inflammation; macromolecular and organelle dysfunction such as changes in proteins, carbohydrates, lipids, mitochondria and DNA; and increased senescent cell burden including senescence in adult stem cell populations. These processes are inter-related in that interventions targeting one also attenuate the others. For example, senescent cells accumulate with ageing and at sites of pathogenesis in chronic diseases (Kirkland 2016; Krishnamurthy et al. 2004) and reducing senescent cell burden can lead to reduced inflammation, decreased macromolecular dysfunction, and enhanced function of progenitors (Tchkonia et al. 2013; Xu et al. 2015; Zhu et al. 2014).

Although there is compelling evidence to support a cell autonomous mechanism for ageing with time-dependent accumulation of stochastic damage to cells, organelles and macromolecules, it is also clear from heterochronic parabiosis (Conboy et al. 2005, 2013; Conboy and Rando 2012; Loffredo et al. 2013; Sinha et al. 2014; Rebo et al. 2016) and serum transfer (Villeda et al. 2014; Rebo et al. 2016) studies that cell non-autonomous mechanisms also play important roles in suppressing or driving degenerative changes that arise as the consequence of spontaneous, stochastic damage. For example, using heterochronic parabiosis, it was demonstrated that factors in young blood rejuvenate certain cell types and tissues in old mice (Conboy et al. 2005, 2013; Conboy and Rando 2012; Loffredo et al. 2013; Sinha et al. 2014; Rebo et al. 2016). Conversely, factors in old blood can drive ageing of certain cell types and tissues in young mice.

Cellular Senescence

Senescence is a cell fate that involves loss of proliferative potential of normally replication-competent cells with associated resistance to cell death through apoptosis and generally increased metabolic activity. Frequently, senescent cells develop a senescence-associated secretory phenotype (SASP) that entails increased release of pro-inflammatory cytokines and chemokines, tissue-damageing proteases, factors that can impact stem and progenitor cell function, hemostatic factors, and growth factors (Tchkonia et al. 2013). Markers of senescent cells include increases in expression of the cell cycle regulators, $p16^{INK4A}$ and $p21^{Cip1}$, of SASP factors, increased senescence-associated β-galactosidase (SA-βgal) activity, senescent-associated distension of satellites (SADS), and telomere-associated DNA damage foci (TAFs), among others. Clearance of senescent cells in genetic model systems (INK-ATTAC and 3MR mice) or treating mice with novel senolytics extended healthspan and improved age-related diseases (Zhu et al. 2015; Fuhrmann-Stroissnigg et al. 2017; Roos et al. 2016; Childs et al. 2016; Schafer et al. 2017; Jeon et al. 2017; Ogrodnik et al. 2017). Thus, the increase in cellular senescence that occurs with ageing appears to play a major role in driving life-limiting, age-related diseases (Tchkonia et al. 2013; LeBrasseur et al. 2015; Palmer et al. 2015; Kirkland and Tchkonia 2015, 2017; Kirkland et al. 2017). As discussed below, many of the key signaling pathways implicated in ageing modulate senescence.

Conserved Pathways in Longevity

Studies in model organisms have demonstrated that the rate of ageing and the frequency and severity of age-related pathologies are influenced by conserved genetic pathways. For example, insulin/IGF-1 signaling (IIS) is a key pathway that regulates ageing in eukaryotes and there is overwhelming evidence that dampening IIS increases lifespan and healthspan in diverse species including nematodes, flies, and mice (Johnson et al. 2013; Lopez-Otin et al. 2013; O'Neill et al. 2012; Partridge et al. 2011; Cypser et al. 2006). The fact that two human longevity-associated IGF1R alleles have been identified that cause functional impairments in IGF-1 signaling clearly demonstrates that evolutionary conservation of IIS reduction contributes to longevity in humans (Tazearslan et al. 2011; Suh et al. 2008). Also, in humans, genome-wide association studies have demonstrated that genetic variations in the transcription factor FOXO3a locus are robustly associated with longevity (Broer et al. 2015; Deelen et al. 2011). Furthermore, FOXO3 has been one of the most replicated longevity-associated genes by numerous "candidate" association studies in diverse populations (Ziv and Hu 2011; Broer et al. 2015). Finally, pharmacologic and genetic studies in models of organismal ageing have clearly implicated TOR (target of rapamycin) as a key regulator of ageing. Inhibition of TOR in rodents, worms and flies leads to an extension in lifespan.

In addition to insulin/IGF-1 and TOR signaling and FOXO3A, the transcription factors NF-κB and Nrf2 have been implicated in modulating ageing and age-related diseases in model systems. Overall, these observations implicating certain cell signaling pathways in ageing suggest that it is the type and extent of cell signaling in response to external and internal stimuli that regulates healthspan and lifespan. In this review, several of these key signaling pathways important for ageing including the DNA damage response, IGF-1 signaling, NF-κB, FOXO3A, mTOR and Nrf2 will be discussed. However, due to space limitations, this review will focus primarily on the IKK/NF-κB and Nrf2/Keap pathways as examples of signaling pathways known to modulate ageing and age-related disease.

Signaling Pathways and Ageing

DNA Damage Response

Signaling induced by DNA damage has been implicated as a key factor in driving cellular senescence and ageing. In response to DNA damage, in particular double strand breaks (DSBs), the PI-3-like kinase ATM is activated by the MRE11-RAD50-NBS1 complex. Auto-phosphorylation of ATM at Ser1981 enhances its kinase activity, leading to phosphorylation of histone H2A variant H2AX (γH2AX) in nucleosomes surrounding damaged sites and recruiting more ATM and other repair factors (Ditch and Paull 2012; Shiloh and Ziv 2013). Additional DDR proteins including KRAB-associated protein-1 (KAP1) and checkpoint kinase 2 (CHK2) are phosphorylated by ATM kinase, promoting DNA repair, cell-cycle arrest, apoptosis and/or senescence (Soutoglou and Misteli 2008; Polo and Jackson 2011). ATM also phosphorylates the transcription factor p53 that regulates the G1/S checkpoint and apoptosis, increasing its stability and activity. ATM also increases the stability of p53 through the phosphorylation of MDM2, blocking its ability to ubiquitinate p53 to target it for degradation. In addition, ATM also phosphorylates the NEMO/IKKγ the regulatory subunit of the IκB kinase IKK, which then activates the transcription factor NF-κB (see below). ATM phosphorylates NEMO at Ser85, which in turn induces sumoylation and mono-ubiquitination of NEMO at Lys277 and 309 (McCool and Miyamoto 2012). These post-translational modifications eventually lead to the nuclear export of the ATM-NEMO complex to the cytoplasm where it associates with ubiquitin and SUMO-1 modified RIP1 and TAK1, activating the catalytic IKKα and subunits (Wu et al. 2006; Biton and Ashkenazi 2011). This DNA damage/ATM mediated activation of IKK then leads to an increase in NF-κB transcriptional activity.

NF-κB

Originally discovered as a transcriptional activator of immunoglobin production in B cells (Sen and Baltimore 1986), nuclear factor-κB (NF-κB) has been shown to be ubiquitously present in almost all cell types and is a key regulator of diverse biological processes, including innate and adaptive immunity, stress responses, apoptosis, and differentiation (Zhang et al. 2017). Normal activation of NF-κB is vital for many physiological functions, whereas its abnormal activation can cause a variety of pathological conditions. The persistent or chronic activation of NF-κB signaling has been well documented in many inflammatory and age-related diseases such as septic shock, arthritis, atherosclerosis, diabetes, sarcopenia, neurodegeneration (Alzheimer's and Parkinson's), and cancers (Baker et al. 2011; Amiri and Richmond 2005). Of note, constitutive NF-κB activation is not only associated with cellular senescence and organismal ageing, but also drives the ageing process (Salminen and Kaarniranta 2009; Osorio et al. 2016; Tilstra et al. 2011).

The DNA-binding activity of NF-κB complex is significantly elevated in many tissues of older rodents compared with younger ones, thereby linking the role of NF-κB and ageing (Helenius et al. 1996a, b; Korhonen et al. 1997; Spencer et al. 1997; Poynter and Daynes 1998). The elevated NF-κB activity in older animals promotes the production of inflammatory cytokines (*e.g.* IL-6 and TNF-α), contributing to many pathological disorders associated with ageing. Bioinformatic analysis using motif mapping demonstrated that NF-κB was the transcription factor most associated with mammalian ageing in multiple human and mouse tissues (Adler et al. 2007). NF-κB hyper-activation not only associates with ageing, but also directly induces senescent phenotypes (Bernard et al. 2004; Seitz et al. 2000; Zhi et al. 2011). For instance, overexpression of c-Rel in normal young keratinocytes induced senescent cellular phenotypes including decreased proliferation, apoptosis resistance, enlargement and polynucleation (Bernard et al. 2004). Moreover, acute genetic blockade of NF-κB signaling in epidermis of old mice reduced the expression of age-associated genes and reverted many features of ageing to that observed in young mice (Adler et al. 2007). Convincingly, administration of dietary antioxidants or PPAR-α agonists could correct the dysregulated NF-κB activation and concomitant expression of inflammatory cytokines (Spencer et al. 1997; Poynter and Daynes 1998).

NF-κB transcription factors refer to a collection of homo- or heterodimeric complexes formed by five structurally related proteins including p65 (RelA), RelB, c-Rel, p50, and p52, encoded respectively by RELA, RELB, REL, NFKB1, and NFKB2 (Hoffmann and Baltimore 2006). Generally, NF-κB can be activated *via* three pathways in different cell types: the canonical, non-canonical, and atypical pathways (Fig. 9.1) (Perkins 2007). The canonical or classical NF-κB signaling pathway can be initiated by physiological stimuli through cytokine receptors (*e.g.* TNFR and IL-1R), pattern-recognition receptors (*e.g.* TLRs), antigen receptors (*e.g.* TCR and BCR), and growth receptors. Ligand binding of these receptors relays the signal transduction into cytoplasm by recruiting distinct adaptor proteins. Subsequent signaling cascades then converge on the IκB kinase (IKK) complex formed by two catalytic subunits, IKKα and IKKβ, and a regulatory subunit IKKγ (NEMO). Activated IKK complex phosphorylates IκB protein whose function is to sequester NF-κB dimer in the cytoplasm, leading to polyubiquitylation and subsequent degradation of IκB by the 26S proteasome. As a result, NF-κB dimers, predominantly p65/p50 and c-Rel/p50, are liberated and translocate to the nucleus to induce gene expression (Perkins 2007). This pathway is essential for innate immunity, inflammation, cell proliferation and apoptosis.

The non-canonical or alternative signaling pathway depends on the activation of NF-κB inducing kinase (NIK) and its substrate IKKα homodimer (Sun 2011). In resting cells, NIK is maintained at an extremely low level due to its continuous degradation by E3 ligases and proteasome. Activation of this pathway can be initiated by ligation of limited receptors, including LTβR, BAFFR, CD40, RANK, TNFR2 and Fn14, leading to inhibition of NIK degradation. Accumulated NIK phosphorylates its downstream kinase IKKα and mediates the proteosomal process-

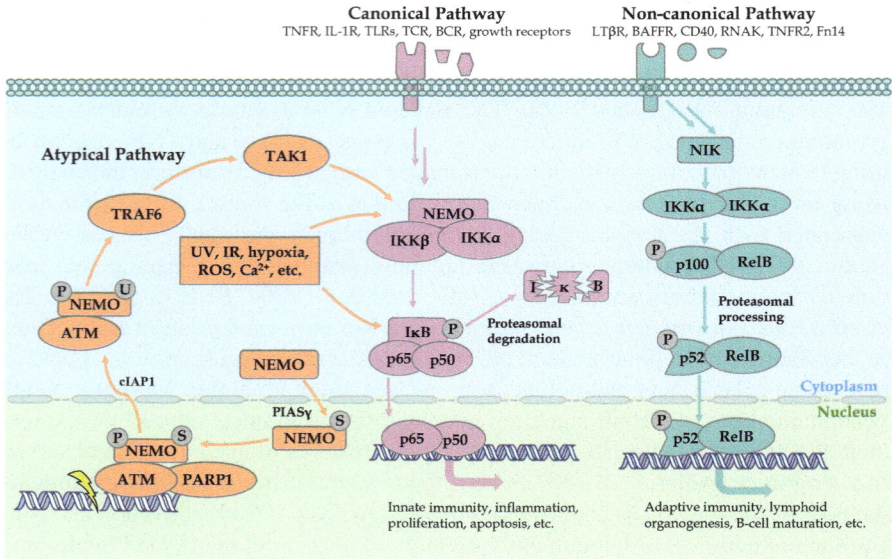

Fig. 9.1 The NF-κB signaling pathways. The canonical pathway can be activated by a variety of stimuli via different membrane receptors, eventually converging on the IKK complex. The non-canonical pathway is dependent on NIK and IKKα homodimer. The atypical pathway can be activated by a variety of stress stimuli. Abbreviations: *NEMO* NF-κB essential modulator, *TNFR* tumor necrosis factor receptor, *IL-1R* interleukin-1 receptor, *TLRs* Toll-like receptors, *TCR* T-cell receptor, *BCR* B-cell receptor, *LTβR* lymphotoxin β receptor, *BAFFR* B-cell activating factor receptor, *RANK* receptor activator of NF-κB, *PARP-1* Poly (ADP-ribose) polymerase 1, *ATM* ataxia telangiectasia mutated kinase, *TRAF* TNF receptor-associated factor, *TAK1* TGF-β activated kinase 1

ing of p100 to generate p52, followed by nuclear translocation of the RelB/p52 active heterodimer. This pathway is important for regulation of adaptive immunity, secondary lymphoid organogenesis, and maturation of B-cells.

NF-κB activation can also be induced via an atypical pathway in response to physical, genotoxic, oxidative or organelle stresses, such as ultraviolet (UV) radiation, ionizing radiation (IR), reactive oxygen species (ROS), hypoxia, dysfunctional mitochondria or endoplasmic reticulum, and secondary messenger calcium (Kriete and Mayo 2009; Siomek 2012). The atypical pathway also includes the DNA damage response discussed above. These stress stimuli activate NF-κB either IKK-dependently or IKK-independently or both. In some circumstances, the activity of IKK or IκB proteins involved in these pathways is different from that of canonical and non-canonical pathways. For example, UV radiation triggers IκBα phosphorylation at C-terminal sites (Kato et al. 2003), whereas hydrogen peroxide leads to phosphorylation at tyrosine residue 42 of IκBα (Schoonbroodt et al. 2000), rather than serines 32/36 in the canonical pathway. In particular, as discussed above, DNA damage response evokes a nuclear-to-cytoplasmic signaling pathway to activate

NF-κB (Miyamoto 2011). This inside-out signaling pathway enables cells to maintain homeostasis and survive under genotoxic stress.

NF-κB can be activated by a wide range of external, internal, and environmental cues, consequently its activation regulates the expression of a multitude of genes involved in many biological processes. Thus, the NF-κB system features a bow-tie structure which integrates diverse upstream input signals into varied downstream output responses (Fig. 9.2) (Halsey et al. 2007). The input signals of the NF-κB system include both exogenous and endogenous stimuli, many of which are danger, damage or survival signals such as viral and pathogenic assaults, DNA damage, and growth factors. These stimuli can be recognized by either membrane or cytoplasmic receptors and then the signaling is relayed to specific adapter proteins. Under normal conditions, the majority of these genes orchestrate biological processes primarily involved in cell protection, repair, and growth, such as fighting against infections, T cell maturation, DNA damage repair, triggering apoptosis to suppress tumorigenesis, facilitating tissue heal after injury. However, aberrant NF-κB activation leads to detrimental results including many age-related diseases and ageing *per se*. Therefore, NF-κB is considered a pleiotropic transcription factor whose activation is beneficial in early life to control normal physiological functions and guarantee successful growth and survival of organisms, but its hyper-activation becomes deleterious in later life by driving ageing and many age-related diseases.

The NF-κB system is tightly regulated both cell autonomously and cell non-autonomously. To elicit effective physiological functions especially inflammatory responses, NF-κB inherently encodes gene effectors to potentiate and amplify its activation in a feedforward manner. Besides, NF-κB activation also programs default feedback loops to ensure its automatic termination by producing negative effectors such as anti-inflammatory cytokines, decoy receptors, microRNAs that inhibit the signaling pathways, inhibitory proteins IκBα and IκBε (Ruland 2011; Renner and Schmitz 2009). In addition, the NF-κB system is closely coordinated with several collateral pathways by positive or negative mechanisms (Oeckinghaus et al. 2011). For example, pro-growth and pro-ageing factors positively promote the NF-κB signaling including insulin/IGF-1 and mTOR pathways. In contrast, longevity factors can negatively suppress the NF-κB activation and delay the ageing process, such as SIRT1, SIRT6, FoxO3a, Wnt4, Nrf2, and p53. Additionally, life-style interventions, especially caloric restriction and exercise, have been found to inhibit NF-κB activity and extend animal lifespan (López-Lluch and Navas 2016; Dolinsky and Dyck 2014).

IGF-1

IGF-1 signaling is a key pathway that regulates ageing in eukaryotes with reduced IGF-1 signaling shown to increase lifespan and healthspan in nematodes, flies, and mice (Johnson et al. 2013; Lopez-Otin et al. 2013; O'Neill et al. 2012; Partridge

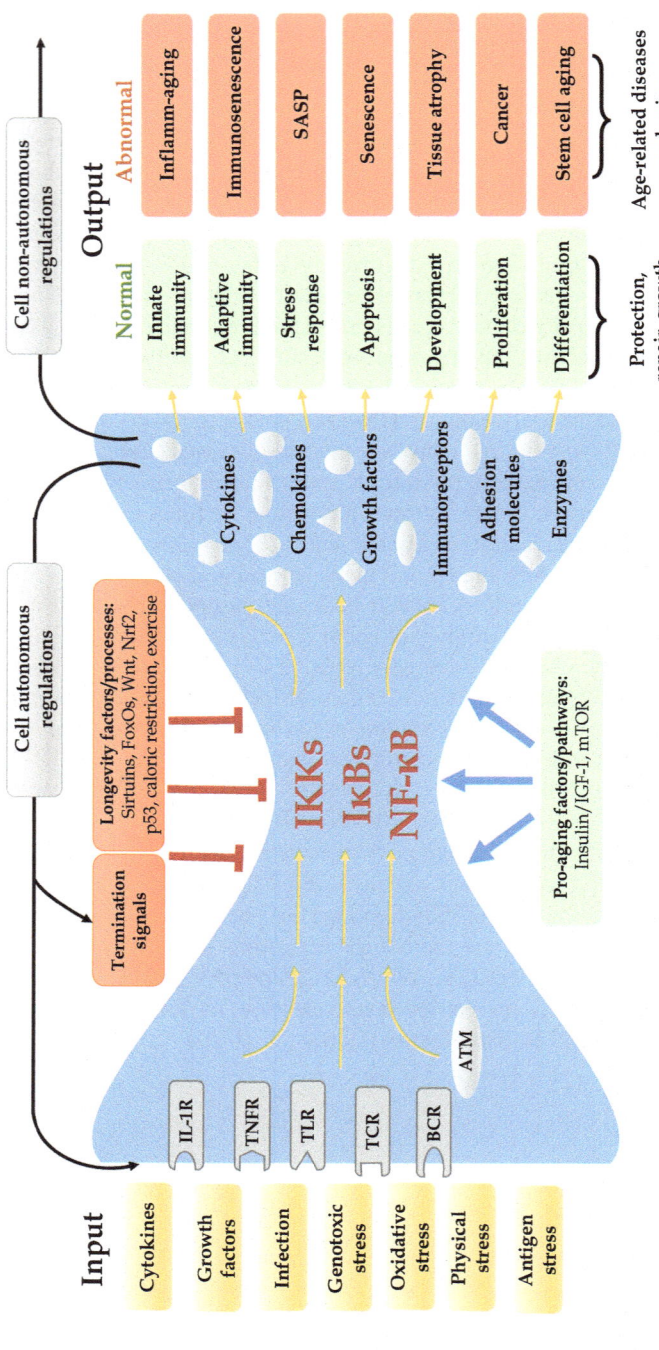

Fig. 9.2 The bow-tie structure of the NF-κB system. A wide range of input signals initiate the upstream NF-κB signaling cascades which converge on the central IKK-IκB-NF-κB hub, eventually resulting in various gene products for diverse output responses. The NF-κB system is also tightly regulated by multiple feedback loops and collateral pathways

et al. 2011; Cypser et al. 2006). In addition, suppression of IGF-1 signaling is also the primary mechanism by which caloric restriction extends lifespan. Furthermore, two human longevity-associated IGF1R alleles have been identified that cause functional impairments in IGF-1 signaling (Tazearslan et al. 2011; Suh et al. 2008). IGF-1 is a mitogenic peptide produced primarily by the liver in response to growth hormone (GH). Circulating IGF-1 is sequestered in high-affinity complexes by interactions with a family of six structurally related binding proteins termed IGF-binding proteins, termed IGFBP-1 to −6. IGF-1 binds to the IGF-1R to initiate a complex intracellular signal cascades, including phosphorylation of the insulin receptor substrate (IRS) molecules, the PI3K-Akt pathway, and the MAPK signaling cascade. These pathways further modulate the activities of mTOR as well as altering the cell localization of FOXO3A. Activation of these signaling pathways by IGF-1 to its receptor leads to promotion of cell survival, proliferation and differentiation. IGF-1 also protects cells against apoptosis in response to genotoxic stress. In contrast, chronic genotoxic stress causes a systemic reduction in circulating IGF-1, possibly mediated by p53, which inhibits expression of IGF-1R and IGF-II while increasing the expression of IGFBP-3.

Nuclear Erythroid-2-p45-Related Factor 2 (Nrf-2)

Constant exposure to oxidants and electrophiles, derived from endogenous enzymatic processes or through exogenous exposure to oxidative toxicants, enhances our susceptibility to oxidative stress (Beal 2002). To combat the harmful effects of oxidative stress, humans have evolved a vigorous antioxidant system in order to maintain redox homeostasis. Equilibrium of oxidant-antioxidant balance can be disrupted by increasing oxidants or loss of antioxidant buffering capacity resulting in oxidative stress (Zhang et al. 2015). This disruption of redox homeostasis is not infrequent as oxidant production and antioxidant buffering capacity can change in response to metabolic changes, pathophysiologic conditions, and environmental exposure (Beal 1995; Jacob et al. 2013). Oxidative stress is quite harmful to cellular survival and tissue homeostasis and is implicated in various age-related chronic diseases (cardiovascular, neurodegenerative disorders, diabetes, and cataracts) as well as cancer (Beal 1995, 2002; Jacob et al. 2013; Butterfield et al. 2001; Emerit et al. 2004). The free radical theory of ageing (FRTA), first described by Denham Harman in 1956, posits that macromolecules are damaged by endogenously-generated reactive oxidative species which contributes to overall ageing (Harman 1956). While the FRTA is controversial and has been challenged and modified over the years, the crux of it relies upon perturbations in redox homeostasis and accumulated damage of macromolecules causes degenerative ageing (Harman 1992; Vina et al. 2013; Stuart et al. 2014).

The nuclear erythroid-2-p45-related factor 2 (Nrf2) transcription factor, encoded by the *NFE2L2* gene, regulates the basal and inducible expression of multiple antioxidant and stress response proteins (Chan et al. 1995; Itoh et al. 1999; Moi et al.

1994; Venugopal and Jaiswal 1996). Nrf2 contains seven functional domains, Neh1-Neh7, of which Neh2 is the major regulatory domain (Zhang et al. 2015). Nrf2 was originally identified over two decades ago to be a regulator of β-globin expression (Moi et al. 1994). In 1996, the first evidence of Nrf2 as a regulator of antioxidant enzymes was discovered, was shown to regulate antioxidants when Venugopal and colleagues showed it to be an activator of antioxidant-response element (ARE) in the promotor of and regulator of the induction of NQO-1 in response to oxidative stress (Venugopal and Jaiswal 1996). Later other Nrf2 targets including catalase, superoxide dismutases, regulators of glutathione metabolism, and peroxiredoxins would go on to be discovered (Zhang et al. 2015).

In addition to regulating expression of target genes, Nrf2 regulates expression of itself with two AREs flanking the *NFE2L2* promotor region, inducing a feed forward effect upon activation of Nrf2 (Jaramillo and Zhang 2013). Further regulating Nrf2 and its antioxidant responses is Kelch-like ECH-associated protein 1 (Keap1) that interacts with Nrf2 through its ETGE and DLG motifs in the Neh2 domain and sequesters Nrf2 in the cytoplasm during unstressed conditions (Zhang et al. 2004; McMahon et al. 2006; Tong et al. 2006). Keap1 also acts as part of the Keap1-Cul3-E3 ubiquitin ligase complex to promote ubiquitination of seven lysine residues in the Neh2 domain of Nrf2 and eventual proteosomal degradation of the protein (Zhang et al. 2004). During oxidative stress, oxidants or electrophilic stress induces conformational changes in Keap1, thus interrupting its interactions with Nrf2 (Zhang et al. 2015; Jaramillo and Zhang 2013). This allows for the translocation of Nrf2 to the nucleus where it dimerizes with Maf family members to bind to AREs within the regulatory regions of antioxidant and stress response genes (Fig. 9.3) (Rushmore et al. 1991; Wasserman and Fahl 1997; Jaramillo and Zhang 2013).

Nrf2-deficient mice are viable and phenotypically normal but vulnerable to oxidative stress and have a greater incidence of cancer (Chan et al. 1996). However, Nrf2-deficient mice have an abbreviated lifespan due to the development of anemia and hyperkeratinosis in the esophagus and forestomach (Lee et al. 2004; Pearson et al. 2008; Wakabayashi et al. 2003; Leiser and Miller 2010). Gain of function studies have proven successful with overexpression of the worm homolog of Nrf2, SKN-1, increasing lifespan (Tullet et al. 2008). Studies of Keap1 heterozygosity in flies increased Nrf2 activity and oxidative stress response, but lifespan extension only occurs in males (Sykiotis and Bohmann 2008). Studies into the effect of enhanced Nrf2 activity on longevity in mammals have been impeded by the lack of viability of the Keap1-deficient mice (Wakabayashi et al. 2003). Future studies would be aided by conditional or tissue-specific overexpression of Nrf2 or deletion of Keap1.

Total and nuclear Nrf2 levels were shown to decrease with age in tissues of multiple species: Drosophila, mice, rats and Rhesus macaque with age (Landis et al. 2012; Hochmuth et al. 2011; Rahman et al. 2013; Ungvari et al. 2011; Duan et al. 2009; George et al. 2009; Shih and Yen 2007; Kim and Nel 2005). Alterations in other proteins besides Nrf2 can cause age-related loss of Nrf2 activity. For example, it was reported that mRNA levels of Nrf2 and its target genes were unchanged with

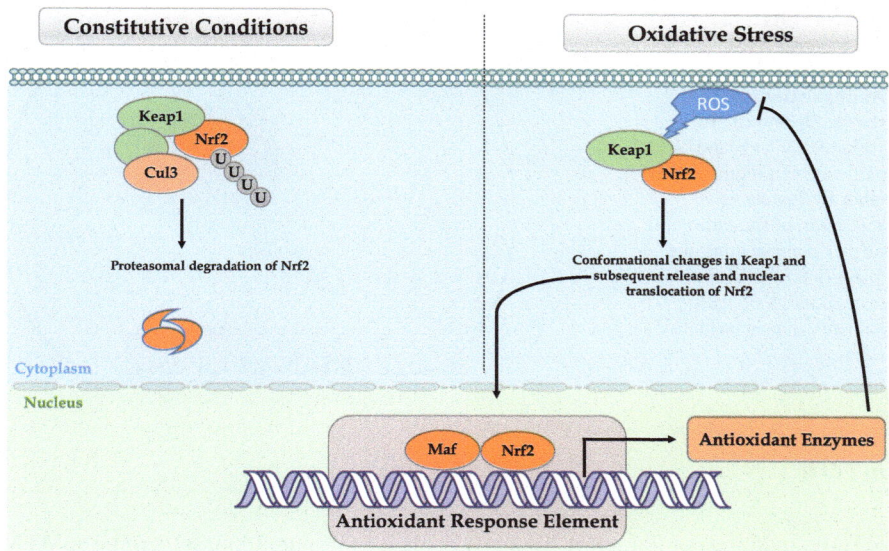

Fig. 9.3 The Nrf2 antioxidant response pathway. Under constitutive conditions Nrf2 is seques-tered in the cytosol and bound to Keap1 which mediates ubiquitin-dependent proteosomal degra-dation of Nrf2. During oxidative stress, Keap1 undergoes conformational changes in response to oxidative damage releasing Nrf2 and allowing its translocation into the nucleus where it can pro-mote transcription of antioxidant genes through antioxidant response elements contained within them

age in flies but the response to oxidative stress was lost in old flies (Rahman et al. 2013). Careful analysis of Nrf2 signaling and maximum lifespan potential in naked mole-rats and nine other rodent species found that longer lived rodent species had enhanced Nrf2 activity. This activity was not linked exclusively to Nrf2 abundance but rather to species differences in Keap1-regulation of Nrf2 (Lewis et al. 2015).

Further complicating the analysis of age-related changes in Nrf2 activity is the complex regulation of downstream antioxidant genes that can give divergent results (Zhang et al. 2015). For example, expression of Nrf2 target HO-1 is also regulated by NF-κB and HIF-1a signaling (Fig. 9.4). While Nrf2 activity decreases with age, both NF-κB and HIF-1a signaling are activated with age (Lavrovsky et al. 2000; Kang et al. 2005). This poses a challenge when analyzing age-related changes in Nrf2 and its targets. Many questions regarding Nrf2 and ageing are still left to be answered such as: 1) the mechanism(s) that cause a loss of Nrf2 signaling in response to oxidative stress with ageing; 2) can this age-related deficit in Nrf2 sig-naling be reverted and to what extent; and 3) how the age-related loss of Nrf2 sig-naling influence cellular senescence?

Fig. 9.4 Interplay of the NF-κB and Nrf2 transcription factors in response to oxidative stress. During oxidative stress the NF-κB and Nrf2 pathways antagonize each other by impairing activation of the other, thus further complicating the dissection of the relative contributions of each pathway during oxidative stress

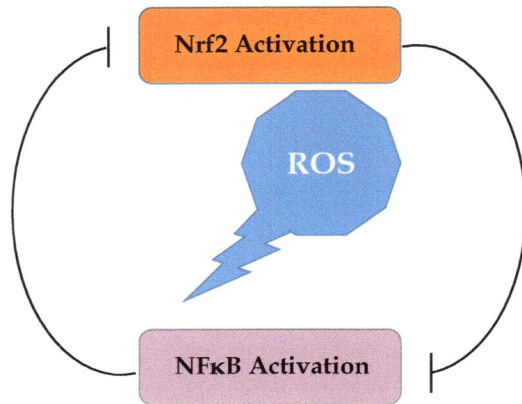

mTOR

mTOR (mammalian Target of Rapamycin) is a PI3-related kinase, similar to ATM that phosphorylates Ser-Thr residues of proteins. It has been implicated in ageing by the fact that treatment with the mTOR inhibitor rapamycin extends lifespan in multiple model systems of ageing. Also, dominant negative forms of TOR or its downstream effector S6K1, or overexpression of the upstream inhibitors of TOR, TSC1 and TSC2, extend lifespan in certain models of ageing. There are two mTOR complexes: 1) mTORC1, comprised of mTOR, mLST8 (GßL) and Raptor, and 2) mTORC2, containing mTOR, mLST8 and Rictor. mTOR, in particular mTORC1, serves to integrate multiple signals to regulate growth including responses to growth factors, stress, nutrients, and energy. mTOR is downstream of IGF-1R with IGF-1 activation of AKT leading to phosphorylation and inactivation of the TSC1–TSC2 complex, an inhibitor of mTOR signaling. Depriving cells of nutrients, such as leucine, results in rapid dephosphorylation of mTOR substrates including S6K1 and 4E-BP1 to inhibit translation. Conversely, addition of nutrients results in rapid phosphorylation of these targets by mTOR. In contrast to IGF-1 signaling, DNA damage leads to inhibition of mTOR *via* p53, AMPK and TSC1–TSC2-dependent mechanism(s). Thus, at least in certain tissue and cell types, the proliferation signals of mTOR are dampened in response to genotoxic stress and with ageing, likely as a protective mechanism, similar to IGF-1.

FOXO3A

The transcription factor FOXO3, a member of the FOXO (Forkhead box O) transcription factor, in mammalian systems and similar to DAF16 in the worm, is implicated in regulating lifespan in variety of species (Hesp et al. 2015; Martins et al. 2016; Webb and Brunet 2014). Interestingly, non-coding variants in FOXO3 have been identified by multiple groups as contributing to lifespan and likely healthspan in humans (Li et al. 2009; Pawlikowska et al. 2009). FOXO3 also protects certain stem/progenitor cells from stress (Renault et al. 2009; Morris et al. 2015), reducing ROS in part through upregulation of anti-oxidants such as catalase and SOD2 as well as promoting autophagy. Suppression of FOXO3A activity has been shown to promotes senescence in culture (Kyoung Kim et al. 2005) and *in vivo* (Yao et al. 2012). Interestingly, FOXO3 is also a target for phosphorylation by IKK which reduces its stability (Liu et al. 2012).

FOXO3 is a transcription factor regulated by modulation of its intracellular localization, translocating in and out of the nucleus *via* AKT phosphorylation and possibly by acetylation and methylation (Monsalve and Olmos 2011). What remains unclear is why DAF-16/FOXO3A nuclear localization and thus activity cannot be sustained with age. Chronic nuclear localization of DAF-16 is observed in several long-lived mutants, suggesting that sustaining DAF-16 activity after genotoxic insult could be beneficial (Kenyon et al. 1993). DAF-16/FOXO3A sub-cellular localization and activity is coordinated *via* several post-translational modifications, including phosphorylation and acetylation, and is regulated by multiple pathways. AMP-activated protein kinase (AMPK) phosphorylates and activates FOXO3, linking metabolism and stress responses (Lutzner et al. 2012). The stress-activated kinase p38 MAPK also phosphorylates FOXO3 in response to genotoxic agents, leading to its nuclear localization (Ho et al. 2012). Conversely, inflammation-induced phosphorylation of FOXO3A by IKK leads to its cytoplasmic sequestration (Hu et al. 2004). DAF-16/FOXO3A activity is also regulated by changes in acetylation status. A reduction in the deacetylation of DAF-16/FOXO3A by sirtuins (Hashimoto et al. 2010) prevents its nuclear localization and activity (Daitoku et al. 2011). Levels of NAD+, which is required for sirtuin activity, decline with age (Imai and Guarente 2016) and are reduced in DNA repair deficient animals (Fang et al. 2014). Moreover, supplementation of NAD+ can extend the lifespan of DNA repair deficient and wild-type worms (Fang et al. 2016). Therefore, changes in ATP levels, MAPK signaling, NF-κB activation (Tilstra et al. 2012; Adler et al. 2007) or changes in acetylation status could all contribute to the deactivation of FOXO3 upon persistent DNA damage.

Summary

Extensive genetic and pharmacologic analysis has identified multiple pathways that contribute to slowing or accelerating ageing in model organisms. These pathways include the insulin/IGF-1, mTOR, Nrf2, FOXO3a and NF-κB. Several of these pathways have more recently been identified in humans as important for exceptional longevity. Interestingly, many of these pathways are part of the organismal response to genotoxic, oxidative and inflammatory stresses. These pathways also modulate different cell fates in response to stress such as senescence and apoptosis as well as modulate mitochondrial dysfunction and inflammation. Thus, these pathways represent key therapeutic targets for extending healthspan and possibly lifespan in humans.

Acknowledgements This work was supported by the NIH grants P01-AG043376 (PDR, LJN), U19-AG056278 (PDR, LJN, YS) and the Glenn/AFAR (LJN).

References

Adler AS, Sinha S, Kawahara TLA, Zhang JY, Segal E, Chang HY (2007) Motif module map reveals enforcement of ageing by continual NF-κB activity. Genes Dev 21(24):3244–3257. https://doi.org/10.1101/gad.1588507

Amiri KI, Richmond A (2005) Role of nuclear factor-κ B in melanoma. Cancer Metastasis Rev 24(2):301–313. https://doi.org/10.1007/s10555-005-1579-7

Baker RG, Hayden MS, Ghosh S (2011) NF-κB, inflammation, and metabolic disease. Cell Metab 13(1):11–22. https://doi.org/10.1016/j.cmet.2010.12.008

Beal MF (1995) Ageing, energy, and oxidative stress in neurodegenerative diseases. Ann Neurol 38(3):357–366. https://doi.org/10.1002/ana.410380304

Beal MF (2002) Oxidatively modified proteins in ageing and disease. Free Radic Biol Med 32(9):797–803

Bernard D, Gosselin K, Monte D, Vercamer C, Bouali F, Pourtier A, Vandenbunder B, Abbadie C (2004) Involvement of Rel/Nuclear Factor-κB transcription factors in keratinocyte senescence. Cancer Res 64(2):472–481. https://doi.org/10.1158/0008-5472.Can-03-0005

Biton S, Ashkenazi A (2011) NEMO and RIP1 control cell fate in response to extensive DNA damage via TNF-α feedforward signaling. Cell 145(1):92–103. https://doi.org/10.1016/j.cell.2011.02.023

Broer L, Buchman AS, Deelen J, Evans DS, Faul JD, Lunetta KL, Sebastiani P, Smith JA, Smith AV, Tanaka T, Yu L, Arnold AM, Aspelund T, Benjamin EJ, De Jager PL, Eiriksdottir G, Evans DA, Garcia ME, Hofman A, Kaplan RC, Kardia SL, Kiel DP, Oostra BA, Orwoll ES, Parimi N, Psaty BM, Rivadeneira F, Rotter JI, Seshadri S, Singleton A, Tiemeier H, Uitterlinden AG, Zhao W, Bandinelli S, Bennett DA, Ferrucci L, Gudnason V, Harris TB, Karasik D, Launer LJ, Perls TT, Slagboom PE, Tranah GJ, Weir DR, Newman AB, van Duijn CM, Murabito JM (2015) GWAS of longevity in CHARGE consortium confirms APOE and FOXO3 candidacy. J Gerontol A Biol Sci Med Sci 70(1):110–118. https://doi.org/10.1093/gerona/glu166

Butterfield DA, Howard BJ, LaFontaine MA (2001) Brain oxidative stress in animal models of accelerated ageing and the age-related neurodegenerative disorders, Alzheimer's disease and Huntington's disease. Curr Med Chem 8(7):815–828

Campisi J, Vijg J (2009) Does damage to DNA and other macromolecules play a role in ageing? If so, how? J Gerontol A Biol Sci Med Sci 64(2):175–178. https://doi.org/10.1093/gerona/gln065

Chan JY, Cheung MC, Moi P, Chan K, Kan YW (1995) Chromosomal localization of the human NF-E2 family of bZIP transcription factors by fluorescence in situ hybridization. Hum Genet 95(3):265–269

Chan K, Lu R, Chang JC, Kan YW (1996) NRF2, a member of the NFE2 family of transcription factors, is not essential for murine erythropoiesis, growth, and development. Proc Natl Acad Sci U S A 93(24):13943–13948

Childs BG, Baker DJ, Wijshake T, Conover CA, Campisi J, van Deursen JM (2016) Senescent intimal foam cells are deleterious at all stages of atherosclerosis. Science 354(6311):472–477. https://doi.org/10.1126/science.aaf6659

Conboy IM, Rando TA (2012) Heterochronic parabiosis for the study of the effects of ageing on stem cells and their niches. Cell Cycle 11(12):2260–2267. https://doi.org/10.4161/cc.20437

Conboy IM, Conboy MJ, Wagers AJ, Girma ER, Weissman IL, Rando TA (2005) Rejuvenation of aged progenitor cells by exposure to a young systemic environment. Nature 433(7027):760–764. https://doi.org/10.1038/nature03260

Conboy MJ, Conboy IM, Rando TA (2013) Heterochronic parabiosis: historical perspective and methodological considerations for studies of ageing and longevity. Aging Cell 12(3):525–530. https://doi.org/10.1111/acel.12065

Cypser JR, Tedesco P, Johnson TE (2006) Hormesis and ageing in Caenorhabditis elegans. Exp Gerontol 41(10):935–939. https://doi.org/10.1016/j.exger.2006.09.004

Daitoku H, Sakamaki J, Fukamizu A (2011) Regulation of FoxO transcription factors by acetylation and protein-protein interactions. Biochim Biophys Acta 1813(11):1954–1960. https://doi.org/10.1016/j.bbamcr.2011.03.001

Deelen J, Beekman M, Uh HW, Helmer Q, Kuningas M, Christiansen L, Kremer D, van der Breggen R, Suchiman HE, Lakenberg N, van den Akker EB, Passtoors WM, Tiemeier H, van Heemst D, de Craen AJ, Rivadeneira F, de Geus EJ, Perola M, van der Ouderaa FJ, Gunn DA, Boomsma DI, Uitterlinden AG, Christensen K, van Duijn CM, Heijmans BT, Houwing-Duistermaat JJ, Westendorp RG, Slagboom PE (2011) Genome-wide association study identifies a single major locus contributing to survival into old age; the APOE locus revisited. Aging cell 10(4):686–698. https://doi.org/10.1111/j.1474-9726.2011.00705.x

Ditch S, Paull TT (2012) The ATM protein kinase and cellular redox signaling: beyond the DNA damage response. Trends Biochem Sci 37(1):15–22. https://doi.org/10.1016/j.tibs.2011.10.002

Dolinsky V, Dyck J (2014) Experimental studies of the molecular pathways regulated by exercise and resveratrol in heart, skeletal muscle and the vasculature. Molecules 19(9):14919–14947

Duan W, Zhang R, Guo Y, Jiang Y, Huang Y, Jiang H, Li C (2009) Nrf2 activity is lost in the spinal cord and its astrocytes of aged mice. In Vitro Cell Dev Biol Anim 45(7):388–397. https://doi.org/10.1007/s11626-009-9194-5

Emerit J, Edeas M, Bricaire F (2004) Neurodegenerative diseases and oxidative stress. Biomed Pharmacother 58(1):39–46

Fang EF, Scheibye-Knudsen M, Brace LE, Kassahun H, SenGupta T, Nilsen H, Mitchell JR, Croteau DL, Bohr VA (2014) Defective mitophagy in XPA via PARP-1 hyperactivation and NAD(+)/SIRT1 reduction. Cell 157(4):882–896. https://doi.org/10.1016/j.cell.2014.03.026

Fang EF, Kassahun H, Croteau DL, Scheibye-Knudsen M, Marosi K, Lu H, Shamanna RA, Kalyanasundaram S, Bollineni RC, Wilson MA, Iser WB, Wollman BN, Morevati M, Li J, Kerr JS, Lu Q, Waltz TB, Tian J, Sinclair DA, Mattson MP, Nilsen H, Bohr VA (2016) NAD(+) replenishment improves lifespan and healthspan in ataxia telangiectasia models via mitophagy and DNA repair. Cell Metab 24(4):566–581. https://doi.org/10.1016/j.cmet.2016.09.004

Fuhrmann-Stroissnigg H, Ling YY, Zhao J, McGowan SJ, Zhu Y, Brooks RW, Grassi D, Gregg SQ, Stripay JL, Dorronsoro A, Corbo L, Tang P, Bukata C, Ring N, Giacca M, Li X, Tchkonia T, Kirkland JL, Niedernhofer LJ, Robbins PD (2017) Identification of HSP90 inhibitors as a novel class of senolytics. Nat Commun 8(1):422. https://doi.org/10.1038/s41467-017-00314-z

George L, Lokhandwala MF, Asghar M (2009) Exercise activates redox-sensitive transcription factors and restores renal D1 receptor function in old rats. Am J Physiol Renal Physiol 297(5):F1174–F1180. https://doi.org/10.1152/ajprenal.00397.2009

Halsey TA, Yang L, Walker JR, Hogenesch JB, Thomas RS (2007) A functional map of NFκB signaling identifies novel modulators and multiple system controls. Genome Biol 8(6):R104. https://doi.org/10.1186/gb-2007-8-6-r104

Harman D (1956) Ageing: a theory based on free radical and radiation chemistry. J Gerontol 11(3):298–300

Harman D (1992) Free radical theory of ageing. Mutat Res 275(3–6):257–266

Hashimoto T, Horikawa M, Nomura T, Sakamoto K (2010) Nicotinamide adenine dinucleotide extends the lifespan of Caenorhabditis elegans mediated by sir-2.1 and daf-16. Biogerontology 11(1):31–43. https://doi.org/10.1007/s10522-009-9225-3

Helenius M, Hänninen M, Lehtinen SK, Salminen A (1996a) Ageing-induced Up-regulation of nuclear binding activities of oxidative stress responsive NF-kB transcription factor in mouse cardiac muscle. J Mol Cell Cardiol 28(3):487–498. https://doi.org/10.1006/jmcc.1996.0045

Helenius M, Hänninen M, Lehtinen SK, Salminen A (1996b) Changes associated with ageing and replicative senescence in the regulation of transcription factor nuclear factor-kappa B. Biochem J 318(Pt 2):603–608

Hesp K, Smant G, Kammenga JE (2015) Caenorhabditis elegans DAF-16/FOXO transcription factor and its mammalian homologs associate with age-related disease. Exp Gerontol 72:1–7. https://doi.org/10.1016/j.exger.2015.09.006

Ho KK, McGuire VA, Koo CY, Muir KW, de Olano N, Maifoshie E, Kelly DJ, McGovern UB, Monteiro LJ, Gomes AR, Nebreda AR, Campbell DG, Arthur JS, Lam EW (2012) Phosphorylation of FOXO3a on Ser-7 by p38 promotes its nuclear localization in response to doxorubicin. J Biol Chem 287(2):1545–1555. https://doi.org/10.1074/jbc.M111.284224

Hochmuth CE, Biteau B, Bohmann D, Jasper H (2011) Redox regulation by Keap1 and Nrf2 controls intestinal stem cell proliferation in Drosophila. Cell Stem Cell 8(2):188–199. https://doi.org/10.1016/j.stem.2010.12.006

Hoeijmakers JH (2009) DNA damage, ageing, and cancer. N Engl J Med 361(15):1475–1485. https://doi.org/10.1056/NEJMra0804615

Hoffmann A, Baltimore D (2006) Circuitry of nuclear factor kappaB signaling. Immunol Rev 210:171–186. https://doi.org/10.1111/j.0105-2896.2006.00375.x

Hu MC, Lee DF, Xia W, Golfman LS, Ou-Yang F, Yang JY, Zou Y, Bao S, Hanada N, Saso H, Kobayashi R, Hung MC (2004) IkappaB kinase promotes tumorigenesis through inhibition of forkhead FOXO3a. Cell 117(2):225–237

Imai SI, Guarente L (2016) It takes two to tango: NAD(+) and sirtuins in ageing/longevity control. NPJ Ageing Mech Dis 2:16017. https://doi.org/10.1038/npjamd.2016.17

Itoh K, Wakabayashi N, Katoh Y, Ishii T, Igarashi K, Engel JD, Yamamoto M (1999) Keap1 represses nuclear activation of antioxidant responsive elements by Nrf2 through binding to the amino-terminal Neh2 domain. Genes Dev 13(1):76–86

Jacob KD, Noren Hooten N, Trzeciak AR, Evans MK (2013) Markers of oxidant stress that are clinically relevant in ageing and age-related disease. Mech Ageing Dev 134(3–4):139–157. https://doi.org/10.1016/j.mad.2013.02.008

Jaramillo MC, Zhang DD (2013) The emerging role of the Nrf2-Keap1 signaling pathway in cancer. Genes Dev 27(20):2179–2191. https://doi.org/10.1101/gad.225680.113

Jeon OH, Kim C, Laberge RM, Demaria M, Rathod S, Vasserot AP, Chung JW, Kim DH, Poon Y, David N, Baker DJ, van Deursen JM, Campisi J, Elisseeff JH (2017) Local clearance of senescent cells attenuates the development of post-traumatic osteoarthritis and creates a pro-regenerative environment. Nat Med 23:775–781. https://doi.org/10.1038/nm.4324

Johnson SC, Rabinovitch PS, Kaeberlein M (2013) mTOR is a key modulator of ageing and age-related disease. Nature 493(7432):338–345. https://doi.org/10.1038/nature11861

Kang MJ, Kim HJ, Kim HK, Lee JY, Kim DH, Jung KJ, Kim KW, Baik HS, Yoo MA, Yu BP, Chung HY (2005) The effect of age and calorie restriction on HIF-1-responsive genes in aged liver. Biogerontology 6(1):27–37. https://doi.org/10.1007/s10522-004-7381-z

Kenyon C, Chang J, Gensch E, Rudner A, Tabtiang R (1993) A C. elegans mutant that lives twice as long as wild type. Nature 366(6454):461–464

Kim HJ, Nel AE (2005) The role of phase II antioxidant enzymes in protecting memory T cells from spontaneous apoptosis in young and old mice. J Immunol 175(5):2948–2959

Kirkland JL (2016) Translating the science of ageing into therapeuticiInterventions. Cold Spring Harb Perspect Med 6(3):a025908. https://doi.org/10.1101/cshperspect.a025908

Kirkland JL, Tchkonia T (2015) Clinical strategies and animal models for developing senolytic agents. Exp Gerontol 68:19–25. https://doi.org/10.1016/j.exger.2014.10.012

Kirkland JL, Tchkonia T (2017) Cellular senescence: a translational perspective. EBioMedicine 21:21–28. https://doi.org/10.1016/j.ebiom.2017.04.013

Kirkland JL, Tchkonia T, Zhu Y, Niedernhofer LJ, Robbins PD (2017) The clinical potential of senolytic drugs. J Am Ger Soc 65:2297–2301

Kirkwood TB (2008) Understanding ageing from an evolutionary perspective. J Intern Med 263(2):117–127. https://doi.org/10.1111/j.1365-2796.2007.01901.x

Korhonen P, Helenius M, Salminen A (1997) Age-related changes in the regulation of transcription factor NF-κB in rat brain. Neurosci Lett 225(1):61–64. https://doi.org/10.1016/S0304-3940(97)00190-0

Krishnamurthy J, Torrice C, Ramsey MR, Kovalev GI, Al-Regaiey K, Su L, Sharpless NE (2004) Ink4a/Arf expression is a biomarker of ageing. J Clin Invest 114(9):1299–1307. https://doi.org/10.1172/JCI22475

Kyoung Kim H, Kyoung Kim Y, Song IH, Baek SH, Lee SR, Hye Kim J, Kim JR (2005) Down-regulation of a forkhead transcription factor, FOXO3a, accelerates cellular senescence in human dermal fibroblasts. J Gerontol A Biol Sci Med Sci 60(1):4–9

Landis G, Shen J, Tower J (2012) Gene expression changes in response to ageing compared to heat stress, oxidative stress and ionizing radiation in Drosophila melanogaster. Ageing (Albany NY) 4(11):768–789. https://doi.org/10.18632/ageing.100499

Lavrovsky Y, Chatterjee B, Clark RA, Roy AK (2000) Role of redox-regulated transcription factors in inflammation, ageing and age-related diseases. Exp Gerontol 35(5):521–532

LeBrasseur NK, Tchkonia T, Kirkland JL (2015) Cellular senescence and the biology of ageing, disease, and frailty. Nestle Nutr Inst Workshop Ser 83:11–18. https://doi.org/10.1159/000382054

Lee JM, Chan K, Kan YW, Johnson JA (2004) Targeted disruption of Nrf2 causes regenerative immune-mediated hemolytic anemia. Proc Natl Acad Sci U S A 101(26):9751–9756. https://doi.org/10.1073/pnas.0403620101

Leiser SF, Miller RA (2010) Nrf2 signaling, a mechanism for cellular stress resistance in long-lived mice. Mol Cell Biol 30(3):871–884. https://doi.org/10.1128/MCB.01145-09

Lewis KN, Wason E, Edrey YH, Kristan DM, Nevo E, Buffenstein R (2015) Regulation of Nrf2 signaling and longevity in naturally long-lived rodents. Proc Natl Acad Sci U S A 112(12):3722–3727. https://doi.org/10.1073/pnas.1417566112

Li Y, Wang WJ, Cao H, Lu J, Wu C, Hu FY, Guo J, Zhao L, Yang F, Zhang YX, Li W, Zheng GY, Cui H, Chen X, Zhu Z, He H, Dong B, Mo X, Zeng Y, Tian XL (2009) Genetic association of FOXO1A and FOXO3A with longevity trait in Han Chinese populations. Hum Mol Genet 18(24):4897–4904. https://doi.org/10.1093/hmg/ddp459

Liu F, Xia Y, Parker AS, Verma IM (2012) IKK biology. Immunol Rev 246(1):239–253. https://doi.org/10.1111/j.1600-065X.2012.01107.x

Loffredo FS, Steinhauser ML, Jay SM, Gannon J, Pancoast JR, Yalamanchi P, Sinha M, Dall'Osso C, Khong D, Shadrach JL, Miller CM, Singer BS, Stewart A, Psychogios N, Gerszten RE, Hartigan AJ, Kim MJ, Serwold T, Wagers AJ, Lee RT (2013) Growth differentiation factor 11 is a circulating factor that reverses age-related cardiac hypertrophy. Cell 153(4):828–839. https://doi.org/10.1016/j.cell.2013.04.015

López-Lluch G, Navas P (2016) Calorie restriction as an intervention in ageing. J Physiol 594(8):2043–2060. https://doi.org/10.1113/JP270543

Lopez-Otin C, Blasco MA, Partridge L, Serrano M, Kroemer G (2013) The hallmarks of ageing. Cell 153(6):1194–1217. https://doi.org/10.1016/j.cell.2013.05.039

Lutzner N, Kalbacher H, Krones-Herzig A, Rosl F (2012) FOXO3 is a glucocorticoid receptor target and regulates LKB1 and its own expression based on cellular AMP levels via a positive autoregulatory loop. PLoS One 7(7):e42166. https://doi.org/10.1371/journal.pone.0042166

Martins R, Lithgow GJ, Link W (2016) Long live FOXO: unraveling the role of FOXO proteins in ageing and longevity. Ageing Cell 15(2):196–207. https://doi.org/10.1111/acel.12427

McCool KW, Miyamoto S (2012) DNA damage-dependent NF-κB activation: NEMO turns nuclear signaling inside out. Immunol Rev 246(1):311–326

McMahon M, Thomas N, Itoh K, Yamamoto M, Hayes JD (2006) Dimerization of substrate adaptors can facilitate cullin-mediated ubiquitylation of proteins by a "tethering" mechanism: a two-site interaction model for the Nrf2-Keap1 complex. J Biol Chem 281(34):24756–24768. https://doi.org/10.1074/jbc.M601119200

Moi P, Chan K, Asunis I, Cao A, Kan YW (1994) Isolation of NF-E2-related factor 2 (Nrf2), a NF-E2-like basic leucine zipper transcriptional activator that binds to the tandem NF-E2/AP1 repeat of the beta-globin locus control region. Proc Natl Acad Sci U S A 91(21):9926–9930

Monsalve M, Olmos Y (2011) The complex biology of FOXO. Curr Drug Targets 12(9):1322–1350

Morris BJ, Willcox DC, Donlon TA, Willcox BJ (2015) FOXO3: a major gene for human longevity--a mini-review. Gerontology 61(6):515–525. https://doi.org/10.1159/000375235

O'Neill C, Kiely AP, Coakley MF, Manning S, Long-Smith CM (2012) Insulin and IGF-1 signalling: longevity, protein homoeostasis and Alzheimer's disease. Biochem Soc Trans 40(4):721–727. https://doi.org/10.1042/BST20120080

Oeckinghaus A, Hayden MS, Ghosh S (2011) Crosstalk in NF-κB signaling pathways. Nat Immunol 12:695–708. https://doi.org/10.1038/ni.2065

Ogrodnik M, Miwa S, Tchkonia T, Tiniakos D, Wilson CL, Lahat A, Day CP, Burt A, Palmer A, Anstee QM, Grellscheid SN, Hoeijmakers JHJ, Barnhoorn S, Mann DA, Bird TG, Vermeij WP, Kirkland JL, Passos JF, von Zglinicki T, Jurk D (2017) Cellular senescence drives age-dependent hepatic steatosis. Nat Commun 8:15691. https://doi.org/10.1038/ncomms15691

Osorio FG, Soria-Valles C, Santiago-Fernández O, Freije JMP, López-Otín C (2016) NF-κB signaling as a driver of ageing. Int Rev Cell Mol Biol 326:133–174. https://doi.org/10.1016/bs.ircmb.2016.04.003

Palmer AK, Tchkonia T, LeBrasseur NK, Chini EN, Xu M, Kirkland JL (2015) Cellular senescence in type 2 diabetes: a therapeutic opportunity. Diabetes 64(7):2289–2298. https://doi.org/10.2337/db14-1820

Partridge L, Alic N, Bjedov I, Piper MD (2011) Ageing in drosophila: the role of the insulin/Igf and TOR signalling network. Exp Gerontol 46(5):376–381. https://doi.org/10.1016/j.exger.2010.09.003

Passos JF, von Zglinicki T, Kirkwood TB (2007) Mitochondria and ageing: winning and losing in the numbers game. BioEssays: News and Reviews in Molecular, Cellular and Developmental Biology 29(9):908–917. https://doi.org/10.1002/bies.20634

Pawlikowska L, Hu D, Huntsman S, Sung A, Chu C, Chen J, Joyner AH, Schork NJ, Hsueh WC, Reiner AP, Psaty BM, Atzmon G, Barzilai N, Cummings SR, Browner WS, Kwok PY, Ziv E, Study of Osteoporotic F (2009) Association of common genetic variation in the insulin/IGF1 signaling pathway with human longevity. Aging Cell 8(4):460–472. https://doi.org/10.1111/j.1474-9726.2009.00493.x

Pearson KJ, Lewis KN, Price NL, Chang JW, Perez E, Cascajo MV, Tamashiro KL, Poosala S, Csiszar A, Ungvari Z, Kensler TW, Yamamoto M, Egan JM, Longo DL, Ingram DK, Navas P, de Cabo R (2008) Nrf2 mediates cancer protection but not prolongevity induced by caloric restriction. Proc Natl Acad Sci U S A 105(7):2325–2330. https://doi.org/10.1073/pnas.0712162105

Perkins ND (2007) Integrating cell-signalling pathways with NF-[kappa]B and IKK function. Nat Rev Mol Cell Biol 8(1):49–62

Polo SE, Jackson SP (2011) Dynamics of DNA damage response proteins at DNA breaks: a focus on protein modifications. Genes Dev 25(5):409–433. https://doi.org/10.1101/gad.2021311

Poynter ME, Daynes RA (1998) Peroxisome proliferator-activated receptor α activation modulates cellular redox status, represses nuclear factor-κB signaling, and reduces inflammatory cytokine production in ageing. J Biol Chem 273(49):32833–32841. https://doi.org/10.1074/jbc.273.49.32833

Rahman MM, Sykiotis GP, Nishimura M, Bodmer R, Bohmann D (2013) Declining signal dependence of Nrf2-MafS-regulated gene expression correlates with ageing phenotypes. Aging Cell 12(4):554–562. https://doi.org/10.1111/acel.12078

Rebo J, Mehdipour M, Gathwala R, Causey K, Liu Y, Conboy MJ, Conboy IM (2016) A single heterochronic blood exchange reveals rapid inhibition of multiple tissues by old blood. Nat Commun 7:13363. https://doi.org/10.1038/ncomms13363

Renault VM, Rafalski VA, Morgan AA, Salih DA, Brett JO, Webb AE, Villeda SA, Thekkat PU, Guillerey C, Denko NC, Palmer TD, Butte AJ, Brunet A (2009) FoxO3 regulates neural stem cell homeostasis. Cell Stem Cell 5(5):527–539. https://doi.org/10.1016/j.stem.2009.09.014

Renner F, Schmitz ML (2009) Autoregulatory feedback loops terminating the NF-kappaB response. Trends Biochem Sci 34(3):128–135. https://doi.org/10.1016/j.tibs.2008.12.003

Roos CM, Zhang B, Palmer AK, Ogrodnik MB, Pirtskhalava T, Thalji NM, Hagler M, Jurk D, Smith LA, Casaclang-Verzosa G, Zhu Y, Schafer MJ, Tchkonia T, Kirkland JL, Miller JD (2016) Chronic senolytic treatment alleviates established vasomotor dysfunction in aged or atherosclerotic mice. Aging Cell 15:973–977. https://doi.org/10.1111/acel.12458

Ruland J (2011) Return to homeostasis: downregulation of NF-κB responses. Nat Immunol 12(8):709–714

Rushmore TH, Morton MR, Pickett CB (1991) The antioxidant responsive element. Activation by oxidative stress and identification of the DNA consensus sequence required for functional activity. J Biol Chem 266(18):11632–11639

Salminen A, Kaarniranta K (2009) NF-κB signaling in the ageing process. J Clin Immunol 29(4):397–405. https://doi.org/10.1007/s10875-009-9296-6

Schafer MJ, White TA, Iijima K, Haak AJ, Ligresti G, Atkinson EJ, Oberg AL, Birch J, Salmonowicz H, Zhu Y, Mazula DL, Brooks RW, Fuhrmann-Stroissnigg H, Pirtskhalava T, Prakash YS, Tchkonia T, Robbins PD, Aubry MC, Passos JF, Kirkland JL, Tschumperlin DJ, Kita H, LeBrasseur NK (2017) Cellular senescence mediates fibrotic pulmonary disease. Nat Commun 8:14532. https://doi.org/10.1038/ncomms14532

Seitz CS, Deng H, Hinata K, Lin Q, Khavari PA (2000) Nuclear factor κB subunits induce epithelial cell growth arrest. Cancer Res 60(15):4085–4092

Sen R, Baltimore D (1986) Inducibility of κ immunoglobulin enhancer-binding protein NF-κB by a posttranslational mechanism. Cell 47(6):921–928. https://doi.org/10.1016/0092-8674(86)90807-x

Shih PH, Yen GC (2007) Differential expressions of antioxidant status in ageing rats: the role of transcriptional factor Nrf2 and MAPK signaling pathway. Biogerontology 8(2):71–80. https://doi.org/10.1007/s10522-006-9033-y

Shiloh Y, Ziv Y (2013) The ATM protein kinase: regulating the cellular response to genotoxic stress, and more. Nat Rev Mol Cell Biol 14(4):197–210

Sinha M, Jang YC, Oh J, Khong D, Wu EY, Manohar R, Miller C, Regalado SG, Loffredo FS, Pancoast JR, Hirshman MF, Lebowitz J, Shadrach JL, Cerletti M, Kim MJ, Serwold T, Goodyear LJ, Rosner B, Lee RT, Wagers AJ (2014) Restoring systemic GDF11 levels reverses age-related dysfunction in mouse skeletal muscle. Science 344(6184):649–652. https://doi.org/10.1126/science.1251152

Soutoglou E, Misteli T (2008) Activation of the cellular DNA damage response in the absence of DNA lesions. Science 320(5882):1507–1510. https://doi.org/10.1126/science.1159051

Spencer NF, Poynter ME, Im SY, Daynes RA (1997) Constitutive activation of NF-kappa B in an animal model of ageing. Int Immunol 9(10):1581–1588. https://doi.org/10.1093/intimm/9.10.1581

Stuart JA, Maddalena LA, Merilovich M, Robb EL (2014) A midlife crisis for the mitochondrial free radical theory of ageing. Longev Healthspan 3(1):4. https://doi.org/10.1186/2046-2395-3-4

Suh Y, Atzmon G, Cho MO, Hwang D, Liu B, Leahy DJ, Barzilai N, Cohen P (2008) Functionally significant insulin-like growth factor I receptor mutations in centenarians. Proc Natl Acad Sci U S A 105(9):3438–3442. https://doi.org/10.1073/pnas.0705467105

Sykiotis GP, Bohmann D (2008) Keap1/Nrf2 signaling regulates oxidative stress tolerance and lifespan in drosophila. Dev Cell 14(1):76–85. https://doi.org/10.1016/j.devcel.2007.12.002

Tazearslan C, Huang J, Barzilai N, Suh Y (2011) Impaired IGF1R signaling in cells expressing longevity-associated human IGF1R alleles. Aging cell 10:551–554. https://doi.org/10.1111/j.1474-9726.2011.00697.x

Tchkonia T, Zhu Y, van Deursen J, Campisi J, Kirkland JL (2013) Cellular senescence and the senescent secretory phenotype: therapeutic opportunities. J Clin Invest 123(3):966–972. https://doi.org/10.1172/JCI64098

Tilstra JS, Clauson CL, Niedernhofer LJ, Robbins PD (2011) NF-κB in ageing and disease. Ageing Dis 2(6):449–465

Tilstra JS, Robinson AR, Wang J, Gregg SQ, Clauson CL, Reay DP, Nasto LA, St Croix CM, Usas A, Vo N, Huard J, Clemens PR, Stolz DB, Guttridge DC, Watkins SC, Garinis GA, Wang

Y, Niedernhofer LJ, Robbins PD (2012) NF-kappaB inhibition delays DNA damage-induced senescence and ageing in mice. J Clin Invest 122(7):2601–2612. https://doi.org/10.1172/JCI45785

Tong KI, Katoh Y, Kusunoki H, Itoh K, Tanaka T, Yamamoto M (2006) Keap1 recruits Neh2 through binding to ETGE and DLG motifs: characterization of the two-site molecular recognition model. Mol Cell Biol 26(8):2887–2900. https://doi.org/10.1128/MCB.26.8.2887-2900.2006

Tullet JM, Hertweck M, An JH, Baker J, Hwang JY, Liu S, Oliveira RP, Baumeister R, Blackwell TK (2008) Direct inhibition of the longevity-promoting factor SKN-1 by insulin-like signaling in C. elegans. Cell 132(6):1025–1038. https://doi.org/10.1016/j.cell.2008.01.030

Ungvari Z, Bailey-Downs L, Gautam T, Sosnowska D, Wang M, Monticone RE, Telljohann R, Pinto JT, de Cabo R, Sonntag WE, Lakatta EG, Csiszar A (2011) Age-associated vascular oxidative stress, Nrf2 dysfunction, and NF-{kappa}B activation in the nonhuman primate Macaca mulatta. J Gerontol A Biol Sci Med Sci 66(8):866–875. https://doi.org/10.1093/gerona/glr092

Venugopal R, Jaiswal AK (1996) Nrf1 and Nrf2 positively and c-Fos and Fra1 negatively regulate the human antioxidant response element-mediated expression of NAD(P)H:quinone oxidoreductase1 gene. Proc Natl Acad Sci U S A 93(25):14960–14965

Vijg J, Campisi J (2008) Puzzles, promises and a cure for ageing. Nature 454(7208):1065–1071. https://doi.org/10.1038/nature07216

Villeda SA, Plambeck KE, Middeldorp J, Castellano JM, Mosher KI, Luo J, Smith LK, Bieri G, Lin K, Berdnik D, Wabl R, Udeochu J, Wheatley EG, Zou B, Simmons DA, Xie XS, Longo FM, Wyss-Coray T (2014) Young blood reverses age-related impairments in cognitive function and synaptic plasticity in mice. Nat Med 20(6):659–663. https://doi.org/10.1038/nm.3569

Vina J, Borras C, Abdelaziz KM, Garcia-Valles R, Gomez-Cabrera MC (2013) The free radical theory of ageing revisited: the cell signaling disruption theory of ageing. Antioxid Redox Signal 19(8):779–787. https://doi.org/10.1089/ars.2012.5111

Wakabayashi N, Itoh K, Wakabayashi J, Motohashi H, Noda S, Takahashi S, Imakado S, Kotsuji T, Otsuka F, Roop DR, Harada T, Engel JD, Yamamoto M (2003) Keap1-null mutation leads to postnatal lethality due to constitutive Nrf2 activation. Nat Genet 35(3):238–245. https://doi.org/10.1038/ng1248

Wasserman WW, Fahl WE (1997) Functional antioxidant responsive elements. Proc Natl Acad Sci U S A 94(10):5361–5366

Webb AE, Brunet A (2014) FOXO transcription factors: key regulators of cellular quality control. Trends Biochem Sci 39(4):159–169. https://doi.org/10.1016/j.tibs.2014.02.003

Wu ZH, Shi Y, Tibbetts RS, Miyamoto S (2006) Molecular linkage between the kinase ATM and NF-kappaB signaling in response to genotoxic stimuli. Science 311(5764):1141–1146. https://doi.org/10.1126/science.1121513

Xu M, Palmer AK, Ding H, Weivoda MM, Pirtskhalava T, White TA, Sepe A, Johnson KO, Stout MB, Giorgadze N, Jensen MD, LeBrasseur NK, Tchkonia T, Kirkland JL (2015) Targeting senescent cells enhances adipogenesis and metabolic function in old age. eLife 4:e12997. https://doi.org/10.7554/eLife.12997

Yao H, Chung S, Hwang JW, Rajendrasozhan S, Sundar IK, Dean DA, McBurney MW, Guarente L, Gu W, Ronty M, Kinnula VL, Rahman I (2012) SIRT1 protects against emphysema via FOXO3-mediated reduction of premature senescence in mice. J Clin Invest 122(6):2032–2045. https://doi.org/10.1172/JCI60132

Zhang DD, Lo SC, Cross JV, Templeton DJ, Hannink M (2004) Keap1 is a redox-regulated substrate adaptor protein for a Cul3-dependent ubiquitin ligase complex. Mol Cell Biol 24(24):10941–10953. https://doi.org/10.1128/MCB.24.24.10941-10953.2004

Zhang H, Davies KJ, Forman HJ (2015) Oxidative stress response and Nrf2 signaling in ageing. Free Radic Biol Med 88 (Pt B):314–336. https://doi.org/10.1016/j.freeradbiomed.2015.05.036

Zhang Q, Lenardo MJ, Baltimore D (2017) 30 Years of NF-κB: a blossoming of relevance to human pathobiology. Cell 168(1):37–57. https://doi.org/10.1016/j.cell.2016.12.012

Zhi H, Yang L, Kuo Y-L, Ho Y-K, Shih H-M, Giam C-Z (2011) NF-κB hyper-activation by HTLV-1 tax induces cellular senescence, but can be alleviated by the viral anti-sense protein HBZ. PLoS Pathog 7(4):e1002025. https://doi.org/10.1371/journal.ppat.1002025

Zhu Y, Armstrong JL, Tchkonia T, Kirkland JL (2014) Cellular senescence and the senescent secretory phenotype in age-related chronic diseases. Curr Opin Clin Nutr Metab Care 17(4):324–328. https://doi.org/10.1097/MCO.0000000000000065

Zhu Y, Tchkonia T, Pirtskhalava T, Gower AC, Ding H, Giorgadze N, Palmer AK, Ikeno Y, Hubbard GB, Lenburg M, O'Hara SP, LaRusso NF, Miller JD, Roos CM, Verzosa GC, LeBrasseur NK, Wren JD, Farr JN, Khosla S, Stout MB, McGowan SJ, Fuhrmann-Stroissnigg H, Gurkar AU, Zhao J, Colangelo D, Dorronsoro A, Ling YY, Barghouthy AS, Navarro DC, Sano T, Robbins PD, Niedernhofer LJ, Kirkland JL (2015) The Achilles' heel of senescent cells: from transcriptome to senolytic drugs. Ageing Cell 14(4):644–658. https://doi.org/10.1111/acel.12344

Ziv E, Hu D (2011) Genetic variation in insulin/IGF-1 signaling pathways and longevity. Ageing Res Rev 10(2):201–204. https://doi.org/10.1016/j.arr.2010.09.002

Chapter 10
Skin Changes During Ageing

Frédéric Bonté, Dorothée Girard, Jean-Christophe Archambault, and Alexis Desmoulière

Abstract The skin provides the primary protection for the body against external injuries and is essential in the maintenance of general homeostasis. During ageing, resident cells become senescent and the extracellular matrix, mainly in the dermis, is progressively damaged affecting the normal organization of the skin and its capacity for repair. In parallel, extrinsic factors such as ultraviolet irradiation, pollution, and intrinsic factors such as diabetes or vascular disease can further accelerate this phenomenon. Indeed, numerous mechanisms are involved in age-induced degradation of the skin and these also relate to non-healing or chronic wounds in the elderly. In particular, the generation of reactive oxygen species seems to play a major role in age-related skin modifications. Certainly, targeting both the hormonal status of the skin or its surface nutrition can slow down age-induced degradation of the skin and improve healing of skin damage in the elderly. Skin care regimens that prevent radiation and pollution damage, and reinforce the skin surface and its microbiota are among the different approaches able to minimize the effects of ageing on the skin.

Keywords Wound healing · Myofibroblast · Extracellular matrix · Oestrogen · Reactive oxygen species · Pollution · Microbiota

F. Bonté · J.-C. Archambault
LVMH Recherche, Saint Jean de Braye, France
e-mail: fredericbonte@research.lvmh-pc.com; jcarchambault@research.lvmh-pc.com

D. Girard
Faculties of Medicine and Pharmacy, Department of Physiology, University of Limoges, Myelin Maintenance and Peripheral Neuropathies (EA 6309), Limoges, France

Institut de Recherche Biomédicale des Armées, INSERM UMRS-MD 1197, Centre de Transfusion Sanguine des Armées, Clamart, France
e-mail: dorothee.girard@inserm.fr

A. Desmoulière (✉)
Faculties of Medicine and Pharmacy, Department of Physiology, University of Limoges, Myelin Maintenance and Peripheral Neuropathies (EA 6309), Limoges, France
e-mail: alexis.desmouliere@unilim.fr

© Springer Nature Singapore Pte Ltd. 2019
J. R. Harris, V. I. Korolchuk (eds.), *Biochemistry and Cell Biology of Ageing: Part II Clinical Science*, Subcellular Biochemistry 91,
https://doi.org/10.1007/978-981-13-3681-2_10

Abbreviations

ADSC   adipose tissue-derived stem cell
AGEs   advanced glycation end-products
AMPK   adenosine monophosphate-activated protein kinase
AP-1   activator protein-1
CR   caloric restriction
CTGF   connective tissue growth factor
DEJ   dermal-epidermal junction
ECM   extracellular matrix
HSD   hydroxysteroid dehydrogenase
IL   interleukin
IR   infrared radiation
MB   methylene blue
MMP   matrix metalloproteinase
MSC   mesenchymal stem cell
NT   neurotrophin
PDGF   platelet-derived growth factor
ROS   reactive oxygen species
SOD   superoxide dismutase
TGF-β1   transforming growth factor-β1
TIMPs   tissue inhibitors of metalloproteinases
TNF-α   tumour necrosis factor-α
UV   ultraviolet

Introduction

The skin is the largest and most visible organ of the human body. Its main role is to permit us to live in our environment, protecting our body from dehydration and preventing xenobiotic penetration of the organism. The skin is often also a mirror of our health and of what is going on inside the body. Many internal disorders can affect skin properties and with time, modifications of skin appearance can affect self-esteem and quality of life, including our social life. It is well known that patients with extensive burn injuries, who can now survive thanks to recent improvements in medical care, often present with psychological problems that can lead, in many cases to suicide, due to the psychological effects of skin scarring and bodily disfigurement (Jain et al. 2017). Aged skin is characterized biologically by modifications of the epidermis and of the dermal-epidermal junction (DEJ), and by global degradation of the dermal extracellular matrix (ECM), where collagen and elastin fibres are particularly affected (Binic et al. 2013). As detailed below, skin ageing can be intrinsic and/or extrinsic, caused by physiological and environmental factors, respectively. Intrinsic skin ageing illustrates the naturally occurring modifications of the skin with age, leading to wrinkles and skin dryness. Extrinsic skin ageing arrives earlier and is due, for example, to exposure to sunlight and/or air or even to indoor pollution, to life style and also can be due to a lack of balanced nutrition. It can already be underlined that in most cases of skin ageing, oxidative damage caused by overproduction of free radicals plays a major role. In this review, the normal architecture of the skin and skin modifications that appear during ageing are described. Factors influencing skin ageing are also discussed and the deleterious role of skin ageing on wound healing processes is emphasised. Finally, several skin-care strategies that may help to avoid or reduce skin ageing will be outlined.

Overall Organization of the Skin

The specific structural and cellular organization of the skin is adapted to ensure its three major functions as primary protective barrier, in thermoregulation, and in sensory perception (Fig. 10.1).

The Epidermis

The epidermis is the outer layer of the skin, in direct contact with the external environment. It consists of a keratinized stratified squamous epithelium which undergoes continuous renewal. Like all epithelia it is not vascularized. Its thickness

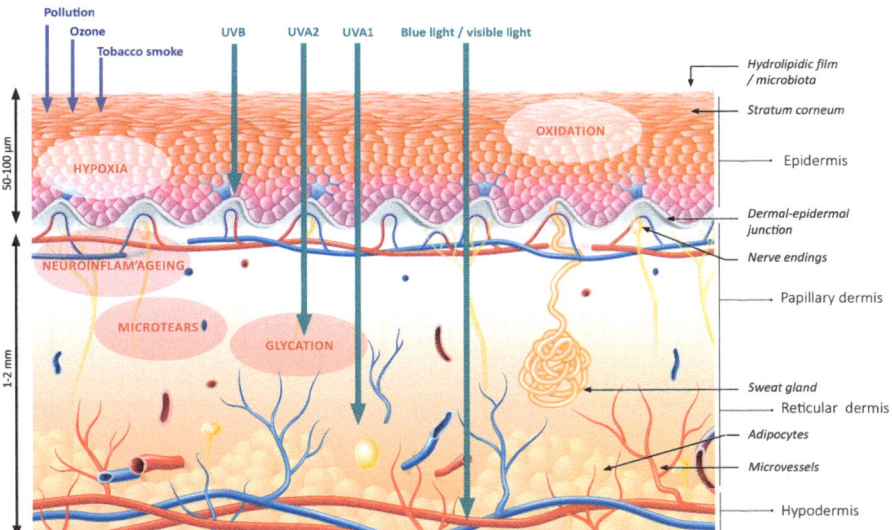

Fig. 10.1 Environmental factors and main biological stresses involved in skin ageing. © LVMH Recherche

ranges, depending on the body location, from 50 to 75 μm on the eyelids to 600 μm on palms and soles (Wong et al. 2016). The epidermis is primarily composed of keratinocytes that are located in distinguishable layers. The deepest layer, sitting on the DEJ, is the basal layer or *stratum basale*. It represents the proliferative or stem cell compartment, a single layer of keratinocyte stem cells which give rise to progenitors and keratinocytes that progressively migrate to the more superficial layers. Basal keratinocytes display characteristic cuboidal shapes with elliptical nuclei. Above the *stratum basale*, the spinous layer or *stratum spinosum* represents 5–15 layers of polygonal keratinocytes, in the lower layers, that are flattened in the upper layers, and are connected to each other by specific anchorage structures called desmosomes (Stahley et al. 2016). Then, the granular cell layer or *stratum granulosum* represents one to three layers made of flattened granular keratinocytes which contains keratohyalin granules and lamellar bodies. Exclusively in thick epidermis, a thin transitional layer of translucent cells called the *stratum lucidum* can also be observed. Finally, the outermost epidermal layer is the *stratum corneum* also termed the horny layer. It is composed of 5–15 layers of non-viable but biochemically active, large, flat, polyhedral cells named corneocytes embedded in an organized lipid matrix (Eckhart et al. 2013). The normal cell turnover rate, proceeding upward from the *stratum basale* through the spinous and granular layers to the *stratum corneum*, is about 28 days. Other cell types found in the epidermis are melanocytes (about one melanocyte for 36 keratinocytes) which are specialized in producing melanin pigments involved in both skin colour and protecting against ultraviolet (UV) (Singh et al. 2017). Langerhans cells, which are mobile and have dendrite-like processes, are antigen presenting cells with a role in the epidermal immune defence (Clayton et al. 2017). In addition, Merkel cells, which are sensory receptors that show close interaction with epidermal nerve endings, are also specifically present within the basal layer (Maksimovic et al. 2014).

The Dermal-Epidermal Junction (DEJ)

The DEJ is located below the epidermis. It is in contact with the basal layer of keratinocytes and allows the anchorage of the epidermis to the upper dermis. It has a role in polarizing the keratinocytes from the basal layer and acts as an interface for nutrition exchange and signalling between the dermal and epidermal compartments. The DEJ is composed of four acellular layers. The upper layer is composed of keratinocyte plasma membranes of the basal layer forming specific adhesion structures with the underlying ECM, called hemidesmosomes. Below this, the *lamina lucida* is crossed by anchoring filaments of laminins (5–7 nm in diameter). These filaments interact with α6β4 integrins located in keratinocyte hemidesmosomes and generate strong attachment structures (Schneider et al. 2007; Tsuruta et al. 2011). These anchoring filaments are also attached to the underlying *lamina densa*. The *lamina densa* is mainly composed of type IV and V collagens and laminins. It is also attached to anchoring plaques located deeper within the fibrillar zone of the upper dermis via type VII collagen anchoring fibrils which are 20–60 nm thick. Thus, the *lamina densa* constitutes an intermediate anchorage zone both for the anchoring filaments and the anchoring fibrils arising from the dermis.

The Dermis

The dermis represents connective tissue that provides tensile strength and resilience properties to the skin. Depending on the body location, the dermis contains more or less numerous skin appendages such as sweat glands and hair follicles associated with sebaceous glands, and its thickness can range from 0.3 mm on the eyelids to 4 mm on the back (Wong et al. 2016). The dermis has another important role in supporting the epidermis and for this purpose, it is richly vascularized with blood and lymphatic vessels arising from the hypodermis underneath. This dense vascular network is essential to sustain both dermal homeostasis and provide all the nutrients necessary to the epidermis (Lugo et al. 2011). The dermis is composed of two distinguishable layers. The upper layer, termed papillary dermis, is thin and composed of loose type I and III collagen and elastin fibres. It is in contact with the DEJ where it forms specific invaginations called dermal papillae. The papillary dermis possesses high cellular density with fibroblasts producing and modulating the ECM and immune cells such as macrophages, dendritic cells and mast cells involved in host defence. The reticular dermis underneath is thicker, less cellular and consists of dense bundles of type I collagen fibres. The ECM plays a critical role in maintaining the structural integrity of the dermis but also in mediating cell-cell and cell-matrix interactions necessary for cellular activity. Examples of specific signalling pathways include integrins and transforming growth factor-β1 (TGF-β1) (Desmoulière et al. 1993; Watt and Fujiwara 2011).

The Hypodermis

The hypodermis is the deepest and thickest layer of the skin located below the reticular dermis. It provides anchorage of the skin to the underlying fascia of bones and muscles. The structural organization of the hypodermis consists of a loose connective tissue of collagen and elastin containing adipocyte lobules and a dense network of blood vessels. Thus, the hypodermis helps protect against mechanical shock, provides thermal insulation while also assuring energy metabolism and fatty acid storage (Klein et al. 2007; Alexander et al. 2015). The hypodermis is also a source of hormones and growth factors. In addition, it is a well-studied reservoir of adult mesenchymal stem cells known as adipose-derived stem cells (ADSCs) which possess multipotent properties (Klar et al. 2017). Other cell types such as resident fibroblasts, involved in the maintenance of the ECM, and immune cells (mainly macrophages) are also present.

Skin Innervation

The skin is densely innervated by both peripheral sensory and autonomic innervation. Autonomic nerve fibres, mainly sympathetic, constitute a minority of cutaneous nerve fibres. These are restricted to the dermis, innervating blood and lymphatic vessels, erector pili muscles, and sweat glands. Thus, dermal autonomic innervation is essential in the regulation of body temperature by acting on skin blood flow and sweat gland function (Blessing et al. 2016). Sensory innervation is present throughout the hypodermis, dermis and epidermis with a higher density in glabrous skin. Sensory fibres detect and transmit signals (*e.g.* heat, pH changes, pressure) via dorsal root ganglia and the spinal cord to specific areas of the central nervous system leading in the sensory perception of pain, touch, itch or burn (Roosterman et al. 2006). Nerves fibres are associated in the dermis with Schwann cells which may or may not produce a myelin sheath. Therefore, sensory fibres are classified into three categories depending on their diameter and conduction velocity: thick myelinated Aβ with the fastest conduction speed, intermediate myelinated Aδ and unmyelinated C fibres also called free nerve endings with the lowest conduction velocities. Aβ fibres are mainly mechanoreceptors detecting pressure or stretch. Aβ fibres are found within the dermis associated with specific structures such as Ruffini, Pacinian and Meissner's corpuscles. Another specific sensory structure consists of the association of epidermal Merkel cells and Aβ fibres (Zimmerman et al. 2014). Aδ and C fibres are primarily involved in nociception and the perception of temperature changes in both the dermis and epidermis. It should be noted that only free-endings of C fibres can cross the DEJ to innervate the epidermis.

Skin Microbiota

The skin, and more precisely the epidermis, possesses its own commensal microbiota (Byrd et al. 2018). The term microbiota refers to the microorganism flora ranging from bacteria, fungi, to viruses living in symbiosis within the superficial cell layers of the *stratum corneum* and skin appendages such as hair follicles, sebaceous glands and sweat glands (Prescott et al. 2017). It is estimated that one billion bacteria inhabit each square centimetre of skin (Grice et al. 2008). The biodiversity of resident skin microorganisms varies depending on the body site, specific cutaneous microenvironment (*e.g.* pH, moisture, temperature, presence of sebum) and depends on the external environment where a transient flora can temporarily arise. These microorganisms are in close dialogue with the host's immune system and are not pathogenic in physiological conditions. However, in case of damage to the integrity of the epidermal barrier, an imbalance in microbiota or the presence of pathogens can lead to cutaneous inflammatory disorders such as acne, psoriasis and atopic dermatitis (Belkaid and Segre 2014; Naik et al. 2015).

Modifications of the Skin During Ageing

The skin changes as it ages and this ageing is complex as it involves tissues with different levels of specialization (Gragnani et al. 2014). Changes occur at all levels of the skin structure, thus modifying its appearance. On the skin surface, these changes are reflected in a duller micro-relief, a thinner epidermis, some cutaneous dryness, more visible pores, small actinic keratoses, redness, and uneven pigmentation (Law et al. 2017). On a more macroscopic level, wrinkles, fine lines, dark spots, tissue ptosis, loss of elasticity and, depending on the area of the body, undesired hair growth and stretch marks (striae) can be observed with ageing (Fig. 10.2). The occurrence and severity of these changes in skin appearance and function are even more visible on photo-exposed areas such as the face. Although there are phenotypic variations in the emergence of signs of ageing due to gender, lifestyle and ethnic origin, these are merely the result of variability in the intrinsic and extrinsic causes of ageing (Dumas et al. 2005; Lee et al. 2014; Voegeli et al. 2015; Vashi et al. 2016). Due to their multifactorial nature, there is a high level of variability in cutaneous data linked to ageing, making it virtually impossible to generalise the causes (Bergler-Czop and Miziotek 2017; Callaghan et al. 2017; Tobin 2017). Most scales for the classification of skin ageing combine the simultaneous grading of wrinkles (Table 10.1), telangiectasia, pigmentation spots and firmness. In fact, all of these parameters and the sagging of features allow the human eye to visually determine age.

Fig. 10.2 Face skin of an aged woman (95 years old). *A*: eye area; *B*: elastosis; *C*: ptosis, *D*: unwanted hair. © FB

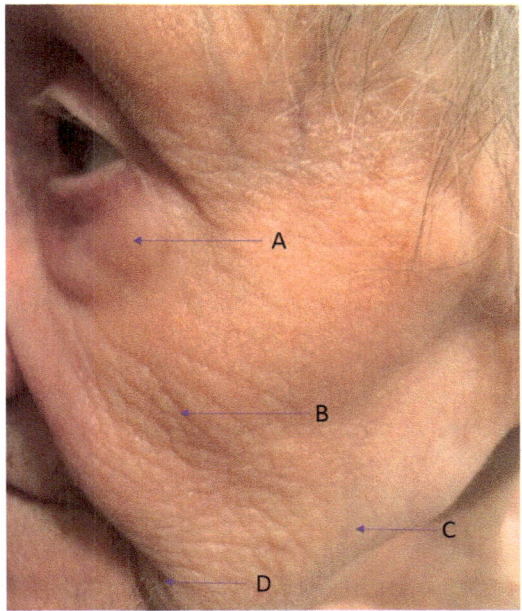

Table 10.1 Main types of wrinkles on the face

Type of wrinkles	Characterisation	Location	Origin
Atrophic	Fine, almost parallel lines	Sun protected area	Collagen atrophy
Elastotic	Hypertrophic and compact	Sun exposed area, cheeks, upper lip, neck	Actinic elastosis
Expressional	Orientation according facial muscles forces, permanent	Forehead, glabella lines crow's feet	Musculocutaneous hypodermal connective tissue
Gravitational	Loss of tone, folding, sagging	Nasolabial folds, cheeks, jowls	Gravitational forces hypodermal fibrous network

Visible Changes

The face is particularly impacted as it suffers from the effects of gravity, exposure to light radiation and pollution, redistribution of superficial and deep fat, and repeated muscular contractions related to facial expressions. All of these age-related phenomena gradually lead to changes in the facial contour making skin changes more perceptible. The extent and characteristics of facial wrinkles differ between men and women, in terms of their location and depth. These differences are the result of both hormonal variations and lifestyle differences (Luebberding et al. 2014).

A gradual decline in skin cell renewal combined with a lower level of sebaceous gland secretion amplifies the disruption of the horny layer. Corneocyte cohesion is more uneven and there are disruptions in differentiation mechanisms, promoting the accumulation of dead cells which gives to the skin a slight xerotic appearance and causes the pores to become dilated. In terms of micro-relief, this then leads to a significant change in interactions between visible light and the skin and thus in the skin appearance. Highly visible actinic keratosis "scales" represent keratinocyte dysplasia resulting from chronic exposure to UV radiation. They are mainly found on the face, ears, scalp and hands and 90% occur in subjects over the age of 50 years. Milia (milk spots or fat granules) in the eye contour area are very often observed in women. These unsightly white bumps are caused by the accumulation of lipids and dead cells.

Wrinkles and pigmentation problems are two of the most visible changes. Surface wrinkles are fine and shallow but are often numerous. They make the skin look slightly creased. They are especially found in areas of facial movement (such as the eye area, mouth, upper lip, *etc.*). They originate in the upper layers of the skin and are the result of dehydration and fragility of the DEJ (Rinnerthaler et al. 2015). Permanent wrinkles take root deep down in the dermis and are caused by a decrease in collagen and changes in the ECM. Thus, the skin becomes less resistant and is more easily marked on the surface. Deep wrinkles are at first visible only when smiling or frowning and almost disappear when the face is at rest. Over time, they gradually become established and change the facial expression (Fig. 10.2). They are the result of repeated muscle contraction messages sent to the skin via the skin-muscle interface: the dermal-muscular junction. Over time, these three types of wrinkles become superimposed. Surprisingly, the viscoelasticity of facial skin is thought to be asymmetric in women after menopause (Piérard et al. 2014).

Colour disorders are caused by melanocyte pigmentation disorders with or without post-inflammatory or microcirculatory mediated reactions. Eumelanin is a nitrogen-containing polymer made up of semiquinone units whose colour ranges from brown to black. It provides photo-protection and is mainly found in subjects with dark photo-types. Pheomelanin is dominant in photo-type I or II fair-skinned individuals with red or blond hair. Its sulfur-containing units, such as benzothiazine (cysteinyldopa), give it its red to yellow colour. Solar lentigos or age spots are brownish irregular pigmented macules due mainly to chronic sun exposure. Pigmentation disorders, especially the formation of pigmentation spots on photo-exposed areas, are due to reduction in the natural elimination of melanin-filled cells during keratinocyte turnover and the uneven or inadequate (lentigo with pendulum melanocytes) distribution of melanosomes (Noblesse et al. 2006; Warrick et al. 2017). There is now a lot of knowledge regarding the various mechanisms involved in pigmentation disorders and age spot formation explained by the accumulation of melanin, and the disruption of normal processing of melanin upwards from the basal layer (Choi et al. 2017). It is recognised that after the age of 30, the number of melanocytes decreases by 6–8% every decade, which explains why skin colour is lighter with age however with some differences depending on ethnicity (Chien et al. 2016). This causes not only hypopigmentation but also a decline in the capacity for photo-protection. Over the years, the yellow component deepens while there are no changes in the red component. This results in a dull, greyish complexion. Women, as they age, are generally highly concerned about their complexion, which is often dull, and their skin tone, which is uneven. Other pigmentation problems are caused by lipofuscin which is a brown pigment produced by the oxidation of molecular debris of lipid and protein origin within lysosomes. This pigment accumulates mainly in pre-senescent or senescent fibroblasts. These dark spots are mainly found on the cheeks, chest and hands (Skoczyńska et al. 2017).

Periorbital hyperpigmentation is often observed. This is genetically induced and also caused by active melanocytes in the dermis giving this area a grey or blue-grey appearance. Post-inflammatory hyperpigmentation is often observed in this thin and highly exposed area (Sarkar et al. 2016).

Telangiectasias are permanently dilated blood vessels or micro-vessels caused by capillary fragility. They are more commonly observed in areas most susceptible to temperature fluctuations such as the convex areas of the face (nose, cheeks, forehead, chin, *etc.*).

Functional Changes

A number of functional skin changes are observed in parallel with the emergence and progression of the visible signs of ageing.

While the thickness of the *stratum corneum* may increase slightly with age, the epidermis thins, and the dermal tissue and hypodermis provide less protection against mechanical shock. The skin is gradually weakened. The enlargement of visible pores on the surface of the face is due to the disruption of differentiation (formation of the cornified envelope), the presence of unsaturated fatty acids such as oleic acid in sebum, and a reduced amount of collagen which promotes pore dilation (Wong et al. 2016). With age, due to a decrease in the secretion of sebaceous and sweat glands, the skin also becomes drier. Ageing skin is not as well protected on the surface due to the decline in the surface hydrolipid film combined with a slightly higher pH and due to the action of exogenous factors such as the chronic overuse of detergents and soaps (Two et al. 2016). These physiological changes potentially modify the cutaneous ecology and thus disrupt the balance of cutaneous microbiota (Prescott et al. 2017). For example, a study recently demonstrated significant changes in the spatial distribution of 38 species of resident bacteria between young and older groups, on four skin sites: forehead, forearm, cheek and scalp. Recent research into the cutaneous microbiota, or resident flora, demonstrated that the microbiota is strongly present in the different compartments and appendages of the skin. Its composition – the ratio between Firmicutes, Bacteroidetes, Proteobacteria and Actinobacteria – vary depending on the area of the body (dry, moist or sebaceous) and this biodiversity is essential to adapt to the ambient environment within short time periods (Shibagaki et al. 2017). Its link to the immune system (via toll-like receptors) is also essential for the control of environmental pathogens and helps prevent pro-inflammatory diseases. Recent studies suggest that direct microbiota-nervous system interaction or microbiota metabolites can elicit biological reactions in skin and even in other body sites. It is well established that skin fungal communities profoundly shift during puberty, linked to modifications in sebaceous gland activation and sebum composition. It seems possible that local alteration in skin nutrients (proteins, lipids, sweat exosomes), due to the ageing process, could also influence the composition and function of the microbiota. In case of dysbiosis, pyruvate dehydrogenase dysregulation, which is normally essential for skin homeostasis, could affect the generation of cellular energy and thus contribute to skin 'tiredness' (Chen et al. 2018).

It is currently acknowledged that skin barrier functions and hydration can show minor changes with age. It is also widely recognized that skin biomechanical properties, firmness, elasticity and shock absorption, decrease with age. Thinner and drier skin causes itching in elderly subjects. Pruritus developing in elderly people is a complex reflex involving transient receptor potential vanilloid type 1, 3 and 4 channels on sensory nerves, serotonin, histamine, G protein receptors, proteases, and toll-like receptors. This chronic itching interferes with quality of life and leads to significant changes in the skin integrity, with the appearance of scratches and even wounds (Gouin et al. 2017). Cutaneous thermoregulation and thermal-resistance properties also decline with age. Pain thresholds increase with age; this reduced tactile acuity promotes burns and traumas on the skin in which repair mechanisms are also less and less effective.

Underlying Biological Changes

In general, a distinction is made between intrinsic or chronological ageing and extrinsic ageing, which is caused by environmental factors. Intrinsic ageing is the result of several factors: genetic, metabolic, hormonal, nutritional, *etc.* The genetic factor essentially relies on disruptions in the function or regulation of gene expression and the accumulation of daily stresses in the biological sense of the term (Rocquet and Bonté 2002). General decreased cellular activity can also be related to the start of cellular senescence or a decrease in cellular bioenergy due to mitochondrial dysfunction, through a decline in the recycling of defective mitochondria, or mitophagy (Childs et al. 2017; Ritschka et al. 2017). Simply by removing senescent cells or defective organelles, new tissue production is almost automatically stimulated. The regulation of lymph, blood and nutrient fluids and exchanges between the various skin compartments is controlled by the vascular and nervous systems which together are key factors in skin ageing and in the skin's response to attacks. Ageing or senescent phenotypes of endothelial cells may also increase stress factors, redox imbalance and the loss of replicative capacity, destroying the integrity of the endothelium and impairing micro-vessel angiogenesis (Zhang et al. 2014).

In the horny layer, the composition of the cornified envelope of the corneocytes drastically changes due to modifications in the expression profiles of the genes that encode its main components. The epidermal calcium gradient in the *stratum granulosum*, which regulates its genes, collapses during ageing, causing changes in gene expression profiles. In the epidermis, differentiation and junction proteins are also reduced. This is true for integrins (markers of the basal layer) and involucrin (marker of cellular differentiation), thus explaining the disruption of epidermal differentiation processes. A decrease in the level of aquaporin-3 (an aquaglyceroporin responsible for transporting water and small solutes such as glycerol) is also observed with age; this phenomenon is aggravated by chronic photo-exposure. The decline in the number or function of epidermal stem and progenitor cells or the major changes in their environment (called a stem cell niche), can accelerate the ageing phenotype of elderly skin (Naik et al. 2017). Similarly, hypoxia is much more pronounced in elderly human skin, especially in photo-exposed areas (Rezvani et al. 2011). Eccrine sweat glands appear to also be anchor points for the epithelium and modification of sweat exosomes with age could also interfere with skin immune homeostasis (Wu and Liu 2018). Sweat glands, as follicle appendages, are deeply colonized by microbiota which find a source of nutrients there and can move bidirectionally, either from or to the surface, and release important biological messages into the skin.

Elderly skin is characterised by the disappearance of the dermal papillae, skin atrophy, and disruption of the ECM of the dermis. There is a flattening of the DEJ and thus a decrease in the surface area for contact between the epidermis and dermis. Moreover, collagen types IV and VII, the main components of the anchoring plaques and fibrils, are reduced (Le Varlet et al. 1998; Langton et al. 2016). This is combined with some degradation of the dermal tissue which shows fragmented collagen bundles, a decrease in the overall level of collagen (mainly I and III) due to decreased fibroblast collagen production, and an increase in matrix metalloprotein-ases (MMPs), added to a decrease in the contractile activity of myofibroblasts. The highly organised architecture of the elastic microfibrils, containing elastin and fibrillin (glycoprotein), is also modified. It is thought that fragmentation of the fibre network causes fine surface wrinkles, the loss of fibrillin is responsible for perma-nent wrinkles, and deposits of poorly-organised elastin cause deep wrinkles (solar elastosis) (Zhang et al. 2017).

A decrease in hyaluronic acid, proteoglycans and glycosaminoglycans illustrates a loss of tone in the ECM and disrupts cell-matrix relations. During ageing, there are changes in the balances of the various cell populations in the epidermis and dermis, particularly in the dendritic cells and cells involved in inflammatory and immune phenomena (Zegarska et al. 2017; Pilkington et al. 2018). A histological comparison of biopsies taken on a wrinkled area of the face, with chronic photo-exposure (pre-auricular area), and a photo-protected retro-auricular control area from the same subjects demonstrated a disruption of cell balance in the skin: some cells migrated to the deepest layers to leave the skin while other cells entered the skin. To do so, they must cross the DEJ, an area already recognised as fragile in wrinkled photo-aged skin. A cutaneous imbalance in favour of cells with pro-inflammatory and proteolytic activity was thought to be responsible for major dis-ruption and collagen degradation, causing the skin to age and producing signs of photo-ageing such as wrinkles, loss of firmness, *etc.* There was also a loss of the skin's defensive cells (Langerhans cells, langerin expressing cells) and an epidermal gain of dendritic cells other than Langerhans cells (CD1A expressing cells which do not express langerin). This increase in the number of dendritic cells was correlated with the increased arrival of functional dermal precursors in the photo-exposed epi-dermis. These were CD1A, CD11B, CD14, CD36 and CCR6-expressing functional dendritic cells which do not express langerin. The photo-exposed skin shows char-acteristics of chronic inflammation with a dermal gain of macrophages, mast cells and CD4-expressing T cells and an epidermal gain of epidermal dendritic cells and CD4-expressing T cells. There was no biological inflammation or increase in the mRNA of inflammatory (interleukin-1α or IL-1α, tumour necrosis factor-α or TNF-α) or anti-inflammatory (IL-4 and IL-10, TGF-β) cytokines, which explained the absence of macroscopic clinical signs (Bosset et al. 2002, 2003). In addition, neu-rotransmitters and neuropeptides secreted by different skin cells can also influence, in an endocrine or paracrine manner, the skin ageing processes (Pincelli and Bonté 2004; Elewa et al. 2013). For example, nerve growth factor is involved in UVB irradiated keratinocyte apoptosis. The neurotrophin (NT) receptor p75NTR coun-terbalances the Trk receptors in neurotrophic functions. It also plays a critical role in keratinocyte stem cell transition to their progeny as well as in epidermal differen-tiation and induced apoptosis. The functional neurotrophin network (NT3, 4, 5), with distinct protective or not protective behavior after UVB or UVA keratinocyte irradiation, influences epidermal homeostasis and regenerative processes (Marconi et al. 2003; Pincelli 2017).

Aggravating Factors

Glycation and Oxidative Stress

Glycation is one of the fundamental mechanisms involved in ageing. Moreover, while glycation products occur in healthy subjects and contribute to the body ageing, long-term hyperglycaemia in diabetic subjects greatly increases these products, triggering toxic mechanisms that have a more serious impact on health. Glycation is a non-enzymatic reaction that occurs between a carbohydrate and a molecule with a free amino group, such as a protein. This irreversible and cumulative reaction takes place spontaneously in the body. All of the resulting metabolic intermediates are then reactive. Advanced glycation end-products (AGEs) accumulate in the ECM of the dermis. They affect long-lived proteins of the ECM such as collagen and elastin (Fig. 10.1). Glycation modifies the physical properties of the skin, rendering it more rigid and less elastic. It is important to note that the receptors for advanced glycation end-products are concentrated in the skin, especially expressed on keratinocytes, fibroblasts, and dendritic cells. Their activation contributes to changes in cellular activity and the production of cytokines and growth factors. AGEs in the epidermis can also disrupt the migration and proliferation functions of keratinocytes, thus impairing skin repair. In addition, the accumulation of AGEs can have a direct or indirect effect on skin pigmentation and its optical qualities. The activities of proteasome enzymes are reduced following their glycation, thus decreasing the process for the destruction of abnormal or oxidized proteins (Fournet et al. 2018). Glycation seems to contribute to a decline in the activity of enzymes that naturally protect against oxidative stress such as catalase and superoxide dismutase, leading to an increase in the formation of signs of ageing (Birch-Machin and Bowman 2016).

Hormones and Menopause

Hormonal factors are significant for the skin quality of women during and after menopause. In fact, oestrogen deficiency is responsible, in combination with other factors, for several cutaneous phenomena (Wilkinson and Hardman 2017). Thus, during menopause, the skin is drier, thinner, less firm, less elastic and may be more sensitive (Verdier-Sévrain et al. 2006; Falcone et al. 2017). These cutaneous changes associated with menopause are due to the wide distribution of hormone receptors. Indeed, the skin contains oestrogen and androgen receptors. Oestrogen receptors are located in keratinocytes, fibroblasts, sebaceous glands, sweat glands, hair follicles and the blood vessels of the skin. Hormonal co-receptors, such as for Klotho protein, are involved in endocrine fibroblast growth factor family signalling. Progesterone receptors are found in melanocytes and play a major role in the formation of dark spots, whereas androgen receptors are located in keratinocytes, hair follicles and sebaceous glands. Hyperandrogenism, hair loss, hirsutism, acne and grey hair are common signs observed in women after menopause. Men tend to have oilier skin, resulting in the formation of dilated pores and unsightly blackheads. In certain areas (eyebrows, nasal cavity, ear canal), there is an increase in hair growth which may or may not coexist with alopecia.

Ultraviolet and Infrared Radiation

It has been recognised for the last few decades that UV radiation has harmful effects on the skin and its components. The sun rays are made up of photons, which are powerful energy particles. The light spectrum includes visible light and infrared radiation in addition to UV radiation which is divided into UVA (415–315 nm) and UVB (315–280 nm) radiation and can penetrate the skin to varying degrees. Ninety-five percent of the UV rays reaching the skin are UVA rays and 5% are UVB rays.

UVB rays mainly penetrate the surface layers of the skin (10% reach the superficial dermis) and cause sunburn (Fig. 10.1). A single episode of sunburn is capable of inducing immunosuppression that can last almost 14 days. UVB radiation mainly affects the keratinocytes and also promotes the depletion of Langerhans cells, for the most part (>50%) 4–7 days after sunburn. It also stimulates melanin production and is responsible for tanning. It further induces epidermal thickening and can induce DNA lesions and mutations. UV rays act by changing the electronic structure of DNA bases, promoting chemical reactions between the bases, especially thymine (cyclobutane derivative) dimerisation. Additionally, UVB rays heavily influence the genes that regulate the circadian cycle in keratinocytes (Matsui et al. 2016).

Furthermore, UVB rays induce the secretion of epidermal cytokines, such as TNF-α and IL-10, which play a major role in immunosuppression. UV radiation also activates numerous cellular receptors for cytokines and growth factors such as the epidermal growth factor receptor, the TNF-α receptor, the platelet-activating factor receptor, the insulin receptor, the IL-1 receptor and the platelet-derived growth factor (PDGF) receptor. This simultaneous increase in cytokines and their receptors induces multiple signalling pathways, entailing an increase in the expression of p21Ras, a kinase regulated by the extracellular signal-regulated kinase (or ERK), c-Jun amino-terminal kinase (or JNK) and p38. The increase in c-Jun, in conjunction with c-Fos, leads to an increase in the activation of activator protein-1 (AP-1). UV irradiation thus activates transcription factors such as AP-1 as well as nuclear factor-kappa B (or NF-κB). AP-1 inhibits collagen expression and increase the expression of MMPs, leading to increased degradation of matrix proteins, thus contributing to and aggravating the formation of cutaneous wrinkles.

UVA rays have less energy and penetrate the deep layers of the skin including the dermis, where they act on the connective tissue and the endothelial cells of blood vessels. A distinction is currently made between UVA1 radiation (340–400 nm), a key player in photo-ageing, and UVA2 radiation (315–340 nm). They act to intensify the darkening of the melanin pigment preformed via UVB rays and can act as photo-sensitizers causing allergic reactions. They act indirectly by way of free radicals that attack proteins, membrane lipids and DNA, both nuclear and mitochondrial. Free radicals are highly unstable molecules with a missing electron that destabilise surrounding molecules (causing damage) by 'stealing' their electrons. UVA rays are poorly absorbed by DNA bases but they can excite cellular chromophores such as melanin, porphyrins, riboflavin and aromatic amino acids. In the dermis, they induce micro-tearing of collagen fibres. The overall rate of lesions formed in DNA following UVB irradiation is approximately 156 lesions/cell/J.cm^{-2} whereas it is 0.024 lesions/cell/J.cm^{-2} after UVA irradiation (Mouret et al. 2006). The capacity of UVB rays to generate thymine dimers is thus 6000 times greater than that of UVA rays. Visible light and infrared radiation activate MMP-1 and -9, modulate gene expression, and can also contribute to the occurrence or aggravation of dermal ageing (Greaves 2016). It is important to keep in mind that frequent and lasting exposure to UV radiation can promote the occurrence of seborrhoeic keratosis or precancerous lesions such as actinic keratosis. Blue light is an integral part of visible light and is emitted by electronic devices used on a daily basis such as smartphones, computers, TVs, LED lights, *etc.* The toxicity of digital blue light to retinal photoreceptors has been recently demonstrated and scientists have studied its action on the skin. Blue light (412, 419, 426 nm) generates free radicals, slows keratinocyte proliferation, and has a differentiation-promoting effect (Nakashima et al. 2017). The mitochondrial DNA 4977-bp common deletion is very widespread in elderly human skin. Accumulation of this deletion impairs the mitochondrial respiratory chain complex balance and may contribute to the 'imflam'ageing' phenomenon (Weiland et al. 2018). Induction of the common deletion, by UV rays, in human fibroblasts is combined with a decrease in oxygen absorption and ATP content and with an increase in MMP-1, while natural tissue inhibitors of these MMPS (termed TIMPs) are not increased. This imbalance is recognised as playing a major role in cutaneous photo-ageing. Infrared radiation (IR) includes IR-A (760–1440 nm), IR-B (1440–3000 nm) and IR-C (3000 nm). This radiation is not as strong as UV radiation but accounts for more than half of the solar energy that reaches the skin (versus 7% for UV rays). IR-A (near infrared) is the type of IR with the highest level of penetration compared to IR-B and -C, which affect only the surface of the skin, and can penetrate the dermis more deeply than UVA rays. Similarly to UV radiation, IR-A first leads to an increase in levels of reactive oxygen species (ROS) by disrupting the electron transport chain required for ATP production. The skin responds by increasing the secretion of ferritin, an intracellular antioxidant, and in the long term, IR-A is related to a decrease in epidermal proliferation and Langerhans cell density.

Blue light induces oxidative stress similar to that of UVA rays, mainly affecting proteins (by carbonylation) and mitochondria, reduces skin autofluorescence, and acts through the production of superoxide radicals more than through that of singlet oxygen. Blue light can also have positive effects and can be used in the event of inflammatory diseases related to acute disruption of the skin barrier, in psoriasis vulgaris and atopic dermatitis. Blue light, a shorter wavelength of the visible light spectrum, leads to sustained melanogenesis activity via calcium flux in melanocytes, and thus may partially explain part of the pigmentation that occurs with visible light (Regazetti et al. 2018).

Environment, Pollution, Lifestyle

As early as 2010, Krutmann and colleagues demonstrated that exposure to atmospheric pollution correlated significantly with signs of extrinsic cutaneous ageing, especially dark spots and also, to a lesser extent, wrinkles (Vierkötter et al. 2010). An increase in soot and particles from traffic (475 kg per year and per square kilometre) was associated with a 20% increase in pigmentation spots on the forehead and cheeks. Outdoor air pollution related to human activities and indoor pollution in homes are made up of several types of pollutants: particulate matter (broken down into both fine and coarse particles), gases (O_3, CO_2, CO, SO_2, NO_2) and volatile organic compounds. Many of these pollutants act by activating the aryl hydrocarbon receptor. According to studies on Caucasian and Asian skin, these pollutants promote the formation of dark spots on the skin, the oxidation of surface lipids, and the formation of wrinkles (Ding et al. 2017; Puri et al. 2017). More generally, the term 'exposome' is now associated with cutaneous ageing (Krutmann et al. 2017). Studies of human skin explants suggest that volatile organic compounds inhibit the proteasome, promote apoptosis, alter mitochondrial membrane potential, induce a decrease in Lon protease, stimulate the production of free radicals and induce lipid peroxidation (Dezest et al. 2017). Lon is a protease that has been shown to degrade proteins oxidised by oxidation in the mitochondrial matrix, a function similar to that of the 20S proteasome in the cytoplasm, and as being an essential factor for the repair of mitochondrial DNA (Ngo and Davis 2007; Petropoulos and Friguet 2006). The role of volatile organic compounds therefore appears to be a new co-factor in cutaneous ageing. The use of tanning beds and thus the exposure to UV rays as well as frequent swimming in chlorinated water can also generate stress and/or strip the skin's hydrolipidic film. Atmospheric ozone is produced as a result of chemical reactions, under the influence of solar radiation, primarily between nitrogen oxides and volatile organic compounds (hydrocarbons, solvents) present due to industrial activity. It is a very powerful oxidant that affects the superficial layers of the skin and is involved in reducing the level of natural antioxidants in the epidermis such as vitamin E and vitamin C. Lipid ozonisation products, which are small and highly diffusible molecules, cross the surface layers of the skin to reach the dermis, act as second messengers via lipid peroxidation, and generate an inflammatory response. In particular, there is an induction of heat shock proteins 27 and 70 and heme oxygenase-1, which are stress proteins activated by numerous endogenous and exogenous stimuli. Moreover, a potential pro-inflammatory role of ozone and its derivatives is related to an increase in levels of cyclooxygenase-2 and inducible nitric oxide synthase and the activation of NF-κB. Over the past years, there has thus been a combination of factors (UV rays, particulate matter, heat, soot, *etc.*) interacting with solar radiation to generate oxidative phenomena that have a significant impact on the skin.

Tobacco, Sugar, Alcohol and Xenobiotics

Smoking is a major environmental factor in premature skin ageing. *In vivo* studies have shown that tobacco smoke extract alters collagen production and increases the production of tropoelastin and MMPs which degrade the ECM, and it also causes an abnormal production of elastosis material in the dermis. Moreover, tobacco smoke induces ROS involved in premature skin ageing and the appearance of a greyish complexion. An epigenetic analysis of various ethnic groups in relation to their exposure to smoking proved that ethnic differences in DNA methylation can provide further insight into the molecular pathways involved in diseases caused by smoking in certain populations (Elliott et al. 2014; Jamal et al. 2017). The increased formation of wrinkles in smokers seems to be simultaneously of molecular origin, through the activation of MMPs and the increase in elastosis, and of behavioural origin, due to repeated pursing of the lips and contraction of the facial muscles, partly explaining the formation of wrinkles around the mouth (Heusèle et al. 2010). Studies have also shown that young children exposed to cigarette smoke are more likely to develop atopic eczema than non-exposed children.

In addition to damaging collagen, a high-sugar diet can promote the formation of glycation products and deactivate natural antioxidant enzymes. People suffering from diabetes often show early signs of cutaneous ageing and poor wound healing in the skin. Depending on the management of their disease, diabetic subjects can have up to 50 times the amount of glycation products in their skin compared to non-diabetic individuals.

The consumption of alcohol has a dehydrating effect on the body and causes the peripheral blood vessels to dilate beneath the skin surface. While a study on a cohort of twins demonstrated that regular alcohol consumption is sometimes associated with lower photo-ageing scores, a recent study significantly linked alcohol consumption to wrinkles (Martires et al. 2009; Hamer et al. 2017).

Elderly subjects are often prescribed several medications to treat their chronic diseases. In the USA, more than 2/3 of people over the age of 65 have high blood pressure, more than 25% have type-2 diabetes, and 17% take anti-psychotics. Anti-hypertensive, anti-diabetic, anti-depressant and anti-inflammatory medications are thus widely prescribed and can trigger, alone or in combination, cutaneous reactions that further weaken the skin of elderly subjects for example causing increased sensitivity to the sun, atrophy, epidermal thickening, rashes… (Skandalis et al. 2011) (Fig. 10.3).

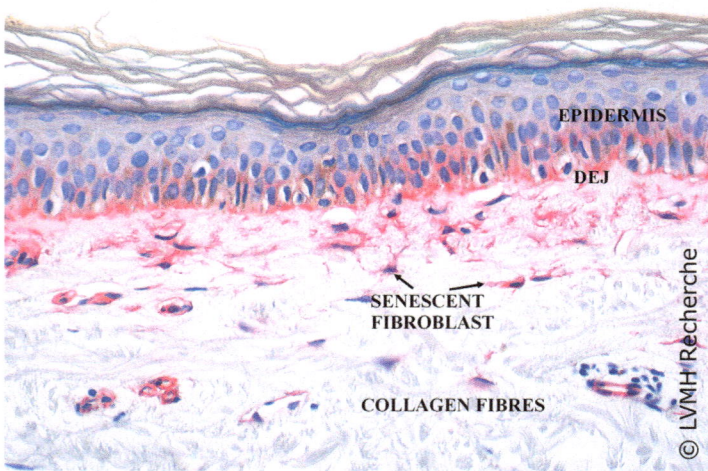

Fig. 10.3 Accumulation of senescent fibroblasts in aged skin. In human skin fibroblasts, senescence results in reduced extracellular matrix and increased matrix metalloproteinase production. In aged skin, reduced mechanical tension induces progressive fibroblast collapse and a loss of attachment to the extracellular matrix. *DEJ* dermal-epidermal junction. ×400. © LVMH Recherche

Effect of Ageing on Wound Healing

Normal Wound Healing Mechanisms

Granulation Tissue Formation and Myofibroblastic Differentiation

Normal tissue repair generally includes a number of overlapping phases. After injury, there is an early inflammatory step which is characterized by haemorrhage and clotting. The next phase allows the development of the granulation tissue involving fibroblasts which invade the wound and commence replacing the provisional matrix (essentially containing fibrin) with a more mature wound matrix rich in collagen type III. As the granulation tissue step proceeds, fibroblasts begin to acquire a new phenotype with prominent microfilament bundles. These typical myofibroblasts have been shown to develop a smooth muscle-like phenotype, and are responsible for wound contraction. Indeed, myofibroblasts express α-smooth muscle actin, the isoform typical of contractile vascular smooth muscle cells. Lastly, in the resolution phase of healing, the cellularity of granulation tissue decreases with a considerable loss of various cell types, including myofibroblasts, by apoptosis (Desmoulière et al. 1995). The signal for this cell death is unknown but may be

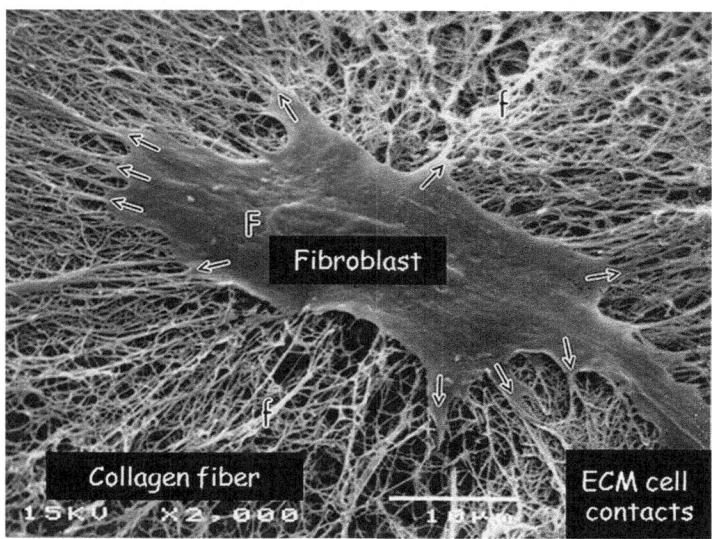

Fig. 10.4 Contractile (myo)fibroblast (F) attached to collagen fibres (f) (*in vitro*). In aged skin, fragmentation of the extracellular matrix (ECM) due to the action of matrix metalloproteinases impairs the structural integrity of the dermis. Fibroblasts cannot attach to fragmented collagen. Loss of attachment prevents fibroblasts from receiving mechanical information from their supporting substrate and they collapse. Mechanical tension is critical for maintenance of a normal balanced production of extracellular matrix and extracellular matrix-degrading enzymes. © LVMH Recherche

related to reductions in the concentrations of local trophic factors or to modifications in myofibroblast adhesion to the ECM (Fig. 10.4). Inappropriate delay of apoptosis, and thus increased survival of myofibroblasts activated during the healing process, may be a factor which leads to pathological situations (such as hypertrophic scars) and to excessive scarring. Indeed, the transient acquisition of the myofibroblast phenotype is beneficial for normal tissue repair processes but persistence of myofibroblasts results in tissue stiffening and deformation.

The growth factor that has the most powerful effect on stimulating conversion of fibroblasts to a myofibroblast phenotype is TGF-β1 (Desmoulière et al. 1993). In the presence of ED-A fibronectin, TGF-β1 stimulates full differentiation into α-smooth muscle actin-expressing myofibroblasts which then results in both increased collagen deposition but also an increase in generation of contractile forces. Certainly, TGF-β plays a major role in both physiological and pathological wound healing (Kiritsi and Nyström 2017). Other growth factors also have effects on myofibroblast proliferation and/or differentiation. These include growth factors such as connective tissue growth factor (CTGF), granulocyte macrophage colony stimulating factor, PDGF and endothelin-1 (for review, see Darby et al. 2016).

Mechanical Forces

It is well established that mechanical stress plays a fundamental role in the regulation of connective tissue as well as in tissue repair and regeneration and it has led to an increase in research linked to mechanobiology (Carver and Goldsmith 2013; Wong et al. 2011a). Stretching of the skin, for example, has been shown to stimulate myofibroblast activity. Indeed, it has been clearly shown that mechanically stressing dermal wounds in mice (which usually do not develop hypertrophic scars) causes hypertrophic scarring, presumably by decreasing the rate of apoptosis in myofibroblasts (Aarabi et al. 2007). Obviously, the ECM is involved in this dialogue between the cells and their mechanical microenvironment (Chiquet et al. 2009) (Fig. 10.4). Numerous *in vitro* studies have been conducted to evaluate the responsiveness of (myo)fibroblasts to stretch and to determine the molecular mechanisms involved (Nishimura et al. 2007). The efficacy of negative pressure therapy which has been shown to be effective for treating both acute and chronic wounds is due to different mechanisms, including reduction of tissue oedema, increase in blood supply, reduction in incidence of infection but also, certainly, to mechanical stimulation of granulation tissue formation via promotion of cell proliferation, increased angiogenesis, local release of growth factors and induction of myofibroblastic differentiation. Interestingly, it has been shown that a shorter healing time decreases the probability of developing a hypertrophic scar (Cubison et al. 2006). In addition, mechanical forces induce something similar to a chronic inflammatory state, through immune-dependent activation of both local and systemic cell populations, leading to pathological scar development (Wong et al. 2011b). Recently, it has also been shown that mechanical stretching of the tissue contributes to the development of fibrosis via mechanotransductional activation of TGF-β1 in human pulmonary fibrosis (Froese et al. 2016). Stretching of the lung ECM, due to breathing or mechanical ventilation which is used to support breathing of patients with lung injury or disease, may contribute to TGF-β1-related disease progression. Since it is not possible to stop breathing to avoid mechanical TGF-β1 activation from established fibrotic tissue, therapeutic intervention may however be possible by targeting the cellular side of mechanical TGF-β1 activation. Furthermore, in mechanically active organs, such as the lungs and skin, the role of TGF-β1 in excessive scarring and fibrosis is clearly crucial and the mechanisms involved in TGF-β1 activation constitute an important therapeutic target.

In addition, after extensive skin damage, such as post burn wounds, scar retraction often occurs and obviously this must be, if possible, avoided. It is important therefore to inhibit the processes leading to this pathological situation which appears to be mainly due to excessive contraction of myofibroblasts which fail to disappear by apoptosis. In this case, a skin substitute, particularly the dermal part of the substitute, may be an efficient contraction-blocker. Indeed, it has been suggested that contraction and regeneration are mutually antagonistic processes (Yannas et al. 2017). Using a specific ECM (such as collagen) scaffold, a drastic reduction in myofibroblast density has been observed and the regeneration processes was promoted. Recently, a specific mechanomodulatory ('embrace') device has been developed to reduce the mechanical forces on healing in surgical incisions (Longaker et al. 2014). The authors demonstrated that the 'embrace' device significantly reduced scarring following abdominoplasty surgery. Lastly, it is now well accepted that many of the methods for physical scar management, including compression therapy, negative pressure therapy, adhesive tape, and occlusive dressing therapy, function through mechanotransduction mechanisms (Yagmur et al. 2010).

Effect of Age on Skin Wound Repair

The elderly population is growing and in parallel, the clinical and socioeconomic burdens of non-healing cutaneous wounds are also increasing. Wound healing is slower in older adults and may also be affected by concomitant diseases, including diabetes mellitus or vascular disease. Additionally, as already discussed, exogenous factors such as nutritional status, dehydration, venous insufficiency or smoking may also reduce the healing capacity in the case of skin injuries.

The capability of skin to heal or to repair damage decreases with age, in correlation with increased cellular senescence (Gosain and DiPietro 2004; Blume-Peytavi et al. 2016). Age-associated senescence alters homeostasis via different mechanisms, amongst these, impaired platelet function and PDGF expression can particularly decrease skin healing capacity. As mentioned above, the aged wound environment contains pro-inflammatory cytokines and MMPs while the expression of TIMPs is also decreased. Consequently, excessive proteolysis associated with fibroblast senescence limits normal ECM deposition (Fig. 10.3). In addition, aged fibroblasts lose their responsiveness to TGFβ-1 and CTGF, leading to reduced collagen deposition. Senescence perturbs myofibroblastic differentiation with a reduction of α-smooth muscle actin expression, thus impairing wound contraction, and with altered ECM remodelling capacities. Finally, the inflammatory profile of the wound bed is altered with excessive immune cell recruitment. Re-epithelialisation is also impaired, showing reduced keratinocyte proliferation and abnormal keratinocyte terminal differentiation.

Numerous studies report alterations in the repair process of aged skin and suggest general factors associated with old age that might impair wound healing (for review, see Sgonc and Gruber 2013). Gould et al. recently discussed research questions that could guide future studies concerning age-associated changes in chronic wound healing. They underlined that in fact, the basic biology underlying chronic wounds and the influence of age-associated changes on wound healing, remain poorly understood (Gould et al. 2015). Novel, though not exhaustive, approaches to better understand age-related wound healing defects will be discussed below based on recently published studies.

Age-related alterations in collagen fibrils impairs skin structure and function and creates a tissue microenvironment that promotes age-related delayed skin wound healing. Quan and Fischer recently reviewed cellular mechanisms that give rise to self-perpetuating, collagen fibril fragmentation inducing a specific age-associated dermal microenvironment which leads to functional impairment of the skin (Quan and Fisher 2015). The DEJ is a zone that is rich in different ECM components and is particularly involved in skin homeostasis (see above). After partial thickness wounds, the DEJ is restored, in young adults, within 4 weeks. In contrast, the DEJ is deficient in elderly individuals (Fisher and Rittié 2017). It has been suggested that improvement of DEJ restoration during wound healing in the elderly could ameliorate repair quality.

In humans, hair follicles and eccrine sweat glands that are present around the wound bed are sources of keratinocytes which can proliferate to form the new epidermis. In aged *vs.* young skin, keratinocyte outgrowth is reduced leading to a delay in wound closure and to a thinner repaired epidermis. The defect of keratinocyte proliferation can be due to an impaired dialogue of these cells with the damaged ECM that is observed in aged skin (Rittié et al. 2016). This study underlines the importance of the skin appendages for the repair of human wounds. In addition, at the wound-edge, aged keratinocytes display a reduced capacity for proliferation and migration. Furthermore, in aged keratinocytes, communication with immune cells is impaired, decreasing their efficacy in restoration of skin barrier function after wounding, underlining that perturbations in epithelial-immune crosstalk can explain, at least in part, the age-related decline in wound-repair (Keyes et al. 2016).

It is now well accepted that wound healing is impaired by ROS through destructive oxidation of intracellular proteins, lipids and nucleic acids. Moor and colleagues reported that the deficiencies in antioxidant pathways in aged rats, leads to impaired wound healing and prolonged inflammation (Moor et al. 2014). Intracellular superoxide dismutase 1 (SOD1) regulates ROS levels and plays a major role for tissue homoeostasis. Wounds in aged animals displayed reduced myofibroblast differentiation and delayed wound healing, consistent with a decrease in the *in vitro* capacity for fibroblast-myofibroblast transition following oxidative stress. Young fibroblasts, with normal SOD1 expression, exhibit increased phosphorylation of ERK in response to elevated ROS, allowing the fibroblast-myofibroblast transition that is necessary for normal skin healing. In contrast, aged fibroblasts with reduced SOD1 expression display a reduced capacity to modulate intracellular ROS (Fujiwara et al. 2017). Age-associated wound healing impairments are correlated with fibroblast dysfunction due to decreased SOD1 expression and subsequent dysregulation of intracellular ROS. Fujiwara et al. (2017) suggested that a new therapeutic approach in the treatment of chronic non-healing wounds in the aged population could be strategies that target these mechanisms.

An interesting new approach concerns the identification of microRNAs which may negatively impact wound repair (Fahs et al. 2015). In unwounded aged skin (versus unwounded younger skin), the level of miR-200c was found to be elevated and overexpression of miR-200c in human *ex vivo* wounds was found to delay re-epithelialisation while modulation of miR-200c expression in keratinocytes *in vitro* revealed inhibitory effects of miR-200c on cell migration (Aunin et al. 2017). Interestingly, anti-miR-200c treatment in an *in vitro* model of wound healing accelerated wound closure, through an upregulation of genes controlling cell migration.

This study identified miR-200c to be a critical determinant inhibiting cell migration during skin repair after injury and may contribute to age-associated alterations in wound repair.

Finally, the last but not least point that we will discuss, concerns hormonal equilibrium. It is clear that after menopause, skin ageing increases. Recent studies reveal crucial insights into sex steroid changes with age, accounting for many of the physiological and architectural features associated with pathological age-related cutaneous healing. However, the link between sex steroid-deficiency and skin ageing is still not clearly understood. The sex steroid most widely discussed as an intrinsic contributor to skin ageing and pathological healing is 17β oestradiol (or oestrogen), although many others are involved. Oestrogen deficiency following menopause results in atrophic skin changes and acceleration of skin ageing (Thornton 2013). In addition, oestrogen deficiency is detrimental to many wound-healing processes, including inflammation and granulation tissue development, while exogenous oestrogen treatment largely reverses these effects (for review, see Wilkinson and Hardman 2017). It is also well accepted that many of the adverse effects associated with cutaneous ageing can also be induced by glucocorticoids (Tiganescu et al. 2011). Within target tissues, cortisol concentrations are regulated at the pre-receptor level by isozymes of 11β-hydroxysteroid dehydrogenase (11β-HSD) and 11β-HSD type 1 converts inactive cortisone to cortisol. 11β-HSD type 1 levels have been shown to increase with age (Tiganescu et al. 2011). Thus, the age-associated increase in dermal 11β-HSD type 1, which induces increased local glucocorticoid activation, may contribute to the adverse changes in skin morphology and function associated with ageing. Interestingly, blockade of the age-related increase in 11β-HSD type 1 activity may promote improved structural and functional properties in ageing skin (Tiganescu et al. 2013).

Skin Care and Ageing

Lifestyle plays a major role in the emergence of signs of ageing and the progression of cutaneous ageing. It is essential to avoid prolonged unprotected sun exposure, to stay away from polluted environments and pollutants, avoid temperature stresses, and to pay attention to diet. A suitable skincare routine can be used to protect the skin and provide it with the means to repair itself. Thoroughly cleansing the skin with gentle surfactants, thoroughly moisturising the skin, and protecting it from the sun and pollution are some basic actions. Cleansing with lotions is a way to eliminate substances that clog pores, traces of pollution, and dead cells, thus leaving the skin clean and receptive to other treatments, especially moisturising ones (Verdier-Sévrain and Bonté 2007). For more targeted actions that are intended to slow down, conceal or repair signs of ageing, specific skincare products can be used (Ganceviciene et al. 2012; Herman et al. 2013). To conceal uneven pigmentation or highlight certain areas of the face, cosmetic or make-up products should be used sparingly. Cosmetic products contain a number of active substances. To prevent ageing, the use of chemical and mineral UVA and UVB filters effectively keeps radiation from penetrating the skin (Flament et al. 2017). It is possible to strengthen the filtering properties of cosmetic products by optimising mixtures of filters and adding natural oils or extracts containing molecules that absorb UV rays (Korać and Khambholja 2011).

To combat the effects of pollution and external stresses, antioxidants are also commonly added. Antioxidants are compounds that provide protection against damage induced by free radicals and attacks by electrophilic compounds. The harmful effects of free radicals on cells depend on their concentration and speed and therefore their distance of action. They act either directly through molecular attacks or as second messengers in various redox pathways. They can also have a positive effect when involved in stress-coping phenomena. Many of the antioxidant substances used in anti-ageing cosmetics are chemical molecules (vitamin C, vitamin E, glutathione, coenzyme Q10) or molecules of plant origin such as carotenoids, flavonoids (flavones, flavanones, anthocyanins, catechins), stilbenoids (resveratrol, imbricatin), and polyphenols (ellagic acid, gallic acid, chlorogenic acid, tannins). Complex plant extracts (tea, orchids, skullcap, liquorice, ruscus) also genuinely combat cutaneous micro-inflammation that promotes ageing or even fibroblast senescence (Draelos 2010; Bonté et al. 2011; Bonté and Beauchef 2013). Other types of molecules, including retinol, retinyl palmitate, monoterpene glycosides such as paeoniflorin, oligosaccharides, hydrolysed fucoidans, ectoine, ginsenosides, ergothioneine, triterpenoidic saponins and sapogenols have been used for various anti-ageing effects. Innovative beauty products are generally actually prebiotics, ingredients that promote the growth of beneficial microorganisms. More recently, products containing probiotics (mainly lysates of Lactobacillus, Bifidobacterium, Vitreoscilla, Aquaphilus dolomiae) have appeared on the market with the aim of strengthening the regulation of microbiota equilibrium, supporting skin barrier function and preventing erythema (Guéniche et al. 2009, 2010). Because it is known that microbiota communicate with skin cells, feeding or activating by skincare ingredients of our skin microbiota help it to release beneficial signals for regeneration of aged skin. The relationship between the microbiota and inter-individual variation of skin status and response to cosmetic interventions, emphasizes the importance of future personalized cosmetic approaches. A growing number of molecules from marine flora resources (cynanobacteria, macro and microalgae) have recently generated attention due to their chemical structural diversity and biological potentials (Brunt and Burgess 2018).

To help smooth the skin microrelief, the use of alpha or beta hydroxyl acids, whether natural or grafted with long-chain fatty acids, or fruit acids (malic, citric acids) has demonstrated real efficacy, since their highly acidic pH promotes local desquamation. A number of glycols (glycerol, propylene glycol and butylene glycol) act as keratin plasticisers and promote skin suppleness and hydration. Plant oils, rich in polyunsaturated fatty acids, tocotrienols and ceramides or their precursors, help to strengthen and "nourish" the skin barrier on the surface. Hyaluronic acids of various molecular weights, with a high capacity for swelling in the presence of water, induce a smoothing effect. Currently, more or less invasive methods in aesthetic medicine (injections, peels, lasers, *etc.*) can be used to combat signs of ageing. These treatments should be used in combination with cosmetic products.

Conclusion

To better understand age-related problems in wound healing and the mechanisms involved in the development of chronic wounds, it is important to comprehensively understand the entire course of wound healing and scar formation, including events occurring very early, that is, at the time of wounding. This point has been recently well-emphasised by Kwon and Gurtner (2017) in comments concerning the study of Butzelaar and colleagues (Butzelaar et al. 2016). Additionally, mechanical stimuli (as discussed above) can prolong acute inflammation via activation of the cellular mechanotransduction pathways (Wong et al. 2011c). In the same vein, Pakshir and Hinz (2018) have underlined that the dialogue between myofibroblasts, macrophages and the ECM in which the mechanical microenvironment is essential for the development of normal healing. Miscommunication can result in either insufficient (chronic) or exacerbated (fibrotic) repair (Pakshir and Hinz 2018). It is interesting to point out that the same types of mechanisms are involved to repair the micro-tearings of the skin which appear within the skin with age. It has been observed that the elderly are often undernourished, accelerating age-related wound healing impairment and explaining, at least in part, the development of chronic non-healing wounds. We can note that hyaluronic acid, which is at high concentration in fetal skin and which is responsible at least in part for the scarless healing of fetal wounds, improves dermal healing of aged damaged skin (Damodarasamy et al. 2014). The authors propose that hyaluronic acid has potential clinical utility for improving cutaneous wound healing in ageing. Indeed, some dressings containing hyaluronic acid that are used on damaged skin are currently available. It is also interesting to underline that various cosmetic anti-ageing products contain hyaluronic acid.

Dietary supplementation with balanced essential amino acids could also serve as a strategy to accelerate wound healing and may therefore be a simple but pivotal therapeutic approach (Corsetti et al. 2017). Pharmacologically, several caloric restriction (CR) mimetics (in particular resveratrol and metformin) have recently been shown to retard ageing and alleviate age-related pathological changes in various experimental models (Vaiserman et al. 2016). Locally applied metformin and resveratrol improves vascularization of the wound bed, this effect being attributed to stimulation of adenosine monophosphate-activated protein kinase (AMPK) pathway, a key mediator in wound healing (Zhao et al. 2017). As a therapeutic approach, local treatment using metformin and resveratrol to prevent age-related AMPK suppression and inhibition of angiogenesis in wound beds could be considered. In aged skin, topically applied metformin and resveratrol were shown to restore cell viability in wound beds, with metformin showing more prominent anti-ageing effects. These observations open new therapeutic approaches by using CR-based anti-ageing pharmacology in wound healing, including prevention of pressure ulcers (bedsores) particularly in the elderly.

At the conclusion of this review, we will propose a number of cosmetic dermatologic approaches developed to reduce age-related skin damages or to improve the skin repair process in the elderly. Prevention, protection and repair are three key words to consider in taking care of the skin.

A healthy lifestyle is the first line of prevention to avoid the accelerated appearance of signs of ageing in the skin which can be reinforced by using non-aggressive cleansers and lotions.

It is crucial to protect the skin from the damaging effects of UV rays and to counteract the deleterious effects of oxidative stress. A complementary method is to use dedicated skin care to reduce and repair the effects of oxidative stress by scavenging free radicals, to improve skin vitality, to increase skin hydration and thickness, to promote skin elastin and collagen synthesis, and to protect the ECM of the skin by inhibiting enzymatic degradation by MMPs. Depending on the signs of ageing to be treated, specialized products, including anti-wrinkle, firming, anti-spot products or even more invasive treatments are proposed to consumers. One of the most basic actions to take care of the skin is to hydrate it by using moisturizing products containing glycols (which act as plasticizers of keratins), active ingredients that stimulate inward water flux and hyaluronic acid as a water fixing agent. Molecules of hyaluronic acid are able to penetrate more or less quickly and easily into the skin depending on their molecular weight. This explains why many skincare products contain mixtures of hyaluronic acid with different molecular weights (ranging from a few thousand Daltons to millions of Daltons) to target the different layers of the skin. To obtain plumper, more toned and therefore brighter skin, it is possible to use beads or capsules based on hyaluronic acid. By ensuring perfect hydration, re-epithelialization and cell migration will also be promoted. Hyaluronic acid is often used in injections in aesthetic medicine, in combination with autologous fat grafting, for facial rejuvenation and contouring. In terms of cosmetics, if a primer can create a protective barrier on the skin, the use of a concealer or a foundation will help to cover any imperfections, blemishes or any other marks.

We will conclude by saying that for an individual, his or her skin health is essential for their well-being and their comfort in both their professional and personal surroundings. In addition, for society, improving the rate of healing in non-healing wounds of the elderly must be, taking into account the cost and poor quality of life engendered by this disease, a primary objective.

References

Aarabi S, Bhatt KA, Shi Y et al (2007) Mechanical load initiates hypertrophic scar formation through decreased cellular apoptosis. FASEB J 21(12):3250–3261

Alexander CM, Kasza I, Yen CL et al (2015) Dermal white adipose tissue: a new component of the thermogenic response. J Lipid Res 56(11):2061–2069

Aunin E, Broadley D, Ahmed MI et al (2017) Exploring a role for regulatory miRNAs in wound healing during ageing: involvement of miR-200c in wound repair. Sci Rep 7(1):3257. https://doi.org/10.1038/s41598-017-03331-6

Belkaid Y, Segre JA (2014) Dialogue between skin microbiota and immunity. Science 346(6212):954–959

Bergler-Czop B, Miziotek B (2017) Aging-how do we know? Acta Dermatoveneol Croat 25(1):50–56

Binic I, Lazarevic V, Ljubenovic M et al (2013) Skin ageing: natural weapons and strategies. Evid Based Complement Alternat Med 2013:1–10. https://doi.org/10.1155/2013/827248

Birch-Machin MA, Bowman A (2016) Oxidative stress and ageing. Br J Dermatol 175(Suppl 2):26–29

Blessing W, McAllen R, McKinley M (2016) Control of the cutaneous circulation by the central nervous system. Compr Physiol 6(3):1161–1197

Blume-Peytavi U, Kottner J, Sterry W et al (2016) Age-associated skin conditions and diseases: current perspectives and future options. Gerontologist 56(Suppl 2):S230–S242

Bonté F, Beauchef G (2013) Antioxidative phytochemicals and skin care. In: Kuang H-X (ed) Phytochemicals: occurrence, nature, health effects and antioxidant properties. Nova Science Publishers, Hauppauge, pp 189–217

Bonté F, Simmler C, Lobstein A et al (2011) Action of an extract of Vanda coerulea on the senescence of skin fibroblasts. Ann Pharm Fr 69(3):177–181

Bosset S, Barré P, Chalon A et al (2002) Skin ageing: clinical and histopathologic study of permanent and reducible wrinkles. Eur J Dermatol 12(3):247–252

Bosset S, Bonnet-Duquennoy M, Barre P et al (2003) Decreased expression of keratinocyte beta1 integrins in chronically sun-exposed skin in vivo. Br J Dermatol 148(4):770–778

Brunt EG, Burgess JG (2018) The promise of marine molecules as cosmetic active ingredients. Int J Cosmet Sci 40(1):1–165

Butzelaar L, Schooneman DP, Soykan EA et al (2016) Inhibited early immunologic response is associated with hypertrophic scarring. Exp Dermatol 25(10):797–804

Byrd AL, Belkaid Y, Segre JA (2018) The human skin microbiome. Nat Rev Microbiol 16(3):143–155

Callaghan DJ, Singh B, Reddy KK (2017) Skin aging in individuals with skin of color. In: Vashi NA, Maibach HI (eds) Dermatoanthropology of ethnic skin and hair. Springer, Basel, pp 389–403

Carver W, Goldsmith EC (2013) Regulation of tissue fibrosis by the biomechanical environment. Biomed Res Int 2013:101979. https://doi.org/10.1155/2013/101979

Chen YE, Fischbach MA, Belkaid Y (2018) Skin microbial-host interactions. Nature 553(7689):427–436

Chien AL, Suh J, Cesar SSA et al (2016) Pigmentation in African American skin decreases with skin aging. J Am Acad Dermatol 75(4):782–787

Childs BG, Gluscevic M, Baker DJ et al (2017) Senescent cells: an emerging target for diseases of ageing. Nat Rev Drug Discov 16(10):718–735

Chiquet M, Gelman L, Lutz R et al (2009) From mechanotransduction to extracellular matrix gene expression in fibroblasts. Biochim Biophys Acta 1793(5):911–920

Choi W, Yin L, Smuda C et al (2017) Molecular and histological characterization of age spots. Exp Dermatol 26(3):242–248

Clayton K, Vallejo AF, Davies J et al (2017) Langerhans cells-programmed by the epidermis. Front Immunol 8:1676. https://doi.org/10.3389/fimmu.2017.01676

Corsetti G, Romano C, Pasini E et al (2017) Diet enrichment with a specific essential free amino acid mixture improves healing of undressed wounds in aged rats. Exp Gerontol 96:138–145

Cubison TCS, Pape SA, Parkhouse N (2006) Evidence for the link between healing time and the development of hypertrophic scars (HTS) in paediatric burns due to scald. Burns 32(8):992–999

Damodarasamy M, Johnson RS, Bentov I et al (2014) Hyaluronan enhances wound repair and increases collagen III in aged dermal wounds. Wound Rep Reg 22(4):521–526

Darby IA, Zakuan N, Billet F et al (2016) The myofibroblast, a key cell in normal and pathological tissue repair. Cell Mol Life Sci 73(6):1145–1157

Desmoulière A, Geinoz A, Gabbiani F et al (1993) Transforming growth factor-beta 1 induces alpha-smooth muscle actin expression in granulation tissue myofibroblasts and in quiescent and growing cultured fibroblasts. J Cell Biol 122(1):103–111

Desmoulière A, Redard M, Darby I et al (1995) Apoptosis mediates the decrease in cellularity during the transition between granulation tissue and scar. Am J Pathol 146(1):56–66

Dezest M, Le Bechec M, Chavatte L et al (2017) Oxidative damage and impairment of protein quality control system in keratinocytes exposed to a volatile organic compounds cocktail. Sci Rep 7(1):107707. https://doi.org/10.1038/s41598-017-11088-1

Ding A, Yang Y, Zhao Z et al (2017) Indoor PM2.5 exposure affects skin aging manifestation in a Chinese population. Sci Rep 7(1):15329. https://doi.org/10.1038/s41598-017-15295-8

Draelos ZD (ed) (2010) Cosmetic dermatology. Products and procedures. Wiley Blackwell, Oxford

Dumas M, Langle S, Noblesse E et al (2005) Histological changes in the histology of Japanese skin with aging. Int J Cos Sci 27(1):47–50

Eckhart L, Lippens S, Tschachler E et al (2013) Cell death by cornification. Biochim Biophys Acta 1833(12):3471–3480

Elewa R, Makrantonaki E, Zouboulis CC (2013) Neuropeptides and skin aging. Horm Mol Biol Clin Investig 16(1):29–33

Elliott HR, Tillin T, McArdle WL et al (2014) Differences in smoking associated DNA methylation patterns in South Asians and Europeans. Clin Epigenetics 6(1):4. https://doi.org/10.1186/1868-7083-6-4

Fahs F, Bi X, Yu FS et al (2015) New insights into microRNAs in skin wound healing. IUBMB Life 67(12):889–896

Falcone D, Richters RJH, Uzunbajakava NE et al (2017) Sensitive skin and the influence of female hormone fluctuations: results from a cross-sectional digital survey in the Dutch population. Eur J Dermatol 27(1):42–48

Fisher G, Rittié L (2017) Restoration of the basement membrane after wounding: a hallmark of young human skin altered with aging. J Cell Commun Signal 12:401–411. https://doi.org/10.1007/s12079-017-0417-3

Flament F, Gautier B, Benize AM et al (2017) Seasonally-induced alterations of some facial signs in Caucasian women and their changes induced by a daily application of a photo-protective product. Int J Cosmet Sci 39(6):664–675

Fournet M, Bonté F, Desmoulière A (2018) Glycation damage: a possible hub for major pathophysiological disorders and aging. Aging Dis 9(5):880–900

Froese AR, Shimbori C, Bellaye PS et al (2016) Stretch-induced activation of transforming growth factor-β1 in pulmonary fibrosis. Am J Respir Crit Care Med 194(1):84–96

Fujiwara T, Dohi T, Maan ZN et al (2017) Age-associated intracellular superoxide dismutase deficiency potentiates dermal fibroblast dysfunction during wound healing. Exp Dermatol 4. https://doi.org/10.1111/exd.13404

Ganceviciene R, Liakou AI, Theodoridis A et al (2012) Skin anti-aging strategies. Dermatoendocrinol 4(3):308–319

Gosain A, DiPietro LA (2004) Aging and wound healing. World J Surg 28(3):321–326

Gouin O, L'Herondelle K, Lebonvallet N et al (2017) TRPV1 and TRPA1 in cutaneous neurogenic and chronic inflammation: pro-inflammatory response induced by their activation and their sensitization. Protein Cell 8(9):644–661

Gould L, Abadir P, Brem H et al (2015) Chronic wound repair and healing in older adults: current status and future research. J Am Geriatr Soc 63(3):427–438

Gragnani A, Mac Cornick S, Chominski V et al (2014) Review of major theories of skin aging. Adv Aging Res 3(4):265–284

Greaves AJ (2016) The effects of narrowbands of visible light upon some skin disorders: a review. Int J Cosmet Sci 38(4):325–345

Grice EA, Kong HH, Renaud G et al (2008) A diversity profile of the human skin microbiota. Genome Res 18(7):1043–1050

Guéniche A, Philippe D, Bastien P et al (2009) Probiotics for photoprotection. Dermatoendocrinology 1(5):275–279

Guéniche A, Benyacoub J, Philippe D et al (2010) Lactobacillus paracasei CNCM I-2116 (ST11) inhibits substance P-induced skin inflammation and accelerates skin barrier function recovery in vitro. Eur J Dermatol 20(6):731–737

Hamer MA, Pardo LM, Jacobs LC et al (2017) Lifestyle and physiological factors associated with facial wrinkling in men and women. J Invest Dermatol 137(8):1692–1699

Herman J, Rost-Roszkowska M, Skotnicka-Graca U (2013) Skin care during the menopause period: noninvasive procedures for beauty studies. Postepy Dermatol Alergol 30(6):388–395

Heusèle C, Cantin H, Bonté F (2010) Lips and lipsticks. In: Draelos ZD (ed) Cosmetic dermatology: products and procedures. Wiley Blackwell, Oxford, pp 184–190

Jain M, Khadilkar N, De Sousa A (2017) Burn-related factors affecting anxiety, depression and self-esteem in burn patients: an exploratory study. Ann Burns Fire Disasters 30(1):30–34

Jamal B, Bokhari A, Aljahdali B (2017) The effect of smoking in facial aging among females in Saudi Arabia. Clin Res Dermatol Open Access 4(2):1–4

Keyes BE, Liu S, Asare A et al (2016) Impaired epidermal to dendritic T cell signaling slows wound repair in aged skin. Cell 167(5):1323–1338

Kiritsi D, Nyström A (2017) The role of TGFβ in wound healing pathologies. Mech Ageing and Dev 172:51–58. https://doi.org/10.1016/j.mad.2017.11.004

Klar AS, Zimoch J, Biedermann T (2017) Skin tissue engineering: application of adipose-derived stem cells. Biomed Res Int 2017:9747010–9747012. https://doi.org/10.1155/2017/9747010

Klein J, Permana PA, Owecki M et al (2007) What are subcutaneous adipocytes really good for? Exp Dermatol 16(1):45–70

Korać RR, Khambholja KM (2011) Potential of herbs in skin protection from ultraviolet radiation. Pharmacogn Rev 5(10):164–173

Krutmann J, Bouloc A, Sore G et al (2017) The skin aging exposome. J Dermatol Sci 85(3):152–161

Kwon SH, Gurtner GC (2017) Is early inflammation good or bad? Linking early immune changes to hypertrophic scarring. Exp Dermatol 26(2):133–134

Langton AK, Halai P, Griffiths CE et al (2016) The impact of instrinsic ageing on the protein composition of the dermal-epidermal junction. Mech Ageing Dev 156:14–16

Law MH, Medland SE, Zhu G et al (2017) Genome-wide association shows that pigmentation genes play a role in skin aging. J Invest Dermatol 137(9):1887–1894

Le Varlet B, Chaudagne C, Saunois A et al (1998) Age-related functional and structural changes in human dermo-epidermal junction components. J Investig Dermatol Symp Proc 3(2):172–179

Lee E, Kim S, Lee J et al (2014) Ethnic differences in objective and subjective skin irritation response: an international study. Skin Res Technol 20(3):265–269

Longaker MT, Rohrich RJ, Greenberg L et al (2014) A randomized controlled trial of the embrace advanced scar therapy device to reduce incisional scar formation. Plast Reconstr Surg 134(3):536–546

Luebberding S, Krueger N, Kerscher M (2014) Age-related changes in male skin: quantitative evaluation of one hundred and fifty male subjects. Skin Pharmacol Physiol 27(1):9–17

Lugo LM, Lei P, Andreadis ST (2011) Vascularization of the dermal support enhances wound re-epithelialization by in situ delivery of epidermal keratinocytes. Tissue Eng Part A 17(5–6):665–675

Maksimovic S, Nakatani M, Baba Y et al (2014) Epidermal Merkel cells are mechanosensory cells that tune mammalian touch receptors. Nature 509(7502):617–621

Marconi A, Terracina M, Fila C et al (2003) Expression and function of neurotrophins and their receptors in cultured human keratinocytes. J Invest Dermatol 121(6):1515–1521

Martires KJ, Fu P, Polster AM et al (2009) Factors that affect skin aging: a cohort-based survey on twins. Arch Dermatol 145(12):1375–1379

Matsui MS, Pelle E, Dong K et al (2016) Biological rhythms in the skin. Int J Mol Sci 17(6). https://doi.org/10.3390/ij.ms1700801

Moor AN, Tummel E, Prather JL et al (2014) Consequences of age on ischemic wound healing in rats: altered antioxidant activity and delayed wound closure. Age (Dordr) 36(2):733–748

Mouret S, Baudouin C, Charveron M et al (2006) Cyclobutane pyrimidine dimers are predominant DNA lesions in whole human skin exposed to UVA radiation. Proc Natl Acad Sci U S A 103(37):13765–13770

Naik S, Bouladoux N, Linehan JL et al (2015) Commensal-dendritic-cell interaction specifies a unique protective skin immune signature. Nature 520(7545):104–108

Naik S, Larsen SB, Gomez NC et al (2017) Inflammatory memory sensitizes skin epithelial stem cells to tissue damage. Nature 550(7677):475–480

Nakashima Y, Ohta S, Wolf A (2017) Blue light-induced oxidative stress in live skin. Free Radic Biol Med 108:300–310

Ngo JK, Davis KJ (2007) Importance of the lon protease in mitochondrial maintenance and the significance of declining lon in aging. Ann N Y Acad Sci 1119:78–87

Nishimura K, Blume P, Ohgi S et al (2007) Effect of different frequencies of tensile strain on human dermal fibroblast proliferation and survival. Wound Rep Reg 15(5):646–656

Noblesse E, Nizard C, Cario-André M et al (2006) Skin ultrastructure in senile lentigo. Skin Pharmacol Physiol 19(2):95–100

Pakshir P, Hinz B (2018) The big five in fibrosis: macrophages, myofibroblasts, matrix, mechanics, and miscommunication. Matrix Biol 68–69:81–93. https://doi.org/10.1016/j.matbio.2018.01.019

Petropoulos I, Friguet B (2006) Maintenance of proteins and aging: the role of oxidized protein repair. Free Radic Res 40(12):1269–1276

Piérard GE, Hermanns-Lê T, Gaspard U et al (2014) Asymmetric facial skin viscoelasticity during climacteric aging. Clin Cosmet Investig Dermatol 7:111–118

Pilkington SM, Ogden S, Eaton LH et al (2018) Lower levels of interleukin-1β gene expression are associated with impaired Langerhans' cell migration in aged human skin. Immunology 153(1):60–70

Pincelli C (2017) p75 Neurotrophin receptor in the skin: beyond its neurotrophic function. Front Med (Lausanne) 4:22. https://doi.org/10.3389/fmed.2017.00022

Pincelli C, Bonté F (2004) The 'beauty' of skin neurobiology. J Cosmet Dermatol 2(3–4):195–198

Prescott SL, Larcombe DL, Logan AC et al (2017) The skin microbiome: impact of modern environments on skin ecology, barrier integrity, and systemic immune programming. World Allergy Organ J 10(1):29. https://doi.org/10.1186/s40413-017-0160-5

Puri P, Nandar SK, Kathuria S et al (2017) Effects of air pollution on the skin: a review. Indian J Dermatol Venereol Leprol 83(4):415–423

Quan T, Fisher GJ (2015) Role of age-associated alterations of the dermal extracellular matrix microenvironment in human skin aging: a mini-review. Gerontology 61(5):427–434

Regazetti C, Sormani L, Debayle D et al (2018) Melanocytes sense blue light and regulate pigmentation through opsine3. J Invest Dermatol 138(1):171–178

Rezvani HR, Ali N, Serrano-Sanchez M et al (2011) Loss of epidermal hypoxia-inducible factor-1α accelerates epidermal aging and affects re-epithelialization in human and mouse. J Cell Sci 124(Pt24):4172–4183

Rinnerthaler M, Bischof J, Streubel MK et al (2015) Oxidative stress in aging human skin. Biomol Ther 5(2):545–589

Ritschka B, Storer M, Mas A et al (2017) The senescence-associated secretory phenotype induces cellular plasticity and tissue regeneration. Genes Dev 31(2):172–183

Rittié L, Farr EA, Orringer JS et al (2016) Reduced cell cohesiveness of outgrowths from eccrine sweat glands delays wound closure in elderly skin. Aging Cell 15:842–852

Rocquet C, Bonté F (2002) Molecular aspects of skin aging: recent data. Acta Dermatoven APA 11(3):71–94

Roosterman D, Goerge T, Schneider SW et al (2006) Neuronal control of skin function: the skin as a neuroimmunoendocrine organ. Physiol Rev 86(4):1309–1379

Sarkar R, Ranjan R, Garg S et al (2016) Periorbital hyperpigmentation: a comprehensive review. J Clin Aesthet Dermatol 9(1):49–55

Schneider H, Mühle C, Pacho F (2007) Biological function of laminin-5 and pathogenic impact of its deficiency. Eur J Cell Biol 86(11–12):701–717

Sgonc R, Gruber J (2013) Age-related aspects of cutaneous wound healing: a mini-review. Gerontology 59(2):159–164

Shibagaki N, Suda W, Clavaud C et al (2017) Aging-related changes in the diversity of women's skin microbiomes associated with oral bacteria. Sci Rep 7(1):10567. https://doi.org/10.1038/s41598-017-10834-9

Singh SK, Baker R, Sikkink SK et al (2017) E-cadherin mediates ultraviolet radiation- and calcium-induced melanin transfer in human skin cells. Exp Dermatol 26(11):1125–1133

Skandalis K, Spirova M, Gaitanis G et al (2011) Drug induces bullous pemphigoid in diabetes mellitus patients receiving dipeptidyl peptidase-IV inhibitors plus metformin. J Eur Acad Dermatol Venereol 26(2):249–253

Skoczyńska A, Budzisz E, Trznadel-Grodzka E et al (2017) Melanin and lipofuscin as hallmarks of skin aging. Postepy Dermatol Alergol 34(2):97–103

Stahley SN, Bartle EI, Atkinson CE et al (2016) Molecular organization of the desmosome as revealed by direct stochastic optical reconstruction microscopy. J Cell Sci 129(15):2897–2904

Thornton MJ (2013) Estrogens and aging skin. Dermatoendocrinology 5(2):264–270

Tiganescu A, Walker EA, Hardy RS et al (2011) Localization, age- and site-dependent expression, and regulation of 11β-hydroxysteroid dehydrogenase type 1 in skin. J Invest Dermatol 131(1):30–36

Tiganescu A, Tahrani AA, Morgan SA et al (2013) 11β-Hydroxysteroid dehydrogenase blockade prevents age-induced skin structure and function defects. J Clin Invest 123(7):3051–3060

Tobin DJ (2017) Introduction to skin aging. J Tissue Viability 26(1):37–46

Tsuruta D, Hashimoto T, Hamill KJ et al (2011) Hemidesmosomes and focal contact proteins: functions and cross-talk in keratinocytes, bullous diseases and wound healing. J Dermatol Sci 62(1):1–7

Two AM, Nakatsuji T, Kotol PF et al (2016) The cutaneous microbiome and aspects of skin antimicrobial defense system resist acute treatment with topical skin cleansers. J Invest Dermatol 136(10):1950–1954

Vaiserman AM, Lushchak OV, Koliada AK (2016) Anti-aging pharmacology: promises and pitfalls. Ageing Res Rev 31:9–35

Vashi NA, de Castro Maymone MB, Kundu RV (2016) Aging differences in ethnic skin. J Clin Aesthet Dermatol 9(1):31–38

Verdier-Sévrain S, Bonté F (2007) Skin hydration: a review on its molecular mechanisms. J Cosmetic Dermatol 6(2):75–82

Verdier-Sévrain S, Bonté F, Gilchrest B (2006) Biology of estrogens in skin: implications for skin aging. Exp Dermatol 15(2):83–94

Vierkötter A, Schikowski T, Ranft U et al (2010) Airborne particle exposure and extrinsic skin aging. J Invest Dermatol 130(12):2719–2726

Voegeli R, Rawlings AV, Seroul P et al (2015) A novel continuous colour mapping approach for visualization of facial skin hydration and transepidermal water loss for four ethnic groups. Int J Cosmet Sci 37(6):595–605

Warrick E, Duval C, Nouveau S et al (2017) Morphological and molecular characterization of actinic lentigos reveals alterations of the dermal extracellular matrix. Br J Dermatol 177(6):1619–1632

Watt FM, Fujiwara H (2011) Cell-extracellular matrix interactions in normal and diseased skin. Cold Spring Harb Perspect Biol 3(4). https://doi.org/10.1101/cshperspect.a005124

Weiland D, Brachvogel B, Horning-Do HT et al (2018) Imbalance of mitochondrial respiratory chain complexes in the epidermis induces severe skin inflammation. J Invest Dermatol 138(1):132–140

Wilkinson HN, Hardman MJ (2017) The role of estrogen in cutaneous ageing and repair. Maturitas 103:60–64

Wong VW, Akaishi S, Longaker MT et al (2011a) Pushing back: wound mechanotransduction in repair and regeneration. J Invest Dermatol 131(11):2186–2196

Wong VW, Paterno J, Sorkin M et al (2011b) Mechanical force prolongs acute inflammation via T-cell-dependent pathways during scar formation. FASEB J 25(12):4498–4510

Wong VW, Rustad KC, Akaishi S et al (2011c) Focal adhesion kinase links mechanical force to skin fibrosis via inflammatory signaling. Nat Med 18(1):148–152

Wong R, Geyer S, Weninger W et al (2016) The dynamic anatomy and patterning of skin. Exp Dermatol 25(2):92–98

Wu CX, Liu ZF (2018) Proteomic profiling of sweat exosome suggests its involvement in skin immunity. J Invest Dermatol 138(1):89–97

Yagmur C, Akaishi S, Ogawa R, Guneren E (2010) Mechanical receptor-related mechanisms in scar management: a review and hypothesis. Plast Reconstr Surg 126(2):426–434

Yannas IV, Tzeranis DS, So PTC (2017) Regeneration of injured skin and peripheral nerves requires control of wound contraction, not scar formation. Wound Repair Regen 25(2):177–191

Zegarska B, Pietkun K, Giemza-Kucharska P et al (2017) Changes of Langerhans cells during skin ageing. Postepy Dermatol Alergol 34(3):260–267

Zhang C, Zhen YZ, Lin YJ et al (2014) KNDC1 knockdown protects human umbilical vein endothelial cells from senescence. Mol Med Rep 10(1):82–88

Zhang J, Hou W, Feng S et al (2017) Classification of facial wrinkles among Chinese women. J Biomed Res 31(2):108–115

Zhao P, Sui BD, Liu N et al (2017) Anti-aging pharmacology in cutaneous wound healing: effects of metformin, resveratrol, and rapamycin by local application. Aging Cell 16(5):1083–1093

Zimmerman A, Bai L, Ginty DD (2014) The gentle touch receptors of mammalian skin. Science 346(6212):950–954

Chapter 11
Connective Tissue and Age-Related Diseases

Carolyn Ann Sarbacher and Jaroslava T. Halper

Abstract We begin this chapter by describing normal characteristics of several pertinent connective tissue components, and some of the basic changes they undergo with ageing. These alterations are not necessarily tied to any specific disease or disorders, but rather an essential part of the normal ageing process. The general features of age-induced changes, such as skin wrinkles, in selected organs with high content of connective or soft tissues are discussed in the next part of the chapter. This is followed by a section dealing with age-related changes in specific diseases that fall into at least two categories. The first category encompasses common diseases with high prevalence among mostly ageing populations where both genetic and environmental factors play roles. They include but may not be limited to atherosclerosis and coronary heart disease, type II diabetes, osteopenia and osteoporosis, osteoarthritis, tendon dysfunction and injury, age-related disorders of spine and joints. Disorders where genetics plays the primary role in pathogenesis and progression include certain types of progeria, such as Werner syndrome and Hutchinson-Gilford progeria belong to the second category discussed in this chapter. These disorders are characterized by accelerated signs and symptoms of ageing. Other hereditary diseases or syndromes that arise from mutations of genes encoding for components of connective tissue and are less common than diseases included in the first group will be discussed briefly as well, though they may not be directly associated with ageing, but their connective tissue undergoes some changes compatible with ageing. Marfan and Ehlers-Danlos syndromes are primary examples of such disorders. We will probe the role of specific components of connective tissue and extracellular matrix if not in each of the diseases, then at least in the main representatives of these disorders.

Keywords Ageing and senescence · Collagens · Proteoglycans · Elastin and elastic fibres · Fibrillins · Fibronectin · Laminins · Skin ageing · Bone disorders ·

C. A. Sarbacher · J. T. Halper (✉)
Department of Pathology, College of Veterinary Medicine, The University of Georgia and AU/UGA Medical Partnership, Athens, GA, USA
e-mail: jhalper@uga.edu

© Springer Nature Singapore Pte Ltd. 2019
J. R. Harris, V. I. Korolchuk (eds.), *Biochemistry and Cell Biology of Ageing: Part II Clinical Science*, Subcellular Biochemistry 91,
https://doi.org/10.1007/978-981-13-3681-2_11

Tendon problems · Ageing of menisci and intervertebral discs · Cardiovascular ageing · Kidney ageing · Diabetes · Hutchinson-Guilford progeria syndrome · Werner syndrome · Ehlers-Danlos syndrome · Marfan syndrome · Loeys-Dietz syndrome

Introduction

Ageing affects all components of an organism including those of connective tissues. Such changes can be part of "normal" or inevitable ageing, or they are exaggerated qualitatively or quantitatively (out of proportion), and then they become part of a disease or a syndrome. For example, we all are (or will be) familiar with skin wrinkles, a normal ageing process due to changes in subcutaneous hydration, fat content and collagen structure. Osteoporosis and muscle atrophy, alopecia are among some other "wonders" of old age. Such changes usually progress slowly and become apparent later in life. However, in certain conditions, of which progerias are the most egregious example, such changes are accelerated due to specific mutations in genes that remain otherwise intact in people without these disorders. Many processes are responsible for ageing and will not be discussed here unless they are relevant to components of connective tissue and extracellular matrix (ECM). As part of the ageing process various components of the ECM, the substance produced by cells of connective tissue, undergo increase in synthesis, e.g., of fibronectin due to increased expression of its gene, or post-synthetic modifications (e.g., proteolytic degradation). Changes in cell matrix interactions are also altered in ageing connective tissues (Robert and Labat-Robert 2000). It is also the case that connective tissues undergo qualitative and quantitative changes in ageing organs and that such changes might be difficult to distinguish from pathological processes such as fibrosis (Schafer et al. 2018). Those "normal" changes in ageing connective tissue include, but are not limited to, upregulation of an elastase, modification of laminin and collagens in addition to upregulation of fibronectin (Labat-Robert 2003). Fibrosis can progress from minor changes compatible with ageing organ to an end stage disease (e.g., in chronic kidney disease, liver cirrhosis). Cell senescence is a hallmark of ageing and chronic disease alike (Munoz-Espin and Serrano 2014; Schafer et al. 2018).

Senescent cells are identifiable by their morphology (Sharpless and Sherr 2015) and the presence of cellular markers which differ from healthy cells, such as p16, p21 and SA-β-Gal (Ding et al. 2001; Melk et al. 2004). Both p16 and p21 are cyclin-independent-kinase (CDK) inhibitors which induce cell cycle arrest and enhance telomere shortening (Bolignano et al. 2014). These anti-proliferative processes are accompanied by decreased expression of growth factors directly promoting cell proliferation, such as epidermal growth factor (EGF), insulin-like growth factor (IGF) 1, and vascular endothelial growth factor (VEGF) (Bolignano et al. 2014). However, pathological fibrous tissues differ in their composition from seemingly similar appearing ageing tissues, at least under a light microscope. Liver fibrosis (and cirrhosis by extension) is probably the most egregious example of the pathological aberrations in the composition of the fibrotic tissue (Ricard-Blum et al. 2017).

Further description of these alterations is beyond the scope of this chapter, so interested readers are encouraged to turn to a recent review by Ricard-Blum et al. (2017). Brief description of selected connective tissue components follows as they are instrumental to ageing process and our understanding of this process.

Selected Connective Tissue Components

Collagens

Twenty eight known types of collagen form the skeleton, or rather the bulk of ECM. These structural proteins are composed as trimers containing at least one collagenous and at least one non-collagenous domain. Each collagen type consists of at least one unique α chain, so one type of collagen can be a homotrimer of 3 identical α-chains, or a heterotrimer of different chains (of which at least one is an α-chain) (Mienaltowski and Birk 2014). All of these chains are composed of repeating triplets of amino acids where glycine is placed at the third position. Any other amino acid can occupy the other two intervening positions and determine the specific type of the chain, and thus of the collagen. The glycine placement is crucial for the formation of the triple helix structure characteristic for collagens (Cole et al. 2018). Post-translational modifications of collagens, particularly the formation of covalent cross-links of collagen fibrils by lysyl oxidase, are essential to collagen stability, including proper assembly and resistance to proteolytic degradation (Ricard-Blum et al. 2017; Svensson et al. 2018). Enzymatically derived covalent immature cross-links are converted into mature trivalent cross-links which can be transformed into even more mature cross-links with non-enzymatic reaction of sugar with the lysyl and arginine residues in the triple helix. This formation of so called advanced glycation end-products (AGEs) is generally associated with ageing, and diabetes (Couppe et al. 2009). Transforming growth factor β (TGFβ) is perhaps the most potent known stimulant of collagen production, an interaction essential to events in ageing processes as well as in fibrosis, a pathological process (Halper 2010).

Collagens can be divided into distinct groups based on their suprastructural organization. The group of fibrillary collagens, of which types I-III are its most prominent members, is perhaps most pertinent to this chapter. Types I and III collagens form the backbone of tendons and ligaments, type II collagen is prominent in the cartilage. Types V and XI are considered "minor" collagens co-assembling with types I-III in tendons and ligaments (Mienaltowski and Birk 2014). Type V collagen can also be found in other tissues, such as skin and bone, its deficiency due to a mutation leads to classic type of Ehlers-Danlos syndrome (see below for more details) (Malfait and De Paepe 2014). Fibrillary collagens are secreted as procollagens. Once they are processed into active collagens by various proteinases, collagen molecules self-assembly into fibrils. The fibril formation and assembly is at least partially regulated by certain proteoglycans (see below). Cross-linking, described briefly above, is especially important for the stability of fibrillary collagens. The formation of AGEs leads to increased stiffness of tissues, for example of tendons, in old age.

Type VI collagen forms an extensive network with collagen fibrils in the form of beaded microfibrils, broad bands and hexagonal networks (Mienaltowski and Birk 2014). Its interactions with several types of collagen and proteoglycans, fibronectin and other glycoproteins of the ECM are essential to successful integration of ECM and connective tissue components, including cells (Kielty and Grant 2002). Mutations in the gene for type VI collagen lead to less characterized forms of muscular dystrophy and joint contractions involving tendons (Bushby et al. 2014; Lamande and Bateman 2017).

Proteoglycans

Proteoglycans are complex molecules consisting of a protein core with one or more glycosaminoglycan chains attached to it. They play important roles in many cellular functions, including, but not limited to, assembly and maintenance of ECM, cell proliferation, interactions with growth factors and tumour cell growth (Halper 2014). Proteoglycans can be divided into two groups of small and large compounds. The group of small proteoglycans, also called small leucine-rich proteoglycans (SLRPs), consists of small core proteins (~40 kD) with one or two dermatan/chondroitin sulfate or several keratan sulfate chains attached (Yoon and Halper 2005). Some members of the SLRPs, e.g., decorin, biglycan, fibromodulin and lumican, bind to several types of collagen and exert influence in the assembly of collagen fibrils, and other ECM functions (Zhang et al. 2006; Schaefer and Iozzo 2008). Deficiency in one or more of them has impact on proper function of collagen and ECM, and processes associated with ageing, such as the progeroid type of Ehlers-Danlos syndrome (see below), and premature ageing or weakening of tendon and skin in decorin knockout mice (Danielson et al. 1997).

Aggrecan and versican are the main representatives of the large proteoglycan group. Their non-covalent binding to hyaluronan is mediated by a link protein binding to the large core protein (Yoon and Halper 2005). Numerous chondroitin and keratan sulfate chains bind to the aggrecan core protein. This helps the ECM containing aggrecan to attract water and acquire resilience. This becomes particularly important in cartilage and other tissues such tendons and in the walls of atherosclerotic blood vessels (Ström et al. 2004). With ageing and remodelling tendons acquire more aggrecan in their fibrocartilage portion (Vogel et al. 1994). Versican is present in the dermis, media of the aorta and tendons. Similarly to aggrecan, versican is important for tendon remodelling (Samiric et al. 2004). This is of particular importance in various tendinopathies and their healing (Scott et al. 2007; Parkinson et al. 2011). As we will see later the presence of versican in the dermis plays a role in the skin ageing process.

Elastin and Elastic Fibres

Elastin is the insoluble polymer of soluble tropoelastin with a very slow turnover lasting literally an entire human lifetime. It is the main component of elastic fibres. The high content of hydrophobic amino acids, especially valine, proline and glycine, makes elastin elastic and very durable. Alternating hydrophilic domains contain many lysine residues that stabilize elastic microfibrils by cross-linking (Halper and Kjaer 2014). Lysyl oxidase mediates not only the formation of cross-links of collagen fibrils but also of elastic fibre cross-links that stabilize these fibres, prevent excessive elasticity and enable deposition of ECM among elastic fibres (Liu et al. 2004; Cole et al. 2018). Fibrillin microfibrils are the other major constituent of these fibres and a mutation in a gene for microfibrils plays a major role in pathogenesis of Marfan syndrome by association of microfibrils with TGFβ (see below). Elastic fibres provide elastic recoil and resilience to many tissues, most notably to aorta and ligaments (Halper and Kjaer 2014). Elastin also participates in cell adhesion, migration, survival and differentiation (Muiznieks et al. 2010; Kielty 2006). Because elastic fibres are synthesized mostly during tissue development, older tissues contain fewer and fragmented elastic fibres than younger ones. This phenomenon contributes to increase in stiffness in ageing arteries and tendons (Wagenseil and Mecham 2012; Kostrominova and Brooks 2013).

Fibrillins

Another group of three large extracellular proteins (~350 kD) is the fibrillins. As mentioned above, they form the core of microfibrils in elastic tissue where they interact with tropoelastin to form elastic fibrils and fibres, and also with integrins in the ECM of various tissues. Fibrillins participate in constructing scaffolds in elastic tissues and contribute to structural integrity of many organs (Halper and Kjaer 2014). Mutations in the fibrillin gene *FBN1* lead to impaired assembly in a disorder named Marfan syndrome. Because fibrillins contain TGFβ-binding sites as well, increased TGFβ signalling contributes to pathophysiology of Marfan syndrome (Wheeler et al. 2014). See below for more discussion of Marfan syndrome as related to premature ageing.

Fibronectin

Fibronectin, a large glycoprotein (molecular weight of 230–270 kD), is present both in the ECM and plasma in the form of numerous isoforms. Its ability to bind simultaneously to cell membrane receptors (integrins) and to components of the ECM (collagens, proteoglycans, focal adhesion molecules) makes it an ideal agent to regulate assembly of fibrillary collagens, thrombospondin-I and microfibrils, including fibrillin-I (Halper and Kjaer 2014). This enables fibronectin to impact ageing processes and connective tissue in disorders marked by accelerated ageing, such as diabetic basement membrane (Asselot-Chapel et al. 1996). Ageing-induced changes in fibronectin include not only its upregulation in plasma and tissues, but also generation of fibronectin fragments, and inhibition of endothelial cell growth and damage of cartilage by some of the fragments (Homandberg et al. 1985, 1998; Labat-Robert 2003).

Laminins

The presence of laminins, a family of at least 19 large multidomain glycoproteins (M.W. 500–800 kD) is limited to the basement membrane where they adhere to cells via binding to integrins, dystrophoglycan, or sulfated glycolipids. This enables laminins to instruct cells in adhesion, differentiation, migration, stability of phenotype and resistance to apoptosis (Halper and Kjaer 2014). Up-regulation of laminin, type IV collagen and fibronectin is held responsible for thickening of basement membranes with ageing and in diabetes (Asselot-Chapel et al. 1996; Labat-Robert 2003).

Connective Tissue Ageing in Selected Organs

Skin Ageing

Some of the most obvious and visible signs of ageing are skin wrinkles and sagginess, a result of numerous changes occurring both at the epidermis in the form of epidermal atrophy, and in the ECM of the dermis (Lawker 1979; Amano 2016). Eighty percent of the dermal collagen is represented by type I collagen, and the remaining 20% mostly by type III collagen (Cole et al. 2018). The main change in aged skin is decrease in both of these fibrillar collagens as their production by ageing skin fibroblasts goes down. The fibre bundles also become thinner, and fragmented as the skin ages. These changes are exaggerated in sun or UV-damaged skin (Varani et al. 2006). The accumulation of AGEs contributes to increased tissue stiffness and interferes with collagen binding with both hyaluronan and decorin (Cole et al. 2018). AGEs are produced by insertion of sugar (usually glucose) between collagen molecules, thus interfering with coupling of amino acid chains on adjacent proteins (Gautieri et al. 2017). The production of AGEs becomes particularly more significant in diabetes (see below).

Though type V collagen constitutes only a very small proportion in the dermis and other connective tissues, it plays an important regulatory role. A genetic defect in this collagen is responsible for the major form of Ehlers-Danlos syndrome (discussed in the last part of this chapter) (Malfait and De Paepe 2014).

Reports by Carrino and colleagues indicate that the patterns of expression of decorin and versican, the most abundant proteoglycans in the skin, change with age (Carrino et al. 2000, 2003, 2011). Because decorin binds to type I collagen and is instrumental for correct assembly of type I collagen fibrils (Reed and Iozzo 2003; Yoon and Halper 2005) any changes in decorin structure or conformation may have significant impact on collagen assembly, its biomechanical properties, including decreased tensile strength of the skin (Danielson et al. 1997) and ultimately wrinkle formation (Amano 2016; Yasui et al. 2013). With ageing, a catabolic fragment of decorin, named decorunt, accumulates in the skin at the expense of intact decorin. Because decorunt lacks the binding site for collagen, the authors postulate that the decrease in proper binding of decorin to collagen fibrils results in impaired collagen fibrillogenesis, and thus in reduced tensile strength of skin, similarly to decorin-deficient mice (Danielson et al. 1997; Carrino et al. 2003). Both decorin and decorunt were identified in the ECM of the adult dermis where they were co-localized with collagen fibrils. However, whereas decorin immune-stained the entire thickness of the dermis, staining for decorunt was mostly present in the inner dermis (Carrino et al. 2003). Because of its fibrillary structure and its abundancy, type I collagen is the most commonly incorporated collagen type in a variety of anti-ageing and anti-wrinkling creams produced by the cosmetics industry (Avila Rodriguez et al. 2018). In addition, upregulation of matrix metalloproteinases 1, 2 and 3 (MMPs 1–3) has been recorded to occur in ageing skin (Hornebeck 2003; Dyer and Miller 2018). Current knowledge indicates that MMP activity is stimulated by oxidative metabolism and oxidative events associated more with solar UV radiation rather than with the ageing process itself (Cole et al. 2018).

As it has already been mentioned, versican is the other major skin proteoglycan. Fetal skin contains the highest amount of versican, its content drops off postnatally (Carrino et al. 2011). Because versican co-localizes with elastic fibres, its high content in fetal skin should not be surprising (Bernstein et al. 1995; Zimmermann et al. 1994). Versican V3 splice variant, a form of versican lacking chondroitin sulfate, stimulates elastogenesis. When present (as a component of versican) chondroitin sulfate inhibits the formation of elastic fibres (Wight et al. 2014). Elastic fibres are important for skin resilience and structural changes in these fibres, in part due to decrease in fibulin-5 (a molecule associating tropoelastin with microfibrils), have been documented in ageing skin (Amano 2016).

Hyaluronan is another molecule playing a role in keeping ageing at bay. Hyaluronan (sometimes called hyaluronic acid) is a large linear non-sulfated glycosaminoglycan composed of numerous disaccharides of N-acetyl-glucosamine and glucuronic acid units (Anderegg et al. 2014). It is not bound covalently to any core protein, though it connects to aggrecan and versican via a link protein or a proteoglycan tandem repeat attached to aggrecan or versican G1 domain (Yoon and Halper 2005; Wu et al. 2005). It is quite abundant in the dermis and its effect as an anti-ageing component and agent is due to its hygroscopic properties (Maytin 2016). In other words, this very large molecule retains large quantities of water and keeps skin well hydrated (and without wrinkles). Its presence reduces friction between collagen fibres (Dyer and Miller 2018). Its decreased presence during ageing or sun-induced damage leads to fragile skin accompanied by frequent wounding, delayed wound healing with senile purpura, stellate pseudo-scars and skin atrophy (Kaya and Saurat 2007).

Dermal changes occurring with ageing have major impact on biomechanical parameters of the skin with increased stiffness and decreased ability to recoil is well documented (Lynch et al. 2017). This biomechanical deterioration can be attributed to changes just described in the above paragraphs: a decrease in collagen, elastin and proteoglycan content, an increase in glycation cross-linking of collagen fibres and matrix proteins, and fragmentation of collagen fibres to name at least the most prominent ones (Waller and Malbach 2005; Lynch et al. 2017). Because the half-life of elastin, and to lesser degree of collagen, is so long, age-dependent degradation of elastin and collagen fibres leads to loss of dermal elasticity and impaired function of dermal fibroblasts (Cole et al. 2018). The stimulation of inflammatory processes by fragmented collagen may accelerate not just ageing changes in the skin but also may contribute to overall ageing of the whole organisms upon entering of inflammatory mediators into the blood circulation and reaching other organs (Cole et al. 2018).

Age-Related Bone Disorders

Osteoarthritis has been considered to be a paradigm of degenerative disease affecting ageing bone and joints. However, recent findings are rapidly revising this notion with our expanding knowledge of the role of inflammatory mediators in the progression of this rather complex disorder (Favero et al. 2015). Injury, wear and tear of ageing, and perhaps hereditary factors are the initiators of osteoarthritis. The first signs of damage are limited to the subchondral bone as a consequence of repetitive loading in the form of microfractures in the bone. Subsequent remodelling and repair lead to activation of secondary ossification centres, followed by bone resorption and bone formation, i.e., by rapid bone turnover (Cucchiarini et al. 2016). The next step is usually characterized by changes compatible with hypertrophic scar formation, such as increased water content secondary to glycosaminoglycan loss (and leading to softening of the articular cartilage), and an increase in type II collagen and proteoglycan (though the increase in proteoglycans is encountered only in the very beginning of the process) (Favero et al. 2015; Madry et al. 2012). Stimulation of production of inflammatory mediators, proteinases and stress response factors is the next step, followed by cartilage loss. Further inflammation localized to the synovium is induced by fragments of type II collagen (Saito et al. 2002), and can lead to further degradation of the cartilage, including its fibrillation, thickening of the subchondral bone, formation of osteophytes, synovitis, degeneration of ligament and hypertrophy of the joint capsule (Goldring and Goldring 2011; Robinson et al. 2016). The participation of synovium implicates the involvement of the entire joint, and not just the articular cartilage in osteoarthritis as proposed originally. Most likely it is inflammation rather than simple "degeneration playing the primary role in the joint damage" as degenerative changes are usually visualized by radiography only after inflammatory markers are identified (Robinson et al. 2016). In contrast to rheumatoid arthritis, the inflammation of osteoarthritis is low-grade and, unlike in rheumatoid arthritis, primarily innate with only minor adaptive part (Haseeb and Haqqi 2013). Reduced bone turnover with an increase in bone deposition is known as subchondral sclerosis and is characteristic of the end-stage osteoarthritis (Kellgren and Lawrence 1957; Burr and Gallant 2012; Cucchiarini et al. 2016). Perhaps pertinent to our discussion of osteoporosis and cardiovascular calcification below is the finding that late stage osteoblasts secrete abnormal type I collagen $\alpha 1$ homotrimer which has less affinity for calcium than the normal $\alpha 1/\alpha 2$ heterodimer (Couchourel et al. 2009). Though the new discoveries on pathogenesis of osteoarthritis are fascinating as they integrate both degenerative and remodelling changes in the bone with inflammatory processes a more detailed discussion is beyond the scope of this chapter. However, two reviews are recommended for interested readers (Cucchiarini et al. 2016; Robinson et al. 2016).

As humans mature (and age) it is natural for bone loss to start as early as 30 years of age. Bone loss occurs in all parts of the skeleton, but usually it starts in the trabecular (cancellous) bone which also suffers the greatest rate of loss (Boskey and Imbert 2017). With age, trabecular bone loses both number of trabeculae and its overall thickness. There is a difference between the sexes: whereas the loss of the number of trabeculae is more prominent in bones of women, thinning of trabecular bone is more characteristic for men (Farr and Khosla 2015). Cortical bone cross-sectional area decreases because endosteal resorption is not compensated for by periosteal apposition. Men lose cortical bone at a relatively low rate of 3–4% per decade which can begin as early as 30 years of age, but more usually it starts by 40 years of age. The rate of loss stays consistent throughout life if the individual remains healthy and injury free. Women on the other hand have an increase in cortical bone loss around menopause (Grynpass et al. 1989). This rate increases to 9% per decade and will decelerate starting around 65 years of age (Grynpass et al. 1989; Kanis et al. 1994; Raisz 2005; Riggs et al. 1981). This is accompanied by decrease in bone strength. This is determined by diminishing mineralization of the bone, the amount of microdamage that accumulates, and the number of cross-links that form between collagen fibres (Saito and Marumo 2010). As bone strength decreases with age the risk of bone fracture increases.

Osteoporosis is a condition thought to be an extreme form of bone ageing that affects the entire skeleton. It is characterized by low bone mass, microarchitectural deterioration, increased bone fragility, factors that increase fracture risk (Kanis et al. 1994). There are multiple mechanisms that can play a role in osteoporosis, including an imbalance of resorption and formation, decreased mitogenic response to IGF-1, and a decreased ability of mesenchymal stem cells (MSCs) to differentiate into osteogenic lineage (Raisz 2005; Rodríguez et al. 2000). The imbalance could be due to increased activity of osteoclasts, decreased synthesis of type I collagen, or increased protease activity. Cells that contribute to osteoporosis produce less collagen and TGFβ. The collagen provides structure and strength to the bone. TGFβ down-regulates expression of MMPs which degrade the ECM so that the less TGFβ is produced by osteoporotic osteoblasts the more MMPs are available, thus increasing the degradation rate of the ECM. The result is that osteoporotic cells cannot maintain a healthy ECM. They also have an increased likelihood of differentiating into adipocytes (Rodríguez et al. 2000). Calcium deficiency can also contribute to osteoporosis and can be caused by decreased intake, impaired absorption by the body, a deficiency of vitamin D secondary to hypothyroidism (Raisz 2005).

The collagen that is produced by osteoporotic cells can be abnormal due to a polymorphism of the SP1 binding site in the *COL1A1* gene (Viguet-Carrin et al. 2006; Young 2003). Normally three polypeptide chains form collagen, two α1 chains and one α2 chain (Young 2003). Together they form a helix made of Gly-X-Y repeats. The X is often proline and Y hydroxyproline. Glycine is at the turn because it is a small enough amino acid to fit the space (Viguet-Carrin et al. 2006; Young 2003). The polymorphism of the Sp1 binding site affects transcription to create abnormal collagen. It increases the amount of α1 chains relative to the amount of α2 chains (Viguet-Carrin et al. 2006). This forms collagen helixes made entirely of α1 chains weakening the structure (Young 2003). This could be one plausible explanation of pathology of osteoporotic fractures (Viguet-Carrin et al. 2006).

There are two conditions that often lead to the development of osteoporosis: post-menopause and diabetes. In postmenopausal women increased bone resorption is the main reason for bone loss (Raisz 2005). Bone loss accelerates between the ages of 51 and 65 then decelerates after the age of 65 (Riggs et al. 1981). A deficiency in estrogen decreases the activity of lysyl oxidase which is a copper metalloenzyme involved in the formation of collagen cross-links (Viguet-Carrin et al. 2006). Cross-links are essential for the stability of bone and they also provide a framework for the tissue. The decrease in lysyl oxidase activity decreases the number of cross-links between collagen fibres decreasing the stability of the bone. Along with the decrease in cross-links there is increased collagen synthesis of abnormal quality, possibly due to the polymorphism of the Sp1 binding site. This abnormal collagen further weakens the bone (Viguet-Carrin et al. 2006). There is decreased proliferation of MSCs and increased turnover rates of bone tissue (Viguet-Carrin et al. 2006; Rodríguez et al. 2000). This contributes to the increased resorption rate because there are not enough MSCs to differentiate and replace osteoblasts to build up the ECM. In diabetes the strength and toughness of the bone is decreased even more. There is a decrease in enzymatic cross-links and an increase in AGE cross -links. Normally, there are more enzymatic cross-links in bone than AGE cross-links. AGE cross-links occur in hyperglycemia and oxidative stress which are both observed in diabetes. Enzymatic cross-links are beneficial to both mineralization and strength of bone while AGE cross-links are theorized to decrease mechanical and biological functions of bone. The increase of AGE cross-links in diabetes is associated with decreased bone strength and fracture risk is even higher with decreased bone mineral density (Saito and Marumo 2010).

Increased rate of bone fractures may occur in several conditions which must be distinguished from osteoporosis. One of them is osteomalacia in adults and rickets in children, both resulting from insufficient mineralization due to vitamin D deficiency, and/or due to calcium/phosphate metabolism. It is the increased incidence of bone fractures in osteomalacia which can be mistaken for osteoporosis. However, soft bones due to defective bone mineralization, hypophosphatemia or hypocalcemia are clearly distinguishable from osteoporosis by imageing (Boskey and Imbert 2017). Two other disorders may be confused with osteoporosis. These two entities are inherited and they are osteogenesis imperfecta and osteopetrosis. Their occurrence since childhood with distinguishing clear radiographic signs should signal that they are not osteoporotic in nature in spite of increased incidence of bone fractures (Boskey and Imbert 2017). Osteogenesis imperfecta or brittle bone disease is the result of faulty synthesis, folding, and transport of type I collagen, affecting all tissues containing type I collagen, and expressed most prominently in bones and teeth because of their high content of type I collagen (Marini and Blissett 2013). The last condition, osteopetrosis, is characterized by genetic defects in bone remodelling leading to accumulation of calcified cartilage and appearance of old bone (Bargman et al. 2012). The presence of calcified cartilage and failure to correct microcracks in the bone provide explanation for fractures in this disease (Boskey and Imbert 2017).

In addition, a newish hypothesis has emerged which connects pathophysiology and pathogenesis of osteoporosis with age-related vascular calcification (Lampropoulos et al. 2012; Zeng et al. 2017). This hypothesis is based on temporal association between the appearance of primary osteoporosis and vascular calcification arising in atherosclerosis, hypertension, diabetes, vascular lesions, chronic renal disease and ageing (Paloian and Giachelli 2014; Shimizu et al. 2014). According to Zeng and colleagues distressed mechanical transduction weakens and damages bone tissue structure (i.e., expressed as breakage of collagen fibres) leading eventually to osteoporosis (Zeng et al. 2017). By the same token it is thought that vascular mechanical stress, i.e., hemodynamics, induces degradation of certain molecules in blood vessels which is followed by calcification of the intima and media (Tsao et al. 2014; Zeng et al. 2017). Normal bones respond to mechanical loading with reconstruction of bone tissue, i.e., with remodelling (Zeng et al. 2017). However, in osteoporosis degradation of bones rather than reconstruction prevails mainly due to increased MMP and aggreganase activities which lead to the degradation of bone matrix (Geissler et al. 2015).

Tendons and Related Structures

The dry weight of tendon consists almost exclusively of connective tissue, primarily of type I collagen, several other types of collagen, proteoglycans (mostly decorin and some aggrecan), glycoproteins and cells (Yoon and Halper 2005; Zhang and Wang 2015). Tenocytes are the most identifiable cells in tendons, though it is their precursors, tendon stem cells that are responsible for the production of collagen fibres and ECM, and the maintenance of the integrity of tendons and their repair

(Zhang and Wang 2015). Tendon injuries are common and their prevalence increases with ageing population participating more in sports. The general conclusion is that ageing is a risk factor for tendon injuries and tendinopathies both in people and in horses, another species also suffering from high rate of tendon problems (Birch et al. 2016). As pointed out above, the degree of cross-links among collagen fibrils increases with age and that should provide tendons with more stiffness, however, studies on human and equine tendons have been equivocal. Increase, no change or even decrease in material stiffness or modulus in tested tendons in different species, including humans, have all been described in many papers. Whether this is because different tendons from different species were compared (e.g., patellar tendons with Achilles, people vs horses vs rodents), because of decline in associated muscle mass in which tendons would attempt to compensate, or because of the role of other mechanical parameters (e.g., hysteresis and fatigue) is not clear. For example, human Achilles tendons and equine superficial digital flexor tendons are subjected to a high number loading and unloading cycles, and not all tendons undergo this stress (Birch et al. 2016). The composition of tendons may undergo changes with age as well. Certain changes, particularly those induced by loading and unloading are caused by MMPs whose expression is stimulated by cyclical stress. An increase in MMP2 and MMP9, more pronounced in ageing tendons, leads to degradation of cartilage oligomeric matrix protein (COMP), and this contributes to ultimate tensile stress loss (Dudhia et al. 2007). Whereas tendon fascicles have the tendency to decrease in diameter in an ageing equine superficial digital flexor tendon (Gillis et al. 1997), and to compensate for this decrease by increasing the amount of ECM to preserve the overall diameter of the tendon that is not the case in human patellar tendon. In the human patellar tendon, lower collagen content accompanied with enhanced mature cross-links was observed in specimens from old people (Couppe et al. 2009). Inconsistent or variable changes in sulphated glycosaminoglycans were observed in different tendons as well (Birch et al. 2016). For example, glycosaminoglycan level decreases with age in the human supraspinatus tendon, but not in the biceps tendon (Riley et al. 1994). However, some findings common to different types of ageing tendons include increased ratio of the nucleus to cytoplasm, and enhanced adipose tissue deposition. This is accompanied usually by decreased vascularization and impaired matrix integrity (Chambers et al. 2007), though it does appear that there is a decline in ageing matrix turnover, especially that of collagen (Birch et al. 2016). It has been noted that tendon stem cells are present in lower numbers in ageing tendons, so this would explain the increased tendency to injury and a diminished ability of ageing tendons to heal (Zhang and Wang 2010). Recent studies indicate that this situation can be ameliorated by exercise (Zhang and Wang 2015). At least in some cases an increase in type I collagen content was identified post exercise (Langberg et al. 2001).

Ageing Changes in the Meniscus

Tears of menisci, structures positioned between the femoral condyles and tibial plateau, are among the most common knee injuries. The incidence of tears increases sharply with age. Water comprises about 72% of total weight of meniscus. Its dry components consist mostly of ECM and, to lesser degree of cells. Collagen (primarily type I) forms 70% of dry weight, proteoglycans 17%, the rest is distributed among non-collagenous proteins, elastin, glycoproteins and DNA. Though adult meniscus is vascularized only in the periphery, the foetal meniscus is fully vascularized, and the presence of vasculature gradually diminishes with postnatal growth and development (Tsujii et al. 2017). The heavy duties of joint stabilization, load transmission, shock absorption and lubrication take their toll on the rather small menisci and would explain the wear leading seemingly to relatively frequent tears as people age. Grossly, with age the meniscus becomes transformed from translucent smooth structure to opaque, dark yellow structure (Pauli et al. 2011), It is thought that the change in colour is mainly result of glycation (Vo et al. 2016). At a microscopic level progression from a fibroblastic phenotype to chondrocytes has been observed with increasing age, together with decreased cellularity and increased Safranin-O staining (Vo et al. 2016). Though total collagen content remains stable through adulthood (Ingman et al. 1974), distribution of individual types of collagen changes with age. Type II collagen was found to be limited to the tip of the inner zone in the newborn, but dispersed in older menisci. Type IV collagen was identified in the perivascular location of the newborn meniscus, but it completely disappeared during adulthood, whereas type I collagen appears to be uniformly present throughout the meniscus during the entire life (Melrose et al. 2005). These changes were accompanied by increasing advanced glycation end-products, leading to increased tissue stiffness similar to changes in ageing tendons, thus making these and related tissues prone to biomechanical damage and failure (Tsujii et al. 2017). The overall proteoglycan content of the meniscus remains stable during adulthood, though changes in individual proteoglycans were observed, including increase in decorin and aggrecan (McAlinden et al. 2001), but decrease in perlecan (Melrose et al. 2005). And, just like in most other tissues and organs calcium depositions have been identified in menisci of older people with or without osteoarthritis (Tsujii et al. 2017). The result of these changes in composition is increased vulnerability of the meniscus to trauma due to repetitive loading or microtrauma (Tsujii et al. 2017). Just as in other tissues and organs, cellular senescence is an integral part of meniscus ageing (Tsujii et al. 2017).

Ageing of Intervertebral Discs

Intervertebral discs are similar to menisci in their proclivity to "degeneration" and age-related wear and tear due to load transmission and shock absorption. However, disc composition is different: the central gelatinous nucleus pulposus consists of a loose network of type II collagen and elastin fibres embedded in proteoglycan aggregates. This makes the nucleus well equipped to withstand and distribute compressive loads (Roughley and Alini 2002). The peripheral annulus fibrosus encloses the gelatinous nucleus pulposus, and its composition of highly organized lamellae of type I collagen fibres allows the annulus to restrain swelling and tensile forces of the nucleus pulposus during bending and twisting (Vo et al. 2016). Age-related changes can occur in the disc fairly early in life and they include tissue fissures of varying numbers and sizes as well as formation of granular debris and neovascularization of disc tissue which in its healthy condition is largely avascular (Boos et al. 2002). Though the differences between an aged and a degenerated disc might not be clear as they resemble each other, they are two distinguishable entities: whereas a degenerated disc occurs usually as the result of a trauma or another pathological event, it occurs as an isolated incident affecting in many cases only one disc. In contrast, an ageing disc does not age in isolation but as a part of a systemic problem affecting all discs of somewhat older individuals, together with other age-related changes, such as osteoporosis and osteopenia of vertebrae that often lead to compression fractures (Ferguson and Steffen 2003) and more pronounced and now degenerative disc problems (Harada et al. 1998). Other structures of the spinal system are affected as well, including spinal ligaments, facet cartilage, and even ageing muscle which accumulates fatty deposits (Vo et al. 2016).

The main age-related changes in the ECM affect disc proteoglycans, and, to lesser degree, collagens. Proteolytic activity leads to transformation of aggregates of aggrecan and versican to non-aggregated form with shorter glycosaminoglycan chains and lower content of link protein, and to increase of hyaluronan in the disc (Roughley 2004; Vo et al. 2016). Loss of glycosaminoglycan chains was also observed for small proteoglycans decorin and biglycan (Roughley et al. 1993). As described above most of the disc collagen is of fibrillary types I and II. The latter becomes degraded with age (Hollander et al. 1996). It should not come as a surprise that aged discs contain elevated levels of certain catabolic enzymes, such as MMP3, MMP 13, ADAMTS-4 and ADAMTS-5 that facilitate the ECM changes (Le Maitre et al. 2007; Sztrolovics et al. 1997; Xu et al. 2014; Zhao et al. 2011; Tian et al. 2013). Loss of hydration which happens mostly in the nucleus pulposus is the consequence of degradation of proteoglycans, and that induces other deleterious events, including cell clustering and changes in cellular interaction with the pericellular matrix. This has at least two effects: biomechanical impairment and formation of AGEs (Sivan et al. 2006).

Many other changes and damageing processes were described contributing to underperforming ageing discs: from cellular senescence to DNA damage by environmental factors or defects in repair mechanism, reactive oxygen species, abnormal mechanical loading and nutritional stress to name just some of the factors (Vo et al. 2016).

Cardiovascular System and Ageing

Atherosclerosis is generally considered a scourge of old age, though its beginnings can be traced to very young age. Healthy intima in a healthy young individual is thin layer of endothelial cells covered by glycocalyx on their luminal surface (Alphonsus and Rodseth 2014; Halper 2018). Endothelial cells are supported by basement membrane and the underlying internal elastic lamina (Halper 2018). However, with ageing and just living, a repeating cycle of endothelial injury and repair leads to thickening of tunica intima that may progress to plaque formation, and to atherosclerosis. Injured endothelial cells are replaced by healthy cells differentiating from bone marrow derived vascular progenitor cells and smooth muscle cells migrating from the tunica media (Head et al. 2017). With ageing disturbances in blood flow pattern accompanied by low shear stress leading to irregular arrangement of endothelial cells become more frequent. This is followed by vasoconstriction and abnormal endothelial cell proliferation and injury (Zeng et al. 2017). Frequently this is preceded and/or accompanied by dyslipidemia. The repair capacity of bone marrow derived vascular progenitor cells declines (Rauscher et al. 2003), and atherosclerotic plaques start appearing (Karra et al. 2005). Macrophages and macrophage-derived foam cells accumulate within a growing plaque and release cytokines, such as tumour necrosis factor- α, and interleukins IL-1β and IL-4 which stimulate the production of MMPs (Chistiakov et al. 2013). As a response to MMP initiation of remodelling and weakening of the ECM around or in the vicinity of vascular smooth muscle cells, these cells undergo stretching and migration to the plaque (Rudijanto 2007). Eventually vascular smooth muscle cells switch from their usual contractile phenotype (expressing α-smooth muscle actin and secreting elastin) to synthetic phenotype marked by collagen synthesis (Rudijanto 2007; Halper 2018). Eventually, this newly synthesized collagen is incorporated into fibrous ECM deposited in the fibrous plaque (Zhang et al. 2016). This is accompanied by vascular calcification that temporally corresponds to progression of osteoporosis described above (Zeng et al. 2017). Some authors connect the calcium loss from bones with deposits in the cardiovascular system, including peripheral arteries (Pennisi et al. 2004) and cardiac valves (Raggi et al. 2000).

Ageing of Kidneys

Some degree of interstitial fibrosis and glomerulosclerosis occurs as part of normal ageing with only a minimal or mild impact on kidney function. However, with increased longevity reduced kidney function leading to chronic kidney disease has been noted in a substantial portion of the elderly population (Valentijn et al. 2017). Advanced kidney fibrosis and glomerulosclerosis appear the same regardless of their origin, due to ageing or chronic kidney disease. Multiple and complex processes and signalling pathways, be it proinflammatory/fibrotic signalling, loss of renoprotective factors, vascular pathophysiology or oxidative stress lead to the same end state (O'Sullivan et al. 2017). The pathway of cellular senescence to fibrosis is the most relevant to this chapter, and it will be the only pathway involved in ageing discussed here. What is the current understanding of cellular senescence, at least in the context of chronic kidney disease? Recent investigations indicate that this process is regulates by the Klotho/FGF23 system (Lu and Hu 2017). The *Klotho* gene was first described in transgenic mice which happened to have a mutation in a gene named *Klotho*. These mice exhibit symptoms and signs associated with early ageing and dying at the age of 8–9 weeks. Klotho is highly expressed in the kidney and brain, and less in other organs (Kuro-o et al. 1997). It is a co-receptor for fibroblast growth factor 23 (FGF23) with FGF receptors. FGF23, produced in bones, maintains mineral homeostasis. FGF23 preserves normal serum level of phosphate, inhibits parathyroid hormone secretion and reduces level of active vitamin D in serum (Consortium 2000). The decline of serum and urine Klotho level is followed by rise of FGF23 in serum and these changes are considered early biomarkers for chronic kidney disease and by extension predictors for development of cardiovascular disease, including aortic and coronary calcifications (Lu and Hu 2017). Though neither marker is a component of the ECM, their effect on ageing, including stimulation of the fibrotic process cannot be underestimated. Moreover, a disturbance in Klotho/FGF23 system may explain the possible causal connection between osteoporosis and calcification in accompanying cardiovascular disease discussed earlier in the chapter. This mechanism provides an explanation why in some elderly people renal fibrotic changes are accompanied by structural pathology of renal vasculature, usually as part of systemic atherosclerosis, and hypertension-induced changes, such as medial hypertrophy of small arteries (Bolignano et al. 2014). As pointed out above, the contribution of the Klotho/FGF23 system cannot be underestimated (Lu and Hu 2017).

The emergence of senescent cells, prominent in ageing kidney, is usually attributed to a permanent cell cycle arrest, either in G1-phase or G2-phase. A variety of stimuli, or events can trigger senescence – repeated cell division, telomere shortening, oxidative stress, and perhaps most important for our topic: growth factors, such as TGFβ and connective tissue growth factor (CCN2/CTGF) which are expressed by cells arrested in G2-phase and are powerful fibrogenic agents, stimulating production of collagen, fibronectin and laminin (O'Sullivan et al. 2017; Yang et al. 2010). Cytokines such as IL-4 and IL-3 participate in the fibrotic process as well (Ding et al. 2001). These peptides are assisted in this process by p53, the protein product of a well characterized suppressor gene, which induces permanent cell cycle arrest in irreversibly damaged cells (O'Sullivan et al. 2017). Healthy kidney in a younger person is protected against these ageing stimuli by Klotho.

Though senescent cells are most common in the renal cortex where they originate presumably from proximal tubule cells, they have been found in the medulla as well. The number of tubules decreases with age and remaining tubules may undergo scarring, dilatation and have hyaline material deposition in the basement membrane (Bolignano et al. 2014). Other renal cells, such as glomerular and interstitial can become senescent as well (Yang et al. 2010; Bonvetre 2014). Pericytes and endothelial cells of renal vasculature contribute to renal fibrosis through production of TGFβ, and upregulation of Wnt signalling (Halper 2018) During our life-time the number of functional glomeruli also decreases with a concomitant rise in sclerotic (=fibrotic) glomeruli. Most of these fibrotic glomerular changes are due to the inability of podocytes to proliferate (Smeets et al. 2009; Steffes et al. 2001) and are accompanied by glomerular basement membrane thickening (Anderson and Brenner 1986) and mesangial expansion (Musso and Oreopoulos 2011).

Diabetes

Diabetes, especially type 2 diabetes, is a condition that can serve as a model of accelerated ageing in multiple organs and systems. Hyperglycemia leads to modification of various proteins by AGEs which then activate pathways responsible for vascular damage (Madonna et al. 2018). In addition to the glycation process other factors promote accelerated atherosclerosis, microvascular disease, diabetic cardiomyopathy, chronic kidney disease, all conditions associated with ageing, and other features, usually not associated with ageing, such as retinopathy and peripheral neuropathy. Other factors are obesity, insulin resistance and hyperinsulinemia (Madonna et al. 2018). Current hypothese explain the pathophysiology of diabetic macroangiopathy by pointing to reduced number or at least loss of function of progenitor cells and to impaired mobilization of these cells from bone marrow (mobilopathy) (DiPersio 2011). These progenitor cells are instrumental to repair and replacement of damaged endothelium so their loss of function in diabetes negatively affects the repair of vascular wall (Fadini et al. 2007). The deterioration of progenitor cells, an inevitable part of ageing (see above), is enhanced or accelerated in diabetes, and appears at an earlier age. This is proportional to haemoglobin A1C levels (Madonna et al. 2018). The consequences in diabetic patients are manifold: vascular calcification with large necrotic cores, inflammation and extensive calcification in the tunica media of large and medium size arteries, similar but perhaps more prominent than in ageing (Tesauro et al. 2017). The presence of AGEs and their binding to their receptors (RAGEs) on the surface of vascular cells, monocytes/macrophages and other cell types is the driving force behind initiation of the vascular changes both in ageing and, more prominently, diabetes (Lopez-Diez et al. 2016). Binding of AGEs to soluble RAGEs in plasma adds to the vascular damage (Schmidt 2015). In addition, AGEs on their own (i.e., without binding to their receptors) participate in the formation of cross-links with collagen fibres, and thus thickening of basement membrane and biomechanical signs of ageing (Lopez-Diez et al. 2016). AGEs also promote thickening of walls of large vessels, hypertension, loss of pericytes and growth factor-induced angiogenesis and vascularization (Eelen et al. 2015). The role of laminin in thickening of diabetic basement membranes was described above (Labat-Robert 2003).

Distinct Diseases of Accelerated Ageing

Progeria and Ageing

Hutchinson-Guilford Progeria Syndrome (HGPS) is an entity in a group of progeria disorders that have abnormal lamin A, otherwise known as progerin, accumulation. It is a rare disorder, dominantly inherited, that causes premature ageing. The rate of ageing is increased sevenfold in affected individuals (Chawla et al. 2017). It is dominantly inherited, and only 1 in 8 million infants are born with the syndrome. Males are slightly more affected than females with a ratio of 1.5:1, respectively (Mounkes and Stewart 2004; Sarkar and Shinto 2001). Symptoms begin to appear around the first year as infants are born clinically normal (Chawla et al. 2017; Tariq et al. 2017). Many of the symptoms are morphological with stunted growth, disproportionate mandible, thin and large calvarium, narrow thoracic cage, thinner long bones with enlarged epiphysis, coxa valga deformity, poor muscle development, alopecia, decreased joint mobility, prominent eyes, and loss of subcutaneous fat. Other patients require more testing for increased hyaluronic acid levels, hyperlipidemia, and osteopenia among other signs (Chawla et al. 2017; Mounkes and Stewart 2004; Sarkar and Shinto 2001). HGPS is a segmental progeroid syndrome meaning that it reproduces the ageing process only in part. Those with HGPS do not have an increased likelihood to develop neoplasms or cataracts and there is no cognitive degeneration (Mounkes and Stewart 2004). Despite this HGPS is used as a model to study ageing. The average lifespan is 13 years and most deaths occur due to cardio-vascular abnormalities (Chawla et al. 2017).

This syndrome is caused by a mutation in the LMNA gene encoding for lamin A, affecting its structure and function (Chawla et al. 2017; Tu et al. 2016). Lamins are type V intermediate filaments that underlie the nuclear envelope to form the nuclear lamina (Charar and Gruenbaum 2017; Mounkes and Stewart 2004; Tariq et al. 2017; Vidak and Foisner 2016). The nuclear lamina supports the nuclear envelope to resist mechanical stress (Tariq et al. 2017). It is involved with the regulation of gene expression and DNA synthesis by both directly and indirectly associating with chromatin (Mounkes and Stewart 2004). The lamina also plays an important role in apoptosis (Tariq et al. 2017). There are two types of lamins coded for by the LMNA gene; type A and type B. They share similar structures, but opposite functions (Mounkes and Stewart 2004). Type B makes the nuclear lamina flexible and type A creates the rigidity (Vidak and Foisner 2016). There are two type A lamins, termed lamin A and lamin C. They are formed by alternate splicing of LMNA gene on chromosome 1 (8, 10) (Tariq et al. 2017; Vidak and Foisner 2016). The difference between the two is that lamin A has an additional 98 amino acids on its C terminus and lamin C has 6 amino acids that are unique to its C terminus (Charar and Gruenbaum 2017).

Normally lamin A is dansylated at the C terminus which helps to target it to the nuclear membrane. Once there the farnesyl group is cleaved by ZMPSTE24 endo-protease allowing lamin A to adhere to the nuclear lamina (Cao et al. 2011). In

HGPS a splicing defect at exon II deletes 50 amino acids from the carboxy-terminal globular domain (Mounkes and Stewart 2004). The deletion of 50 amino acids includes the cleavage site for ZMPSTE24 (Cao et al. 2011). Without the cleavage site lamin A stays farnesylated and is then referred to as progerin. Progerin accumulates in the nuclear envelope instead of adhering to the nuclear lamina.

Defective lamins such as progerin lead to fragile nuclei which can lead to cell death or senescence. Normal lamin A accounts for the rigidity of the nuclear lamina supporting the nuclear envelope under mechanical stress (Mounkes and Stewart 2004). Progerin is unable to adhere to the lamina and this weakens the nucleus. This leads to cell death which would explain cardiac and skeletal pathologies, decreased muscle mass and subcutaneous fat in HGPS patients. Progerin disrupts other cellular processes as well. The nuclear lamina acts as an anchoring point for signalling molecules and transcription factors (Vidak and Foisner 2016). DNA and chromosomes during mitosis are anchored there. Defects in mitosis occur in cells containing progerin including cytokinesis delay, abnormal chromosome segregation and binucleation (Cao et al. 2011). The lack of lamina A in the lamina disorders impairs processes such as gene expression and DNA synthesis. It also can upset the cell signalling to the nucleus (Mounkes and Stewart 2004).

There are certain cellular phenotypes that can be observed in HGPS cells caused by the accumulation of progerin. The most common are thickening of the nuclear lamina, loss of heterochromatin, abnormal histone methylation, gene misregulation, and nuclear blebbing (Cao et al. 2011; Tariq et al. 2017; Vidak and Foisner 2016). Interestingly these phenotypes are secondary to changes in gene expression and delayed DNA repair. Persistent unchecked damage will trigger p53 to activate inducing senescence seen frequently in these cells (Vidak and Foisner 2016).

Senescence is an inevitable occurrence in the normal ageing process. Eventually all cells, except for cancerous cells, become too damaged or worn out to continue cell division. Increased amounts of progerin are believed to be correlated with senescence. HGPS is a useful model to study ageing because it shares similar symptoms, but also because progerin is found in the cell of normal individuals (Cao et al. 2011), though the accumulation of progerin begins at later ages and at a much slower rate. The same cellular phenotypes seen in HGPS cells are also observed in cells of older individuals.

There is no consensus on how progerin may induce senescence yet, but it may have a connection to the shortening of telomeres. Cells with large amounts of progerin have very short to no telomeres both in HGPS and normal individuals. In immortalized cells progerin is absent. This suggests that there is a correlation between progerin production and telomere length. It is possible that capping proteins on the telomeres, when cleaved off, signal spliceosomes to production of progerin instead of lamin A (Cao et al. 2011). Another theory is that progerin impedes with the breakdown of the nuclear envelope during prophase (Moiseeva et al. 2015). It has already been determined that progerin disrupts the nuclear lamina and the nuclear envelope so this is entirely possible (Mounkes and Stewart 2004). This delay or halt during prophase could cause the cell to take an alternate pathway skipping the apoptotic checkpoints in the normal cell cycle and entering senescence instead (Moiseeva et al. 2015).

The presence of lipofuscin deposition in kidneys, brain, adrenal glands, liver, testes, and heart is quite prominent and it is consistent with ageing. Affected children develop atherosclerosis, arteriosclerosis of small vessels, and prominent adventitial fibrosis with increasing deposition of progerin within coronary arteries (Olive et al. 2010). The media of arteries is mostly devoid of smooth muscle cells (Stehbens et al. 1999) which are replaced by fibrosis (Olive et al. 2010). Similarly, increase in thick collagen fibres and fibrosis is present in the dermis and subcutaneous tissue (Rork et al. 2014). The atherosclerosis is usually rapidly progressive, often fatal and many times accompanied by congestive heart failure.

Theories for possible therapies to address progerin production include correcting progerin by cutting out the farnesyl group, reducing the levels of progerin though mechanisms to remove aggregated and misfolded proteins, and targeting the cellular phenotypes (Vidak and Foisner 2016). One group of researchers conducted a clinical trial on children with HGPS using farnesyltransferase inhibitors (FTIs). Previous studies in HGPS mouse models had indicated treatment with FTIs could improve cardiovascular pathologies, bone morphologies, weight loss, and life expectancy. The FTIs prevent farnesylation of progerin by reversibly binding to farnesyltransferase CAAX binding site. The HGPS children treated with lonafarnin for at least 2 years experienced improvements in weight gain, bone development, hearing, and cardiovascular abnormalities (Gordon et al. 2012).

Werner Syndrome

Werner syndrome (WS) is an autosomal recessive segmental progeroid syndrome (Chen et al. 2003; Mounkes and Stewart 2004). It causes premature ageing in individuals similar to HGPS, but with distinctive differences. WS has a later onset of symptoms with diagnosis occurring at ages 30–40 (Ding and Shen 2008). Similar symptoms to HGPS include fat loss, alopecia, and atrophy of muscle. Characteristic symptoms are an early onset of cataracts, diabetes, premature greying of hair, and increased risk of neoplasms (Mounkes and Stewart 2004). These are symptoms that do not occur in HGPS patients. Also, different from HGPS is the average lifespan of these individuals, which is in the late 40s rather than in the teens (Mounkes and Stewart 2004). Similar to patients with HGPS, skin and subcutaneous tissue of the extremities shows epidermal thinning and dermal fibrosis with or without collagen hyalinization. However, the presence of diabetes mellitus, arteriosclerosis, hypogonadism, and an increase in cancer is clinically more significant than the skin changes. The most frequent causes of death are cancer and myocardial dysfunction (Oshimo et al. 2017).

A mutation in the WRN gene that codes for a RecQ DNA helicase-exonuclease is typically what causes WS. There are 70 identifiable mutations in the WRN gene that render the protein unstable. The types of mutation include stop codons, small indels and splicing mutations. Normally the WRN gene interacts with DNA in double-stranded break repairs and at multiple points during DNA synthesis. Cells taken from WS patients have limited replicative ability and are genomically unstable which coincides with clinical signs of premature ageing and neoplasm development (Oshimo et al. 2017).

A small subset of WS patients do not have a mutation in the WRN gene (Mounkes and Stewart 2004). Patients without this mutation are considered to have an atypical variation of WS (Chen et al. 2003). About 15% of WS cases have a mutation in the LMNA gene affecting lamin A. Patients with the LMNA mutation have symptom onset in their 20s rather than their 30s. The mutation is a missense mutation causing a substitution of amino acids in either the heptad repeats of the lamin rod domain or in the N-terminal globular domain (Mounkes and Stewart 2004). Several amino acids have the possibility of being substituted. This disturbs intermolecular interactions within lamin and possibly its dimer structure to an extent (Chen et al. 2003). Like in HGPS the change in lamin A causes abnormal nuclear phenotypes. Nuclei of fibroblasts are deformed by lobulation, invaginations and hypertrophy (Doh et al. 2009). It is possible the clinical signs of these two syndromes differ, but their nuclear phenotypes are similar because each mutation affects different interacts of lamin A (Chen et al. 2003).

Disorders of Soft Tissues with Signs of Ageing

Ehlers-Danlos Syndrome

Ehlers-Danlos syndrome (EDS) is classified into several types based on the underlying biochemical defect and/or gene mutation (Malfait and De Paepe 2014). As not all of them give the appearance of premature ageing, only those with "aged or ageing" phenotype will be discussed here. The classic type of Ehlers-Danlos is usually the result of a mutation in the gene for type V (or type I) collagen. The main manifestations are skin hyperextensibility, muscle weakness, atrophic scars without a history of trauma and joint hypermobility. The skin problems may also include smooth velvety appearance, easy bruising and difficulties with wound healing. Excess skin over the eyelids, the presence of skin scars on the forehead and pale skin may appear the person look prematurely aged (Malfait and De Paepe 2014). Muscle weakness has been described in other types of EDS. Excessive wrinkling and thinning of skin of hands and feet ("acrogeria") gives the appearance of an old person in a vascular type of EDS (Germain 2007). Though persons with vascular type EDS are at high risk from rupture of large arteries or of rupture of hollow internal organs in fairly young age, the aged appearance is the most obvious feature. The underlying cause is a mutation in the *COL3A1* gene, encoding type III collagen.

Patients with kyphoscoliotic EDS present with kyphoscoliosis, muscle weakness and fragile skin with bruising and poor healing. The usual defect is a mutation in the gene encoding lysyl hydroxylase, an enzyme crucial for proper cross-linking of collagen fibrils (Yeowell et al. 2005; Giunta et al. 2005).

However, it is the rare form of EDS, the so called progeroid type, which is phenotypically most aligned with changes attributed to ageing or progeria. Homozygous mutation in *BSGALT7*, a gene encoding β-1, 4 galactosyltransferase is responsible for this phenotype (Quentin et al. 1990). Similar phenotype was described in other rare forms of EDS with a mutation in enzymes participating in the glycosylation of proteoglycans such as decorin and biglycan (e.g., galactosyltransferase I or galactosyltransferase II) (Seidler 2006; Miyake et al. 2014).

Marfan Syndrome

The primary defect of Marfan syndrome is a mutation in *FBN1* which leads to multiple secondary disruptions because of complexity of fibrillin-1 binding to other modulators of connective tissue function and because of variability in expression and phenotypes (Cook and Ramirez 2014). A mutation in *FBN1* is typically identified in more than 90% of patients with Marfan syndrome (Wheeler et al. 2014). The site of mutation is highly variable, up to 1800 mutations in *FBN1* have been documented. The mutations affecting TGFβ-binding site lead to up-regulation of this growth factor. This leads to up-regulation of TGFβ signalling, and some of the pathology associated with Marfan (Wheeler et al. 2014; Meester et al. 2017). A disruption of the binding site for versican within the fibrillin usually leads to a severe, but less common form of Marfan, so called neonatal Marfan syndrome, characterized by early expression and severe pathology of the cardiovascular system (Wu et al. 2005; Cook and Ramirez 2014). Characteristically, multiple organ systems are involved, including cardiovascular, skeletal systems, and eyes, to name just a few of them. Most signs and symptoms of Marfan syndrome are not associated with ageing or with disorders with some ageing symptomatology, but some are. Dilatation and dissection of the aorta occurs at younger age and more commonly it affects the thoracic aorta rather than the abdominal aorta in older patients with atherosclerosis. The histopathology is significant for disorganization and fragmentation of elastic fibres of the media rather thickened intima and the presence of atherosclerotic plaque and aneurysm associated with atherosclerosis. Likewise osteoarthritic changes and osteopenia, including scoliosis and thoracolumbar kyphosis may be reminiscent of bone and/or joint pathology of old or older age, but for associated features like disproportionately long tubular bones and ligament laxity (Cook and Ramirez 2014; Meester et al. 2017). Also, the development of cataracts and glaucoma are more common in Marfan and in older population. The combination of young age at presentation, tall stature and some unusual features involving bones (e.g., dolichocephaly, arachnodactyly) should alert the examining doctor or other medical professionals that one is not dealing with accelerated atherosclerosis, a skeletal or ocular disorder associated more directly with ageing.

Loeys-Dietz Syndrome

Because the main defect in this syndrome lies in mutations affecting genes encoding for TGFβ receptors or for a Smad molecule mediating TGFβ signalling pathway, some of the signs or symptoms resemble those found in Marfan (Van Laer et al. 2014). Cardiovascular presentation is quite prominent and consists of aortic dilatation and dissection with rupture, and tortuous arteries in the head and neck. Among the complicated skeletal presentations, scoliosis and osteopenia are observed as signs related to ageing and the frail skin with easy bruising and dystrophic scars is also reminiscent of ageing integument (Loeys et al. 2006; Van Laer et al. 2014; Meester et al. 2017).

References

Alphonsus CS, Rodseth RN (2014) The endothelial glycocalyx: a review of the vascular barrier. Anaesthesia 69:777–784

Amano S (2016) Characterization and mechanisms of photoageing-related structures changes in skin. Damages of basement membrane and dermal structure. Exp Dermatol 25:14–19

Anderegg U, Simon JC, Averbeck M (2014) More than just a filler – the role of hyaluronan for skin homeostasis. Exp Dermatol 23:295–303

Anderson S, Brenner BM (1986) Effects of ageing on the renal glomerulus. Am J Med 80:435–442

Asselot-Chapel C, Borchiellini C, Labat-Robert J, Kern P (1996) Expression of laminin and type IV collagen by basement membrane-producing EHS tumours in streptozotocin-induced diabetic mice in vivo modulation by low molecular weight heparin fragments. Biochem Pharmacol 52:1695–1701

Avila Rodriguez MI, Rodriguez Barroso LG, Sanchez ML (2018) Collagen: a review on its sources and potential cosmetic applications. J Cosmet Dermatol 17:20–26

Bargman R, Posham R, Boskey A, Carter E, DiCarlo E, Verdelis K, Raggio C, Pleshko N (2012) High- and low-dose OPG-Fc cause osteopetrosis-like changes in infant mice. Pediatr Res 72:495–501

Bernstein EF, Fisher LW, Li K, LeBaron RG, Tan EM, Uitto J (1995) Differential expression of the versican and decorin genes in photoaged and sun-protected skin. Comparison by immunohistochemical and northern analyses. Lab Investig 72:662–669

Birch HL, Peffers MJ, Clegg PD (2016) Influence of ageing on tendon homeostasis. Adv Exp Med Biol 920:247–260

Bolignano D, Mattace-Raso F, Sijbrands EJG (2014) The ageing kidney revisited: a systematic review. Ageing Res Rev 14:65–80

Bonvetre JV (2014) Maladaptive proximal tubule repair: cell cycle arrest. Nephron Clin Pract 127:61–64

Boos N, Weissbach S, Rohrbach H, Christoph Weiler C, Spratt KF, Nerlich AG (2002) Classification of age-related changes in lumbar intervertebral discs: 2002 Volvo Award in basic science. Spine (Phila Pa 1976) 27:2631–2644

Boskey AL, Imbert L (2017) Bone quality changes associated with ageing and disease: a review. Ann N Y Acad Sci 1410:93–106

Burr DB, Gallant MA (2012) Bone remodelling in osteoarthritis. Nat Rev Rheumatol 8:665–673

Bushby KMD, Collins JE, Hicks D (2014) Collagen type VI myopathies. Adv Exp Med Biol 802:185–199

Cao K, Blair CD, Faddah DA, Kieckhaefer JE, Olive M, Erdos MR, Nabel EG, Collins FS (2011) Progerin and telomere dysfunction collaborate to trigger cellular senescence in normal human fibroblasts. J Clin Invest 121:2833–2844

Carrino DA, Sorrell JM, Caplan AI (2000) Age-related changes in the proteoglycans of human skin. Arch Biochem Biophys 373:91–101

Carrino DA, Onnerfjord P, Sandy JD, Cs-Szabo G, Scott PG, Sorrell JM, Heinegard D, Caplan AI (2003) Age-related changes in the proteoglycans of human skin. Specific cleavage of decorin to yield a major catabolic fragment in adult skin. J Biol Chem 278:17566–17572

Carrino DA, Calabro A, Darr AB, Dours-Zimmermann MT, Sandy JD, Zimmermann DR, Sorrell JM, Hascall VC, Caplan AI (2011) Age-related differences in human skin proteoglycans. Glycobiology 21:257–268

Chambers SM, Shaw CA, Gatza C, Fisk CJ, Donehower LA, Goodell MA (2007) Ageinghematopoietic stem cells decline in function and exhibit epigenetic dysregulation. PLoS Biol 5:e201

Charar C, Gruenbaum Y (2017) Lamins and metabolism. Clin Sci 131:105–111

Chawla GS, Agrawal PM, Dhok A (2017) Progeria: an extremely unusual disorder. Skelet Radiol 46:1149–1153

Chen L, Lee L, Kudlow BA, Dos Santos HG, Sletvold O, Shafeghati Y, Botha EG, Garg A, Hanson NB, Martin GM, Mian IS, Kennedy BK, Oshima J (2003) LMNA mutations in atypical Werner's syndrome. Lancet 362:440–445

Chistiakov DA, Sobenin IA, Orekhov AN (2013) Vascular extracellular matrix in atherosclerosis. Cardiol Rev 21:270–288

Cole MA, Quan T, Voorhees JJ, Fisher GJ (2018) Extracellular matrix regulation of fibroblast function: redefining our perspective on skin ageing. J cell Commun Signal 12:35–43

Consortium ADHR (2000) Autosomal dominant hypophosphaetemic rickets is associated with mutations in FGF23. Nat Genet 26:345–348

Cook JR, Ramirez F (2014) Clinical, diagnostic, and therapeutic aspects of the Marfan syndrome. Adv Exp Med Biol 802:77–94

Couchourel D, Aubry I, Delalandre A, Lavigne M, Martel-Pelletier J, Pelletier JP, Lajeunesse D (2009) Altered mineralization of human osteoarthritic osteoblasts is attributable to abnormal type I collagen production. Arthritis Rheum 60:1438–1450

Couppe C, Hansen P, Kongsgaard M, Kovanen V, Suetta C, Aagaard P, Kjaer M, Magnusson SP (2009) Mechanical properties and collagen cross-linking of the patellar tendon in old and young men. J Appl Physiol 107:880–886

Cucchiarini M, de Girolamo L, Filardo G, Oliveira M, Orth P, Pape D, Reboul P (2016) Basic science of osteoarthritis. J Exp Orthop 3:22

Danielson KG, Baribault H, Holmes DF, Graham H, Kadler KE, Iozzo RV (1997) Targeted disruption of decorin leads to abnormal collagen fibril morphology and skin fragility. J Cell Biol 136:729–743

Ding SL, Shen CY (2008) Model of human ageing: recent findings on Werner's and Hutchinson-Gilford progeria syndromes. Clin Interv Ageing 3:431–444

Ding G, Franki N, Kapasi AA, Reddy K, Gibbons N, Singhal PC (2001) Tubular cell senescence and expression of TGF-beta1 and p21(WAF1/CIP1) in tubulointerstitial fibrosis of ageingrats. Exp Mol Pathol 70:43–53

Dipersio JF (2011) Diabetic stem cell "mobilopathy". N Engl J Med 365:2536–2538

Doh YJ, Kim HK, Jung ED, Choi SH, Kim JG, Kim BW, Lee IK (2009) Novel LMNA gene mutation in a patient with atypical Werner's syndrome. Korean J Intern Med 24:68–72

Dudhia J, Scott CM, Draper ERC, Heinegaard D, Pitsillides AA, Smith RK (2007) Ageingenhances a mechanically-induced reduction in tendon strength by an active process involving matrix metalloproteinase activity. Ageing Cell 6:547–556

Dyer JM, Miller RA (2018) Chronic skin fragility: current concepts in the pathogenesis, recognition and management of dermatoporosis. J Clin Aesthet Dermatol 11:13–18

Eelen G, de Zeeuw P, Simons M, Carmeliet P (2015) Endothelial cell metabolism in normal and diseased vasculature. Circ Res 116:1231–1244

Fadini GP, Agostini C, Sartore S, Avogaro S (2007) Endothelial progenitor cells in the natural history of atherosclerosis. Atherosclerosis 194:46–54

Farr JN, Khosla S (2015) Skeletal changes through the lifespan – from growth to senescence. Nat Rev Endocrinol 11:513–521

Favero M, Ramonda R, Goldring MB, Goldring SR, Punzi L (2015) Early knee osteoarthritis. RMD Open 1:e000062

Ferguson SJ, Steffen T (2003) Biomechanics of the ageingspine. Eur Spine J 12:S97–S103

Gautieri A, Passini FS, Silvan U, Guizar-Sicairos M, Carimati G, Volpi P, Moretti M, Schoenhuber H, Redaelli A, Berli M, Snedeker JG (2017) Advanced glycation end-products: mechanics of aged collagen from molecule to tissue. Matrix Biol 59:95–108

Geissler JR, Bajaj D, Fritton JC (2015) American Society of Biomechanics Journal of Biomechanics Award 2013: cortical bone tissue mechanical quality and biological mechanisms possibly underlying atypical fractures. J Biomech 48:883–894

Germain DP (2007) Ehlers-Danlos syndrome type IV. Orphanet J Rare Dis 2:32

Gillis C, Pool RR, Meagher DM, Stover SM, Reiser K, Willits N (1997) Effect of maturation and ageing on the histomorphometric and biochemical characteristics of equine superficial digital flexor tendon. Am J Vet Res 58:425–430

Giunta C, Randolph A, Steinmann B (2005) Mutation analysis of the PLOD1 gene: an efficient multistep approach to the molecular diagnosis of the kyphoscoliotic type of Ehlers-Danlos syndrome (EDS VIa). Mol Genet Metab 86:269–276

Goldring MB, Goldring SR (2011) Inflammation in osteoarthritis. Curr Opin Rheumatol 23:471–478

Gordon LB, Kleinman ME, Miller DT, Neuberg DS, Giobbie-Hurder A, Gerhard-Herman M, Smoot LB, Gordon CM, Cleveland R, Snyder BD, Fligor B, Bishop WR, Statkevich P, Regen A, Sonis A, Riley S, Ploski C, Correia A, Quinn N, Ullrich NJ, Nazarian A, Liang MG, Huh SY, Schwartzman A, Kieran MW (2012) Clinical trial of a farnesyltransferase inhibitor in children with Hutchinson-Gilford progeria syndrome. Proc Natl Acad Sci U S A 109:16666–16671

Grynpass MA, Huckell B, Pritzker KP, Hancock RG, Kessler MJ (1989) Bone mineral and osteoporosis in ageing rhesus monkey. P R Health Sci J 8:197–204

Halper J (2010) Growth factors as active participants in carcinogenesis: a perspective. Vet Pathol 47:77–97

Halper J (2014) Proteoglycans and diseases of soft tissues. Adv Exp Med Biol 802:49–58

Halper J (2018) Basic components of vascular connective tissue and extracellular matrix. Adv Pharmacol 81:95–127

Halper J, Kjaer M (2014) Basic components of connective tissues and extracellular matrix: elastin, fibrillin, fibulins, fibrinogen, fibronectin, laminin, tenascins and thrombospondins. Adv Exp Med Biol 802:31–47

Harada A, Okuizumi H, Miyagi N, Genda E (1998) Correlation between bone mineral density and intervertebral disc degeneration. Spine (Phila Pa 1976) 23:857–661

Haseeb A, Haqqi TM (2013) Immunopathogenesis of osteoarthritis. Clin Immunol 146:185–196

Head T, Daunert S, Goldschmidt-Clermont PJ (2017) The ageing risk and atherosclerosis: a fresh look at arterial homeostasis. Front Genet 8:1–11

Hollander AP, Heathfield TF, Liu JJ, Pidoux I, Roughley PJ, Most JS, Poole AR (1996) Enhanced denaturation of the a1(II) chains of type-II collagen in normal adult human intervertebral discs compared with femoral articular cartilage. J Orthop Res 14:61–66

Homandberg GA, Williams JE, Grant D, Schumacher B, Eisenstein R (1985) Heparin-binding fragments are potent inhibitors of endothelial cell growth. Am J Pathol 120:327–332

Homandberg GA, Wen C, Hui F (1998) Cartilage damaging activity of fibronectin fragments derived from cartilage and synovial fluid. Osteoarthr Cartil 6:231–244

Hornebeck W (2003) Down-regulation of tissue inhibitor matrix metalloproteinase-1 (TIMP-1) contributes to matrix degradation and impaired cell growth and survival. Pathol Biol 51:569–573

Ingman AM, Ghosh P, Taylor TK (1974) Variation of collagenous and non-collagenous proteins of human knee joint menisci with age and degeneration. Gerontologia 20:212–223

Kanis JA, Melton LJ 3rd, Christiansen C, Johnston CC, Khaltaev N (1994) The diagnosis of osteoporosis. J Bone Miner Res 9:1137–1141

Karra R, Vemullapalli S, Dong C, Herderick EE, Song X, Slosek K, Nevins JR, West M, Goldschmidt-Clermont PJ, Seo D (2005) Molecular evidence for arterial repair in atherosclerosis. Proc Natl Acad Sci U S A 102:16789–16794

Kaya G, Saurat JH (2007) Dermatoporosis: a chronic cutaneoud insufficiency. Dermatol 215:284–294

Kellgren JH, Lawrence JS (1957) Rodiological assessment of osteo-arthrosis. Ann Rheum Dis 16:494–502

Kielty CM (2006) Elastic fibres in health and disease. Expert Rev Mol Med 8:1–23

Kielty C, Grant ME (2002) The collagen family: structure, assembly, and organization in the extracellular matrix. In: Royce PM, Steinmann B (eds) Connective tissue and its heritable disorders. Wiley-Liss, New York

Kostrominova TY, Brooks SV (2013) Age-related changes in structure and extracellular matrix protein expression levels in rat tendons. Age 35:2203–2214

Kuro-o, M., Y. Matsumura, H. Aizawa, H. Kawaguchi, T. Suga, T. Utsugi, Y. Ohyama, M. Kurabayashi, T. Kaname, E. Kumek, H. Iwasakik, A. Iida, .T Shiraki-Iida, S. Nishikawa, R. Nagai, and Y. Nabeshima. 1997. Mutation of the mouse klotho gene leads to a syndrome resembling ageing, Nature, 380: 45–51

Labat-Robert J (2003) Age-dependent remodeling of connective tissue: role of fibronectin and laminin. Pathol Biol (Paris) 51:563–568

Lamande SR, Bateman JF (2018) Collagen VI disorders: insights on form and function in the extracellular matrix and beyond. Matrix Biol 71–72:348–367

Lampropoulos CE, Papaioannou L, D'Cruz DP (2012) Osteoporosis – a risk factor for cardiovascular disease? Nat Rev Rheumatol 8:587–598

Langberg H, Rosendal L, Kjaer M (2001) Training-induced changes in peritendinous type I collagen turnover determined by microdialyis in humans. J Physiol 534:297–302

Lawker RM (1979) Structural alterations in exposed and unexposed aged skin. J Invest Dermatol 73:59–66

Le Maitre CL, Freemont AJ, Hoyland JA (2007) Accelerated cellular senescence in degenerate intervertebral discs: a possible role in the pathogenesis of intervertebral disc degeneration. Arthritis Res Ther 9:R45

Liu XQ, Zhao Y, Gao JG, Pawlyk B, Starcher B, Spencer JA, Yanagisawa H, Zuo J, Li TS (2004) Elastic fibre homeostasis requires lysyl oxidase-like I protein. Nat Genet 36:178–182

Loeys BL, Schwarze U, Holm T, Callewaert BL, Thomas GH, Pannu H, De Backer JF, Oswald GL, Symoens S, Manouvrier S, Roberts AE, Faravelli F, Greco MA, Pyeritz RE, Milewicz DM, Coucke PJ, Cameron DE, Braverman AC, Byers PH, De Paepe AM, Dietz HC (2006) Aneurysm syndromes caused by mutations in the TGF-beta receptor. N Engl J Med 355:788–798

Lopez-Diez R, Shekhtman A, Ramasamy R, Schmidt AM (2016) Cellular mechanisms and consequences of glycation in atherosclerosis and obesity. Biochim Biophys Acta 1862:2244–2252

Lu X, Hu MC (2017) Klotho/FGF23 axis in chronic kidney disease and cardiovascular disease. Kidney Dis 3:15–21

Lynch B, Bonod-Bidaud C, Ducourthial G, Affagard JS, Bancelin S, Psilodimitrakopoulos S, Ruggiero F, Allain JM, Schanne-Klein MC (2017) How ageing impacts skin biomechanics: a multiscale study in mice. Sci Rep 7:13750

Madonna R, Pieragostini D, Balistreri CR, Rossi C, Geng YJ, Del Boccio P, De Caterina R (2018) Diabetic macroangiopathy: pathogenetic insights and novel therapeutic approaches with focus on high glucose-mediated vascular damage. Vasc Pharmacol 107:27–34

Madry H, Luyten FP, Facchini A (2012) Biological aspects of early osteoarthritis. Knee Surg Sports Traumatol Arthrosc 20:407–420

Malfait F, De Paepe A (2014) The Ehlers-Danlos syndrome. Adv Exp Med Biol 802:129–143

Marini JC, Blissett AR (2013) New genes in bone development: what's new in osteogenesis imperfecta. J Clin Endocrinol Metab 98:3095–3103

Maytin EV (2016) Hyaluronan: more than just a wrinkle filler. Glycobiology 26:553–559

McAlinden A, Dudhia J, Bolton MC, Lorenzo P, Heinegaard D, Bayliss MT (2001) Age-related changes in the synthesis and mRNA expression of decorin and aggrecan in human meniscus and articular cartilage. Osteoarthr Cartil 9:33–41

Meester J, Verstraeten A, Schepers D, Alaerts M, Van Laer L, Loeys BL (2017) Differences in manifestations of Marfan syndrome, Ehlers-Danlos syndrome, and Loeys-Dietz syndrome. Ann Cardiothorac Surg 6:582–594

Melk A, Schmidt BMW, Takeuchi O, Sawitzki B, Rayner DC, Halloran PF (2004) Expression of p16INK4a and other cell cycle regulator and senescence associated genes in ageinghuman kidney. Kidney Int 65:510–520

Melrose J, Smith S, Cake M, Read R, Whitelock J (2005) Comparative spatial and temporal location of perlecan, aggrecan, and type I, II, and IV collagen in the ovine meniscus: an ageing study. Histochem Cell Biol 124:225–235

Mienaltowski MJ, Birk DE (2014) Structure, physiology, and biochemistry of collagens. Adv Exp Med Biol 802:5–29

Miyake N, Kosho T, Matsumoto N (2014) Ehlers-Danlos syndrome associated with glycosaminoglycan abnormalities. Adv Exp Med Biol 802:145–159

Moiseeva O, Lessard F, Acevedo-Aquino M, Vernier M, Tsantrizos YS, Ferbeyre G (2015) Mutant lamin A links prophase to a p53 independent senescence program. Cell Cycle 14:2408–2421

Mounkes LC, Stewart CL (2004) Ageing and nuclear organization: lamins and progeria. Curr Opin Cell Biol 16:322–327

Muiznieks LD, Weiss AS, Keeley FW (2010) Structural disorder and dynamics of elastin. Biochem Cell Biol 88:239–250

Munoz-Espin D, Serrano M (2014) Cellular senescence: from physiology to pathology. Nat Rev Mol Cell Biol 15:482–496

Musso CG, Oreopoulos DG (2011) Ageing and physiological changes of the kidneys including changes in glomerular filtration rate. Nephron Physiol 119(Suppl 1):1–5

O'Sullivan ED, Hughes J, Ferenbach DA (2017) Renal ageing: causes and consequences. J Am Soc Nephrol 28:407–420

Olive M, Harten I, Mitchell R, Beers JK, Djabali K, Cao K, Erdos MR, Blair C, Funke B, Smoot L, Gerhard-Herman M, Machan JT, Kutys R, Virmani R, Collins FS, Wight TN, Nabel EG, Gordon LB (2010) Cardiovascular pathology in Hutchinson-Gilford progeria: correlation with the vascular pathology of ageing. Arterioscler Throm Vasc Biol 30:2301–2309

Oshimo J, Sidorova JM, Monnat RJJ (2017) Werner syndrome: clinical features, pathogenesis and potential therapeutic interventions. Ageing Res Rev 33:105–114

Paloian NJ, Giachelli CM (2014) A current understanding of vascular calcification in CKD. Am J Physiol Renal Physiol 307:F891–F900

Parkinson J, Samiric T, Ilic MZ, Cook J, Handley CJ (2011) Involvement of proteoglycans in tendinopathy. J Musculoskelet Neuronal Interact 11:86–93

Pauli C, Grogan SP, Patil S, Otsuki S, Hasegawa A, Koziol J, Lotz MK, D'Lima DD (2011) Macroscopic and histopathologic analysis of human knee menisci in ageing and osteoarthritis. Osteoarthr Cartil 19:1132–1141

Pennisi P, Signorelli SS, Riccobene S, Celotta G, Di Pino L, La Malfa T, Fiore CE (2004) Low bone density and abnormal bone turnover in patients with atherosclerosis of peripheral vessels. Osteoporos Int 15:389–395

Quentin E, Gladen A, Rodén L, Kresse H (1990) A genetic defect in the biosynthesis of dermatan sulfate proteoglycan: galactosyl-transferase I deficiency in fibroblasts from a patient with progeroid syndrome. Proc Natl Acad Sci U S A 87:1342–1346

Raggi P, Callister TQ, Lippolis MJ, Russo DJ (2000) Is mitral valve prolapse due to cardiac entrapment in the chest cavity? A CT view. Chest 117:636–642

Raisz LG (2005) Pathogenesis of osteoporosis: concepts, conflicts, and prospects. J Clin Invest 115:3318–3325

Rauscher FM, Goldschmidt-Clermont PJ, Davis BH, Wang T, Gregg D, Ramaswami P, Pippen AM, Annex BH, Dong C, Taylor DA (2003) Ageing, progenitor cells exhaustion, and athero-sclerosis. Circulation 108:457–463

Reed CC, Iozzo RV (2003) The role of decorin in collagen fibrillogenesis and skin homeostasis. Glycoconjug J 19:249–255

Ricard-Blum S, Baffet G, Theret N (2018) Molecular and tissue alterations of collagens in fibrosis. Matrix Biol 68–69:122–149

Riggs BL, Wahner HW, Dunn WL, Mazess RB, Offord KP, Melton LJ 3rd (1981) Differential changes in bone mineral density of the appendicular and axial skeleton with ageing: relation-ship to spinal osteoporosis. J Clin Invest 67:328–335

Riley GP, Harrall RL, Constant CR, Chard MD, Cawston TE, Hazleman BL (1994) Glycosaminoglycans of human rotator cuff tendons: changes with age and in chronic rotator cuff tendinitis. Ann Rheum Dis 53:367–376

Robert L, Labat-Robert J (2000) Ageing of connective tissues: from genetic to epigenetic mecha-nisms. Biogerontology 1:123–131

Robinson WH, Lepus CM, Wang Q, Raghu H, Mao R, Lindstrom TM, Sokolove J (2016) Low-grade inflammation as a key mediator of the pathogenesis of osteoarthritis. Nat Rev Rheumatol 12:580–592

Rodríguez JP, Montecinos L, Ríos S, Reyes P, Martínez J (2000) Mesenchymal stem cells from osteoporotic patients produce a type I collagen-deficient extracellular matrix favoring adipo-genic differentiation. J Cell Biochem 79:557–565

Rork JF, Huang JT, Gordon LB, Kleinman M, Kieran MW, Liang MG (2014) Initial cutaneous manifestations of Hutchinson-Gilford progeria syndrome. Pediatr Dermatol 31:196–202

Roughley PJ (2004) Biology of intervertebral disc ageing and degeneration: involvement of the extracellular matrix. Spine (Phila Pa 1976) 29:2691–2699

Roughley PJ, Alini MAJ (2002) The role of proteoglycans in ageing, degeneration and repair of the intervertebral disc. Biochem Soc Trans 30:869–874

Roughley PJ, White RJ, Magny MC, Liu JH, Pearce RH, Mort JS (1993) Non-proteoglycan forms of biglycan increase with age in human articular cartilage. Biochem J 295:421–426

Rudijanto A (2007) The role of vascular smooth muscle cells in the pathogenesis of atherosclero-sis. Acta Med Indones 39:86–93

Saito M, Marumo K (2010) Collagen cross-links as a determinant of bone quality: a possible explanation for bone fragility in ageing, osteoporosis, and diabetes mellitus. Osteoporos Int 21:195–214

Saito I, Koshino T, Nakashima K, Uesugi M, Saito T (2002) Increased cellular infiltrate in inflam-matory synovia of osteoarthritic knees. Osteoarthr Cartil 10:156–162

Samiric T, Ilic MZ, Handley CJ (2004) Characterization of proteoglycans and their catabolic prod-ucts in tendons and explant cultures of tendon. Matrix Biol 23:127–140

Sarkar PK, Shinto RA (2001) Hutchinson-Guilford progeria syndrome. Postgrad Med J 77:312–317

Schaefer L, Iozzo RV (2008) Biological functions of the small leucine-rich proteoglycans: from genetics to signal transduction. J Biol Chem 283:21305–21309

Schafer MJ, Haak MJ, Tschumperlin DJ, LeBrasseur NK (2018) Targeting senescent cells in fibro-sis: pathology, paradox, and practical considerations. Curr Rheum Rep 20:3

Schmidt AM (2015) Soluble RAGEs – prospects for treating and tracking metabolic and inflam-matory disease. Vasc Pharmacol 72:1–8

Scott A, Lian O, Roberts CR, Cook JL, Handley CJ, Bahr R, Samiric T, Ilic MZ, Parkinson J, Hart DA, Duronio V, Khan KM (2007) Increased versican content is associated with tendino-sis pathology in the patellar tendon of athletes with jumper's knee. Scand J Med Sci Sports 8:427–435

Seidler DG, Faiyaz-Ul-Haque M, Hansen U, Yip GW, Zaidi SHE, Teebi AS, Kiesel L, Götte M (2006) Defective glycosylation of decorin and biglycan, altered collagen stucture, and abnormal phenotype of the skin fibroblasts of an Ehlers-Danlos syndrome patient carrying the novel Arg270Cys substitution in galactosyltransferase I (β4GalT-7). J Mol Med 84:583–594

Sharpless NE, Sherr CJ (2015) Forging a signature of in vivo senescence. Nat Rev Cancer 15:397–408

Shimizu H, Nakagami H, Morishita R (2014) Bone metabolism and cardiovascular function update. Cross link of hypertension, bone loss and vascular calcification – common backgrounds in renin angiotensin system with anti-ageing aspect. Clin Calcium 24:53–62

Sivan SS, Tsitron E, Wachtel E, Roughley P, Sakkee N, van der Ham F, Degroot J, Maroudas A (2006) Age-related accumulation of pentosidine in aggrecan and collagen from normal and degenerate human intervertebral discs. Biochem J 399:29–35

Smeets B, Uhlig S, Fuss A, Mooren F, Wetzels JF, Floege J, Moeller MJ (2009) Tracing the origin of glomerular extracapillary lesions from parietal epithelial cells. J Am Soc Nephrol 20:2604–2615

Steffes MW, Schmidt D, McCrery R, Basgen JM (2001) Glomerular cell number in normal subjects and in type 1 diabetic patients. Kidney Int 59:2104–2113

Stehbens WE, Wakefield SJ, Gilbert-Barness E, Olson RE, Ackerman J (1999) Histological and ultrastructural features of atherosclerosis in progeria. Cardiovasc Pathol 8:29–39

Ström Å, Ahlqvist E, Franzén A, Heinegård D, Hultgårdh-Nilsson A (2004) Extracellular matrix components in atherosclerotic arteries of Apo E/LDL receptor deficient mice: an immunohistochemical study. Histol Histopathol 19:337–347

Svensson RB, Smith ST, Moyer PJ, Magnusson SP (2018) Effects of maturation and advanced glycation on tensile mechanics of collagen fibrils from rat tail and Achilles tendons. Acta Biomater 70:270–280

Sztrolovics R, Alini M, Roughley PJ, Mort JS (1997) Aggrecan degradation in human intervertebral disc and articular cartilage. Biochem J 326:235–241

Tariq Z, Zhang H, Chia-Liu A, Shen Y, Gete Y, Xiong ZM, Tocheny C, Campanello L, Wu D, Losert W, Cao K (2017) Lamin A and microtubules collaborate to maintain nuclear morphology. Nucleus 8:433–446

Tesauro M, Mauriello A, Rovella V, Annicchiarico-Petruzzelli M, Cardillo C, Melino G, Di Daniele N (2017) Arterial ageing: from endothelial dysfunction to vascular calcification. J Intern Med 281:471–482

Tian Y, Yuan W, Fujita N, Wang J, Shapiro IM, Risbud MV (2013) Inflammatory cytokines associated with degenerative disc disease control aggrecanase-1 (ADAMTS-4) expression in nucleus pulposus cells through MAPK and NF-κB. Am J Pathol 182:2310–2321

Tsao CW, Pencina KM, Massaro JM, Benjamin EJ, Levy D, Vasan RS, Hoffmann U, O'Donnell CJ, Mitchell GF (2014) Cross-sectional relations of arterial stiffness, pressure pulsatility, wave reflection, and arterial calcification. Arterioscler Thromb Vasc Biol 34:2495–2500

Tsujii A, Nakamura N, Horibe S (2017) Age-related changes in the knee meniscus. Knee 24:1262–1270

Tu Y, Sánchez-Iglesias S, Araújo-Vilar D, Fong LG, Young SG (2016) LMNA missense mutations causing familial partial lipodystrophy do not lead to an accumulation of prelamin A. Nucleus 7:512–521

Valentijn FA, Falke LL, Nguyen TQ, Goldschmeding R (2017) Cellular senescence in the ageing and diseased kidney. J Cell Commun Signal 12:69–82

Van Laer L, Dietz H, Loeys B (2014) Loeys-Dietz syndrome. Adv Exp Med Biol 802:95–105

Varani J, Dame MK, Rittie L, Fligiel SE, Kang S, Fisher GJ, Voorhees JJ (2006) Decreased collagen production in chronologically aged skin: roles of age-dependent alteration in fibroblast function and defective mechanical stimulation. Am J Pathol 168(6):1861–1868

Vidak S, Foisner R (2016) Molecular insights into the premature ageing disease progeria. Histochem Cell Biol 145:401–417

Viguet-Carrin S, Garnero P, Delmas PD (2006) The role of collagen in bone strength. Osteoporos Int 17:319–336

Vo NV, Hartman RA, Patil PR, Risbud MV, Kletsas D, Iatridis JC, Hoyland JA, Le Maitre CL, Sowa GA, Kang JD (2016) Molecular mechanisms of biological ageing in intervertebral discs. J Orthop Res 34:1289–12306

Vogel KG, Sandy JD, Pogany G, Robbins JR (1994) Aggrecan in bovine tendon. Matrix Biol 14:171–179

Wagenseil JE, Mecham RP (2012) Elastin in large artery stiffness and hypertension. J Cardiovasc Transl Res 5:264–273

Waller JM, Malbach HI (2005) Age and skin structure and function, a quantitative approach (I): blood flow, pH, thickness, and ultrasound echogenicity. Skin Res Technol 11:221–235

Wheeler JB, Ikonomidis JS, Jones JA (2014) Connective tissue disorders and cardiovascular complications: the indomitable role of transforming growth factor-beta signalling. Adv Exp Med Biol 802:107–127

Wight TN, Kang I, Merrilees MJ (2014) Versican and the control of inflammation. Matrix Biol 35:152–161

Wu YJ, La Pierre DP, Wu J, Yee AJ, Yang BB (2005) The interaction of versican with its binding partners. Cell Res 15:483–494

Xu H, Qiang M, Bin X, Liu G, Zhao J (2014) Expression of matrix metalloproteinases is positively related to the severity of disc degeneration and growing age in the East Asian lumbar disc herniation patients. Cell Biochem Biophys 70:1219–1225

Yang L, Besschetnova TY, Brooks CR, Shah JV, Bonventre JV (2010) Epithelial cell cycle arrest in G2/M mediates kidney fibrosis after injury. Nat Med 16:535–543

Yasui T, Yonetsu M, Tanaka R, Tanaka Y, Fukushima S, Yamashita T, Ogura Y, Hirao T, Murota H, Araki T (2013) In vivo observation of agerelated structural changes of dermal collagen in human facial skin using collagen-sensitive second harmonic generation microscope equipped with 1250-nm mode-locked Cr:Forsterite laser. J Biomed Opt 18:31108

Yeowell HN, Walker LC, Neumann LM (2005) An Ehlers-Danlos syndrome type VIA patient with cystic malformation of the meninges. Eur J Dermatol 15:353–358

Yoon JH, Halper J (2005) Tendon proteoglycans: biochemistry and function. J Musculoskelet Neuronal Interact 5:22–34

Young MF (2003) Bone matrix proteins: their function, regulation, and relationship to osteoporosis. Osteoporos Int 14:535–542

Zeng Y, Wu J, He X, Li L, Liu X, Liu X (2017) Mechanical microenvironment regulation of age-related disease involving degeneration of human skeletal and cardiovascular systems. Prog Biophys Mol Biol. https://doi.org/10.1016/j.pbiomolbio.2017.09.022. (in press)

Zhang J, Wang JH (2010) Characterization of differential properties of rabbit tendon stem cells and tenocytes. BMC Musculoskelet Disord 11:10

Zhang J, Wang HC (2015) Moderate exercise mitigates the detrimental effects on tendon stem cells. PLoS One 10(6):e0130454

Zhang G, Ezura Y, Chervoneva I, Robinson PS, Beason DP, Carine ET, Soslowsky LJ, Iozzo RV, Birk DE (2006) Decorin regulates assembly of collagen fibrils and acquisition of biomechanical properties during tendon development. J Cell Biochem 98:1436–1449

Zhang YN, Xie BD, Sun L, Chen W, Jiang SL, Liu W, Bian F, Tian H, Li RK (2016) Phenotypic switching of vascular smooth muscle cells in the 'normal region' of aorta from atherosclerosis patients is regulated by miR-145. J Cell Mol Med 20:1049–1061

Zhao CQ, Zhang YH, Jiang SD, Li H, Jiang LS, Dai LY (2011) ADAMTS-5 and intervertebral disc degeneration: the results of tissue immunohistochemistry and in vitro cell culture. J Orthop Res 29:718–725

Zimmermann DR, Dours-Zimmermann MT, Schubert M, Bruckner-Tuderman L (1994) Versican is expressed in the proliferating zone of the epidermis and in association with the elastic network of the dermis. J Cell Biol 124:817–825

Chapter 12
Potential Cellular and Biochemical Mechanisms of Exercise and Physical Activity on the Ageing Process

Mark Ross, Hannah Lithgow, Lawrence Hayes, and Geraint Florida-James

Abstract Exercise in young adults has been consistently shown to improve various aspects of physiological and psychological health but we are now realising the potential benefits of exercise with advancing age. Specifically, exercise improves cardiovascular, musculoskeletal, and metabolic health through reductions in oxidative stress, chronic low-grade inflammation and modulating cellular processes within a variety of tissues. In this this chapter we will discuss the effects of acute and chronic exercise on these processes and conditions in an ageing population, and how physical activity affects our vasculature, skeletal muscle function, our immune system, and cardiometabolic risk in older adults.

Keywords Ageing · Exercise · Physical activity · Cardiovascular disease · Endothelial function · Tissue regeneration · Skeletal muscle · Atrophy · Sarcopenia · Insulin resistance · Immunology · Bone health · Oxidative stress

Physical Activity in the Elderly and Non Communicable Disease

Advancing age is associated with increased risk of non-communicable disease (NCD), such as cardiovascular disease (CVD), type 2 diabetes mellitus (T2DM), and cancer (Lozano et al. 2012). Using mathematical modelling, Lozano et al. (2012) suggested that there is a 39% increase in the incidence of deaths attributable to NCD as a direct consequence of the ageing process. Healthcare provision, and healthcare insurance costs are significant as we age due to the debilitating effects of such diseases, whilst epidemiological evidence strongly suggests that we become

M. Ross (✉) · H. Lithgow · G. Florida-James
School of Applied Science, Edinburgh Napier University, Edinburgh, Scotland, UK
e-mail: M.Ross@napier.ac.uk; H.Lithgow@napier.ac.uk; G.Florida-James@napier.ac.uk

L. Hayes
Active Ageing Research Group, University of Cumbria, Lancaster, UK
e-mail: Lawrence.Hayes@cumbria.ac.uk

© Springer Nature Singapore Pte Ltd. 2019
J. R. Harris, V. I. Korolchuk (eds.), *Biochemistry and Cell Biology of Ageing: Part II Clinical Science*, Subcellular Biochemistry 91,
https://doi.org/10.1007/978-981-13-3681-2_12

more inactive as we age (Hall et al. 2017), further increasing the risk of NCD incidence, morbidity and mortality in this population (Lee et al. 2012; Stenholm et al. 2016). Insufficient physical activity in the older population is itself associated with muscle mass loss/atrophy and sarcopenia (Evans 2010), T2DM (Amati et al. 2009), CVD (Wannamethee et al. 1998), and increased risk of infection (Leveille et al. 2000). This is estimated to contribute to $65.7bn worth of healthcare costs per annum worldwide (Torjesen 2016), equivalent to the gross domestic product of Costa Rica in a single year. Conversely, increasing physical activity levels in the older population is linked with enhanced cognitive and physical function, improved cardiovascular health measures (Carlsson et al. 2016), and a reduced T2DM risk (de Souto Barreto et al. 2017), resulting in improved quality of life.

Physical activity and exercise can stimulate a host of changes at the molecular, cellular, and tissue level, which translates to improved physical, as well as psychological health. In the following sections we will explore the physiological effects of exercise, the benefits of exercise to the older population, detailing molecular, cellular and tissue-level effects and help explain the health benefits of exercise and physical activity.

The Ageing Cardiovascular System and Physical Activity/ Inactivity

The cardiovascular system (CVS) is essential for the delivery of oxygen and nutrients to every cell in the body, the removal of waste products, such as carbon dioxide, lactate and ammonia, and also works to help the immune system fight infection through distributing leukocytes to sites of infection. As we age, various aspects of our CVS change. Our heart undergoes structural changes, as do our blood vessels, which makes it difficult for the CVS to perform its roles efficiently. Ageing itself is strongly associated with a high risk of CVD morbidity and mortality (Lozano et al. 2012), as a result of increasing incidence of stroke, myocardial infarction (MI) and heart failure (HF). Therefore maintaining the performance of our CVS is key, not only for longevity but also health in longevity.

Ageing and Vascular Function: Role of Exercise and Physical Activity

Our blood vessels are key structures within our body which regulate blood flow to all tissues of the body, and the ability of our vasculature to do so, is termed 'vascular function'. The cells of the inner lining of all blood vessels are the endothelial cells. These cells are crucial in regulating blood flow via production and release of vasoactive substances such as nitric oxide (NO) (Furchgott and Zawadzki 1980). NO subsequently diffuses across to the surrounding vascular smooth muscle cells (VSMCs) and stimulate these cells to relax via Ca^{2+} active re-uptake by the sarcoplasmic reticulum. The relaxation causes a widening of the diameter of the blood vessel, thus allowing increased blood flow to tissues distal to the vessel. This predominantly occurs at the arteriolar level, rather than the artery or capillary level, due to the relative ratio of VSMCs to endothelial cells. We can assess vascular/endothelial function through a technology called 'flow-mediated dilatation', or FMD, which is the use of ultrasound technology to determine changes in vascular diameter (typically the brachial or femoral arteries) in response to an increase in flow after a period of ischaemia or occlusion. The subsequent shear stress, after occlusion is removed, results in an increase in NO production by the endothelium (Chistiakov et al. 2017), and so FMD has been validated to be a measure of endothelial, NO-dependent vasodilation (Green 2005). Studies to date have found significant relationships between endothelial function/FMD scores and cardiovascular-related mortality, with poorer scores and lower levels of vasodilation being predictive of earlier mortality (Green et al. 2011). Unfortunately, with advancing age, we display significant reductions in endothelial function, as demonstrated in several studies (Black et al. 2008, 2009; Muller-Delp 2006; Soucy et al. 2006; Taddei et al. 2001). Potential causes include age-related elevations in oxidative stress, which may uncouple endothelial NO synthase (eNOS), which is required for NO production from its precursor, L-arginine. Aged vascular tissue exhibit greater production of superoxide ($O_2 \cdot -$) anions (Chrissobolis and Faraci 2008; Hamilton et al. 2001; Mayhan et al. 2008) which may contribute to the uncoupling of eNOS. The role of oxidants in the age-related reductions in endothelial function were confirmed in a study by Eskurza et al. (2004). In this study, young and old sedentary adults were assessed for vascular function. They confirmed that vascular function was reduced in the older group, but that an acute dose of ascorbic acid (vitamin C, a powerful antioxidant) reversed this effect, so much so that there was no longer a significant difference in vascular function between the two age groups.

Interestingly, the study by Eskurza et al. (2004) also included an older, endurance trained group. Vascular function between the young group and the endurance trained older group were not different from one another, indicating a potentially powerful role for exercise and physical activity to prevent or at least attenuate age-related vascular dysfunction. The potential for exercise and physical activity to do this, as indicated by this cross-sectional study, has been confirmed by longitudinal studies in both young (Birk et al. 2012; Rakobowchuk et al. 2008) and older adults (Black et al. 2008, 2009).

Cardiovascular Regeneration and Repair with Ageing and Exercise

The human body has the remarkable ability for endogenous regeneration, through its own stem and progenitor cell network. Stem cells, located in specific tissues, or from the bone marrow, contribute to tissue repair and growth. The walls of the heart contain c-kit[+] cardiac stem/progenitor cells (Ellison et al. 2013; Renko et al. 2018), which have been shown to differentiate into myocardial cells under stimulation *in vitro* and *in vivo* (Ellison et al. 2013). Ageing influences the function of these cardiac stem cells (Castaldi et al. 2017), with reduction in stemness of cardiac progenitor cells (CPC), impairments in differentiation into myocardial cells, and failure to secrete vital paracrine factors in response to stimulation in animal models (Castaldi et al. 2017). Aged mice also display CPCs expressing greater levels of senescent markers such as p27kip1, p53 and p19ARF, and subsequent loss of CPCs due to apoptosis (Torella et al. 2004). Unfortunately, due to the invasive nature of CPC isolation, characterization and functional assessment, human data are lacking.

Interestingly, exercise training in animals activates c-kit[+] and Sca1[+] cardiac progenitor cells, which may contribute to left ventricular physiological hypertrophy, a response that appears to be dose-dependent (Xiao et al. 2014). Mice that underwent physical training displayed greater number of c-Kit[+]Lin[−] cells than sedentary controls, potentially due to increased survival or increased proliferation of cardiac resident progenitors (Leite et al. 2015). It is possible that the increase in cardiac workload leads to increased cellular activation of these CPCs (Urbanek et al. 2003), which in turn would support the subsequent physiological hypertrophy observed with exercise training in humans. There is however, a lack of research in this area, which opens up opportunities for exciting future work to determine if exercise can be used to stimulate cardiac repair after ischaemic events in patients and in the elderly.

Bone marrow-derived, or tissue-resident endothelial progenitor cells (EPCs) contribute to the regeneration and growth of the vascular endothelium (Asahara et al. 1997, 1999). These cells may or may not differentiate into mature endothelial cells, but they do also have the ability to secrete pro-angiogenic factors, such as vascular endothelial growth factor (VEGF) and interleukin-8 (IL-8) to support endothelial cell turnover and replication (Hur et al. 2004). Unfortunately, they circulate in such small numbers, within the region of 0.001–0.01% of all circulating mononuclear cells (Case et al. 2007). Despite this, their circulating number has been related to vascular function (Bruyndonckx et al. 2014) and mortality risk, with lower progenitor cell numbers associated with impairments in peripheral arterial tonometry and greater risk of mortality and morbidity in humans (Patel et al. 2015). Several studies have observed lower circulating EPCs in older humans compared to their younger counterparts (Ross et al. 2018; Thijssen et al. 2006), independent of other cardiometabolic risk factors (Ross et al. 2018). EPC function and survival are also affected by ageing, with older adults displaying a greater number of apoptotic EPCs than younger individuals (Kushner et al. 2011). Additionally, these cells display functional deficits, such as secretion of pro-angiogenic cytokines and growth factors (Kushner et al. 2010). Together, these data show that ageing-associated increased vascular and mortality risk may be partly due to loss of EPC number and/ or function. Additionally, Xia et al. (2012a) treated mouse ischemic hind limbs with human EPCs from young and old donors and they showed that cells from young donors homed to the site of ischemia, helped to promote vascular repair, and recover blood flow more so than sham delivery. Interestingly, they also found that EPCs from older individuals lacked this ability, which the researchers associated with an inability of these EPCs to migrate *in vitro*, which in turn was shown to be associated with impaired intracellular CXCR4:JAK-2 signaling (Xia et al. 2012a, b).

Single bouts of exercise have a remarkable ability to mobilize these progenitor cells from peripheral tissues, such as the bone marrow, into the circulation in the post-exercise recovery period (Ross et al. 2014; Van Craenenbroeck et al. 2008), even in older adults, despite an attenuated response (Ross et al. 2018). The mobilization of such progenitor cells are accompanied by elevations in circulating VEGF (Ross et al. 2014; Van Craenenbroeck et al. 2010b; Wang et al. 2014), granulocyte colony-stimulating factor (G-CSF) (Ross et al. 2014) and stromal-derived factor-1α (SDF-1α) (Van Craenenbroeck et al. 2010b; Wang et al. 2014), which are thought to act as chemoattractive factors. The response of EPCs to acute exercise is both time and intensity-dependent (Laufs et al. 2005). Studies investigating the effect of regular exercise training on circulating EPCs provide mixed results with regards to outcomes. Most (Choi et al. 2014; Hoetzer et al. 2007; Laufs et al. 2004; Manfredini et al. 2009; Sarto et al. 2007; Schlager et al. 2011; Sonnenschein et al. 2011; Steiner et al. 2005; Van Craenenbroeck et al. 2010a; Xia et al. 2012a), but not all studies (Luk et al. 2012; Thijssen et al. 2006) demonstrate either an improvement in EPC number (due to increased mobilization or enhanced survival) or function with regular exercise training. In an elegant study, Xia et al. (2012a) demonstrated that 12 weeks physical exercise training in older populations can restore the age-related impairment in EPC function. The researchers transplanted human EPCs (young and old donors, before and after exercise training) into mice that had undergone femoral artery ligation. Their data concur with their earlier finding that EPCs from older adults displayed reduced neovascularization and ability to recover blood flow in ischemic hind limb in mice (Xia et al. 2012b), but exercise training resulted in improved vascular repair capability, and recovery of blood flow.

The current evidence strongly suggest that exercise has a strong positive benefit for the cardiovascular system in ageing populations, through its effects on improving vascular function via increasing NO bioavailability, angiogenesis, and both cardiac (c-kit[+] CPC activation and survival) and vascular (improving EPC number and function) repair mechanisms (Fig. 12.1).

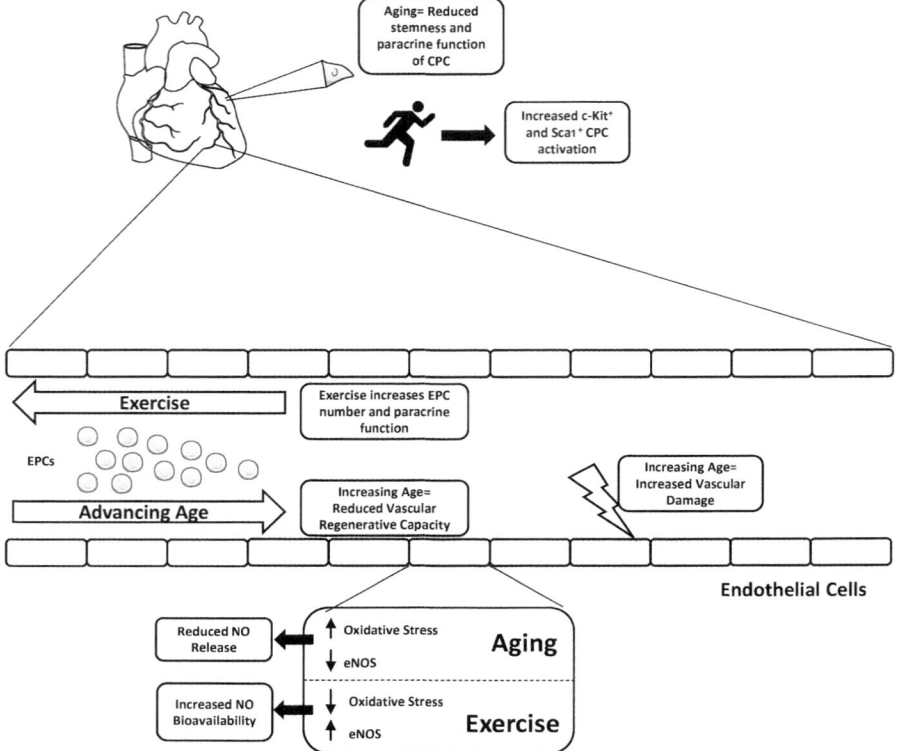

Fig. 12.1 Exercise-associated effects on the ageing cardiovascular system. Vascular endothelial health improvements are seen via improvements in nitric oxide (NO) bioavailability, increased number and function of endothelial progenitor cells (EPC) and activation of cardiac regenerative progenitor cells (CPC)

Musculoskeletal Health and Function with Healthy Ageing

One important change associated with biological ageing is the reduced ability of our skeletal muscles to exert force (or torque) around a joint. Age-associated dysfunction of the muscular system, termed 'sarcopenia', is defined as a syndrome characterised by progressive loss of muscle mass and strength. This results in an increased risk of adverse outcomes such as physical disability, inferior quality of life, and mortality (Cruz-Jentoft et al. 2010; Delmonico et al. 2007). Therefore, the European Working Group on Sarcopenia in Older People (EWGSOP) proposed that diagnosis requires evidence of reduced muscle mass, and either low muscle strength, or low physical performance. Recently, it has been observed that reduced muscle power (dynapenia) occurs faster than force or mass losses and may be more predictive of functional impairment (Manini and Clark 2012), as many tasks of daily living require us to exert force over a short space of time, e.g. when we stand from a chair.

Various interacting tissues, including connective, nervous, skeletal and muscular, determine measurable force and power *in vivo*. These systems do not operate in isolation, so a holistic view of force production and transmission is required. For example, nerve conduction velocity, motor unit recruitment, and firing frequency all influence force via recruitment of muscle, and decline with age (Kommalage and Gunawardena 2013). However, skeletal muscle is our most important organ for generating force and power, and therefore this section will focus on the biology of ageing muscle.

Causes of Age-Associated Muscle Deterioration

Several theories are proposed to explain our reduced muscular capacity with age. Alterations to contractile characteristics, namely decreased twitch speed and a shift in fibre type (from fast to slow) are observed in the elderly which reduces rate of force development. Decreased anabolic hormone production, increased proinflammatory cytokines, and protein turnover imbalances attenuate the ability of aged muscle to regenerate (which leads to atrophy and therefore reduced muscle mass). Importantly, these mechanisms that occur with advancing age, are exacerbated (or even detected in isolation) by physical inactivity. However, masters athletes do display a younger phenotype than age-matched sedentary counterparts, which results in lower incidence of frailty and dependency. As such, masters athletes may be considered as a model of successful ageing (Hawkins et al. 2003). This hypothesis is supported by masters athletes presenting greater relative lean mass, and muscle power than sedentary counterparts, thus suggesting chronic exercise (even aerobic in nature) preserves muscle mass into later life (Hayes et al. 2013). The following sections will discuss how each mechanism may cause muscle deterioration, and how exercise may mediate these mechanisms, with evidence from human studies.

Ageing and Reduction in Anabolic Hormones: Influence of Exercise

As we age, less anabolic hormones are released into circulation to interact with muscular receptors, to exert muscle-building effects (Sipila et al. 2013). This theory of muscle ageing is supported by cell culture experiments (Deane et al. 2013), but also administering older adults testosterone and observing increased muscle mass and strength (Atkinson et al. 2010; Frederiksen et al. 2012; Smith et al. 2014). Although supraphysiologic doses of anabolic hormones increase muscle mass, the effect of lifelong exercise or physical fitness on naturally occurring anabolic hormones is unclear. Ari et al. (2004) reported higher testosterone levels in masters athletes compared with sedentary counterparts. However, this finding is not ubiquitous (Hayes et al. 2015). Several studies inducting sedentary individuals onto an exercise programme do see an increase in 'anabolic' hormones, which accompanies increases in lean muscle mass (Hayes et al. 2017; Herbert et al. 2017a). What is evident however, is that a threshold level of metabolic stress may be required to induce hormone changes, as Khoo et al. (2013) noted greater increases in testosterone following high volume, compared to low volume training. Similarly, Herbert et al. (2017a) reported increased insulin-like growth factor-1 (IGF-1) following high intensity training, but not following low intensity training in previously sedentary older men.

Endocrinology is a complex discipline, with hormones exerting multiple actions, which confounds our ability to draw conclusions about whether age-related hormone changes are to blame for muscle deterioration. For example, IGF-1 may be increased post-exercise compared to pre-exercise, but testosterone, cortisol, myostatin, and growth hormone may not be different, so we cannot say for sure that an individual is in a greater anabolic state than before exercise. Similarly, a hormone in circulation may be increased post-exercise, but unless the hormone is bioavailable (i.e. not bound to a carrier), it cannot exert a cellular effect. The hormone is also reliant upon receptors within the muscle to commence a downstream signalling cascade, resulting in transcription and translation of muscle protein. As such, we are some distance from understanding the endocrinology of ageing and the effect exercise can exert.

Inflammatory Cytokines: Effect of Ageing and Exercise

As we age, we experience increased systemic inflammation. We now know elevated inflammatory cytokines negatively correlate with muscle mass and strength in the elderly. Cytokines are small secreted proteins released by cells, which permit interaction and communication between cells (Zhang and An 2007). Rodent and cell culture experiments have demonstrated that inflammatory cytokines directly impair expression of muscle-specific transcription factors, ultimately inhibiting protein synthesis (Otis et al. 2014; Strle et al. 2004; Tidball 2017). More evidence on a human level for the *inflamm-ageing* hypothesis is provided by Aguirre et al. (2014) who reported significant correlations between knee flexor strength and interleukin-6 (IL-6) and C-reactive protein (CRP), both inflammatory cytokines, in frail, obese, older adults. Furthermore, Mikkelsen et al. (2013) measured muscle size and strength, maximal oxygen uptake, but also inflammatory cytokines in old runners, young runners, and age-matched untrained individuals. CRP and IL-6 were higher in older groups, but lower in trained groups compared to untrained groups. It therefore appears age increases inflammation, but exercise may exert an anti-ageing effect.

Whilst increased low grade inflammation in the elderly is commonly observed, ageing reduces cytokines that contribute to local recruitment of immune cells responsible for muscle remodelling. Therefore ageing may reduce the adaptive response of skeletal muscle to exercise by reducing inflammatory cytokines

(Hamada et al. 2005). For example, Hamada et al. (2005) observed an increase in systemic inflammation (demonstrated by elevations in CRP), yet lower local exercise-induced inflammation (demonstrated by reduced transcripts for CD18, IL-1β, IL-6, tumour necrosis factor-α [TNF-α], and transforming growth factor-ß1 [TGF-β1] in muscle biopsies) in older adults compared to younger adults. To date, the effect of training status on exercise-induced inflammatory cytokine response in older adults is unexamined.

Ageing-Associated Effects on Skeletal Muscle Protein Turnover

Regardless of the precise contribution of each of the above factors to muscle deterioration, reduced muscle mass ultimately results from an imbalance between muscle protein synthesis (MPS) and muscle protein breakdown (MPB). Amino acid-based feeding increased net protein balance, via increased MPS, and reduced MPB (Moore et al. 2015). Exercise exerts a synergistic effect on MPS and ultimately net protein balance, with resistance exercise most potent (Devries et al. 2015; Francaux et al. 2016). MPS in response to amino acids (Cuthbertson et al. 2005; Volpi et al. 2000) and resistance exercise (Kumar et al. 2009) is reduced in aged muscle compared to young muscle tissue, and is termed *anabolic resistance* (Rennie 2009). It is worth noting however, that reduced MPS is not always observed in older adults (Koopman et al. 2009; Symons et al. 2011). Therefore, blaming chronological age for anabolic resistance, rather than physical inactivity (often associated with advanced age), may have led to type I error in cross-section comparison. Breen et al. (2013) suggested that inactivity induces anabolic resistance as they observed 2 weeks reduced physical activity resulted in decreased lean leg mass by ~4%, and postprandial MPS by 26% in ~72 year olds. Furthermore, Symons et al. (2011) reported MPS increased in young and old adults to the same extent following resistance exercise and protein ingestion. In the sole study investigating MPS in masters athletes, masters and young triathletes completed 30 min downhill running to induce muscle damage, MPS was lower in the masters triathletes compared to the younger triathletes, which resulted in poorer subsequent cycling performance in the masters athletes (Doering et al. 2016).

Due to this disruption in net protein balance, older individuals likely require greater protein intake to maintain muscle mass and function (Phillips 2015; Phillips et al. 2016). This is often difficult as older individuals have a lower appetite, and protein is the most satiating of the macronutrients. Therefore, it has been suggested that pragmatic supplementation may be necessary to optimise health (Phillips 2015).

Physical inactivity and ageing-induced reductions in muscle function increase our likelihood of sarcopenia or dynapenia. Therefore, applied studies that demonstrate improved physical function may have the greatest practical application. For example, Fiatrione and colleagues (1990) reported improved strength (175%), lean leg mass (9%), and gait speed (47%) in nonagenarian women following resistance

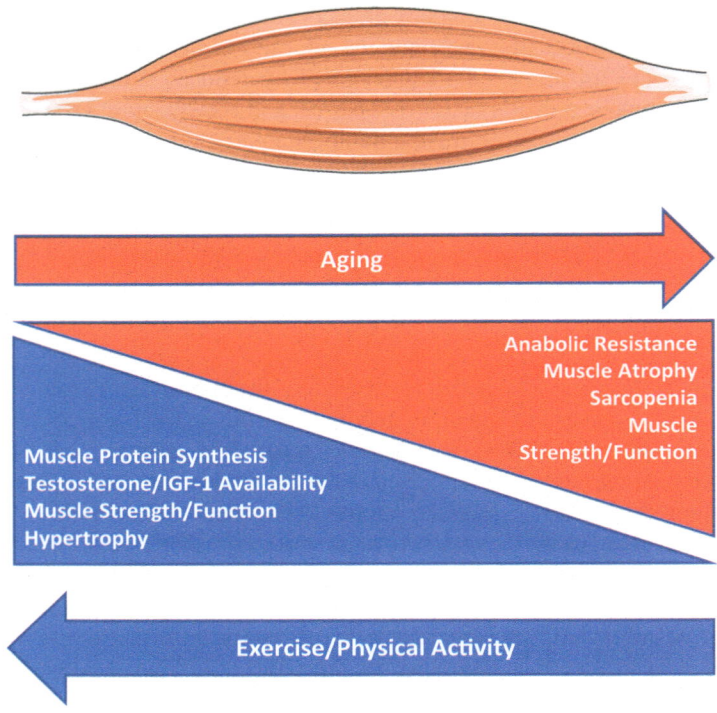

Fig. 12.2 Ageing-associated changes in skeletal muscle function and the role of physical activity and exercise to counteract the deleterious ageing effects

exercise, which demonstrates great muscle plasticity into old age. More recently, high intensity interval training (HIIT) has shown some promise for increasing muscle power in older adults (~65 years) (Hayes et al. 2017; Herbert et al. 2017b; Sculthorpe et al. 2015). Yet, the efficacy and safety of this model in the old-old (85+ years) is still untested (Fig. 12.2).

Skeletal Health in Older Adults: Role of Vitamin D and Exercise

Ageing leads to anabolic impairments in skeletal muscle and bone, which in turn impacts musculoskeletal mass and strength leading to the prevalence of conditions such as sarcopenia, dynapenia, low bone mineral density, low bone strength, risk of falls and fractures, and increased mortality rates in the elderly (Janssen et al. 2002; Lexell 1995; Mitchell et al. 2012; Newman et al. 2006; Visser et al. 2005). Even in the concept of 'healthy ageing' there is a progressive decline in skeletal muscle size and quality, characterised by muscle structure, fibre type, and mechanical function (Lexell 1995). The Health, Ageing and Body Composition (Health ABC) study found that muscle strength is a more important contributor to muscle quality than quantity/mass (Newman et al. 2006). Furthermore, the Health ABC study reported than even gaining or maintaining muscle mass with ageing does not prevent the ageing-induced decline in strength (Delmonico et al. 2009). As with the decline in muscle health, older adults become more susceptible to age-related detrimental effects on bone health, exhibiting a loss of bone mineral and bone strength. This common age-associated physiological environment stems from a decrease in osteoblast (bone cell that secretes the substance of bone) activity, an increase in osteoclast (bone cell that mediates bone destruction) activity, and diminished bone marrow stem cell differentiation potential (Khosla and Riggs 2005). As a result, both bone mass and strength are reduced, and play a role in the pathophysiology of conditions such as osteoporosis (Gallagher 1990). In light of this, the prevalence and impact of musculoskeletal health conditions pose a global threat to 'healthy ageing' (Briggs et al. 2016).

Vitamin D status and metabolism is associated with numerous negative skeletal consequences affecting both bone and muscle, such as reduced bone mineral density (BMD), sarcopenia and dynapenia, osteomalacia (marked softening of bones), and impaired calcium absorption (Holick 2007) which further exacerbates the negative impact of ageing itself on skeletal health. A large proportion of the global population are vitamin D deficient for the twin reasons of not meeting recommended intake guidelines and the climate restricting sufficient dermatological metabolism of vitamin D. The primary source of vitamin D is from direct skin exposure to Ultra Violet B (UVB) rays from the sun initiating the conversion of pre-vitamin D (7-dehydrocholersterol) to vitamin D_3 (cholecalciferol) (Holick and Chen 2008), which is inherently dependent on climate and weather and thus latitude and season. Vitamin D concentrations are inversely linked with advancing age (Chapuy et al. 1983), with evidence suggesting that ageing affects the cutaneous capacity for the initial metabolic conversion in the vitamin D pathway, and the concentration and expression of subsequent vitamin D metabolites, such as the vitamin D binding protein (DBP) and the vitamin D receptor (VDR) (Bischoff-Ferrari et al. 2004a). The ligand-activated VDR, expressed in skeletal muscle as well as most other tissues, is a strong mediator of mRNA transcription and thus protein synthesis (Bischoff et al. 2001; Simpson et al. 1985). The expression of VDR and the post-transcriptional regulation of VDR and can be affected by ageing (Coleman et al. 2016). As a result, evidence has suggested that a lack of vitamin D in an ageing population may affect skeletal muscle mass and strength and thus induce a risk of falls and immobility.

It is generally accepted that vitamin D availability in combination with calcium regulation, beneficially affects musculoskeletal health, particularly bone strength and quality (Bischoff-Ferrari et al. 2004b; Bischoff et al. 2003; Dawson-Hughes et al. 1997). This is notably evident following courses of vitamin D supplementation, often in combination with calcium (Dawson-Hughes et al. 1997; Jackson et al. 2006; Tang et al. 2007). During the ageing process there is a decline in the intestinal absorption of calcium, which may be predetermined by the bioavailability of the active form of vitamin D $(1,25(OH)_2D_3)$ (Veldurthy et al. 2016), which declines with advancing age. Vitamin D stimulates the production of calcium-binding protein (CBP) in the intestine to facilitate the absorption of calcium. Vitamin D is also a regulator of cell growth and maturation, particularly of osteoblasts (bone cells), and mediates the function of white blood cells such as macrophages and activated T- and B-lymphocytes, which modulate the immune system.

Although not conclusive, research on the effect of exercise and vitamin D metabolism in older adults has indicated that mechanical stress such as exercise and strength training can alter the expression and action of key vitamin D metabolites and increase skeletal muscle mass and strength (Aly et al. 2016; Makanae et al. 2015). This may be due to alterations in vitamin D signalling, which has been found to influence skeletal muscle protein synthesis and turnover (Ceglia and Harris 2013), that can be manipulated by exercise and physical activity. A lower vitamin D status has been associated with a decline in musculoskeletal strength, which becomes increasingly prevalent as age advances. Investigations *in vitro* have reported the bioavailable form of vitamin D, $1,25(OH)_2D_3$, stimulates key cellular pathways of muscle growth and differentiation, acting primarily through the action of VDR, to induce myogenesis (Ceglia et al. 2013; Garcia et al. 2011). Currently it is uncertain if the effect of vitamin D on skeletal health is association or causation.

Whilst recent advancements in physiological imaging and molecular biology provide insight into the mechanisms underpinning muscle and skeletal ageing, loss of function, and frailty, we are still some way from conclusive evidence to suggest which cause is dominant. What we know, is the above causes occur simultaneously, and are often interlinked. To conclude, ageing has a significant deleterious effect on our skeletal muscle and our bones, such as loss of muscle mass and function, as well as reduced integrity of our skeletal tissue. The evidence strongly suggests that exercise may attenuate the ageing-related impairments, by modulating muscle anabolism, and potentially vitamin D metabolism, which is an exciting new area of research.

The Elderly Immune System and Changes with Exercise

Immune Cell Senescence and Ageing

Human immunosenescence is the canopy term used to refer to the gradual deterioration of the immune system and function attributed to advancing age. The complex process of ageing negatively impacts the innate and adaptive immune system and their functional capacity, therefore compromising the ability of the host to elicit an effective immune response to fight (ever-evolving) invading pathogens or prevent the development of a pro-inflammatory environment. The innate and adaptive immune systems are differentially affected by ageing, whereby innate immunity appears to be better preserved while adaptive immunity exhibits age-dependent depreciation.

Immunological parameters that impact health and mortality, creating the immune risk profile (Pawelec et al. 2001), become exhausted with the ageing process. The functionality of the components of the adaptive immune system can become exhausted, specifically the main matured cells involved: bone marrow cells (B cells) and thymus lymphocytes (T cells) and their subsets. The primary lymphocyte subpopulation, CD3$^+$ T cells can be divided into CD4$^+$ and CD8$^+$ subsets, which exhibit helper and cytotoxic functions. In particular, CD8$^+$ T cells are affected by age, inducing the development of an inverted CD4:CD8 T cell ratio and thus contributing to immune incompetence.

Thymic Atrophy with Ageing

Age-dependent regression of the thymus, thymic atrophy, defined as the loss of thymic mass, induces a decline in the output of naïve T cells. Therefore, as age advances fewer T cells are developed and exported into the vascular pool (Lazuardi et al. 2005), directly impacting on the peripheral T cell repertoire and altering white blood cell subset diversity, and thus the cells that are circulated to the target tissues.

There is an increase in the proportion of T cells expressing markers associated with senescence delineating T cell subpopulations from naïve T cells (recent thymic products with no proliferative history) to exhausted senescent T cells (not so recent poorly proliferative cells that exhibit severe functional abnormalities). These markers are primarily used to identify T cell subpopulations, but may also be used to provide insight into T cell differentiation, activation, and functional status. The combination of markers can be utilised to define naïve T cells turnover and loss of naïve T cells, assessing proliferative history. Ageing can also restrict the T cell receptor (TCR) repertoire. T cell receptors are complex integral membrane proteins that are responsible for recognising antigens that are bound to the major histocompatibility complex (MHC). A diminished TCR pool reduces the capacity for T cells to identify specific bound antigens and illicit a distinct and critical immune response.

Exercise and Immunosenescence

The beneficial effect of exercise became apparent in the early work of David Neiman in the 1990s who demonstrated that individuals who exercise are at less risk of upper respiratory tract infections (URTI) (Nieman 1994), which are a major cause of visits to and treatment from a physician. However, there is a hyperbolic relationship between intensity and volume of exercise and the risk of URTIs, suggesting that excessive or too intense exercise can be detrimental to effective immunity by supressing immune function (Malm 2006). There are both acute and chronic effects of exercise on immune function. In response to an acute bout of exercise, one of the major changes that occurs is a change in the number of leukocytes (Robson et al. 1999), with a biphasic response induced. The redeployment of lymphocytes from tissues or the blood vessel wall with exercise consists of an initial increase, known as lymphocytosis, that is followed by a significant transient drop in lymphocyte number, known as lymphocytopenia. Immediately upon cessation of exercise the rise in lymphocyte and neutrophil number usually precedes a reduction to below baseline levels, creating a pocket period of reduced immune protection, known as exercise-induced immunosuppression. Each of the individual cell types respond differently to exercise as they all perform different tasks to achieve sound immune function, however it is the Natural Killer (NK) cells and the cytotoxic T cells that display the largest response (Simpson et al. 2006). Exercise-induced immunosuppression can also be altered by cytokines, the signalling molecules of the immune system. Circulating concentrations of cytokines have numerous responsibilities and roles in the inflammatory profile and protection against pathogens, directly and indirectly. Ageing is recognised to strongly affect the redeployment of lymphocytes with particular subsets not mobilised in the bloodstream: although the relative numbers of T cells are similar between young and old, it is the absolute numbers that change. The process of age-associated deterioration of immune cells and thus the operation of the system, is referred to as immunosenescence (Castle 2000). This causes a rise in senescent T cells that are mobilised and thus circulate around the body unable to play an efficient role in immune function and protection (Simpson et al. 2012). This age-related accumulation of senescent T cells lowers the naïve T cell stock and can increase host infection risk. This is also due to older individuals having less naïve and low differentiated cells in the circulation and peripheral tissues for redeployment (Provinciali et al. 2009). Exercise can potentially override the age-related impairments in T cell subset redeployment, specifically CD8[+] T cells (Spielmann et al. 2014), as aerobic fitness level, achieved through regular exercise, is inversely associated with the proportion of senescent T cells, with the relationship withstanding adjustment for age (Spielmann et al. 2011). Recently, Duggal et al. (2018) reported that individuals above the age of 55 years, regular exercisers had significantly fewer senescent T-cells than less active individuals, together suggesting that regular exercise can alleviate the deleterious effect of ageing on the immune system.

Programmed cell death, or apoptosis, is an important mechanism in the mediation of the immune response, serving as a key role in the removal of damaged, infected, exhausted or redundant cells. This orchestrated system then allows for alterations in the proportion of cells that make up the bloodstream repertoire of T cells. Acute bouts of exercise have been shown to induce increases in both senescent and naïve T cells, and elevate apoptotic lymphocytes (Mooren and Kruger 2015). Since ageing induces an accumulation of senescent T cells, it is imperative for effective immune function to induce apoptosis in specific cell types, preferentially the older less functional cells, to allow for naïve T cells to be exported into the circulation, favourably altering the bloodstream repertoire. Exercise has been associated with an increase in apoptotic cells, although the mechanisms are not yet fully under-

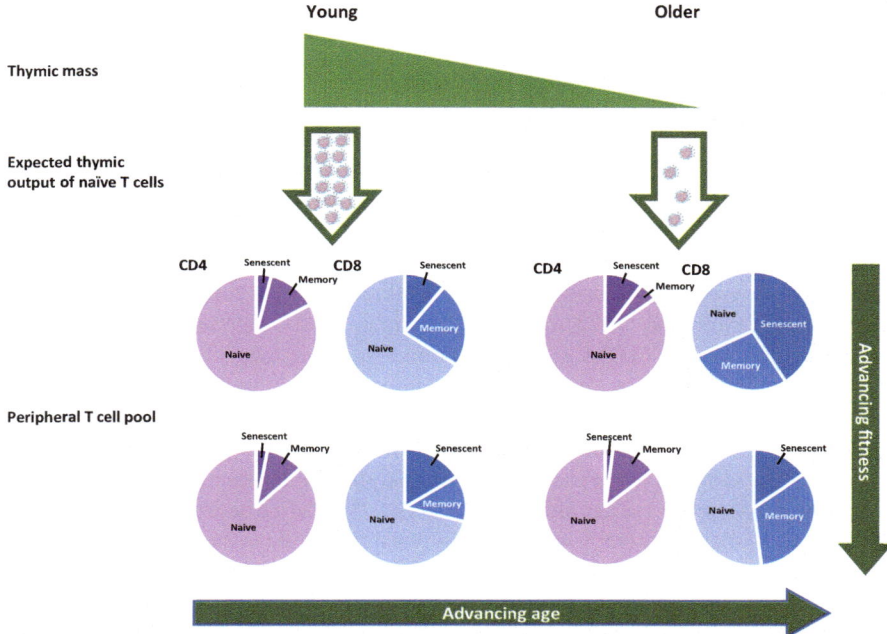

Fig. 12.3 Ageing-related T-lymphocyte changes and influence of maintaining physical fitness. Thymic output of T-cells is reduced with age partially due to reductions in thymic mass, leading to an expansion of senescent T-cell pool, which can be offset by regular exercise and fitness. (Adapted from Spielmann et al. 2011)

stood. In addition, despite the modality of exercise, there is no evidence to suggest that lymphocyte-apoptosis contributes to exercise-induced lymphocytopenia (Simpson et al. 2007).

In summary, immunosenescence contributes to increased susceptibility to illness in the ageing population, which, as evident by the IRP studies, also strongly affects mortality risk. However, exercise can orchestrate a protective effect on immune function (Fig. 12.3), through phenotypic and functional changes in the T cell component of the immune profile.

Endocrine System, Ageing and Physical Activity/Inactivity

Incidence and Prevalence of Diabetes Mellitus

Diabetes is a global health problem, costing national health services millions of dollars annually. Diabetes is a serious chronic disease, classified into type 1 diabetes mellitus (T1DM) and type 2 diabetes mellitus (T2DM). The latest prevalence stats from Diabetes UK in 2016, report that almost 3.6 million people suffer from diabetes across the UK, with an additional 1 million likely to have undiagnosed T2DM, based on the Diabetes Prevalence Model 2016. Worldwide there are an expected 422 million people currently with Diabetes, with global trends indicating the incidence rate is on the rise (NCD Risk Factor Collaboration 2016).

The initial pathophysiological events in the development of diabetes are insulin resistance, high blood glucose levels (hyperglycaemia), and impaired beta cell function (Kahn 2003). Beta cells are insulin-producing cells located in the islets of Langerhans in the pancreas. Progressive dysfunction and degeneration of beta cells is the main cause of both forms of diabetes, although a beta cell deficit is prevalent in T2DM (Butler et al. 2003), it is more variable than T1DM (Cnop et al. 2005; Kloppel et al. 1985). T1DM is defined as insulin-dependent diabetes mellitus and requires medical monitoring and management in order to maintain euglycaemia (normal blood glucose concentration) (ADA 2014). The immune system attacks beta cells often inducing beta cell apoptosis (Butler et al. 2003), seizing the secretion of insulin and exposing the body to a hyperglycaemic state, since insulin is the hormone responsible for the uptake of glucose. T2DM is defined as non-insulin-dependent diabetes whereby the body is resistant to the insulin that is secreted (ADA 2014), developing a physiological state known as insulin resistance or reduced insulin sensitivity (Abdul-Ghani et al. 2006; Kahn 2003). A lack of physical activity and exercise and a poor diet can lead to T2DM (Pan et al. 1997), suggesting it is a lifestyle-induced disease, and hence merits further consideration here.

Role of Physical Activity and Exercise on Improving Insulin Sensitivity in Older Adults

Diabetes is very common in adults over the age of 65, with a decrease in insulin sensitivity observed with advancing age (Basu et al. 2003; Gumbiner et al. 1989). Age-related changes, such as reduced physical activity, changes in diet, and undesirable changes in body composition, i.e. reduced muscle mass and increased fat mass, can affect glucose tolerance (Coon et al. 1992; Shimokata et al. 1991). A healthy lifestyle can reverse the detrimental effects on glucose metabolism, however most interventions focus on prevention rather than treatment, as diabetes is difficult to fully reverse. Although, it has been found that regular exercise and a physically active lifestyle can help attenuate the usual decline in insulin secretion and glucose tolerance associated with ageing (Kahn et al. 1990; Lithgow and Leggate 2018; Mann et al. 2014; Short et al. 2003), current evidence is inconclusive. Older people, including those who are frail and/or weak, have been shown to benefit from endurance and resistance training, which can prevent age-related loss of muscle mass and strength, defined as sarcopenia and dynapenia. Muscle tissue has been identified as a major regulator of glucose homeostasis and tolerance (Petersen et al. 2007). The initial step paramount for cellular glucose utilisation is the transport of glucose across the cell membrane into the matrix of the cell by the action of insulin, thus preventing hypo- and hyperglycaemia (Kahn 2003). The sensitivity of cells to the action of insulin may thus determine the rate at which glucose is cleared from the (Utriainen et al. 1998). Since muscle tissue stores glucose, primarily as glycogen, the muscle mass has an available supply of glucose to maintain homeostasis in the case of hypoglycaemia. However, only during muscle activity can skeletal muscle glycogen breakdown provide a source of glucose (Jensen et al. 2011), as it is converted to lactate and then into blood glucose.

In addition, exercise training and maintaining a physically active lifestyle can be beneficial in promoting loss of excess abdominal or visceral adiposity that accumulates with an energy imbalance. In turn, this can result in alleviating insulin resistance (Gastaldelli et al. 2002; Goodpaster et al. 2003). Obesity is associated with a low-grade chronic inflammatory response, resulting from the secretion and activation of some pro-inflammatory cytokines/adipokines and respective pathways (Spranger et al. 2003). Adipocytes exhibit properties shared by immune cells, mainly pro-inflammatory cytokine production, such as IL-6, TNF-), and CRP, which can influence insulin production. Therefore, if older adults perform regular exercise and maintain a healthy body composition, the daily control over blood glucose levels may prevent the onset of a chronic hyperglycaemic state.

In addition to altering body composition, exercise can also influence insulin sensitivity on a cellular basis (Henriksen 2002). Exercise upregulates the demand on hepatic and skeletal muscle metabolism to provide fuel for the mechanical stress induced. During exercise, the production, regulation, and uptake of glucose is mediated by the glucose transporter type 4 (GLUT4) through insulin-controlled pathways (Park et al. 2015), with excess glucose contributing to glycogen stores if not metabolically required. Insulin secretion is inhibited during exercise and thus the body relies on hepatic and skeletal muscle tissue cells being sufficiently sensitive to insulin to maintain glucose homeostasis. Regular exercise can improve the efficiency of this mechanism, and as a result can improve insulin sensitivity of cells. Continuing to exercise throughout the life span, particularly in older adult or elderly ages, can delay the onset or reduce the risk of insulin resistance (Goodyear and Kahn 1998) and thus diabetes.

Oxidative Stress: An Ageing Problem, an Exercise Solution?

Free radicals, and reactive oxygen species (ROS) can be generated within the body by various metabolic pathways and enzymes, such as mitochondrial complexes in the electron transport chain (ETC), cytochrome P450, xanthine oxidase and nicotinamide dinucleotide phosphate (NADPH) oxidase (Murphy 2009). Oxidative stress occurs when free radical or ROS production exceeds the body's antioxidant capacity, leading to unchecked effects of these reactive molecules and compounds on tissues, such as DNA modifications, damage to lipids, proteins and other macromolecules. The accumulation of oxidative stress has been purported to lead to the ageing associated tissue dysfunction. This 'free radical theory of ageing' (Harman 1956) hypothesizes that this elevated exposure to oxidative stress damages macromolecules, impairing antioxidant and repair mechanisms, which leads to the deleterious effects on tissues (Sallam and Laher 2016). Indeed ageing is associated with elevated levels of oxidative stress in various tissues in the body such as skeletal muscle (Bejma and Ji 1999), the heart, brain (Navarro et al. 2004) and the vascular tree (Luttrell et al. 2013). Specifically, advanced age is linked with defective mitochondria which itself results from reduction in cytochrome C oxidase activity (Navarro et al. 2004). This mitochondrial dysfunction leads to greater escape of generated electrons which can stimulate oxidative damage. Oxidative stress may play a role in processes such as inflammation (Chung et al. 2009), sarcopenia (Jackson 2016), and insulin resistance (Paneni et al. 2015). Whilst there is plethora of evidence to show that lowering oxidative stress promotes tissue function (Eskurza et al. 2004; Taddei et al. 2001; Tatchum-Talom and Martin 2004), there is some evidence to challenge the free radical theory of ageing, with studies showing that increasing antioxidant capacity in mice fails to extend lifespan (Chen et al. 2004), indicating that lowering oxidative stress may promote tissue function without affecting longevity.

Exercise and physical activity modulates some of the deleterious side-effects of ageing, and is known to be protective against oxidative stress-associated conditions, including CVD, diabetes (de Souto Barreto et al. 2017), and cancer (Celis-Morales et al. 2017). However, with acute exercise, due to the elevated oxygen consumption ($\dot{V}O_2$), there is an enhanced leakage of superoxide ($O_2\cdot-$) from the ETC (Wang et al. 2015), leading to an imbalance between ROS production and antioxidant capacity. This overproduction of $O_2\cdot-$ though, acts as an important redox signal for regular exercise-induced adaptations (Cobley et al. 2015; Margaritelis et al. 2018; Webb et al. 2017). Several studies in human ageing populations report reductions in plasma or urine markers of oxidative stress with endurance training or regular aerobic exercise (e.g. Thiobarbituric Acid Reactive Substances; TBARS, lipid peroxidation, $O_2\cdot-$) (Ghosh et al. 2011; Jessup et al. 2003; Karolkiewicz et al. 2009) or an improvement in antioxidant capacity (upregulation of antioxidant enzymes, such as superoxide dismutase; SOD, and catalase) (Done and Traustadottir 2016; Johnson et al. 2015; Takahashi et al. 2013). Resistance exercise may also confer some benefits, with some studies reporting positive effects on oxidative stress biomarkers and antioxidant capacity (Bloomer et al. 2008; Bobeuf et al. 2011; Parise et al. 2005b; Vincent et al. 2006). There is however, contrasting evidence to show lack of efficacy of exercise training to modulate some oxidative stress biomarkers (Parise et al. 2005a). These differences are due to variety of biomarkers of oxidative stress and damage, as well as antioxidant capacity, and as yet, due to the rapid appearance and subsequent disappearance of ROS and free radicals, measurement is difficult, and often requires downstream markers (Cobley et al. 2017).

Physical inactivity itself promotes the elevation of basal ROS and oxidative stress (Bar-Shai et al. 2008; Pierre et al. 2016). Animal models of physical inactivity show that skeletal muscle from immobilized limbs in mice produce higher levels of $O_2\cdot-$ and hydrogen peroxide (H_2O_2) than mobilized limbs (Cannavino et al. 2014; Talbert et al. 2013; Xu et al. 2010). In cross-sectional studies comparing active vs.

inactive animals, lipid peroxidation and protein damage levels in skeletal muscle are elevated in sedentary vs. active rodent models (Figueiredo et al. 2009; Rosa et al. 2005). In humans, one study showed that 2 weeks of unilateral limb immobilization in old men resulted in greater H_2O_2 production and mitochondrial leakage than the mobilized limb, however this returned to normal after a period of exercise training, suggesting that exercise may be able to counteract the pro-oxidant effect of inactivity. Further studies show that inactive older individuals display greater levels of oxidative stress biomarkers than trained age-matched controls (Santos-Parker et al. 2017). Together, these animal and human models of inactivity show that sedentary behaviours promote localised ROS production, which may have significant effects on tissue function, compromising the health of older individuals. Considering the positive effect of regular physical aerobic and/or resistance exercise, physical activity should be promoted to counteract the negative effects of both ageing and inactivity.

Conclusion

The exact 'dose' of exercise to promote healthy ageing and longevity is still unknown, and unlikely to be described in the near future due to the varying effects that manipulating the time, intensity and frequency of exercise has on our cells and tissues. However, what is known is that exercise acts as a powerful, health-promoting, stimulus. Its ability to positively benefit a wide variety of cells, tissues and organs means it can be regarded as a potent anti-ageing therapeutic intervention. The evidence available shows that physical activity and exercise can reduce NCD risk, improve cardiovascular, immune and muscle function, leading to improved quality of life in our ever increasing ageing population.

References

Abdul-Ghani MA, Tripathy D, DeFronzo RA (2006) Contributions of beta-cell dysfunction and insulin resistance to the pathogenesis of impaired glucose tolerance and impaired fasting glucose. Diabetes Care 29:1130–1139

ADA (2014) Diagnosis and classification of diabetes mellitus. Diabetes Care 37(Suppl 1):S81–S90

Aguirre LE, Jan IZ, Fowler K et al (2014) Testosterone and adipokines are determinants of physical performance, strength, and aerobic fitness in frail, obese, older adults. Int J Endocrinol 2014:507395

Aly YE, Abdou AS, Rashad MM et al (2016) Effect of exercise on serum vitamin D and tissue vitamin D receptors in experimentally induced type 2 diabetes mellitus. J Adv Res 7:671–679

Amati F, Dubé JJ, Coen PM et al (2009) Physical inactivity and obesity underlie the insulin resistance of ageing. Diabetes Care 32:1547–1549

Ari Z, Kutlu N, Uyanik BS et al (2004) Serum testosterone, growth hormone, and insulin-like growth factor-1 levels, mental reaction time, and maximal aerobic exercise in sedentary and long-term physically trained elderly males. Int J Neurosci 114:623–637

Asahara T, Murohara T, Sullivan A et al (1997) Isolation of putative progenitor endothelial cells for angiogenesis. Science 275:964–966

Asahara T, Masuda H, Takahashi T et al (1999) Bone marrow origin of endothelial progenitor cells responsible for postnatal vasculogenesis in physiological and pathological neovascularization. Circ Res 85:221–228

Atkinson RA, Srinivas-Shankar U, Roberts SA et al (2010) Effects of testosterone on skeletal muscle architecture in intermediate-frail and frail elderly men. J Gerontol A Biol Sci Med Sci 65A:1215–1219

Bar-Shai M, Carmeli E, Ljubuncic P et al (2008) Exercise and immobilization in ageing animals: the involvement of oxidative stress and NF-κB activation. Free Rad Biol Med 44:202–214

Basu R, Breda E, Oberg AL et al (2003) Mechanisms of the age-associated deterioration in glucose tolerance: contribution of alterations in insulin secretion, action, and clearance. Diabetes 52:1738–1748

Bejma J, Ji LL (1999) Ageing and acute exercise enhance free radical generation in rat skeletal muscle. J Appl Physiol 87:465–470

Birk GK, Dawson EA, Atkinson C et al (2012) Brachial artery adaptation to lower limb exercise training: role of shear stress. J Appl Physiol 112:1653–1658

Bischoff HA, Borchers M, Gudat F et al (2001) In situ detection of 1,25-dihydroxyvitamin D3 receptor in human skeletal muscle tissue. Histochem J 33:19–24

Bischoff HA, Stahelin HB, Dick W et al (2003) Effects of vitamin D and calcium supplementation on falls: a randomized controlled trial. J Bone Miner Res 18:343–351

Bischoff-Ferrari HA, Borchers M, Gudat F et al (2004a) Vitamin D receptor expression in human muscle tissue decreases with age. J Bone Miner Res 19:265–269

Bischoff-Ferrari HA, Dietrich T, Orav EJ et al (2004b) Positive association between 25-hydroxy vitamin D levels and bone mineral density: a population-based study of younger and older adults. Am J Med 116:634–639

Black MA, Green DJ, Cable NT (2008) Exercise prevents age-related decline in nitric-oxide-mediated vasodilator function in cutaneous microvessels. J Physiol 586:3511–3524

Black MA, Cable NT, Thijssen DH et al (2009) Impact of age, sex, and exercise on brachial artery flow-mediated dilatation. Am J Physiol Heart Circ Physiol 297:H1109–H1116

Bloomer RJ, Schilling BK, Karlage RE et al (2008) Effect of resistance training on blood oxidative stress in Parkinson disease. Med Sci Sports Exerc 40:1385–1389

Bobeuf F, Labonte M, Dionne IJ et al (2011) Combined effect of antioxidant supplementation and resistance training on oxidative stress markers, muscle and body composition in an elderly population. J Nutr Health Ageing 15:883–889

Breen L, Stokes KA, Churchward-Venne TA et al (2013) Two weeks of reduced activity decreases leg lean mass and induces "anabolic resistance" of myofibrillar protein synthesis in healthy elderly. J Clin Endocrinol Metab 98:2604–2612

Briggs AM, Cross MJ, Hoy DG et al (2016) Musculoskeletal health conditions represent a global threat to healthy ageing: a report for the 2015 World Health Organization World Report on ageing and health. Gerontologist 56:S243–S255

Bruyndonckx L, Hoymans VY, Frederix G et al (2014) Endothelial progenitor cells and endothelial microparticles are independent predictors of endothelial function. J Pediatr 165:300–305

Butler AE, Janson J, Bonner-Weir S et al (2003) Beta-cell deficit and increased beta-cell apoptosis in humans with type 2 diabetes. Diabetes 52:102–110

Cannavino J, Brocca L, Sandri M et al (2014) PGC1-α over-expression prevents metabolic alterations and soleus muscle atrophy in hindlimb unloaded mice. J Physiol 592:4575–4589

Carlsson AC, Arnlov J, Sundstrom J et al (2016) Physical activity, obesity and risk of cardiovascular disease in middle-aged men during a median of 30 years of follow-up. Eur J Prev Cardiol 23:359–365

Case J, Mead LE, Bessler WK et al (2007) Human CD34+AC133+VEGFR-2+ cells are not endothelial progenitor cells but distinct, primitive hematopoietic progenitors. Exp Hematol 35:1109–1118

Castaldi A, Dodia RM, Orogo AM et al (2017) Decline in cellular function of aged mouse c-kit(+) cardiac progenitor cells. J Physiol 595:6249–6262

Castle SC (2000) Clinical relevance of age-related immune dysfunction. Clin Infect Dis 31:578–585

Ceglia L, Harris SS (2013) Vitamin D and its role in skeletal muscle. Calcif Tissue Int 92:151–162

Ceglia L, Niramitmahapanya S, da Silva Morais M et al (2013) A randomized study on the effect of vitamin D(3) supplementation on skeletal muscle morphology and vitamin D receptor concentration in older women. J Clin Endocrinol Metab 98:E1927–E1935

Celis-Morales CA, Lyall DM, Welsh P et al (2017) Association between active commuting and incident cardiovascular disease, cancer, and mortality: prospective cohort study. BMJ 357:j1456

Chapuy MC, Durr F, Chapuy P (1983) Age-related changes in parathyroid hormone and 25 hydroxycholecalciferol levels. J Gerontol 38:19–22

Chen X, Liang H, Van Remmen H et al (2004) Catalase transgenic mice: characterization and sensitivity to oxidative stress. Arch Biochem Biophys 422:197–210

Chistiakov DA, Orekhov AN, Bobryshev YV (2017) Effects of shear stress on endothelial cells: go with the flow. Acta Physiol (Oxf) 219:382–408

Choi J, Moon K, Jung S et al (2014) Regular exercise training increases the number of endothelial progenitor cells and decreases homocysteine levels in healthy peripheral blood. Kor J Physiol Pharmacol 18:163–168

Chrissobolis S, Faraci FM (2008) The role of oxidative stress and NADPH oxidase in cerebrovascular disease. Trends Mol Med 14:495–502

Chung HY, Cesari M, Anton S et al (2009) Molecular inflammation: underpinnings of ageing and age-related diseases. Ageing Res Rev 8:18–30

Cnop M, Welsh N, Jonas JC et al (2005) Mechanisms of pancreatic beta-cell death in type 1 and type 2 diabetes: many differences, few similarities. Diabetes 54(Suppl 2):S97–S107

Cobley JN, McHardy H, Morton JP et al (2015) Influence of vitamin C and vitamin E on redox signaling: implications for exercise adaptations. Free Radic Biol Med 84:65–76

Cobley JN, Close GL, Bailey DM et al (2017) Exercise redox biochemistry: conceptual, methodological and technical recommendations. Redox Biol 12:540–548

Coleman LA, Mishina M, Thompson M et al (2016) Age, serum 25-hydroxyvitamin D and vitamin D receptor (VDR) expression and function in peripheral blood mononuclear cells. Oncotarget 7:35512–35521

Collaboration NRF (2016) Worldwide trends in diabetes since 1980: a pooled analysis of 751 population-based studies with 4·4 million participants. Lancet 387:1513–1530

Coon PJ, Rogus EM, Drinkwater D et al (1992) Role of body fat distribution in the decline in insulin sensitivity and glucose tolerance with age. J Clin Endocrinol Metab 75:1125–1132

Cruz-Jentoft AJ, Baeyens JP, Bauer JM et al (2010) Sarcopenia: European consensus on definition and diagnosis: report of the European Working Group on sarcopenia in older people. Age Ageing 39:412–423

Cuthbertson D, Smith K, Babraj J et al (2005) Anabolic signaling deficits underlie amino acid resistance of wasting, ageing muscle. FASEB J 19:422–424

Dawson-Hughes B, Harris SS, Krall EA et al (1997) Effect of calcium and vitamin D supplementation on bone density in men and women 65 years of age or older. N Engl J Med 337:670–676

de Souto Barreto P, Cesari M, Andrieu S et al (2017) Physical activity and incident chronic diseases: a longitudinal observational study in 16 European countries. Am J Prev Med 52:373–378

Deane CS, Hughes DC, Sculthorpe N et al (2013) Impaired hypertrophy in myoblasts is improved with testosterone administration. J Ster Biochem Mol Biol 138:152–161

Delmonico MJ, Harris TB, Lee JS et al (2007) Alternative definitions of sarcopenia, lower extremity performance, and functional impairment with ageing in older men and women. J Am Geriatr Soc 55:769–774

Delmonico MJ, Harris TB, Visser M et al (2009) Longitudinal study of muscle strength, quality, and adipose tissue infiltration. Am J Clin Nutr 90:1579–1585

Devries MC, Breen L, Von Allmen M et al (2015) Low-load resistance training during step-reduction attenuates declines in muscle mass and strength and enhances anabolic sensitivity in older men. Physiol Rep 3:e12493

Doering T, Jenkins D, Reaburn P et al (2016) Lower integrated muscle protein synthesis in masters compared with younger athletes. Med Sci Sports Exerc 48:1613–1618

Done AJ, Traustadottir T (2016) Aerobic exercise increases resistance to oxidative stress in sedentary older middle-aged adults. A pilot study. Age (Dordr) 38:505–512

Duggal NA, Pollock RD, Lazarus NR et al (2018) Major features of immunesenescence, including reduced thymic output, are ameliorated by high levels of physical activity in adulthood. Ageing Cell 17:e12750

Ellison G, Vicinanza C, Smith A et al (2013) Adult c-kitpos cardiac stem cells are necessary and sufficient for functional cardiac regeneration and repair. Cell 154:827–842

Eskurza I, Monahan KD, Robinson JA et al (2004) Effect of acute and chronic ascorbic acid on flow-mediated dilatation with sedentary and physically active human ageing. J Physiol 556:315–324

Evans WJ (2010) Skeletal muscle loss: cachexia, sarcopenia, and inactivity. Am J Clin Nutr 91:1123S–1127S

Fiatarone MA, Marks EC, Ryan ND et al (1990) High-intensity strength training in nonagenarians: effects on skeletal muscle. JAMA 263:3029–3034

Figueiredo PA, Powers SK, Ferreira RM et al (2009) Impact of lifelong sedentary behavior on mitochondrial function of mice skeletal muscle. J Gerontol A Biol Sci Med Sci 64A:927–939

Francaux M, Demeulder B, Naslain D et al (2016) Ageing reduces the activation of the mTORC1 pathway after resistance exercise and protein intake in human skeletal muscle: potential role of REDD1 and impaired anabolic sensitivity. Nutrients 8:47

Frederiksen L, Højlund K, Hougaard DM et al (2012) Testosterone therapy increased muscle mass and lipid oxidation in ageing men. Age 34:145–156

Furchgott R, Zawadzki J (1980) The obligatory role of endothelial cells in the relaxation of arterial smooth muscle by acetylcholine. Nature 288:373–376

Gallagher JC (1990) The pathogenesis of osteoporosis. Bone Miner 9:215–227

Garcia LA, King KK, Ferrini MG et al (2011) 1,25(OH)2vitamin D3 stimulates myogenic differentiation by inhibiting cell proliferation and modulating the expression of promyogenic growth factors and myostatin in C2C12 skeletal muscle cells. Endocrinology 152:2976–2986

Gastaldelli A, Miyazaki Y, Pettiti M et al (2002) Metabolic effects of visceral fat accumulation in type 2 diabetes. J Clin Endocrinol Metab 87:5098–5103

Ghosh S, Lertwattanarak R, Lefort N et al (2011) Reduction in reactive oxygen species production by mitochondria from elderly subjects with normal and impaired glucose tolerance. Diabetes 60:2051–2060

Goodpaster BH, Krishnaswami S, Resnick H et al (2003) Association between regional adipose tissue distribution and both type 2 diabetes and impaired glucose tolerance in elderly men and women. Diabetes Care 26:372–379

Goodyear LJ, Kahn BB (1998) Exercise, glucose transport, and insulin sensitivity. Annu Rev Med 49:235–261

Green D (2005) Point: flow-mediated dilation does reflect nitric oxide-mediated endothelial function. J Appl Physiol 99:1233–1234

Green DJ, Jones H, Thijssen D et al (2011) Flow-mediated dilation and cardiovascular event prediction. Hypertension 57:363–369

Gumbiner B, Polonsky KS, Beltz WF et al (1989) Effects of ageing on insulin secretion. Diabetes 38:1549–1556

Hall KS, Cohen HJ, Pieper CF et al (2017) Physical performance across the adult life span: correlates with age and physical activity. J Gerontol A Biol Sci Med Sci 72:572–578

Hamada K, Vannier E, Sacheck JM et al (2005) Senescence of human skeletal muscle impairs the local inflammatory cytokine response to acute eccentric exercise. FASEB J 19:264–266

Hamilton CA, Brosnan MJ, McIntyre M et al (2001) Superoxide excess in hypertension and ageing: a common cause of endothelial dysfunction. Hypertension 37:529–534

Harman D (1956) Ageing: a theory based on free radical and radiation chemistry. J Gerontol 11:298–300

Hawkins SA, Wiswell RA, Marcell TJ (2003) Exercise and the master athlete – a model of successful ageing? J Gerontol A Biol Sci Med Sci 58:M1009–M1011

Hayes LD, Grace FM, Sculthorpe N et al (2013) Does chronic exercise attenuate age-related physiological decline in males? Res Sports Med 21:343–354

Hayes LD, Sculthorpe N, Herbert P et al (2015) Resting steroid hormone concentrations in lifetime exercisers and lifetime sedentary males. Ageing Male 18:22–26

Hayes LD, Herbert P, Sculthorpe NF et al (2017) Exercise training improves free testosterone in lifelong sedentary ageing men. Endocr Connect 6:306–310

Henriksen EJ (2002) Invited review: effects of acute exercise and exercise training on insulin resistance. J Appl Physiol 93:788–796

Herbert P, Hayes LD, Sculthorpe N et al (2017a) High-intensity interval training (HIIT) increases insulin-like growth factor-I (IGF-I) in sedentary ageing men but not masters' athletes: an observational study. Ageing Male 20:54–59

Herbert P, Hayes LD, Sculthorpe NF et al (2017b) HIIT produces increases in muscle power and free testosterone in male masters athletes. Endocr Connect 6:430–436

Hoetzer GL, Van Guilder GP, Irmiger HM et al (2007) Ageing, exercise, and endothelial progenitor cell clonogenic and migratory capacity in men. J Appl Physiol 102:847–852

Holick MF (2007) Vitamin D deficiency. N Engl J Med 357:266–281

Holick MF, Chen TC (2008) Vitamin D deficiency: a worldwide problem with health consequences. Am J Clin Nutr 87:1080S–1086S

Hur J, Yoon C-H, Kim H-S et al (2004) Characterization of two types of endothelial progenitor cells and their different contributions to neovasculogenesis. Arterioscler Thromb Vasc Biol 24:288–293

Jackson MJ (2016) Reactive oxygen species in sarcopenia: should we focus on excess oxidative damage or defective redox signalling? Mol Asp Med 50:33–40

Jackson RD, LaCroix AZ, Gass M et al (2006) Calcium plus vitamin D supplementation and the risk of fractures. N Engl J Med 354:669–683

Janssen I, Heymsfield SB, Ross R (2002) Low relative skeletal muscle mass (sarcopenia) in older persons is associated with functional impairment and physical disability. J Am Geriatr Soc 50:889–896

Jensen J, Rustad PI, Kolnes AJ et al (2011) The role of skeletal muscle glycogen breakdown for regulation of insulin sensitivity by exercise. Front Physiol 2:112

Jessup JV, Horne C, Yarandi H et al (2003) The effects of endurance exercise and vitamin E on oxidative stress in the elderly. Biol Res Nurs 5:47–55

Johnson ML, Irving BA, Lanza IR et al (2015) Differential effect of endurance training on mitochondrial protein damage, degradation, and acetylation in the context of ageing. J Gerontol A Biol Sci Med Sci 70:1386–1393

Kahn SE (2003) The relative contributions of insulin resistance and beta-cell dysfunction to the pathophysiology of type 2 diabetes. Diabetologia 46:3–19

Kahn SE, Larson VG, Beard JC et al (1990) Effect of exercise on insulin action, glucose tolerance, and insulin secretion in ageing. Am J Phys 258:E937–E943

Karolkiewicz J, Michalak E, Pospieszna B et al (2009) Response of oxidative stress markers and antioxidant parameters to an 8-week aerobic physical activity program in healthy, postmenopausal women. Arch Gerontol Ger 49:e67–e71

Khoo J, Tian H-H, Tan B et al (2013) Comparing effects of low- and high-volume moderate-intensity exercise on sexual function and testosterone in obese men. J Sex Med 10:1823–1832

Khosla S, Riggs BL (2005) Pathophysiology of age-related bone loss and osteoporosis. Endocrinol Metab Clin N Am 34:1015–1030. xi

Kloppel G, Lohr M, Habich K et al (1985) Islet pathology and the pathogenesis of type 1 and type 2 diabetes mellitus revisited. Surv Synth Pathol Res 4:110–125

Kommalage M, Gunawardena S (2013) Influence of age, gender, and sidedness on ulnar nerve conduction. J Clin Neurophysiol 30:98–101

Koopman R, Walrand S, Beelen M et al (2009) Dietary protein digestion and absorption rates and the subsequent postprandial muscle protein synthetic response do not differ between young and elderly men. J Nutr 139:1707–1713

Kumar V, Selby A, Rankin D et al (2009) Age-related differences in the dose–response relationship of muscle protein synthesis to resistance exercise in young and old men. J Physiol 587:211–217

Kushner E, Van Guilder G, MacEneaney O et al (2010) Ageing and endothelial progenitor cell release of proangiogenic cytokines. Age Ageing 39:268–272

Kushner EJ, MacEneaney OJ, Weil BR et al (2011) Ageing is associated with a proapoptotic endothelial progenitor cell phenotype. J Vasc Res 48:408–414

Laufs U, Werner N, Link A et al (2004) Physical training increases endothelial progenitor cells, inhibits neointima formation, and enhances angiogenesis. Circulation 109:220–226

Laufs U, Urhausen A, Werner N et al (2005) Running exercise of different duration and intensity: effect on endothelial progenitor cells in healthy subjects. Eur J Cardio Prev Rehab 12:407–414

Lazuardi L, Jenewein B, Wolf AM et al (2005) Age-related loss of naive T cells and dysregulation of T-cell/B-cell interactions in human lymph nodes. Immunology 114:37–43

Lee I-M, Shiroma EJ, Lobelo F et al (2012) Effect of physical inactivity on major non-communicable diseases worldwide: an analysis of burden of disease and life expectancy. Lancet 380:219–229

Leite CF, Lopes CS, Alves AC et al (2015) Endogenous resident c-Kit cardiac stem cells increase in mice with an exercise-induced, physiologically hypertrophied heart. Stem Cell Res 15:151–164

Leveille SG, Gray S, Lacroix AZ et al (2000) Physical inactivity and smoking increase risk for serious infections in older women. J Am Geriatr Soc 48:1582–1588

Lexell J (1995) Human ageing, muscle mass, and fiber type composition. J Gerontol A Biol Sci Med Sci 50 Spec No:11–16

Lithgow H, Leggate M (2018) The effect of a single bout of high intensity intermittent exercise on glucose tolerance in non-diabetic older adults. Int J Exerc Sci 11:95–105

Lozano R, Naghavi M, Foreman K et al (2012) Global and regional mortality from 235 causes of death for 20 age groups in 1990 and 2010: a systematic analysis for the global burden of disease study 2010. Lancet 380:2095–2128

Luk T-H, Dai Y-L, Siu C-W et al (2012) Effect of exercise training on vascular endothelial function in patients with stable coronary artery disease: a randomized controlled trial. Eur J Prev Cardiol 19:830–839

Luttrell M, Seawright J, Wilson E et al (2013) Effect of age and exercise training on protein: protein interactions among eNOS and its regulatory proteins in rat aortas. Eur J Appl Physiol 113:2761–2768

Makanae Y, Ogasawara R, Sato K et al (2015) Acute bout of resistance exercise increases vitamin D receptor protein expression in rat skeletal muscle. Exp Physiol 100:1168–1176

Malm C (2006) Susceptibility to infections in elite athletes: the S-curve. Scand J Med Sci Sports 16:4–6

Manfredini F, Rigolin GM, Malagoni AM et al (2009) Exercise training and endothelial progenitor cells in haemodialysis patients. J Int Med Res 37:534–540

Manini TM, Clark BC (2012) Dynapenia and ageing: an update. J Gerontol A Biol Sci Med Sci 67A:28–40

Mann S, Beedie C, Balducci S et al (2014) Changes in insulin sensitivity in response to different modalities of exercise: a review of the evidence. Diabetes Metab Res Rev 30:257–268

Margaritelis NV, Theodorou AA, Paschalis V et al (2018) Adaptations to endurance training depend on exercise-induced oxidative stress: exploiting redox interindividual variability. Acta Physiol (Oxf) 222:e12898

Mayhan WG, Arrick DM, Sharpe GM et al (2008) Age-related alterations in reactivity of cerebral arterioles: role of oxidative stress. Microcirculation 15:225–236

Mikkelsen UR, Couppé C, Karlsen A et al (2013) Life-long endurance exercise in humans: circulating levels of inflammatory markers and leg muscle size. Mech Age Dev 134:531–540

Mitchell WK, Williams J, Atherton P et al (2012) Sarcopenia, dynapenia, and the impact of advancing age on human skeletal muscle size and strength; a quantitative review. Front Physiol 3:260

Moore DR, Churchward-Venne TA, Witard O et al (2015) Protein ingestion to stimulate myofibrillar protein synthesis requires greater relative protein intakes in healthy older versus younger men. J Gerontol A Biol Sci Med Sci 70:57–62

Mooren FC, Kruger K (2015) Apoptotic lymphocytes induce progenitor cell mobilization after exercise. J Appl Physiol 119:135–139

Muller-Delp JM (2006) Ageing-induced adaptations of microvascular reactivity. Microcirculation 13:301–314

Murphy MP (2009) How mitochondria produce reactive oxygen species. Biochem J 417:1–13

Navarro A, Gomez C, López-Cepero JM et al (2004) Beneficial effects of moderate exercise on mice ageing: survival, behavior, oxidative stress, and mitochondrial electron transfer. Am J Physiol Regul Integr Comp Physiol 286:R505–R511

Newman AB, Kupelian V, Visser M et al (2006) Strength, but not muscle mass, is associated with mortality in the health, ageing and body composition study cohort. J Gerontol A Biol Sci Med Sci 61:72–77

Nieman DC (1994) Exercise, infection, and immunity. Int J Sports Med 15(Suppl 3):S131–S141

Otis JS, Niccoli S, Hawdon N et al (2014) Pro-inflammatory mediation of myoblast proliferation. PLoS One 9:e92363

Pan XR, Li GW, Hu YH et al (1997) Effects of diet and exercise in preventing NIDDM in people with impaired glucose tolerance. The Da Qing IGT and Diabetes Study. Diabetes Care 20:537–544

Paneni F, Costantino S, Cosentino F (2015) Role of oxidative stress in endothelial insulin resistance. World J Diabetes 6:326–332

Parise G, Brose AN, Tarnopolsky MA (2005a) Resistance exercise training decreases oxidative damage to DNA and increases cytochrome oxidase activity in older adults. Exp Gerontol 40:173–180

Parise G, Phillips SM, Kaczor JJ et al (2005b) Antioxidant enzyme activity is up-regulated after unilateral resistance exercise training in older adults. Free Rad Biol Med 39:289–295

Park D-R, Park K-H, Kim B-J et al (2015) Exercise ameliorates insulin resistance via Ca^{2+} signals distinct from those of insulin for GLUT4 translocation in skeletal muscles. Diabetes 64:1224–1234

Patel RS, Li Q, Ghasemzadeh N et al (2015) Circulating CD34+ progenitor cells and risk of mortality in a population with coronary artery disease. Circ Res 116:289–297

Pawelec G, Ferguson FG, Wikby A (2001) The SENIEUR protocol after 16 years. Mech Ageing Dev 122:132–134

Petersen KF, Dufour S, Savage DB et al (2007) The role of skeletal muscle insulin resistance in the pathogenesis of the metabolic syndrome. PNAS 104:12587–12594

Phillips SM (2015) Nutritional supplements in support of resistance exercise to counter age-related sarcopenia. Adv Nutr (Bethesda, MD) 6:452–460

Phillips SM, Chevalier S, Leidy HJ (2016) Protein "requirements" beyond the RDA: implications for optimizing health. Appl Physiol Nutr Metab 41:565–572

Pierre N, Appriou Z, Gratas-Delamarche A et al (2016) From physical inactivity to immobilization: dissecting the role of oxidative stress in skeletal muscle insulin resistance and atrophy. Free Radic Biol Med 98:197–207

Provinciali M, Moresi R, Donnini A et al (2009) Reference values for CD4+ and CD8+ T lymphocytes with naive or memory phenotype and their association with mortality in the elderly. Gerontology 55:314–321

Rakobowchuk M, Tanguay S, Burgomaster KA et al (2008) Sprint interval and traditional endurance training induce similar improvements in peripheral arterial stiffness and flow-mediated dilation in healthy humans. Am J Physiol Regul Integr Comp Physiol 295:R236–R242

Renko O, Tolonen A-M, Rysä J et al (2018) SDF1 gradient associates with the distribution of c-Kit+ cardiac cells in the heart. Sci Rep 8:1160

Rennie MJ (2009) Anabolic resistance: the effects of ageing, sexual dimorphism, and immobilization on human muscle protein turnover. Appl Physiol Nutr Metab 34:377–381

Robson PJ, Blannin AK, Walsh NP et al (1999) Effects of exercise intensity, duration and recovery on in vitro neutrophil function in male athletes. Int J Sports Med 20:128–135

Rosa EF, Silva AC, Ihara SSM et al (2005) Habitual exercise program protects murine intestinal, skeletal, and cardiac muscles against ageing. J Appl Physiol 99:1569–1575

Ross MD, Wekesa AL, Phelan JP et al (2014) Resistance exercise increases endothelial progenitor cells and angiogenic factors. Med Sci Sports Exerc 46:16–23

Ross MD, Malone EM, Simpson R et al (2018) Lower resting and exercise-induced circulating angiogenic progenitors and angiogenic T cells in older men. Am J Physiol Heart Circ Physiol 314:H392–H402

Sallam N, Laher I (2016) Exercise modulates oxidative stress and inflammation in ageing and cardiovascular diseases. Oxidative Med Cell Longev 2016:7239639

Santos-Parker JR, Strahler TR, Vorwald VM et al (2017) Habitual aerobic exercise does not protect against micro- or macrovascular endothelial dysfunction in healthy estrogen-deficient post-menopausal women. J Appl Physiol 122:11–19

Sarto P, Balducci E, Balconi G et al (2007) Effects of exercise training on endothelial progenitor cells in patients with chronic heart failure. J Cardiac Fail 13:701–708

Schlager O, Giurgea A, Schuhfried O et al (2011) Exercise training increases endothelial progenitor cells and decreases asymmetric dimethylarginine in peripheral arterial disease: a randomized controlled trial. Atherosclerosis 217:240–248

Sculthorpe N, Herbert P, Grace FM (2015) Low-frequency high-intensity interval training is an effective method to improve muscle power in lifelong sedentary ageing men: a randomized controlled trial. J Am Geriatr Soc 63:2412–2413

Shimokata H, Muller DC, Fleg JL et al (1991) Age as independent determinant of glucose tolerance. Diabetes 40:44–51

Short KR, Vittone JL, Bigelow ML et al (2003) Impact of aerobic exercise training on age-related changes in insulin sensitivity and muscle oxidative capacity. Diabetes 52:1888–1896

Simpson RU, Thomas GA, Arnold AJ (1985) Identification of 1,25-dihydroxyvitamin D3 receptors and activities in muscle. J Biol Chem 260:8882–8891

Simpson RJ, Florida-James GD, Whyte GP et al (2006) The effects of intensive, moderate and downhill treadmill running on human blood lymphocytes expressing the adhesion/activation molecules CD54 (ICAM-1), CD18 (beta2 integrin) and CD53. Eur J Appl Physiol 97:109–121

Simpson RJ, Florida-James GD, Whyte GP et al (2007) Apoptosis does not contribute to the blood lymphocytopenia observed after intensive and downhill treadmill running in humans. Res Sports Med 15:157–174

Simpson RJ, Lowder TW, Spielmann G et al (2012) Exercise and the ageing immune system. Ageing Res Rev 11:404–420

Sipila S, Narici M, Kjaer M et al (2013) Sex hormones and skeletal muscle weakness. Biogerontology 14:231–245

Smith GI, Yoshino J, Reeds DN et al (2014) Testosterone and progesterone, but not estradiol, stimulate muscle protein synthesis in postmenopausal women. J Clin Endocrinol Metab 99:256–265

Sonnenschein K, Horváth T, Mueller M et al (2011) Exercise training improves in vivo endothelial repair capacity of early endothelial progenitor cells in subjects with metabolic syndrome. Eur J Cardio Prev Rehab 18:406–414

Soucy KG, Ryoo S, Benjo A et al (2006) Impaired shear stress-induced nitric oxide production through decreased NOS phosphorylation contributes to age-related vascular stiffness. J Appl Physiol 101:1751–1759

Spielmann G, McFarlin BK, O'Connor DP et al (2011) Aerobic fitness is associated with lower proportions of senescent blood T-cells in man. Brain Behav Immun 25:1521–1529

Spielmann G, Bollard CM, Bigley AB et al (2014) The effects of age and latent cytomegalovirus infection on the redeployment of CD8+ T cell subsets in response to acute exercise in humans. Brain Behav Immun 39:142–151

Spranger J, Kroke A, Mohlig M et al (2003) Inflammatory cytokines and the risk to develop type 2 diabetes: results of the prospective population-based European prospective investigation into cancer and nutrition (EPIC)-Potsdam Study. Diabetes 52:812–817

Steiner S, Niessner A, Ziegler S et al (2005) Endurance training increases the number of endothelial progenitor cells in patients with cardiovascular risk and coronary artery disease. Atherosclerosis 181:305–310

Stenholm S, Koster A, Valkeinen H et al (2016) Association of physical activity history with physical function and mortality in old age. J Gerontol A Biol Sci Med Sci 71:496–501

Strle K, Broussard SR, McCusker RH et al (2004) Proinflammatory cytokine impairment of insulin-like growth factor I-induced protein synthesis in skeletal muscle myoblasts requires ceramide. Endocrinology 145:4592–4602

Symons TB, Sheffield-Moore M, Mamerow MM et al (2011) The anabolic response to resistance exercise and a protein-rich meal is not diminished by age. J Nutr Health Ageing 15:376–381

Taddei S, Virdis A, Ghiadoni L et al (2001) Age-related reduction of NO availability and oxidative stress in humans. Hypertension 38:274–279

Takahashi M, Miyashita M, Kawanishi N et al (2013) Low-volume exercise training attenuates oxidative stress and neutrophils activation in older adults. Eur J Appl Physiol 113:1117–1126

Talbert EE, Smuder AJ, Min K et al (2013) Immobilization-induced activation of key proteolytic systems in skeletal muscles is prevented by a mitochondria-targeted antioxidant. J Appl Physiol 115:529–538

Tang BM, Eslick GD, Nowson C et al (2007) Use of calcium or calcium in combination with vitamin D supplementation to prevent fractures and bone loss in people aged 50 years and older: a meta-analysis. Lancet 370:657–666

Tatchum-Talom R, Martin DS (2004) Tempol improves vascular function in the mesenteric vascular bed of senescent rats. Can J Physiol Pharmacol 82:200–207

Thijssen DH, Vos JB, Verseyden C et al (2006) Haematopoietic stem cells and endothelial progenitor cells in healthy men: effect of ageing and training. Ageing Cell 5:495–503

Tidball JG (2017) Regulation of muscle growth and regeneration by the immune system. Nat Rev Immunol 17:165–178

Torella D, Rota M, Nurzynska D et al (2004) Cardiac stem cell and myocyte ageing, heart failure, and insulin-like growth factor-1 overexpression. Circ Res 94:514–524

Torjesen I (2016) Global cost of physical inactivity is estimated at $67.5bn a year. BMJ 354:i4187

Urbanek K, Quaini F, Tasca G et al (2003) Intense myocyte formation from cardiac stem cells in human cardiac hypertrophy. PNAS 100:10440–10445

Utriainen T, Takala T, Luotolahti M et al (1998) Insulin resistance characterizes glucose uptake in skeletal muscle but not in the heart in NIDDM. Diabetologia 41:555–559

Van Craenenbroeck EM, Vrints CJ, Haine SE et al (2008) A maximal exercise bout increases the number of circulating CD34+/KDR+ endothelial progenitor cells in healthy subjects. Relation with lipid profile. J Appl Physiol 104:1006–1013

Van Craenenbroeck E, Hoymans V, Beckers P et al (2010a) Exercise training improves function of circulating angiogenic cells in patients with chronic heart failure. Bas Res Cardiol 105:665–676

Van Craenenbroeck EM, Beckers PJ, Possemiers NM et al (2010b) Exercise acutely reverses dysfunction of circulating angiogenic cells in chronic heart failure. Eur Heart J 31:1924–1934

Veldurthy V, Wei R, Oz L et al (2016) Vitamin D, calcium homeostasis and ageing. Bone Res 4:16041

Vincent HK, Bourguignon C, Vincent KR (2006) Resistance training lowers exercise-induced oxidative stress and homocysteine levels in overweight and obese older adults. Obesity 14:1921–1930

Visser M, Goodpaster BH, Kritchevsky SB et al (2005) Muscle mass, muscle strength, and muscle fat infiltration as predictors of incident mobility limitations in well-functioning older persons. J Gerontol A Biol Sci Med Sci 60:324–333

Volpi E, Mittendorfer B, Rasmussen BB et al (2000) The response of muscle protein anabolism to combined hyperaminoacidemia and glucose-induced hyperinsulinemia is impaired in the elderly. J Clin Endocrinol Metab 85:4481–4490

Wang J-S, Lee M-Y, Lien H-Y et al (2014) Hypoxic exercise training improves cardiac/muscular hemodynamics and is associated with modulated circulating progenitor cells in sedentary men. Int J Cardiol 170:315–323

Wang P, Li CG, Qi Z et al (2015) Acute exercise induced mitochondrial H(2)O(2) production in mouse skeletal muscle: association with p(66Shc) and FOXO3a signaling and antioxidant enzymes. Oxidative Med Cell Longev 2015:536456

Wannamethee SG, Shaper AG, Walker M (1998) Changes in physical activity, mortality, and incidence of coronary heart disease in older men. Lancet 351:1603–1608

Webb R, Hughes MG, Thomas AW et al (2017) The ability of exercise-associated oxidative stress to trigger redox-sensitive signalling responses. Antioxidants (Basel) 6:63

Xia W-H, Li J, Su C et al (2012a) Physical exercise attenuates age-associated reduction in endothelium-reparative capacity of endothelial progenitor cells by increasing CXCR4/JAK-2 signaling in healthy men. Ageing Cell 11:111–119

Xia WH, Yang Z, Xu SY et al (2012b) Age-related decline in reendothelialization capacity of human endothelial progenitor cells is restored by shear stress. Hypertension 59:1225–1231

Xiao J, Xu T, Li J et al (2014) Exercise-induced physiological hypertrophy initiates activation of cardiac progenitor cells. Int J Clin Exp Pathol 7:663–669

Xu X, C-N C, Arriaga EA et al (2010) Asymmetric superoxide release inside and outside the mitochondria in skeletal muscle under conditions of ageing and disuse. J Appl Physiol 109:1133–1139

Zhang J-M, An J (2007) Cytokines, inflammation and pain. Int Anesth Clin 45:27–37

Chapter 13
Health Benefits of Anti-aging Drugs

Veronika Piskovatska, Olha Strilbytska, Alexander Koliada, Alexander Vaiserman, and Oleh Lushchak

Abstract Aging, as a physiological process mediated by numerous regulatory pathways and transcription factors, is manifested by continuous progressive functional decline and increasing risk of chronic diseases. There is an increasing interest to identify pharmacological agents for treatment and prevention of age-related disease in humans. Animal models play an important role in identification and testing of anti-aging compounds; this step is crucial before the drug will enter human clinical trial or will be introduced to human medicine. One of the main goals of animal studies is better understanding of mechanistic targets, therapeutic implications and side-effects of the drug, which may be later translated into humans. In this chapter, we summarized the effects of different drugs reported to extend the lifespan in model organisms from round worms to rodents. Resveratrol, rapamycin, metformin and aspirin, showing effectiveness in model organism life- and healthspan extension mainly target the master regulators of aging such as mTOR, FOXO and PGC1α, affecting autophagy, inflammation and oxidative stress. In humans, these drugs were demonstrated to reduce inflammation, prevent CVD, and slow down the functional decline in certain organs. Additionally, potential anti-aging pharmacologic agents inhibit cancerogenesis, interfering with certain aspects of cell metabolism, proliferation, angioneogenesis and apoptosis.

Keywords Aging · Lifespan · Healthspan · Model organisms · Anti-aging drugs

V. Piskovatska
Clinic for Heart Surgery, University Clinic of Martin Luther University, Halle, Germany
e-mail: veronika.piskovatska@uk-halle.de

O. Strilbytska · O. Lushchak (✉)
Biochemistry and Biotechnology Department, Vasyl Stefanyk Precarpathian National University, Ivano-Frankivsk, Ukraine
e-mail: olehl@pu.if.ua

A. Koliada · A. Vaiserman
D.F. Chebotarev Institute of Gerontology, NAMS, Kyiv, Ukraine
e-mail: vaiserman@geront.kiev.ua

© Springer Nature Singapore Pte Ltd. 2019
J. R. Harris, V. I. Korolchuk (eds.), *Biochemistry and Cell Biology of Ageing: Part II Clinical Science*, Subcellular Biochemistry 91,
https://doi.org/10.1007/978-981-13-3681-2_13

Introduction

During the last few centuries, advances in medicine and pharmacology allowed mankind to substantially increase human lifespan. Dramatic increase in the proportion of individuals >60 years is observed worldwide – from 9.2% in 1990 to 11.7% in 2013, with expected 21.1% (>2 billion) in 2050 (Sander et al. 2015). However, age-related disease prevalence and costs, associated with cardiovascular diseases, cancer, type 2 diabetes mellitus, cognitive impairment and dementia in elderly are overwhelming for many well-developed healthcare systems. Elderly patients are often experiencing an everyday struggle with multiple chronic conditions and disability. All of the above mentioned shifts focus on healthspan/quality of life prolongation, not just life extension (Hansen and Kennedy 2016). Non-pharmacological approaches to delay age-related decline including diet and physical activity have been extensively studied and have proved to provide some life-extending and disease-preventing benefits (Di Daniele et al. 2017; McPhee et al. 2016). Strategies of life- and healthspan extension with administration of drugs and their combinations remain tempting. Recent research, concentrated on prolonging life of model organisms and humans, is mainly focused on well-known pharmacological agents, as well as food-derived substances (Vaiserman et al. 2016; Vaiserman and Lushchak 2017a). Some of the FDA-approved substances target one or several mechanisms, associated with age-related cellular and molecular dysfunctions. Aspirin, statins, rapalogs, metformin, as well as lisinopril, propranolol, caloric restriction and exercise target compounds contributing the same age-related cellular dysfunction, known as senescence-associated secretory phenotype (SASP). SASP seems to be a hallmark of age-related diseases, being a cause of functional decline (Blagosklonny 2017). Each of the above mentioned agents blocks different parts of the mechanism contributing to dysfunction and further organ damage. However, a large part of the mechanisms mediating the anti-aging properties of medications remain essentially unknown (Vaiserman and Lushchak 2017b). In the present chapter, molecular targets and the anti-aging effects of aspirin, statins, metformin and rapamycin will be described. We will look into data obtained in different model organisms and assess the evidence obtained in clinical trials, addressing gaps, restrictions and limitations of current research.

Anti-aging pharmacology seems to be an extremely promising and challenging field, facing multiple constrains in different aspects. A lack of tangible, measurable, biomarkers of human aging appears to be one of the cornerstone issues. Surrogate endpoints, like cardiovascular disease or cognitive impairment or mortality are used in clinical trials, thus the introduction of alternative, easily measurable biomarkers with high predictive value is of extreme importance. The same problem affects biomarkers applied to animal models; it remains unclear whether these might be efficiently translated into human medicine. Many side effects, potentially occurring in humans cannot be evaluated in model organisms. Further investigation on optimal beginning of drug exposure need to be conducted, while some medications might have a certain therapeutic window of effect, when drug exposure is the most beneficial. The question of dosing also remains unsolved, while some of the actions seem to have a dose-dependent effect, strongly varying in dependence on individual metabolizing activity (Burd et al. 2016).

Some of the drugs might undergo repurposing, like aspirin, have emerged over the course of the last century, and further attempts are being made to introduce aspirin and metformin as adjuvant anti-cancer treatments and/or cancer prevention medication (Yue et al. 2014).

Food-derived substances with evidence for prolonging lifespan and preventing age-related functional decline and disease, represent an expanding research area with promising perspectives for development of new preventive and/or treatment strategies. Difficulties in this field include absent recommended daily allowance for humans, extremely high variability in content of these substances in different products, depending on the area of cultivation, type of processing etc. Furthermore, many of the antioxidant substances have failed to demonstrate benefits in life extension or disease prevention when tested in model organisms. Perhaps, the synergetic effects of multiple whole food compounds remain underestimated by researchers. Many reviews concentrate on the idea that targeting reactive oxygen species as primary aging-driving mechanism is wrong (Gruber and Halliwell 2017).

Lifespan Extension in Model Organisms

Aging is the physiological process that is characterized by the loss of normal organ function caused by damage accumulation in cells and tissues (Fontana et al. 2007). Longevity can be modulated by alterations in age-related genes. Moreover, lifespan might be extended using some drugs. Discovery of chemicals that can delay aging and extend lifespan is one of the most promising potential ways to improve the quality of life in older age. Today, many of the prolongevity drugs are effective at relatively low concentrations from 5 to 200 mg/kg of body weight (Hayashi and McMahon 2002). However, it is important to know that every compound may possess some side effects. Furthermore, many drugs may extend lifespan by so called hormetic effect: i.e. the conditions, where relatively toxic substances may have beneficial effects.

There is a continuous need of optimal model system to discover potential anti-aging properties and evaluate the effect on healthspan. Ideally, model system should maximally replicate ageing processes in humans, including genes and signaling pathways that exhibit high conservation. More often the researchers use simple model organisms such as nematodes, fruit flies and rodents. Most studies aimed to investigate the anti-ageing drugs were performed by using invertebrate models, which are considered as useful models for human disease exploration and are widely used for discovering potential anti-ageing agents (Markaki and Tavernarakis 2010; Millburn et al. 2016; Ugur et al. 2016). Pathways controlling lifespan and aging are partially conserved in a wide range of species, from yeast to humans (Bitto et al. 2015; Fontana et al. 2010). In this part, we collected and summarized data obtained in invertebrate and rodent models indicating the anti-aging potential medicines. Aspirin, rapamycin, resveratrol and metformin were effective to extend the lifespan

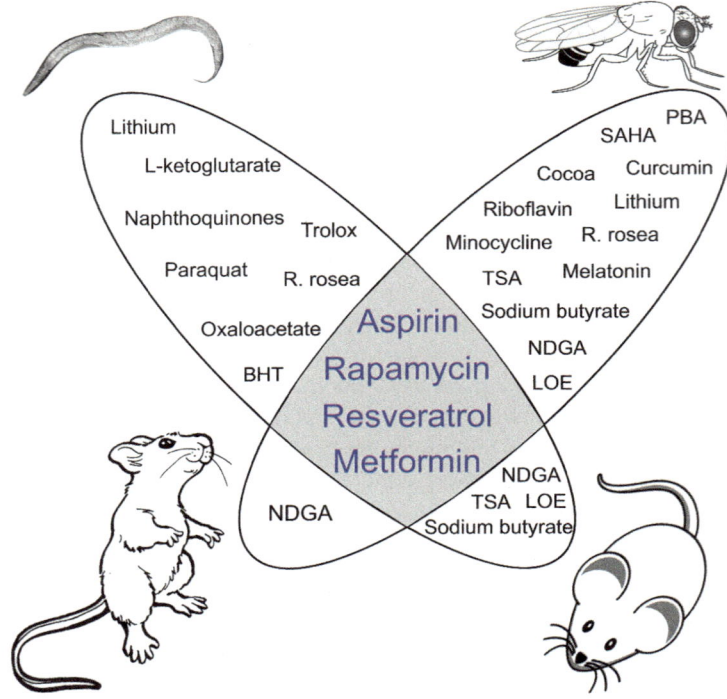

Fig. 13.1 Pharmaceuticals with lifespan-extending properties in different animal models. Drugs extending lifespan of nematode *C. elegans* belong to antioxidants, metabolites, natural compounds, kinase inhibitors. Many agents were shown to extend the lifespan in fruit fly *D. melanogaster*. Some drugs have significant effects on healthspan in several rodent models, however direct effects on lifespan were not demonstrated. Commonly used drugs, including aspirin, rapamycin, resveratrol and metformin, have been found to delay aging and improve overall health in all discussed animal models. *PBA* phenylbutyrate, *BHT* D-β-hydroxybutyrate, *SAHA* suberoylanilide hydroxamic acid, *TSA* Trichostatin A, *NDGA* nordihydroguaiaretic acid, *LOE Ludwigia octovalvis* extract

in four model organisms. Many more drugs, such as Trichostatin A (TSA), sodium butyrate, lithium, Nordihydroguaiaretic acid (NDGA) as well as plant extracts from *Rhodiola rosea* or *Ludwigia octovalvis* have extended the lifespan in more than one model (Fig. 13.1).

Round Worm *Caenorhabditis elegans*

To date, many drugs extending nematode *C. elegans* lifespan have been discovered. They mostly belong to antioxidants, metabolites, natural compounds and kinase inhibitors. Interestingly, these compounds affect the signaling pathways involved in lifespan regulation. Similar mechanisms involved in modulation of lifespan in nematodes and mammals were shown for aspirin, rapamycin, metformin, resveratrol (Fig. 13.1). Thus, this worm is an excellent model system to test prolongevity properties of pharmacological interventions. Besides short lifespan, small size and amenability to genetic manipulations, many genetic pathways affecting aging are conserved in *C. elegans*.

Antioxidants are the most studied class of anti-aging compound (Harman 1972; Melov et al. 2000). Vitamin E (tocopherol) was shown to extend *C. elegans* lifespan (Harrington and Harley 1988; Ishii et al. 2004). Moreover, the α-tocopherol derivative trolox also demonstrated a positive effect on nematode survival (Benedetti et al. 2008). These results support the oxidative damage theory, with the central concept of molecular damage caused by reactive oxygen species (ROS) affect aging (Harman 1956). Interestingly, there are also some controversial results, which revealed lifespan shortening effect under EUK-8 and EUK-134 supplementation (Kim et al. 2008). However, vitamin C, which is the most powerful antioxidant, had no effect on nematode lifespan (Harrington and Harley 1988). The theory of hormesis claims that low doses of stressful agents activate a stress response and as a result improved longevity. Indeed, ROS generating compounds such as naphthoquinones extend nematode lifespan (Hunt et al. 2011). Furthermore, administration of low doses of paraquat or rotenone increased the lifespan in *C. elegans* (Lee et al. 2010a, b) (Figs. 13.2 and 13.3).

Aging is strongly affected by metabolism. Besides dietary restriction, some pharmacological perturbation of metabolism showed beneficial effects on longevity. Several studies have demonstrated increased nematode lifespan by metabolic intermediates. Williams and colleagues showed that oxaloacetate supplementation

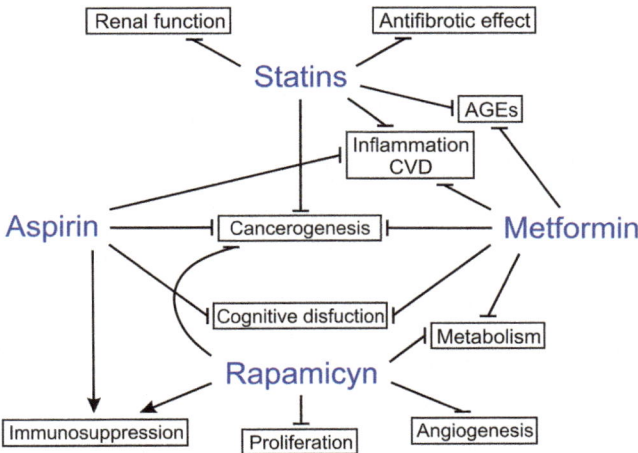

Fig. 13.2 Aging-associated processes affected by statins, aspirin, metformin and rapamycin. Schematic representation of the main physiological processes and diseases targeted by life-extending drugs. *CVD* cardiovascular disease, *AGEs* advanced glycation end-products

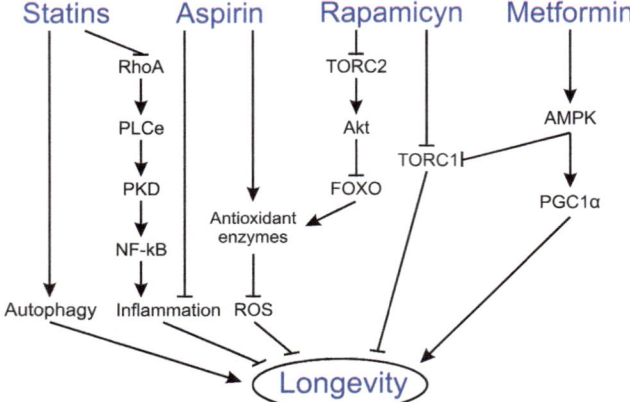

Fig. 13.3 Signaling pathways and mechanisms mediating the lifespan-extending effects of drugs. An association between autophagy induction and lifespan extension has been revealed in various experimental models. Both statins and aspirin decrease inflammation. The ability of aspirin and rapamycin to regulate expression of genes encoding antioxidant enzymes was elicited. Rapamycin inhibits TORC1 that regulates longevity by phosphorylation of 4EBP and S6K. Metformin acts via AMPK-activator to inhibit TORC1 or activate PGC1α. *RhoA* Ras homolog gene family member A, *PLCe* Phospholipase C Epsilon, *PKD* protein kinase D, *NF-kB* nuclear factor k-light-chain-enhancer of activated B cells, *ROS* reactive oxygen species, *TORC1/2* target of rapamycin complex 1/2, *Akt* Protein kinase B, *FOXO* Forkhead box protein O, *AMPK* AMP-activated protein kinase, *PGC1α* Peroxisome proliferator-activated receptor γ coactivator 1α

extended lifespan of *C. elegans* (Williams et al. 2009). Also, α-ketoglutarate, a tricarboxylic acid cycle intermediate, extended the lifespan of adult *C. elegans* (Chin et al. 2014). Furthermore, α-ketoglutarate as a key metabolite of TCA, acts as a messenger of dietary restriction in aging control. Interestingly, α-ketoglutarate affects the nematode lifespan through inhibiting the TOR signaling pathway (Chin et al. 2014). Malate, as a TCA metabolite, increased lifespan and thermotolerance in *C. elegans*. Additionally, fumarate and succinate also extend nematode lifespan. These compounds realized their effect through the activation of conserved stress response pathways.

Metabolomic analysis demonstrated increased content of glutamine, valine, and isoleucine in long-lived daf-2 mutant worms (Edwards et al. 2015). This suggested that amino acid food supplementation can slightly extend nematode lifespan. Moreover, a lifespan extension effect under histidine and tryptophan was accompanied with endoplasmic reticulum (ER) stress response activation; proline and tryptophan increase thermotolerance; tryptophan supplementation increases proteotoxicity (Edwards et al. 2015).

Metformin is widely known for its anti-hyperglycemic properties and thus used for type 2 diabetes (T2D) treatment. The effect of metformin treatment on *C. elegans* aging was investigated by Onken and Driscoll (2010). This study demonstrated improved nematode healthspan, locomotor activity and median lifespan under metformin administration. Interestingly, this biguanide acts in a dietary restriction manner and activates an oxidative stress response. Furthermore, it affects lifespan via AMPK, LKB1, and SKN-1 but is independent of insulin signaling (Onken and Driscoll 2010). Metformin also mediates DR and oxidative stress pathways in a lifespan extending effect. Hence, metfomin may be a promising pharmacological intervention, with a complex beneficial impact on physiology.

Many kinase inhibitors are known to extend *C. elegans* lifespan, including rapamycin (Ye et al. 2014). A life-extending effect of rapamycin was shown in a number of studies from yeast to mammals. It is known that rapamycin realize its action through inhibiting the target of rapamycin (TOR) signaling pathway. TOR is essential for growth processes and is associated with disease and aging. Rapamycin triggers TOR inhibition which in turn leads to activation protective genes by SKN-1/ Nrf and DAF-16/FoxO, and it enhances stress resistance and longevity (Robida-Stubbs et al. 2012). TOR inhibition or rapamycin treatment promotes longevity at least in part by reducing mRNA translation (Bjedov et al. 2010; Kapahi et al. 2010; Zid et al. 2009).

Potent cyclooxygenase-2 (COX-2) inhibitor, celecoxib extends *C. elegans* lifespan and delays the age-associated physiological changes, such as decline of locomotor activity (Ching et al. 2011). Celecoxib as a nonsteroidal drug is often used to treat pain and inflammation. The lifespan extension by celecoxib depends on the activity of DAF-16, the FOXO transcription factor known to regulate development and longevity downstream of insulin signaling (Lin et al. 1997; Ching et al. 2011). It was suggested that celecoxib might extend lifespan by inhibiting the kinase activity of 3-phosphoinositide-dependent kinase-1 (PDK-1) (Ching et al. 2011). Furthermore, celecoxib extends the lifespan of animals with reduced food uptake and mitochondrial respiration (Ching et al. 2011).

A *C. elegans* lifespan extension effect was also shown for histone deacylase (HDAC) inhibitors (Pasyukova and Vaiserman 2017). Ketone body D-beta-hydroxybutyrate (D-βHB), an endogenous and specific inhibitor of class I HDACs, has important implications for the pathogenesis and treatment of metabolic, neuro-degenerative, and other aging-related pathological conditions (Edwards et al. 2014; Newman and Verdin 2014). In addition, the supplementation with D-βHB enhanced thermotolerance, prevented glucose toxicity, decreased α-synuclein aggregation (a sign of Parkinson's disease) and delayed amyloid β toxicity (a sign of Alzheimer's disease). Lithium is frequently used for bipolar disorder treatment; it also extend *C. elegans* lifespan via an epigenetic mechanism by altering expression of genes encoding nucleosome-associated functions (McColl et al. 2008). Longevity caused by lithium treatment is achieved by modulating histone methylation and chromatin structure.

Aspirin was found to have many beneficial effects on physiological traits and is often used to treat pain and inflammation. Aspirin treatment extended the nematode lifespan and improved stress resistance (Wan et al. 2013; Ayyadevara et al. 2013). Data also suggest that aspirin may act in a DR-like manner (Wan et al. 2013). Furthermore, it attenuates ROS amounts by triggering the expression of genes encoding antioxidant enzymes, such as catalase, superoxide dismutase and glutatione-*S*-transferase (Ayyadevara et al. 2013).

Numerous natural compounds including blueberry polyphenols, curcumin, quercetin, *ginkgo biloba* extracts, and resveratrol were shown to extend *C. elegans* lifespan and increase resistance to various stresses (Wu et al. 2002; Wilson et al. 2006; Pietsch et al. 2009). The plant polyphenol resveratrol is well characterized natural compound with annotated lifespan-extending properties. Lifespan extension by resveratrol is associated with NAD^+-dependent histone deacetylase, SIR-2.1 (Bass et al. 2007). The effect of resveratrol on lifespan in *C. elegans* could indicate induction of phase 2 drug detoxification or activation of AMP kinase (Bass et al. 2007). Interestingly, resveratrol did not extend nematode lifespan under normal condition, but induced longevity phenotype under condition of oxidative stress (Chen et al. 2013). *Rhodiola rosea* extract was also found as a promising anti-aging remedy. The lifespan-extending effect was observed in nematodes treated with *R. rosea* extract (Wiegant et al. 2009). This effect was associated with increased stress resistance. The suggested molecular mechanism of *R. rosea* is hormesis (Wiegant et al. 2009), due to the toxic effect at high doses that causes additional stress (Mattson 2008).

Some antidepressants showed positive effects on nematode longevity (Rangaraju et al. 2015). The atypical antidepressant, mianserin, induced oxidative stress resistance and extended *C. elegans* lifespan, but typical antidepressant fluoxetine had no impact on nematode physiology (Rangaraju et al. 2015).

Conclusively, drugs and compounds of different classes and origins were shown to extend the lifespan of worms evolving different pathways and mechanisms. However, there is a need to confirm the effects on other model organisms.

The Fruit Fly *Drosophila melanogaster*

Invertebrate model systems, including *Drosophila melanogaster*, are essential for better understanding the genetic pathways that control aging. Furthermore, the fruit fly is proved to be valuable in testing chemical compounds that may influence longevity. The *Drosophila* model is characterized by the existence of complex behavioral phenotypes and availability of several models of human age-related diseases. Furthermore, *Drosophila* experiments might be conducted in demographic cages, which allow one to take into account any bio-demographic effect on lifespan.

Numerous studies describe the effect of antioxidant dietary supplementation on *Drosophila* longevity. It was shown that an increased median and maximum fly lifespan could be induced by the antioxidant and glutathione precursor N-acetylcysteine (Brack et al. 1997). Vitamin supplementation with known antioxidants, also has lifespan extending effects. Vitamin E or α-tocopherol, was found to suppress several phenotypes in *Drosophila* models of neurodegenerative diseases like taupathies (Dias-Santagata et al. 2007), Parkinson disease (Wang et al. 2006), or pantothenate-kinase-associated neurodegeneration (Yang et al. 2005). Vitamin E extends lifespan of flies with deficiencies in oxidative stress response such as hyperoxia or in Cu/Zn superoxide dismutase-deficient (SOD1-deficient) flies under normoxia (Bahadorani et al. 2008). An extended lifespan increased reproduction and catalase activity, with decreasing level of lipofuscin level shown under riboflavin supplementation. It acts via pathways related to oxidative stress (Zou et al. 2017). The lifespan-extending effect of melatonin was found to prevent oxidative damage to the fly tissues and slow down aging (Bonilla et al. 2002). The optimal dosage of NDGA (nordihydroguaiaretic acid) shown to extend the lifespan of flies was 100 µg/ml (Miquel et al. 1982).

Protein homeostasis plays an important role in aging and age-related pathologies. Curcumin prevents protein aggregation and increases lifespan in *D. melanogaster*. Its effects are associated with enhanced stress resistance (Lee et al. 2010a, b) by increased superoxide dismutase activity (Suckow and Suckow 2006; Shen et al. 2013). It was demonstrated, that curcumin acts as scavenger for the superoxide anion, hydroxyl radical (at high concentration) and nitric oxide (Kunchandy and Rao 1990; Brouet and Ohshima 1995).

Beneficial effects of lithium suggest its use for treatment of age-related diseases, mostly by hermetic mechanisms. Therapeutic concentration of lithium for mood disorders is close to the toxic doses (Lucanic et al. 2013). Lithium in low-doses extended fly lifespan by inhibition of glycogen synthase kinase-3 (GSK-3) and activation of the transcription factor nuclear factor erythroid 2-related factor (NRF-2) (Castillo-Quan et al. 2016). Higher doses of the drug, however, were found to reduce the lifespan.

Inhibitors of the histone deacetylases (HDACs) have been proposed as a promising type of therapeutic drugs able to modulate aging (Pasyukova and Vaiserman 2017). A lifespan extending effect in *Drosophila* was shown for 4-phenylbutyrate (PBA) treatment (Kang et al. 2002). PBA acts by histone (de)acetylation, a primary mechanism of epigenetic regulation (Vaiserman 2011). The effect of PBA is accompanied by changes in the acetylation level of histones H3 and H4 and transcriptional changes in significant set of genes (Pasyukova and Vaiserman 2017). Also, lifespan extending effects were shown for sodium butyrate (Zhao et al. 2005; McDonald et al. 2013; Vaiserman et al. 2013a, b) with annotated HDAC inhibition activity and shown to influence the processes of cell growth, differentiation and apoptosis (Buommino et al. 2000; Khan and Jena 2014). An increase of median and maximum *Drosophila* lifespan was observed when food was supplemented with Trichostatin A (TSA) (Tao et al. 2004; Zhao et al. 2005). This effect is characterized by induction of terminal differentiation, cell cycle arrest and also apoptosis in various cancer cell lines, thereby inhibiting tumorogenesis. The longevity phenotype induced by TSA is associated with the improvement of the cell stress resistance and locomotor activity. Suberoylanilide hydroxamic acid (SAHA) treatment resulted in a decreased mortality rate and extended longevity (McDonald et al. 2013). All these compounds affect several pathways involved in the regulation of gene expression patterns associated with healthy aging.

In general, the aging process is associated with inflammation. Non-steroidal anti-inflammatory drugs extended the lifespan in various model organisms, with demonstrated anticancer effects, tumor suppression and apoptosis stimulation (Poole et al. 2004; Danilov et al. 2015). Furthermore, NSAIDs increase locomotor activity and stress resistance in *Drosophila*. Interestingly, they have both anti- and pro-oxidant properties, dependent on concentration (Danilov et al. 2015). Antioxidant properties are realized by their radical scavenging activity and membrane-stabilizing action (Danilov et al. 2015).

Aspirin was found to suppress aging process by interfering with oxidant production, cytokine response processes, and by blocking glycooxidation reactions (Phillips and Leeuwenburgh 2004). Prolonged *Drosophila* lifespan and improved healthspan was observed following aspirin administration (Song et al. 2017). The molecular mechanism of aspirin action still remains to be determined, but it influences the metabolism of amino acids, carbohydrates and urea.

The ability of resveratrol to extend the lifespan is conserved from yeast to mammals. Resveratrol treatment at a concentration of 400 μM was shown to extend mean lifespan in *Drosophila* when fed high-fat diet (Wang et al. 2013). Fly prolongevity phenotype was associated with downregulation of genes in aging-related pathways, including antioxidant peroxiredoxins and insulin-like peptides (Wang et al. 2013). As in worms, the lifespan extension was related to NAD^+-dependent histone deacetylase, Sir2 in *Drosophila* (Bass et al. 2007).

The antibiotic minocycline with anti-inflammatory, antioxidant and neuroprotective properties increases *Drosophila* survival and improves overall healthspan (Oxenkrug et al. 2012). Minocycline, as a key metabolite in the formation of kynurenine from tryptophan (KYN-TRP metabolism), is a promising candidate drug for delaying aging and treatment of aging-associated disorders.

TOR inhibition delays the aging process and increases lifespan in fruit flies (Kapahi et al. 2004; Luong et al. 2006). It was demonstrated that rapamycin causes a longevity phenotype via TOR inhibition, which in turn, modulates autophagy and translation. Increased resistance to starvation and paraquat treatment is accompanied by the lifespan-extending effect by rapamycin (Bjedov et al. 2010). Interestingly, rapamycin also reduces fecundity and increases lipid level in the fly body (Bjedov et al. 2010). The analysis of the rapamycin treatment under dietary restriction revealed slightly extended lifespan that had already been maximized by DR (Bjedov et al. 2010). The potential mechanism of rapamycin action is associated with its anticancer effects (Blagosklonny 2012a, b), induction of autophagy (Perluigi et al. 2015), anti-inflammatory effect (Araki et al. 2011).

Pharmacological interventions that mimic the effects of DR delay the onset of several age-related diseases in laboratory animals. A lifespan extending effect of metformin treatment in *C. elegans* was previously claimed (Onken and Driscoll 2010). However, it did not increase lifespan in either male or female flies (Slack et al. 2012). Moreover, higher metformin doses can be toxic to flies. Interestingly, metformin acts via AMPK to decrease body lipid stores (Slack et al. 2012).

R. rosea root extract as a promising anti-aging natural compound has been extensively used to protect against stress and to improve overall healthspan. The controversial studies showed that lifespan extension by *R. rosea* is independent (Schriner et al. 2013) or dependent of dietary composition (Gospodaryov et al. 2013). It delays age-related decline of physical activity and increases stress resistance. The mechanism of *R. rosea* action is still unknown, but it was demonstrated that it acts independently from TOR, IS and SIR2 (Schriner et al. 2013).

Cocoa, as a natural anti-aging compound, derived from *Theobroma cacao* increased lifespan in *D. melanogaster* and demonstrated antioxidant properties (Bahadorani and Hilliker 2008). Cocoa acts as an antioxidant under hyperoxia or during Cu/Zn-superoxide dismutase-deficiency (Bahadorani and Hilliker 2008). Moreover, it is involved in a heavy metal chelation process but may act as a pro-oxidant, when the level of oxidative stress is extremely high (Bahadorani and Hilliker 2008).

Rosemary (*Rosmarinus officinalis L.*) is known for its powerful antioxidant activity, antibacterial and hepatoprotective properties. Rosemary extract causes longevity phenotype in *Drosophila*, that is associated with increased superoxide dismutase and catalase activities (Wang et al. 2017a, b). *Ludwigia octovalvis* (LOE) is a rich source of antioxidants, including polyphenol compounds, phytosterols and squalene, extended lifespan of fruit fly fed regular or high-calorie diet (Lin et al. 2014; Wang et al. 2015). LOE attenuated age-related cognitive decline in fruit flies (Lin et al. 2014) and was shown to have anti-bacterial and anti-cancer activities (Chang et al. 2004).

Testing drugs with anti-aging properties, using *Drosophila* as a model system, is one of the most promising ways to understand the pharmacology of aging and lifespan extension. Rodent aging pharmaceuticals identification is the next step to investigate promising anti-aging therapies for humans.

The House Mouse *Mus musculus*

Mice have become a favorite model system for testing interventions in aging. The numerous of mouse advantages include short generation time, genetic proximity to humans (99% of human genes have their homologues in mouse), availability of strains and relatively small size. Mouse age-related models ensure a comprehensive tool for pharmaceutic compounds to affect longevity and genes implicated in the effects, to provide important general information about the genetic basis of ageing.

Beneficial effects of antioxidant supplements seem certain, however, lifespan extension were generally not described. Indeed, it was demonstrated that vitamin E is effective in suppressing phenotypes found in mouse models of human tauopathy (Nakashima et al. 2004) or Down syndrome (Lockrow et al. 2009). Furthermore, it eliminates the increased oxidative stress observed in these mice models. Hence, vitamin E is considered as a powerful antioxidant with anti-aging effect.

The National Institute on Aging Interventions Testing Program (ITP) examined the effects of compounds which are suggested to increase lifespan and prevent age-related disease in genetically heterogeneous mice. Significantly extended lifespan under nordihydroguaiaretic acid (NDGA) and aspirin supplementation for male mice was revealed by Strong et al. (2008). Aspirin, as a non-steroidal anti-inflammatory drug demonstrates anti-thrombotic and anti-oxidant properties (Shi et al. 1999; Vane 2000; Weissmann 1991). Aspirin activates the NF-kB signaling pathway and induces apoptosis in two in vivo models of human colorectal cancer (Stark et al. 2007). NDGA has anti-oxidant and anti-inflammatory properties, moreover it demonstrates life-extension properties in metazoans (Wood et al. 2004). NDGA has been shown to prevent neuronal death and cognitive deficits which occurred under forebrain ischemia/reperfusion injury (Shishido et al. 2001). NDGA has an anti-cancer activity by affecting 5-lipoxygenase (Nony et al. 2005).

Anti-diabetic biguanides are considered as the most promising among pharmacological treatments of aging. This class of anti-aging compounds inhibits fatty acid oxidation, gluconeogenesis in the liver, increases the availability of insulin receptors and reduces excretion of glucocorticoid metabolites (Dilman 1994; Muntoni 1999). Metformin causes physiological and anti-aging effects similar to caloric restriction. Chronic metformin treatment enhanced mean and maximum lifespan of SHR mice, decreased body weight and slowed down the age-related switch-off of estrous function (Anisimov et al. 2008). However, metformin is not able to prevent tumor formation (Anisimov et al. 2008) and is toxic at high doses (Martin-Montalvo et al. 2013). The molecular mechanism of metformin action is by increasing activation of adenosine monophosphate-activated protein kinase (AMPK), which is involved in maintaining energy balance (Martin-Montalvo et al. 2013).

Positive effects of HDAC inhibitors on life- and health-span have been reproduced in numerous model organisms, including rodents. For instance, mice fed a high fat diet and additionally treated with SB (sodium butyrate), showed reduced obesity and insulin resistance (Gao et al. 2009). Also, SB induced insulin sensitivity and reduced adiposity in obese mice. The lifespan extension effect in mice by SB treatment is accompanied with functional improvement of myocardial function and also attenuation of cardiac hypertrophy and increased angiogenesis in myocardium (Chen et al. 2015). It is noteworthy that the level of superoxide dismutase was significantly enhanced in SB-treated diabetic mice (Chen et al. 2015). SB was also shown to be effective in treatment of neurodegenerative diseases which are associated with the aging process. Mice with neurodegenerative disease phenotype treated with SB demonstrated ameliorated defects in histone acetylation, thereby substantially improving motor performance and extended mean lifespan. A positive effect was observed under TSA administration, which resulted in increased mean survival time following the treatment (Yoo and Ko 2011).

Resveratrol is widely known for its properties to extend the lifespan of yeast, worms, and flies (Bhullar and Hubbard 2015). It is the best studied anti-aging drug, without known toxicity, with the ability to treat and counteract a number of age-related diseases such as cancer, Alzheimer's disease, and diabetes (Baur and Sinclair 2006; Hubbard and Sinclair 2014). It is interesting to note that resveratrol does not extend the lifespan of healthy mice (Pearson et al. 2008; Strong et al. 2013). However, resveratrol's lifespan extending effects were shown for metabolically compromised mice fed a high-calorie diet (Baur et al. 2006). It improved mitochondrial number, locomotor activity and increased insulin sensitivity. Resveratrol supplementation might be beneficial against environmental toxins, pathogens and radiation. Indeed, resveratrol restored renal microcirculation and extend lifespan in mice with kidney injury (Holthoff et al. 2011). Additionally, a protecting effect of resveratrol on the tubular epithelium by scavenge reactive nitrogen species has been shown (Holthoff et al. 2012). It also improves longevity and prevents tumor formation in mice subjected to ionizing radiation (Oberdoerffer et al. 2008). A DR-mimicing effect of resveratrol treatment in lower organisms has been shown. Furthermore, resveratrol induces gene expression patterns that parallel those induced by DR (Pearson et al. 2008). Resveratrol-treated animals are characterized by a generalized reduction in oxidative stress and inflammation, which are consistent features of the DR effect. Besides improving insulin sensitivity and increasing survival in mice, resveratrol improves cardiovascular function, bone density, and motor coordination (Pearson et al. 2008).

TOR signaling inhibition was shown to extend lifespan in invertebrate models. A lifespan extension effect was demonstrated in mice of both sexes, which were treated late in life (aged 20 months) (Harrison et al. 2009; Miller et al. 2011). Furthermore, late rapamycin treatment reverses the age-related heart dysfunction and also resulted in beneficial skeletal, motor and behavioral changes (Flynn et al. 2013).

A significant body of evidence suggests an anti-aging effect for *Ludwigia octovalvis* (LOE) in mice. Indeed, it was demonstrated that this natural compound attenuated age-related cognitive decline in senescence-accelerated-prone 8 (SAMP8) mice (Lin et al. 2014). Hence, LOE can be proposed as a potential anti-aging compound for attenuating oxidative damage and activating AMPK-related pathways (Lin et al. 2014).

The Rat *Rattus norvegicus*

In addition to mice, rats are used extensively in studies related to aging. Early studies demonstrated a prolongevity effect following Vitamin E supplementation in male rats, which were reared on a high fat diet (Porta et al. 1980). This phenotype was associated with reduction of incidence of malignant neoplasms, but there was no influence on the incidence or severity of chronic nephropathy, which was developed in all rats (Porta et al. 1980).

Metformin, as a CR mimetic, was shown to extend lifespan in Fischer-344 rats, mostly by prevention of tumor formation (Smith et al. 2010). Furthermore, it significantly reduced body weight and adipose tissue (Muzumdar et al. 2008). However, CR has a stronger impact on median and maximum lifespan as compared to metformin treatment (Smith et al. 2010). It worth noting, however, that an effective dose for Fisher-344 rats was about tenfold higher than the maximum daily dose used in human treatment (Ma et al. 2007).

Low resveratrol (LR) doses and red wine (RW) supplementation both improved vascular function and aerobic capacity, moreover, they decrease markers of senescence (P53, P16) in rats. However, these experimental rats did not live longer (da Luz et al. 2012). The potential molecular mechanism of RW and LR action involve ubiquitous NAD⁺-dependent protein deacetylases. Resveratrol protects against oxidative stress via the Nrf2 pathway and, in turn, attenuates mortality in obese rats. Both acute and chronic resveratrol treatment improves post-ischemic cerebral perfusion in rats (Ritz et al. 2008a, b). Additionally, down-regulation of inducible NO synthase (iNOS) and up-regulation of vasorelaxant eNOS were observed in treated rats.

Rapamycin, as a potent and specific mTOR inhibitor, is widely used for treatment of renal cell carcinoma and mantle cell lymphoma, moreover, it is often tested in clinical trials as a therapeutic compound to cure various cancer types (Konings et al. 2009; Dancey 2010). Systemic chronic administration by rapamycin causes impaired glucose homeostasis in type 2 diabetes (Deblon et al. 2012). Furthermore, rapamycin prevents excessive body weight gain, fat accumulation and hepatic steatosis, however, it also leads to insulin resistance and glucose intolerance in rats fed a high-fat diet (Deblon et al. 2012). The molecular mechanism, underlying these effects is realized through insulin-related signaling pathways. Interestingly, rapamycin prevents spontaneous retinopathy in senescence-accelerated OXYS rats (Kolosova et al. 2012). Hence, rapamycin can be suggested as having therapeutic potential for treatment and prevention of age-associated pathologies.

The longevity phenotype under NDGA supplementation in flies (Miquel et al. 1982) and mice (Strong et al. 2008) was previously reported. Furthermore, there is evidence concerning the anti-aging effect of NDGA in rats (Buu-Hoi and Ratsimamang 1959). Interestingly, this longevity phenotype has been associated with enhanced glucose clearance, reduced triglycerides and insulin sensitivity in a diabetic rat model (Reed et al. 1999). The molecular mechanism of NDGA action are related to its ability to block fatty acid synthesis in adipocytes, through inhibition of fatty acid synthase and lipoprotein lipase (Li et al. 2005; Park and Pariza 2001).

Numerous studies have demonstrated the highly conserved effect of aspirin on lifespan and healthspan in different models. There was no direct impact of aspirin supplementation on rat survival. However, aspirin normalized blood pressure in rats with hypertension phenotype (Tuttle et al. 1988).

Conclusions from Animal Models

Testing chemical compounds for their ability to slow down aging in multiple species may support their usefulness in treating and retarding age-related disease in humans. Indeed, the main rationale for testing these compounds in animal models is to introduce them into clinical research and, in turn, improvement of human health and longevity. Model organisms, including *C. elegans, D. melanogaster* and rodents are important for understanding signaling pathways and genes related to aging, and the possibility to alter them with drugs or supplements. Here we have reviewed the publications showing that many compounds have a highly conserved effect on healthspan and lifespan across model organisms. Furthermore, the molecular mechanism of their action and longevity pathways involved in ageing control are also often conserved. The reasonable question is now to ask is how we can extrapolate these data to humans? First of all, ageing is characterized by the occurrence of age-related diseases. For this reason, potential anti-aging compounds, which have a positive effect in preventing or treating some age-related disorders in animal models, should be thoroughly investigated in clinical trials. Secondly, it is important to emphasise possible side effect and any hormetic action of many pharmaceutical compounds. The optimal concentration differs in every animal model, and also should be individually assessed in each patient.

Most of the age-related diseases are characterized by accumulation of oxidative damage to macromolecules. Antioxidants such as N-acetylcysteine (NAC) and vitamin E prevent oxidative stress and in this way modulate longevity. Interestingly, the natural compound cocoa imparts its antioxidant properties to extend organismal lifespan. The formation of molecular aggregates is the major feature of numerous age-associated diseases including Parkinson's, Alzheimer's and Huntington's disease. Hence, formation of protein aggregates is an important indicator of aging. Both lithium and curcumin prevent protein aggregation and produce prolongevity phenotypes. Using *C. elegans* and *D. melanogaster* as model systems it was demonstrated that the pharmacological approach to maintain protein homeostasis might be promising to prevent pathology caused by age-related disease and extend lifespan. The best studied and robust means of extending the lifespan of model organisms is dietary intervention termed dietary restriction (DR). TOR and insulin signaling pathways have been shown to be important for modulating lifespan in animal models.

Hence, preclinical studies on *in vitro* models have established a large number of promising pharmaceutical compounds with proven anti-aging properties. Moreover, the possible direct or indirect molecular targets and mechanisms of action mediating these chemicals have been established. Chemical screens in animal models are particularly promising for the development of possible drugs for humans.

Beneficial Health Effects in Humans

Metformin

Metformin is inexpensive, safe and widely prescribed glucose-lowering drug, being the first-line treatment for patients with T2D, it was proven to effectively reduce risk of cardiovascular diseases and death (Chamberlain et al. 2017; Palmer et al. 2016). In older patients, use of metformin is accompanied by reduced risk of hypoglycaemia and non-fatal cardiovascular events, compared to other antidiabetic drugs (Schlender et al. 2017). Apart from its direct hypoglycemic effects and prevention of target-organ damage in T2D patients it has shown numerous advantageous effects in patients with diverse conditions such as impaired glucose tolerance (Hostalek et al. 2015), obesity (Bouza et al. 2012; Siskind et al. 2016), metabolic syndrome (Zimbron et al. 2016), polycystic ovary syndrome (PCOS) (Patel and Shah 2017) and nonalcoholic fatty liver disease (Li et al. 2013).

Metabolic Effects of Metformin

Numerous pleiotropic metabolic effects of the hypoglycemic drug metformin have emerged during its clinical use. Metformin exerts its glucose-lowering effect via AMPK, which is also involved in the lipid metabolism regulation. AMPK phosphorylates and in this way inactivates acetyl-Co-A carboxylase, which plays essential role in the synthesis of fatty acids. One thousand and five hundred milligram per day metformin monotherapy has been shown to reduce total cholesterol, triglycerides, LDL-C and VLDL-C, increasing HDL-C levels (Garimella et al. 2016). A small study aimed to evaluate the effects of metformin on lipid peroxidation in T2D patients assigned to metformin, gliclazide or diet, showed a significant increase in activity of antioxidant enzymes in erythrocytes, along with malondialdehyde reduction in the metformin group compared to diet alone. These results suggested that administration of metformin might decrease oxidative stress in T2D subjects (Memisogullari et al. 2008).

Direct influence of metformin on insulin resistance, combined with leptin reduction and GLP-1-mediated lipolytic and anorectic effects contribute to weight loss in T2D patients and non-diabetic individuals (Rojas and Gomes 2013). Six months of metformin treatment in overweight and obese, mostly insulin-resistant patients resulted in significant weight loss in insulin-resistant group, comparing to untreated controls (Seifarth et al. 2013).

A systemic review and meta-analysis by Björkhem-Bergman and colleagues revealed that metformin, compared to placebo, caused a significant weight reduction in adults and children, treated with atypical antipsychotic medication (Björkhem-Bergman et al. 2011). Metformin was also shown to reduce amounts of advanced glycation end-products (AGEs) in patients with T2D and PCOS (Diamanti-Kandarakis et al. 2007; Haddad et al. 2016). AGEs contribute to cellular senescence and are linked to target-organ damage in T2D, neurodegeneration, inflammation and oncogenesis (Ott et al. 2014; Yamagishi et al. 2012). Metformin seems to reduce formation of AGEs through its hypoglycemic action and additionally by down-regulating expression of cellular receptors to AGEs, thus preventing activation of downstream signaling targets (Ishibashi et al. 2012).

Cardiovascular Effects of Metformin

Metformin is the only anti-diabetic drug, shown to reduce microvascular outcome, most likely due to its miscellaneous effects beyond glycemic control (Rojas and Gomes 2013). Sub-analysis of obese patients from one of the largest trials, the United Kingdom Prospective Diabetes study (UKPDS), intensively treated with metformin, experienced a 33% reduction of myocardial infarction risk, compared to conventionally treated patients (American Diabetes Association 2002). During the 10 years follow-up a sustainable reduction in microvascular risk and reduction of risk of myocardial infarction and death from any cause was observed among overweight patients. These effects were thought to be exerted due to pleiotropic effects of metformin, not just due to glycemic action alone (Holman et al. 2008).

Papanas and coauthors concluded that metformin might also be beneficial for patients with heart failure, improving the 2-year survival. These authors have suggested that potential prevention of cardiac fibrosis could be mediated by AMPK-dependent mechanisms (Papanas et al. 2012). Experimental data also shows that metformin ameliorates endothelial function, phosphorylating eNOS and stimulating release of NO (Eriksson and Nyström 2015).

However, according to the recent meta-analysis performed by S.J. Griffin and colleagues, summarizing the reports from 13 trials (with 2079 individuals with type 2 diabetes allocated to metformin and a similar number to comparison groups), there is no certainty as to whether metformin reduces the risk of cardiovascular disease. In this review metformin reduced risk of all-cause mortality by up to 16% and at the same time, increased risk of stroke by up to 48%. The authors underlined the fact that cardiovascular endpoint data from studies of metformin is derived from a small studies with quite specified categories of patients – relatively young, overweight or obese, North American and Northern European – with poorly controlled diabetes, lacking evidence from older adults with HbA1c less than 8%, people of diverse ethnic group and geographical origin (Griffin et al. 2017).

Effects of Metformin on Inflammation

Chronic low-grade inflammation is an important pathogenetic mechanism involved in aging (Franceschi and Campisi 2014). Targeting inflammatory mechanisms has a promising contribution to human longevity and prevention of age-associated diseases (Fougère et al. 2017). Experimental data obtained from human cells shows that metformin inhibits IL-1β-induced release of the pro-inflammatory cytokines IL-6 and IL-8 in human vascular smooth muscle cells (SMCs), macrophages, and endothelial cells (ECs) in a dose-dependent manner. This study also demonstrated reduction in activation and nuclear translocation of nuclear factor-κB (NF-κB) in SMCs under metformin treatment, as well as suppression of pro-inflammatory phosphokinases Akt, p38, and Erk activation (Isoda et al. 2006).

Several studies have demonstrated anti-inflammatory pleiotropic effect of metformin in diabetic patients. A study by Chen et al. (2016) showed that compared to other antidiabetic drugs (gliclazide, acarbose, or repaglinide), metformin significantly reduced levels of proinflammatory cytokines (IL-6, TNF-α) in serum and MCP-1 in urine of T2D patients. These effects were time- and dose-dependent. The authors concluded that metformin reduces inflammatory responses in the systemic circulation and urine, thereby contributing to its beneficial effects on type 2 diabetes.

In a population cohort study, involving 3575 naïve T2D patients, Cameron and coauthors showed that compared to sulfonylurea treatment, metformin reduced the mean neutrophil to lymphocyte ratio, an inflammation marker and predictor of all-cause mortality and cardiac events (Cameron et al. 2016).

Anticarcinogenous Properties

There is an established increased risk of certain types of cancer in T2D patients. Numerous studies and meta-analyses showed an association between T2D and an increased risk of liver, pancreas, endometrial, colorectal, breast, and bladder cancer (Giovannucci et al. 2010). This association might have a causal reason, explained by hyperglycemia, insulin resistance and hyperinsulinemia, however the high prevalence of confounding factors such as adiposity in T2D patients could also have a potential influence on cancer incidence in this specific category of patients (Tsilidis et al. 2015).

Meta-analysis of studies, comparing metformin with other drugs in diabetic patients, demonstrated a 30% lower cancer incidence of all cancer types in metformin-allocated patients (Decensi et al. 2010). A recent systematic review assessed cancer risk and cancer mortality in 12 randomized controlled trials (21,595 patients) and 41 observational studies (1,029,389 patients). Metformin intake was associated with reduction of the cancer mortality risk, as well as risk of any cancer for 35% and 31%, respectively (Franciosi et al. 2013). Also, Zhang and coauthors demonstrated the preventive effect of metformin towards liver cancer. In T2D patients use of metformin was associated with 62% reduction in the estimated risk of liver cancer and 70% risk reduction for hepatocellular carcinoma (Zhang et al. 2012).

Li et al. (2017) analysed nine retrospective cohort studies and two RCTs for potential effects of metformin on survival of pancreatic cancer patients. Results showed significant improvement in survival of patients receiving metformin compared to controls. However, effects of metformin were insignificant in patients with advanced disease stages. Observational studies showed a reduced incidence of endometrial cancer (EC) in metformin-treated diabetic patients and improved the overall survival of patients with EC (Tang et al. 2017). Meireles et al. (2017) performed a systematic review and meta-analysis of studies assessing potential metformin effects in patients with endometrial hyperplasia and EC. Use of metformin was associated with reversion of atypical endometrial hyperplasia to a normal endometrium. There was also a significant decrease in expression of cell proliferation biomarkers. Metformin-treated EC patients had a higher overall survival compared to non-metformin users and non-diabetic patients. The authors suggested that EC patients might benefit from addition of metformin to standard treatment, considering the evidence of reversing atypical hyperplasia, cell proliferation biomarkers reduction and overall survival improvement.

A recent analysis by Hou and colleagues, including seven studies with 7178 participants, evaluated the impact of metformin treatment on the occurrence of colorectal adenoma (CRA) in T2D patients. Metformin therapy correlated with a significant decrease in the risk of CRA in the T2D patients, with a 27% reduction in comparison to T2D treatment without metformin (Hou et al. 2017). Other meta-analysis showed improved overall survival in colorectal cancer patients, however no improvement in cancer survival (Meng et al. 2017).

The protective effect of metformin against breast cancer (BC) was summarized by Col et al. (2012) in postmenopausal diabetic women. Meta-analysis of 11 studies, including 5464 BC patients with diabetes (2760 patients who had received metformin and 2704 patients who had not) showed that metformin use was associated with a 47% decreased risk of death from all causes in BC patients with diabetes, as well as with reduced cancer-related mortality. After adjusting patients for hormonal receptor expression, metformin showed improvement in overall survival by 65% (Xu et al. 2015). Recently, metformin demonstrated its potential role as an additional cancer-treatment option in non-diabetic BC patients, by unveiling indirect insulin-dependent effects of intervention. Women with newly diagnosed, treatment-naïve, early-stage BC were recruited for participation in the study, regardless of tumor subtype. Patients were administered 500 mg of metformin three times daily for about 2 weeks after diagnostic core biopsy until the moment of surgery. Tumor biopsies were collected prior to metformin administration and after the surgery. Immunohistochemical analysis of tumors demonstrated the reduction in PKB/Akt and ERK1/2 phosphorylation, decreased insulin receptor expression in the tumor, reduction in PI3K and Ras-MAPK signaling following metformin administration. Researchers claim that fasting insulin levels and insulin receptor expression by the tumor cells could possibly stratify patients to further allocate them for metformin treatment (Dowling et al. 2015). Analysis by Yu et al. suggests that metformin use appears to be associated with a significant reduction in the cancer risk and biochemical recurrence of prostate cancer (Yu et al. 2014).

A retrospective cohort study by Tseng C.-H., pointed out that metformin significantly reduces gastric cancer risk, especially when the cumulative treatment duration is more than ~2 years (Tseng 2016). Meta-analysis of cohort studies revealed that the risk of gastric cancer among patients with T2D is lower in metformin-users, compared to those who were not treated with metformin (Zhou et al. 2017).

However, data from different meta-analysis on cancer incidence and mortality among metformin users is often conflicting due to diverse analysis methodology, heterogeneity of studies, absence of cancer data in some studies and short follow up time (Stevens et al. 2012). Due to potential causal interpretation of findings and observational nature of studies, often included into meta-analysis, this is identified as another limitation (Col et al. 2012). Even when considering the epidemiologically proven anticarcinogenous effects of metformin, the exact mechanisms of tumor suppression remain essentially unknown. Researchers suggest the potential direct effect on cancer cells to be via AMPK-mediated mechanisms and mammalian target of rapamycin (mTOR) inhibition (Col et al. 2012). Other possible mechanisms include inhibition of the HER2 and NF-κB signaling pathways (Lei et al. 2017). There is an ongoing discussion and research as to whether metformin should be introduced into cancer treatment as a strategy to improve survival, however the question whether non-diabetic cancer patients will surely benefit from this treatment remains open.

Prevention of Frailty in Elderly Patients

Frailty is a complex geriatric syndrome, associated with increased risk of death in elderly patients, and could be potentially prevented or influenced by metformin. Comparison of metformin-treated patients vs. metformin-naïve showed reduction of frailty risk and comorbidity in elderly T2D patients (Sumantri et al. 2014). Metformin-treated patients, included in this study also demonstrated better muscle strength and body balance characteristics. Previously, a study by Musi et al. (2002) described enhanced AMPK phosphorylation and glucose uptake in muscle tissue of metformin-treated diabetes subjects, which might partially explain better frailty indexes under metformin treatment. Another study by Gore and colleagues demonstrated enhanced muscle protein anabolism of metformin in intensive care patients with severe burns. Patients treated with metformin groups experienced increased fractional synthetic rate of muscle protein and the greater net rate of phenylalanine deposition into the leg, compared to placebo (Gore et al. 2005). The cohort study by Wang et al. (2014) suggested that metformin could be associated with reduced mortality mediated by reducing the onset of frailty older adults with T2DM. Compared to the sulfonylureas, metformin treatment was significantly associated with decreased incidence of frailty in the studied cohort. Some success was demonstrated metformin-associated prevention of osteoporosis in experimental models (Gao et al. 2010; Mai et al. 2011; Tolosa et al. 2013). However, a 12-week study by S.K. Hegasy failed to show improvement of bone-turnover markers in metformin-treated postmenopausal diabetic women, compared to the study baseline (Hegazy 2015). Perhaps, studies with larger numbers of subjects and longer-follow-up are required to clearly establish effects of metformin on osteogenesis and bone tissue loss in susceptible elderly populations.

Metformin and Cognitive Function

Currently there is conflicting evidence about the effects of metformin on cognitive decline in elderly individuals. Certain studies, involving diabetic patients, emphasized the protective role of metformin against cognitive decline, others, on contrary claimed that exposure to metformin contributes to neurodegeneration, Parkinson and Alzheimer disease. Contradicting results of these studies are explained by prevalent comorbidity of T2D patients, necessity to prescribe more than one medication and practical impossibility to evaluate the exclusive effect of metformin alone. Cognitive decline could also potentially arise from other concomitant conditions, not necessarily stemming from pharmacological intervention of any sort. Cross-sectional study by D. Hervás and colleagues showed that the use of metformin was associated with better cognitive function in patients with Huntington's disease (Hervás et al. 2017).

Longitudinal multivariate analysis in the population-based Singapore Longitudinal Aging Study showed a significant inverse association of metformin use and cognitive impairment in T2D patients, controlling for age, education, diabetes duration, fasting blood glucose, vascular and non-vascular risk factors. This study also showed a linear trend of the lowest risk for cognitive decline under use of metformin for longer than 6 years, in cross-sectional and longitudinal analysis. The authors of this study concluded that long-term metformin treatment in T2D patients might reduce the risk of dementia and cognitive impairment (Ng et al. 2014).

In a cohort study of Taiwan's National Health Insurance Research Database, 4651 patients were recruited in the metformin cohort and a comparable number of non-metformin controls, by using propensity score matching. During the 12 year follow-up it was demonstrated that metformin users had a higher risk of Parkinson's disease, along with an increased risk of all-cause dementia, risk of Alzheimer's disease, and vascular dementia. Time and dose-dependent effects were observed for occurrence of Parkinson's disease and dementia (Kuan et al. 2017).

Undoubtedly, there is a strong need for further large-scale prospective controlled trials, focusing on cognitive function of patients allocated to metformin. Recently, several promising studies, evaluating the exclusive role of metformin, as an anti-aging agent with emphasis on age-related diseases in humans have been designed and launched. The Metformin in Longevity Study (MILES) aims to determine whether 1700 mg/daily metformin can potentially restore gene expression in elderly with impaired glucose tolerance (https://clinicaltrials.gov/ct2/show/NCT02432287?term=metformin&cond=Ageing&rank=2). The Targeting Aging with Metformin (TAME) trial is designed as a first randomized controlled clinical study to evaluate metformin as an anti-aging drug. Primary trial outcome is the time until occurrence of any of aging-related multimorbidity composite (coronary heart disease, stroke, congestive heart failure, peripheral arterial disease, cancer, T2D, cognitive impairment, mortality) (Newman et al. 2016) (https://www.afar.org/natgeo/).

Rapamycin and Rapalogs

Rapamycin (sirolimus) is an mTOR inhibitor, widely used for immunosuppression for organ transplant recipients and cancer patients (Blagosklonny 2017). Some of the common adverse reactions observed upon administration of rapamycin in clinical trials include hypertriglyceridaemia, hypertension, hypercholesterolaemia, creatinine increase, urinary tract infection, anaemia and thrombocytopaenia. Currently, several rapalogs with pharmacokinetic properties superior to rapamycin and reduced immunosuppressive effects are being tested as immunosuppressant or treatment options for advanced solid tumors (Xie et al. 2016).

After numerous animal studies, rapamycin is widely discussed as a medicamentous intervention to increase healthspan and longevity in humans. Potential highlights of its use in prevention of certain age-related diseases and conditions should be obtained from follow-up studies, performed on cohorts, where rapamycin is used according to its primary indication – as immunosuppressant or cytostatic drug.

Evidence from clinical trials suggests that continuous use of mTOR inhibitors in transplant patients increases risk of diabetes. Data from the United States Renal Data System, evaluating association of sirolimus use in 20,124 adult kidney recipients without diabetes, concludes that patients on sirolimus are at increased risk of new-onset diabetes, compared to subjects on other immunosuppression schemes. This risk did not depend on the immunosuppressant combination, in which mTOR inhibitor was initially prescribed to the patient (Johnston et al. 2008). However, the exclusive role of rapamycin in onset of diabetes is very difficult to evaluate. On the contrary, patients after kidney transplantation, converted to rapalog everolimus, experienced reduced risks of diabetes mellitus in meta-analysis of RCTs (Liu et al. 2017a, b). Risk of diabetes in post-transplant patients is also negotiable due to potential investment of other medications into pathophysiology of abnormal glucose metabolism and initial risk factors of the given population. It remains unclear whether chronic use of rapamycin might result in hyperglycemia and diabetes in healthy individuals, thus future studies are warranted (Blagosklonny 2012a, b). Another known side effect of rapalogs, occurring in 40–75% of patients is dyslipidemia (Kurdi et al. 2018). Rapamycin altered levels of serum lipids in patients with autosomal-dominant polycystic kidney disease (ADPKD), increasing serum total cholesterol, triglycerides and LDL-C, however without influencing HDL-C (Liu et al. 2014a, b). Several authors suggest that combined use of mTOR inhibitor with medications capable of controling adverse effects of rapalogs, such as statins and metformin might be a promising combined formula to control age-associated diseases (Martinet et al. 2014; Blagosklonny 2017).

Despite certain pro-atherogenic effects, local application of sirolimus is widely used for revascularization interventions in patients with coronary artery diseases. Sirolimus-eluting stents (SES) demonstrated to reduce the short-, long- and overall-term risk of target lesion revascularization (TLR) and target vessel revascularization, as restenosis, major adverse cardiac events (MACE), overall-term risk of myocardial infarction in randomized controlled trials, comparing SES with paclitaxel-eluted scaffolds (Zhang et al. 2014). Another study, including population of 2877 patients, who underwent stenting with polymer-free sirolimus-coated scaffolds, demonstrated favorable rates of TLR and MACE reduction (Krackhardt et al. 2017).

mTOR inhibitors can cause regression of atherosclerosis with subsequent artery lumen enlargement and plaque regression. Thr underlying mechanisms of these interventions are not precisely understood; Martinet and coauthors hypothesize that these outcomes arise from cell proliferation suppression, modulation of authophagy, cell survival and cholesterol efflux. Currently, there is no evidence that systemic administration of mTOR inhibitor might be beneficial in dyslipidemia and atherosclerosis prevention. On the contrary, systemic mTOR inhibitor use is associated with dyslipidemia and hyperglycemia, risk factors, contributing to plaque formation and destabilization (Martinet et al. 2014).

Inhibitors of mTOR showed certain positive effects in transplant recipients in several collaborative studies and meta-analyses. Based on data from 6867 patients in 21 randomized trials, it was found that sirolimus use was associated with a 40% reduction of malignancy risk after kidney or combined pancreatic and kidney transplantation. This meta-analysis also showed 56% reduction of non-melanoma skin cancer (NMSC) risk, compared to non-sirolimus immunosuppression. These authors concluded that a cancer-protective effect was even more obvious in those patients who converted to sirolimus treatment after an established regimen of immunosuppression; this intervention resulted in reduction of NMSC risk and other cancers. However, sirolimus administration, both de novo and after switch from other immunosuppression agent, was associated with an increased risk of death, compared to controls (Knoll et al. 2014).

The Collaborative Transplant Study involved 78,146 patients after kidney transplantation, receiving either mTOR inhibitor, or non-mTOR-inhibitor immunosuppressant demonstrated that kidney transplant recipients, receiving mTOR inhibitor de novo have had reduced incidence of NMSC, however no influence on other cancers was found (Opelz et al. 2016).

Many oncogenic pathways, including the Ras/Raf/MEK/ERK pathway and the phosphoinositide 3-kinase (PI3K)/AKT (PKB) pathway are linked to mTOR signalling. Approximately 70% of human tumors harbor gain-of-function mutations in oncogenes (i.e. PI3K, AKT, or Ras) and/or loss-of-function mutations in tumor suppressors (i.e. PTEN, LKB1 or TSC1/2), resulting in mTORC1 hyperactivation (Forbes et al. 2011; Li et al. 2014). Inhibition of mTOR appears to be a promising approach in oncology, as this molecular target is involved in cell proliferation, metabolism, angiogenesis, survival, and is involved in cancer development and immune microenvironment modulation (Cash et al. 2015). mTOR inhibitors (everolimus, serolimus or temsirolimus) were also shown to improve the effects of hormonal therapy and the outcome in patients with metastatic luminal breast cancer. It was suggested that mTOR inhibition might affect hormone sensitivity of the tumor cells (Rotundo et al. 2016).

A recent Cochrane review and meta-analysis revealed beneficial effects of rapamycin and rapalogs on tuberous sclerosis – rare multisystem disease with formation of benign tumors and neurological disorders. Results of 3 placebo-controlled studies with a total of 263 participants demonstrated, that administration of everolimus achieved a 50% reduction in the size of sub-ependymal giant cell astrocytoma and renal angiomyolipoma. Sasongko and coauthors concluded that use of rapalogs in clinical practice is supported by significant evidence, as the benefits outweigh the risks, as risk of adverse events among treated patients was the same as in those who received no treatment (Sasongko et al. 2016).

One of the main concerns regarding chronic use of rapalogs is related to suppression of the immune system and potential high susceptibility to infection, including opportunistic infections, fatal infections, sepsis (https://www.rxlist.com/rapamune-drug.htm). Meta-analysis of RCTs, evaluating sirolimus for treatment of ADPKD showed a slight increase of the rate of infection (generally aphtous stomatitis and pharyngitis), yet its use was not associated with induction of severe infections (Liu et al. 2014a, b). Recent meta-analysis revealed that treatment with mTOR inhibitors in patients with lymphangioleiomyomatosis, a chronic destructive cystic lung disease, does not appear to increase the incidence of respiratory infections. On the contrary, sirolimus and everolimus demonstrated a trend towards reduction of respiratory infection risk, compared to placebo. The authors underline the point that additional studies are warranted to determine the mechanism of the potential protective effect of mTOR inhibitors, which may involve numerous mechanisms in the respiratory tract, immune system and microbiome (Courtwright et al. 2017).

It is also necessary to underline the fact that rapamycin is used for immunosuppression only in specific categories of transplant recipients. I.e., the FDA issued a black box warning towards use of sirolimus in liver transplant patients. De novo use of sirolimus in post-liver transplant patients was associated with high incidence of hepatic artery thrombosis and decreased patient and graft survival. Massoud and Wiesner (2012) reported a controversial role of rapamycin in liver transplant recipients, showing potential benefit of its use in liver transplant patients with hepatocellular carcinoma (HCC) (sirulimus appears to increase recurrence-free survival for liver recipients due to HCC) (Chinnakotla et al. 2009). They underlined the need for further evaluation of anti-neoplastic and anti-viral effects of sirolimus, potentially favorable for certain categories of liver recipients (Massoud and Wiesner 2012). Sirolimus previously demonstrated suppression of hepatitis C recurrence in a small cohort of liver transplantation candidates (Wagner et al. 2010). A later study by Yanik and colleagues showed that HCC recurrence and cancer-specific mortality rates were lower in patients prescribed with sirolimus, but not statistically significant. This study also demonstrated more favorable outcomes in patients older than 55 years, while younger patients had worse outcomes, including all-cause mortality, HCC recurrence and cancer-specific mortality (Yanik et al. 2016).

Evaluating the effects of rapamycin on aging, its biomarkers and age-related diseases in humans appears to be a very challenging task. This drug is used in very specific patient populations – transplant recipients, who often experience multimorbidity, initial functional decline in many physiological parameters along with the strong need to take several medications to control immune response against the graft and provide proper nephroprotection (Lamming et al. 2013). Better prognosis and rate of certain outcomes in these subjects could be also explained by direct immunosuppressive effects of rapamycin, preventing graft rejection and preserving renal function, thus reducing the impact of renal- dependent mechanisms. It is necessary to highlight the fact that most of the data regarding pleiotropic actions of rapalogs come from small, retrospective, case-control studies. A pilot randomized control trial, establishing feasibility of rapamycin in a small group of overall healthy volunteers aged 70–95 years, showed that 1 mg of this agent can be safely used in short-term settings. The reported findings on cognitive, physical performance and immune changes are rather occasionally anecdotal than consistent, and do not enable one to draw a firm conclusions. Larger trials with longer treatment duration are warranted for further analysis, interpretation and possible practical translation (Kraig et al. 2018).

An additional large concern in case of chronic rapamycin administration in women of reproductive age is its potential embryo- and fetotoxicity. No proper controlled studies have been conducted in pregnant women, thus the FDA marks this agent as pregnancy category C, recommending avoiding pregnancy and nursing during rapamycin therapy (https://www.rxlist.com/rapamune-drug.htm). Use of rapamycin in healthy individuals to control mechanisms involved into aging remains a widely discussible and negotiable intervention, considering the balance between potential risks and beneficial outcome. Nevertheless, toxicity issues might be successfully addressed by targeted drug delivery, allowing rapamycin to be introduced as a senolytic drug to specific cell types or tissues. These technologies are already being developed and tested in preclinical settings (Gholizadeh et al. 2017; Thapa et al. 2017).

Aspirin

Aspirin (acetylsalicylic acid, ASA) is one of the top-prescribed medications worldwide. This agent made a historical entry from being originally developed as analgesic and antipyretic drug; nowadays is predominantly used for primary and secondary cardiovascular prevention. Yet, many of its complementary effects regarding age-related diseases are still to be unveiled (Desborough and Keeling 2017). Aspirin might be one of the most appropriate agents to be appointed as a promising anti-aging drug due to low costs and simplicity of treatment and well-investigated multifaceted properties regarding cardiovascular disease (CVD) and cancer.

Aspirin and CVD

Low-dose aspirin is an anti-platelet therapy with established clinically benefit in secondary prevention of cardiovascular diseases. The anti-platelet effect of aspirin results from COX-1 acetylation with subsequent inhibited production of thromboxane A_2. Growing evidence suggests, that aspirin-mediated acetylation might play additional, non-COX-dependent role in thrombosis prevention, together with anti-inflammatory and anti-tumor effects of this drug (Ornelas et al. 2017; Warner et al. 2011).

Currently, ASA is recommended to patients presenting with ST-elevation myocardial infarction (STEMI) and non-STEMI patients. A convincing body of evidence suggests that aspirin provides a beneficial reduction of CVD mortality and new CVD events (Ittaman et al. 2014). In non-STEMI, ASA is recommended to be prescribed to all patients without contraindications, using an initial loading dose (150–300 mg), followed by maintenance dose 75–100 mg daily for a long term, regardless whether an invasive or non-invasive treatment strategy was selected

(Roffi et al. 2016). Aspirin administration is recommended indefinitely in all patients with STEMI (Ibanez et al. 2018).

Meanwhile, the role of low-dose aspirin in primary prevention remains unclear (Brotons et al. 2015). Recent meta-analysis of 9 trials, where aspirin was used for primary cardiovascular prevention and involved 100,076 patients, showed that long-term aspirin in comparison to placebo or no aspirin reduced myocardial infarction, ischemic stroke, and all-cause mortality. Yet, this intervention increased risk of hemorrhagic complications – hemorrhagic stroke, major bleeding, and gastrointestinal bleeding. Raju et al. (2016) concluded that this prevention strategy always requires an evaluation of the balance between the potential benefit and harm of long-term ASA prescription; the shown reduction of all-cause mortality is a favorable fact deserving to be taken into account. Yet a high risk of bleeding, especially intracerebral hemorrhage remains the major concern for different patient cohorts, despite potentially favorable cardiovascular outcome. The clinical decision about prescription of low-dosed ASA in primary prevention remains complicated and requires evaluation of risk-benefit ratio for each individual. Additionally, selection of correct dose, coated/non-coated form of the drug, considering concomitant conditions and medications, possible twice-a-day dosing in patients with increased platelet turnover are among variables to be considered and further researched (Leggio et al. 2017).

Anti-cancer Effects of Aspirin

A large body of evidence exists on certain roles of chronic ASA administration in cancer prevention and cancer survival. Massive-scale meta-analysis, involving 23 RCTs on low-dose aspirin and non-vascular deaths reported significant reduction of cancer deaths; a preventive effect was observed after 4 years of aspirin intake (Mills et al. 2012). Results of meta-analyses suggest that aspirin might serve as a preventive treatment for breast cancer (Lu et al. 2017; Luo et al. 2012; Zhong et al. 2015a, b), prostate (Huang et al. 2014; Liu et al. 2014a, b), pancreatic (Zhang et al. 2015) and gastric cancer (Huang et al. 2017; Kong et al. 2016). Protective effects were not prominent in some cancer types, however these findings could have different explanations. For instance, meta-analysis by Hochmuth et al. (2016) demonstrated protective effect of ASA against non-small cell lung cancer with strong heterogeneity. Researchers assume that aspirin potentially prevents lung cancer, but only in certain patient populations, while others do not benefit.

Chronic use of aspirin was shown to provide certain benefits for patients with established cancer diagnosis, improving overall survival, reducing risk of cardiovascular events and in some cases influencing cancer-related survival, and slowing the rate of metastasis. Meta-analyses of trials regarding aspirin use in patients with different types of cancer often provide conflicting results due to variability in research methodology and large heterogeneity of available studies. E.g., meta-analysis of observational studies by Zhong et al. (2015a, b) reported a small if any effect of aspirin intake on survival of breast cancer patients. Another meta-analysis, demonstrated decreased rates of breast cancer specific mortality, all-cause mortality and metastasis in aspirin and non-steroidal anti-inflammatory drug (NSAID) users. Beneficial effects of breast cancer survival were observed only if treatment was initiated after diagnosis, not before (Huang et al. 2015).

Aspirin intake was inversely related with prostate-cancer-specific mortality, according to meta-analysis by Li et al. (2014). A systematic review and meta-analysis by P.C. Elwood et al. concluded that low-dose aspirin is beneficial agent for adjuvant treatment of cancer. It was shown to reduce mortality in colon cancer, specifically in tumors, expressing *PIK3CA*. Probable and possible benefits were also demonstrated for breast cancer and prostate cancer patients. The authors emphasise that there is large heterogeneity among analyzed studies, with a lack of adequately planned and controlled randomized trials for less common types of cancers. Apart from that, reduction of vascular events and suppression of metastatic growth is promising evidence, allowing clinicians to discuss and incorporate aspirin as additional anti-cancer treatment into patients' treatments (Elwood et al. 2016).

The most remarkable chemopreventive and disease-modifying effects of ASA were observed in case of colorectal cancer (CRC). Current evidence even inspired US preventive services task force to recommend low-dose aspirin for primary prevention of CVD and CRC in adults aged 50–59, having ≥10% 10-year risk of CVD and are not at increased risk of bleeding (Chubak et al. 2015; Patrignani and Patrono 2016).

Meta-analysis by Ye et al. (2013) showed that low-dose (75–325 mg daily) regular (two to seven times a week) aspirin treatment lasting more than 5 years provides effective CRC risk reduction. Evidence from studies investigating CVD primary and secondary prevention suggested that ASA administration reduces the incidence of CRC, and CRC-mortality 10 years after treatment initiation (Chubak et al. 2016). Aspirin intake provides survival benefits for CRC-patients. Findings from meta-analyses of studies, evaluating aspirin use in patients with CRC demonstrated benefit in overall survival only in cases when aspirin was administered after cancer diagnosis. Meta-analyses suggest that post-diagnosis use of aspirin might be beneficial in CRC patients with positive expression of *COX-2* and *PIK3CA* mutated tumors, reducing overall mortality in *PIK3CA* mutated cancers by 29% (Li et al. 2015; Paleari et al. 2016).

Emillson and colleagues examined whether ASA intake might be an effective alternative to available CRC screening methods. Network meta-analysis compared efficacy of low-dose aspirin vs. flexible sigmoidoscopy or guaiac-based fecal occult blood test in reduction of colorectal cancer incidence and mortality. Low-dose aspirin seemed to be as effective as screening tools in colorectal cancer prevention, with effects more visible for malignancies with proximal colon localization. Randomized controlled trials are warranted to make a definite conclusion about possible colorectal cancer chemoprevention with ASA (Emillson et al. 2017).

ASA could be used for CRC prevention in susceptible populations, e.g. hereditary cancers like Lynch syndrome. The randomized trial CAPP2 (Cancer Prevention Programme) involved 861 participants with Lynch syndrome. Patients received 600 mg aspirin or placebo for up to 4 years. Intake of aspirin resulted in substantial (around 60%) decrease of cancer incidence (Burn et al. 2011). Results from 1858 participants of Colon Cancer Family Registry support evidence that aspirin is effective to reduce risk of CRC in *MMR* gene mutation carriers (Ait Ouakrim et al. 2015). Aspirin could be pa romising agent for secondary chemoprevention in CRC patients who already have experienced an intervention due to confirmed CRC diagnosis. Meta-analysis of 15 RCTs, evaluating 10 different candidate chemoprevention agents among individuals with previous colorectal neoplasia, showed that non-aspirin NSAIDs were the most effective agents for prevention of advanced metachronous neoplasia. Low-dose aspirin, being second in efficacy, also demonstrated the most favorable safety profile. According to the results of this analysis, low-dose aspirin with superior risk/benefit profile might be considered for secondary colorectal cancer chemoprevention in patients with previous colorectal neoplasia (Dulai et al. 2016).

Anti-neoplastic effects of ASA include mechanisms related to COX-1 and COX-2 inhibition, however an increasing body of evidence also supports the hypothesis about non-COX mechanisms. COX-2 inhibition is considered to be crucial in prevention of colorectal neoplasia (Chubak et al. 2015). Discussion about possible mechanisms and pathways involved into carcinogenesis inhibition include inhibition of IκB kinase β, preventing activation of NF-κB, inhibition of extracellular-signal-regulated kinase (ERK) and Wnt/ β-catenin pathway (Alfonso et al. 2014). Aspirin was also shown to activate AMPK, which further inhibits activity of mTORC1 (Din et al. 2012; Lamming et al. 2013).

It is necessary to emphasise that future studies, evaluating the role of aspirin as a chemopreventive and/or adjuvant cancer treatment are warranted. Many studies report beneficial effects of ASA only if intake was initiated after cancer diagnosis, dose and treatment duration might also matter and vary in each case. Cancer prevention in specific patient cohorts at high risk of neoplasia (e.g. genetic cancers) is an increasingly interesting direction for future research.

Anti-inflammatory Properties of Aspirin

Direct COX-inhibition mediated mechanisms and indirect modulation of NF-kB pathway, along with inhibition of IL-7 release control the anti-inflammatory properties of ASA (Ornelas et al. 2017). Limited, but promising evidence exists about the role of aspirin in prevention of sepsis – life-threatening condition, often affecting elderly. An individual patient data meta-analysis with propensity matching showed 7–12% mortality risk reduction in sepsis patients, taking aspirin prior to sepsis onset (Trauer et al. 2017). The aspirin to Inhibit SEPSIS (ANTISEPSIS) trial is substudy of ASPirin in Reducing Events in the Elderly (ASPREE) to be finished in 2018; it is expected to answer the questions as to whether low-dose aspirin reduces sepsis-related mortality and sepsis related hospital admissions in elderly (Eisen et al. 2017).

Aspirin and Cognitive Function

Antiplatelet effects could provide potential benefits in neuroprotection, reducing impairments occurring as a result of small neurovascular lesions. However, data regarding possible buffering of cognitive decline among aspirin users is not that optimistic. Specific influence of ASA on brain white matter lesions (WML) was evaluated in patients from Women's Health Initiative Memory Study of Magnetic Resonance Imaging study. There was no significant difference between WML volumes among aspirin users and non-users (Holcombe et al. 2017). Furthermore, research, conducted in patients with Alzheimer's disease concludes that ASA use does not provide any additional therapeutic benefit; furthermore, due to increased risk of intracerebral hemorrhage, this exposes patients to high risk of additional cognitive loss (Thoonsen et al. 2010). Recent meta-analysis by Veronese et al. (2017) involving data from 36,196 patients does not confirm protective effect of aspirin towards cognitive decline in older age. Pooled data from RCTs and observational studies showed that use of low-dose aspirin was not associated with significantly better global cognition, or onset of dementia or cognitive impairment. Results of an ongoing the ASPREE trial (ASPirin in Reducing Events in the Elderly), assessing role of aspirin in maintenance of disability-free and dementia-free life in a healthy population of elderly is expected to be finished by the end of 2018. Use of sophisticated neurovisualization techniques in this study could be advantageous for better understanding of prevention of microvascular dementia with ASA. Trial is also aimed to evaluate whether potential benefits outweigh the risks in this specific population (McNeil et al. 2017).

Statins

Statins represent a heterogeneous group of pharmacological agents, mediating their lipid-lowering effects by 3-Hydroxy-3-methylglutaryl coenzyme A reductase inhibition. Statins are the most commonly used prescription drugs for the treatment of dyslipidemia due to their well-established cholesterol-lowering properties and reduction of cardiovascular events and mortality (Collins et al. 2016). Apart from their primary mechanism of action, involving hepatic cholesterol synthesis inhibition, statins exert multiple pleiotropic effects, independently of LDL-C lowering mechanisms. Statins inhibit synthesis of certain substances – farnesyl pyrophosphate, geranylgeranyl pyrophosphate, isopentanyl adenosine, dolichols and polyisoprenoid side chains of ubiquinone, heme A and nuclear lamins, classified as isoprenoid intermediates and playing key role in activation of numerous intracellular signaling proteins. Reduction of circulating isoprenoid intermediates affects activity of Ras and Ras-like proteins (Rho, Rab, Rac, Ral, and Rap). Thus, statins have anti-inflammatory, immunomodulatory, antioxidant, antiproliferative effects, stabilize atherosclerotic plaques and prevent aggregation of platelets; each of these effects are cholesterol-independent and mediated via isoprenoid-dependent signaling pathways (Kavalipati et al. 2015; Oesterle et al. 2017). These additional properties along with long-term use safety and cost-efficiency promote potential repurposing of statins for treatment and prevention of multiple age-related diseases and conditions.

Cardiovascular Effects of Statins

An impressive body of evidence from multiple high-quality randomized clinical trials confirms that statins effectively reduce total cholesterol, LDL-C, risk of acute coronary syndrome, stroke, venous thromboembolic disease, and death (Chou et al. 2016; Fulcher et al. 2015; Taylor et al. 2013).

Meta-analysis of 92 placebo-controlled and active-comparator trials demonstrated that statins are affective for both primary and secondary cardiovascular prevention as a class, significantly reducing major coronary events and all-cause mortality. Among atorvastatin, fluvastatin, lovastatin, pravastatin, rosuvastatin, and simvastatin, consistently strong evidence of benefit was provided for use of atorvastatin and simvastatin (Naci et al. 2013).

There is a gap in guidelines related to primary cardiovascular prevention with statins in elderly patients, due to the fact that most of the data on primary prevention was derived from patients at age 40–75. Huge concerns are related to potential side-effects of these agents in older patients, however, and many health professional address this issue and emphasize that potential benefits often outweigh risk of undesired side-effects (Mortensen and Falk 2018; Otto 2016).

Additional Benefits in CVD Prevention in T2DM Patients

Atorvastatin was shown to reduce the amount of glyceraldehyde-derived advanced glycation end-products in patients with acute myocardial infarction, thus, being important evidence of potential cardiovascular protection via reduction of AGE-RAGE signalling and oxidative stress (Shimomura et al. 2016). Significant decrease in serum AGEs was observed after 3 and 6 months of simvastatin treatment in elderly with hyperlipidaemia (Yang et al. 2016). Another study confirms that statins mediate decrease of plasma AGEs, contributing to plaque stabilization and lack of its progression in patients with acute coronary syndrome. Fukushima et al. (2013) showed that pitavastatin, but not atorvastatin significantly reduced serum levels of AGEs.

Anti-inflammatory Effects

Statins demonstrate convincing clinical benefit by reduction of inflammatory markers, making the use of statin for the treatment of chronic inflammatory diseases very appealing (Gilbert et al. 2017). Possible mechanism of anti-inflammatory effects of statins could be related to Ras-prenylation inhibition, resulting in beneficial reduction of C-reactive protein (CRP), including hs-CRP. Some studies confirm hs-CRP as being a non-conventional cardiovascular risk factor and predictor for the clinical outcome (Joo 2012; Kitagawa et al. 2017; Ridker 2016). Meta-analysis of statin use after revascularization marked beneficial modulation of systemic inflammatory markers in a group of statin-treated patients, followed by reduction of risk for postoperative atrial fibrillation (An et al. 2017). Recent meta-analysis of RCTs, evaluating statin use among patients with rheumatoid arthritis showed additional benefits of atorvastatin for management of RA. Atorvastatin was shown to decrease the levels of inflammatory markers – C-reactive protein and erythrocyte sedimentation rate, along with decrease in DAS28 score. This meta-analysis demonstrated superior anti-inflammatory properties of atorvastatin, compared to simvastatin (Li et al. 2018). Chronic low-grade inflammation, thus supporting aging, might be targeted with this group of agents.

Potential Anti-infectious Properties

Conflicting evidence exists regarding potential effects of statins as agents that reduce the risk of infectious diseases. Some of them, being substances of fungal origin, are widely discussed to have beneficial antimicrobial effects, also claimed as potential agents to fight microbial antibiotic resistance. Simvastatin is even considered to be repurposed as a novel adjuvant antibiotic (Ko et al. 2017). Among patients with drug-treated T2DM statins were associated with a reduced risk of infections (Pouwels et al. 2016). However, large meta-analyses showed no effect of statins on the risk of infections or deaths, related to infections (Deshpande et al. 2015; van den Hoek et al. 2011).

There is also a wide discussion related to utilization of statins as antiviral agents, including treatment of dangerous infections, like influenza (Mehrbod et al. 2014) and HIV (Kelesidis, 2012). Meta-analysis of 16 homogeneous studies showed that statins as adjuvant therapy, improved sustained virological response rate by 31% in patients with chronic hepatitis C infection, compared with those who obtained antiviral treatment alone. HCV-infected patients, receiving statins also had reduced risks of HCC, cirrhosis and mortality (Zheng et al. 2017).

Anticarcinogenous Properties of Statins

Numerous meta-analyses have addressed the potential role of statins in chemoprevention of cancer. Thus, it was shown, that use of statins can reduce the risk of colorectal (8–12%), gastric (27–44%), hematological (19%), liver (37–42%), oesophageal (14–28%), ovarian (21%) and prostate cancer (7%) (Undela et al. 2017).

Apart from lipid-lowering actions, which could potentially contribute to repression of tumor growth, statins interfere with several pathways which modulate carcinogenesis. Statins demonstrate multi-level anti-cancer functions, modulating divergent signaling, cell-adhesion, epithelial-mesenchymal transition, DNA replication, angiogenesis, tumor-associated macrophages (Kavalipati et al. 2015; Papanagnou et al. 2017). In experimental studies, different statins showed capabilities to induce apoptosis, cause cytostatic and antiproliferative effects, augment anticarcinogenous properties of standard chemotherapies and attenuate cancer cell migration and invasion. All of the above mentioned properties are of increased importance for potential drug repurposing and use of statins as chemopreventive or adjuvant agents (Papanagnou et al. 2017). Potential and problems and existing gaps in current evidence are related to excessive amount of fragmentary research and data obtained from different models and cancer cell lines, with a lack of direct comparison between different statin molecules.

A systematic review and network meta-analysis by Zhou and colleagues, based on 87,127 patient data from observational studies, indicates that statin treatment is associated with decreased incidence of hepatocellular carcinoma (HCC). Among seven different statins, fluvastatin appeared to provide superior benefit in HCC risk reduction (Zhou et al. 2016). Also, a recent meta-analysis by Yi et al. (2017), demonstrated a statistically significant association between statin use and liver cancer risk reduction. This associated was dose dependent and was observed both in Asian and Caucasian patient subgroups (Yi et al. 2017). Another meta-analysis of 24 studies showed that statin users experienced a significantly decreased risk for developing primary liver cancer, compared with statin non-users, and the risk reduction was more evident in case of rosuvastatin administration. Subgroup analyses revealed greater risk reduction with statins in high-risk patients versus non-high-risk populations (Zhong et al. 2016).

Two recent meta-analyses, involving observational studies assessing statin use and colorectal cancer (CRC) prognosis showed that prescription of statins before and after CRC diagnosis showed a beneficial association between statin use and reduction in all-cause mortality and cancer-specific mortality (Cai et al. 2015; Ling et al. 2015).

In meta-analysis of 7 studies involving 5449 patients with endocrine-related gynecologic cancers, statin use was associated with improved overall survival. Current work also provided evidence on improved disease-specific survival and progression-free survival in statin users with endometrial and ovarian cancers (Xie et al. 2017).

In different pre-clinical studies on breast cancer cell lines, statins have demonstrated a great potential to increase apoptosis, improve radiosensitivity, suppressing cell proliferation, invasion and metastatic dissemination (Van Wyhe et al. 2017). A recent large meta-analysis concluded that use of statins is associated with reduced breast cancer mortality, including both breast cancer-specific and all-cause mortality. Beneficial effects varied in different types of statins – lipophilic statins (lovastatin, simvastatin, fluvastatin, cerivastatin) showed a strong protective function in breast cancer patients, whereas hydrophilic (pravastatin, rosuvastatin, atorvastatin) only slightly improved all-cause mortality (Liu et al. 2017a, b). Another meta-analysis reports improved recurrence-free survival in breast cancer patients, prescribed with lipophilic statins, specifically simvastatin (Manthravadi et al. 2016).

Undoubtely, thoroughly planned studies, assessing perspective placement of statins as adjuvant agents and radio- or chemosensitizers are warranted. Considering lipophilic/hydrophilic properties, targeted efficacy of certain agents against specific cancer types, and preferred duration of exposure might be of substantial importance in oncology.

Antifibrotic Action of Statins

Fibrosis occurring due to chronic tissue inflammation and its retardation under statin use was intensively investigated in patients with liver diseases. Six cohort studies, including 38,951 cases of cirrhosis in 263,573 patients with hepatitis B or C, were included into meta-analysis. Use of statins was associated with a significant 42% reduction in the risk of cirrhosis, being dose-dependent and more pronounced in patients from Asian countries (Wang et al. 2017a, b). Results of another meta-analysis of statin effects in patients with chronic liver disease (CLD) suggest that such drug intervention might delay progression of liver fibrosis and prevent hepatic decompensation in cirrhosis, and reduce all-cause mortality in CLD patients (Kamal et al. 2017). Meta-analysis by Kim et al. (2017) showed that statin use was associated with 46% lower risk of hepatic decompensation and 46% lower mortality in patients with liver cirrhosis. According to data, obtained in RCTs, statin use was associated with 27% lower risk of variceal bleeding or progression of portal hypertension.

Effects of Statins on Renal Function

Apart from benefits in reduction of MACE and all-cause mortality in patients with chronic kidney disease (CKD) (Messow and Isles 2017), statins are continuously reported to reduce the rate of kidney functional decline. Results of meta-analysis evaluating renal outcomes in patients with CKD, demonstrated slower decline of estimated glomerular filtration rate (eGFR) in statin users, compared to controls.

This effect was observed only in case of high-intensity statin treatment; statin intervention did not reduce proteinuria in CKD patients (Sanguankeo et al. 2015). Meta-analysis of data from 11 RCTs, involving 543 diabetic kidney disease patients demonstrated that statins were associated with beneficial reduction of albuminuria, however no influence on eGFR or total proteinuria was observed (Qin et al. 2017). A systematic review and meta-analysis of RCTs from 143,888 non-dialysis CKD patients reports modest reduction in proteinuria and rate of eGFR decline and no benefit in prevention of kidney failure events (Su et al. 2016). Recent survival meta-analysis demonstrated improved patient and graft survival among kidney transplant recipients receiving statins. However, there is still lack of studies evaluating outcomes in renal transplant and dialysis patients or providing head-to-head comparison of different statins (Rostami et al. 2017).

Major limitations of routine statin use as potential anti-aging formula includes the fact that most of the data was obtained from patients at initially high risk of cardiovascular disease or presenting with existing cardiovascular problems. There is still scarce evidence from particularly non-white populations and elderly patients. Long-term outcomes of prolonged treatment, lasting over several decades and its possible risks, remain unknown (Otto 2016).

Problems of Multimorbidity and Polypharmacy in the Elderly: Confounding Effects of Combined Treatment Strategies in Patients with Several Chronic Age-Associated Diseases

Research related to drugs, promoting healthy aging and preventing age-associated diseases and conditions is surrounded by numerous ethical considerations and limitations. For sure, the most desirable setting is life-long use of potential anti-ageing substances in healthy individuals. Assessment of those anti-aging and lifespan-prolonging properties in humans would require a lifetime to determine (Moskalev et al. 2017). Furthermore, as for any type of medications, anti-aging drugs do have side effects, so lifelong prescriptions of those in healthy individuals makes clinical trials unacceptable (Blagosklonny 2009). However, medications which have demonstrated their confident benefits and a reliable safety profile in long-term use, might be potentially evaluated in primary prevention studies. For example, an ongoing RCT STAREE (statins therapy for reducing events in elderly) randomized individuals aged 70 and older without prior cardiovascular disease, dementia, diabetes, or a life-limiting illness to atorvastatin or placebo, with results expected in 2019, will evaluate the role of statins in mortality and functional status (https://clinicaltrials.gov/ct2/show/NCT02099123). The statins are a quite representative group of medications, surrounded by a large number of constraints and considerations regarding prescription in populations of elderly patients. Evidence of statin long-term use in patients aged 75 and older is scarce, due to an insufficient number of patients having been included into clinical trials (advanced age is often an exclusion criterion) (Leya and Stone 2017); unfortunately, existing data about statin use in people aged 85 and older is insufficient to shape guidelines about rational and safe statin use in this category of patients (Orkaby et al. 2017).

Another complicated aspect of establishing the preventive role of certain agents towards age-associated diseases includes conflicting results, obtained in variable patient populations. A review by Islam et al. (2017) concluded that observational studies, claiming a breast cancer preventive effect of statins could not c establish a certain preventive causal role of these agents. These authors underline the importance of possibly unmeasured confounding variables, having an established effect on risk of cancer, i.e. obesity, physical activity, diet, tobacco and alcohol consumption. Furthermore, there is current discussion that some of the cancer trends in developed countries are a reflection of the wide chronic use of medications with potential chemopreventive properties. Thus, even data from well-designed randomized controlled trials does not consider all of the individual aspects of health and treatment and might not justify certain preventive properties of prescribed agents (Gronich and Rennert 2013).

All of the data about anti-aging properties of certain pharmacologic agents were obtained from large groups of patients, treated for one or more diseases by one certain medication. In these studies confounding and controversial results might be partially explained by multimorbidity of the included patients. So, on one hand, multimorbidity is a desirable inclusion criteria, because aging is accompanied by a number of health-deteriorating conditions and it is necessary to evaluate potential effects of certain medication on several conditions (i.e. improvement of survival of T2DM patients with certain cancers in case of chronic metformin use). On the other hand, multimorbidity is a synonym of multiple medication prescription, when it is impossible to assess individual action of each substance per se, thus providing researchers with confounding and inconclusive data (Fabbri et al. 2015). Many chronic conditions in elderly patients require combined treatment strategies and the use of polypill strategies, improving compliance and outcome is nowadays not rare (Hedner et al. 2016; Lafeber et al. 2016). This problem is especially large in population of elderly patients, as different studies show that more than one third of elderly patients are receiving five and more prescription drugs at once, which is often accompanied by intake of one or more over the counter drugs or dietary supplements (Maher et al. 2014).

Study by Johanna Jyrkka and colleagues established that excessive polypharmacy in elderly patients, aged 80 years and over, defined as concomitant consumption of ten and more prescription medications is an indicator of 5-year mortality (Jyrkkä et al. 2009). This issue might be addressed by the fact that certain medications with anti-aging properties affect several disease at once, thus clustering of several diseases with common underlying pathophysiological mechanisms into triades or groups might be beneficial for research and clinical practice (Schäfer et al. 2014; Violan et al. 2014).

A Cochrane review by Smith et al. (2016) underlines the fact that there is still lack of solid definitions for multimorbidity and related concepts such as comorbidity, complexity, frailty, and vulnerability. This leads to misclassification of patients, especially when considering the heterogeneity of multimorbidity, and might provide false conclusions about the effects of intervention. RCTs often include age and concomitant disease limitations, which does not allow clinicians to precisely depict the whole spectrum of effects for potential anti-aging substances (Zulman et al. 2011). Studies evaluating drug effects in chronic anti-aging agent users could also be biased by self-reporting of patients, often experiencing cognitive decline, lack of compliance due to the cost of treatment or just partial adherence to recommendations (Yap et al. 2016).

Additionally, the population of older adults has a long story of inconsistent prescription patterns, including under-prescription of certain agents with side effects (i.e. statins) (Orkaby et al. 2017). Clinicians often hesitate to use drug with known side effects, being additionally concerned about potential drug-drug interactions, which have essentially not been studied in elderly individuals (Strandberg et al. 2014).

Some life-span prolonging agents, affecting various pro-aging targets, often demonstrate controversial or negative results when tested in vivo. An appropriate example of this phenomenon is co called "antioxidant paradox": reactive oxygen species and oxygen radicals are involved into process of aging and pathogenesis of age-related diseases, however dietary supplementation with large doses of antioxidants have no preventive or therapeutic effect. Biswas (2016) underlined the fact that unsuccessful attempts to reduce oxidative stress in humans by antioxidant supplementation is partially explained by lack of reliable biomarkers, which could be used to measure redox status in humans. On the other hand, preferential targeting of ROS is possibly harmful, especially in case of cancer, when ROS production plays an important role in malignant cell apoptosis induction (Biswas 2016). Still, many of the crucial mechanisms of aging remain undiscovered and untargeted, and selective targeted of ROS, with very limited or nearly no success, is a good demonstration of our limited knowledge in this field (Halliwell 2013).

Another translational problem is based on the fact that we still do not have reliable biomarkers of aging, easily assessable in patients and allowing one to draw conclusions about the definitive anti-aging effect of intervention (Blagosklonny 2009; Moskalev et al. 2016). Basically, available clinically relevant evidence from clinical trials is based on the outcome of certain age-related disease prevention treatments, when their onset and progression had already happened, and not on the earlier markers of aging and age-related functional decline itself.

Conclusions and Perspectives

Because of their widespread used as medications, some FDA approved drugs are effective to prolong both life- and healthspan. Their effects are mostly associated with triggering the pathways shown to regulate lifespan in various experimental models. Unfortunately, many drugs were effective only in worms and flies and much less in rodents. The reason for this phenomenon is their differences in physiology and additional factors not taken into account. However, the use of combined treatments with an already beneficial compound may give significant steps forward, leading to the extension of human lifespan and decreasing age-related pathologies.

References

Ait Ouakrim D, Dashti SG, Chau R, Buchanan DD, Clendenning M, Rosty C, Winship IM, Young JP, Giles GG, Leggett B, Macrae FA, Ahnen DJ, Casey G, Gallinger S, Haile RW, Le Marchand L, Thibodeau SN, Lindor NM, Newcomb PA, Potter JD, Baron JA, Hopper JL, Jenkins MA, Win AK (2015) Aspirin, ibuprofen, and the risk of colorectal cancer in Lynch syndrome. J Natl Cancer Inst 107(9). pii: djv170

Alfonso L, Ai G, Spitale RC, Bhat GJ (2014) Molecular targets of aspirin and cancer prevention. Br J Cancer 111(1):61–67

American Diabetes Association (2002). Implications of the United Kingdom prospective diabetes study. Diabetes Care 25(suppl 1):s28–s32. https://www.clinicaltrials.gov/ct2/show/NCT0243 2287?term=metformin&cond=Ageing&rank=2, https://www.afar.org/natgeo/, https://www.clinicaltrials.gov/ct2/show/NCT02099123

An J, Shi F, Liu S, Ma J, Ma Q (2017) Preoperative statins as modifiers of cardiac and inflammatory outcomes following coronary artery bypass graft surgery: a meta-analysis. Interact Cardiovasc Thorac Surg 25:958–965

Anisimov VN, Berstein LM, Egormin PA, Piskunova TS, Popovich IG, Zabezhinski MA, Tyndyk ML, Yurova MV, Kovalenko IG, Poroshina TE, Semenchenko AV (2008) Metformin slows down ageing and extends life span of female SHR mice. Cell Cycle 7(17):2769–2773

Araki K, Ellebedy AH, Ahmed R (2011) TOR in the immune system. Curr Opin Cell Biol 23(6):707–715

Ayyadevara S, Bharill P, Dandapat A, Hu C, Khaidakov M, Mitra S, Shmookler Reis RJ, Mehta JL (2013) Aspirin inhibits oxidant stress, reduces age-associated functional declines, and extends lifespan of Caenorhabditis elegans. Antioxid Redox Signal 18(5):481–490

Bahadorani S, Hilliker AJ (2008) Cocoa confers life span extension in Drosophila melanogaster. Nutr Res 28(6):377–382

Bahadorani S, Bahadorani P, Phillips JP, Hilliker AJ (2008) The effects of vitamin supplementation on Drosophila life span under normoxia and under oxidative stress. J Gerontol A Biol Sci Med Sci 63(1):35–42

Bass TM, Weinkove D, Houthoofd K, Gems D, Partridge L (2007) Effects of resveratrol on lifespan in Drosophila melanogaster and Caenorhabditis elegans. Mech Ageing Dev 128(10):546–552

Baur JA, Sinclair DA (2006) Therapeutic potential of resveratrol: the in vivo evidence. Nat Rev Drug Discov 5(6):493–506

Baur JA, Pearson KJ, Price NL, Jamieson HA, Lerin C, Kalra A, Prabhu VV, Allard JS, Lopez-Lluch G, Lewis K, Pistell PJ, Poosala S, Becker KG, Boss O, Gwinn D, Wang M, Ramaswamy S, Fishbein KW, Spencer RG, Lakatta EG, Le Couteur D, Shaw RJ, Navas P, Puigserver P, Ingram DK, de Cabo R, Sinclair DA (2006) Resveratrol improves health and survival of mice on a high-calorie diet. Nature 444(7117):337–342

Benedetti MG, Foster AL, Vantipalli MC, White MP, Sampayo JN, Gill MS, Olsen A, Lithgow GJ (2008) Compounds that confer thermal stress resistance and extended lifespan. Exp Gerontol 43(10):882–891

Bhullar KS, Hubbard BP (2015) Lifespan and healthspan extension by resveratrol. Biochim Biophys Acta 1852(6):1209–1218

Biswas SK (2016) Does the interdependence between oxidative stress and inflammation explain the antioxidant paradox? Oxidative Med Cell Longev 2016:5698931

Bitto A, Wang AM, Bennett CF, Kaeberlein M (2015) Biochemical genetic pathways that modulate ageing in multiple species. Cold Spring Harb Perspect Med 5(11). pii: a025114

Bjedov I, Toivonen JM, Kerr F, Slack C, Jacobson J, Foley A, Partridge L (2010) Mechanisms of life span extension by rapamycin in the fruit fly *Drosophila melanogaster*. Cell Metab 11(1):35–46

Björkhem-Bergman L, Asplund AB, Lindh JD (2011) Metformin for weight reduction in non-diabetic patients on antipsychotic drugs: a systematic review and meta-analysis. J Psychopharmacol 25(3):299–305

Blagosklonny MV (2009) Validation of anti-ageing drugs by treating age-related diseases. Ageing (Albany NY) 1:281–288

Blagosklonny MV (2012a) Rapalogs in cancer prevention: anti-ageing or anticancer? Cancer Biol Ther 13(14):1349–1354

Blagosklonny MV (2012b) Once again on rapamycin-induced insulin resistance and longevity: despite of or owing to. Ageing (Albany NY) 4(5):350–358

Blagosklonny MV (2017) From rapalogs to anti-ageing formula. Oncotarget 8(22):35492–35507

Bonilla E, Medina-Leendertz S, Díaz S (2002) Extension of life span and stress resistance of *Drosophila melanogaster* by long-term supplementation with melatonin. Exp Gerontol 37(5):629–638

Bouza C, López-Cuadrado T, Gutierrez-Torres LF, Amate JM (2012) Efficacy and safety of metformin for treatment of overweight and obesity in adolescents: an updated systematic review and meta-analysis. Obes Facts 5:753–765

Brack C, Bechter-Thüring E, Labuhn M (1997) N-acetylcysteine slows down ageing and increases the life span of *Drosophila melanogaster*. Cell Mol Life Sci 53(11–12):960–966

Brotons C, Benamouzig R, Filipiak KJ, Limmroth V, Borghi CA (2015) Systematic review of aspirin in primary prevention: is it time for a new approach? Am J Cardiovasc Drugs 15(2):113–133

Brouet I, Ohshima H (1995) Curcumin, an anti-tumour promoter and anti-inflammatory agent, inhibits induction of nitric oxide synthase in activated macrophages. Biochem Biophys Res Commun 206(2):533–540

Buommino E, Pasquali D, Sinisi AA, Bellastella A, Morelli F, Metafora S (2000) Sodium butyrate/retinoic acid costimulation induces apoptosis-independent growth arrest and cell differentiation in normal and ras-transformed seminal vesicle epithelial cells unresponsive to retinoic acid. J Mol Endocrinol 24(1):83–94

Burd CE, Gill MS, Niedernhofer LJ, Robbins PD, Austad SN, Barzilai N, Kirkland JL (2016) Barriers to the preclinical development of therapeutics that target ageing mechanisms. J Gerontol A Biol Sci Med Sci 71(11):1388–1394

Burn J, Gerdes AM, Macrae F, Mecklin JP, Moeslein G, Olschwang S, Eccles D, Evans DG, Maher ER, Bertario L, Bisgaard ML, Dunlop MG, Ho JW, Hodgson SV, Lindblom A, Lubinski J, Morrison PJ, Murday V, Ramesar R, Side L, Scott RJ, Thomas HJ, Vasen HF, Barker G, Crawford G, Elliott F, Movahedi M, Pylvanainen K, Wijnen JT, Fodde R, Lynch HT, Mathers JC, Bishop DT, CAPP2 Investigators (2011) Long-term effect of aspirin on cancer risk in carriers of hereditary colorectal cancer: an analysis from the CAPP2 randomised controlled trial. Lancet 378(9809):2081–2087

Buu-Hoi NP, Ratsimamanga AR (1959) Retarding action of nordihydroguiaiaretic acid on ageing in the rat. C R Seances Soc Biol Fil 153:1180–1182. (in French)

Cai H, Zhang G, Wang Z, Luo Z, Zhou X (2015) Relationship between the use of statins and patient survival in colorectal cancer: a systematic review and meta-analysis. PLoS One 10(6):e0126944

Cameron AR, Morrison VL, Levin D, Mohan M, Forteath C, Beall C, McNeilly AD, Balfour DJ, Savinko T, Wong AK, Viollet B, Sakamoto K, Fagerholm SC, Foretz M, Lang CC, Rena G (2016) Anti-inflammatory effects of metformin irrespective of diabetes status. Circ Res 119(5):652–665

Cash H, Shah S, Moore E, Caruso A, Uppaluri R, Van Waes C, Allen C (2015) mTOR and MEK1/2 inhibition differentially modulate tumor growth and the immune microenvironment in syngeneic models of oral cavity cancer. Oncotarget 6(34):36400–36417

Castillo-Quan JI, Li L, Kinghorn KJ, Ivanov DK, Tain LS, Slack C, Kerr F, Nespital T, Thornton J, Hardy J, Bjedov I, Partridge L (2016) Lithium promotes longevity through GSK3/NRF2-dependent hormesis. Cell Rep 15(3):638–650

Chamberlain JJ, Herman WH, Leal S, Rhinehart AS, Shubrook JH, Skolnik N, Kalyani RR (2017) Pharmacologic therapy for type 2 diabetes: synopsis of the 2017 American Diabetes Association Standards of Medical Care in Diabetes. Ann Intern Med 166:572–578

Chang CI, Kuo CC, Chang JY, Kuo YH (2004) Three new oleanane-type triterpenes from *Ludwigia octovalvis* with cytotoxic activity against two human cancer cell lines. J Nat Prod 67(1):91–93

Chen W, Rezaizadehnajafi L, Wink M (2013) Influence of resveratrol on oxidative stress resistance and life span in *Caenorhabditis elegans*. J Pharm Pharmacol 65(5):682–688

Chen Y, Du J, Zhao YT, Zhang L, Lv G, Zhuang S, Qin G, Zhao TC (2015) Histone deacetylase (HDAC) inhibition improves myocardial function and prevents cardiac remodeling in diabetic mice. Cardiovasc Diabetol 14:99

Chen W, Liu X, Ye S (2016) Effects of metformin on blood and urine pro-inflammatory mediators in patients with type 2 diabetes. J Inflamm (London, England) 13:34

Chin RM, Fu X, Pai MY, Vergnes L, Hwang H, Deng G, Diep S, Lomenick B, Meli VS, Monsalve GC, Hu E, Whelan SA, Wang JX, Jung G, Solis GM, Fazlollahi F, Kaweeteerawat C, Quach A, Nili M, Krall AS, Godwin HA, Chang HR, Faull KF, Guo F, Jiang M, Trauger SA, Saghatelian A, Braas D, Christofk HR, Clarke CF, Teitell MA, Petrascheck M, Reue K, Jung ME, Frand AR, Huang J (2014) The metabolite α-ketoglutarate extends lifespan by inhibiting ATP synthase and TOR. Nature 510(7505):397–401

Ching TT, Chiang WC, Chen CS, Hsu AL (2011) Celecoxib extends *C. elegans* lifespan via inhibition of insulin-like signaling but not cyclooxygenase-2 activity. Ageing. Cell 10(3):506–519

Chinnakotla S, Davis GL, Vasani S, Kim P, Tomiyama K, Sanchez E, Onaca N, Goldstein R, Levy M, Klintmalm GB (2009) Impact of sirolimus on the recurrence of hepatocellular carcinoma after liver transplantation. Liver Transpl 15(12):1834–1842

Chou R, Dana T, Blazina I, Daeges M, Jeanne TL (2016) Statins for prevention of cardiovascular disease in adults: evidence report and systematic review for the US Preventive Services Task Force. JAMA 316(19):2008–2024

Chubak J, Kamineni A, Buist DSM, Anderson ML, Whitlock EP (2015) Aspirin use for the prevention of colorectal cancer. Agency for Healthcare Research and Quality (US), Rockville

Chubak J, Whitlock EP, Williams SB, Kamineni A, Burda BU, Buist DS et al (2016) Aspirin for the prevention of cancer incidence and mortality: systematic evidence reviews for the U.S. Preventive Services Task Force. Ann Intern Med 164:814–825

Col NF, Ochs L, Springmann V, Aragaki AK, Chlebowski RT (2012) Metformin and breast cancer risk: a meta-analysis and critical literature review. Breast Cancer Res Treat 135:639–646

Collins R, Reith C, Emberson J, Armitage J, Baigent C, Blackwell L, Blumenthal R, Danesh J, Smith GD, DeMets D, Evans S, Law M, MacMahon S, Martin S, Neal B, Poulter N, Preiss D, Ridker P, Roberts I, Rodgers A, Sandercock P, Schulz K, Sever P, Simes J, Smeeth L, Wald N, Yusuf S, Peto R (2016) Interpretation of the evidence for the efficacy and safety of statin therapy. Lancet 388(10059):2532–2561

Courtwright AM, Goldberg HJ, Henske EP, El-Chemaly S (2017) The effect of mTOR inhibitors on respiratory infections in lymphangioleiomyomatosis. Eur Respir Rev 26(143):160004

da Luz PL, Tanaka L, Brum PC, Dourado PM, Favarato D, Krieger JE, Laurindo FR (2012) Red wine and equivalent oral pharmacological doses of resveratrol delay vascular ageing but do not extend life span in rats. Atherosclerosis 224(1):136–142

Dancey J (2010) mTOR signaling and drug development in cancer. Nat Rev Clin Oncol 7(4):209–219

Danilov A, Shaposhnikov M, Shevchenko O, Zemskaya N, Zhavoronkov A, Moskalev A (2015) Influence of non-steroidal anti-inflammatory drugs on *Drosophila melanogaster* longevity. Oncotarget 6(23):19428–19444

Deblon N, Bourgoin L, Veyrat-Durebex C, Peyrou M, Vinciguerra M, Caillon A, Maeder C, Fournier M, Montet X, Rohner-Jeanrenaud F, Foti M (2012) Chronic mTOR inhibition by rapamycin induces muscle insulin resistance despite weight loss in rats. Br J Pharmacol 165(7):2325–2340

Decensi A, Puntoni M, Goodwin P, Cazzaniga M, Gennari A, Bonanni B, Gandini S (2010) Metformin and cancer risk in diabetic patients: a systematic review and meta-analysis. Cancer Prev Res (Phila) 3(11):1451–1461

Desborough MJR, Keeling DM (2017) The aspirin story – from willow to wonder drug. Br J Haematol 177(5):674–683

Deshpande A, Pasupuleti V, Rothberg MB (2015) Statin therapy and mortality from sepsis: a meta-analysis of randomized trials. Am J Med 128(4):410–417

Di Daniele N, Noce A, Vidiri MF, Moriconi E, Marrone G, Annicchiarico-Petruzzelli M, D'Urso G, Tesauro M, Rovella V, De Lorenzo A (2017) Impact of Mediterranean diet on metabolic syndrome, cancer and longevity. Oncotarget 8(5):8947–8979

Diamanti-Kandarakis E, Alexandraki K, Piperi C, Aessopos A, Paterakis T, Katsikis I, Panidis D (2007) Effect of metformin administration on plasma advanced glycation end product levels in women with polycystic ovary syndrome. Metabolism 56(1):129–134

Dias-Santagata D, Fulga TA, Duttaroy A, Feany MB (2007) Oxidative stress mediates tau-induced neurodegeneration in *Drosophila*. J Clin Invest 117(1):236–245

Dilman VM (1994) Ageing, rate of ageing and cancer. A search for preventive treatment. Ann N Y Acad Sci 719:454–455

Din FV, Valanciute A, Houde VP, Zibrova D, Green KA, Sakamoto K, Alessi DR, Dunlop MG (2012) Aspirin inhibits mTOR signaling, activates AMP-activated protein kinase, and induces autophagy in colorectal cancer cells. Gastroenterology 142(7):1504–15.e3

Dowling RJ, Niraula S, Chang MC, Done SJ, Ennis M, McCready DR, Leong WL, Escallon JM, Reedijk M, Goodwin PJ, Stambolic V (2015) Changes in insulin receptor signaling underlie neoadjuvant metformin administration in breast cancer: a prospective window of opportunity neoadjuvant study. Breast Cancer Res BCR 17(1):32

Dulai PS, Singh S, Marquez E, Khera R, Prokop LJ, Limburg PJ, Gupta S, Murad MH (2016) Chemoprevention of colorectal cancer in individuals with previous colorectal neoplasia: systematic review and network meta-analysis. BMJ 355:i6188

Edwards C, Canfield J, Copes N, Rehan M, Lipps D, Bradshaw PC (2014) D-beta-hydroxybutyrate extends lifespan in *C. elegans*. Ageing (Albany NY) 6(8):621–644

Edwards C, Canfield J, Copes N, Brito A, Rehan M, Lipps D, Brunquell J, Westerheide SD, Bradshaw PC (2015) Mechanisms of amino acid-mediated lifespan extension in *Caenorhabditis elegans*. BMC Genet 16:8

Eisen DP, Moore EM, Leder K et al (2017) AspiriN To Inhibit SEPSIS (ANTISEPSIS) randomised controlled trial protocol. BMJ Open 7(1):e013636

Elwood PC, Morgan G, Pickering JE et al (2016) Aspirin in the treatment of cancer: reductions in metastatic spread and in mortality: a systematic review and meta-analyses of published studies. Ali R, ed. PLoS ONE 11(4):e0152402

Emilsson L, Holme Ø, Bretthauer M, Cook N, Buring JE, Løberg M, Adami HO, Sesso HD, Gaziano MJ, Kalager M (2017) Systematic review with meta-analysis: the comparative effectiveness of aspirin vs. screening for colorectal cancer prevention. Aliment Pharmacol Ther 45(2):193–204

Eriksson L, Nyström T (2015) Antidiabetic agents and endothelial dysfunction – beyond glucose control. Basic Clin Pharmacol Toxicol 117:15–25

Fabbri E, Zoli M, Gonzalez-Freire M, Salive ME, Studenski SA, Ferrucci L (2015) Ageing and multimorbidity: new tasks, priorities, and frontiers for integrated gerontological and clinical research. J Am Med Dir Assoc 16(8):640–647

Flynn JM, O'Leary MN, Zambataro CA, Academia EC, Presley MP, Garrett BJ, Zykovich A, Mooney SD, Strong R, Rosen CJ, Kapahi P, Nelson MD, Kennedy BK, Melov S (2013) Late-life rapamycin treatment reverses age-related heart dysfunction. Ageing Cell 12(5):851–862

Fontana L, Villareal DT, Weiss EP, Racette SB, Steger-May K, Klein S, Holloszy JO, Washington University School of Medicine CALERIE Group (2007) Calorie restriction or exercise: effects on coronary heart disease risk factors. A randomized, controlled trial. Am J Physiol Endocrinol Metab 293(1):E197–E202

Fontana L, Partridge L, Longo VD (2010) Extending healthy life span – from yeast to humans. Science 328(5976):321–326

Forbes SA, Bindal N, Bamford S, Cole C, Kok CY, Beare D, Jia M, Shepherd R, Leung K, Menzies A, Teague JW, Campbell PJ, Stratton MR, Futreal PA (2011) COSMIC: mining complete cancer genomes in the Catalogue of Somatic Mutations in Cancer. Nucleic Acids Res 39(Database issue):D945–D950

Fougère B, Boulanger E, Nourhashémi F, Guyonnet S, Cesari M (2017) Chronic inflammation: accelerator of biological ageing. J Gerontol A Biol Sci Med Sci 72(9):1218–1225

Franceschi C, Campisi J (2014) Chronic inflammation (inflammaging) and its potential contribution to age-associated diseases. J Gerontol A Biol Sci Med Sci 69(Suppl 1):S4–S9

Franciosi M, Lucisano G, Lapice E, Strippoli GFM, Pellegrini F, Nicolucci A (2013) Metformin therapy and risk of cancer in patients with type 2 diabetes: systematic review. PLoS One 8(8):e71583

Fukushima Y, Daida H, Morimoto T, Kasai T, Miyauchi K, Yamagishi S, Takeuchi M, Hiro T, Kimura T, Nakagawa Y, Yamagishi M, Ozaki Y, Matsuzaki M, JAPAN-ACS Investigators (2013) Relationship between advanced glycation end products and plaque progression in patients with acute coronary syndrome: the JAPAN-ACS sub-study. Cardiovasc Diabetol 12:5

Fulcher J, O'Connell R, Voysey M, Emberson J, Blackwell L, Mihaylova B, Simes J, Collins R, Kirby A, Colhoun H, Braunwald E, La Rosa J, Pedersen TR, Tonkin A, Davis B, Sleight P, Franzosi MG, Baigent C, Keech A (2015) Efficacy and safety of LDL-lowering therapy among men and women: meta-analysis of individual data from 174,000 participants in 27 randomised trials. Lancet 385(9976):1397–1405

Gao Z, Yin J, Zhang J, Ward RE, Martin RJ, Lefevre M, Cefalu WT, Ye J (2009) Butyrate improves insulin sensitivity and increases energy expenditure in mice. Diabetes 58(7):1509–1517

Gao Y, Li Y, Xue J, Jia Y, Hu J (2010) Effect of the anti-diabetic drug metformin on bone mass in ovariectomized rats. Eur J Pharmacol 635:231–236

Garimella S, Seshayamma V, Rao HJ, Kumar S, Kumar U, Saheb SH (2016) Effect of metformin on lipid profile of type II diabetes. Int J Intg Med Sci 3(11):449–453

Gholizadeh S, Visweswaran GRR, Storm G, Hennink WE, Kamps JAAM, Kok RJ (2017) E-selectin targeted immunoliposomes for rapamycin delivery to activated endothelial cells. Int J Pharm 548(2):759–770

Gilbert R, Al-Janabi A, Tomkins-Netzer O, Lightman S (2017) Statins as anti-inflammatory agents: a potential therapeutic role in sight-threatening non-infectious uveitis. Porto Biomed J 2(2):33–39

Giovannucci E, Harlan DM, Archer MC, Bergenstal RM, Gapstur SM, Habel LA, Pollak M, Regensteiner JG, Yee D (2010) Diabetes and cancer: a consensus report. Diabetes Care 33:1674–1685

Gore DC, Wolf SE, Sanford A, Herndon DN, Wolfe RR (2005) Influence of metformin on glucose intolerance and muscle catabolism following severe burn injury. Ann Surg 241(2):334–342

Gospodaryov DV, Yurkevych IS, Jafari M, Lushchak VI, Lushchak OV (2013) Lifespan extension and delay of age-related functional decline caused by *Rhodiola rosea* depends on dietary macronutrient balance. Longev Healthspan 2(1):5

Griffin SJ, Leaver JK, Irving GJ (2017) Impact of metformin on cardiovascular disease: a meta-analysis of randomised trials among people with type 2 diabetes. Diabetologia 60(9):1620–1629

Gronich N, Rennert G (2013) Beyond aspirin-cancer prevention with statins, metformin and bisphosphonates. Nat Rev Clin Oncol 10(11):625–642

Gruber J, Halliwell B (2017) Approaches for extending human healthspan: from antioxidants to healthspan pharmacology. Essays Biochem 61(3):389–399

Haddad M, Knani I, Bouzidi H, Berriche O, Hammami M, Kerkeni M (2016) Plasma levels of pentosidine, carboxymethyl-lysine, soluble receptor for advanced glycation end products, and metabolic syndrome: the metformin effect. Dis Markers 2016:6248264

Halliwell B (2013) The antioxidant paradox: less paradoxical now? Br J Clin Pharmacol 75(3):637–644

Hansen M, Kennedy BK (2016) Does longer lifespan mean longer healthspan? Trends Cell Biol 26(8):565–568

Harman AD (1956) Ageing: a theory based on free radical and radiation chemistry. J Gerontol 11(3):298–300

Harman D (1972) Free radical theory of ageing: dietary implications. Am J Clin Nutr 25(8):839–843

Harrington LA, Harley CB (1988) Effect of vitamin E on lifespan and reproduction in *Caenorhabditis elegans*. Mech Ageing Dev 43(1):71–78

Harrison DE, Strong R, Sharp ZD, Nelson JF, Astle CM, Flurkey K, Nadon NL, Wilkinson JE, Frenkel K, Carter CS, Pahor M, Javors MA, Fernandez E, Miller RA (2009) Rapamycin fed late in life extends lifespan in genetically heterogeneous mice. Nature 460(7253):392–395

Hayashi S, McMahon AP (2002) Efficient recombination in diverse tissues by a tamoxifen-inducible form of Cre: a tool for temporally regulated gene activation/inactivation in the mouse. Dev Biol 244(2):305–318

Hedner T, Kjeldsen SE, Narkiewicz K, Oparil S (2016) The polypill: an emerging treatment alternative for secondary prevention of cardiovascular disease. Blood Press 25(5):276–279

Hegazy SK (2015) Evaluation of the anti-osteoporotic effects of metformin and sitagliptin in postmenopausal diabetic women. J Bone Miner Metab 33(2):207–212

Hervás D, Fornés-Ferrer V, Gómez-Escribano AP, Sequedo MD, Peiró C, Millán JM, Sequedo MD, Peiró C, Millán JM, Vázquez-Manrique RP (2017) Metformin intake associates with better cognitive function in patients with Huntington's disease. PLoS One 12(6):e0179283

Hochmuth F, Jochem M, Schlattmann P (2016) Meta-analysis of aspirin use and risk of lung cancer shows notable results. Eur J Cancer Prev 25(4):259–268

Holcombe A, Ammann E, Espeland MA, Kelley BJ, Manson JE, Wallace R, Robinson J (2017) Chronic use of aspirin and total white matter lesion volume: results from the women's health initiative memory study of magnetic resonance imageing study. J Stroke Cerebrovasc Dis 26(10):2128–2136

Holman RR, Paul SK, Bethel MA, Matthews DR, Neil HAW (2008) 10-year follow-up of intensive glucose control in type 2 diabetes. N Engl J Med 359:1577–1589

Holthoff JH, Wang Z, Seely KA, Gokden N, Mayeux PR (2012) Resveratrol improves renal microcirculation, protects the tubular epithelium, and prolongs survival in a mouse model of sepsis-induced acute kidney injury. Kidney Int 81(4):370–378

Hostalek U, Gwilt M, Hildemann S (2015) Therapeutic use of metformin in prediabetes and diabetes prevention. Drugs 75(10):1071–1094

Hou YC, Hu Q, Huang J, Fang JY, Xiong H (2017) Metformin therapy and the risk of colorectal adenoma in patients with type 2 diabetes: a meta-analysis. Oncotarget 8(5):8843–8853

Huang TB, Yan Y, Guo ZF, Zhang XL, Liu H, Geng J, Yao XD, Zheng JH (2014) Aspirin use and the risk of prostate cancer: a meta-analysis of 24 epidemiologic studies. Int Urol Nephrol 46(9):1715–1728

Huang XZ, Gao P, Sun JX, Song YX, Tsai CC, Liu J, Chen XW, Chen P, Xu HM, Wang ZN (2015) Aspirin and nonsteroidal anti-inflammatory drugs after but not before diagnosis are associated with improved breast cancer survival: a meta-analysis. Cancer Causes Control 26(4):589–600

Huang X, Chen Y, Wu J, Zhang X, Wu CC, Zhang CY, Sun SS, Chen WJ (2017) Aspirin and non-steroidal anti-inflammatory drugs use reduce gastric cancer risk: a dose-response meta-analysis. Oncotarget 8(3):4781–4795

Hubbard BP, Sinclair DA (2014) Small molecule SIRT1 activators for the treatment of ageing and age-related diseases. Trends Pharmacol Sci 35(3):146–154

Hunt PR, Son TG, Wilson MA, Yu QS, Wood WH, Zhang Y, Becker KG, Greig NH, Mattson MP, Camandola S, Wolkow CA (2011) Extension of lifespan in *C. elegans* by naphthoquinones that act through stress hormesis mechanisms. PLoS One 6(7):e21922

Ibanez B, James S, Agewall S, Antunes MJ, Bucciarelli-Ducci C, Bueno H, Caforio ALP, Crea F, Goudevenos JA, Halvorsen S, Hindricks G, Kastrati A, Lenzen MJ, Prescott E, Roffi M, Valgimigli M, Varenhorst C, Vranckx P, Widimský P, ESC Scientific Document Group (2018) 2017 ESC guidelines for the management of acute myocardial infarction in patients presenting with ST-segment elevation: the Task Force for the management of acute myocardial infarction in patients presenting with ST-segment elevation of the European Society of Cardiology (ESC). Eur Heart J 39(2):119–177

Ishibashi Y, Matsui T, Takeuchi M, Yamagishi S (2012) Metformin inhibits advanced glycation end products (AGEs)-induced renal tubular cell injury by suppressing reactive oxygen species generation via reducing receptor for AGEs (RAGE) expression. Horm Metab Res 44(12):891–895

Ishii N, Senoo-Matsuda N, Miyake K, Yasuda K, Ishii T, Hartman PS, Furukawa S (2004) Coenzyme Q10 can prolong *C. elegans* lifespan by lowering oxidative stress. Mech Ageing Dev 125(1):41–46

Islam MM, Yang HC, Nguyen PA, Poly TN, Huang CW, Kekade S, Khalfan AM, Debnath T, Li YJ, Abdul SS (2017) Exploring association between statin use and breast cancer risk: an updated meta-analysis. Arch Gynecol Obstet 296(6):1043–1053

Isoda K, Young JL, Zirlik A, MacFarlane LA, Tsuboi N, Gerdes N, Schönbeck U, Libby P (2006) Metformin inhibits proinflammatory responses and nuclear factor-κB in human vascular wall cells. Arterioscler Thromb Vasc Biol 26:611–617

Ittaman SV, VanWormer JJ, Rezkalla SH (2014) The role of aspirin in the prevention of cardiovascular disease. Clin Med Res 12(3–4):147–154

Johnston O, Rose CL, Webster AC, Gill JS (2008) Sirolimus is associated with new-onset diabetes in kidney transplant recipients. J Am Soc Nephrol JASN 19(7):1411–1418

Joo SJ (2012) Anti-inflammatory effects of statins beyond cholesterol lowering. Korean Circ J 42(9):592–594

Jyrkkä J, Enlund H, Korhonen MJ, Sulkava R, Hartikainen S (2009) Polypharmacy status as an indicator of mortality in an elderly population. Drugs Ageing 26(12):1039–1048

Kamal S, Khan MA, Seth A, Cholankeril G, Gupta D, Singh U, Kamal F, Howden CW, Stave C, Nair S, Satapathy SK, Ahmed A (2017) Beneficial effects of statins on the rates of hepatic fibrosis, hepatic decompensation, and mortality in chronic liver disease: a systematic review and meta-analysis. Am J Gastroenterol 112(10):1495–1505

Kang HL, Benzer S, Min KT (2002) Life extension in *Drosophila* by feeding a drug. Proc Natl Acad Sci U S A 99(2):838–843

Kapahi P, Zid BM, Harper T, Koslover D, Sapin V, Benzer S (2004) Regulation of lifespan in *Drosophila* by modulation of genes in the TOR signaling pathway. Curr Biol 14(10):885–890

Kapahi P, Chen D, Rogers AN, Katewa SD, Li PW, Thomas EL, Kockel L (2010) With TOR, less is more: a key role for the conserved nutrient-sensing TOR pathway in ageing. Cell Metab 11(6):453–465

Kavalipati N, Shah J, Ramakrishan A, Vasnawala H (2015) Pleiotropic effects of statins. Indian J Endocrinol Metabol 19(5):554–562

Kelesidis T (2012) Statins as antiviral and anti-inflammatory therapy in HIV infection. Virol Mycol 1:e102

Khan S, Jena G (2014) Sodium butyrate, a HDAC inhibitor ameliorates eNOS, iNOS and TGF-β1-induced fibrogenesis, apoptosis and DNA damage in the kidney of juvenile diabetic rats. Food Chem Toxicol 73:127–139

Kim J, Takahashi M, Shimizu T, Shirasawa T, Kajitha M, Kanayama A, Miyamoto Y (2008) Effects of a potent antioxidant, platinum nanoparticle, on the lifespan of *Caenorhabditis elegans*. Mech Ageing Dev 129(6):322–331

Kim RG, Loomba R, Prokop LJ, Singh S (2017) Statin use and risk of cirrhosis and related complications in patients with chronic liver diseases: a systematic review and meta-analysis. Clin Gastroenterol Hepatol 15(10):1521–1530.e8

Kitagawa K, Hosomi N, Nagai Y, Kagimura T, Ohtsuki T, Origasa H, Minematsu K, Uchiyama S, Nakamura M, Matsumoto M, J-STARS Investigators (2017) Reduction in high-sensitivity C-reactive protein levels in patients with ischemic stroke by statin treatment: Hs-CRP Sub-Study in J-STARS. J Atheroscler Thromb 24(10):1039–1047

Knoll GA, Kokolo MB, Mallick R, Beck A, Buenaventura CD, Ducharme R, Barsoum R, Bernasconi C, Blydt-Hansen TD, Ekberg H, Felipe CR, Firth J, Gallon L, Gelens M, Glotz D, Gossmann J, Guba M, Morsy AA, Salgo R, Scheuermann EH, Tedesco-Silva H, Vitko S, Watson C, Fergusson D (2014) Effect of sirolimus on malignancy and survival after kidney transplantation: systematic review and meta-analysis of individual patient data. BMJ 349:g6679

Ko HHT, Lareu RR, Dix BR, Hughes JD (2017) Statins: antimicrobial resistance breakers or makers?. Flores-Valdez MA. PeerJ 5:e3952

Kolosova NG, Stefanova NA, Muraleva NA, Skulachev VP (2012) The mitochondria-targeted antioxidant SkQ1 but not N-acetylcysteine reverses ageing-related biomarkers in rats. Ageing (Albany NY) 4(10):686–694

Kong P, Wu R, Liu X et al (2016) The effects of anti-inflammatory drug treatment in gastric cancer prevention: an update of a meta-analysis. J Cancer 7(15):2247–2257

Konings IR, Verweij J, Wiemer EA, Sleijfer S (2009) The applicability of mTOR inhibition in solid tumors. Curr Cancer Drug Targets 9(3):439–450

Krackhardt F, Kočka V, Waliszewski MW, Utech A, Lustermann M, Hudec M, Studenčan M, Schwefer M, Yu J, Jeong MH, Ahn T, Wan Ahmad WA, Boxberger M, Schneider A, Leschke M (2017) Polymer-free sirolimus-eluting stents in a large-scale all-comers population. Open Heart 4:e000592

Kraig E, Linehan LA, Liang H, Romo TQ, Liu Q, Wu Y, Benavides AD, Curiel TJ, Javors MA, Musi N, Chiodo L, Koek W, Gelfond JAL, Kellogg DL (2018) A randomized control trial to establish the feasibility and safety of rapamycin treatment in an older human cohort: immunological, physical performance, and cognitive effects. Exp Gerontol. pii: S0531-5565(17)30913-0

Kuan YC, Huang KW, Lin CL, Hu CJ, Kao CH (2017) Effects of metformin exposure on neurodegenerative diseases in elderly patients with type 2 diabetes mellitus. Prog Neuropsychopharmacol Biol Psychiatry 79(Pt B):77–83

Kunchandy E, Rao MNA (1990) Oxygen radical scavenging activity of curcumin. Int J Pharm 58:237–240

Kurdi A, Martinet W, De Meyer GRY (2018) mTOR inhibition and cardiovascular diseases: dyslipidemia and atherosclerosis. Transplantation 102(2S Suppl 1):S44–S46

Lafeber M, Spiering W, Visseren FLJ, Grobbee DE (2016) Multifactorial prevention of cardiovascular disease in patients with hypertension: the cardiovascular polypill. Curr Hypertens Rep 18:40

Lamming DW, Ye L, Sabatini DM, Baur JA (2013) Rapalogs and mTOR inhibitors as anti-ageing therapeutics. J Clin Invest 123(3):980–989

Lee KS, Lee BS, Semnnani S, Avanesian A, Um CY, Jeon HJ, Seong KM, Yu K, Min KJ, Jafari M (2010a) Curcumin extends life span, improves health span, and modulates the expression of age-associated ageing genes in *Drosophila melanogaster*. Rejuvenation Res 13(5):561–570

Lee SJ, Hwang AB, Kenyon C (2010b) Inhibition of respiration extends *C. elegans* life span via reactive oxygen species that increase HIF-1 activity. Curr Biol 20:2131–2136

Leggio M, Bendini MG, Caldarone E, Lombardi M, Severi P, D'Emidio S, Stavri DC, Armeni M, Bravi V, Mazza A (2017) Low-dose aspirin for primary prevention of cardiovascular events in patients with diabetes: benefit or risk? Diabetes Metab 44(3):217–225

Lei Y, Yi Y, Liu Y, Liu X, Keller ET, Qian CN, Zhang J, Lu Y (2017) Metformin targets multiple signaling pathways in cancer. Chin J Cancer 36:17

Leya M, Stone NJ (2017) Statin prescribing in the elderly: special considerations. Curr Atheroscler Rep 19(11):47

Li BH, Ma XF, Wang Y, Tian WX (2005) Structure-activity relationship of polyphenols that inhibit fatty acid synthase. J Biochem (Tokyo) 138:679–685

Li Y, Liu L, Wang B, Wang J, Chen D (2013) Metformin in non-alcoholic fatty liver disease: a systematic review and meta-analysis. Biomed Rep 1(1):57–64

Li J, Kim SG, Blenis J (2014) Rapamycin: one drug, many effects. Cell Metab 19(3):373–379

Li P, Wu H, Zhang H, Shi Y, Xu J, Ye Y, Xia D, Yang J, Cai J, Wu Y (2015) Aspirin use after diagnosis but not prediagnosis improves established colorectal cancer survival: a meta-analysis. Gut 64(9):1419–1425

Li X, Li T, Liu Z, Gou S, Wang C (2017) The effect of metformin on survival of patients with pancreatic cancer: a meta-analysis. Sci Rep 7:5825

Li GM, Zhao J, Li B, Zhang XF, Ma JX, Ma XL, Liu J (2018) The anti-inflammatory effects of statins on patients with rheumatoid arthritis: a systemic review and meta-analysis of 15 randomized controlled trials. Autoimmun Rev 17(3):215–225

Lin K, Dorman JB, Rodan A, Kenyon C (1997) daf-16: an HNF-3/forkhead family member that can function to double the life-span of *Caenorhabditis elegans*. Science 278:1319–1322

Lin WS, Chen JY, Wang JC, Chen LY, Lin CH, Hsieh TR, Wang MF, Fu TF, Wang PY (2014) The anti-ageing effects of *Ludwigia octovalvis* on *Drosophila melanogaster* and SAMP8 mice. Age (Dordr) 36(2):689–703

Ling Y, Yang L, Huang H, Hu X, Zhao C, Huang H, Ying Y (2015) Prognostic significance of statin use in colorectal cancer: a systematic review and meta-analysis. Medicine (Baltimore) 94(25):e908

Liu Y, Chen J-Q, Xie L, Wang J, Li T, He Y, Gao Y, Qin X, Li S (2014a) Effect of aspirin and other non-steroidal anti-inflammatory drugs on prostate cancer incidence and mortality: a systematic review and meta-analysis. BMC Med 12:55

Liu YM, Shao YQ, He Q (2014b) Sirolimus for treatment of autosomal-dominant polycystic kidney disease: a meta-analysis of randomized controlled trials. Transplant Proc 46(1):66–74

Liu B, Yi Z, Guan X, Zeng YX, Ma F (2017a) The relationship between statins and breast cancer prognosis varies by statin type and exposure time: a meta-analysis. Breast Cancer Res Treat 164(1):1–11

Liu J, Liu D, Li J, Zhu L, Zhang C, Lei K, Xu Q, You R (2017b) Efficacy and safety of everolimus for maintenance immunosuppression of kidney transplantation: a meta-analysis of randomized controlled trials. PLoS One 12(1):e0170246

Lockrow J, Prakasam A, Huang P, Bimonte-Nelson H, Sambamurti K, Granholm AC (2009) Cholinergic degeneration and memory loss delayed by vitamin E in a Down syndrome mouse model. Exp Neurol 216(2):278–289

Lu L, Shi L, Zeng J, Wen Z (2017) Aspirin as a potential modality for the chemoprevention of breast cancer: a dose-response meta-analysis of cohort studies from 857,831 participants. Oncotarget 8(25):40389–40401

Lucanic M, Lithgow GJ, Alavez S (2013) Pharmacological lifespan extension of invertebrates. Ageing Res Rev 12(1):445–458

Luo T, Yan HM, He P, Luo Y, Yang YF, Zheng H (2012) Aspirin use and breast cancer risk: a meta-analysis. Breast Cancer Res Treat 131(2):581–587

Luong N, Davies CR, Wessells RJ, Graham SM, King MT, Veech R, Bodmer R, Oldham SM (2006) Activated FOXO-mediated insulin resistance is blocked by reduction of TOR activity. Cell Metab 4(2):133–142

Ma TC, Buescher JL, Oatis B, Funk JA, Nash AJ, Carrier RL, Hoyt KR (2007) Metformin therapy in a transgenic mouse model of Huntington's disease. Neurosci Lett 411:98–103

Maher RL, Hanlon JT, Hajjar ER (2014) Clinical consequences of polypharmacy in elderly. Expert Opin Drug Saf 13(1):57–65

Mai QG, Zhang ZM, Xu S, Lu M, Zhou RP, Zhao L, Jia CH, Wen ZH, Jin DD, Bai XC (2011) Metformin stimulates osteoprotegerin and reduces RANKL expression in osteoblasts and ovariectomized rats. J Cell Biochem 112:2902–2909

Manthravadi S, Shrestha A, Madhusudhana S (2016) Impact of statin use on cancer recurrence and mortality in breast cancer: a systematic review and meta-analysis. Int J Cancer 139(6):1281–1288

Markaki M, Tavernarakis N (2010) Modeling human diseases in *Caenorhabditis elegans*. Biotechnol J 5(12):1261–1276

Martinet W, De Loof H, De Meyer GR (2014) mTOR inhibition: a promising strategy for stabilization of atherosclerotic plaques. Atherosclerosis 233(2):601–607

Martin-Montalvo A, Mercken EM, Mitchell SJ, Palacios HH, Mote PL, Scheibye-Knudsen M, Gomes AP, Ward TM, Minor RK, Blouin MJ, Schwab M, Pollak M, Zhang Y, Yu Y, Becker KG, Bohr VA, Ingram DK, Sinclair DA, Wolf NS, Spindler SR, Bernier M, de Cabo R (2013) Metformin improves healthspan and lifespan in mice. Nat Commun 4:2192

Massoud O, Wiesner RH (2012) The use of sirolimus should be restricted in liver transplantation. J Hepatol 56(1):288–290

Mattson MP (2008) Hormesis and disease resistance: activation of cellular stress response pathways. Hum Exp Toxicol 27(2):155–162

McColl G, Killilea DW, Hubbard AE, Vantipalli MC, Melov S, Lithgow GJ (2008) Pharmacogenetic analysis of lithium-induced delayed ageing in *Caenorhabditis elegans*. J Biol Chem 283(1):350–357

McDonald, Maizi BM, Arking R (2013) Chemical regulation of mid- and late-life longevities in *Drosophila*. Exp Gerontol 48:240–249

McNeil JJ, Woods RL, Nelson MR, Murray AM, Reid CM, Kirpach B, Storey E, Shah RC, Wolfe RS, Tonkin AM, Newman AB, Williamson JD, Lockery JE, Margolis KL, Ernst ME, Abhayaratna WP, Stocks N, Fitzgerald SM, Trevaks RE, Orchard SG, Beilin LJ, Donnan GA, Gibbs P, Johnston CI, Grimm RH, ASPREE Investigator Group (2017) Baseline characteristics of participants in the ASPREE (ASPirin in Reducing Events in the Elderly) study. J Gerontol A Biol Sci Med Sci 72(11):1586–1593

McPhee JS, French DP, Jackson D, Nazroo J, Pendleton N, Degens H (2016) Physical activity in older age: perspectives for healthy ageing and frailty. Biogerontology 17:567–580

Mehrbod P, Omar AR, Hair-Bejo M, Haghani A, Ideris A (2014) Mechanisms of action and efficacy of statins against influenza. Biomed Res Int 2014:872370

Meireles CG, Pereira SA, Valadares LP, Rêgo DF, Simeoni LA, Guerra ENS, Lofrano-Porto A (2017) Effects of metformin on endometrial cancer: systematic review and meta-analysis. Gynecol Oncol 147(1):167–180

Melov S, Ravenscroft J, Malik S, Gill MS, Walker DW, Clayton PE, Wallace DC, Malfroy B, Doctrow SR, Lithgow GJ (2000) Extension of life-span with superoxide dismutase/catalase mimetics. Science 289(5484):1567–1569

Memisogullari R, Gümüştekin K, Dane Ş, Akçay F (2008) Effect of preinjury supplementation of selenium or vitamin E on lipid peroxidation and antioxidant enzymes in burn injury. Turk J Med Sci 38(6):545–548

Meng F, Song L, Wang W (2017) Metformin improves overall survival of colorectal cancer patients with diabetes: a meta-analysis. J Diabetes Res 2017:5063239

Messow CM, Isles C (2017) Meta-analysis of statins in chronic kidney disease: who benefits? QJM 110(8):493–500

Millburn GH, Crosby MA, Gramates LS, Tweedie S, FlyBase Consortium (2016) FlyBase portals to human disease research using *Drosophila* models. Dis Model Mech 9(3):245–252

Miller RA, Harrison DE, Astle CM, Baur JA, Boyd AR, de Cabo R, Fernandez E, Flurkey K, Javors MA, Nelson JF, Orihuela CJ, Pletcher S, Sharp ZD, Sinclair D, Starnes JW, Wilkinson JE, Nadon NL, Strong R (2011) Rapamycin, but not resveratrol or simvastatin, extends life span of genetically heterogeneous mice. J Gerontol A Biol Sci Med Sci 66(2):191–201

Mills EJ, Wu P, Alberton M, Kanters S, Lanas A, Lester R (2012) Low-dose aspirin and cancer mortality: a meta-analysis of randomized trials. Am J Med 125(6):560–567

Miquel J, Fleming J, Economos AC (1982) Antioxidants, metabolic rate and ageing in *Drosophila*. Arch Gerontol Geriatr 1(2):159–165

Mortensen MB, Falk E (2018) Primary prevention with statins in the elderly. J Am Coll Cardiol 71(1):85–94

Moskalev A, Chernyagina E, Tsvetkov V, Fedintsev A, Shaposhnikov M, Krut'ko V, Zhavoronkov A, Kennedy BK (2016) Developing criteria for evaluation of geroprotectors as a key stage toward translation to the clinic. Ageing Cell 15(3):407–415

Moskalev A, Chernyagina E, Kudryavtseva A, Shaposhnikov M (2017) Geroprotectors: a unified concept and screening approaches. Ageing Dis 8(3):354–363

Muntoni S (1999) Metformin and fatty acids. Diabetes Care 22(1):179–180

Musi N, Hirshman MF, Nygren J, Svanfeldt M, Bavenholm P, Rooyackers O, Zhou G, Williamson JM, Ljunqvist O, Efendic S, Moller DE, Thorell A, Goodyear LJ (2002) Metformin increases AMP-activated protein kinase activity in skeletal muscle of subjects with type 2 diabetes. Diabetes 51(7):2074–2081

Muzumdar R, Allison DB, Huffman DM, Ma X, Atzmon G, Einstein FH, Fishman S, Poduval AD, McVei T, Keith SW, Barzilai N (2008) Visceral adipose tissue modulates mammalian longevity. Ageing Cell 7(3):438–440

Naci H, Brugts JJ, Fleurence R, Tsoi B, Toor H, Ades AE (2013) Comparative benefits of statins in the primary and secondary prevention of major coronary events and all-cause mortality: a network meta-analysis of placebo-controlled and active-comparator trials. Eur J Prev Cardiol 20(4):641–657

Nakashima K, Ishihara T, Yokota O, Terada S, Trojanowski JQ, Lee VM, Kuroda S (2004) Effects of alpha-tocopherol on an animal model of tauopathies. Free Radic Biol Med 37(2):176–186

Newman JC, Verdin E (2014) β-hydroxybutyrate: much more than a metabolite. Diabetes Res Clin Pract 106:173–181

Newman JC, Milman S, Hashmi SK, Austad SN, Kirkland JL, Halter JB, Barzilai N (2016) Strategies and challenges in clinical trials targeting human ageing. J Gerontol Ser A Biol Med Sci 71(11):1424–1434

Ng TP, Feng L, Yap KB, Lee TS, Tan CH, Winblad B (2014) Long-term metformin usage and cognitive function among older adults with diabetes. J Alzheimers Dis 41(1):61–68

Nony PA, Kennett SB, Glasgow WC, Olden K, Roberts JD (2005) 15SLipoxygenase- 2 mediates arachidonic acid-stimulated adhesion of human breast carcinoma cells through the activation of TAK1, MKK6, and p38 MAPK. J Biol Chem 280:31413–31419

Oberdoerffer P, Michan S, McVay M, Mostoslavsky R, Vann J, Park SK, Hartlerode A, Stegmuller J, Hafner A, Loerch P, Wright SM, Mills KD, Bonni A, Yankner BA, Scully R, Prolla TA, Alt FW, Sinclair DA (2008) SIRT1 redistribution on chromatin promotes genomic stability but alters gene expression during ageing. Cell 135(5):907–918

Oesterle A, Laufs U, Liao JK (2017) Pleiotropic effects of statins on the cardiovascular system. Circ Res 120(1):229–243

Onken B, Driscoll M (2010) Metformin induces a dietary restriction-like state and the oxidative stress response to extend C. elegans healthspan via AMPK, LKB1, and SKN-1. PLoS One 5(1):e8758

Opelz G, Unterrainer C, Süsal C, Döhler B (2016) Immunosuppression with mammalian target of rapamycin inhibitor and incidence of post-transplant cancer in kidney transplant recipients. Nephrol Dial Transplant 31(8):1360–1367

Orkaby AR, Gaziano JM, Djousse L, Driver JA (2017) Statins for primary prevention of cardiovascular events and mortality in older men. J Am Geriatr Soc 65(11):2362–2368. 15

Ornelas A, Zacharias-Millward N, Menter DG, Davis JS, Lichtenberger L, Hawke D, Hawk E, Vilar E, Bhattacharya P, Millward S (2017) Beyond COX-1: the effects of aspirin on platelet biology and potential mechanisms of chemoprevention. Cancer Metastasis Rev 36(2):289–303

Ott C, Jacobs K, Haucke E, Navarrete Santos A, Grune T, Simm A (2014) Role of advanced glycation end products in cellular signaling. Redox Biol 2:411–429

Otto CM (2016) Statins for primary prevention of cardiovascular disease. BMJ 355:i6334

Oxenkrug G, Navrotskaya V, Vorobyova L, Summergrad P (2012) Minocycline effect on life and health span of Drosophila melanogaster. Ageing Dis 3(5):352–359

Paleari L, Puntoni M, Clavarezza M, DeCensi M, Cuzick J, DeCensi A (2016) PIK3CA mutation, aspirin use after diagnosis and survival of colorectal cancer. A systematic review and meta-analysis of epidemiological studies. Clin Oncol (R Coll Radiol) 28(5):317–326

Palmer SC, Mavridis D, Nicolucci A, Johnson DW, Tonelli M, Craig JC, Maggo J, Gray V, De Berardis G, Ruospo M, Natale P, Saglimbene V, Badve SV, Cho Y, Nadeau-Fredette AC, Burke M, Faruque L, Lloyd A, Ahmad N, Liu Y, Tiv S, Wiebe N, Strippoli GF (2016) Comparison of clinical outcomes and adverse events associated with glucose-lowering drugs in patients with type 2 diabetes: a meta-analysis. JAMA 316:313–324

Papanagnou P, Stivarou T, Papageorgiou I, Papadopoulos GE, Pappas A (2017) Marketed drugs used for the management of hypercholesterolemia as anticancer armament. Oncol Targets Ther 10:4393–4411

Papanas N, Maltezos E, Mikhailidis DP (2012) Metformin and heart failure: never say never again. Expert Opin Pharmacother 13(1):1–8

Park Y, Pariza MW (2001) Lipoxygenase inhibitors inhibit heparin-releasable lipoprotein lipase activity in 3T3-L1 adipocytes and enhance body fat reduction in mice by conjugatedlinoleic acid. Biochim Biophys Acta 1534:27–33

Pasyukova EG, Vaiserman AM (2017) HDAC inhibitors: a new promising drug class in anti-ageing research. Mech Ageing Dev 166:6–15

Patel R, Shah G (2017) Effect of metformin on clinical, metabolic and endocrine outcomes in women with polycystic ovary syndrome: a meta-analysis of randomized controlled trials. Curr Med Res Opin 33(9):1545–1557

Patrignani P, Patrono C (2016) Aspirin and cancer. J Am Coll Cardiol 68(9):967–976

Pearson KJ, Baur JA, Lewis KN, Peshkin L, Price NL, Labinskyy N, Swindell WR, Kamara D, Minor RK, Perez E, Jamieson HA, Zhang Y, Dunn SR, Sharma K, Pleshko N, Woollett LA, Csiszar A, Ikeno Y, Le Couteur D, Elliott PJ, Becker KG, Navas P, Ingram DK, Wolf NS, Ungvari Z, Sinclair DA, de Cabo R (2008) Resveratrol delays age-related deterioration and mimics transcriptional aspects of dietary restriction without extending life span. Cell Metab 8(2):157–168

Perluigi M, Di Domenico F, Butterfield DA (2015) mTOR signaling in ageing and neurodegeneration: at the crossroad between metabolism dysfunction and impairment of autophagy. Neurobiol Dis 84:39–49

Phillips T, Leeuwenburgh C (2004) Lifelong aspirin supplementation as a means to extending life span. Rejuvenation Res 7:243–251

Pietsch K, Saul N, Menzel R, Stürzenbaum SR, Steinberg CE (2009) Quercetin mediated lifespan extension in Caenorhabditis elegans is modulated by age-1, daf-2, sek-1 and unc-43. Biogerontology 10(5):565–578

Poole JC, Thain A, Perkins ND, Roninson IB (2004) Induction of transcription by p21Waf1/Cip1/Sdi1: role of NFkappaB and effect of non-steroidal anti-inflammatory drugs. Cell Cycle 3(7):931–940

Porta EA, Joun NS, Nitta RT (1980) Effects of the type of dietary fat at two levels of vitamin E in Wistar male rats during development and ageing. I. Life span, serum biochemical parameters and pathological changes. Mech Ageing Dev 13(1):1–39

Pouwels KB, Widyakusuma NN, Bos JHJ, Hak E (2016) Association between statins and infections among patients with diabetes: a cohort and prescription sequence symmetry analysis. Pharmacoepidemiol Drug Saf 25:1124–1130

Qin X, Dong H, Fang K, Lu F (2017) The effect of statins on renal outcomes in patients with diabetic kidney disease: a systematic review and meta-analysis. Diabetes Metab Res Rev 33(6)

Raju N, Sobieraj-Teague M, Bosch J, Eikelboom JW (2016) Updated meta-analysis of aspirin in primary prevention of cardiovascular disease. Am J Med 129(5):35–36

Rangaraju S, Solis GM, Andersson SI, Gomez-Amaro RL, Kardakaris R, Broaddus CD, Niculescu AB 3rd, Petrascheck M (2015) Atypical antidepressants extend lifespan of Caenorhabditis elegans by activation of a non-cell-autonomous stress response. Ageing Cell 14(6):971–981

Reed MJ, Meszaros K, Entes LJ, Claypool MD, Pinkett JG, Brignetti D, Luo J, Khandwala A, Reaven GM (1999) Effect of masoprocol on carbohydrate and lipid metabolism in a rat model of type II diabetes. Diabetologia 42:102–106

Ridker PM (2016) A test in context: high-sensitivity C-reactive protein. J Am Coll Cardiol 67(6):712–723

Ritz MF, Curin Y, Mendelowitsch A, Andriantsitohaina R (2008a) Acute treatment with red wine polyphenols protects from ischemia-induced excitotoxicity, energy failure and oxidative stress in rats. Brain Res 1239:226–234. A

Ritz MF, Ratajczak P, Curin Y, Cam E, Mendelowitsch A, Pinet F, Andriantsitohaina R (2008b) Chronic treatment with red wine polyphenol compounds mediates neuroprotection in a rat model of ischemic cerebral stroke. J Nutr 138:519–525

Robida-Stubbs S, Glover-Cutter K, Lamming DW, Mizunuma M, Narasimhan SD, Neumann-Haefelin E, Sabatini DM, Blackwell TK (2012) TOR signaling and rapamycin influence longevity by regulating SKN-1/Nrf and DAF-16/FoxO. Cell Metab 15(5):713–724

Roffi M, Patrono C, Collet JP, Mueller C, Valgimigli M, Andreotti F, Bax JJ, Borger MA, Brotons C, Chew DP, Gencer B, Hasenfuss G, Kjeldsen K, Lancellotti P, Landmesser U, Mehilli J, Mukherjee D, Storey RF, Windecker S, Baumgartner H, Gaemperli O, Achenbach S, Agewall S, Badimon L, Baigent C, Bueno H, Bugiardini R, Carerj S, Casselman F, Cuisset T, Erol Ç, Fitzsimons D, Halle M, Hamm C, Hildick-Smith D, Huber K, Iliodromitis E, James S, Lewis BS, Lip GY, Piepoli MF, Richter D, Rosemann T, Sechtem U, Steg PG, Vrints C, Luis Zamorano J, Management of Acute Coronary Syndromes in Patients Presenting without Persistent ST-Segment Elevation of the European Society of Cardiology (2016) 2015 ESC guidelines for the management of acute coronary syndromes in patients presenting without persistent ST-segment elevation: task force for the management of acute coronary syndromes in patients presenting without persistent ST-segment elevation of the European Society of Cardiology (ESC). Eur Heart J 37(3):267–315

Rojas LBA, Gomes MB (2013) Metformin: an old but still the best treatment for type 2 diabetes. Diabetol Metab Syndr 5:6

Rostami Z, Moteshaker Arani M, Salesi M, Safiabadi M, Einollahi B (2017) Effect of statins on patients and graft survival in kidney transplant recipients: a survival meta-analysis. Iran J Kidney Dis 11(5):329–338

Rotundo MS, Galeano T, Tassone P, Tagliaferri P (2016) mTOR inhibitors, a new era for metastatic luminal HER2-negative breast cancer? A systematic review and a meta-analysis of randomized trials. Oncotarget 7(19):27055–27066

RxList: The Internet Drug Index. Rapamune (Sirolimus, Last reviewed on RxList: 4/21/2017). https://www.rxlist.com/rapamune-drug.htm

Sander M, Oxlund B, Jespersen A, Krasnik A, Mortensen EL, Westendorp RG, Rasmussen LJ (2015) The challenges of human population ageing. Age Ageing 44(2):185–187

Sanguankeo A, Upala S, Cheungpasitporn W, Ungprasert P, Knight EL (2015) Effects of statins on renal outcome in chronic kidney disease patients: a systematic review and meta-analysis. PLoS One 10(7):e0132970

Sasongko TH, Ismail NFD, Zabidi-Hussin ZAMH (2016) Rapamycin and rapalogs for tuberous sclerosis complex. Cochrane Database Syst Rev 7:CD011272

Schäfer I, Kaduszkiewicz H, Wagner H-O, Schön G, Scherer M, van den Bussche H (2014) Reducing complexity: a visualisation of multimorbidity by combining disease clusters and triads. BMC Public Health 14:1285

Schlender L, Martinez YV, Adeniji C, Reeves D, Faller B, Sommerauer C, Al Qur'an T, Woodham A, Kunnamo I, Sönnichsen A, Renom-Guiteras A (2017) Efficacy and safety of metformin in the management of type 2 diabetes mellitus in older adults: a systematic review for the development of recommendations to reduce potentially inappropriate prescribing. BMC Geriatr 17(Suppl 1):227

Schriner SE, Lee K, Truong S, Salvadora KT, Maler S, Nam A, Lee T, Jafari M (2013) Extension of *Drosophila* lifespan by *Rhodiola rosea* through a mechanism independent from dietary restriction. PLoS One 8(5):e63886

Seifarth CB, Schehler H, Schneider J (2013) Effectiveness of metformin on weight loss in nondiabetic individuals with obesity. Exp Clin Endocrinol Diabetes 121:27–31

Shen LR, Xiao F, Yuan P, Chen Y, Gao QK, Parnell LD, Meydani M, Ordovas JM, Li D, Lai CQ (2013) Curcumin-supplemented diets increase superoxide dismutase activity and mean lifespan in Drosophila. Age (Dordr) 35(4):1133–1142

Shi X, Min D, Zigang D, Fei C, Jiangping Y, Suwei W, Stephen S, Leonard Vince C, Val V (1999) Antioxidant properties of aspirin: characterization of the ability of aspirin to inhibit silica-induced lipid peroxidation, DNA damage, NF-+¦B activation, and TNF-+¦ production. Mol Cell Biochem 199:93–102

Shimomura M, Oyama J, Takeuchi M, Shibata Y, Yamamoto Y, Kawasaki T, Komoda H, Kodama K, Sakuma M, Toyoda S, Inoue Y, Mine D, Natsuaki M, Komatsu A, Hikichi Y, Yamagishi S, Inoue T, Node K (2016) Acute effects of statin on reduction of angiopoietin-like 2 and glyceraldehyde-derived advanced glycation end-products levels in patients with acute myocardial infarction: a message from SAMIT (Statin for Acute Myocardial Infarction Trial). Heart Vessel 31(10):1583–1589

Shishido Y, Furushiro M, Hashimoto S, Yokokura T (2001) Effect of nordihydroguaiaretic acid on behavioral impairment and neuronal cell death after forebrain ischemia. Pharmacol Biochem Behav 69(3–4):469–474

Siskind DJ, Leung J, Russell AW, Wysoczanski D, Kisely S (2016) Metformin for clozapine associated obesity: a systematic review and meta-analysis. PLoS One 11(6):e01562085

Slack C, Foley A, Partridge L (2012) Activation of AMPK by the putative dietary restriction mimetic metformin is insufficient to extend lifespan in *Drosophila*. PLoS One 7(10):e47699

Smith DL Jr, Elam CF Jr, Mattison JA, Lane MA, Roth GS, Ingram DK, Allison DB (2010) Metformin supplementation and life span in Fischer-344 rats. J Gerontol A Biol Sci Med Sci 65(5):468–474

Smith SM, Wallace E, O'Dowd T, Fortin M (2016) Interventions for improving outcomes in patients with multimorbidity in primary care and community settings. Cochrane Database Syst Rev 3:CD006560

Song C, Zhu C, Wu Q, Qi J, Gao Y, Zhang Z, Gaur U, Yang D, Fan X, Yang M (2017) Metabolome analysis of effect of aspirin on *Drosophila* lifespan extension. Exp Gerontol 95:54–62

Stark LA, Reid K, Sansom OJ, Din FV, Guichard S, Mayer I, Jodrell DI, Clarke AR, Dunlop MG (2007) Aspirin activates the NF-kappaB signalling pathway and induces apoptosis in intestinal neoplasia in two in vivo models of human colorectal cancer. Carcinogenesis 28:968–976

Stevens RJ, Ali R, Bankhead CR, Bethel MA, Cairns BJ, Camisasca RP, Crowe FL, Farmer AJ, Harrison S, Hirst JA, Home P, Kahn SE, McLellan JH, Perera R, Plüddemann A, Ramachandran A, Roberts NW, Rose PW, Schweizer A, Viberti G, Holman RR (2012) Cancer outcomes and all-cause mortality in adults allocated to metformin: systematic review and collaborative meta-analysis of randomised clinical trials. Diabetologia 55(10):2593–2603

Strandberg TE, Kolehmainen L, Vuorio A (2014) Evaluation and treatment of older patients with hypercholesterolemia: a clinical review. JAMA 312:1136–1144

Strong R, Miller RA, Astle CM, Floyd RA, Flurkey K, Hensley KL, Javors MA, Leeuwenburgh C, Nelson JF, Ongini E, Nadon NL, Warner HR, Harrison DE (2008) Nordihydroguaiaretic acid and aspirin increase lifespan of genetically heterogeneous male mice. Ageing Cell 7(5):641–650

Strong R, Miller RA, Astle CM, Baur JA, de Cabo R, Fernandez E, Guo W, Javors M, Kirkland JL, Nelson JF, Sinclair DA, Teter B, Williams D, Zaveri N, Nadon NL, Harrison DE (2013) Evaluation of resveratrol, green tea extract, curcumin, oxaloacetic acid, and medium-chain triglyceride oil on life span of genetically heterogeneous mice. J Gerontol A Biol Sci Med Sci 68(1):6–16

Su X, Zhang L, Lv J, Wang J, Hou W, Xie X, Zhang H (2016) Effect of statins on kidney disease outcomes: a systematic review and meta-analysis. Am J Kidney Dis 67(6):881–892

Suckow BK, Suckow MA (2006) Lifespan extension by the antioxidant curcumin in *Drosophila*. Int J Biomed Sci 2(4):401–404

Sumantri S, Setiati S, Purnamasari D, Dewiasty E (2014) Relationship between metformin and frailty syndrome in elderly people with type 2 diabetes. Acta Med Indones 46(3):183–188

Tang YL, Zhu LY, Li Y, Yu J, Wang J, Zeng XX, Hu KX, Liu JY, Xu JX (2017) Metformin use is associated with reduced incidence and improved survival of endometrial cancer: a meta-analysis. Biomed Res Int 2017:5905384

Tao LJ, Sun H, Zhao YM, Yuan ZG, Li XX, Huang BQ (2004) Trichostatin A extends the lifespan of *Drosophila melanogaster* by elevating hsp22 expression. Acta Biochim Biophys Sin (Shanghai) 36:618–622

Taylor F, Huffman MD, Macedo AF, Moore TH, Burke M, Davey Smith G, Ward K, Ebrahim S (2013) Statins for the primary prevention of cardiovascular disease. Cochrane Database Syst Rev 1:CD004816

Thapa RK, Nguyen HT, Jeong JH, Kim JR, Choi HG, Yong CS, Kim JO (2017) Progressive slow-down/prevention of cellular senescence by CD9-targeted delivery of rapamycin using lactose-wrapped calcium carbonate nanoparticles. Sci Rep 7:43299

Thoonsen H, Richard E, Bentham P, Gray R, van Geloven N, De Haan RJ, Van Gool WA, Nederkoorn PJ (2010) Aspirin in Alzheimer's disease: increased risk of intracerebral hemorrhage: cause for concern? Stroke 41(11):2690–2692

Tolosa MJ, Chuguransky SR, Sedlinsky C, Schurman L, McCarthy AD, Molinuevo MS, Cortizo AM (2013) Insulin-deficient diabetes-induced bone microarchitecture alterations are associated with a decrease in the osteogenic potential of bone marrow progenitor cells: preventive effects of metformin. Diabetes Res Clin Pract 101:177–186

Trauer J, Muhi S, McBryde ES, Al Harbi SA, Arabi YM, Boyle AJ, Cartin-Ceba R, Chen W, Chen YT, Falcone M, Gajic O, Godsell J, Gong MN, Kor D, Lösche W, McAuley DF, O'Neal HR Jr, Osthoff M, Otto GP, Sossdorf M, Tsai MJ, Valerio-Rojas JC, van der Poll T, Violi F, Ware L, Widmer AF, Wiewel MA, Winning J, Eisen DP (2017) Quantifying the effects of prior acetylsalicylic acid on sepsis-related deaths: an individual patient data meta-analysis using propensity matching. Crit Care Med 45(11):1871–1879

Tseng CH (2016) Metformin reduces gastric cancer risk in patients with type 2 diabetes mellitus. Ageing (Albany NY) 8(8):1636–1649

Tsilidis KK, Kasimis JC, Lopez DS, Ntzani EE, Ioannidis JP (2015) Type 2 diabetes and cancer: umbrella review of meta-analyses of observational studies. BMJ 350:g7607

Tuttle RS, Yager J, Northrup N (1988) Age and the antihypertensive effect of aspirin in rats. Br J Pharmacol 94(3):755–758

Ugur B, Chen K, Bellen HJ (2016) *Drosophila* tools and assays for the study of human diseases. Dis Model Mech 9(3):235–244

Undela K, Shah CS, Mothe RK (2017) Statin use and risk of cancer: an overview of meta-analyses. World J Meta-Anal 5(2):41–53

Vaiserman AM (2011) Hormesis and epigenetics: is there a link? Ageing Res Rev 10(4):413–421

Vaiserman A, Lushchak O (2017a) Implementation of longevity-promoting supplements and medications in public health practice: achievements, challenges and future perspectives. J Transl Med 15(1):160

Vaiserman AM, Lushchak OV (2017b) Anti-ageing drugs: where are we and where are we going? RSC Drug Discov Ser 2017(57):1–10

Vaiserman A, Koliada AK, Koshel' NM, Simonenko AV, Pasiukova EG (2013a) Effect of the histone deacetylase inhibitor sodium butyrate on the viability and lifespan in *Drosophila melanogaster*. Adv Gerontol 3:30–34

Vaiserman et al (2013b) Determination of geroprotective potential of sodium butyrate in *Drosophila melanogaster*: long-term effects. Adv Gerontol 26:111–116. In Russian

Vaiserman AM, Lushchak OV, Koliada AK (2016) Anti-ageing pharmacology: promises and pitfalls. Ageing Res Rev 31:9–35

van den Hoek HL, Bos WJ, de Boer A, van de Garde EM (2011) Statins and prevention of infections: systematic review and meta-analysis of data from large randomised placebo controlled trials. BMJ 343:d7281

Van Wyhe RD, Rahal OM, Woodward WA (2017) Effect of statins on breast cancer recurrence and mortality: a review. Breast Cancer Targets Ther 9:559–565

Vane SJ (2000) Aspirin and other anti-inflammatory drugs. Thorax 55:35–39

Veronese N, Stubbs B, Maggi S, Thompson T, Schofield P, Muller C, Tseng PT, Lin PY, Carvalho AF, Solmi M (2017) Low-dose aspirin use and cognitive function in older age: a systematic review and meta-analysis. J Am Geriatr Soc 65(8):1763–1768

Violan C, Foguet-Boreu Q, Flores-Mateo G, Salisbury C, Blom J, Freitag M, Glynn L, Muth C, Valderas JM (2014) Prevalence, determinants and patterns of multimorbidity in primary care: a systematic review of observational studies. PLoS One 9(7):e102149

Wagner D, Kniepeiss D, Schaffellner S, Jakoby E, Mueller H, Fahrleitner-Pammer A, Stiegler P, Tscheliessnigg KH, Iberer F (2010) Sirolimus has a potential to influent viral recurrence in HCV positive liver transplant candidates. Int Immunopharmacol 10(8):990–993

Wan QL, Zheng SQ, Wu GS, Luo HR (2013) Aspirin extends the lifespan of *Caenorhabditis elegans* via AMPK and DAF-16/FOXO in dietary restriction pathway. Exp Gerontol 48(5):499–506

Wang D, Qian L, Xiong H, Liu J, Neckameyer WS, Oldham S, Xia K, Wang J, Bodmer R, Zhang Z (2006) Antioxidants protect PINK1-dependent dopaminergic neurons in *Drosophila*. Proc Natl Acad Sci U S A 103(36):13520–13525

Wang C, Wheeler CT, Alberico T, Sun X, Seeberger J, Laslo M, Spangler E, Kern B, de Cabo R, Zou S (2013) The effect of resveratrol on lifespan depends on both gender and dietary nutrient composition in *Drosophila melanogaster*. Age (Dordr) 35(1):69–81

Wang CP, Lorenzo C, Espinoza SE (2014) Frailty attenuates the impact of metformin on reducing mortality in older adults with type 2 diabetes. J Endocrinol Diabetes Obes 2(2):1031

Wang HL, Sun ZO, Rehman RU, Wang H, Wang YF, Wang H (2017a) Rosemary extract-mediated lifespan extension and attenuated oxidative damage in *Drosophila melanogaster* fed on high-fat diet. J Food Sci 82(4):1006–1011

Wang Y, Xiong J, Niu M, Chen X, Gao L, Wu Q, Zheng K, Xu K (2017b) Statins and the risk of cirrhosis in hepatitis B or C patients: a systematic review and dose-response meta-analysis of observational studies. Oncotarget 8(35):59666–59676

Warner TD, Nylander S, Whatling C (2011) Anti-platelet therapy: cyclo-oxygenase inhibition and the use of aspirin with particular regard to dual anti-platelet therapy. Br J Clin Pharmacol 72(4):619–633

Weissmann G (1991) Aspirin. Sci Am 264:84–90

Wiegant FA, Surinova S, Ytsma E, Langelaar-Makkinje M, Wikman G, Post JA (2009) Plant adaptogens increase lifespan and stress resistance in *C. elegans*. Biogerontology 10(1):27–42

Williams DS, Cash A, Hamadani L, Diemer T (2009) Oxaloacetate supplementation increases lifespan in *Caenorhabditis elegans* through an AMPK/FOXO-dependent pathway. Ageing Cell 8(6):765–768

Wilson MA, Shukitt-Hale B, Kalt W, Ingram DK, Joseph JA, Wolkow CA (2006) Blueberry polyphenols increase lifespan and thermotolerance in *Caenorhabditis elegans*. Ageing Cell 5(1):59–68

Wood JG, Rogina B, Lavu S, Howitz K, Helfand SL, Tatar M, Sinclair D (2004) Sirtuin activators mimic caloric restriction and delay ageing in metazoans. Nature 430:686–689

Wu Z, Smith JV, Paramasivam V, Butko P, Khan I, Cypser JR, Luo Y (2002) *Ginkgo biloba* extract EGb 761 increases stress resistance and extends life span of *Caenorhabditis elegans*. Cell Mol Biol (Noisy-le-Grand) 48(6):725–731

Xie J, Wang X, Proud CG (2016) mTOR inhibitors in cancer therapy. F1000Research 5:F1000. Faculty Rev-2078

Xie W, Ning L, Huang Y, Liu Y, Zhang W, Hu Y, Lang J, Yang J (2017) Statin use and survival outcomes in endocrine-related gynecologic cancers: a systematic review and meta-analysis. Oncotarget 8(25):41508–41517

Xu H, Chen K, Jia X, Tian Y, Dai Y, Li D, Xie J, Tao M, Mao Y (2015) Metformin use is associated with better survival of breast cancer patients with diabetes: a meta-analysis. Oncologist 20(11):1236–1244

Yamagishi S, Maeda S, Matsui T, Ueda S, Fukami K, Okuda S (2012) Role of advanced glycation end products (AGEs) and oxidative stress in vascular complications in diabetes. Biochim Biophys Acta 1820(5):663–671

Yang Y, Wu Z, Kuo YM, Zhou B (2005) Dietary rescue of fumble–a *Drosophila* model for pantothenate-kinase-associated neurodegeneration. J Inherit Metab Dis 28:1055–1064

Yang ZX, Wang YZ, Jia BB, Mao GX, Lv YD, Wang GF, Yu H (2016) Downregulation of miR-146a, cyclooxygenase-2 and advanced glycation end-products in simvastatin-treated older patients with hyperlipidemia. Geriatr Gerontol Int 16(3):322–328

Yanik EL, Chinnakotla S, Gustafson SK, Snyder JJ, Israni AK, Segev DL, Engels EA (2016) Effects of maintenance immunosuppression with sirolimus after liver transplant for hepatocellular carcinoma. Liver Transpl 22(5):627–634

Yap AF, Thirumoorthy T, Kwan YH (2016) Medication adherence in the elderly. J Clin Gerontol Geriatr 7(20):64–67

Ye X, Fu J, Yang Y, Chen S (2013) Dose-risk and duration-risk relationships between aspirin and colorectal cancer: a meta-analysis of published cohort studies. PLoS One 8(2):e57578

Ye X, Linton JM, Schork NJ, Buck LB, Petrascheck M (2014) A pharmacological network for lifespan extension in *Caenorhabditis elegans*. Ageing Cell 13(2):206–215

Yi C, Song Z, Wan M, Chen Y, Cheng X (2017) Statins intake and risk of liver cancer: a dose–response meta-analysis of prospective cohort studies. Castiella. A, ed. Medicine 96(27):e7435

Yoo Y-E, Ko C-P (2011) Treatment with trichostatin A initiated after disease onset delays disease progression and increases survival in a mouse model of amyotrophic lateral sclerosis. Exp Neurol 231:147–159

Yu H, Yin L, Jiang X, Sun X, Wu J, Tian H, Gao X, He X (2014) Effect of metformin on cancer risk and treatment outcome of prostate cancer: a meta-analysis of epidemiological observational studies. Medeiros R, ed. PLoS ONE 9(12):e116327

Yue W, Yang CS, DiPaola RS, Tan XL (2014) Repurposing of metformin and aspirin by targeting AMPK-mTOR and inflammation for pancreatic cancer prevention and treatment. Cancer Prev Res 7(4):388–397

Zhang ZJ, Zheng ZJ, Shi R, Su Q, Jiang Q, Kip KE (2012) Metformin for liver cancer prevention in patients with type 2 diabetes: a systematic review and meta-analysis. J Clin Endocrinol Metab 97(7):2347–2353

Zhang X, Xie J, Li G, Chen Q, Xu B (2014) Head-to-head comparison of sirolimus-eluting stents versus paclitaxel-eluting stents in patients undergoing percutaneous coronary intervention: a meta-analysis of 76 studies. PLoS One 9(5):e97934

Zhang YP, Wan YD, Sun YL, Li J, Zhu RT (2015) Aspirin might reduce the incidence of pancreatic cancer: a meta-analysis of observational studies. Sci Rep 5:15460

Zhao Y, Sun H, Lu J, Li X, Chen X, Tao D, Huang W, Huang B (2005) Lifespan extension and elevated hsp gene expression in *Drosophila* caused by histone deacetylase inhibitors. J Exp Biol 208:697–705

Zheng YX, Zhou PC, Zhou RR, Fan XG (2017) The benefit of statins in chronic hepatitis C patients: a systematic review and meta-analysis. Eur J Gastroenterol Hepatol 29(7):759–766

Zhong S, Chen L, Zhang X, Yu D, Tang J, Zhao J (2015a) Aspirin use and risk of breast cancer: systematic review and meta-analysis of observational studies. Cancer Epidemiol Biomark Prev 24(11):1645–1655

Zhong S, Zhang X, Chen L, Ma T, Tang J, Zhao J (2015b) Association between aspirin use and mortality in breast cancer patients: a meta-analysis of observational studies. Breast Cancer Res Treat 150(1):199–207

Zhong GC, Liu Y, Ye YY, Hao FB, Wang K, Gong JP (2016) Meta-analysis of studies using statins as a reducer for primary liver cancer risk. Sci Rep 6:26256

Zhou YY, Zhu GQ, Wang Y, Zheng JN, Ruan LY, Cheng Z, Hu B, Fu SW, Zheng MH (2016) Systematic review with network meta-analysis: statins and risk of hepatocellular carcinoma. Oncotarget 7(16):21753–21762

Zhou XL, Xue WH, Ding XF, Li LF, Dou MM, Zhang WJ, Lv Z, Fan ZR, Zhao J, Wang LX (2017) Association between metformin and the risk of gastric cancer in patients with type 2 diabetes mellitus: a meta-analysis of cohort studies. Oncotarget 8(33):55622–55631

Zid BM, Rogers AN, Katewa SD, Vargas MA, Kolipinski MC, Lu TA, Benzer S, Kapahi P (2009) 4E-BP extends lifespan upon dietary restriction by enhancing mitochondrial activity in *Drosophila*. Cell 139(1):149–160

Zimbron J, Khandaker GM, Toschi C, Jones PB, Fernandez-Egea E (2016) A systematic review and meta-analysis of randomised controlled trials of treatments for clozapine-induced obesity and metabolic syndrome. Eur Neuropsychopharmacol 26(9):1353–1365

Zou YX, Ruan MH, Luan J, Feng X, Chen S, Chu ZY (2017) Anti-ageing effect of riboflavin via endogenous antioxidant in fruit fly *Drosophila melanogaster*. J Nutr Health Ageing 21(3):314–319

Zulman DM, Sussman JB, Chen X, Cigolle CT, Blaum CS, Hayward RA (2011) Examining the evidence: a systematic review of the inclusion and analysis of older adults in randomized controlled trials. J Gen Intern Med 26(7):783–790

Chapter 14
Visual Defects and Ageing

Sergio Claudio Saccà, Carlo Alberto Cutolo, and Tommaso Rossi

Abstract Many diseases are related to age, among these neurodegeneration is particularly important. Alzheimer's disease Parkinson's and Glaucoma have many common pathogenic events including oxidative damage, Mitochondrial dysfunction, endothelial alterations and changes in the visual field. These are well known in the case of glaucoma, less in the case of neurodegeneration of the brain. Many other molecular aspects are common, such as the role of endoplasmic reticulum autophagy and neuronal apoptosis while others have been neglected due to lack of space such as inflammatory cytokine or miRNA. Moreover, the loss of specific neuronal populations, the induction of similar mechanisms of cell injury and the deposition of protein aggregates in specific anatomical areas are very similar events between these diseases. Intracellular and/or extracellular accumulation of protein aggregates is a key feature of many neurodegenerative disorders. The existence of abnormal protein aggregates has been documented in the RGCs of glaucomatous patients such as the anomalous Tau protein or the β-amyloid accumulations. Intra-cell catabolic processes also appear to be common in both glaucoma and neurodegeneration. They also help us to understand how the basis between these diseases is common and how the visual aspects can be a serious problem for those who are affected.

Keywords Glaucoma · Alzheimer's disease · Parkinson's disease · Oxidative stress · Mitochondria · Endothelial dysfunction · Trabecular meshwork · Autophagy · Visual field

S. C. Saccà (✉) · T. Rossi
Department of Head/Neck Pathologies, St Martino Hospital, Ophthalmology Unit, Genoa, Italy
e-mail: sergio.sacca@hsanmartino.it; tommaso.rossi@hsanmartino.it

C. A. Cutolo
Department of Neuroscience, Rehabilitation, Ophthalmology,
Genetics and Maternal and Child Science, University of Genoa,
Policlinico San Martino Hospital, Eye Clinic Genoa, Genoa, Italy

© Springer Nature Singapore Pte Ltd. 2019
J. R. Harris, V. I. Korolchuk (eds.), *Biochemistry and Cell Biology of Ageing:
Part II Clinical Science*, Subcellular Biochemistry 91,
https://doi.org/10.1007/978-981-13-3681-2_14

Abbreviations

AC	Anterior chamber
AD	Alzheimer's disease
AF	Actin microfilaments
AH	Aqueous humour
AMD	Age related macular degeneration
ECM	Extracellular matrix
ER	Endoplasmic reticulum
GSH	Glutathione
IF	Intermediate filaments
JCT	Juxtacanalicular connective tissue
MT	Microtubules
mtDNA	Mitochondrial DNA
NO	Nitric oxide
ONH	Optic nerve head
PD	Parkinson's disease
PKC	Protein kinases
POAG	Primary open-angle glaucoma
RGCs	Retinal ganglion cells
RNFL	Retinal nerve fibre layer
ROS	Reactive oxygen species
ROS	Reactive oxygen species
SC	Schlemm's canal
SOD	Superoxide dismutase
TM	Trabecular meshwork
UPR	Unfolded protein response
UV	Ultraviolet rays

Introduction

Health is defined by the World Health Organization as a state of complete physical, mental and social well-being and not just the absence of disease or infirmity. This state, which should be physiological, is undermined by ageing events. According to Flatt (2012), old age consists of age-progressive decline in intrinsic physiological function, involving tissues, organs and apparatuses, thus leading to an increase in age-specific mortality. Against this background, quality of life assumes significant implications from both the social and the economic points of view. Indeed, ageing combines all the main forms of neurodegenerative diseases that lead to irreversible damage to cognitive, motor and behavioral functions. Furthermore, proper cognitive functioning is essential to independent living and successful ageing (Valentijn et al. 2005).

In this article, we obviously cannot examine the cognitive functions of all the sensorineural organs that play a significant role in the quality of ageing. Nor can we consider all the diseases that impair the quality of life and of ageing. Rather, we will focus our attention on sight and ageing, with particular regard to visual-field defects and ocular ageing.

Eye Diseases and Their Impact on the Visual Field

Many individual features affect vision in ageing subjects; these are personal differences and environmental factors that may have an impact on the brain and ocular structures. For example, the density of the crystalline lens increases with age, although this condition does not constitute a true cataract (Xu et al. 1997), and pupillary miosis influences retinal illumination (Loewenfeld 1979). Although these are not true pathologies, they are able to affect spatial contrast sensitivity (Owsley 2011) and consequently also the visual field (Korth et al. 1989). One of the most serious diseases that involve vision and the eye is age-related macular degeneration (AMD), which is characterized by progressive degeneration of the retinal pigment epithelium and the photoreceptors immersed in it. The consequence of this disease is that the patient progressively loses central vision and with it the central visual field. AMD chiefly affects the cells targeted by the visual signal, rather than the output cells of the retina, i.e. the retinal ganglion cells (RGCs). In the eye, the photoreceptors are a hundred times more than ganglion cells. This implies that the visual signal generated by the photoreceptors is compressed and conveyed to the ganglion cells, and along the neurons to the visual cortex in the central nervous system (Meister 1996; Meister and Berry 1999). Therefore, although a visual field defect occurs in AMD, the structures affected are different from those involved in neurodegeneration or glaucoma. Indeed, in AMD, retrograde trans-synaptic degeneration of the retinal ganglion cell layer is one of the mechanisms contributing to permanent disability following visual post-geniculate pathway injury (Keller et al. 2014; Mitchell et al. 2015). Perimetry, a method for evaluating the spectrum of AMD severity, appears to be valid technique for assessing retinal sensitivity in AMD when colloid bodies/drusen >125 μm in diameter are present, but before the development of late AMD (Luu et al. 2013), while microperimetry provides information beyond that of visual acuity and contrast sensitivity in the functional assessment of AMD (Cassels et al. 2018).

However, numerous anatomic changes that occur in the eye with age are able to alter vision. In the aged eye, as well as in the brain, and therefore in neurodegenerative disease, normal antioxidant defense mechanisms decline, which increases the vulnerability of tissues to the effects of oxidative stress (Finkel and Holbrook 2000). For example, exposure to ultraviolet-A sunlight induces major modifications in the epithelial cells of the cornea (Zinflou and Rochette 2017), while in human corneal endothelial cells, a significant increase in DNA oxidative damage occurs (Joyce et al. 2009). In many studies, this type of damage has been assessed by measuring the levels of 8-hydroxy-2′-deoxyguanosine (8-OH-dG), an indicator of oxidative DNA damage (Ames and Gold 1991). In the cell, when DNA is damaged events occur, which may include: DNA repair, cell cycle delay, entry into senescence, or induction of apoptosis (Toussaint et al. 2002; Zgheib et al. 2005). DNA-induced stress inhibits proliferation and induces cellular senescence by activating specific pathways (von Zglinicki et al. 2005), including those that cause intracellular oxidative stress, such as that of hydrogen peroxide (Erusalimsky and Skene 2009) or the mitochondrial damage that increases it (Agarwal and Sohal 1994). One of the indicators of the presence of senescent cells in tissues is senescence-associated-ß-galactosidase (SA-β-gal), although it should be noted that it is not specific, and that today several markers are used to identify senescent cells. For example, the changes in heterochromatin formation (Narita et al. 2003) or telomere length associated to DNA damage (Hewitt et al. 2012), or the presence of cyclin-dependent kinase inhibitors p21, p16 (Baker et al. 2016) have been used to type cells from ocular tissues. Moreover, it has been shown that senescent cells produce changes in metabolism, the epigenome and gene expression (de Magalhães and Passos 2017). Furthermore, these cells have a peculiar secretome profile, called Senescence-Associated Secretory Phenotype (SASP), which can affect the microenvironment of the tissues by modifying the production of growth factors, extracellular matrix (ECM) -degrading proteins and also pro-inflammatory cytokines and chemokines (Coppé et al. 2008). Many types of cells acquire senescent phenotypes with age: keratinocytes, endothelial cells, lymphocytes, smooth vascular muscle cells, and others (Rodier and Campisi 2011). This pattern is also seen in the pathogenesis of glaucoma, in which there is progressive decay of the trabecular meshwork and its cells, which are markedly aged (Saccà et al. 2016a, b). While senescence cannot cause proliferation loss in non-divisive cells, such as endothelial cells or neurons, these cells undergo major conversion to SASP (Salminen et al. 2011). Therefore, as their numbers dwindle, these cells lose their ability to proliferate and assume the SASP phenotype; moreover, subcellular damage occurs through cellular senescence, which results in tissue malfunction and then in macroscopically visible manifestations of ageing (Bhatia-Dey et al. 2016). In addition, the morphology of the optic nerve, retinal nerve fiber layer and maculae are reported to vary across racial groups and with age (Girkin et al. 2011). In humans, from 500 to 7000 axons are lost per year (Parikh et al. 2007), and a sharp decline in RNFL (Retinal Nerve Fiber Layer) occurs after 50 years of age, as has been observed in histological studies (Johnson et al. 1987). Furthermore, all tests of vision show a deterioration of performance with increasing age (Rudolph and Frisén 2007) as does the visual field test. Figure 14.1 shows the normal range of the visual field and of the Retinal Nerve Fiber Layer (NRFL). From what has been said so far, it emerges that old age, albeit not in itself a disease, is manifested as a functional decay that involves many factors, both physical, such as senile miosis or lens changes, and psychophysical, such as reflexes or perception. In older adults, Owsley (2016) has identified three visual features that can be seriously impaired by the development of common eye conditions and diseases of ageing. The first is "spatial contrast sensitivity", where "contrast sensitivity" is the ability of the eye to recognize different shades of the same color in two adjacent areas. While

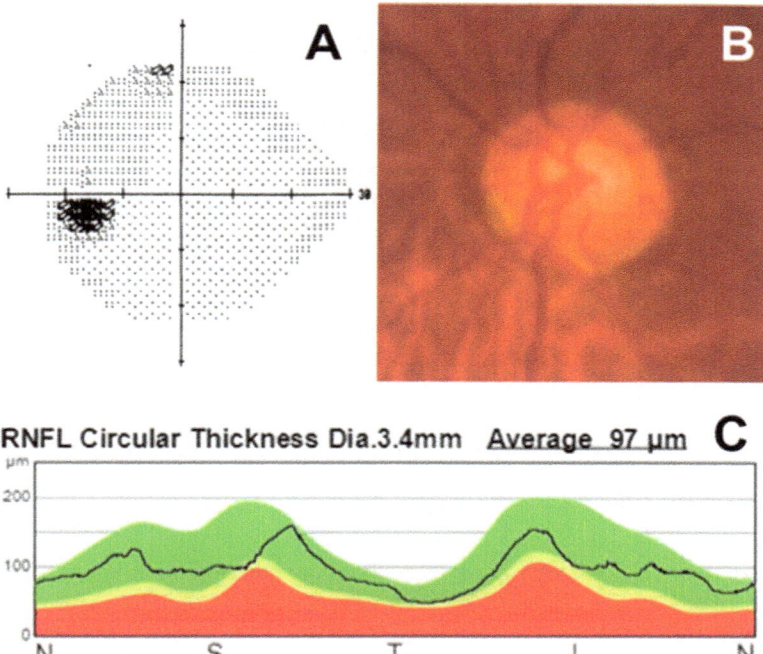

Fig. 14.1 Normal optic nerve (left eye) (**a**) Normal visual field represented on a grayscale map. Darker areas indicate lower sensitivities, while lighter areas indicate higher sensitivities. This graphical representation allows field loss to be interpreted easily and is usually used to demonstrate vision changes to the patient. (**b**) Normal optic disc with a healthy neuro-retinal rim. (**c**) Retinal nerve fiber layer (RNFL) measured by optical coherence tomography over a 3.4-mm diameter circle centered on the optic nerve head. The green area is the 5th–95th percentile by age, the yellow area is the 1st–5th percentile, and the red area is below the 1st percentile. In this case, all the measurements (continuous black line) are in the green area

normally very high, this sensitivity declines in some diseases affecting the transparency of the optical apparatus and in some diseases of the optic nerve. The second is "scotopic vision"; this is the kind of vision that enables us to see in poor light, and which corresponds to the ability of the photoreceptors to adapt to darkness. This function also declines with age. Moreover, older adults with severe dark adaptation delays are more likely to have several risk factors for AMD (Owsley et al. 2014). The third feature is "visual processing speed"; this refers to the speed at which visual information is processed automatically, i.e. without the subject having to focus intentionally on a specific visual target. This third visual function slows down with old age; this is one of the most inhibiting phenomena connected with human ageing (Birren and Fisher 1991), being related to activities such as driving a car or normal mobility.

Globally, of the 7.33 billion people alive in 2015, an estimated 36.0 million were blind. Moreover, the number of people suffering from moderate or severe visual impairment increased from 159.9 million in 1990 to 216.6 million in 2015 (Bourne et al. 2017). Subjects aged 50 years and older account for 65% and 82% of the visually impaired and blind, respectively. The main causes of visual impairment are uncorrected refractive errors (43%) followed by cataract (33%); the leading cause of blindness is cataract (51%) (Pascolini and Mariotti 2012). Blindness caused by AMD is estimated at 5% (Zetterberg 2016), though AMD has become the most common cause of blindness in high-income countries (Bourne et al. 2014). In addition, diseases such as AMD and glaucoma have probably become relatively common causes of visual impairment and blindness worldwide (Pascolini and Mariotti 2012).

In this chapter, however, we will deal only with visual field defects that may occur during age-related illnesses, such as neurodegeneration and glaucoma, which have the same pathogenic root.

Visual Field and Neurodegeneration

The human eye is constantly exposed to sunlight and artificial lighting. Exogenous sources of ROS, such as UV light, visible light, ionizing radiation, chemotherapy, and environmental toxins contribute to oxidative damage to eye tissues. UV rays, besides being able to directly damage ocular tissue, can induce oxidative stress in irradiated cells through the production of ROS and riboflavin by activating tryptophan and porphyrin, which in turn can activate cellular oxygen (Ikehata and Ono 2011). Oxidative stress can be defined as an imbalance between the production of ROS and the antioxidant capacity of the cell. Long-term exposure to these exogenous sources of ROS puts the ageing eye at considerable risk, owing to the pathological consequences of oxidative stress. As all the ocular structures, from the tear film to the retina, undergo oxidative stress over time, the antioxidant defenses of each tissue are called upon to safeguard against degenerative ocular pathologies. The ocular surface and the cornea protect by the eyelid tissues but they are significantly exposed to oxidative stress of environmental origin (Saccà et al. 2013). Indeed, ultraviolet rays modulate the expression of antioxidants and proinflammatory mediators by interacting with corneal epithelial cells (Black et al. 2011). The decay of antioxidant defenses in these tissues is clinically manifested as pathologies, including pterygium (Balci et al. 2011), corneal dystrophy (Choi et al. 2009) and Fuch's endothelial dystrophy (Jurkunas et al. 2010). The crystalline is highly susceptible to oxidative damage in ageing because its cells and intracellular proteins are not replaced, thus providing the basis for cataractogenesis. H_2O_2 is the main oxidant involved in the formation of cataracts and in damage to the DNA of the lens and membrane pump systems, thus leading to the loss of vitality of epithelial cells and death due to necrotic and apoptotic mechanisms (Spector 1995).

The trabecular meshwork is the anterior chamber (AC) tissue which allows the drainage of the aqueous humor. It is particularly susceptible to mitochondrial oxidative damage (Izzotti et al. 2009), which affects its endothelium. Its ensuing malfunction leads to an increase in intraocular pressure (IOP) and the onset of glaucoma (Saccà et al. 2016b).

As occurs in several neurodegenerative diseases, the normal antioxidant defense mechanisms of the eye decline with ageing; this increases the vulnerability of the eye to the deleterious effects of oxidative damage, as has also been seen with regard to the brain (Finkel and Holbrook 2000).

It is believed that mitochondrial free radicals are among the main causes of mitochondrial DNA damage (mtDNA). Several studies have found high levels of 8-hydroxy-2′-deoxyguanosine (8-OHdG), a biomarker of mtDNA oxidative damage, in the aged brain (Agarwal and Sohal 1994) and trabecular meshwork (Saccà and Izzotti 2014). High levels of 8-OHdG have also been found post-mortem in both the nuclear DNA (nDNA) and mtDNA of the brains of elderly subjects (Mecocci et al. 1993). Similarly, in glaucoma patients, damage to the trabecular mtDNA is greater than in healthy subjects. This increased sensitivity to mtDNA oxidative damage may be due to a lack of mtDNA repair mechanisms, lack of protection by the histone proteins, and the fact that mtDNA is located near the inner mitochondrial membrane, where reactive oxygen species are generated (Mecocci et al. 1993; Barja 2004). Neurodegeneration is characterized by a chronic and selective process of neuronal apoptosis and typically affects a specific neuronal system bilaterally. In Alzheimer's disease (AD), for example, neurons in the medial temporal lobe and the limbic system are affected, while in Parkinson's Disease (PD) the motor neurons are affected. In both cases, these neurons undergo apoptosis.

What patients with neurodegeneration have in common with glaucoma patients, in addition to old age, is neuronal loss, which leads to neurological changes in the visual field. In patients with AD, there is extensive loss of ganglion cells in the central retina (Blanks et al. 1996) and a volume reduction of the optic nerve (Kusbeci et al. 2015) (Fig. 14.2), where macular volume reduction reflects neuronal loss and is related to the severity of cognitive impairment (Iseri et al. 2006). Moreover, in patients with PD, the temporal retinal nerve fiber layer (RNFL) (Fig. 14.3) and central-parafoveal macular layers are reduced. Patients with greater RNFL damage tend to have more severe symptoms and, consequently, a lower quality of life

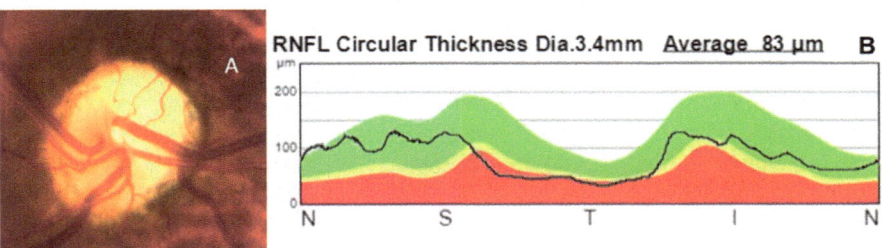

Fig. 14.2 Optic disc of a 67-year-old man affected by initial Alzheimer's disease. The papilla appears paler than normal (**a**). The retinal nerve fiber layer (RNFL), measured by optical coherence tomography along a 3.4-mm diameter circle centered on the optic nerve head, shows a broad part below the 1st percentile (**b**)

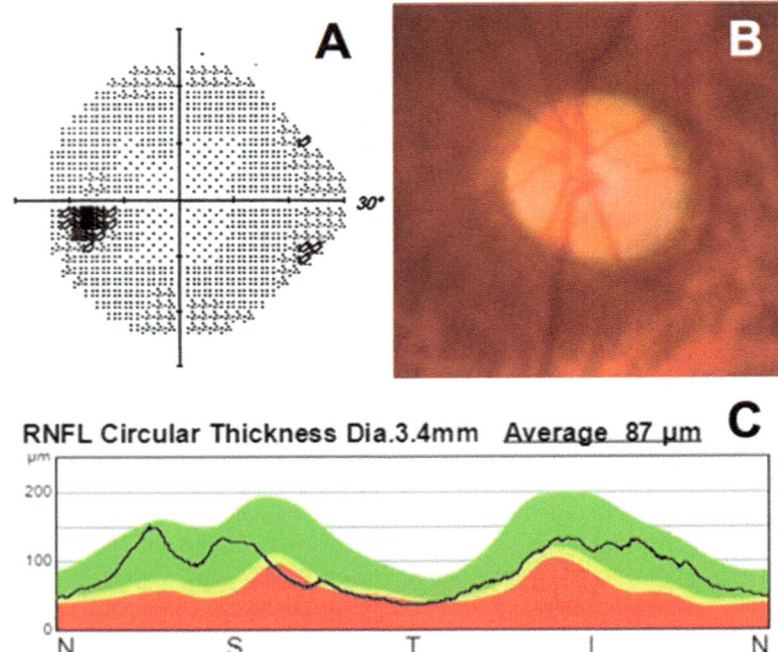

Fig. 14.3 Optic nerve and visual field in a patient affected by Parkinson's disease (left eye) (**a**) The visual field shows areas of decreased sensitivity. (**b**) The optic nerve head appears within normal limits. However, OCT measurement of the retinal nerve fiber layers (**c**) shows a significant reduction in the temporal sector. In neurological diseases, such as Parkinson's disease and Alzheimer's disease, visual field examination can only be performed in the early stages of the disease, as good collaboration on the part of the patient is necessary. Conversely, OCT can be performed even in more advanced stages, as the acquisition time is limited to a few seconds

(Mendoza-Santiesteban et al. 2017). The molecular mechanisms that determine neurodegeneration and diseases such as glaucoma and AMD are similar, in that they display the same pattern: first oxidative stress and inflammation, then mitochondrial dysfunction, and finally cellular apoptosis.

These diseases have many other molecular aspects in common, such as autophagy; this is a metabolic process through which cells degrade cytoplasmic substrates through their lysosomes, and participates in the basal turnover of long-lived proteins and organelles. There are three different types of autophagy: macroautophagy, chaperone-mediated autophagy, and microautophagy, all of which are indispensable to cellular homeostasis (Plaza-Zabala et al. 2017). Indeed, dysfunctional mitochondria appear to be selectively removed by means of autophagy (Li et al. 2012). By contrast, non-selective autophagy occurs when mitochondria are digested by lysosomes to obtain nutrients and energy, and plays an important role in promoting neuronal health and survival (Plaza-Zabala et al. 2017). Mitochondria play a vital role in several phases of autophagy, from initial autogenous biogenesis and autophagic regulation through beclin-1 to autophagy-mediated cell death (Rubinsztein et al.

2012). In the normal human brain, autophagy diminishes with ageing (Rubinsztein et al. 2011). Indeed, insufficient protective autophagy accelerates both ageing and AD pathology, and is possibly caused by defects in autophagosome fusion with lysosomes (Barnett and Brewer 2011). In the nervous system, autophagy is associated with the maintenance of the normal balance between the formation and degradation of cellular proteins. As neurons are susceptible to the accumulation of aggregated or damaged cytosolic compounds or membranes, their survival depends on autophagy (Tooze and Schiavo 2008). Autophagy can inhibit proinflammatory signaling by eliminating dysfunctional mitochondria (Saitoh and Akira 2010) and can also inhibit the activation of the NLRP3 inflammasome – the caspase-1 activation complex required for the production of interleukin-1β – by removing permeabilizing or ROS-producing mitochondria (Zhou et al. 2011). These anti-inflammatory effects of autophagy are useful both to brain health and in neurodegenerative processes and pathological ageing, which are often accompanied by chronic inflammation (Rubinsztein et al. 2011). Mutations in Presenilin-1, which is a protein of the gamma secretase complex, and which plays an important role in the generation of amyloid beta, are associated with the onset of AD, reducing lysosomal acidification and inducing a blockade of autophagy flow in the fibroblasts of AD patients (Lee et al. 2010). Phenotypes indicative of defective autophagy are also found in PD, in which protein aggregation and mitochondrial damage occurs (Kim et al. 2017). Moreover, the native form of α-Synuclein, which is a presynaptic neuronal protein involved in PD pathogenesis, is degraded by means of chaperone-mediated autophagy (Cuervo et al. 2004). Thus, autophagy plays an important role in cellular homeostasis in neurodegenerative diseases, too. Moreover, mitochondria participate in the formation of autophosomes; indeed, the membranes of these organelles are transiently shared (Hailey et al. 2010). The main factors influencing autophagy are the perturbation and starvation of Ca^{++} homeostasis. Signals from Ca^{++} perturbations might be integrated into the unfolded protein response (UPR) and the activation of endoplasmic reticulum (ER) stress (Kania et al. 2015). In PD, prolonged stimulation of mitochondrial oxidative phosphorylation occurs, owing to an excess of Ca^{++}; this causes oxidative stress, which impairs mitochondrial function, increases mitophagy and induces high α-synuclein expression (Surmeier et al. 2016). Ca^{++} dysregulation may involve several mechanisms, since the ER, the lysosomes and the Golgi apparatus, as well as the mitochondria, are important intracellular Ca^{++} reservoirs (Patel and Muallem 2011; Kilpatrick et al. 2013). An accumulation of unfolded or misfolded proteins in the ER leads to stress conditions. To mitigate such circumstances, stressed cells activate a homeostatic intracellular signaling network to restore normal cell function by halting protein translation, degrading misfolded proteins, and activating the signaling pathways that lead to the increased production of the molecular chaperones involved in protein folding. If these molecular conditions are not restored, the UPR is directed towards apoptosis. Although ER stress and autophagy can function independently, they are dynamically interconnected and ER stress can either stimulate or inhibit autophagy (Rashid et al. 2015). ER stress can also promote the NF-kB activation associated to inflammatory pathways (Muriach et al. 2014) and activate cellular inflammatory pathways, which, in turn impair cellular functions and induce mitochondrial changes and, finally, cell apoptosis (Rao et al. 2004). ER stress also causes cellular accumulation of the ROS associated to oxidative stress (Cullinan and Diehl 2006), which, in turn, can reciprocally promote ER stress, thus creating a vicious circle.

Visual Field and Glaucoma

Glaucoma is a neurodegenerative disease characterized by progressive optic atrophy, which is the result of the death of retinal ganglion cells due to apoptosis induced by the inhibition of cell survival. This disease causes visual field alterations, and can be acute or chronic, depending on whether the corneal iris angle is open or closed. The opening of this angle enables the aqueous humor (AH) to circulate. Produced by the ciliary body, the AH passes through the pupil into the AC and flows through the trabecular meshwork (TM).

Trabecular Meshwork Organization

For a long time it was believed that AH outflow was a passive phenomenon, and that the TM was a structure similar to a filter. This is not so. The TM consists of endothelial cells that are immersed in their matrix to form a network of beams, in which the matrix occupies the spaces between the beams (Tian et al. 2000) (Fig. 14.4). TM cells respond to mechanical stretching by increasing ECM turnover and reducing total versican mRNA levels (Keller et al. 2007). As is obvious, AH outflow is an active phenomenon, and is regulated by the cells of the TM. Indeed, the TM is a true organ made up of endothelial cells that are able to increase or reduce AH outflow in accordance with the requirements of the eye (Fig. 14.5). The TM incorporates two barriers: the first is constituted by trabecular meshwork endothelial cells that are immersed in the AH of the AC; the second is formed by the endothelial cells that line the anterior wall of Schlemm's canal (SC). The first controls the second through a cytokine system, which modulates the permeability of the endothelial cells of the SC (Alvarado et al. 2005a). These cytokines include IL1a, IL1b and IL8 (Alvarado and Shifera 2010). In reality, interactions between these barriers occur in both directions and involve relationships within both barriers. Indeed, the endothelial cells lining the SC, which make up the second barrier, are juxtaposed to juxtacanalicular tissue (JCT); these cells form tight junctions, providing a significant barrier to the passage of fluid (Gong et al. 2001, 2002). They also act as a "control" site, regulating the outflow of AH from the eye by increasing/reducing the permeability of this barrier, and drive a mechanism controlling SC endothelium permeability (Alvarado et al. 2005a, b). By altering the physiological catabolism of the ECM, TM cells are able to change the TM outflow, and hence to regulate aqueous outflow resistance (Aga et al. 2014). In this structure, endothelial cells are organized on the collagen

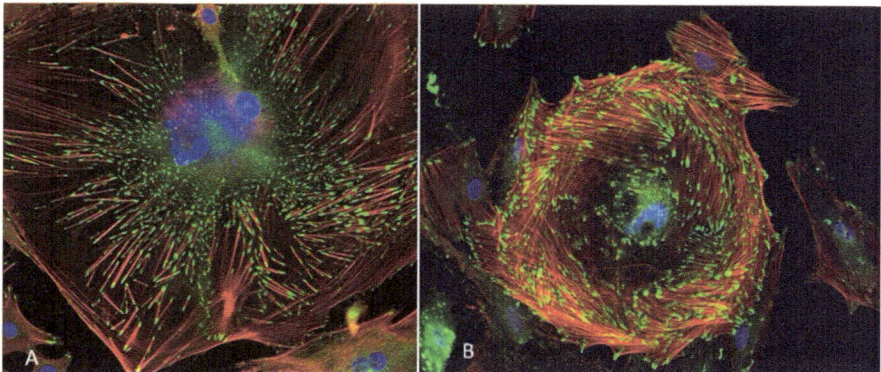

Fig. 14.4 (**a, b**) TM cells were fibronectin-coated under standard culture conditions. Images were generated by means of a GE In Cell 1000 high-content imaging system and were colored for nuclei (blue), actin fibers (red), and focal adhesions (green). As DNA is stained blue, the large clumps of blue just above center are cell nuclei. The red lines are filaments of actin extending throughout the cell, while the green patches at their tips are focal adhesions. These images document the plasticity of these endothelial cells, and the different shapes of the same cell-type demonstrate that these cells have the ability to change shape; the endothelial cells are endowed with an intricate actin network, which extends from the cell wall throughout the cell body to the nucleolus (blue) and enables these cells to assume various shapes. Owing to the functions of their unique and complex metabolism, these endothelial cells are able to change their gene expression in order to preserve their barrier function of the TM. (Saccà et al. 2016b)

beams, maintaining evident cell–ECM focal contact-like structures and adherent-type cell–cell junctions (Tian et al. 2000). The cytoskeletal interactions between JCT and the SC can be regulated by a variety of environmental and cytoplasmic factors, such as the level of extracellular calcium, activation of specific small G-proteins, mechanical tension and hydrostatic pressure (Ye et al.1997), and particular molecular components of tight junctions may help to regulate flow resistance (Underwood et al. 1999). The spaces between JCT cells and ECM fibers contain a ground substance consisting of various proteoglycans and hyaluronan (Gong et al. 1992; Tamm 2009). Therefore, the endothelial cells that form this outflow pathway, besides having the ability to secrete proteins and cytokines, have a cytoskeleton that allows them to change shape and, accordingly, their gene expression (Saccà et al. 2016b). This plasticity is also due to the contractile properties of the ciliary body and the trabecular meshwork; according to the circumstances, these properties are activated through various signal transduction pathways involved in the regulation of smooth muscle contractility (Wiederholt et al. 2000). The TM is thought to be a smooth muscle-like tissue with contractile properties (Lepple-Wienhues et al. 1991). Contraction and relaxation of the cells are thought to regulate aqueous humor outflow (Wiederholt 1998). Indeed, the TM possesses smooth muscle-like properties and is actively involved in the regulation of aqueous humor outflow and intraocular pressure; the TM and ciliary muscle appear to be functional antagonists, with ciliary muscle contraction leading to distension of the TM and subsequent reduction

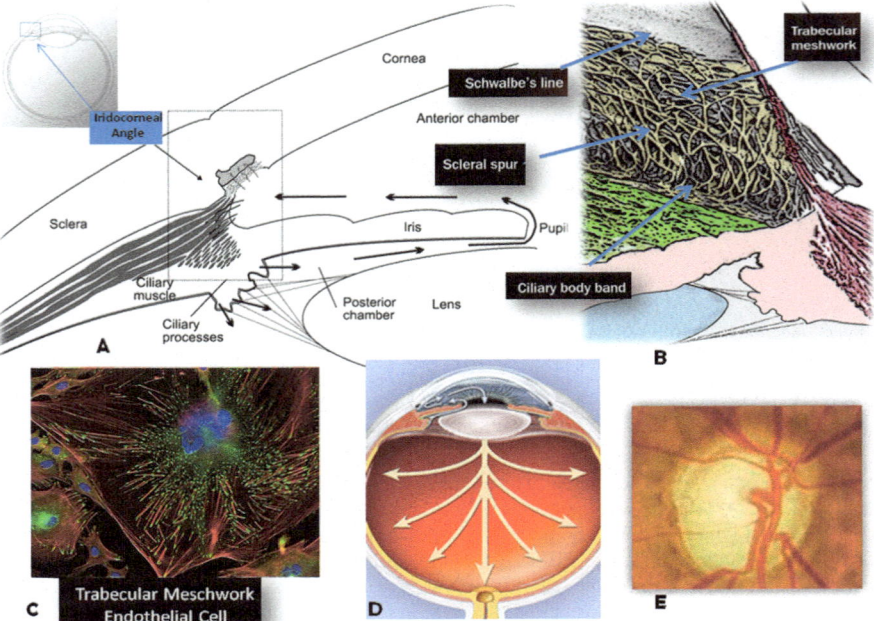

Fig. 14.5 (**a**) The conventional aqueous outflow pathway: from the ciliary process the aqueous humor flows through the pupil into the anterior chamber, where it encounters the trabecular meshwork endothelial cells that line aqueous channels, and then subsequently encounters the endothelial cells that line the lumen of Schlemm's canal. (**b**) Anatomy of the iridocorneal angle with the structures observable on gonioscopy. Schlemm's canal is not normally visible. (**c**) The endothelial cells that delineate the structures of the collagen framework of the trabecular meshwork are equipped with a cytoskeleton and thus are able to change their shape. The cytoskeleton (red filaments) is attached to the nuclear membrane (colored in blue) and can, in milliseconds, send signals to the nucleus in order to alter the expression of genes in an attempt to adapt to biomechanical injury. (**d**) The proteins released by the trabecular-toothed cells in the anterior chamber may become pro-apoptotic signals for the retina, and in particular for retinal ganglion cells, activating apoptosis. (**e**) If glaucoma is not adequately treated, the typical appearance of optic nerve head atrophy, with its characteristic cupping, is observed

in outflow, and with TM contraction leading to the opposite effect (Wiederholt et al. 2000). Similarly, increases or decreases in the volume of TM cells could influence outflow (O'Donnell et al. 1995; Soto et al. 2004). TM cell volume is influenced by the activities of ions (Al-Aswad et al. 1999; Mitchell et al. 2002). Moreover, it is thought that both cellular contractile mechanisms and cell volume regulatory mechanisms are functionally linked (Dismuke et al. 2008), and may be part of the homeostatic mechanisms of the TM whereby outflow is regulated. Furthermore the large-conductance Ca^{2+}-activated K^+ (BK_{Ca}) channel has been shown to regulate TM cell volume and contractility (Wiederholt et al. 2000; Soto et al. 2004) and outflow facility (Soto et al. 2004). The flow from the AC to Schlemm's canal takes place basically through two pathways: a paracellular route through the junctions formed between the endothelial cells (Epstein and Rohen 1991), and a transcellular pathway through intracellular pores (Johnson and Erickson 2000) or giant vacuoles in the cells themselves. SC cells probably have the ability to modulate local pore density and the filtration characteristics of the inner-wall endothelium on the basis of local biomechanical cues (Braakman et al. 2014).

Factors Involved in Glaucoma Pathogenesis

Numerous factors, both metabolic and environmental, can affect the conventional outflow pathway, the main one being its cellularity. Indeed, the decline in the cellularity of human TM endothelial cells is linearly related to age, and plays a major role in glaucoma pathogenesis (Alvarado et al. 1981, 1984). When the capacity of the TM cells to remove aqueous humor decreases, IOP increases (Saccà and Izzotti 2014). The relationship between IOP and optic nerve damage has not yet been elucidated, but it is certain that this relationship is real. Indeed, the only currently recognized anti-glaucoma therapy is that which lowers IOP (Konstas et al. 2015). Therefore, all that is able to influence the functioning of the endothelial cells plays a role in the pathogenetic cascade of glaucoma. Oxidative damage is now recognized as an important step in the pathogenesis of glaucoma. Indeed, it affects the TM cells, and if it is measured in terms of 8-hydroxy-2′-deoxyguanosine (8-OH-dG) levels, the DNA of TM cells in glaucoma patients is seen to be altered in comparison with the healthy TM; moreover, this damage is directly proportional to the IOP level and to the visual field damage (Saccà et al. 2005). Both in ageing and during the course of glaucoma, normal antioxidant defenses decline, which determines an increased susceptibility to the effects of this type of damage (Finkel and Holbrook 2000) The TM is the tissue most sensitive to oxidative radicals in the AC (Izzotti et al. 2009). As in other neurodegenerative diseases, mitochondrial involvement is also seen in glaucoma; in particular, mtDNA damage occurs in the target tissue of POAG, the TM. Such damage is detectable only in the TM, and not in other AC tissues, cornea or iris (Izzotti et al. 2010b). The dysfunction of oxidative phosphorylation secondary to mtDNA mutations can reduce ATP production and the generation of reactive oxygen (ROS). An increase in ROS that exceeds the antioxidant capacity of the tissue results in oxidative stress, contributing to the ageing process through the induction and further progression of cellular senescence. The defective mitochondrial function in the TM cells of patients with POAG renders these cells abnormally vulnerable to Ca^{2+} stress, with subsequent failure of IOP control (He et al. 2008). Conversely, the increased expression of SIRT1 antagonizes the development of oxidative stress-induced premature senescence in human endothelial cells (Ota et al. 2007). The sirtuins are a highly conserved family of nicotinamide adenine dinucleotide (NAD+)-dependent histone deacetylases that help to regulate the lifespan of several organisms and may provide protection against diseases related to oxidative stress-induced ocular damage (Mimura et al. 2013). In the case of glaucoma, this is likely to occur through the interaction of SIRT1 with eNOS

(Ota et al. 2010). Indeed, eNOS activity in HTM cells regulates inflow and outflow pathways (Coca-Prados and Ghosh 2008), and the regulation of eNOS is, in turn, influenced by the activation of Rho GTPase signaling (Shiga et al. 2005) in the AH outflow pathway, which influences actomyosin assembly, cell adhesive interactions and the expression of ECM proteins and cytokines in TM cells in a cascade-like manner (Zhang et al. 2008). It is evident therefore that the mitochondria of TM cells play a crucial role in meeting the high metabolic demand of these cells. Mitochondrial integrity declines with age, aged organelles being morphologically altered and producing more oxidants and less ATP than younger organelles (Shigenaga et al. 1994). The mitochondrial dysfunction that occurs during the course of glaucoma remains of unknown origin. However it is known that mitochondrial DNA deletion, which is correlated with levels of 8-OH-dG (Hayakawa et al. 1992), is far greater in the TM of patients with glaucoma than in controls. This finding is paralleled by a decrease in the number of mitochondria per cell and by cell loss (Izzotti et al. 2010b), with the result that mitochondria bearing this deletion are less efficient in ATP production and release more ROS than intact mitochondria (Izzotti 2009), thereby causing an energy deficit and tissue atrophy (Morris 1990). Moreover, the cytoskeleton-dependent changes in ATP release are correlated with changes in cell volume regulation (Sanderson et al. 2014), and the entire TM cell homeostasis is altered. In glaucomatous TM tissue, the increased number of cells positively stained for senescence-associated-β-galactosidase (SA-β-Gal) activity has been seen to disrupt the local tissue micro-environment through the over-expression of several pro-inflammatory cytokines and the production of ROS (Liton et al. 2005). Oxidative DNA damage in the TM has been significantly correlated with age, and there is a significant relationship between DNA oxidative damage and autophagy activation (Pulliero et al. 2014). Autophagy is a catabolic process involving lysosomal degradation, which has the characteristic of activating a cellular survival mechanism against a variety of stressors. During glaucoma, the TM cells react to IOP increase by triggering autophagy (Porter et al. 2013) even though dysregulation of the autophagic pathway in TM cells occurs (Porter et al. 2015). However, the activation of autophagy must be coupled with the lysosomal system; indeed, concurrently with autophagy, the endoplasmic reticulum (ER) undergoes stress in HTMcs (Li et al. 2011).

Myocilin and Optineurin

Myocilin may play a role in autophagy. Myocilin is a glycoprotein that is normally expressed in both ocular and non-ocular tissues. Although its physiological function is as yet unknown, it may play a role in the development of retinal cell apoptosis (Koch et al. 2014) and in axonal myelination and oligodendrocyte differentiation (Kwon et al. 2014). This protein is assembled in the extracellular space between the ER and Golgi complex through the secretory pathway that is responsible for the synthesis, folding and delivery of cellular proteins; here, it may undergo mutation, leading to myocilin aggregation, ER stress and TM cell toxicity (Stothert et al. 2016). Indeed, the chronic accumulation of misfolded proteins in the ER can facilitate cell death (Gorbatyuk and Gorbatyuk 2013; Zhu and Lee 2014). Furthermore, autophagy contributes to cellular homeostasis through the turnover of mitochondria, ER and peroxisomes (Johansen and Lamark 2011; Wang and Klionsky 2011). Myocilin mutations are typically associated with high IOP; for this reason they have more impact on the HTM cells, while optineurin (OPTN) mutations are associated with normal-tension glaucoma, even though its expression is increased by IOP increases (Vittow and Borras 2002). OPTN plays an important role in regulating several genes, including myocilin (Park et al. 2007). Moreover, OPTN expression is regulated by various cytokines, particularly NF-κB, which can be activated by increased IOP, ageing, vascular diseases and oxidative stress (Saccà and Izzotti 2014). Optineurin can also mediate the removal of protein aggregates through a ubiquitin-independent mechanism that also serves as a substrate for autophagic degradation (Ying and Yue 2016). Optineurin is therefore involved not only in glaucoma but also in other neurodegenerative diseases or in cancer, and again, in age-related macular degeneration (Wang et al. 2014). Myocilin mutations in the MYOC gene are observed in 10–30% of cases of juvenile-onset open-angle glaucoma (Shimizu et al. 2000), and in 2–4% of patients with POAG (Stone et al. 1997). In combination, mutated myocilins may exacerbate sensitivity to oxidative stress (Joe and Tomarev 2010). Furthermore, it is interesting that over-expession of the senescence-related biomarkers SM22 and osteonectin (SPARC) occurs in conditions of oxidative stress. Both proteins are involved in ECM turnover (Nair et al. 2006; Dumont et al. 2000). SPARC over-expression increases IOP in perfused cadaveric human anterior segments as a result of a qualitative change in the juxtacanalicular tissue ECM (Oh et al. 2013). Furthermore, the function of myocilin as an extracellular protein may link ECM molecules such as fibronectin and laminin with matricellular proteins such as SPARC (Aroca-Aguilar et al. 2011). SM22 is an actin-binding protein involved in senescence-associated morphological changes (Nair et al. 2006). Increasing macroautophagy can delay the ageing process and extend tissue longevity (Stothert et al. 2016). In glaucomatous TM cells, the autophagic mechanism is dysregulated, and they show a decreased response to oxidative stress (Porter et al. 2015). Micro RNAs (miRNAs), post-trascriptional regulators of gene expression, are associated with the development of stress-induced premature senescence (SIPS) in the HTM. TGF-β also induces SA-β-Gal activity and increases the miRNA levels of senescence-associated genes (Frippiat et al. 2001), thereby promoting SIPS. The down-regulation of members of the miR-15 and miR-106b families may contribute to some features of senescent cells, such as increased resistance to apoptosis; likewise, the up-regulation of miRNAs, such as miR-182, may contribute to specific changes in gene expression that are associated with the senescence phenotype (Li et al. 2009). It is evident that this situation leads to a functional decay of the TM, in which both oxidative stress and mitochondrial dysfunction induce endothelium dysfunction of trabecular cells (Saccà et al. 2016a). The endothelium of the TM is immersed in the aqueous humor, and its operation is comparable to that of a small vessel. Indeed, the endothelial leukocyte adhesion molecule-1 (ELAM-1) is found in the AH; this oxidatively induced molecule is known as an early marker of atherosclerotic plaque in the vasculature (Eriksson et al. 2001), along with all the other atherosclerotic biomarkers (Saccà et al. 2012).

The Proteome

In the glaucomatous aqueous humor, the proteome is altered and has special characteristics: many proteins expressed at high levels in healthy subjects are reduced in POAG patients; by contrast, other proteins detected at low levels in normal aqueous humor are increased in glaucoma (Table 14.1). Total proteins do not show quantitative differences between POAG patients and controls; the qualitative differences, however, are significant (Izzotti et al. 2010a). These differences can be divided into six groups, the first of which concerns mitochondria, mitochondrial proteins involved in the electron transport chain, trans-membrane transport, protein repair, and the maintenance of mitochondrial integrity. The proteins involved are found in the AH because they come from mitochondria that no longer work, and also because trabecular cells that die release their contents into the medium. The apoptosis that

Table 14.1 POAG-marker proteins in aqueuous humour as detected by Ab microarray (Izzotti et al. 2010a)

Classes of proteins found in glaucomatous aqueous humour	Their pathologic meaning
Mitochondrial damage	The presence of these proteins in AH provides evidence of mitochondrial dysfunction
Apoptosis	These proteins produce induction through the intrinsic (i.e., mitochondrial dependent) pathway and lead endothelial cell death
Cell adhesion defects	This group is related to loss of mechanical integrity in the Trabecular Meshwork as a consequence of reduced cell–cell and cell–matrix adhesion
Protein kinases	This group influence the morphological and cytoskeletal characteristics of Trabecular and Schlemm Canal cells. AKAP 2 plays a critical role in controlling excitability of neurons and K ions
Neuronal proteins	This group includes nestin and optineurin and growth and differentiation factors involved in neurogenesis and neuronal survival
Antioxidant activities	The levels of the antioxidant enzymes are significantly lower in POAG patients than in controls
Vascular proteins	Their presence reflect the endothelial dysfunction, characterized by mechanisms and molecular events that resemble the events that occur during atherosclerosis

Histopathological and ultrastructural functional alterations occurring in Trabecular Meshwork is reflected in the protein expression of trabecular cells that flow in the aqueous humor and highlighted by AH proteome analysis. These accurately describe the cascade of events that first leads to the malfunction of the trabecular meshwork and then to the IOP increase. Upstream of this damage there are all genetic factors favoring this damage. Just because the trabecular meshwork is particularly susceptible to oxidative damage, and oxidative stress is greater in GSTM1-null subjects, this gene should be considered. Indeed, Oxidative attack induces a loss of trabecular meshwork functions: inducing into it mitochondrial damage and degenerative phenomena in the trabecular meshwork, thus increasing intraocular pressure possibly increasing the need for surgery and facilitating the deterioration of clinical glaucoma occurs in ocular tissues during POAG is induced by a variety of mechanisms: mainly mitochondrial damage, but also inflammation, vascular dysregulation, and hypoxia. It is evident that, as the cells gradually undergo apoptosis, their physiological TM functions fail. The barrier function is altered, and this is reflected in the AH by the presence of a group of proteins associated with cell adhesion defects. This group includes chains, junction proteins and cadherins. All the biological events observed in glaucoma are indicative of the progressive functional failure of the TM. Indeed, the glaucomatous AH contains another important group of proteins: those that induce apoptosis. One of these proteins, which seems to be very important and which is expressed more than in healthy subjects, is insulin receptor substrate 1 (IRS-1). This catalyzes the intramolecular autophosphorylation of specific tyrosine residues of the insulin β subunit, further enhancing the binding of tyrosine kinase to the receptor for other protein substrates (White et al. 1985). IRS-1 exerts an important biological function both in metabolic and mitogenic pathways and in signal-activating pathways, including the PI3K pathway and the MAP kinase pathway. It should be remembered that the PI3K gene is related to the system of reporting insulin/IGF-1, one of the mechanisms of regulation of ageing tissue and that the PI3K/AKT/mTOR pathway is an intracellular signaling pathway involved in apoptosis. Indeed, in vascular tissue, activation of the PI3K/Akt/mTOR survival signal pathway and concomitant suppression of the p38 MAPK proapoptotic pathway protects the endothelium against damage caused by oxidative stress, cell migration and/or proliferation and apoptosis/survival (Joshi et al. 2005). The MAPK (originally called ERK) pathway is a chain of proteins in the cell that communicates a signal from a receptor on the surface of the cell to the DNA in the nucleus of the cell. The pathway includes many proteins, which communicate by adding phosphate groups to a neighboring protein, thereby acting as an "on"/"off" switch. When one of the proteins in the pathway is mutated, it can be stuck in the "on" or "off" position, which is a necessary step in the development of many diseases. During glaucoma, the components of this MAP kinase pathway in the TM are dramatically affected by TNF-α, and inhibition of extracellular signal-regulated kinase phosphorylation blocks changes in MMP and tissue inhibitor expression (Alexander and Acott 2003). Furthermore, the mitogen-activated protein MAP kinase superfamily constitutes a signaling pathway that is also active in the posterior segment. Indeed, p38, a member of this superfamily, is involved in the RGC apoptosis that is mediated by glutamate neurotoxicity through NMDA receptors after damage to the optic nerve (Kikuchi et al. 2000). Again, the MAP signaling pathway is probably involved in the induction and/or maintenance of the activated glial phenotype in glaucoma. Because MAPKs are involved in determining the ultimate fate of these cells, their differential activity in neuronal and activated glial cells in the glaucomatous retina may be partly associated with the differential susceptibility of these cell types to glaucomatous injury (Tezel et al. 2003). It is interesting that apoptosis signal-regulating kinase 1 is a MAP kinase involved in neural cell apoptosis after various kinds of oxidative stress (Harada et al. 2006) and its deficiency attenuates neural cell death in normal-tension glaucoma-like pathology in both neural and glial cells, in which the TNF-induced activation of p38 MAPK is suppressed and inducible nitric oxide synthase is produced (Harada et al. 2010). The PI3K pathway also modulates the expression of angiogenic factors, such as vascular endothelial growth factor, nitric oxide and angiopoietins (Karar and Maity 2011). Moreover, IRS-1 plays necessary roles in insulin-signaling pathways leading to the activation of eNOS Ca^{2+}-mediated pathways (Montagnani et al. 2002). Gogg et al. (2009) introduced the concept of "selective" insulin resistance, involving IRS-1 and the PI3kinase pathway, as an underlying factor for the dysregulation of microvascular endothelium, in which MAPK are involved in increasing endothelin (ET)-1levels. This could happen at the molecular level in the TM. Indeed, the levels of pro-inflammatory cytokine TNF-alpha are reported to be significantly higher in the glaucomatous AH (Sawada et al. 2010). TNF-alpha induces a state of insulin resistance in terms of glucose uptake in myocytes because of the activation of pro-inflammatory pathways that impair insulin signaling at the level of the IRS proteins (Lorenzo et al. 2008). This increases contractile dysfunction (Reid and Moylan 2011; Bhatnagar et al. 2010) and mitochondrial ROS, which in turn activate apoptosis signal-regulating kinase 1, which aggravates endothelial cell dysfunction. It is also possible that the diffusion of these proteins in the posterior segment, through the route described by Smith et al. (1986) may reduce retinal Müller cell death (Walker et al. 2012).

Another important group of proteins in the glaucomatous aqueous humour that reflects TM motility damage is that of Protein Kinases (PKC). Indeed, PKC plays an important role both in the regulation of phosphorylation of myosin light chain, which induces cellular contraction and in the dynamics of the cytoskeleton in the Trabecular Meshwork. Furthermore, PKC could affect outflow, by influencing the cell shape (expansion, contraction and morphological changes) in the trabecular and sclerocorneal cells. This group also includes one of the proteins most abundantly expressed in the AH of POAG, i.e. A kinase (PRKA) anchor protein 2. Protein phosphorylation is one of the most important mechanisms of enzyme regulation and signal transduction in eukaryotic cells. cAMP-dependent protein kinase (PKA), one of the first protein kinases discovered, appears to be the main 'read-out' for cAMP to downstream signaling pathways. These downstream substrates include other protein kinases, protein phosphatases, other enzymes and ion channels. Metabolism, gene transcription, ion channel conductivity, cell growth, cell division and actin cytoskeleton rearrangements are modulated by PKA-catalyzed phosphorylation in response to hormonal stimuli (Francis and Corbin 1994; Scott 1991). Localization of the cAMP- PKA and other signaling enzymes is mediated by interaction with A-kinase anchoring proteins (AKAPs). These AKAPs are classified on the basis of their ability to associate with the PKA holoenzyme inside cells (Colledge and Scott 1999); they function as targeting units and tether the kinase to specific subcellular localizations (Huang et al. 1997). The AKAPs are a group of structurally diverse proteins, which have the common function of binding to one or more of the regulatory subunits of cAMP-dependent protein kinase A (PRKA) and confining the holoenzyme to discrete locations within the cell. Activation of PRKA usually results from the binding of cAMP to the R subunits of PRKA, this promotes dissociation and activation of the catalytic subunits, leading to a wide variety of cellular responses. The encoded protein is expressed in endothelial cells, cultured fibroblasts, and osteosarcoma cells. It associates with protein kinases A and C and phosphatase, and serves as a scaffold protein in signal transduction. Cytoskeletal signalling complexes facilitate this process by optimizing the relay of messages from membrane receptors to specific sites on the actin cytoskeleton. AKAP mediated organization of kinases and phosphatases is particularly important for the transduction of signals to the cytoskeleton (Diviani and Scott 2001). Indeed, AKAP-Lbc is a protein kinase A-anchoring protein that also functions as a scaffolding protein to coordinate a Rho signaling pathway (Diviani et al. 2001). The Rho family of small GTPases are the principal transmitters of signals from transmembrane receptors that stimulate actin filament nucleation (Bishop and Hall 2000). The actin cytoskeleton is critical to a variety of essential biological processes in all eukaryotic cells, including the establishment of cell shape and polarity, motility, and cell division (Hall 1998). PRKA activation has been shown to negatively regulate RhoA signaling (Lang et al. 1996; Chen et al. 2005).These signals influence fundamental cell properties such as shape, movement and division (Scott 2003). PRKA 2 is highly enriched in mitochondria (Wang et al. 2001). This protein is a cell growth-related protein. It is well known that from anatomical, physiological and pathological perspectives, the Anterior Chamber (AC) is similar to a vessel and behaves like a vessel, so much so that during the course of POAG, the AC is characterized by mechanisms and molecular events resembling those that occur during atherosclerosis (Saccà et al. 2012). The presence of PRKA2 in AH of the AC is justified on the basis of the endothelial dysfunction that occurs during POAG (Saccà and Izzotti 2014). From a patho-physiological point of view, we think that this is a result of the reduction in TM endothelial cells that occurs in glaucoma (Alvarado et al. 1984), due both to ageing (Alvarado et al. 1981) and to mitochondrial dysfunction (Izzotti et al. 2010b, 2011). Probably, protein which originates from the cytoplasm of endothelial cell is lost into AH, where it acts as a signal to the surviving cells or other tissues. Indeed, oxidative stress is a plausible mechanism for the development of glaucoma, manifesting its effects on the HTM (Izzotti et al. 2003) as an IOP increase (Saccá et al. 2005).The TM endothelium is involved in modulating the permeability of the endothelial barrier and the release of endothelins and nitric oxide. In AH, there are many factors that have a protective role on endothelial cells, such as GSH, which protects anterior segment tissues from high levels of H_2O_2 (Costarides et al. 1991). Unfortunately, the TM is highly susceptible to oxidative injury (Izzotti et al. 2009) so the increased concentration of NO, that occurs in glaucoma (Tsai et al. 2002) reacts with anion superoxide to form peroxynitrite. The production of peroxynitrite is counteracted by the antioxidant defenses and repair systems located in the AC tissues and AH (Saccà et al. 2007). Together, however, NO and peroxynitrite can suppress eNOS expression via the activation of RhoA, hence causing vascular dysfunction (El-Remessy et al. 2010). Moreover, myocilin, which is localized in TM cells (Tamm 2002), has several functions that are mediated by Rho GTPase and cAMP/protein kinase A signaling. Indeed, when moderately over-expressed, myocilin induces a loss of actin stress fibers and focal adhesions (Wentz-Hunter et al. 2004), inhibits the adhesion of human TM cells to ECM proteins, and compromises TM cell-matrix cohesiveness leading to TM damage (Shen et al. 2008). In addition to acting on the motility of the trabecular meshwork, it should be stressed that the AKAP 2 regulates many ion channels, including those of Na$^+$ involving PCK (Bengrine et al. 2007) and K$^+$ channels (Zhang and Shapiro 2012). The TM cells utilize Na-K-Cl cotransport to modulate their intracellular volume, and thus the volume of the paracellular pathways through which aqueous humor may travel (Brandt and O'Donnell 1999). Thus, the role of AKAP 2 could also be related to the maintenance of the endothelial barrier within the anterior chamber. In the course of glaucoma endothelial dysfunction occurs which impairs the functioning of this barrier (Saccà et al. 2015).

Another important protein that is a highly expressed by human endothelial trabecular meshwork cells is the Actin related protein 2/3 complex, subunit 3, 21 kDa (Arp2/3). Human Arp2 and Arp3 are very similar and this protein complex has been implicated in the control of actin polymerization in cells (Welch et al. 1997). Moreover, the complex promotes actin assembly in lamellipodia and may participate in lamellipodial protrusion (Machesky et al. 1997). It is interesting that the Arp2/3 complex is critical for spine and synapse formation in the central nervous system, and this synaptic plasticity, which underlies cognitive functions such as learning and memory is attributed to the reorganization of actin (Wegner et al. 2008). In all probability we discovered the Arp2/3 complex in glaucomatous AH because the TM endothelial barrier is altered during this disease (Saccà et al. 2012), likely owing to oxidative stress (Saccà et al. 2009). Indeed, actin cytoskeleton rearrangements are the basis of many fundamental processes of cell biology such as motility, adhesion, mitosis, endocytosis, and morphogenesis (Moreau et al. 2003). Actin not only facilitates membrane deformation, cytoskeleton remodeling and the formation of vesicles, but also contributes to vesicle movement and targeting within the cell. Indeed, normal regulation of vesicular transport events is essential to cell proliferation and apoptosis, as well as to the maintenance of homeostasis. Deregulation of vesicular transport can lead to decreased capacitive calcium entry, which in turn results in cell apoptosis (Jayadev et al. 1999). Rho GTPases play a key role in vesicle trafficking through their ability to regulate the actin cytoskeleton (Cory et al. 2003). Actin plays multiple roles in vesicle trafficking (Smythe and Ayscough 2006; Lanzetti 2007). Indeed the Arp2/3 complex leads to actin polymerization (Chi et al. 2013). Furthermore, actin assembly regulates the nuclear import of Junction-mediating and regulatory protein which is a regulator of both transcription and actin filament assembly in response to DNA damage (Zuchero et al. 2012).

During glaucoma TM motility is altered (Khurana et al. 2003) and the presence of the Arp 2/3 complex, together with other classes of AH proteins (in the AH) could represent an attempt to restore normal TM motility. This hypothesis fits well with the finding that Protein kinase C levels are significantly increased in the AH of POAG (Izzotti et al. 2010a). Moreover, the presence of Arp 2/3 complex in POAG AH indicates that the endothelium of the TM behaves as a vascular tissue and therefore may change shape. The results obtained by Mao et al. (2011) implicate podosomes in normal development of the iridocorneal angle and the genes influencing podosomes as candidates in glaucoma. Moreover, in various models, the cytoskeletal dynamics underlying these processes have been shown to be driven by small

G-protein members of the Rho family. Indeed, endothelin-1 (ET-1) is known to induce Ca^{2+}-independent contraction of the trabecular meshwork. This contraction involves RhoA and its kinases as intracellular mediators (Renieri et al. 2008). Protein kinase C (PKC) levels are significantly increased in POAG AH influencing cytoskeletal dynamics within the TM (Izzotti et al. 2010a). PCK activation triggers other regulatory mechanisms which in turn influence the shape of TM cells (Khurana et al. 2003).

Importantly, Rho GTPases regulate actin dynamics by acting as molecular switches that transduce signals from activated membrane receptors to cytoskeleton organizers (Van Aelst and D'Souza-Schorey 1997). Podosomes in the machinery required for actin polymerization constitute the localization dedicated to a specific physiological process such as angiogenesis, or vascular permeability, or represent the manifestation of a pathological status, with inevitable consequences on endothelial cell functions (Moreau et al. 2003). In addition, it is important to remember that TM cells appear to sense an actomyosin-derived contractile force and induce ECM synthesis/assembly via Rho GTPase activation (Zhang et al. 2008). Another group of proteins present in the glaucomatous aqueous humor is that of the proteins involved in oxidative stress; in which the reduced expression of antioxidant enzymes SOD and GST worsens the imbalance between ROS and NOS that occurs during the course of glaucoma (Bagnis et al. 2012), aggravating molecular damage.

Lastly, another group of important proteins is that of neuronal proteins. Nestin and the above mentioned Optineurin belong to this group. Nestin, an intermediate filament protein, is thought to be expressed exclusively by neural progenitor cells in the normal brain, and is replaced by the expression of proteins specific for neurons or glia in differentiated cells. Furthermore, nestin is expressed not only in nervous system organs, but also in other organs and tissues such as the retina (Kohno et al. 2006), muscle (Kachinsky et al. 1994; Sejersen and Lendahl 1993), skin (Medina et al. 2006), liver (Forte et al. 2006), pancreas (Ueno et al. 2005), the testes (Frojdman et al. 1997), and others. We found significantly greater amounts of this protein in glaucomatous AH than in controls (Izzotti et al. 2010a). It has been suggested that nestin is expressed in dividing cells during the early stages of development and on differentiation, is downregulated and replaced by tissue-specific IF proteins. Although it remains unclear what factors regulate in vitro and in vivo expression of nestin, this protein is significantly more widespread than was previously thought. Mature neurons in the adult brain express nestin in the lateral ventricle and the dentate gyrus zone (Hendrickson et al. 2011). Furthermore, the fact that nestin expression in the endothelial cells of adult tissues is replenished by angiogenesis and in the endothelium of vascular neoplasms (Shimizu et al. 2006) and cancers (Aihara et al. 2004; Teranishi et al. 2002) suggests that nestin is also a marker of angiogenesis (Suzuki et al. 2010). Indeed nestin is expressed in proliferating endothelial cells and may be useful as a marker protein for neovascularization (Suzuki et al. 2010). This protein also participates in the formation of the cytoskeleton of newly formed endothelial cells (Mokrý et al. 2004). Our study confirmed the presence of nestin in glaucomatous AH, probably due to a response of the TM endothelial cells to injury. In glaucoma a dysfunction of the trabecular meshwork endothelium occurs which manifests itself through a hypertrophy and proliferation at the initial stages (Fine et al. 1981; Kuleshova et al. 2008). In addition, nestin has been found to be closely related to the cell membrane, intercellular junctions, and the nuclear membrane (Djabali 1999; Fuchs and Yang 1999) and is released upon the occurrence of cell damage (Yang et al. 1997; Vaittinen et al. 2001). Accordingly, its presence in AH in normal subjects may be due to physiological damage to TM endothelial cells, as occurs in response to the presence of free radicals in healthy conditions (Izzotti et al. 2009).

It is interesting that in glaucoma, this protein is used later to activate the glia, becoming a precise molecular signal for the retina, optic nerve and central nervous system (Saccà et al. 2016a). Furthermore nestin may be involved in mechanisms such as cell migration, generation of new neurons or glial cells and/or in retinal remodeling (Valamanesh et al. 2013).

In this complex scenario, the presence of nestin is indicative of the disintegration of the trabecular meshwork and its function which is to regulate the outflow of AH from the AC to Schlemm's Canal. Therefore, a hypertonus occurs which in some way also determines optical atrophy. However, the intraocular pressure increase may be the consequence of the dysfunction of the endothelial cells of the trabecular meshwork and the hypertonus its epiphenomenon. The change of the proteoma in glaucoma precisely describes this alteration. The interesting thing is that the same proteins that can be found in the aqueous humor can also be found in the posterior segment (Steely et al. 2000). In order to determine the degree of uveoscleral outflow in the pony, Smith et al. (1986) 1- and 3-microns (diam) microspheres were perfused through the anterior chamber for 60 and 90 min. After 90 min, 1- and 3-microns spheres penetrated the prominent supraciliary space and mixed with the suprachoroidea of the midchoroid. The 1-micron spheres infiltrated tissues more extensively than did the 3-micron spheres, packing into the anterior meshwork and supraciliary space and also moving as far posterior as the suprachoroidea of the peripapillary retina. Therefore, it is more than likely that these proteins may also end up in the posterior segment, if so, it is probable that AKAP 2 plays critical roles in the control of excitability neurons and K^+ transport. Synaptic plasticity is the ability of the connection between two neurons to change in strength in response to either the use or disuse of transmission over synaptic pathways. Synaptic plasticity plays important roles during normal postnatal development in learning and memory and in neurodegenerative diseases such as Alzheimer's. The increase or decrease in synaptic plasticity, and hence in the strength of synaptic transmission, occurs through postsynaptic AMPA-type glutamate receptors. Both actions are induced by activation of NMDA-type glutamate receptors but differ in the level and duration of Ca^{2+} influx through the NMDA receptor and the subsequent engagement of downstream signaling by protein kinases including PKA, PKC and others (Sanderson and Dell'Acqua 2011). Moreover, AKAP controls NMDA and AMPA receptor function and hence synaptic plasticity through its interaction with two Ca^{2+}-binding proteins caldendrin and calmodulin (Sekiguchi et al. 2013). Glutamate is a central nervous system excitatory neurotransmitter and has a central role in the conduction of signals between neurons. However, high extracellular levels of glutamate can induce neuronal cell death by excitotoxicity (Choi 1988). This phenomenon is well known at the level of the retina, and is determined by excessive exposure to the neurotransmitter glutamate or overstimulation of its membrane receptors, leading to neuronal injury or death (Lucas and Newhouse 1957). The ionotropic glutamate receptors have been classified into three major subtypes, AMPA, kainate, and *N*-methyl-D-aspartate (NMDA) receptors, named after their most selective agonist (Watkins et al. 1981). AMPA receptors are responsible for primary depolarization in glutamate-mediated neurotransmission and play key roles in synaptic plasticity. In glaucoma, the role of glutamate excitotoxicity remains unclear, not least because RGCs are relatively resistant to glutamate excitotoxicity in the presence of neurotrophic factors (Ullian et al. 2004).

The most abundantly expressed proteins in POAG explain both TM impairment and how these proteins can in turn be used as signals that spread in the posterior segment in the peri-papillary area. It is therefore conceivable that nestin, which is expressed in the anterior segment to activate stem cells present in the trabecular meshwork (Sacca et al. 2016b), functions in the posterior segment as an activator for glia. By contrast, AKAP 2 which in the posterior segment testifies to the breakdown of TM motility, in the posterior segment could be a signal for the apoptosis of retinal ganglion cells. Figure 14.6 summarizes the molecular events that lead to apoptosis.

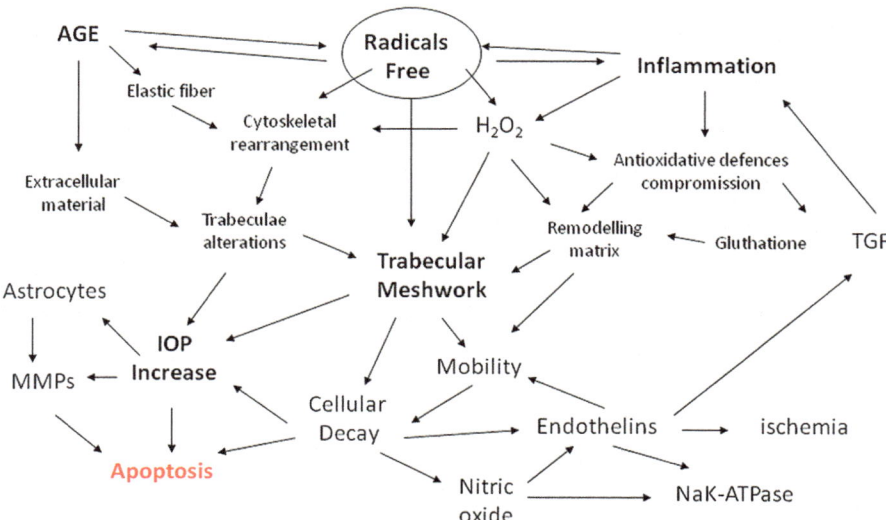

Fig. 14.6 Diagram of the molecular events that lead to the decay of the trabecular meshwork, determining apotosis both in the anterior segment and at the level of the optic nerve head

Apoptosis of Retinal Ganglion Cells in Glaucoma

Apoptosis is the prevalent cause of neuronal cells death in neurodegenerative diseases (Honig and Rosenberg 2000). It is a highly regulated and controlled process characterized by biochemical events that lead to characteristic changes in cell morphology and death. These changes include blebbing, cell shrinkage, nuclear fragmentation, chromatin condensation, chromosomal DNA fragmentation, and global mRNA decrease (Elmore 2007).

Retinal ganglion cells (RCGs) have been demonstrated to die by apoptosis in several models of experimental optic nerve lesions and in human glaucoma, and Cordeiro et al. (2004) showed in vivo apoptotic RCGs death in experimental ocular hypertension in rats. The processes that lead to apoptosis during glaucoma include various stimuli (i.e. mitochondrial dysfunction, inflammation, ischemia, neurotoxicity, neurotrophic factor deprivation, increased intracellular calcium concentration) and this process generally involves two pathways: the intrinsic and extrinsic pathways (Almasieh et al. 2012; Nickells 1999). The intrinsic pathway, also called the mitochondrial pathway, is modulated by intracellular signals produced when cells are stressed and is based on the release of specific proteins from the intermembrane space of mitochondria. Conversely, the extrinsic pathway is modulated by extracellular molecules which bind to cell-surface death receptors, and leads to the production of the death-inducing signaling complex (DISC) (Elmore 2007).

Intrinsic Pathway

Mitochondria are well-known coordination center for apoptosis. Indeed, many pro- or anti-apoptotic signals converge on the mitochondria (Izzotti et al. 2010a, b). When there is an imbalance in mitochondria, owing apoptosis, we observe an increase in the permeability of the mitochondrial membrane and the release of a variety of mediators of apoptosis (Scaffidi et al. 1998). One of the most important molecules involved in apoptosis is cytochrome c which binds in the cytoplasm the apoptotic protease-activating factor-1 (Apaf-1) to form the apoptosome, thereby initiating apoptosis (Qu et al. 2010). It has been shown that experimental optic nerve damage cause early cytochrome c release in RGCs (Cheung et al. 2003). However, in a chronic disease such as glaucoma, the persisting mitochondrial dysfunction is believed to have an important role in the RCG loss. Of note, mitochondrial dysfunction is widely accepted to be one of the most important features of many neurological diseases such as Alzheimer's disease, Parkinson's disease and, glaucoma (Kong et al. 2009; Kumar 2016). Healthy RCG cells have a high metabolic rate and their soma contains a large number of mitochondria, in order to support the elevated demand for energy. In glaucoma, mitochondrial DNA has been found to be damaged; this could reflects a decrease in the respiratory activity and energy production for the cell resulting in dysfunction and possibly apoptosis (Almasieh et al. 2012).

Interestingly, a negative correlation has been observed between age and both ATP levels in the optic nerve and intraocular pressure in mice (Baltan et al. 2010).

Moreover, damage to the mitochondria of RGCs may generate high levels of ROS causing the oxidation of lipids, proteins and mitochondrial DNA and creating a vicious circle that leads to their further deterioration (Izzotti et al. 2010).

The mitogen-activated protein kinases (MAPKs) are a broad family of protein Ser/Thr kinases that convert extracellular signals to series of molecular events, which ultimately result in a cellular response, and are central regulators of many cellular functions (Cargnello and Roux 2011). Among the MAPKs, a subfamily called Janus kinases (JNKs) has been demonstrated to be up-regulated in RGGs in animal models and in human glaucoma (Tezel et al. 2003). Moreover, administration of Janus kinase inhibitors has proved to be protective against RCG apoptosis in experimental ocular hypertension (Sun et al. 2011). Another subfamily of the MAPKs is that of the p38 MAPKs; these are over-expressed in glaucoma and their inhibition has proved to reduce the apoptosis of RGCs (Kikuchi et al. 2000). The fact that these MAPKs have been found to be associated with glaucomatous damage, does not prove their etiologic role. However, they may constitute a target for novel glaucoma therapies.

Bcl-2 family members are another class of proteins involved in the apoptotic process of the RGCs. Among them, the Bcl-XL is the predominant anti-apoptotic protein in the rat retina, and Bax also plays an important role in apoptosis (Elmore 2007). However, there is a lack of evidence that modulation of Bax expression by means of small interfering RNA (siRNA) promotes RCGs survival in animal models (Lingor et al. 2005). Again, identifying novel actors of the RCGs apoptotic pathway may open the door in future to innovative glaucoma therapies.

Extrinsic Pathway

TNF-alpha is the cytokine that has the most important role in the extrinsic pathway of apoptosis. When TNF-alpha binds to TNF-alpha receptors (TNFR1, TNFR2), it is able to initiate a cascade that leads to caspase activation and cell apoptosis (Elmore 2007). Of note, elevated values of TNF-alpha have been found in the brain, cerebrospinal fluid, and serum of patients affected by neurodegenerative diseases such as Alzheimer's disease, Parkinson's disease and, multiple sclerosis (Sriram and O'Callaghan 2007). Indeed, intravitreal injection of TNF-alpha is associated with RGC loss and TNF-alpha levels in optic nerve axons and in the aqueous humor of glaucoma patients level have been seen to increase (Tezel 2008). Although inhibition of TNF-alpha activity in the retina has been seen to result in marked RGC neuroprotection and TNF-alpha inhibitors are currently approved for the treatment of several autoimmune diseases such as rheumatoid arthritis and psoriasis, the role of such inhibitors in the treatment of glaucoma has not been yet demonstrated.

Triggers of RCG Apoptosis

In experimental glaucoma, a model of trans-synaptic degeneration has been described (Gupta and Yücel 2001), whereby damage spreads through synaptic connections, gradually influencing the whole neural pathway, from the retro-laminar portion of the ON to the distal portion at the level of the orbital apex (Bolacchi et al. 2012). Axonal transport is critical to the correct functioning of neurons, while the retrograde transport of neurotrophins may be necessary for the survival of RGCs. One of these neurotrophins is brain-derived neurotrophic factor (BDNF). BDNF is essential for the growth and survival of nerve cells (Sampaio et al. 2017). Conversely, excitotoxicity may also be associated with apoptosis of RGCs (Morrone et al. 2015). Excitotoxicity is the pathological process that leads to the death of neurons exposed to over-activation of the N-methyl-D-aspartate (NMDA) glutamate receptors (Prentice et al. 2015). Excessive activation of the NMDA receptor causes an influx of ions into the cell, in particular Ca^{2+}. The excessive influx of calcium activates enzymes that degrade the cell membranes, cellular proteins, and nucleic acids, and eventually leads to apoptosis (Saccà and Izzotti 2008). It has been observed that the level of glutamate in the vitreous is increased in glaucoma patients and animal models of glaucoma. These data may suggest the involvement of excitotoxicity in RGC apoptosis (Dreyer and Grosskreutz 1997).

Another recently recent hypothesis (Saccà et al. 2016) speculates that pro-apoptotic proteins could pass from the anterior chamber to the posterior segment, becoming biological signals for the retina. Indeed, there is a pathway through which molecules pass from the anterior chamber to the optic nerve head, through the suprachoroidal space (Smith et al. 1986). Hence, these proteins may reach the retina, and in particular the peripapillary area. Indeed, while the nestin expressed in the AC by TM cells attracts stem cells in an attempt to repair the TM, in the retina it could activate glia (Wang et al. 2000; Xue et al. 2006).The failure of optic nerve glia to clear axonal debris may lead to the accumulation of toxic proteins and hence to neurodegeneration. Therefore, the proteins secreted by the damaged TM endothelial cells might become molecular messengers that enable communication between AC structures and the ONH, thereby playing a role in the RGC apoptosis that characterizes glaucomatous optic neuropathy. Translational research is in progress to further elucidate this pathogenic mechanism.

Finally, we must also briefly mention the role that autophagy plays in cell death. Autophagy is a cellular process of lysosomal degradation that is essential for survival, differentiation, development and homeostasis. The relationship between ROS formation and mitochondrial metabolism is essential to cell homeostasis, and therefore also to mitochondria, in that it regulates immune responses and autophagy (Dan Dunn et al. 2015). Mitochondrial dysfunction, as we have seen, plays a fundamental role in neurodegeneration and glaucoma. Damaged mitochondria generate further ROS, especially if mitophagy is insufficient (Pryde et al. 2016). Mitophagy is a process of selective autophagic destruction of mitochondria that have become defective as a result of stress or damage, through autophagy. It is a process of selective destruction of defective mitochondria as a result of stress or damage. Autophagy also takes place constitutively in RGCs. Indeed, acute IOP elevation induces a reduction in markers of autophagy, suggesting a possible role of IOP in disrupting the retinal autophagic mechanism. This supports the notion that autophagy exerts a neuroprotective effect in the retina, and suggests that autophagy dysfunction may have a key role in the neuronal degeneration processes occurring in both glaucoma and Alzheimer disease (Nucci et al. 2013). This dysfunction is influenced by myocilin. The turnover of endogenous myocilin involves the ubiquitin-proteasome and lysosomal pathways. When myocilin is up-regulated or mutated, the ubiquitin-proteasome function is compromised and autophagy is induced (Qiu et al. 2014). Induced autophagy has also been demonstrated in vivo in retinal ganglion cells of transgenic mice with optineurin mutation (Shen et al. 2011). Finally, numerous studies have indicated that autophagy declines with age, and that its induction can promote longevity (Madeo et al. 2010; Morselli et al. 2010). The relevance of the autophagic pathway in the context of glaucoma-associated RGC death and how this contributes to the pathology is still unclear. Furthermore, a number of studies have focused on autophagy as a potential target for pharmacological modulation in order to achieve neuroprotection (Russo et al. 2015).

Conclusions

Some changes in the visual field are age-related rather than disease-related. Indeed, it is known that after the age of 40 years the peripheral visual field deteriorates, probably owing to a progressive decay of the peripheral lamina choriocapillaris (Rutkowski and May 2017). It remains to be ascertained whether these defects have a real impact on the evaluation of the possible presence of disease. VF is the only direct method which allows us to measure the visual function of patients and, therefore, also in the case of glaucoma; it also enables to make a point-to-point map of the retinal function projected in space (Fig. 14.7). Therefore, it allows us to assess disease progression and consequently, the effectiveness of treatment to slow down this progression and the loss of VF. Fortunately, progression is very fast only in 6–13% of cases (Chen 2003). Heijl et al. (2012) have calculated that in only 15% is a progression of the VF defect greater than −1.5 dB/year. Glaucoma damage primarily affects the optic nerve fibers entering the upper and lower poles and coming from the ganglionic temporal retina cells. Extending along the papillomacular beam in an arch-like fashion and respecting the horizontal meridian. The initial damage occurs as a paracentral scotoma prevalently nasal to the blind spot. Automatic perimetry is now the standard examination. Unfortunately, however, the great individual variability of the VF determines the initial examination (Heijl et al. 1989) (not everyone has the skill to do this test well!); consequently, frequent monitoring and/or a long period of time is required in order to accurately detect glaucomatous progression (Gardiner and Crabb 2002; Chauhan et al. 2008). Furthermore, the subjects have no symptoms except in the advanced stages of the disease. Thus, little is

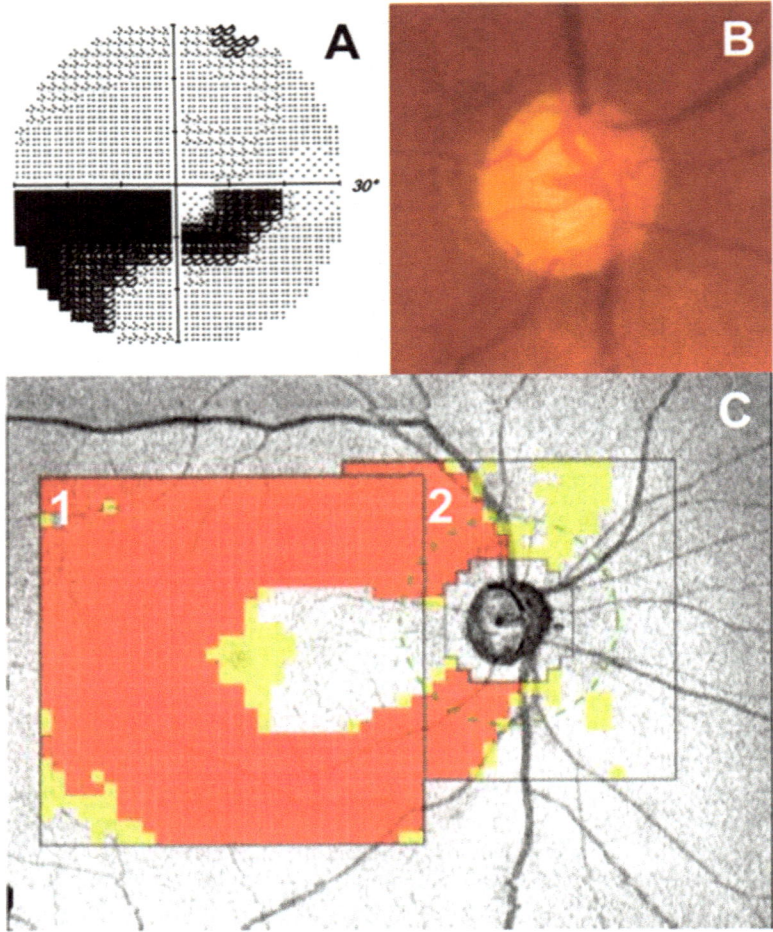

Fig. 14.7 Glaucomatous optic nerve (right eye) (**a**) Dense glaucomatous arcuate scotoma. (**b**) Glaucomatous optic disc with evident narrowing of the superotemporal and inferotemporal neuroretinal rim. (**c**) *Box 1*: Probability map of the ganglion cell layer (GCL) of the retina. Each pixel in the box is color-coded: no color (within the normal limit), yellow (outside 95% of the normal limit), or red (outside 99% of the normal limit). *Box 2* shows the retinal nerve fiber layer (RNFL). The superior arcuate defect of the GCL and RNFL correlates with the inferior defect shown in the visual field (**a**)

known about the precise relationship between functional measurements made in the clinic and the patient's visual disability.

Some VF defects, even advanced VF loss do not compromise the normal quality of life (Glen et al. 2012; Burton et al. 2014); in other cases, however quality of life is seriously impaired even if the disease is mild or moderate (Alqudah et al. 2016). In any case, vision-related quality of life first declines gradually, and then more quickly when functional abilities are significantly compromised (Jones et al. 2017).

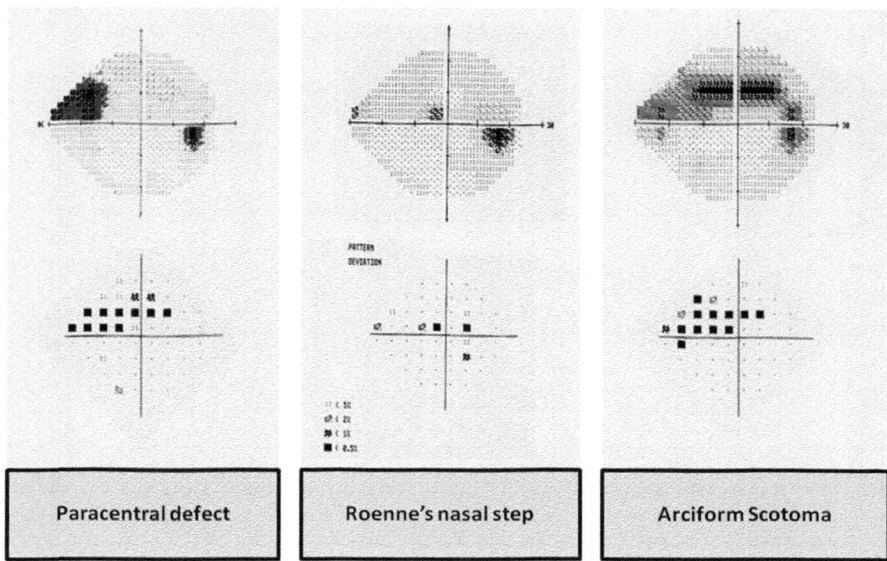

Paracentral defect   **Roenne's nasal step**   **Arciform Scotoma**

Fig. 14.8 Examples of different visual field defects that can occur in different phases of neurode-generative diseases or in glaucoma. These defects may proceed differently in the upper and lower portions of the VF, though the upper VF is more frequently involved in the early stages of glaucoma

It is probable that the degree of quality of life impairment is in strictly dependent once the site of the VF damage (Fig. 14.8). Indeed, a loss of peripheral vision can compromise a person's ability to move around safety in the environment. By contrast, a loss of central vision is a very serious obstacle to visual performance. Indeed, progressive loss of binocular RNFL thickness is associated with longitudinal loss of quality of life, even after adjustment for progressive visual field loss (Gracitelli et al. 2015). The loss of axons that occurs with ageing (Calkins 2013) is about 40% over the lifespan (Neufeld and Gachie 2003); photoreceptors also show a decline, 30% over a lifetime, and while the number of cones remains stable (Curcio et al. 1993), this natural evolution does not affect VF in healthy subjects. This visual decline is probably caused by diminished mitochondrial efficiency (Toescu 2005) with the result that the homeostatic balance of intracellular Ca^{2+} is lost, Na^+/K^+ -ATPase activity decreases, and increased oxidative injury occurs (Toescu and Verkhratsky 2000). From a clinical point of view the visual disability of glaucoma patients can be quantified in a suitable environment by measuring the patient's actual performance of real-life tasks (Wei et al. 2012; Crabb 2016). Indeed, it is now known that many patients with advanced visual field loss, even with preserved visual acuity, have measurable difficulty in reading (Burton et al. 2014). These subjects have difficulty in grasping objects (Kotecha et al. 2009), (Fig. 14.9), and those with a scotoma in the inferior VF are more likely to fall (Kotecha et al. 2012). In addition, these subjects have reduced ability to recognize obstacles while

Fig. 14.9 A person with visual impairment due to neurodegeneration is likely to have vision in which the brain is able to reconstruct the picture seen. However, the details are blurred owing to the VF damage (**a**) compared with normal vision (**b**).Tunnel vision does not adequately describe the visual experience of patients: indeed, it is very tough for a patient to describe his visual defect in terms of "black" or "tunnel", the terms usually used in books to describe these types of visual defects. Affected patients usually use words such as "foggy", "fuzzy", or "unfocused" (Crabb et al. 2013). In this figure, the patient's has a visual defect affecting his upper right side (**a**)

driving (Haymes et al. 2008; Blane 2016), and have difficulty locating everyday objects and recognizing faces (Smith et al. 2011; Glen et al. 2012). Therefore, how do patients who have a VF defect see? Certainly the damaged portion of the VF cannot be replaced by undamaged portions. However, the brain has the ability to complete and to reconstruct the missing VF portions. This means that patients suffering from abnormal VF may not perceive their impaired vision as a problem or a disability (Crabb 2016). However, as in all patients with neurodegenerative diseases, the problem that exists is real and can actually be dangerous for themselves and for others.

Acknowledgements The authors would like to thank Dr. Carmen Laethem for allowing us to publish her beautiful photograph of TM endothelial cells (Fig. 14.4).

References

Aga M, Bradley JM, Wanchu R, Yang YF, Acott TS et al (2014) Differential effects of caveolin-1 and -2 knockdown on aqueous outflow and altered extracellular matrix turnover in caveolin-silenced trabecular meshwork cells. Invest Ophthalmol Vis Sci 55:5497–5509

Agarwal S, Sohal RS (1994) Ageing and protein oxidative damage. Mech Ageing Dev 75:11–19

Aihara M, Sugawara K, Torii S, Hosaka M, Kurihara H et al (2004) Angiogenic endothelium-specific nestin expression is enhanced by the first intron of the nestin gene. Lab Investig 84:1581–1592

Al-Aswad LA, Gong H, Lee D, O'Donnell ME, Brandt JD et al (1999) Effects of Na-K-2Cl cotransport regulators on outflow facility in calf and human eyes in vitro. Invest Ophthalmol Vis Sci 40:1695–1701

Alexander JP, Acott TS (2003) Involvement of the Erk-MAP kinase pathway in TNFalpha regulation of trabecular matrix metalloproteinases and TIMPs. Invest Ophthalmol Vis Sci 44:164–169

Almasieh M, Wilson AM, Morquette B, Cueva Vargas JL, Di Polo A (2012) The molecular basis of retinal ganglion cell death in glaucoma. Prog Retin Eye Res 31:152–181

Alqudah A, Mansberger SL, Gardiner SK, Demirel S (2016) Vision-related quality of life in glaucoma suspect or early glaucoma patients. J Glaucoma 25:629–633

Alvarado JA, Shifera AS (2010) Progress towards understanding the functioning of the trabecular meshwork based on lessons from studies of laser trabeculoplasty. Br J Ophthalmol 94:1417–1418

Alvarado JA, Murphy CG, Polansky JR, Juster R (1981) Agerelated changes in trabecular meshwork cellularity. Invest Ophthalmol Vis Sci 21:714–727

Alvarado JA, Murphy C, Juster R (1984) Trabecular meshwork cellularity in primary open-angle glaucoma and nonglaucomatous normals. Ophthalmology 91:564–579

Alvarado JA, Alvarado RG, Yeh RF, Franse-Carman L, Marcellino GR et al (2005a) A new insight into the cellular regulation of aqueous outflow: how trabecular meshwork endothelial cells drive a mechanism that regulates the permeability of Schlemm's canal endothelial cells. Br J Ophthalmol 89:1500–1505

Alvarado JA, Yeh RF, Franse-Carman L, Marcellino G, Brownstein MJ (2005b) Interactions between endothelia of the trabecular meshwork and of Schlemm's canal: a new insight into the regulation of aqueous outflow in the eye. Trans Am Ophthalmol Soc 103:148–162

Ames BN, Gold LS (1991) Endogenous mutagens and the causes of ageing and cancer. Mutat Res 250:3–16

Aroca-Aguilar JD, Sánchez-Sánchez F, Ghosh S, Fernández-Navarro A, Coca-Prados M et al (2011) Interaction of recombinant myocilin with the matricellular protein SPARC: functional implications. Invest Ophthalmol Vis Sci 52:179–189

Bagnis A, Izzotti A, Centofanti M, Saccà SC (2012) Aqueous humor oxidative stress proteomic levels in primary open angle glaucoma. Exp Eye Res 103:55–62

Baker DJ, Childs BG, Durik M, Wijers ME, Sieben CJ et al (2016) Naturally occurring p16Ink4a-positive cells shorten healthy lifespan. Nature 530:184–189

Balci M, Sahin S, Mutlu FM, Yağci R, Karanci P, Yildiz M (2011) Investigation of oxidative stress in pterygium tissue. Mol Vis 17:443–447

Baltan S, Inman DM, Danilov CA, Morrison RS, Calkins DJ et al (2010) Metabolic vulnerability disposes retinal ganglion cell axons to dysfunction in a model of glaucomatous degeneration. J Neurosci 30:5644–5652

Barja G (2004) Free radicals and ageing. Trends Neurosci 27:595–600

Barnett A, Brewer GJ (2011) Autophagy in ageing and Alzheimer's disease: pathologic or protective? J Alzheimers Dis 25:385–394

Bengrine A, Li J, Awayda MS (2007) The A-kinase anchoring protein 15 regulates feedback inhibition of the epithelial Na+ channel. FASEB J 21:1189–1201

Bhatia-Dey N, Kanherkar RR, Stair SE, Makarev EO, Csoka AB (2016) Cellular senescence as the causal nexus of ageing. Front Genet 7:13

Bhatnagar S, Panguluri SK, Gupta SK, Dahiya S, Lundy RF, Kumar A (2010) Tumor necrosis factor-α regulates distinct molecular pathways and gene networks in cultured skeletal muscle cells. PLoS One 5:e13262

Birren JE, Fisher LM (1991) Ageing and slowing of behavior: consequences for cognition and survival. Nebr Symp Motiv 39:1–37

Bishop AL, Hall A (2000) Rho GTPases and their effector proteins. Biochem J 348:241–255

Black AT, Gordon MK, Heck DE, Gallo MA, Laskin DL, Laskin JD (2011) UVB light regulates expression of antioxidants and inflammatory mediators in human corneal epithelial cells. Biochem Pharmacol 81:873–880

Blane A (2016) Through the looking glass: a review of the literature investigating the impact of glaucoma on crash risk, driving performance, and driver self-regulation in older drivers. J Glaucoma 25:13–21

Blanks JC, Torigoe Y, Hinton DR, Blanks RH (1996) Retinal pathology in Alzheimer's disease. I Ganglion cell loss in foveal/parafoveal retina. Neurobiol Ageing 17:377–384

Bolacchi F, Garaci FG, Martucci A, Meschini A, Fornari M et al (2012) Differences between proximal versus distal intraorbital optic nerve diffusion tensor magnetic resonance imageing properties in glaucoma patients. Invest Ophthalmol Vis Sci 53:4191–4196

Bourne RR, Jonas JB, Flaxman SR, Keeffe J, Leasher J et al (2014) Prevalence and causes of vision loss in high-income countries and in Eastern and Central Europe: 1990–2010. Br J Ophthalmol 98:629–638

Bourne RRA, Flaxman SR, Braithwaite T, Cicinelli MV, Das A et al (2017) Magnitude, temporal trends, and projections of the global prevalence of blindness and distance and near vision impairment: a systematic review and meta-analysis. Lancet Glob Health 5:e888–e897

Braakman ST, Pedrigi RM, Read AT, Smith JA, Stamer WD et al (2014) Biomechanical strain as a trigger for pore formation in Schlemm's canal endothelial cells. Exp Eye Res 127:224–235

Brandt JD, O'Donnell ME (1999) How does the trabecular meshwork regulate outflow? Clues from the vascular endothelium. J Glaucoma 8:328–339

Burton R, Smith ND, Crabb DP (2014) Eye movements and reading in glaucoma: observations on patients with advanced visual field loss. Graefes Arch Clin Exp Ophthalmol 252:1621–1630

Calkins DJ (2013) Age-related changes in the visual pathways: blame it on the axon. Invest Ophthalmol Vis Sci 54:37–41

Cargnello M, Roux PP (2011) Activation and function of the MAPKs and their substrates, the MAPK-activated protein kinases. Microbiol Mol Biol Rev 75:50–83

Cassels NK, Wild JM, Margrain TH, Chong V, Acton JH (2018) The use of microperimetry in assessing visual function in age-related macular degeneration. Surv Ophthalmol 63:40–55

Chauhan BC, Garway-Heath DF, Goni FJ, Rossetti L, Bengtsson B et al (2008) Practical recommendations for measuring rates of visual field change in glaucoma. Br J Ophthalmol 92:569–573

Chen PP (2003) Blindness in patients with treated open-angle glaucoma. Ophthalmology 110:726–733

Chen Y, Wang Y, Yu H, Wang F, Xu W (2005) The cross talk between protein kinase A- and RhoA-mediated signaling in cancer cells. Exp Biol Med (Maywood) 230:731–741

Cheung ZH, Yip HK, Wu W, So KF (2003) Axotomy induces cytochrome c release in retinal ganglion cells. Neuroreport 14:279–282

Chi X, Wang S, Huang Y, Stamnes M, Chen JL (2013) Roles of rho GTPases in intracellular transport and cellular transformation. Int J Mol Sci 14:7089–7108

Choi DW (1988) Glutamate neurotoxicity and diseases of the nervous system. Neuron 1:623–634

Choi SI, Kim TI, Kim KS, Kim BY, Ahn SY et al (2009) Decreased catalase expression and increased susceptibility to oxidative stress in primary cultured corneal fibroblasts from patients with granular corneal dystrophy type II. Am J Pathol 175:248–261

Coca-Prados M, Ghosh S (2008). Functional modulators linking inflow with outflow of aqueous humor. In: Civan M (ed) In the eye's aqueous humour, 2nd edn. Curr Topics Membr 5:123–160. Vol. 62, Elsevier

Colledge M, Scott JD (1999) AKAPs from structure to function. Trends Cell Biol 9:216–221

Coppé JP, Patil CK, Rodier F, Sun Y, Muñoz DP et al (2008) Senescence-associated secretory phenotypes reveal cell-Nonautonomous functions of oncogenic RAS and the p53 tumor suppressor. PLoS Biol e301:6

Cordeiro MF, Guo L, Luong V, Harding G, Wang W et al (2004) Real-time imaging of single nerve cell apoptosis in retinal neurodegeneration. Proc Natl Acad Sci U S A 101:13352–13356

Cory GOC, Cramer R, Blanchoin L, Ridley AJ (2003) Phosphorylation of the WASP-VCA domain increases its affinity for the Arp2/3 complex and enhances actin polymerization by WASP. Mol Cell 11:1229–1239

Costarides AP, Riley MV, Green K (1991) Roles of catalase and the glutathione redox cycle in the regulation of anterior-chamber hydrogen peroxide. Ophthalmic Res 23:284–294

Crabb DP (2016) A view on glaucoma – are we seeing it clearly? Eye (Lond) 30:304–313

Crabb DP, Smith ND, Glen FC, Burton R, Garway-Heath DF (2013) How does glaucoma look?: patient perception of visual field loss. Ophthalmology 120:1120–1126

Cuervo AM, Stefanis L, Fredenburg R, Lansbury PT, Sulzer D (2004) Impaired degradation of mutant α-synuclein by chaperone-mediated autophagy. Science 305:1292–1295

Cullinan SB, Diehl JA (2006) Coordination of ER and oxidative stress signaling: the PERK/Nrf2 signaling pathway. Int J Biochem Cell Biol 38:317–332

Curcio CA, Millican CL, Allen KA, Kalina RE (1993) Ageing of the human photoreceptor mosaic: evidence for selective vulnerability of rods in central retina. Invest Ophthalmol Vis Sci 34:3278–3296

Dan Dunn J, Alvarez LA, Zhang X, Soldati T (2015) Reactive oxygen species and mitochondria: a nexus of cellular homeostasis. Redox Biol 6:472–485

de Magalhães JP, Passos JF (2017) Stress, cell senescence and organismal ageing. Mech Ageing Dev S0047–6374:30078–30077

Dismuke WM, Mbadugha CC, Ellis DZ (2008) NO-induced regulation of human trabecular meshwork cell volume and aqueous humor outflow facility involve the BKCa ion channel. Am J Physiol Cell Physiol 294:C1378–C1386

Diviani D, Scott JD (2001) AKAP signaling complexes at the cytoskeleton. J Cell Sci 114:1431–1437

Diviani D, Soderling J, Scott JD (2001) AKAP-Lbc anchors protein kinase A and nucleates Galpha 12-selective Rho-mediated stress fiber formation. J Biol Chem 276:44247–44257

Djabali K (1999) Cytoskeletal proteins connecting intermediate filaments to cytoplasmic and nuclear periphery. Histol Histopathol 14:501–509

Dreyer EB, Grosskreutz CL (1997) Excitatory mechanisms in retinal ganglion cell death in primary open angle glaucoma (POAG). Clin Neurosci 4:270–273

Dumont P, Burton M, Chen QM, Gonos ES, Frippiat C et al (2000) Induction of replicative senescence biomarkers by sublethal oxidative stresses in normalhuman fibroblast. Free Radic Biol Med 28:361–373

Elmore S (2007) Apoptosis: a review of programmed cell death. Toxicol Pathol 35:495–516

El-Remessy AB, Tawfik HE, Matragoon S, Pillai B, Caldwell RB et al (2010) Peroxynitrite mediates diabetes-induced endothelial dysfunction: possible role of Rho kinase activation. Exp Diabetes Res 247861:2010

Epstein DL, Rohen JW (1991) Morphology of the trabecular meshwork and inner-wall endothelium after cationized ferritin perfusion in the monkey eye. Invest Ophthalmol Vis Sci 32:160–171

Eriksson EE, Xie X, Werr J, Thoren P, Lindbom L (2001) Direct viewing of atherosclerosis in vivo: plaque invasion by leukocytes is initiated by the endothelial selectins. FASEB J 15:1149–1157

Erusalimsky JD, Skene C (2009) Mechanisms of endothelial senescence. Exp Physiol 94:299–304

Fine BS, Yanoff M, Stone RA (1981) A clinicopathologic study of four cases of primary open-angle glaucoma compared to normal eyes. Am J Ophthalmol 91:88–105

Finkel T, Holbrook NJ (2000) Oxidants, oxidative stress and the biology of ageing. Nature 408:239–247

Flatt T (2012) A new definition of ageing? Front Genet 3:148

Forte G, Minieri M, Cossa P, Antenucci D, Sala M et al (2006) Hepatocyte growth factor effects on mesenchymal stem cells: proliferation, migration, and differentiation. Stem Cells 24:23–33

Francis SH, Corbin JD (1994) Structure and function of cyclic nuleotide-dependent protein kinases. Annu Rev Physiol 56:237–272

Frippiat C, Chen QM, Zdanov S, Magalhaes JP, Remacle J et al (2001) Subcytotoxic H2O2 stress triggers a release of transforming growthfactor-beta1, which induces biomarkers of cellular senescence of humandiploid fibroblasts. J Biol Chem 276:2531–2537

Frojdman K, Pelliniemi L, Lendahl U, Virtanen I, Eriksson JE (1997) The intermediate filament protein nestin occurs transiently in differentiating testis of rat and mouse. Differentiation 61:243–249

Fuchs E, Yang Y (1999) Crossroads on cytoskeletal highways. Cell 98:547–550

Gardiner SK, Crabb DP (2002) Frequency of testing for detecting visual field progression. Br J Ophthalmol 86:560–564

Girkin CA, McGwin G Jr, Sinai MJ, Sekhar GC, Fingeret M et al (2011) Variation in optic nerve and macular structure with age and race with spectral-domain optical coherence tomography. Ophthalmology 118:2403–2408

Glen FC, Crabb DP, Smith ND, Burton R, Garway-Heath DF (2012) Do patients with glaucoma have difficulty recognizing faces? Invest Ophthalmol Vis Sci 53:3629–3637

Gogg S, Smith U, Jansson PA (2009) Increased MAPK activation and impaired insulin signaling in subcutaneous microvascular endothelial cells in type 2 diabetes: the role of endothelin-1. Diabetes 58:2238–2245

Gong H, Freddo TF, Johnson M (1992) Age-related changes of sulfated proteoglycans in the normal human trabecular meshwork. Exp Eye Res 55:691–709

Gong H, Overby D, Ruberti J, Freddo T, Johnson M (2001) Human outflow pathway viewed by quick freeze deep etch. Invest Ophthalmol Vis Sci 42:S749

Gong H, Ruberti J, Overby D, Johnson M, Freddo TF (2002) A new view of the human trabecular meshwork using quick-freeze, deep-etch electron microscopy. Exp Eye Res 75:347–358

Gorbatyuk M, Gorbatyuk O (2013) Review: retinal degeneration: focus on the unfolded protein response. Mol Vis 19:1985–1998

Gracitelli CP, Abe RY, Tatham AJ, Rosen PN, Zangwill LM et al (2015) Association between progressive retinal nerve fiber layer loss and longitudinal change in quality of life in glaucoma. JAMA Ophthalmol 133:384–390

Gupta N, Yücel YH (2001) Glaucoma and the brain. J Glaucoma 10:S28–S29

Hailey DW, Rambold AS, Satpute-Krishnan P, Mitra K, Sougrat R et al (2010) Mitochondria supply membranes for autophagosome biogenesis during starvation. Cell 141:656–667

Hall A (1998) Rho GTPases and the actin cytoskeleton. Science 279:509–514

Harada C, Nakamura K, Namekata K, Okumura A, Mitamura Y et al (2006) Role of apoptosis signal-regulating kinase 1 in stress-induced neural cell apoptosis in vivo. Am J Pathol 168:261–269

Harada C, Namekata K, Guo X, Yoshida H, Mitamura Y et al (2010) ASK1 deficiency attenuates neural cell death in GLAST-deficient mice, a model of normal tension glaucoma. Cell Death Differ 17:1751–1759

Hayakawa M, Hattori K, Sugiyama S, Ozawa T (1992) Age associated oxygen damage and mutations in mitochondrial DNA in human hearts. Biochem Biophys Res Commun 189:979–985

Haymes SA, LeBlanc RP, Nicolela MT, Chiasson LA, Chauhan BC (2008) Glaucoma and on-road driving performance. Invest Ophthalmol Vis Sci 49:3035–3041

He Y, Ge J, Tombran-Tink J (2008) Mitochondrial defects and dysfunction in calcium regulation in glaucomatous trabecular meshwork cells. Invest Ophthalmol Vis Sci 49:4912–4922

Heijl A, Lindgren A, Lindgren G (1989) Test–retest variability in glaucomatous visual fields. Am J Ophthalmol 108:130–135

Heijl A, Buchholz P, Norrgren G, Bengtsson B (2012) Rates of visual field progression in clinical glaucoma care. Acta Ophthalmol 91:406–412

Hendrickson ML, Rao AJ, Demerdash ON, Kalil RE (2011) Expression of nestin by neural cells in the adult rat and human brain. PLoS One 6:e18535

Hewitt G, Jurk D, Marques FDM, Correia-Melo C, Hardy T et al (2012) Telomeres are favoured targets of a persistent DNA damage response in ageing and stress-induced senescence. Nat Commun 3:708

Honig LS, Rosenberg RN (2000) Apoptosis and neurologic disease. Am J Med 108:317–330

Huang LJ, Durick K, Weiner JA, Chun J, Taylor SS (1997) D-AKAP2, a novel protein kinase A anchoring protein with a putative RGS domain. Proc Natl Acad Sci U S A 94:11184–11189

Ikehata H, Ono T (2011) The mechanisms of UV mutagenesis. J Radiat Res (Tokyo) 52:115–125

Iseri PK, Altinaş O, Tokay T, Yüksel N (2006) Relationship between cognitive impairment and retinal morphological and visual functional abnormalities in Alzheimer disease. J Neuroophthalmol 26:8–24

Izzotti A (2009) Gene environment interactions in noncancer degenerative diseases. Mutat Res Fundam Mol Mech Mutagen 667:1–3

Izzotti A, Sacca SC, Cartiglia C, De Flora S (2003) Oxidative deoxyribonucleic acid damage in the eyes of glaucoma patients. Am J Med 114:638e646

Izzotti A, Saccà SC, Longobardi M, Cartiglia C (2009) Sensitivity of ocular anterior chamber tissues to oxidative damage and its relevance to the pathogenesis of glaucoma. Invest Ophthalmol Vis Sci 50:5251–5258

Izzotti A, Longobardi M, Cartiglia C, Saccà SC (2010a) Proteome alterations in primary open angle glaucoma aqueous humor. J Proteome Res 9:4831–4838

Izzotti A, Saccà SC, Longobardi M, Cartiglia C (2010b) Mitochondrial damage in the trabecular meshwork of patients with glaucoma. Arch Ophthalmol 128:724–730

Izzotti A, Longobardi M, Cartiglia C, Saccà SC (2011) Mitochondrial damage in the trabecular meshwork occurs only in primary open-angle glaucoma and in pseudoexfoliative glaucoma. PLoS One 6:e14567

Jayadev S, Petranka JG, Cheran SK, Biermann JA, Barrett JC et al (1999) Reduced capacitative calcium entry correlates with vesicle accumulation and apoptosis. J Biol Chem 274:8261–8268

Joe MK, Tomarev SI (2010) Expression of myocilin mutants sensitizes cells to oxidative stress-induced apoptosis: implication for glaucoma pathogenesis. Am J Pathol 176:2880–2890

Johansen T, Lamark T (2011) Selective autophagy mediated by autophagic adapter proteins. Autophagy 7:279e296

Johnson M, Erickson K (2000) Aqueous humor and the dynamics of its flow. In: Albert DM, Jakobiec FA (eds) Principles and practice of ophthalmology. Saunders, Philadelphia, pp 2577–2595

Johnson BM, Miao M, Sadun AA (1987) Age-related decline of human optic nerve axon populations. Age (Omaha) 10:5–9

Jones L, Bryan SR, Crabb DP (2017) Gradually then suddenly? Decline in vision-related quality of life as glaucoma worsens. J Ophthalmol 2017:1621640

Joshi MB, Philippova M, Ivanov D, Allenspach R, Erne P, Resink TJ (2005) T-cadherin protects endothelial cells from oxidative stress-induced apoptosis. FASEB J 19:1737–1739

Joyce NC, Zhu CC, Harris DL (2009) Relationship among oxidative stress, DNA damage, and proliferative capacity in human corneal endothelium. Invest Ophthalmol Vis Sci 50:2116–2122

Jurkunas UV, Bitar MS, Funaki T, Azizi B (2010) Evidence of oxidative stress in the pathogenesis of fuchs endothelial corneal dystrophy. Am J Pathol 177:2278–2289

Kachinsky AM, Dominov JA, Miller JB (1994) Myogenesis and the intermediate filament protein, nestin. Dev Biol 165:216–228

Kania E, Pająk B, Orzechowski A (2015) Calcium homeostasis and ER stress in control of autophagy in cancer cells. Biomed Res Int 2015:352794

Karar J, Maity A (2011) PI3K/AKT/mTOR Pathway in Angiogenesis. Front Mol Neurosci 4:51

Keller KE, Kelley MJ, Acott TS (2007) Extracellular matrix gene alternative splicing by trabecular meshwork cells in response to mechanical stretching. Invest Ophthalmol Vis Sci 48:1164–1172

Keller J, Sánchez-Dalmau BF, Villoslada P (2014) Lesions in the posterior visual pathway promote trans-synaptic degeneration of retinal ganglion cells. PLoS One 9:e97444

Khurana RN, Deng PF, Epstein DL, Vasantha Rao P (2003) The role of protein kinase C in modulation of aqueous humor outflow facility. Exp Eye Res 76:39–47

Kikuchi M, Tenneti L, Lipton SA (2000) Role of p38 mitogen-activated protein kinase in axotomy-induced apoptosis of rat retinal ganglion cells. J Neurosci 20:5037–5044

Kilpatrick BS, Eden ER, Schapira AH, Futter CE, Patel S (2013) Direct mobilization of lysosomal Ca2+ triggers complex Ca2+ signals. J Cell Sci 126:60–66

Kim M, Ho A, Lee JH (2017) Autophagy and human neurodegenerative diseases-A fly's perspective. Int J Mol Sci 18(7):E1596

Koch MA, Rosenhammer B, Koschade SE, Braunger BM, Volz C et al (2014) Myocilin modulates programmed cell death during retinal development. Exp Eye Res 125:41e52

Kohno H, Sakai T, Kitahara K (2006) Induction of nestin, Ki-67, and cyclin D1 expression in Muller cells after laser injury in adult rat retina., *Graefes Arch*. Clin Exptl Ophthalmol 244:90–95

Kong GY, Van Bergen NJ, Trounce IA, Crowston JG (2009) Mitochondrial dysfunction and glaucoma. J Glaucoma 18:93–100

Konstas AG, Katsanos A, Quaranta L, Mikropoulos DG, Tranos PG et al (2015) Twenty-four hour efficacy of glaucoma medications. Prog Brain Res 221:297–318

Korth M, Horn F, Storck B, Jonas JB (1989) Spatial and spatiotemporal contrast sensitivity of normal and glaucoma eyes. Graefes Arch Clin Exp Ophthalmol 227:428–435

Kotecha A, O'Leary N, Melmoth D, Grant S, Crabb DP (2009) The functional consequences of glaucoma for eye-hand coordination. Invest Ophthalmol Vis Sci 50:203–213

Kotecha A, Richardson G, Chopra R, Fahy RTA, Garway-Heath DF et al (2012) Balance control in glaucoma. Invest Ophthalmol Vis Sci 53:7795–7801

Kuleshova ON, Nepomnyashchikh GI, Aidagulova SV, Shvedova EV (2008) Ultrastructure of the endothelium of the drainage system of the eye. Bull Exp Biol Med 145:634–637

Kumar A (2016) Editorial (Thematic selection): mitochondrial dysfunction & neurological disorders. Curr Neuropharmacol 14:565–566

Kusbeci T, Kusbeci OY, Mas NG, Karabekir HS, Yavas G et al (2015) Stereological evaluation of the optic nerve volume in Alzheimer disease. Surgery 26:1683–1686

Kwon HS, Nakaya N, Abu-Asab M, Kim HS, Tomarev SI (2014) Myocilin is involved in NgR1/Lingo-1-mediated oligodendrocyte differentiation and myelination of the optic nerve. J Neurosci 34:5539e5551

Lang P, Gesbert F, Delespine-Carmagnat M, Stancou R, Pouchelet M et al (1996) Protein kinase A phosphorylation of RhoA mediates the morphological and functional effects of cyclic AMP in cytotoxic lymphocytes. EMBO J 15:510–519

Lanzetti L (2007) Actin in membrane trafficking. Curr Opin Cell Biol 19:453–458

Lee JH, Yu WH, Kumar A, Lee S, Mohan PS et al (2010) Lysosomal proteolysis and autophagy require presenilin 1 and are disrupted by Alzheimer-related PS1 mutations. Cell 141:1146–1158

Lepple-Wienhues A, Stahl F, Wiederholt M (1991) Differential smooth muscle-like contractile properties of trabecular meshwork and ciliary muscle. Exp Eye Res 53:33–38

Li G, Luna C, Qiu J, Epstein DL, Gonzalez P (2009) Alterations in microRNA expression in stress-induced cellular senescence. Mech Ageing Dev 130:731–741

Li G, Luna C, Qiu J, Epstein DL, Gonzalez P (2011) Role of miR-204 in the regulation of apoptosis, endoplasmic reticulum stress response, and inflammation in human trabecular meshwork cells. Invest Ophthalmol Vis Sci 52:2999–3007

Li WW, Li J, Bao JK (2012) Microautophagy: lesser-known self-eating. Cell Mol Life Sci 69:1125–1136

Lingor P, Koeberle P, Kügler S, Bähr M (2005) Down-regulation of apoptosis mediators by RNAi inhibits axotomy-induced retinal ganglion cell death in vivo. Brain 128:550–558

Liton PB, Challa P, Stinnett S, Luna C, Epstein DL et al (2005) Cellular senescence in the glaucomatous outflow pathway. Exp Gerontol 40:745–748

Loewenfeld IE (1979) Pupillary changes related to age. In: Thompson HS (ed) Topics in neuroophthalmolgy. Williams & Wilkins, Baltimore, pp 124–150

Lorenzo M, Fernández-Veledo S, Vila-Bedmar R, Garcia-Guerra L et al (2008) Insulin resistance induced by tumor necrosis factor-alpha in myocytes and brown adipocytes. J Anim Sci 86:E94–E104

Lucas DR, Newhouse JP (1957) The toxic effect of sodium L-glutamate on the inner layers of the retina. AMA Arch Ophthalmol 58:193–201

Luu CD, Dimitrov PN, Wu Z, Ayton LN, Makeyeva G et al (2013) Static and flicker perimetry in age-related macular degeneration. Invest Ophthalmol Vis Sci 54:3560–3568

Machesky LM, Reeves E, Wientjes F, Mattheyse FJ, Grogan A et al (1997) Mammalian actin-related protein 2/3 complex localizes to regions of lamellipodial protrusion and is composed of evolutionarily conserved proteins. Biochem J 328:105–112

Madeo F, Tavernarakis N, Kroemer G (2010) Can autophagy promote longevity? Nat Cell Biol 12:842–846

Mao M, Hedberg-Buenz A, Koehn D, John SW, Anderson MG (2011) Anterior segment dysgenesis and early-onset glaucoma in nee mice with mutation of Sh3pxd2b. Invest Ophthalmol Vis Sci 52:2679–2688

Mecocci P, MacGarvey U, Kaufman AE, Koontz D, Shoffner JM et al (1993) Oxidative damage to mitochondrial DNA shows marked age-dependent increases in human brain. Ann Neurol 34:609–616

Medina RJ, Kataoka K, Takaishi M et al (2006) Isolation of epithelial stem cells from dermis by a three-dimensional culture system. J Cell Biochem 98:174–184

Meister M (1996) Multineuronal codes in retinal signaling. Proc Natl Acad Sci U S A 93:609–614

Meister M, Berry MJ 2nd (1999) The neural code of the retina. Neuron 22:435–450

Mendoza-Santiesteban CE, Gabilondo I, Palma JA, Norcliffe-Kaufmann L, Kaufmann H (2017) The retina in multiple system atrophy: systematic review and meta-analysis. Front Neurol 8:206

Mimura T, Kaji Y, Noma H, Funatsu H, Okamoto S (2013) The role of SIRT1 in ocular ageing. Exp Eye Res 116:17–26

Mitchell CH, Fleischhauer JC, Stamer WD, Peterson-Yantorno K, Civan MM (2002) Human trabecular meshwork cell volume regulation. Am J Physiol Cell Physiol 283:C315–C326

Mitchell JR, Oliveira C, Tsiouris AJ, Dinkin MJ (2015) Corresponding ganglion cell atrophy in patients with postgeniculate homonymous visual field loss. J Neuroophthalmol 35:353–359

Mokrý J, Cízková D, Filip S, Ehrmann J, Osterreicher J et al (2004) Nestin expression by newly formed human blood vessels. Stem Cells Dev 13:658–664

Montagnani M, Ravichandran LV, Chen H, Esposito DL, Quon MJ (2002) Insulin receptor substrate-1 and phosphoinositide-dependent kinase-1 are required for insulin-stimulated production of nitric oxide in endothelial cells. Mol Endocrinol 16:1931–1942

Moreau V, Tatin F, Varon C, Génot E (2003) Actin can reorganize into podosomes in aortic endothelial cells, a process controlled by Cdc42 and RhoA. Mol Cell Biol 23:6809–6822

Morris MA (1990) Mitochondrial mutations in neuro-ophthalmological diseases. Rev J Clin Neuroophthalmol 10:159–166

Morrone LA, Rombolà L, Corasaniti MT, Bagetta G, Nucci C et al (2015) Natural compounds and retinal ganglion cell neuroprotection. Prog Brain Res 220:257–281

Morselli E, Maiuri MC, Markaki M, Megalou E, Pasparaki A et al (2010) The life span-prolonging effect of sirtuin-1 is mediated by autophagy. Autophagy 6:186–188

Muriach M, Flores-Bellver M, Romero FJ, Barcia JM (2014) Diabetes and the brain: oxidative stress, inflammation, and autophagy. Oxid Med Cell Longev 2014:102158

Nair RR, Solway J, Boyd DD (2006) Expression cloning identifies transgelin (SM22) as a novel repressor of 92-kDa type IV collagenase (MMP-9) expression. J Biol Chem 281:26424–26436

Narita M, Nuñez S, Heard E, Narita M, Lin AW et al (2003) Rb-Mediated heterochromatin formation and silencing of E2F target genes during cellular senescence. Cell 113:703–716

Neufeld AH, Gachie EN (2003) The inherent, age-dependent loss of retinal ganglion cells is related to the lifespan of the species. Neurobiol Ageing 24:167–172

Nickells RW (1999) Apoptosis of retinal ganglion cells in glaucoma: an update of the molecular pathways involved in cell death. Surv Ophthalmol 43:S151–S161

Nucci C, Martucci A, Mancino R, Cerulli L (2013) Glaucoma progression associated with Leber's hereditary optic neuropathy. Int Ophthalmol 33:75–77

O'Donnell ME, Brandt JD, Curry FR (1995) Na-K-Cl cotransport regulates intracellular volume and monolayer permeability of trabecular meshwork cells. Am J Physiol Cell Physiol 268:C1067–C1074

Oh DJ, Kang MH, Ooi YH, Choi KR, Sage EH, Rhee DJ (2013) Overexpression of SPARC in human trabecular meshwork increases intraocular pressure and alters extracellular matrix. Invest Ophthalmol Vis Sci 54:3309–3319

Ota H, Akishita M, Eto M, Iijima K, Kaneki M et al (2007) Sirt1 modulates premature senescence-like phenotype in human endothelial cells. J Mol Cell Cardiol 43:571–579

Ota H, Eto M, Kano MR, Kahyo T, Setou M et al (2010) Induction of endothelial nitric oxide synthase, SIRT1, and catalase by statins inhibits endothelial senescence through the Akt pathway. Arterioscler Thromb Vasc Biol 30:2205–2211

Owsley C (2011) Ageing and vision. Vis Res 51:1610–1622

Owsley C (2016) Vision and ageing. Annu Rev Vis Sci 2:255–271

Owsley C, Huisingh C, Jackson GR, Curcio CA, Szalai AJ et al (2014) Associations between abnormal rodmediated dark adaptation and health and functioning in older adults with normal macular health. Invest Ophthalmol Vis Sci 55:4776–4789

Parikh RS, Parikh SR, Sekhar GC, Prabakaran S, Babu JG et al (2007) Normal age-related decay of retinal nerve fiber layer thickness. Ophthalmology 114:921–926

Park BC, Tibudan M, Samaraweera M, Shen X, Yue BY (2007) Interaction between two glaucoma genes, optineurin and myocilin. Genes Cells 12:969–979

Pascolini D, Mariotti SP (2012) Global estimates of visual impairment. Br J Ophthalmol 96:614–618

Patel S, Muallem S (2011) Acidic Ca(2+) stores come to the fore. Cell Calcium 50:109–112

Plaza-Zabala A, Sierra-Torre V, Sierra A (2017) Autophagy and microglia: novel partners in neurodegeneration and ageing. Int J Mol Sci 18(3)

Porter K, Nallathambi J, Lin Y, Liton PB (2013) Lysosomal basification and decreased autophagic flux in oxidatively stressed trabecular meshwork cells: implications for glaucoma pathogenesis. Autophagy 9:581–594

Porter K, Hirt J, Stamer WD, Liton PB (2015) Autophagic dysregulation in glaucomatous trabecular meshwork cells. Biochim Biophys Acta 1852:379–385

Prentice H, Modi JP, Wu JY (2015) Mechanisms of neuronal protection against excitotoxicity, endoplasmic reticulum stress, and mitochondrial dysfunction in stroke and neurodegenerative diseases. Oxid Med Cell Longev 2015:964518

Pryde KR, Taanman JW, Schapira AH (2016) A LON-ClpP proteolytic axis degrades complex I to extinguish ROS production in depolarized mitochondria. Cell Rep 17:2522–2531

Pulliero A, Seydel A, Camoirano A, Saccà SC, Sandri M et al (2014) Oxidative damage and autophagy in the human trabecular meshwork as related with ageing. PLoS One 9:e98106

Qiu Y, Shen X, Shyam R, Yue BY, Ying H (2014) Cellular processing of myocilin. PLoS One 9:e92845

Qu J, Wang D, Grosskreutz CL (2010) Mechanisms of retinal ganglion cell injury and defense in glaucoma. Exp Eye Res 91:48–53

Rao RV, Ellerby HM, Bredesen DE (2004) Coupling endoplasmic reticulum stress to the cell death program. Cell Death Differ 11:372–380

Rashid HO, Yadav RK, Kim HR, Chae HJ (2015) ER stress: autophagy induction, inhibition and selection. Autophagy 11:1956–1977

Reid MB, Moylan JS (2011) Beyond atrophy: redox mechanisms of muscle dysfunction in chronic inflammatory disease. J Physiol 589:2171–2179

Renieri G, Choritz L, Rosenthal R, Meissner S, Pfeiffer N et al (2008) Effects of endothelin-1 on calcium independent contraction of bovine trabecular meshwork. Graefes Arch Clin Exp Ophthalmol 246:1107–1115

Rodier F, Campisi J (2011) Four faces of cellular senescence. J Cell Biol 192:547–556

Rubinsztein DC, Mariño G, Kroemer G (2011) Autophagy and ageing. Cell 146:682–695

Rubinsztein DC, Shpilka T, Elazar Z (2012) Mechanisms of autophagosome biogenesis. Curr Biol 22:R29–R34

Rudolph T, Frisén L (2007) Influence of ageing on visual field defects due to stable lesions. Br J Ophthalmol 91:1276–1278

Russo R, Nucci C, Corasaniti MT, Bagetta G, Morrone LA (2015) Autophagy dysregulation and the fate of retinal ganglion cells in glaucomatous optic neuropathy. Prog Brain Res 220:87–105

Rutkowski P, May CA (2017) The peripheral and Central Humphrey visual field – morphological changes during ageing. BMC Ophthalmol 17:127

Saccà SC, Izzotti A (2008) Oxidative stress and glaucoma: injury in the anterior segment of the eye. Prog Brain Res 173:385–407

Saccà SC, Izzotti A (2014) Focus on molecular events in the anterior chamber leading to glaucoma. Cell Mol Life Sci 71:2197–2218

Saccà SC, Pascotto A, Camicione P, Capris P, Izzotti A (2005) Oxidative DNA damage in the human trabecular meshwork: clinical correlation in patients with primary open-angle glaucoma. Arch Ophthalmol 123:458–463

Saccà SC, Izzotti A, Rossi P, Traverso C (2007) Glaucomatous outflow pathway and oxidative stress. Exp Eye Res 84:389–399

Saccà SC, Bolognesi C, Battistella A, Bagnis A, Izzotti A (2009) Gene-environment interactions in ocular diseases. Mutat Res 667:98–117

Saccà SC, Centofanti M, Izzotti A (2012) New proteins as vascular biomarkers in primary open angle glaucomatous aqueous humor. Invest Ophthalmol Vis Sci 53:4242–4253

Saccà SC, Roszkowska AM, Izzotti A (2013) Environmental light and endogenous antioxidants as the main determinants of non-cancer ocular diseases. Mutat Res 752:153–171

Saccà SC, Pulliero A, Izzotti A (2015) The dysfunction of the trabecular meshwork during glaucoma course. J Cell Physiol 230:510–525

Saccà SC, Gandolfi S, Bagnis A, Manni G, Damonte G et al (2016a) From DNA damage to functional changes of the trabecular meshwork in ageing and glaucoma. Ageing Res Rev 29:26–41

Saccà SC, Gandolfi S, Bagnis A, Manni G, Damonte G et al (2016b) The outflow pathway: a tissue with morphological and functional unity. J Cell Physiol 231:1876–1893

Saitoh T, Akira S (2010) Regulation of innate immune responses by autophagy-related proteins. J Cell Biol 189:925–935

Salminen A, Ojala J, Kaarniranta K, Haapasalo A, Hiltunen M et al (2011) Astrocytes in the ageing brain express characteristics of senescence-associated secretory phenotype. Eur J Neurosci 34:3–11

Sampaio TB, Savall AS, Gutierrez MEZ, Pinton S (2017) Neurotrophic factors in Alzheimer's and Parkinson's diseases: implications for pathogenesis and therapy. Neural Regen Res 12:549–557

Sanderson JL, Dell'Acqua ML (2011) AKAP signaling complexes in regulation of excitatory synaptic plasticity. Neuroscientist 17:321–336

Sanderson J, Dartt DA, Trinkaus-Randall V, Pintor J, Civan MM et al (2014) Purines in the eye: recent evidence for the physiological and pathological role of purines in the RPE, retinal neurons, astrocytes, Müller cells, lens, trabecular meshwork, cornea and lacrimal gland. Exp Eye Res 127:270–279

Sawada H, Fukuchi T, Tanaka T, Abe H (2010) Tumor necrosis factor-alpha concentrations in the aqueous humor of patients with glaucoma. Invest Ophthalmol Vis Sci 51:903–906

Scaffidi C, Fulda S, Srinivasan A, Friesen C, Li F, Tomaselli KJ et al (1998) Two CD95 (APO-1/Fas) signaling pathways. EMBO J 17:1675–1687

Scott JD (1991) Cyclic nucleotide-dependent protein kinases. Pharmacol Ther 50:123–145

Scott JD (2003) A-kinase-anchoring proteins and cytoskeletal signalling events. Biochem Soc Trans 31:87–89

Sejersen T, Lendahl U (1993) Transient expression of the intermediate filament nestin during skeletal muscle development. J Cell Sci 106(Pt 4):1291–1300

Sekiguchi F, Aoki Y, Nakagawa M, Kanaoka D, Nishimoto Y et al (2013) AKAP-dependent sensitization of Ca(v) 3.2 channels via the EP(4) receptor/cAMP pathway mediates PGE(2)-induced mechanical hyperalgesia. Br J Pharmacol 168:734–745

Shen X, Koga T, Park BC, SundarRaj N, Yue BY (2008) Rho GTPase and cAMP/protein kinase A signaling mediates myocilin-induced alterations in cultured human trabecular meshwork cells. J Biol Chem 283:603–612

Shen X, Ying H, Qiu Y, Park JS, Shyam R et al (2011) Processing of optineurin in neuronal cells. J Biol Chem 286:3618–3629

Shiga N, Hirano K, Hirano M, Nishimura J, Nawata H et al (2005) Long-term inhibition of RhoA attenuates vascular contractility by enhancing endothelial NO production in an intact rabbit mesenteric artery. Circ Res 96:1014–1021

Shigenaga MK, Hagen TM, Ames BN (1994) Oxidative damage and mitochondrial decay in ageing. Proc Natl Acad Sci U S A 91:10771–10778

Shimizu S, Lichter PR, Johnson AT, Zhou Z, Higashi M et al (2000) Age-dependent prevalence of mutations at the GLC1A locus in primary open-angle glaucoma. Am J Ophthalmol 130:165–177

Shimizu T, Sugawara K, Tosaka M, Imai H, Hoya K et al (2006) Nestin expression in vascular malformations: a novel marker for proliferative endothelium. Neurol Med Chir (Tokyo) 46:111–117

Smith PJ, Samuelson DA, Brooks DE, Whitley RD (1986) Unconventional aqueous humour outflow of microspheres perfused into the equine eye. Am J Vet Res 47:2445–2453

Smith ND, Crabb DP, Garway-Heath DF (2011) An exploratory study of visual search performance in glaucoma. Ophthalmic Physiol Opt 31:225–232

Smythe E, Ayscough KR (2006) Actin regulation in endocytosis. J Cell Sci 119:4589–4598

Soto D, Comes N, Ferrer E, Morales M, Escalada A et al (2004) Modulation of aqueous humor outflow by ionic mechanisms involved in trabecular meshwork cell volume regulation. Invest Ophthalmol Vis Sci 45:3650–3661

Spector A (1995) Oxidative stress-induced cataract: mechanism of action. FASEB J 9:1173–1182

Sriram K, O'Callaghan JP (2007) Divergent roles for tumor necrosis factor-alpha in the brain. J Neuroimmune Pharmacol 2:140–153

Steely HT Jr, English-Wright SL, Clark AF (2000) The similarity of protein expression in trabecular meshwork and lamina cribrosa: implications for glaucoma. Exp Eye Res 70:17–30

Stone EM, Fingert JH, Alward WL, Nguyen TD, Polansky JR et al (1997) Identification of a gene that causes primary open angle glaucoma. Science 275:668–670

Stothert AR, Fontaine SN, Sabbagh JJ, Dickey CA (2016) Targeting the ER-autophagy system in the trabecular meshwork to treat glaucoma. Exp Eye Res 144:38e45

Sun H, Wang Y, Pang IH, Shen J, Tang X et al (2011) Protective effect of a JNK inhibitor against retinal ganglion cell loss induced by acute moderate ocular hypertension. Mol Vis 17:864–875

Surmeier DJ, Schumacker PT, Guzman JD, Ilijic E, Yang B et al (2016) Calcium and Parkinson's disease. Biochem Biophys Res Commun 483:1013–1019

Suzuki S, Namiki J, Shibata S, Mastuzaki Y, Okano H (2010) The neural stem/progenitor cell marker nestin is expressed in proliferative endothelial cells, but not in mature vasculature. J Histochem Cytochem 58:721–730

Tamm ER (2002) Myocilin and glaucoma: facts and ideas. Prog Retin Eye Res 21:395–428

Tamm ER (2009) The trabecular meshwork outflow pathways: structural and functional aspects. Exp Eye Res 88:648–655

Teranishi N, Naito Z, Ishiwata T, Tanaka N, Furukawa K et al (2002) Identification of neovasculature using nestin in colorectal cancer. Int J Oncol 30:593–603

Tezel G (2008) TNF-alpha signaling in glaucomatous neurodegeneration. Prog Brain Res 173:409–421

Tezel G, Chauhan BC, LeBlanc RP, Wax MB (2003) Immunohistochemical assessment of the glial mitogen-activated protein kinase activation in glaucoma. Invest Ophthalmol Vis Sci 44:3025–3033

Tian B, Geiger B, Epstein DL, Kaufman PL (2000) Cytoskeletal involvement in the regulation of aqueous humor outflow. Invest Ophthalmol Vis Sci 41:619–623

Toescu EC (2005) Normal brain ageing: models and mechanisms. Philos Trans R Soc Lond Ser B Biol Sci 360:2347–2354

Toescu EC, Verkhratsky A (2000) Parameters of calcium homeostasis in normal neuronal ageing. J Anat 197(pt 4):563–569

Tooze SA, Schiavo G (2008) Liaisons dangereuses: autophagy, neuronal survival and neurodegeneration. Curr Opin Neurobiol 18:504–515

Toussaint O, Dumont P, Remacle J, Dierick JF, Pascal T et al (2002) Stress-induced premature senescence or stress-induced senescence-like phenotype: one in vivo reality, two possible definitions? Sci World J 2:230–247

Tsai DC, Hsu WM, Chou CK, Chen SJ, Peng CH et al (2002) Significant variation of the elevated nitric oxide levels in aqueous humor from patients with different types of glaucoma. Ophthalmologica 216:346–350

Ueno U, Yamada Y, Watanabe R, Mukai E, Hosokawa M et al (2005) Nestin-positive cells in adult pancreas express amylase and endocrine precursor cells. Pancreas 31:126–131

Ullian EM, Barkis WB, Chen S, Diamond JS, Barres BA (2004) Invulnerability of retinal ganglion cells to NMDA excitotoxicity. Mol Cell Neurosci 26:544–557

Underwood JL, Murphy GM, Chen J, Franse-Carman L, Wood I et al (1999) Glucocorticoids regulate transendothelial fluid flow resistance and formation of intercellular junctions. AmJ Physiol Cell Physiol 277:C330–C342

Vaittinen S, Lukka R, Sahlgren C, Hurme T, Rantanen J et al (2001) The expression of intermediate filament protein nestin as related to vimentin and desmin in regenerating skeletal muscle. J Neuropathol Exp Neurol 60:588–597

Valamanesh F, Monnin J, Morand-Villeneuve N, Michel G, Zaher M et al (2013) Nestin expression in the retina of rats with inherited retinal degeneration. Exp Eye Res 110C:26–34

Valentijn SA, van Boxtel MP, van Hooren SA, Bosma H, Beckers HJ et al (2005) Change in sensory functioning predicts change in cognitive functioning: results from a 6-year follow-up in the maastricht ageing study. J Am Geriatr Soc 53:374–380

Van Aelst L, D'Souza-Schorey C (1997) Rho GTPases and signaling networks. Genes Dev 11:2295–2322

Vittow J, Borras T (2002) Expression of optineurin, a glaucoma-linked gene, is influenced by elevated intraocular pressure. Biochem Biophys Res Commun 298:67–74

von Zglinicki T, Saretzki G, Ladhoff J, d'Adda di Fagagna F, Jackson SP (2005) Human cell senescence as a DNA damage response. Mech Ageing Dev 126:111–117

Walker RJ, Anderson NM, Bahouth S, Steinle JJ (2012) Silencing of insulin receptor substrate-1 increases cell death in retinal Müller cells. Mol Vis 18:271–279

Wang K, Klionsky DJ (2011) Mitochondria removal by autophagy. Autophagy 7:297e300

Wang X, Tay SS, Ng YK (2000) An immunohistochemical study of neuronal and glial cell reactions in retinae of rats with experimental glaucoma. Exp Brain Res 132:476–484

Wang L, Sunahara RK, Krumins A, Perkins G, Crochiere ML et al (2001) Cloning and mitochondrial localization of full-length D-AKAP2, a protein kinase A anchoring protein. Proc Natl Acad Sci U S A 98:3220–3225

Wang L, Cano M, Handa JT (2014) p62 provides dual cytoprotection against oxidative stress in the retinal pigment epithelium. Biochim Biophys Acta 1843:1248e1258

Watkins JC, Davies J, Evans RH, Francis AA, Jones AW (1981) Pharmacology of receptors for excitatory amino acids. Adv Biochem Psychopharmacol 27:263–273

Wegner AM, Nebhan CA, Hu L, Majumdar D, Meier KM, Weaver AM, Webb DJ (2008) N-wasp and the arp2/3 complex are critical regulators of actin in the development of dendritic spines and synapses. J Biol Chem 283:15912–15920

Wei H, Sawchyn AK, Myers JS, Katz LJ, Moster MR et al (2012) A clinical method to assess the effect of visual loss on the ability to perform activities of daily living. Br J Ophthalmol 96:735–741

Welch MD, DePace AH, Verma S, Iwamatsu A, Mitchison TJ (1997) The human Arp2/3 complex is composed of evolutionarily conserved subunits and is localized to cellular regions of dynamic actin filament assembly. J Cell Biol 138:375–384

Wentz-Hunter K, Shen X, Okazaki K, Tanihara H, Yue BY (2004) Overexpression of myocilin in cultured human trabecular meshwork cells. Exp Cell Res 297:39–48

White MF, Maron R, Kahn CR (1985) Insulin rapidly stimulates tyrosine phosphorylation of a Mr-185,000 protein in intact cells. Nat (Lond) 318:183–186

Wiederholt M (1998) Direct involvement of trabecular meshwork in the regulation of aqueous humor outflow. Curr Opin Ophthalmol 9:46–49

Wiederholt M, Thieme H, Stumpff F (2000) The regulation of trabecular meshwork and ciliary muscle contractility. Prog Retin Eye Res 19:271–295

Xu J, Pokorny J, Smith VC (1997) Optical density of the human lens. J Opt Soc Am A Opt Image Sci Vis 14:953–960

Xue LP, Lu J, Cao Q, Hu S, Ding P et al (2006) Müller glial cells express nestin coupled with glial fibrillary acidic protein in experimentally induced glaucoma in the rat retina. Neuroscience 139:723–732

Yang HY, Lieska N, Kriho V, Wu CM, Pappas GD (1997) Subpopulation of reactive astrocytes at the immediate site of cerebral cortical injury. Exp Neuol 146:199–205

Ye W, Gong H, Sit A, Johnson M, Freddo TF (1997) Interendothelial junctions in normal human Schlemm's canal respond to changes in pressure. Invest Ophthalmol Vis Sci 38:2460–2468

Ying H, Yue BYT (2016) Optineurin: the autophagy connection. Exp Eye Res 144:73–80

Zetterberg M (2016) Age-related eye disease and gender. Maturitas 83:19–26

Zgheib O, Huyen Y, DiTullio RA Jr, Snyder A, Venere M et al (2005) ATM signaling and 53BP1. Radiother Oncol 76:119–122

Zhang J, Shapiro MS (2012) Activity-dependent transcriptional regulation of M-Type (Kv7) K(+) channels by AKAP79/150-mediated NFAT actions. Neuron 76:1133–1146

Zhang M, Maddala R, Rao PV (2008) Novel molecular insights into RhoA GTPase-induced resistance to aqueous humor outflow through the trabecular meshwork. Am J Physiol Cell Physiol 295:C1057–C1070

Zhou R, Yazdi AS, Menu P, Tschopp J (2011) A role for mitochondria in NLRP3 inflammasome activation. Nature 469:221–225

Zhu G, Lee AS (2014) Role of the unfolded protein response, GRP78 and GRP94 in organ homeostasis. J Cell Physiol 230:1413e1420

Zinflou C, Rochette PJ (2017) Ultraviolet A-induced oxidation in cornea: characterization of the early oxidation-related events. Free Radic Biol Med 108:118–128

Zuchero JB, Belin B, Mullins RD (2012) Actin binding to WH2 domains regulates nuclear import of the multifunctional actin regulator JMY. Mol Biol Cell 23:853–863

Chapter 15
Hutchinson-Gilford Progeria Syndrome: Challenges at Bench and Bedside

Ray Kreienkamp and Susana Gonzalo

Abstract The structural nuclear proteins known as "lamins" (A-type and B-type) provide a scaffold for the compartmentalization of genome function that is important to maintain genome stability. Mutations in the *LMNA* gene -encoding for A-type lamins- are associated with over a dozen of degenerative disorders termed laminopathies, which include muscular dystrophies, lipodystrophies, neuropathies, and premature ageing diseases such as Hutchinson Gilford Progeria Syndrome (HGPS). This devastating disease is caused by the expression of a truncated lamin A protein named "progerin". To date, there is no effective treatment for HGPS patients, who die in their teens from cardiovascular disease. At a cellular level, progerin expression impacts nuclear architecture, chromatin organization, response to mechanical stress, and DNA transactions such as transcription, replication and repair. However, the current view is that key mechanisms behind progerin toxicity still remain to be discovered. Here, we discuss new findings about pathological mechanisms in HGPS, especially the contribution of replication stress to cellular decline, and therapeutic strategies to ameliorate progerin toxicity. In particular, we present evidence for retinoids and calcitriol (hormonal vitamin D metabolite) being among the most potent compounds to ameliorate HGPS cellular phenotypes in vitro, providing the rationale for testing these compounds in preclinical models of the disease in the near term, and in patients in the future.

Keywords Hutchinson Gilford progeria syndrome · HGPS · Lamins · Premature ageing · Progerin · Retinoids · Calcitriol

R. Kreienkamp · S. Gonzalo (✉)
Edward A. Doisy Department of Biochemistry and Molecular Biology, Doisy Research Center, St Louis University School of Medicine, St. Louis, MO, USA
e-mail: ray.kreienkamp@health.slu.edu; sgonzalo@slu.edu

© Springer Nature Singapore Pte Ltd. 2019
J. R. Harris, V. I. Korolchuk (eds.), *Biochemistry and Cell Biology of Ageing: Part II Clinical Science*, Subcellular Biochemistry 91,
https://doi.org/10.1007/978-981-13-3681-2_15

Introduction

To the study of normal ageing, Hutchinson-Gilford Progeria Syndrome (HGPS) is certainly an outlier. As a severe premature ageing disease, patients develop alopecia, bone and joint abnormalities, subcutaneous fat loss, and severe atherosclerosis, all before their teenage years. Patients ultimately die at an average age of 14.6 years from myocardial infarction or stroke as a result of rapidly progressive atherosclerosis (Ullrich and Gordon 2015). Fortunately, this disease is extremely rare, with an estimated 350–400 children worldwide. Yet, since the mutation driving its pathophysiology was discovered in 2003, it has been the subject of an ever-growing volume of research. This is not only because treatment is desperately needed to help these patients, but it is thought that studying the complexities of this fascinating disease might reveal new insights into the normal ageing process. This is corroborated by the finding that progerin, the toxic protein driving disease pathology, is also found in the fibroblasts and vascular smooth muscle cells from old individuals (Dahl et al. 2006, McClintock et al. 2007). Importantly, progerin is upregulated in the hearts of dilated cardiomyopathy patients, where its expression correlates with left ventricular remodeling (Messner et al. 2018). This suggests that progerin might contribute to the progression of cardiovascular disease with age. Here, we review the clinical manifestations of HGPS, underlying cellular drivers of this disease, and emerging therapies for treating patients.

Hutchinson-Gilford Progeria Syndrome

HGPS results from the disruption of the nuclear lamina, a key nuclear structure for innumerate cellular processes, by a *de novo* single-base substitution within the *LMNA* exon 11 (c.1824C>T) (De Sandre-Giovannoli et al. 2003, Eriksson et al. 2003). This mutation activates a cryptic splice site, leading to an in-frame deletion of 50 amino acids near the C-terminus of prelamin A. This prevents proper post-translational processing of prelamin A to lamin A and leaves a permanently farnesylated and carboxymethylated toxic product called "progerin." This mutant form of lamin A acts in a dominant fashion to induce a variety of abnormalities in nuclear processes, which eventually lead to cellular and organismal decline. Although c.1824C>T remains the most frequent mutation in HGPS patients, other mutations in the *LMNA* gene have also been reported that result in increased usage of the cryptic splice site.

HGPS patients are seemingly normal at birth, but quickly begin showing symptoms of their underlying disease. Skin alterations are often among the first manifestations of HGPS. Though manifestations can present with differing degrees of severity, typical alterations include areas of discoloration, stippled pigmentation, and tightened areas that restrict movement. Sclerodermoid changes, which give the skin a dimpled appearance with varying pigmentation, frequently appear over the abdomen and lower extremities (Rork et al. 2014). By 1 year of age, patients often present with failure to thrive, alopecia, circumoral cyanosis, prominent scalp veins, and decreased range of motion (Merideth et al. 2008, Ullrich and Gordon 2015). They develop a distinct progeroid appearance. Often remaining less than four feet tall and 30 kg, a decreased and linear rate of weight gain prevents growth comparable to age matched peers (Gordon et al. 2007; Kieran et al. 2007). Patients begin to lose cranial hair around 10 months of age, with progression to almost complete alopecia with time (Rork et al. 2014). HGPS patients also have distinct craniofacial characteristics, developing micrognathia, prominent eyes, and a beaked nose (Kieran et al. 2007, Domingo et al. 2009). Prominent forehead scalp veins and perioral cyanosis become evident, both likely the result of decreased subcutaneous fat (Rork et al. 2014). Patients also have multiple dental abnormalities, including both lack of teeth as well as dental crowding, which can manifest as double rows of teeth (Gordon et al. 2007, Domingo et al. 2009). Middle ear abnormalities and aberrations in the ear canal also lead to low-frequency hearing loss in many patients (Guardiani et al. 2011).

HGPS is a "segmental ageing disease," since some features of normal ageing are present, whereas other features are notably absent. The liver, kidneys, lungs, and gastrointestinal tract are relatively spared in these patients (Kieran et al. 2007; Ullrich and Gordon 2015). However, others cell and tissue types, such as those of mesenchymal origin, are particularly susceptible to progerin-induced cellular defects, causing notable fat and bone abnormalities in HGPS patients (McClintock et al. 2007; Merideth et al. 2008; Zhang et al. 2011). HGPS patients develop lipodystrophy as well as bone and joint abnormalities consistent with skeletal dysplasia (Gordon et al. 2011). A profound loss of subcutaneous fat is readily apparent in examining these patients. Loss of fat in some body areas, such as the feet, can lead to discomfort and often requires supportive therapies (Gordon et al. 2014b; Ullrich and Gordon 2015). Interestingly, levels of body fat did not correlate with onset of menarche, which girls with this condition often experience in spite of lack of other pubertal features (Greer et al. 2017). Bone problems for these patients include small clavicles, thin ribs, and acroosteolysis. Patients exhibit reduced bone mineral density with accentuated demineralization at the end of long bones. Avascular necrosis is also present, including at the femoral head, likely resulting from vascular compromise (Cleveland et al. 2012). Interestingly, fracture incidence among HGPS patients is not increased compared to the general population, though HGPS patients are more susceptible to skull fractures. This is likely the result of disrupted bone formation in the skull. Patent anterior and posterior fontanels can persist in patients as old as 9 years of age, and these patients often also have widened calvarial sutures and a thin calvarium (Ullrich and Gordon 2015).

The most significant problems in HGPS are the cardiovascular complications, which underlie patient death. Patients develop severe and progressive atherosclerosis, eventually leading to myocardial ischemia, infarction, and stroke (Stehbens et al. 1999). Patients also develop readily evident left ventricular diastolic dysfunction, which increases with age (Prakash et al. 2018). Left ventricular hypertrophy and systolic dysfunction are also observed, which are more evident in older patients.

Cardiac manifestations include increased afterload and angina (Ullrich and Gordon 2015). Remarkably, it is estimated that 50% of children have radiographically detectable strokes by the age of eight, and infarcts were common on imageing studies of patients between 5 and 10 years of age (Silvera et al. 2013). Most of these strokes are often clinically silent. This suggests that cardiovascular problems are present well before the end of life contributing to both morbidity and mortality.

The atherosclerosis that develops in HGPS has some important differences from the normal ageing population, although calcification, inflammation, and plaque rupture are present in both HGPS and normal ageing. Interestingly, HGPS patients do not develop hypercholesterolemia or increased serum high-sensitivity C-reactive protein, which are often seen with cardiovascular disease in the normal population (Stehbens et al. 1999; Olive et al. 2010). Additionally, vessels have a more complete fibrosis throughout the vessel wall, as arteries and veins show marked adventitial fibrosis with a dense rim of collagen. This complete stiffening of the wall leads to many measurable changes in the vasculature. Patients can become hypertensive, and some patients also have elongated QT intervals by EKG (Merideth et al. 2008; Gerhard-Herman et al. 2012). Carotid-femoral pulse wave velocity is dramatically elevated, indicating an increase in arterial stiffness. Patients also have abnormally echodense vascular walls by ultrasound, thought to correspond to a dramatically thickened fibrotic matrix. In these patients, as well as mouse models of disease, there is a striking depletion of vascular smooth muscle cells from the media, even in the outermost lamellar units adjacent to the adventitia, that is replaced by proteoglycans and collagen (Varga et al. 2006; Osorio et al. 2011; Gerhard-Herman et al. 2012; Villa-Bellosta et al. 2013). This is likely due to the extreme sensitivity of vascular smooth muscle cells to progerin expression.

The vascular abnormalities present in HGPS can compromise the nervous system as well. In the absence of vascular disease, the nervous system is relatively spared due to the fact that both lamin A and progerin expression is limited in the nervous system by miR-9 (Jung et al. 2012). HGPS patients show no evidence of memory or cognitive challenges often associated with the normal ageing process and have normal cognition. However, many HGPS patients experience neurological symptoms such as headaches of migraine-type quality, muscular weakness, or seizures as a result of impaired blood flow and diseased vasculature (Ullrich and Gordon 2015).

Cellular Disruptions

For the plethora of pathologies comprising this disease, it is surprising that it all results from a single nucleotide substitution and a deleterious protein. Progerin's toxic cellular effects are substantial and caused primarily by alterations in genome function and integrity. Hallmarks of progerin-expressing cells include nuclear morphological abnormalities, changes in chromatin organization, DNA damage, telomere shortening, and premature senescence (Goldman et al. 2004; Prokocimer et al.

2013; Gonzalo and Kreienkamp 2015; Gonzalo et al. 2017). Despite enormous progress in recent years identifying cellular processes altered by progerin, we still lack a clear picture of the molecular mechanisms whereby progerin expression causes all these cellular phenotypes.

Nuclear morphological abnormalities are probably the most robust marker of HGPS patient-derived fibroblasts, and the phenotype that is most often ameliorated by therapeutic strategies (Capell et al. 2005) (Fig. 15.1). HGPS nuclei appear big, dysmorphic, and full of protrusions throughout that are accompanied by nuclear lamina thickening and disorganization of nuclear pore complexes and chromatin (Goldman et al. 2004; Kubben et al. 2015; Kreienkamp et al. 2016). A dosage-dependent effect of progerin inducing morphological nuclear defects has been reported, which is exacerbated with continuous proliferation (Chojnowski et al. 2015). Thus, strategies that lower levels of progerin have shown improvement of HGPS cells in vitro and in mouse models of disease in vivo. This is the case of anti-sense oligonucleotides (ASO) targeting lamin A/progerin production (Scaffidi and Misteli 2005; Osorio et al. 2011; Bridgeman et al. 2017), compounds such as rapamycin and everolimus that increase clearance of progerin *via* autophagy (DuBose et al. 2018), and MG132 that induces progerin nucleocytoplasmic translocation and progerin clearance through macroautophagy (Harhouri et al. 2017). In addition, the E3 ubiquitin ligase Smurf2 directly binds, ubiquitinates, and negatively regulates expression of lamin A/progerin, which in HGPS cells reduces nuclear deformability (Borroni et al. 2018). Furthermore, nuclei from HGPS cells exhibit increased nuclear stiffness and impaired mechanotransduction (Dahl et al. 2006; Verstraeten et al. 2008); phenotypes that are thought to have a big impact on tissues such as bone, skeletal muscle, heart, and vessels that are subjected to significant mechanical stress (Prokocimer et al. 2013; Dobrzynska et al. 2016).

Epigenetic changes are characteristic of HGPS cells, including alterations in DNA methylation (Osorio et al. 2010; Heyn et al. 2013), post-translational modifications of histones -mainly H3K9me3, H3K27me3, and H4K20me3- (Columbaro et al. 2005; Scaffidi and Misteli 2005; Shumaker et al. 2006), expression levels of chromatin-modifying activities such as the NURD complex (Pegoraro et al. 2009), and miRNAs (Arancio et al. 2014; Frankel et al. 2018). Interestingly, these chromatin changes and expression of progerin are also observed in cells from old individuals, suggesting their implication in physiological ageing (Dahl et al. 2006). Special attention has been given lately to the consequences of loss of function of SIRT6 in progerin-expressing cells. This sirtuin family member has histone deacetylation and mono-ADP ribosylation activities and has been associated with genomic instability and accelerated ageing similar to HGPS (Ghosh et al. 2015).

Deregulated gene expression is another hallmark of HGPS, which resembles the gene expression profile of disorders affecting mesodermal and mesenchymal cell lineages. Functional categories more often found differentially expressed in HGPS cells include transcription factors (Duband-Goulet et al. 2011) and extracellular matrix (ECM) components (Csoka et al. 2004), in addition to signaling cascades such as the Wnt pathway (Hernandez et al. 2010; Vidak and Foisner 2016), the retinoblastoma pathway (Marji et al. 2010), Notch signalling (Scaffidi and Misteli

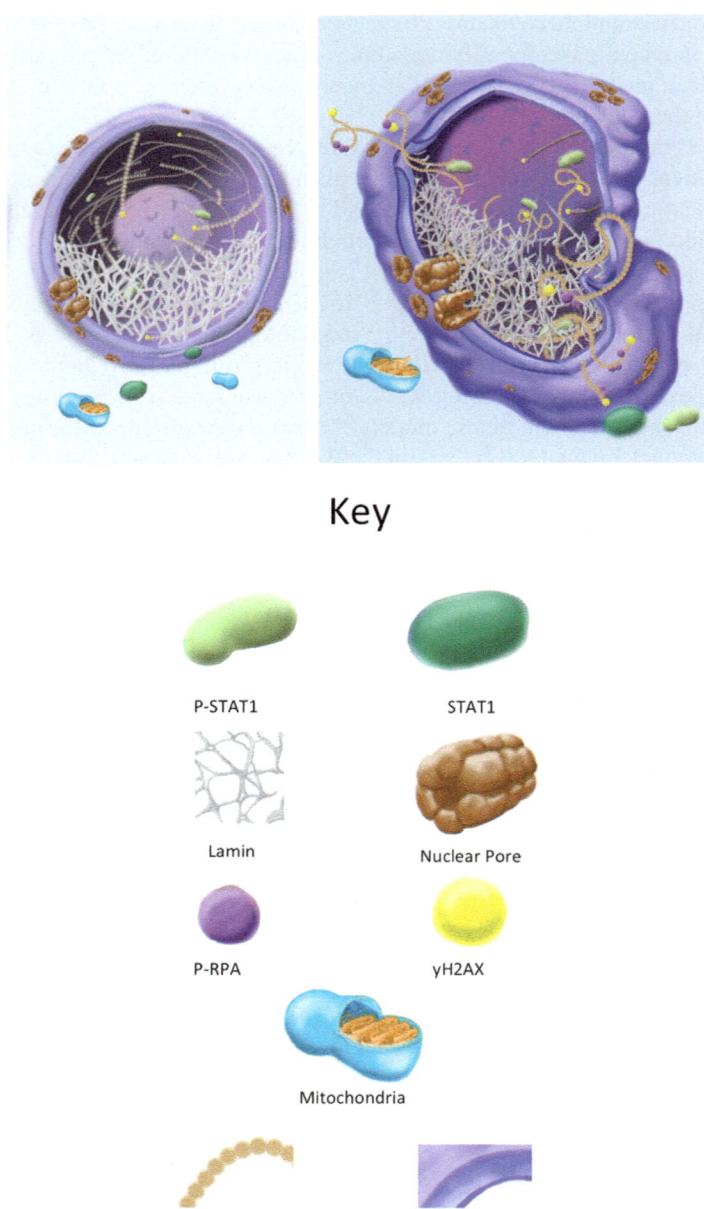

Fig. 15.1 Progerin expression elicits profound alterations in nuclear morphology and genome integrity and function. Compared to normal fibroblasts, HGPS patient-derived fibroblasts exhibit increased nuclear size and nuclear envelope morphological abnormalities characterized by invaginations, protrusions and blebbing. Progerin-expressing cells lose heterochromatin from the nuclear periphery and accumulate DNA damage. In particular, progerin causes increased levels of γH2AX

2008), NFκB inflammatory pathway, and the recently identified, interferon (IFN)-related innate immune responses (Kreienkamp et al. 2018).

DNA repair defects, telomere dysfunction, and genomic instability are amongst the more potent drivers of ageing and malignancy. In HGPS, there is strong evidence for deficiencies in DNA repair, which are characterized by delayed recruitment of DNA repair factors such as 53BP1 and RAD51 to γH2AX-labeled DNA repair foci (Liu et al. 2005) or anomalous accumulation of Xeroderma Pigmentosum group A (XPA) (Liu et al. 2005, 2008), a protein with an important function in nucleotide excision repair (NER), among others. HGPS cells also develop telomere dysfunction, with faster telomere attrition during proliferation that elicits DNA damage and premature senescence (Gonzalo and Kreienkamp 2015). In addition, there is accumulation of ROS due to mitochondrial dysfunction, which was has been linked to impaired NRF2 pathway activity (Kubben et al. 2016). An interesting study has recently found that inhibition of ROCK (rho-associated protein kinase) activity recovers mitochondrial function in HGPS fibroblasts, ameliorating nuclear morphological abnormalities and genomic instability phenotypes (Kang et al. 2017). Many reviews have previously described the phenotypes of genomic instability in progerin-expressing cells (Gonzalo and Kreienkamp 2015; Dobrzynska et al. 2016; Gonzalo et al. 2017). Here, we will focus on newly identified mechanisms underlying genomic instability and their contribution to premature ageing.

Recently, special emphasis has been placed on understanding how progerin affects DNA replication, given that most of the DNA damage that accumulates in cells is generated during replication. Early replication studies using *Xenopus* extracts showed that nuclear lamina disruption causes a marked reduction in DNA replication, concomitant with alteration in the distribution of Proliferating Cell Nuclear Antigen (PCNA) and the Replication Factor Complex (RFC), key factors in the elongation phase of DNA replication (Spann et al. 1997). Another study revealed that altered organization of the nuclear lamina inhibits chain elongation in a dose-dependent manner (Moir et al. 2000). In addition, PCNA has been found to co-localize with A- and B-type lamins in early and late sites of DNA replication, respectively (Moir et al. 1994; Goldberg et al. 1995; Jenkins et al. 1995; Kennedy et al. 2000; Dechat et al. 2008), suggesting a role for lamins in the spatial/temporal organization of replication.

(marker of DNA DSBs), and phosphorylated-RPA (marker of replication stress). The structure of the nuclear lamina is impacted by progerin expression. A-type and B-type lamins form independent networks, and progerin seems to be able to intercalate in both types of networks, eliciting structural alterations that affect nuclear stiffness and stability. In fact, nuclear rupture is common in progerin-expressing cells, with leakage of DNA fragments, chromatin, and other nucleoplasmic proteins into the cytoplasm. Similarly, mitochondrial integrity and function is often compromised in HGPS patient-derived cells. Moreover, progerin causes broad changes in gene expression. Recently, we showed that the transcription factor STAT1 is activated by phosphorylation in progerin-expressing cells, leading to its translocation to the nucleus and the activation of target genes in the interferon (IFN) response. (Graphic illustrations generated by Michael Andrus, BS, St Louis University)

More recently, expression of pre-lamin-A was associated with mono-ubiquitination of PCNA and induction of Pol η, two hallmarks of replication fork stalling (Cobb et al. 2016). It was suggested that pre-lamin-A mitigates the interaction of PCNA with mature lamin-A, eliciting replication fork stalling. In HGPS cells, RFC1 is aberrantly degraded by a serine protease, and the cleavage causes defects in the loading of PCNA and Pol δ onto DNA for replication (Tang et al. 2012). In an unbiased screen of lamin-A- and progerin-interacting proteins by mass spectrometry, progerin was also shown to interact with PCNA more robustly than lamin A (Kubben et al. 2010), and also reported to sequester PCNA away from replicating DNA (Wheaton et al. 2017). These findings support the idea that expression of pre-lamin-A and progerin elicit replication stress by sequestering PCNA away from the replication fork. Consistent with this idea, progerin-expressing cells accumulate XPA at stalled or collapsed replication forks, concomitant with a significant loss of PCNA at the forks (Hilton et al. 2017). Depletion of XPA or progerin restores PCNA at replication forks, while reducing the extent of progerin-induced apoptosis. Therefore, progerin expression seems to alter the binding of key factors to the replication fork, including PCNA and proteins such as XPA that participate in the repair of DNA lesions. Altogether, these findings suggest that alterations in nuclear lamins impact DNA replication and that replication stress could play a major role in the proliferation defects and genomic instability that characterize lamins-defective cells. Despite these findings, our mechanistic understanding of how mutant lamins hinder DNA replication is limited.

Mutant lamins such as progerin could hinder the progression of the replication fork by inducing mis-localization of factors that associate with the replisome -PCNA and RFC-. Lamin dysfunction could also hinder the proper recruitment of replication fork protective factors upon fork stalling, causing replication stress-induced genomic instability. To understand mechanistically how lamin dysfunction affects DNA replication requires the utilization of newly developed techniques such as genome-wide single-molecule replication assays (DNA fiber assays), iPOND (Isolation of Proteins On Nascent DNA) (Sirbu et al. 2011), and electron microscopy (Vindigni and Lopes 2017). Our recent studies performing DNA fiber assays have revealed that progerin expression, but not overexpression of lamin-A, causes a robust phenotype hindering replication (Kreienkamp et al. 2018). Progerin elicits replication stress, characterized by increased replication fork stalling in the absence of drugs that inhibit replication. In addition, we find that stalled replication forks are deprotected and susceptible to MRE11 nuclease-mediated fork degradation. As such, inhibition of MRE11 nuclease rescues replication defects in progerin-expressing cells. Moreover, we find that a variety of compounds known to ameliorate phenotypes of genomic instability in progerin-expressing cells, including vitamin D (Gonzalez-Suarez et al. 2011), all-trans retinoic acid (ATRA) (Swift et al. 2013; Kubben et al. 2015; Pellegrini et al. 2015), remodelin (Larrieu et al. 2014), and the combination of a farnesyltransferase inhibitor (FTI) and rapamycin (Cao et al. 2011; Pellegrini et al. 2015; Gordon et al. 2016) markedly reduce replication stress in progerin-expressing cells (Kreienkamp et al. 2018). Despite the fact that molecular mechanisms underlying the beneficial effects of these drugs rescuing replication stress and genomic instability in progerin-expressing cells remain to be identified, this finding has important implications from a therapeutic perspective and for defining the importance of replication stress to the progeria phenotype.

Interestingly, our recent studies demonstrate that replication stress in progerin-expressing cells not only contribute to genomic instability, but also activate inflammatory responses that contribute to cellular ageing. Replication stress in HGPS patient-derived fibroblasts and progerin-expressing normal fibroblasts is accompanied by accumulation of chromatin at the cytoplasm, upregulation of cytosolic sensors of nucleic acids -cGAS, STING, RIG-I, MDA5, and OASs-, and robust activation of a cell intrinsic interferon (IFN)-like response (Kreienkamp et al. 2018). This IFN-like response, which is regulated by STAT1, contributes to cellular ageing phenotypes such as reduced proliferation and migration capabilities. This finding is important because STAT1 is a notorious regulator of inflammation in immune and vascular cells during atherosclerosis, and an important therapeutic target for cardiovascular disease (Szelag et al. 2016), the main cause of death of HGPS patients. We hypothesize that progerin expression in vascular cells from HGPS patients could recapitulate the STAT1 pathway activation observed in fibroblasts, being a contributor to the decline of vascular cells characteristic of this disease.

Importantly, we showed that the same treatments that ameliorate replication stress -vitamin D, ATRA, remodelin, FTI and rapamycin-, markedly repress the STAT1/IFN-like response (Kreienkamp et al. 2018). We propose that in progerin-expressing cells, DNA damage and replication stress, together with disruption of nuclear integrity, results in accumulation of immunogenic nucleic acids in the cytoplasm, where they activate cytosolic sensors of foreign nucleic acids. This in turn leads to activation of inflammatory pathways such as NFκB and STAT1 that drive an IFN-like response. Defining the causes of this cell-intrinsic IFN response and its consequences for organismal decline in HGPS, as well as the mechanisms whereby different compounds ameliorate this response might reveal ways to reduce its pathological impact in HGPS and in normal ageing, as progerin is expressed in cells from old individuals (Dahl et al. 2006).

Current and Future Therapies

With such a multitude of cellular processes and organ systems affected, developing therapies for HGPS has proven challenging. However, since progerin was identified as the driver behind disease phenotypes, researchers have searched for ways to combat its detrimental effects (Harhouri et al. 2018). Among the first of the drugs that emerged was lonafarnib, a farnesyltransferase inhibitor (FTI) designed to prevent processing of prelamin A to progerin. FTIs reduced nuclear blebbing, nuclear stiffness, rescued heterochromatin organization in HGPS cells, decreasing onset of premature senescence and improving proliferation (Capell et al. 2005; Columbaro et al. 2005; Yang et al. 2005; Verstraeten et al. 2008). Further, FTIs had a remarkable effect in mouse models of disease (Fong et al. 2006; Yang et al. 2006; Varela et al. 2008).

Based on these promising results, a clinical trial was initiated. There, the results were equally compelling. Administration of the FTI lonafarnib for 2 years in HGPS patients improved pulse-wave velocity, carotid artery wall echodensity, and incidence of stroke, headaches, and seizures (Gordon et al. 2012). FTIs increased mean survival by 1.6 years (Gordon et al. 2014a). Recently, it was demonstrated that lonafarnib monotherapy was associated with a lower mortality rate after 2.2 years of follow-up (Gordon et al. 2018). While this drug certainly is the first treatment for HGPS patients, it can be hard for patients to tolerate due to a number of side effects. Therefore, the next wave of therapies, and the possibility for combination therapies, are desperately needed to allow for synergy or even lower dosages for effectiveness.

The next compound that has made its way to patients is everolimus. Everolimus is an analog of rapamycin, which promotes the removal of toxic, insoluble aggregates like progerin by enhanced autophagy (Cao et al. 2011). Everolimus increased proliferative ability and delayed cellular senescence in cell lines, including those without the classical HGPS mutation (DuBose et al. 2018). Based on these studies, a phase I/II dose-escalation clinical trial of everolimus in combination with lonafarnib was initiated in 2015, with results from these studies expected by 2020 (https://clinicaltrials.gov/ct2/show/study/NCT02579044).

Sulforaphane acts in a similar manner to everolimus, and, as such, has shown benefit in vitro by increasing progerin clearance by autophagy (Gabriel et al. 2015). More recent studies have demonstrated that combination of sulforaphane with lonafarnib is toxic, but intermittent treatment of sulforaphane with lonafarnib separately and in repeated cycles rescues HGPS cellular phenotype (Gabriel et al. 2017).

Other therapies are likely next for testing in patients. Remodelin, an inhibitor of N-acetyltransferase-10 (NAT10), increases chromatin compaction while rescuing nuclear morphological abnormalities, proliferation defects, and accumulation of DNA damage characteristic of progerin-expressing cells (Larrieu et al. 2014). A preclinical study recently performed with remodelin treatment revealed an improvement in healthspan in the progeria mice (Balmus et al. 2018), and similar effects by chemical inhibition of NAT10.

The retinoids are among other treatment strategies that now await testing in vivo. The *LMNA* gene promoter contains retinoic acid responsive elements (L-RARE) that downregulate *LMNA* gene expression with all trans retinoic acid (ATRA) treatment (Swift et al. 2013). In HGPS patient-derived fibroblasts, ATRA treatment reduces significantly progerin expression and actually synergizes with rapamycin in downregulating progerin levels, ameliorating a variety of progerin-induced phenotypes (Pellegrini et al. 2015). Retinoids were also identified in a high-throughput, high-content based screening of a library of FDA approved drugs as a class of compounds able to revert cellular HGPS phenotypes (Kubben et al. 2016). Similarly, our studies show that activation of vitamin D receptor signaling by ligand ($1,25\alpha$-dihydroxy-vitamin D_3) binding ameliorates a broad repertoire of phenotypes of HGPS patient-derived cells (Kreienkamp et al. 2016, 2018). Other therapeutic strategies of benefit have included inhibitors of the enzyme responsible for carboxymethylation of the farnesylcysteine of progerin (Ibrahim et al. 2013); the

ROS scavenger N-acetyl cysteine, which reduces the amount of unrepairable DNA damage caused by the increased generation of ROS (Pekovic et al. 2011; Richards et al. 2011; Lattanzi et al. 2012; Sieprath et al. 2012); methylene blue, a mitochondrial-targeting antioxidant (Xiong et al. 2016); or resveratrol, an enhancer of SIRT1 deacetylase activity that alleviates progeroid features (Liu et al. 2012). The plethora of potential therapeutic options is encourageing, since it is likely that the best treatment options for these patients will consist of combination therapy. Combination therapy might allow for synergy among compounds, reducing toxicity owed to lowering the doses of each single compound.

Concluding Remarks

HGPS, with its severity and time-course, is certainly an outlier to the ageing process. As an outlier, though, its study portends value not only for these patients, but also for the normal ageing population. Since 2003, our understanding of both normal and abnormal ageing has greatly increased. In the last few years, the fruits of these studies have become tangible for HGPS patients with the first wave of therapies to help them. It is hoped that the next years will yield further therapies of benefit to these children.

Much remains to be learned. Our understanding of the cardiovascular disease that drives patient death is still quite limited. Our studies are also challenged by the fact that we still lack a suitable animal model for recapitulating well the cardiovascular disease driving human death. The coming years seem destined for advances in our understanding of these critical disease processes. Finding targeted biomarkers for disease remains an important goal, as is identifying other genes that impact disease phenotype. With the increased utilization of next generation sequencing, it is hoped that this technology will also benefit HGPS, particularly with identifying other genetic traits that either potentiate or reduce progerin's toxic effect. Other new technologies, like CRISPR-Cas9, have obvious applications in diseases like HGPS, which now just wait application. Time will certainly reveal new mysteries for disease. And, as we develop ways to help these children live longer, maybe some of these findings might also have relevance for the rest of us in the normal ageing population.

References

Arancio W, Pizzolanti G, Genovese SI, Pitrone M, Giordano C (2014) Epigenetic involvement in Hutchinson-Gilford progeria syndrome: a mini-review. Gerontology 60(3):197–203

Balmus G, Larrieu D, Barros AC, Collins C, Abrudan M, Demir M, Geisler NJ, Lelliott CJ, White JK, Karp NA, Atkinson J, Kirton A, Jacobsen M, Clift D, Rodriguez R, Sanger Mouse Genetics P, Adams DJ, Jackson SP (2018) Targeting of NAT10 enhances healthspan in a mouse model of human accelerated ageing syndrome. Nat Commun 9(1):1700

Borroni AP, Emanuelli A, Shah PA, Ilic N, Apel-Sarid L, Paolini B, Manikoth Ayyathan D, Koganti P, Levy-Cohen G, Blank M (2018) Smurf2 regulates stability and the autophagic-lysosomal turnover of lamin A and its disease-associated form progerin. Ageing Cell 17(2):e12732

Bridgeman VL, Vermeulen PB, Foo S, Bilecz A, Daley F, Kostaras E, Nathan MR, Wan E, Frentzas S, Schweiger T, Hegedus B, Hoetzenecker K, Renyi-Vamos F, Kuczynski EA, Vasudev NS, Larkin J, Gore M, Dvorak HF, Paku S, Kerbel RS, Dome B, Reynolds AR (2017) Vessel co-option is common in human lung metastases and mediates resistance to anti-angiogenic therapy in preclinical lung metastasis models. J Pathol 241(3):362–374

Cao K, Graziotto JJ, Blair CD, Mazzulli JR, Erdos MR, Krainc D, Collins FS (2011) Rapamycin reverses cellular phenotypes and enhances mutant protein clearance in Hutchinson-Gilford progeria syndrome cells. Sci Transl Med 3(89):89ra58

Capell BC, Erdos MR, Madigan JP, Fiordalisi JJ, Varga R, Conneely KN, Gordon LB, Der CJ, Cox AD, Collins FS (2005) Inhibiting farnesylation of progerin prevents the characteristic nuclear blebbing of Hutchinson-Gilford progeria syndrome. Proc Natl Acad Sci U S A 102(36):12879–12884

Chojnowski A, Ong PF, Dreesen O (2015) Nuclear lamina remodelling and its implications for human disease. Cell Tissue Res 360(3):621–631

Cleveland RH, Gordon LB, Kleinman ME, Miller DT, Gordon CM, Snyder BD, Nazarian A, Giobbie-Harder A, Neuberg D, Kieran MW (2012) A prospective study of radiographic manifestations in Hutchinson-Gilford progeria syndrome. Pediatr Radiol 42(9):1089–1098

Cobb AM, Murray TV, Warren DT, Liu Y, Shanahan CM (2016) Disruption of PCNA-lamins A/C interactions by prelamin A induces DNA replication fork stalling. Nucleus 7(5):498–511

Columbaro M, Capanni C, Mattioli E, Novelli G, Parnaik VK, Squarzoni S, Maraldi NM, Lattanzi G (2005) Rescue of heterochromatin organization in Hutchinson-Gilford progeria by drug treatment. Cell Mol Life Sci 62(22):2669–2678

Csoka AB, English SB, Simkevich CP, Ginzinger DG, Butte AJ, Schatten GP, Rothman FG, Sedivy JM (2004) Genome-scale expression profiling of Hutchinson-Gilford progeria syndrome reveals widespread transcriptional misregulation leading to mesodermal/mesenchymal defects and accelerated atherosclerosis. Ageing Cell 3(4):235–243

Dahl KN, Scaffidi P, Islam MF, Yodh AG, Wilson KL, Misteli T (2006) Distinct structural and mechanical properties of the nuclear lamina in Hutchinson-Gilford progeria syndrome. Proc Natl Acad Sci U S A 103(27):10271–10276

De Sandre-Giovannoli A, Bernard R, Cau P, Navarro C, Amiel J, Boccaccio I, Lyonnet S, Stewart CL, Munnich A, Le Merrer M, Levy N (2003) Lamin a truncation in Hutchinson-Gilford progeria. Science 300(5628):2055

Dechat T, Pfleghaar K, Sengupta K, Shimi T, Shumaker DK, Solimando L, Goldman RD (2008) Nuclear lamins: major factors in the structural organization and function of the nucleus and chromatin. Genes Dev 22(7):832–853

Dobrzynska A, Gonzalo S, Shanahan C, Askjaer P (2016) The nuclear Lamina in health and disease. Nucleus 7:233–248

Domingo DL, Trujillo MI, Council SE, Merideth MA, Gordon LB, Wu T, Introne WJ, Gahl WA, Hart TC (2009) Hutchinson-Gilford progeria syndrome: oral and craniofacial phenotypes. Oral Dis 15(3):187–195

Duband-Goulet I, Woerner S, Gasparini S, Attanda W, Konde E, Tellier-Lebegue C, Craescu CT, Gombault A, Roussel P, Vadrot N, Vicart P, Ostlund C, Worman HJ, Zinn-Justin S, Buendia B (2011) Subcellular localization of SREBP1 depends on its interaction with the C-terminal region of wild-type and disease related A-type lamins. Exp Cell Res 317(20):2800–2813

DuBose AJ, Lichtenstein ST, Petrash NM, Erdos MR, Gordon LB, Collins FS (2018) Everolimus rescues multiple cellular defects in laminopathy-patient fibroblasts. Proc Natl Acad Sci U S A 115(16):4206–4211

Eriksson M, Brown WT, Gordon LB, Glynn MW, Singer J, Scott L, Erdos MR, Robbins CM, Moses TY, Berglund P, Dutra A, Pak E, Durkin S, Csoka AB, Boehnke M, Glover TW, Collins FS (2003) Recurrent de novo point mutations in lamin A cause Hutchinson-Gilford progeria syndrome. Nature 423(6937):293–298

Fong LG, Frost D, Meta M, Qiao X, Yang SH, Coffinier C, Young SG (2006) A protein farnesyltransferase inhibitor ameliorates disease in a mouse model of progeria. Science 311(5767):1621–1623

Frankel D, Delecourt V, Harhouri K, De Sandre-Giovannoli A, Levy N, Kaspi E, Roll P (2018) MicroRNAs in hereditary and sporadic premature ageing syndromes and other laminopathies. Aging Cell. 2018 Apr 25:e12766. https://doi.org/10.1111/acel.12766

Gabriel D, Roedl D, Gordon LB, Djabali K (2015) Sulforaphane enhances progerin clearance in Hutchinson-Gilford progeria fibroblasts. Ageing Cell 14(1):78–91

Gabriel D, Shafry DD, Gordon LB, Djabali K (2017) Intermittent treatment with farnesyltransferase inhibitor and sulforaphane improves cellular homeostasis in Hutchinson-Gilford progeria fibroblasts. Oncotarget 8(39):64809–64826

Gerhard-Herman M, Smoot LB, Wake N, Kieran MW, Kleinman ME, Miller DT, Schwartzman A, Giobbie-Hurder A, Neuberg D, Gordon LB (2012) Mechanisms of premature vascular ageing in children with Hutchinson-Gilford progeria syndrome. Hypertension 59(1):92–97

Ghosh S, Liu B, Wang Y, Hao Q, Zhou Z (2015) Lamin A Is an Endogenous SIRT6 Activator and Promotes SIRT6-Mediated DNA Repair. Cell Rep 13(7):1396–1406

Goldberg M, Jenkins H, Allen T, Whitfield WG, Hutchison CJ (1995) Xenopus lamin B3 has a direct role in the assembly of a replication competent nucleus: evidence from cell-free egg extracts. J Cell Sci 108(Pt 11):3451–3461

Goldman RD, Shumaker DK, Erdos MR, Eriksson M, Goldman AE, Gordon LB, Gruenbaum Y, Khuon S, Mendez M, Varga R, Collins FS (2004) Accumulation of mutant lamin A causes progressive changes in nuclear architecture in Hutchinson-Gilford progeria syndrome. Proc Natl Acad Sci U S A 101(24):8963–8968

Gonzalez-Suarez I, Redwood AB, Grotsky DA, Neumann MA, Cheng EH, Stewart CL, Dusso A, Gonzalo S (2011) A new pathway that regulates 53BP1 stability implicates cathepsin L and vitamin D in DNA repair. EMBO J 30(16):3383–3396

Gonzalo S, Kreienkamp R (2015) DNA repair defects and genome instability in Hutchinson-Gilford progeria syndrome. Curr Opin Cell Biol 34:75–83

Gonzalo S, Kreienkamp R, Askjaer P (2017) Hutchinson-Gilford progeria syndrome: a premature ageing disease caused by LMNA gene mutations. Ageing Res Rev 33:18–29

Gordon LB, McCarten KM, Giobbie-Hurder A, Machan JT, Campbell SE, Berns SD, Kieran MW (2007) Disease progression in Hutchinson-Gilford progeria syndrome: impact on growth and development. Pediatrics 120(4):824–833

Gordon CM, Gordon LB, Snyder BD, Nazarian A, Quinn N, Huh S, Giobbie-Hurder A, Neuberg D, Cleveland R, Kleinman M, Miller DT, Kieran MW (2011) Hutchinson-Gilford progeria is a skeletal dysplasia. J Bone Miner Res 26(7):1670–1679

Gordon LB, Kleinman ME, Miller DT, Neuberg DS, Giobbie-Hurder A, Gerhard-Herman M, Smoot LB, Gordon CM, Cleveland R, Snyder BD, Fligor B, Bishop WR, Statkevich P, Regen A, Sonis A, Riley S, Ploski C, Correia A, Quinn N, Ullrich NJ, Nazarian A, Liang MG, Huh SY, Schwartzman A, Kieran MW (2012) Clinical trial of a farnesyltransferase inhibitor in children with Hutchinson-Gilford progeria syndrome. Proc Natl Acad Sci U S A 109(41):16666–16671

Gordon LB, Massaro J, D'Agostino RB Sr, Campbell SE, Brazier J, Brown WT, Kleinman ME, Kieran MW, C. Progeria Clinical Trials (2014a) Impact of farnesylation inhibitors on survival in Hutchinson-Gilford progeria syndrome. Circulation 130(1):27–34

Gordon LB, Rothman FG, Lopez-Otin C, Misteli T (2014b) Progeria: a paradigm for translational medicine. Cell 156(3):400–407

Gordon LB, Kleinman ME, Massaro J, D'Agostino RB Sr, Shappell H, Gerhard-Herman M, Smoot LB, Gordon CM, Cleveland RH, Nazarian A, Snyder BD, Ullrich NJ, Silvera VM, Liang MG, Quinn N, Miller DT, Huh SY, Dowton AA, Littlefield K, Greer MM, Kieran MW (2016) Clinical trial of the protein farnesylation inhibitors lonafarnib, pravastatin, and zoledronic acid in children with Hutchinson-Gilford progeria syndrome. Circulation 134(2):114–125

Gordon LB, Shappell H, Massaro J, D'Agostino RB Sr, Brazier J, Campbell SE, Kleinman ME, Kieran MW (2018) Association of lonafarnib treatment vs no treatment with mortality rate in patients with Hutchinson-Gilford progeria syndrome. JAMA 319(16):1687–1695

Greer MM, Kleinman ME, Gordon LB, Massaro J, D'Agostino RB Sr, Baltrusaitis K, Kieran MW, Gordon CM (2017) Pubertal progression in female adolescents with progeria. J Pediatr Adolesc Gynecol. 2018 Jun 31(3):238–241

Guardiani E, Zalewski C, Brewer C, Merideth M, Introne W, Smith AC, Gordon L, Gahl W, Kim HJ (2011) Otologic and audiologic manifestations of Hutchinson-Gilford progeria syndrome. Laryngoscope 121(10):2250–2255

Harhouri K, Navarro C, Depetris D, Mattei MG, Nissan X, Cau P, De Sandre-Giovannoli A, Levy N (2017) MG132-induced progerin clearance is mediated by autophagy activation and splicing regulation. EMBO Mol Med 9(9):1294–1313

Harhouri K, Frankel D, Bartoli C, Roll P, De Sandre-Giovannoli A, Levy N (2018) An overview of treatment strategies for Hutchinson-Gilford Progeria syndrome. Nucleus 9(1):246–257

Hernandez L, Roux KJ, Wong ES, Mounkes LC, Mutalif R, Navasankari R, Rai B, Cool S, Jeong JW, Wang H, Lee HS, Kozlov S, Grunert M, Keeble T, Jones CM, Meta MD, Young SG, Daar IO, Burke B, Perantoni AO, Stewart CL (2010) Functional coupling between the extracellular matrix and nuclear lamina by Wnt signaling in progeria. Dev Cell 19(3):413–425

Heyn H, Moran S, Esteller M (2013) Aberrant DNA methylation profiles in the premature ageing disorders Hutchinson-Gilford Progeria and Werner syndrome. Epigenetics 8(1):28–33

Hilton BA, Liu J, Cartwright BM, Liu Y, Breitman M, Wang Y, Jones R, Tang H, Rusinol A, Musich PR, Zou Y (2017) Progerin sequestration of PCNA promotes replication fork collapse and mislocalization of XPA in laminopathy-related progeroid syndromes. FASEB J 31(9):3882–3893

Ibrahim MX, Sayin VI, Akula MK, Liu M, Fong LG, Young SG, Bergo MO (2013) Targeting isoprenylcysteine methylation ameliorates disease in a mouse model of progeria. Science 340(6138):1330–1333

Jenkins H, Whitfield WG, Goldberg MW, Allen TD, Hutchison CJ (1995) Evidence for the direct involvement of lamins in the assembly of a replication competent nucleus. Acta Biochim Pol 42(2):133–143

Jung HJ, Coffinier C, Choe Y, Beigneux AP, Davies BS, Yang SH, Barnes RH 2nd, Hong J, Sun T, Pleasure SJ, Young SG, Fong LG (2012) Regulation of prelamin A but not lamin C by miR-9, a brain-specific microRNA. Proc Natl Acad Sci U S A 109(7):E423–E431

Kang HT, Park JT, Choi K, Choi HJC, Jung CW, Kim GR, Lee YS, Park SC (2017) Chemical screening identifies ROCK as a target for recovering mitochondrial function in Hutchinson-Gilford progeria syndrome. Ageing Cell 16(3):541–550

Kennedy BK, Barbie DA, Classon M, Dyson N, Harlow E (2000) Nuclear organization of DNA replication in primary mammalian cells. Genes Dev 14(22):2855–2868

Kieran MW, Gordon L, Kleinman M (2007) New approaches to progeria. Pediatrics 120(4):834–841

Kreienkamp R, Croke M, Neumann MA, Bedia-Diaz G, Graziano S, Dusso A, Dorsett D, Carlberg C, Gonzalo S (2016) Vitamin D receptor signaling improves Hutchinson-Gilford progeria syndrome cellular phenotypes. Oncotarget 7:30018–30031

Kreienkamp R, Graziano S, Coll-Bonfill N, Bedia-Diaz G, Cybulla E, Vindigni A, Dorsett D, Kubben N, Batista LFZ, Gonzalo S (2018) A cell-intrinsic interferon-like response links replication stress to cellular ageing caused by progerin. Cell Rep 22(8):2006–2015

Kubben N, Voncken JW, Demmers J, Calis C, van Almen G, Pinto Y, Misteli T (2010) Identification of differential protein interactors of lamin A and progerin. Nucleus 1(6):513–525

Kubben N, Brimacombe KR, Donegan M, Li Z, Misteli T (2015) A high-content imageing-based screening pipeline for the systematic identification of anti-progeroid compounds. Methods. 2016 Mar 1:96:46–58. https://doi.org/10.1016/j.ymeth.2015.08.024. Epub 2015 Sep 1.

Kubben N, Zhang W, Wang L, Voss TC, Yang J, Qu J, Liu GH, Misteli T (2016) Repression of the antioxidant NRF2 pathway in premature ageing. Cell 165(6):1361–1374

Larrieu D, Britton S, Demir M, Rodriguez R, Jackson SP (2014) Chemical inhibition of NAT10 corrects defects of laminopathic cells. Science 344(6183):527–532

Lattanzi G, Marmiroli S, Facchini A, Maraldi NM (2012) Nuclear damages and oxidative stress: new perspectives for laminopathies. Eur J Histochem 56(4):e45

Liu B, Wang J, Chan KM, Tjia WM, Deng W, Guan X, Huang JD, Li KM, Chau PY, Chen DJ, Pei D, Pendas AM, Cadinanos J, Lopez-Otin C, Tse HF, Hutchison C, Chen J, Cao Y, Cheah KS, Tryggvason K, Zhou Z (2005) Genomic instability in laminopathy-based premature ageing. Nat Med 11(7):780–785

Liu Y, Wang Y, Rusinol AE, Sinensky MS, Liu J, Shell SM, Zou Y (2008) Involvement of xeroderma pigmentosum group A (XPA) in progeria arising from defective maturation of prelamin A. FASEB J 22(2):603–611

Liu B, Ghosh S, Yang X, Zheng H, Liu X, Wang Z, Jin G, Zheng B, Kennedy BK, Suh Y, Kaeberlein M, Tryggvason K, Zhou Z (2012) Resveratrol rescues SIRT1-dependent adult stem cell decline and alleviates progeroid features in laminopathy-based progeria. Cell Metab 16(6):738–750

Marji J, O'Donoghue SI, McClintock D, Satagopam VP, Schneider R, Ratner D, Worman HJ, Gordon LB, Djabali K (2010) Defective lamin A-Rb signaling in Hutchinson-Gilford progeria syndrome and reversal by farnesyltransferase inhibition. PLoS One 5(6):e11132

McClintock D, Ratner D, Lokuge M, Owens DM, Gordon LB, Collins FS, Djabali K (2007) The mutant form of lamin A that causes Hutchinson-Gilford progeria is a biomarker of cellular ageing in human skin. PLoS One 2(12):e1269

Merideth MA, Gordon LB, Clauss S, Sachdev V, Smith AC, Perry MB, Brewer CC, Zalewski C, Kim HJ, Solomon B, Brooks BP, Gerber LH, Turner ML, Domingo DL, Hart TC, Graf J, Reynolds JC, Gropman A, Yanovski JA, Gerhard-Herman M, Collins FS, Nabel EG, Cannon RO 3rd, Gahl WA, Introne WJ (2008) Phenotype and course of Hutchinson-Gilford progeria syndrome. N Engl J Med 358(6):592–604

Messner M, Ghadge SK, Goetsch V, Wimmer A, Dorler J, Polzl G, Zaruba MM (2018) Upregulation of the ageing related LMNA splice variant progerin in dilated cardiomyopathy. PLoS One 13(4):e0196739

Moir RD, Montag-Lowy M, Goldman RD (1994) Dynamic properties of nuclear lamins: lamin B is associated with sites of DNA replication. J Cell Biol 125(6):1201–1212

Moir RD, Spann TP, Herrmann H, Goldman RD (2000) Disruption of nuclear lamin organization blocks the elongation phase of DNA replication. J Cell Biol 149(6):1179–1192

Olive M, Harten I, Mitchell R, Beers JK, Djabali K, Cao K, Erdos MR, Blair C, Funke B, Smoot L, Gerhard-Herman M, Machan JT, Kutys R, Virmani R, Collins FS, Wight TN, Nabel EG, Gordon LB (2010) Cardiovascular pathology in Hutchinson-Gilford progeria: correlation with the vascular pathology of ageing. Arterioscler Thromb Vasc Biol 30(11):2301–2309

Osorio FG, Varela I, Lara E, Puente XS, Espada J, Santoro R, Freije JM, Fraga MF, Lopez-Otin C (2010) Nuclear envelope alterations generate an ageing-like epigenetic pattern in mice deficient in Zmpste24 metalloprotease. Ageing Cell 9(6):947–957

Osorio FG, Navarro CL, Cadinanos J, Lopez-Mejia IC, Quiros PM, Bartoli C, Rivera J, Tazi J, Guzman G, Varela I, Depetris D, de Carlos F, Cobo J, Andres V, De Sandre-Giovannoli A, Freije JM, Levy N, Lopez-Otin C (2011) Splicing-directed therapy in a new mouse model of human accelerated ageing. Sci Transl Med 3(106):106ra107

Pegoraro G, Kubben N, Wickert U, Gohler H, Hoffmann K, Misteli T (2009) Ageing-related chromatin defects through loss of the NURD complex. Nat Cell Biol 11(10):1261–1267

Pekovic V, Gibbs-Seymour I, Markiewicz E, Alzoghaibi F, Benham AM, Edwards R, Wenhert M, von Zglinicki T, Hutchison CJ (2011) Conserved cysteine residues in the mammalian lamin A tail are essential for cellular responses to ROS generation. Ageing Cell 10(6):1067–1079

Pellegrini C, Columbaro M, Capanni C, D'Apice MR, Cavallo C, Murdocca M, Lattanzi G, Squarzoni S (2015) All-trans retinoic acid and rapamycin normalize Hutchinson Gilford progeria fibroblast phenotype. Oncotarget 6(30):29914–29928

Prakash A, Gordon LB, Kleinman ME, Gurary EB, Massaro J, D'Agostino R Sr, Kieran MW, Gerhard-Herman M, Smoot L (2018) Cardiac abnormalities in patients with Hutchinson-Gilford progeria syndrome. JAMA Cardiol 3(4):326–334

Prokocimer M, Barkan R, Gruenbaum Y (2013) Hutchinson-Gilford progeria syndrome through the lens of transcription. Ageing Cell 12(4):533–543

Richards SA, Muter J, Ritchie P, Lattanzi G, Hutchison CJ (2011) The accumulation of un-repairable DNA damage in laminopathy progeria fibroblasts is caused by ROS generation and is prevented by treatment with N-acetyl cysteine. Hum Mol Genet 20(20):3997–4004

Rork, J. F., J. T. Huang, L. B. Gordon, M. Kleinman, M. W. Kieran, M. G. Liang (2014). Initial cutaneous manifestations of Hutchinson-Gilford progeria syndrome. Pediatr Dermatol 31(2): 196–202

Scaffidi P, Misteli T (2005) Reversal of the cellular phenotype in the premature ageing disease Hutchinson-Gilford progeria syndrome. Nat Med 11(4):440–445

Scaffidi P, Misteli T (2008) Lamin A-dependent misregulation of adult stem cells associated with accelerated ageing. Nat Cell Biol 10(4):452–459

Shumaker DK, Dechat T, Kohlmaier A, Adam SA, Bozovsky MR, Erdos MR, Eriksson M, Goldman AE, Khuon S, Collins FS, Jenuwein T, Goldman RD (2006) Mutant nuclear lamin A leads to progressive alterations of epigenetic control in premature ageing. Proc Natl Acad Sci U S A 103(23):8703–8708

Sieprath T, Darwiche R, De Vos WH (2012) Lamins as mediators of oxidative stress. Biochem Biophys Res Commun 421(4):635–639

Silvera VM, Gordon LB, Orbach DB, Campbell SE, Machan JT, Ullrich NJ (2013) Imageing characteristics of cerebrovascular arteriopathy and stroke in Hutchinson-Gilford progeria syndrome. AJNR Am J Neuroradiol 34(5):1091–1097

Sirbu BM, Couch FB, Feigerle JT, Bhaskara S, Hiebert SW, Cortez D (2011) Analysis of protein dynamics at active, stalled, and collapsed replication forks. Genes Dev 25(12):1320–1327

Spann TP, Moir RD, Goldman AE, Stick R, Goldman RD (1997) Disruption of nuclear lamin organization alters the distribution of replication factors and inhibits DNA synthesis. J Cell Biol 136(6):1201–1212

Stehbens WE, Wakefield SJ, Gilbert-Barness E, Olson RE, Ackerman J (1999) Histological and ultrastructural features of atherosclerosis in progeria. Cardiovasc Pathol 8(1):29–39

Swift J, Ivanovska IL, Buxboim A, Harada T, Dingal PC, Pinter J, Pajerowski JD, Spinler KR, Shin JW, Tewari M, Rehfeldt F, Speicher DW, Discher DE (2013) Nuclear lamin-A scales with tissue stiffness and enhances matrix-directed differentiation. Science 341(6149):1240104

Szelag M, Piaszyk-Borychowska A, Plens-Galaska M, Wesoly J, Bluyssen HA (2016) Targeted inhibition of STATs and IRFs as a potential treatment strategy in cardiovascular disease. Oncotarget 7:48788–48812

Tang H, Hilton B, Musich PR, Fang DZ, Zou Y (2012) Replication factor C1, the large subunit of replication factor C, is proteolytically truncated in Hutchinson-Gilford progeria syndrome. Ageing Cell 11(2):363–365

Ullrich NJ, Gordon LB (2015) Hutchinson-Gilford progeria syndrome. Handb Clin Neurol 132:249–264

Varela I, Pereira S, Ugalde AP, Navarro CL, Suarez MF, Cau P, Cadinanos J, Osorio FG, Foray N, Cobo J, de Carlos F, Levy N, Freije JM, Lopez-Otin C (2008) Combined treatment with statins and aminobisphosphonates extends longevity in a mouse model of human premature ageing. Nat Med 14(7):767–772

Varga R, Eriksson M, Erdos MR, Olive M, Harten I, Kolodgie F, Capell BC, Cheng J, Faddah D, Perkins S, Avallone H, San H, Qu X, Ganesh S, Gordon LB, Virmani R, Wight TN, Nabel EG, Collins FS (2006) Progressive vascular smooth muscle cell defects in a mouse model of Hutchinson-Gilford progeria syndrome. Proc Natl Acad Sci U S A 103(9):3250–3255

Verstraeten VL, Ji JY, Cummings KS, Lee RT, Lammerding J (2008) Increased mechanosensitivity and nuclear stiffness in Hutchinson-Gilford progeria cells: effects of farnesyltransferase inhibitors. Ageing Cell 7(3):383–393

Vidak S, Foisner R (2016) Molecular insights into the premature ageing disease progeria. Histochem Cell Biol 145(4):401–417

Villa-Bellosta R, Rivera-Torres J, Osorio FG, Acin-Perez R, Enriquez JA, Lopez-Otin C, Andres V (2013) Defective extracellular pyrophosphate metabolism promotes vascular calcification in a mouse model of Hutchinson-Gilford progeria syndrome that is ameliorated on pyrophosphate treatment. Circulation 127(24):2442–2451

Vindigni A, Lopes M (2017) Combining electron microscopy with single molecule DNA fiber approaches to study DNA replication dynamics. Biophys Chem 225:3–9

Wheaton, K., D. Campuzano, W. Ma, M. Sheinis, B. Ho, G. W. Brown, S. Benchimol (2017) Progerin-induced replication stress facilitates premature senescence in Hutchinson-Gilford progeria syndrome. Mol Cell Biol 37(14)

Xiong ZM, Choi JY, Wang K, Zhang H, Tariq Z, Wu D, Ko E, LaDana C, Sesaki H, Cao K (2016) Methylene blue alleviates nuclear and mitochondrial abnormalities in progeria. Ageing Cell 15(2):279–290

Yang SH, Bergo MO, Toth JI, Qiao X, Hu Y, Sandoval S, Meta M, Bendale P, Gelb MH, Young SG, Fong LG (2005) Blocking protein farnesyltransferase improves nuclear blebbing in mouse fibroblasts with a targeted Hutchinson-Gilford progeria syndrome mutation. Proc Natl Acad Sci U S A 102(29):10291–10296

Yang SH, Meta M, Qiao X, Frost D, Bauch J, Coffinier C, Majumdar S, Bergo MO, Young SG, Fong LG (2006) A farnesyltransferase inhibitor improves disease phenotypes in mice with a Hutchinson-Gilford progeria syndrome mutation. J Clin Invest 116(8):2115–2121

Zhang J, Lian Q, Zhu G, Zhou F, Sui L, Tan C, Mutalif RA, Navasankari R, Zhang Y, Tse HF, Stewart CL, Colman A (2011) A human iPSC model of Hutchinson Gilford Progeria reveals vascular smooth muscle and mesenchymal stem cell defects. Cell Stem Cell 8(1):31–45

Chapter 16
Osteoporosis and the Ageing Skeleton

Terry J. Aspray and Tom R. Hill

Abstract Osteoporosis is a "skeletal disorder characterized by compromised bone strength predisposing a person to an increased risk of fracture" which, in light of demographic change, is becoming an increasing burden on health care worldwide. Increasing age and female gender are associated with the condition, although a wider range of clinical risk factors are being used increasingly to identify those at risk of osteoporosis and its most important sequelae, fracture.

While osteoporosis and fracture have long been associated with women in the post-menopausal age, fracture incidence increases because of the ageing of our population. Interventions to abate the progression of osteoporosis and to prevent fractures must focus on the old and the very old. Evidence associating nutritional factors, particularly calcium and vitamin D are reviewed as are the association of falls risk with fracture and the potential for interventions to prevent falls. Finally, the assessment of frailty in the oldest old, associated sarcopenia and multi-morbidity are considered in the evaluation of fall and fracture risk and the management of osteoporosis in the ninth decade of life and beyond.

Keywords Osteoporosis · Bone mineral density · Hip fracture · Fracture risk assessment · Frailty · Calcium nutrition · Vitamin D · Sarcopenia · Frailty

T. J. Aspray (✉)
NIHR Biomedical Research Centre, Newcastle University, Campus for Ageing and Vitality, Newcastle upon Tyne, UK

Institute of Cellular Medicine, Newcastle University, Newcastle-Upon-Tyne, UK

Institute of Ageing, Newcastle University, Newcastle-Upon-Tyne, UK
e-mail: Terry.Aspray@Newcastle.ac.uk

T. R. Hill
Institute of Cellular Medicine, Newcastle University, Newcastle-Upon-Tyne, UK

Institute of Ageing, Newcastle University, Newcastle-Upon-Tyne, UK

Human Nutrition Research Centre, Newcastle University, Newcastle-Upon-Tyne, UK
e-mail: tom.hill@newcastle.ac.uk

© Springer Nature Singapore Pte Ltd. 2019
J. R. Harris, V. I. Korolchuk (eds.), *Biochemistry and Cell Biology of Ageing: Part II Clinical Science*, Subcellular Biochemistry 91,
https://doi.org/10.1007/978-981-13-3681-2_16

Introduction

Osteoporosis is a "skeletal disorder characterized by compromised bone strength predisposing a person to an increased risk of fracture" (NIH 2001). In this clinical definition, we encounter two key themes. Firstly, the *quality* of the bone is affected and its strength weakened but secondly, and most important in a clinical context, the individual with osteoporosis is at increased risk of fracture. Some of the qualitative features may be observed in skeletal histomorphometry, with osteoporotic bone showing evidence of fewer trabeculae, an overall reduction in trabecular bone volume (see Fig. 16.1) and significant differences in microstructure when compared with normal controls. Differences in bone remodelling (bone turnover) may also be observed, with affected individuals having a low, normal or increased bone turnover, depending on the aetiology of osteoporosis (Steiniche 1995).

Unfortunately, bone histomorphometry is an invasive method of assessment and, although a gold standard diagnostic test for osteoporosis, other non-invasive ways of evaluating the strength of bone are required in practice. Estimates of mineral content by bone mineral density (BMD) correspond to bone strength in vitro (Rudang et al. 2016). Using this method, a beam of x-rays is passed through a skeletal site prone to fracture (spine, hip or wrist) and the attenuation of the x-ray beam is measured using x-rays of differing energies in a dual energy x-ray absorptiometer (DXA), with calibration of the measurement against a bone/soft tissue *phantom* (see Fig. 16.2). In a meta-analysis of prospective studies, BMD at either the spine or hip predict overall fracture risk, with a reduction in BMD by one standard deviation associated with a relative fracture risk of 1.5. Spinal fracture risk was better estimated where the measurement was made at spine with a relative fracture risk of 2.3 and similar findings were seen for hip measurements, predicting hip fracture with a relative risk of 2.6, which are in accordance with results of case-control studies

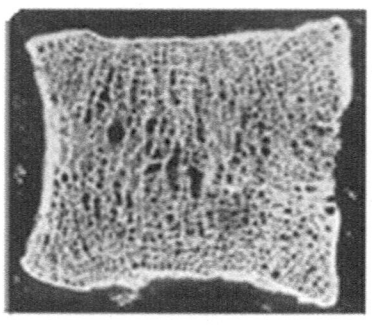
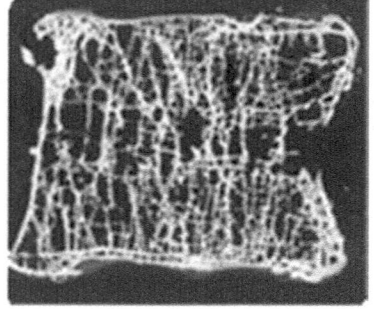

Normal Bone   Osteoporotic Bone

* permission to use image granted by Turner Biomechanics Laboratory.

Fig. 16.1 Two slides comparing the vertebrae of a healthy 37 year old male with a 75 year old female suffering from osteoporosis. (Creative commons licence)

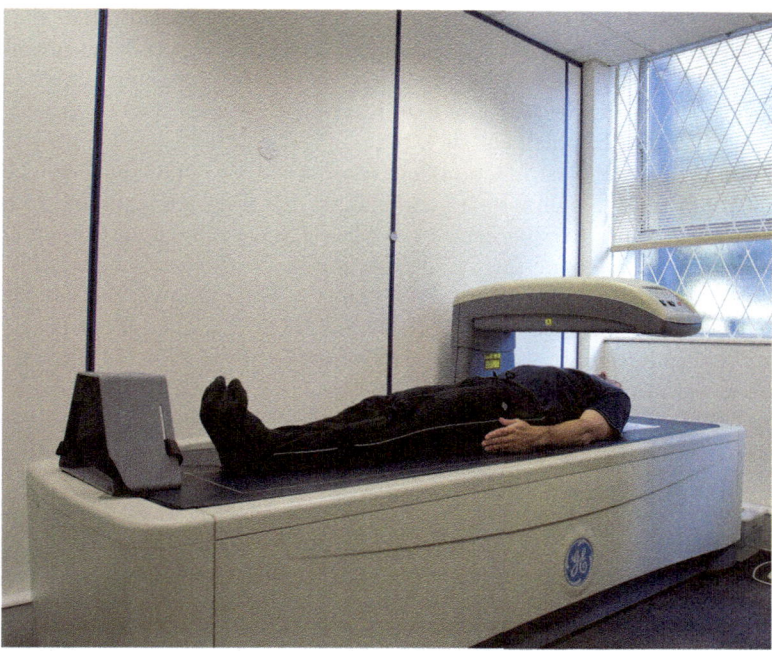

Fig. 16.2 A dual-energy X-ray absorptiometry (DXA) scan being administered. A man lies on the scanner while the arm of the scanner moves over him, taking a full scan of his body tissue density. (Creative commons licence)

(Marshall et al. 1996). However, two individuals of a similar BMD may still have very different likelihoods of sustaining a fracture. This is evident if we consider the child who has not yet achieved peak bone mass and whose BMD is similar to her grandmother, whilst she is at a much lower fracture risk. In vitro studies evaluate a number of parameters to assess bone strength, including elastic modulus, yield stress, yield strain, ultimate stress, ultimate strain, strain energy density, as well as fracture load by fracture force, maximum viscous response, energy at fracture and time to fracture. Notable variations are seen in such studies in the correlation of these parameters to BMD between skeletal sites, trabecular and cortical bone and between the variables themselves at individual sites (Beason et al. 2003; Kemper et al. 2006; Njeh et al. 1997). BMD may also be low in mineralisation disorder, such as seen in nutritional osteomalacia (Bhambri et al. 2006).

There are also other functional concerns to be considered in assessing the contribution of bone quality to osteoporosis. For example, we see that bone stiffness and BMD contribute independently to bone strength (Njeh et al. 1997). Beyond BMD, there have been technical advances in determining bone micro-architecture, using high resolution (HR) computed tomography (CT) at various anatomical sites. This permits the evaluation of bone micro-architecture including trabeculation and the measurement of cancellous bone porosity, which may both contribute to bone strength independent of BMD (Vilayphiou et al. 2016). Large scale clinical trials of alternative bone measurement methodologies such as HR peripheral quantitative CT(HRpQCT) are not available but they may prove to be promising tools for evaluating bone quality in a wider sense than just the areal mean BMD currently measured by DXA. To complement such radiological methods, biochemical tests of bone formation (osteoblast function) and resorption (osteoclast function) are used in clinical practice. It is important to appreciate that there are a number of factors influencing bone turnover markers. In childhood and young adult life, there is an *anabolic* balance reflecting increased bone formation (higher bone-specific alkaline phosphatase or P1NP) with a consolidation of bone turnover throughout adult life, and relatively balanced levels of P1NP and CTX (a marker of osteoclast function) up to middle age. In women, at the beginning of the sixth decade, there is a marked increase in both resorption (CTX) and formation (P1NP) but a relative *uncoupling*, such that the increase in resorption is greater, and results in a decrease in BMD. Increased bone turnover has been identified as a risk factor for osteoporotic fractures independent of BMD, although more recently an influential case control study analysis of the Women's Health Initiative (WHI) study found no evidence that bone turnover markers predicted hip fracture (Crandall et al. 2018). Thus, these biomarkers tend to increase with age, after peak bone mass has been achieved in the late third or early fourth decade of life but, unfortunately, they are not useful at an individual level for estimating fracture risk (Eastell et al. 2018; Gossiel et al. 2014).

Fracture Epidemiology

Throughout life, there are gender disparities in fracture risk. In childhood and younger adult life, fracture incidence in males is greater and peaks in the decade aged 15–24 years with a UK annual incidence approaching 200 per 10,000, whereas fewer than 40 per 10,000 women experience a fracture (Donaldson et al. 1990). However, around the age of 50 years, the baton is passed on to women, whose annual fracture incidence is greater than for men thereafter, exceeding 450 per 10,000 in those aged over 85 years, compared with a rate of approximately 350 per 10,000 in men of this age in the UK (Donaldson et al. 1990). A similar pattern has also been seen elsewhere in the UK (Johansen et al. 1997) and in Australia (Pasco et al. 2015). Osteoporosis is not believed to be a major contributor to fracture risk in childhood and younger adult life, as boys and young men have a higher fracture incidence despite a greater BMD than girls and young women, while typical 'osteoporotic' fractures are rare in either gender before the age of 50 years. However, past the age of 50 years, fractures of the distal forearm, rib, pelvis, humerus, femur and patella increase among women, and rib, pelvis and humerus among men, in addition to fractures of the hip and spine (Pasco et al. 2015). At the age of 50 years, the lifetime risk of having a low trauma (*osteoporotic*) fracture is 53% for women and 21%

for men (van Staa et al. 2001). In older adults, the considerable gender disparity in fracture incidence continues to increase. While low trauma fractures of the wrist, hip and spine steadily increase from the sixth decade onwards, annual hip fracture incidence in women aged over 85 years are around 4%, compared with a risk for men of the same age group of 2% (Johansen et al. 1997). Overall, there is a 10–15 year time lag in fracture risk between men and women, although older men remain at significant risk of osteoporotic fracture in their seventh decade and beyond.

Internationally, there are differences in fracture incidence with ethnic differences observed as well as secular changes in osteoporotic fracture rates. Clearly, some of the differences in osteoporosis and fracture risk relate to demography, as countries with fewer old people will experience fewer fractures. However, age-standardized incidence rates of hip fracture also vary between countries by more than 200-fold in women and 140-fold in men, with age-standardized rates highest in North America and Europe and lowest in Africa. It is anticipated that crude fracture rates will show the greatest proportional increase in Africa, where demographic transition and the anticipated increase in the older adult population at risk of fracture will be seen most. A decline over time in hip fracture rates has been seen in some countries of North America, Europe and Oceania, most notably seen in women. However, hip fracture rates continue to increase in Asia and Latin America. Indicators of health, education and socioeconomic status such as Gross National Income (GNI) per capita, Human Development Index and life expectancy at birth are correlated to hip fracture incidence rates in men and women, even after adjustment for age. This supports the hypothesis that a number of lifestyle factors contribute to osteoporosis and fracture risk in older age and offers the promise that modification of these factors may be a tool for decreasing fracture risk (Cauley et al. 2014; Kannus et al. 1996; Melton 1993).

Risk Factors for Osteoporosis

There are a number of risk factors for osteoporosis, which can be viewed across the life course. Some have already been discussed: BMD, bone turnover (Vilaca et al. 2017) and bone microarchitecture (Vilayphiou et al. 2016), together with skeletal geometry (Leslie et al. 2016) and muscle (Malkov et al. 2015). These effects are primarily on bone itself, mediated through pathophysiology or anatomical effects on bone structure and strength. While fracture risk is greatest in the ninth decade of life and beyond, some of these factors may be effected in younger adult life, childhood or even the uterus, where skeletal size and density increase from early embryogenesis through intrauterine growth to infancy. Genetic and epigenetic effects- at least on BMD, have been identified, including maternal body build, lifestyle and 25(OH)-vitamin D status. These factors might have important effects on developmental plasticity, as the osteoporotic phenotype may be viewed as a product of genotype and the prevailing environment at various stages in life (Holroyd et al. 2012). Specific gene loci are being sought to explore the potential for epigenetic mechanisms to influence BMD (Curtis et al. 2017; Morris et al. 2017; Yu and Wang 2016) and early work has shown that circulating MicroRNAs (miRNAs) have been linked to fragility fracture risk, at least in postmenopausal women with type 2 diabetes mellitus (Heilmeier et al. 2016).

Fracture Risk Assessment

More conventional clinical risk factors have also been identified, which focus on fracture risk assessment rather than the diagnosis of low BMD. These have been reviewed and over recent years adopted in the UK (although not Scotland) by the National Institute for Health and Care Excellence (National Institute for Health and Care Excellence 2012). The guidelines prompt good practice in case finding, using clinical risk factors. They recommend fracture risk assessment using clinical risk factors in women aged 65 years and over and men of 75 years and over with assessment of fracture risk in women and men under these ages if they have a risk factor (see Table 16. 1) but not to routinely assess fracture risk in people under 50 years of age unless they have *major* risk factors (*see* Table 16. 1), as they are unlikely to be at high risk. One of the indications for fracture risk assessment is the presence of a potential cause of secondary osteoporosis. A (non-exhaustive) list of these diseases is presented in Table 16. 2.

When assessing fracture risk, either FRAX or QFracture can be used to calculate a 10 year predicted absolute fracture risk. However, routine measurement of BMD without prior BMD measurement or the risk assessment can be refined by adding in BMD. The QFracture tool is derived from routine GP data on more than two million adults aged 30–85 years (Hippisley-Cox and Coupland 2009). It incorporates many more risk factors than FRAX but cannot use BMD to contribute to fracture risk assessment, since this is not routinely recorded in UK general practice. Unfortunately,

Table 16.1 Clinical risk factors for women aged 65 years and over and men aged 75 years and over

Previous fragility fracture[a]
Oral or systemic glucocorticoids[a]
Untreated early menopause (or male hypogonadism)[a]
History of falls[b]
Family history of hip fracture[b]
Secondary osteoporosis[b]
Low BMI (<18.5 kg/m^2)[b]
Smoking[b]

Alcohol intake[b]
    Women: >14 units/week
    Men: 21 units units/week

[a]These are major risk factors which apply to all ages
[b]These risk factors DO NOT apply to patients aged less than 50 years

Table 16.2 Causes of secondary osteoporosis

Endocrine
Early menopause
Hyperthyroidism
Hyperparathyroidism
Hyperprolactinaemia
Cushing's disease
Diabetes mellitus
Gastrointestinal
Coeliac disease
Inflammatory bowel disease
Malabsorption syndromes e.g. short bowel
Rheumatological
Rheumatoid arthritis
Haematological
Multiple myeloma
Haemoglobinopathies
Mastocytosis
Respiratory
Cystic fibrosis
Chronic obstructive lung disease
Metabolic
Homocystinuria
Chronic kidney disease
Immobility
For example: neurological injury

there are no calibration studies comparing the performance of FRAX and QFracture in predicting fracture incidence. FRAX has been developed, refined and calibrated for a number of nations (Fraser et al. 2011) and there are also a number of other tools worldwide, which have been developed with local and specific populations in mind. These tools vary, with some intended to predict BMD alone, while others focus on specific populations, such as the Garvan tool for fracture risk assessment in older people. Generally, there seems to be little to commend complex tools which incorporate many variables and it remains disappointing that there is so little research comparing tools in their performance, as discussed elsewhere (Aspray 2015).

Osteoporosis in Old Age

Most of the discussion so far in this chapter has considered osteoporosis in its widest context and not focused on those who are at greatest risk: the old. As already highlighted, BMD progressively declines from the fourth decade. The risk of fracture increases progressively into the ninth decade and beyond. Risk factors for fracture include increasing age, female gender and secondary osteoporosis (see

Table 16.2). In addition, there are a number of factors which are of particular relevance to an older population.

Nutrition: Calcium and Vitamin D

Nutritional osteomalacia, and rickets, in children, are the diseases most closely linked to calcium, vitamin D and skeletal health, although the effect of calcium and vitamin D status on the aetiology and treatment of osteoporosis is often highlighted. Evidence in this area is contentious. In the UK, the national diet and nutrition survey (NDNS) found that just 10% of women and 4% of men aged 75 years or over were taking less than the lower reference nutrient intake (LRNI) for calcium, which is a better proportion than seen in children and adults aged 11–64 years (Roberts et al. 2018). However, cases of mineralisation disorder due to osteomalacia are associated with low circulating blood levels of 25(OH) vitamin D, presumably relating to poor intakes and lack of sun exposure. Looking at the population aged 75 years or older in the NDNS survey, only 28% of the recommended nutrient intake (RNI) for vitamin D was obtained from diet and this increased to an average of 53% of RNI, when nutritional supplement sources were included. In the same survey, 11% men and 15% women had a circulating 25(OH) vitamin D concentration less than 25 nmol/L (Roberts et al. 2018), which the UK scientific advisory committee in nutrition (SACN) has confirmed as the threshold for risk to musculoskeletal health (Scientific Advisory Committee on Nutrition 2016). The risks of vitamin D deficiency are discussed in more detail elsewhere (see chapter on "Vitamin D in Biomedical Sciences" volume). However, interest has continued in the role of calcium and vitamin D in osteoporosis and fracture prevention. Epidemiological data from NHANES in North America have shown that lower dietary calcium intakes are associated with lower BMD, although this relationship is only seen in the population with a circulating 25(OH) vitamin D level less than 50 nmol/L (Bischoff-Ferrari et al. 2009b). A Swedish study of 5022 women followed up for 19 years found a non-linear relationship between fracture risk and calcium intake (including supplement). With the third quintile as reference (3.1 first hip fractures per 1000 person-years), the first hip fracture rates highest in the first quintile were 48% greater and the fifth quintile were 13% greater and similar relationships were seen for any fracture, any first fracture and any hip fracture (Warensjo et al. 2011).

Considering evidence from supplementation studies, one randomised controlled trial of calcium supplement in 1471 postmenopausal women showed a significantly higher BMD at the spine and the hip but no difference in fracture risk over 5 years (Reid et al. 2006), whereas the UK RECORD study, which was a placebo controlled trial of secondary fracture prevention comprising supplementation with calcium, vitamin D or both, found no difference in fracture rate between any of the four arms (Grant et al. 2005). The various evidence has been synthesised into a meta-analysis of prospective cohort and randomised controlled trials which showed no relationship between calcium intake, whether food alone or including supplements, and hip fracture (Bischoff-Ferrari et al. 2007). Adherence to long term calcium supplementation may prove difficult and affect outcomes of studies and this has been addressed by a meta-analysis considering the effects of "poor" compliance (<80%) with better compliance (80% or above), which showed a lower fracture rate in the latter than in controls (Tang et al. 2007). Pulling the available data from nutritional studies into a meta-analysis is difficult on a subject such as dietary calcium intake but it is also important to look at the *hard* clinical outcome of fracture. However, Bolland and colleagues concluded that the epidemiological evidence did not support an association between dietary calcium intake and risk of fracture, and that there was no clinical trial evidence that increasing calcium intake from dietary sources prevents fractures and they also concluded that the evidence from calcium dietary supplement trials was inconclusive (Bolland et al. 2015).

Briefly reviewing epidemiological data on vitamin D, results from the NHANES survey in North America have suggested a positive relationship between circulating 25(OH) vitamin D and BMD, across all ranges of 25(OH) vitamin D in the white population. However, this association is strongest in the third to the fifth decades and less conspicuous in older adults, and the effect is not seen in black and hispanic adults, where the gradient is not so steep and tails off as 25(OH) vitamin D levels exceed 80 nmol/L (Bischoff-Ferrari et al. 2004). Comparing this epidemiological with meta-analysis of clinical trial evidence, vitamin D supplementation does not increase BMD, other than at the femoral neck, where there is a small effect (Reid et al. 2014). More recently, a secondary analysis of a clinical trial of vitamin D supplementation in older women did show an effect of supplement on BMD which was confined to participants who started with a low 25(OH) vitamin D level (Macdonald et al. 2018), implying that the rise in BMD may be the treatment of occult osteomalacia. Beyond the potential effect of vitamin D on BMD, fracture prevention remains unproven. While it has been argued that a decrease in fracture incidence in clinical trials has only been shown where a vitamin D dose of at least 800 IU/day (20 ug/day) was used (Bischoff-Ferrari et al. 2012), the Cochrane collaboration meta-analysis failed to confirm any effect of vitamin D monotherapy on fracture prevention, irrespective of dose (Avenell et al. 2014). Beyond the direct effects of calcium and vitamin D on bone and BMD, there are other considerations, including muscle function and falls risk, which will be addressed separately when considering falls.

Falls

A good starting point, when considering falls in older people is to set out our definition. Specifically a fall occurs when an individual "inadvertently comes to rest on the ground, floor or other lower level, excluding an intentional change in position to rest in furniture, wall or other objects" (World Health Organization et al. 2008). Approximately a third of falls are believed to result in a requirement for medical attention (Berry and Miller 2008), with falls resulting in injury, including fractures, as well as longer term health issues such as fear of falling and even death and older age a critically important predictor of adverse outcome (World Health Organization et al. 2008). In the context of osteoporosis, we usual exclude falls from greater than standing height, which are likely to result in greater force being applied to the skeleton and an increased likelihood of fracture, even in the absence of osteoporosis. We talk of *low trauma fractures* of the appendicular skeleton (arms, legs, shoulder and pelvis) which are often associated with a fall, while vertebral fractures may be occult or be recognised as acute severe back pain with no experience of fall or injury. As both falls and osteoporosis are common in older people, it is an important challenge to evaluate their independent risk factors as well as how they can influence one another. A range of models have been suggested for considering falls risk with the WHO recommending these be considered in terms of biological, behavioural, environmental and socio-economic risk factors (World Health Organization et al. 2008). Simplifying that, for this discussion, some are *intrinsic*, such as: chronic diseases, including impaired cognition, arthritis, dizziness and visual impairment; acute illness, such as delirium as well as some age-related changes, including sarcopenia and loss of muscle mass and function. *Extrinsic and environmental* factors may also be important, including the quality of domestic and public lighting, environment risks such as rugs and furniture, as well as the risks posed by pets and other sources of distraction, while simple personal matters such as footwear and access to appropriate walking aids, frames and appliances may also influence an older person's postural stability. Between 20% and 33% of older people fall each year (Peel 2011), with estimates varying greatly between study populations and methods of case finding (e.g. whether using self-reported falls rates, recall or prospective diaries with or without prompts). Older people may fall for a number of reasons, including gait and balance problems, sensory impairments (including vision and neuropathy), environmental hazards or syncope (transient loss of consciousness).

With such a range of mechanisms by which an older person might sustain a fall, interventions to prevent falls and falls-related injury have to be complex. Some of the best evidence comes from multifactorial interventions, targeting a range of culprit mechanisms for falls. The Cochrane review on falls prevention in older people living in the community found that trials which involved a strategy of case finding and onward referral resulted in an overall [95% CI] decrease in falls incidence by 18[5–29]%, whereas active intervention trials were associated with a slightly greater 26[11–39]% decrease. There were a number of effective components of multifactorial intervention most of which also incorporated exercise, including home safety assessment, which decreased falls by 23%, visual assessment (28% fewer falls), home assessment (31% fewer falls), educational intervention (54% fewer falls), ankle exercises (36% fewer falls), vibration therapy (54% fewer falls) and nutritional assessment and intervention associated with a remarkable 81% reduction in falls (Gillespie et al. 2012). It is estimated that 10–15% falls result in serious injuries with 0.2–1.5% of falls resulting in a hip fracture (Peel 2011), so an important clinical question is whether fall prevention can be shown to decrease fracture risk. Exercise interventions do appear to decrease fractures by 66 [37–82]%, albeit based on clinical trials involving only 810 participants (Gillespie et al. 2012). Unfortunately, the evidence for multifaceted interventions to prevent falls and fractures in older people in institutions (e.g. care homes and hospitals) is not as impressive. In certain settings, vitamin D supplementation, probably with calcium, is effective in reducing the rate of falls as may exercise interventions. Multifactorial interventions to decrease the risk of falling in care settings were inconclusive, although targeted interventions can decrease falls in those at risk. (Cameron et al. 2012).

Calcium, Vitamin D and Falls

Suboptimal dietary calcium intakes and poor vitamin D status have been recognized as potential risk factors for falls and fractures. Individual studies vary in their results and so it is left to meta-analysis to synthesise the evidence. Here we also see some inconsistencies with three meta-analyses coming to different conclusions. Bischoff Ferrari concluded that vitamin D supplementation (without calcium) could decrease falls rates, so long as a dose of at least 700 IU (17.5 ug) per day is given (Bischoff-Ferrari et al. 2009a). Another meta-analysis, which included 45,782 older adults, concluded that vitamin D monotherapy was ineffective, while vitamin D with calcium was associated with a 14 [4–27]% decrease in falls incidence, with the greatest evidence of benefit seen in older women (Murad et al. 2011). In a third meta-analysis, intended to evaluate the potential value of performing further trials on vitamin D and falls prevention, 20 randomised controlled trials were included with a total of 29,535 participants. They concluded that supplementation with vitamin D, with or without calcium, did not reduce falls by 15% or more, based on current trial evidence, which is consistent with the conclusions of Murad (Bolland et al. 2014). With a rate of around 30% per annum in the general population aged over 75 years, is a decrease by 15% a significant effect? Here we encounter one of the challenges of intervention in falls research. It seems highly unlikely that those who are at risk of falls can be *cured*, except perhaps in the rare cases of transient loss of consciousness due to a medically treatable cause, such as heart block. When one reviews the distribution of fall frequency in an older population, some people will fall frequently (daily in some cases) while others may be extremely rare fallers. A 15% decrease in falls for a person who falls 300 times a year may not make a major impact or significantly decrease their risk of major injury or fracture. However, to prevent a fall in a person who has a 30% of falling in that year may have a much more significant impact on injury or fracture risk for them.

What about risks of intervention? A couple of vitamin D intervention studies were published over 10 years ago, using high dose vitamin D2. In one study, 100,000 IU were given every 3 months orally to care home residents with the intention of decreasing falls and fractures. Over 10 months of follow up and an average dosing of a little over 300,000 IU of vitamin D, there was a non-significant increase in risk of hip fracture (Law et al. 2006). In another study, giving 300,000 IU vitamin D2 by injection, there was a significant increase in hip fractures seen in community dwelling older people (Smith et al. 2007). These results were mostly ignored, until an Australian study of participants, known to be at high risk of falling, were given 500,000 IU as a once yearly dose to for an average of 3 years. This study showed a very high falls rate in the participants over 3 years, with 73% of those receiving

Table 16.3 Falls observed in recent studies of vitamin D supplementation

Study	Number of participants Sex Mean Age	Intervention	Dose [frequency]	Follow up in months	Fall estimate used	Result: falls rate with Vitamin D	Comments
Sanders (2010)	2317 Females 76 years	Vitamin D3 or Placebo	500,000 IU [yearly]	36	IRR & time to 1st fall	IRR=1.15 [1.02–1.30]	High falls risk group at baseline 50% fell at 1 year Prospective falls questionnaire Early effects post-dose? 10% difference in absolute risk with NNH of 10
Bischoff Ferrari (2016)	200 Males & Females 77 years	Vitamin D3 (2 doses) or D3+ calcifediol	24,000 IU or 60,000 IU or 24,000 IU+300 µg calcifediol [monthly]	12	Compare incidence at 6 & 12 months	More falls at higher doses, 67% vs lower dose, 48% ($p = .048$)	High falls risk group at baseline >60% fell at 1 year Falls NOT primary outcome. Good ascertainment
Smith (2017)	273 Females 66 years	Vitamin D3 or Placebo	Placebo or Low: 400 800 or Mid:1600 2400 3200 or High: 4000 4800 [daily]	12	1+ falls (%)	Placebo as reference: Low (NS) Mid-dose (fewer falls) High dose (NS)	60% fell at 1 year at lower 25OHD 72% in highest quintiles for 25OHD Falls NOT primary outcome. There was a difference in baseline fall history: 68% on low dose 27% on medium dose 100% on higher dose
Kay-Tee Khaw (2017) ViDA	5110 Males & Females 66 years	D3 Vs PBO	100,000 [monthly]	41	time to 1st fall	No effect	Relatively low risk population c.30% fell at 1 year Falls NOT primary outcome of the study No validation for falls (unlike fracture) Fewer completed falls questionnaires than fracture questionnaire

Abbreviations used: incidence rate rate ratio (IRR), number needed to harm (NNH), vitamin D3 (D3), not significant difference (NS)

placebo sustaining at least one fall over this period but 83% in the intervention arm falling. That is an absolute risk increase of 10% and a number needed to harm (NNH) of 10. There was also a (not significant, p = 0.06) increased risk of fracture by 26 [−1 to 59]% (Sanders et al. 2010). Table 16.3 summarises the Sanders study and more recent and relevant studies, including those of Bischoff Ferrari, Kay-Tee Khaw and Smith and Gallagher (Bischoff-Ferrari et al. 2016; Khaw et al. 2017; Smith et al. 2017). They show that there appears to be no benefit of high dose vitamin D supplementation in fall prevention. There is therefore clinical justification for treating vitamin D deficiency, with the expectation of treating the complications of osteomalacia and, for the general population of older people, there is some argument for modest vitamin D supplementation but no benefit and possibly some risk of adverse outcomes, when those at higher risk of falls are given very high doses of vitamin D, for example, in excess of 300,000 IU.

Frailty

Between the ages of 60 and 80 years, fracture risk increases by a factor of 13 but BMD explains only half of fracture risk (De Laet et al. 1997). Other factors are also very important, including previous fragility fracture and gender, as women continue to sustain more fragility fractures than men and, in the broadest sense, frailty may also explain much of the increase in fracture risk seen in the old and very old.

Frailty is a clinical state in which there is an increase in an individual's vulnerability for developing dependency with a possibly increased risk of mortality when exposed to a stressor (Morley et al. 2013). Frailty has been defined in operational terms by Rockwood using an index, which comprises a list of deficits which may be accumulated. Such aspects of health, function and social condition are markers of increasing frailty, as the greater the number of impairments, the frailer the person. The resulting frail phenotype is determined by items identified in multiple domains, reflecting multiple co-morbidities, as represented by a high aggregate score (Rockwood and Mitnitski 2007). By contrast, Fried devised a frailty phenotype, which reflects the domains of impairment seen in frailty. The clinical syndrome is evaluated functionally, using weight loss, muscle weakness, subjective physical exhaustion, slowed walking speed and physical inactivity. Three or more of these five characteristics predicts frailty (Fried et al. 2001).

The impact of frailty, whether evaluated by a frailty index or function assessment, is an increased likelihood of moving into institutional care and, ultimately, an increased mortality (Jones et al. 2005; Abellan van Kan et al. 2008). Considering skeletal health, the Study of Osteoporotic Fracture (SOF) derived a parsimonious frailty index from the available data, similar to that used by Fried but using only three criteria: weight loss of 5% or more in a year, inability to rise from a chair and a reduced energy level using data from the (self reported) geriatric depression scale (GDS) In the older population studied, frailty using the SOF criteria predicted:

- Increased risk of hip fracture in women with an odds ratio (OR) of 1.8
- Increased risk of non-spine fracture in men with an OR of 2.2
- Increased risk of falls and mortality in men (OR = 3.0) and women (OR = 2.4)
- Increased risk of disability in in men (OR = 5.3) and women (OR = 2.2) (Ensrud et al. 2008, 2009).

Sarcopenia

Sarcopenia, defined as both a loss of muscle mass and muscle function is not simply a nutritional disorder, associated with older age (Cruz-Jentoft et al. 2010). Approximately 5–13% of those aged 60–70 years are affected by sarcopenia, increasing to 11–50% for those aged 80 or above (von Haehling et al. 2010) and 20% of patients sustaining a hip fracture also have sarcopenia at presentation (Gonzalez-Montalvo et al. 2016). There is considerable overlap between sarcopenia and frailty, osteoporosis, falls and fracture risk (Landi et al. 2012) as well as a number of other chronic diseases, including insulin resistance and type 2 diabetes (Levine and Crimmins 2012), cardiovascular disease, chronic kidney disease (Honda et al. 2007) and adverse outcomes from cancer (Fearon et al. 2011). Di Monaco and colleagues found considerable overlap between frailty and sarcopenia, in patients sustaining a hip fracture, with 45% having both, 28% sarcopenia alone and only 14% having neither (Di Monaco et al. 2011).

Fracture Risk Assessment in Old Age

Looking at practical aspects of frailty, while a range of clinical risk factors contribute to fracture risk, their significance will vary with age and the presence of comorbidities, including sarcopenia and frailty. FRAX (Kanis et al. 2012) and QFracture (Cummins et al. 2011; Hippisley-Cox and Coupland 2009) can both identify fracture risk, using a range of risk factors which include glucocorticoid therapy, smoking and drinking habits and body mass index for FRAX (and an even longer list for QFracture). However, there are some practical limitations to the application of FRAX (and QFracture), particularly in the very old and frail. Firstly, there is an upper limit on age, with FRAX currently working up to the age of 90 years (QFracture up to 100). Much of the data used to create the FRAX algorithms comes from research studies, where participants will have had to give informed consent. Can we rely on such data to predict fracture risk in older adults with dementia? There are certainly data to suggest a higher prevalence of risk factors and lower BMD in older people with dementia, at least in those living in institutions (Aspray et al. 2006), as well as undertreatment of older adults with dementia (Haasum et al. 2012). There are also pragmatic problems with the potential robustness of data. Can frail older patients (with or without cognitive impairment) remember their previous fracture history or a history of parental fracture (or do we need to depend on imperfect medical record or relatives)? Are measurements of weight and height likely to be valid (and should we use current or adult height)? Weight loss is an important independent predictor of frailty, but it is not considered in FRAX. Previous glucocorticoid usage may be common in an aged population, but the current prescription of high dose glucocorticoids is more likely to be relevant (at any age), as noted by the IOF/ECTS (Lekamwasam et al. 2012). Current smoking and alcohol consumption may be less likely among the very old and the relevance of secondary osteoporosis in this population is unknown. Having highlighted that frailty is important as a cause of fracture risk, it is disappointing that there is no clinical risk factor estimating frailty and, in particular, aspects of the phenotype including falls, dementia, immobility or weight loss in FRAX. While alternative risk assessment tools include falls, such as QFracture, the data quality on which this algorithm is based is dubious, since routine general practice databases in the UK rarely document incident falls although (Hippisley-Cox and Coupland 2012), as discussed elsewhere in this chapter, a third of older adults will experience a fall each year.

Co-morbidities

Much has already been discussed about calcium, vitamin D and falls, although it is perhaps necessary to justify such a large contribution on this topic. The means of assessing falls incidence is critically important to the quality of the research evidence on this topic, particularly when evaluating interventions to decrease risk. However, it cannot be doubted that falls are clinically important in the aetiology of fragility fractures, particular in the frailer population. In one study of people over the age of 90 years, fractures occurred in the context of a fall in 86% of cases (Court-Brown and Clement 2009). Such injuries may be associated with comorbidities, including dementia, chronic kidney disease and diabetes mellitus (Mayne et al. 2010) due to a number of mechanisms, including dysautonomia or the adverse cardiovascular effects of multiple medications used to treat these conditions.

Although data from Holland, already cited (De Laet et al. 1997), suggest that BMD is not the main determinant of fracture risk in old age, there are a number of factors commoner in frail older people which have a negative effect on the skeleton, including immobility, specific diseases, such as cancer myeloma, and their treatment, including anti-androgen for prostate cancer, anti-oestrogens for breast cancer and thiazolidinediones for diabetes mellitus. Dementia is another important factor in older age, with 6.4% of adults over 65 years of age and 28.5% aged over 90 years affected (Lobo et al. 2000). Meta-analysis of epidemiological data suggest that fractures are commoner in older people with dementia living in the community (OR, 2.13) or institutions (OR, 1.88) (Muir et al. 2012), while the incidence of hip fractures in patients with dementia in the UK between 1988 and 2007 was 17.4/1000 patient year (a hazard ratio (HR) of 3.2 for fracture and 1.5 for fracture mortality) (Baker et al. 2011). There are a number of potential effects of dementia treatments, with acetylcholine esterase inhibitors increasing the risk of syncopal episodes, while memantine is associated with a lower risk of falling (Kim et al. 2011). Other treatments, such as sedative drugs, are all too frequently used in the management of patients with dementia, which may be associated with an increase in fracture risk (Finkle et al. 2011).

Moderate to severe chronic kidney disease (CKD) is prevalent in 8% of the adult population (Castro and Coresh 2009) and its incidence increases with age (Van Pottelbergh et al. 2012). It is important to recognise that BMD does not predict osteoporosis or increased fracture risk as effectively in the presence of CKD-associated metabolic bone disease (CKD-MBD) (Moe et al. 2006). Distinct entities exist, including adynamic, hyperparathyroidism, osteitis fibrosa cystica, osteomalacia or mixed uraemic osteodystrophy. The presence of CKD also influences treatment choices, with bisphosphonate best avoided at low glomerular filtration rates due to an increased risk of acute kidney injury.

Polypharmacy is a common, with 57% women in the USA, aged 65 years or older treated with five or more drugs (12% with 10 or more!) (Kaufman et al. 2002). As already discussed, the prescription of many drugs to frail older patients potentiates the risk of adverse effects on falls risk, postural stability, calcium homeostasis and bone health, with diuretics, sedatives, glucocorticoids and proton pump inhibitors frequent culprit medications. Decision-making about treatment may also be impaired, with the patient lacking capacity to make informed choices and balancing risks versus benefits. Frail older people may struggle to identify adverse effects and become dependent on others to help in monitoring their treatment.

For those unfortunate enough to sustain a fracture, surgery may be more difficult due to presence of comorbidities, such as delirium at the time of surgery which is associated with a greater postoperative mortality (Mitchell et al. 2017). Nutrition warrants special mention as active nutritional support with dietary supplements and assisted feeding may help mitigate the risks associated with low body mass index (BMI), a postoperative catabolic state, sarcopenia and immobility. The establishment of such strategies has proven beneficial in randomised controlled trials of postoperative and rehabilitation care to improve nutrition (Duncan et al. 2006), resulting in decreased rates of hospital and nursing home admissions after hip fracture, when targeted with exercise, falls prevention, home safety and polypharmacy. (Singh et al. 2012)

Care Homes

As frailty indices predict, many frail older people are admitted to residential and nursing homes, where their physical state and health needs may be supported. Hospital admission rates are greater for this group, particularly those living in residential care, who have a hospital admission rate of 312/1000 per year compared with an age-matched population rate from home of 190/1000 per year (Godden and Pollock 2001) Fractures are also commoner in care homes, with a relative risk (RR) for fracture of 2.9 [2.5–3.3], equating to 11% p.a. in Residential Care and, for hip fractures, the RR is 3.3 [2.6–4.2], a rate of 3.6% p.a. for residential care homes (Brennan nee Saunders et al. 2003; Godden and Pollock 2001). In one study of 392 care home residents in the UK with a mean age of 85 years, peripheral dual energy X-ray absorptiometry (pDXA) was used to evaluate the prevalence of osteoporosis in residential and nursing home residents (as many were unable to travel). Osteoporosis was present in 69%, with a mean Z-score of −0.96 ± 0.20. However, despite the evidence of high fracture incidence and prevalent osteoporosis, many were left untreated with 2.4–12.6% receiving calcium and vitamin D supplementation and less than 2% receiving bisphosphonate therapy (Aspray and Francis 2006). However, practice is changing and the targeted treatment of older people who are frail and living in residential and nursing homes is likely to have benefits on fracture prevention.

Fracture Summary

While older age is the greatest determinant of fracture risk, other factors, relating to frailty, are also important, particularly with regard to falls and fractures as well as mortality and the likelihood of institutionalised care. In practice, common morbidities associated with frailty include both physical and cognitive impairments, which influence treatment options, response to treatment and outcomes of rehabilitation. Care homes are an important target for the identification of frail older people at risk of falls and fractures, and future interventions should focus on preventing or reversing frailty to prevent fractures. However, there will always be fragility fractures and, in order to optimise outcomes for the frail, we must promote treatment in this group with better adherence from both carers, patients and practitioners, who can be fatalistic about preventative strategies and treatment, so improvements are also needed for rehabilitation in this group.

Conclusion and Future Direction

We have come a long way from the 1980s to 1990s, shifting the clinical assessment and treatment of osteoporosis from concerns about the menopause and early postmenopausal period. Focus has moved from the management of low BMD in women in their sixth decade to the health burden and much greater fracture risk associated with osteoporosis in older women and also men aged 80 years and above. We are learning about bone quality beyond the naïf model of bone mineral density, identified using DXA, as new technologies give more information about microstructure, which contribute independently to bone strength. Epidemiological evidence confirms that non-skeletal factors can tell us more than BMD alone about individuals' fracture risk and simple, practical questionnaire-based tools can be used to evaluate patients.

We still need to target osteoporosis in old age more effectively. The evidence around optimal nutrition is confusing, as it appears that for both calcium and vitamin D, while some may be good, much more is not necessarily better. There are links between falls, muscle function, frailty and multiple comorbidities and it may be possible to intervene to decrease fracture risk, using comprehensive interventions probably incorporated within comprehensive geriatric assessment (CGA) (Jones et al. 2005). However, beyond risk assessment and case finding, there are two areas, not covered here, which require major development. Firstly, we need effective interventions, likely to be pharmacological, to prevent fractures, which are well tolerated with few adverse effects in a frail population. Finally, we have to accept that, even with the most effective treatments available, older people will fall and fracture their bones, and we need to ensure that the potentially devastating experience of surgery and rehabilitation is as good as it can be. Innovations such as the hip fracture database can highlight where outcomes are good and where patient care could be improved (Neuburger et al. 2018; Johansen et al. 2017).

References

Abellan van Kan G, Rolland Y, Bergman H, Morley JE, Kritchevsky SB, Vellas B (2008) The I.A.N.A Task Force on frailty assessment of older people in clinical practice. J Nutr Health Aging 12(1):29–37

Aspray TJ (2015) Fragility fracture: recent developments in risk assessment. Ther Adv Musculoskelet Dis 7(1):17–25. https://doi.org/10.1177/1759720X14564562

Aspray TJ, Francis RM (2006) Fracture prevention in care home residents: is vitamin D supplementation enough? Age Ageing 35(5):455–456. https://doi.org/10.1093/ageing/afl043

Aspray TJ, Stevenson P, Abdy SE, Rawlings DJ, Holland T, Francis RM (2006) Low bone mineral density measurements in care home residents – a treatable cause of fractures. Age Ageing 35(1):37–41. https://doi.org/10.1093/ageing/afj018

Avenell A, Mak JC, O'Connell D (2014) Vitamin D and vitamin D analogues for preventing fractures in post-menopausal women and older men. Cochrane Database Syst Rev 4:CD000227. https://doi.org/10.1002/14651858.CD000227.pub4

Baker NL, Cook MN, Arrighi HM, Bullock R (2011) Hip fracture risk and subsequent mortality among Alzheimer's disease patients in the United Kingdom, 1988–2007. Age Ageing 40(1):49–54. https://doi.org/10.1093/ageing/afq146

Beason DP, Dakin GJ, Lopez RR, Alonso JE, Bandak FA, Eberhardt AW (2003) Bone mineral density correlates with fracture load in experimental side impacts of the pelvis. J Biomech 36(2):219–227

Berry SD, Miller RR (2008) Falls: epidemiology, pathophysiology, and relationship to fracture. Curr Osteoporos Rep 6(4):149–154

Bhambri R, Naik V, Malhotra N, Taneja S, Rastogi S, Ravishanker U, Mithal A (2006) Changes in bone mineral density following treatment of osteomalacia. J Clin Densitom 9(1):120–127. https://doi.org/10.1016/j.jocd.2005.11.001

Bischoff-Ferrari HA, Dietrich T, Orav EJ, Dawson-Hughes B (2004) Positive association between 25-hydroxy vitamin D levels and bone mineral density: a population-based study of younger and older adults. Am J Med 116(9):634–639. https://doi.org/10.1016/j.amjmed.2003.12.029

Bischoff-Ferrari HA, Dawson-Hughes B, Baron JA, Burckhardt P, Li R, Spiegelman D, Specker B, Orav JE, Wong JB, Staehelin HB, O'Reilly E, Kiel DP, Willett WC (2007) Calcium intake and hip fracture risk in men and women: a meta-analysis of prospective cohort studies and randomized controlled trials. Am J Clin Nutr 86(6):1780–1790

Bischoff-Ferrari HA, Dawson-Hughes B, Staehelin HB, Orav JE, Stuck AE, Theiler R, Wong JB, Egli A, Kiel DP, Henschkowski J (2009a) Fall prevention with supplemental and active forms of vitamin D: a meta-analysis of randomised controlled trials. BMJ 339:b3692. https://doi.org/10.1136/bmj.b3692

Bischoff-Ferrari HA, Kiel DP, Dawson-Hughes B, Orav JE, Li R, Spiegelman D, Dietrich T, Willett WC (2009b) Dietary calcium and serum 25-hydroxyvitamin D status in relation to BMD among U.S. adults. J Bone Miner Res 24(5):935–942. https://doi.org/10.1359/jbmr.081242

Bischoff-Ferrari HA, Willett WC, Orav EJ, Lips P, Meunier PJ, Lyons RA, Flicker L, Wark J, Jackson RD, Cauley JA, Meyer HE, Pfeifer M, Sanders KM, Stahelin HB, Theiler R, Dawson-Hughes B (2012) A pooled analysis of vitamin D dose requirements for fracture prevention. N Engl J Med 367(1):40–49. https://doi.org/10.1056/NEJMoa1109617

Bischoff-Ferrari HA, Dawson-Hughes B, Orav EJ, Staehelin HB, Meyer OW, Theiler R, Dick W, Willett WC, Egli A (2016) Monthly high-dose vitamin D treatment for the prevention of functional decline: a randomized clinical trial. JAMA Intern Med 176(2):175–183. https://doi.org/10.1001/jamainternmed.2015.7148

Bolland MJ, Grey A, Gamble GD, Reid IR (2014) Vitamin D supplementation and falls: a trial sequential meta-analysis. Lancet Diabetes Endocrinol 2(7):573–580. https://doi.org/10.1016/S2213-8587(14)70068-3

Bolland MJ, Leung W, Tai V, Bastin S, Gamble GD, Grey A, Reid IR (2015) Calcium intake and risk of fracture: systematic review. BMJ 351:h4580. https://doi.org/10.1136/bmj.h4580

Brennan nee Saunders J, Johansen A, Butler J, Stone M, Richmond P, Jones S, Lyons RA (2003) Place of residence and risk of fracture in older people: a population-based study of over 65-year-olds in Cardiff. Osteoporos Int 14(6):515–519. https://doi.org/10.1007/s00198-003-1404-5

Cameron ID, Gillespie LD, Robertson MC, Murray GR, Hill KD, Cumming RG, Kerse N (2012) Interventions for preventing falls in older people in care facilities and hospitals. Cochrane Database Syst Rev 12:CD005465. https://doi.org/10.1002/14651858.CD005465.pub3

Castro AF, Coresh J (2009) CKD surveillance using laboratory data from the population-based National Health and Nutrition Examination Survey (NHANES). Am J Kidney Dis: Off J Natl Found 53(3 Suppl 3):S46–S55. https://doi.org/10.1053/j.ajkd.2008.07.054

Cauley JA, Chalhoub D, Kassem AM, Fuleihan Gel H (2014) Geographic and ethnic disparities in osteoporotic fractures. Nat Rev Endocrinol 10(6):338–351. https://doi.org/10.1038/nrendo.2014.51

Court-Brown CM, Clement N (2009) Four score years and ten: an analysis of the epidemiology of fractures in the very elderly. Injury 40(10):1111–1114. https://doi.org/10.1016/j.injury.2009.06.011

Crandall CJ, Vasan S, LaCroix A, LeBoff MS, Cauley JA, Robbins JA, Jackson RD, Bauer DC (2018) Bone turnover markers are not associated with hip fracture risk: a case-control study in the women's health initiative. J Bone Miner Res. https://doi.org/10.1002/jbmr.3471

Cruz-Jentoft AJ, Baeyens JP, Bauer JM, Boirie Y, Cederholm T, Landi F, Martin FC, Michel JP, Rolland Y, Schneider SM, Topinkova E, Vandewoude M, Zamboni M, European Working Group on Sarcopenia in Older P (2010) Sarcopenia: European consensus on definition and diagnosis: report of the European working group on sarcopenia in older people. Age Ageing 39(4):412–423. https://doi.org/10.1093/ageing/afq034

Cummins NM, Poku EK, Towler MR, O'Driscoll OM, Ralston SH (2011) clinical risk factors for osteoporosis in Ireland and the UK: a comparison of FRAX and QFractureScores. Calcif Tissue Int 89(2):172–177. https://doi.org/10.1007/s00223-011-9504-2

Curtis EM, Murray R, Titcombe P, Cook E, Clarke-Harris R, Costello P, Garratt E, Holbrook JD, Barton S, Inskip H, Godfrey KM, Bell CG, Cooper C, Lillycrop KA, Harvey NC (2017) Perinatal DNA methylation at CDKN2A is associated with offspring bone mass: findings from the Southampton women's survey. J Bone Miner Res 32(10):2030–2040. https://doi.org/10.1002/jbmr.3153

De Laet CE, van Hout BA, Burger H, Hofman A, Pols HA (1997) Bone density and risk of hip fracture in men and women: cross sectional analysis. BMJ 315(7102):221–225

Di Monaco M, Vallero F, Di Monaco R, Tappero R (2011) Prevalence of sarcopenia and its association with osteoporosis in 313 older women following a hip fracture. Arch Gerontol Geriatr 52(1):71–74. https://doi.org/10.1016/j.archger.2010.02.002

Donaldson LJ, Cook A, Thomson RG (1990) Incidence of fractures in a geographically defined population. J Epidemiol Community Health 44(3):241–245

Duncan DG, Beck SJ, Hood K, Johansen A (2006) Using dietetic assistants to improve the outcome of hip fracture: a randomised controlled trial of nutritional support in an acute trauma ward. Age Ageing 35(2):148–153. https://doi.org/10.1093/ageing/afj011

Eastell R, Pigott T, Gossiel F, Naylor KE, Walsh JS, Peel NFA (2018) Diagnosis of endocrine disease: bone turnover markers: are they clinically useful? Eur J Endocrinol 178(1):R19–R31. https://doi.org/10.1530/EJE-17-0585

Ensrud KE, Ewing SK, Taylor BC, Fink HA, Cawthon PM, Stone KL, Hillier TA, Cauley JA, Hochberg MC, Rodondi N, Tracy JK, Cummings SR (2008) Comparison of 2 frailty indexes for prediction of falls, disability, fractures, and death in older women. Arch Intern Med 168(4):382–389. https://doi.org/10.1001/archinternmed.2007.113

Ensrud KE, Lui LY, Taylor BC, Schousboe JT, Donaldson MG, Fink HA, Cauley JA, Hillier TA, Browner WS, Cummings SR, Study of Osteoporotic Fractures Research G (2009) A comparison of prediction models for fractures in older women: is more better? Arch Intern Med 169(22):2087–2094. https://doi.org/10.1001/archinternmed.2009.404

Fearon K, Strasser F, Anker SD, Bosaeus I, Bruera E, Fainsinger RL, Jatoi A, Loprinzi C, MacDonald N, Mantovani G, Davis M, Muscaritoli M, Ottery F, Radbruch L, Ravasco P, Walsh D, Wilcock A, Kaasa S, Baracos VE (2011) Definition and classification of cancer cachexia: an international consensus. Lancet Oncol 12(5):489–495. https://doi.org/10.1016/S1470-2045(10)70218-7

Finkle WD, Der JS, Greenland S, Adams JL, Ridgeway G, Blaschke T, Wang Z, Dell RM, VanRiper KB (2011) Risk of fractures requiring hospitalization after an initial prescription for zolpidem, alprazolam, lorazepam, or diazepam in older adults. J Am Geriatr Soc 59(10):1883–1890. https://doi.org/10.1111/j.1532-5415.2011.03591.x

Fraser LA, Langsetmo L, Berger C, Ioannidis G, Goltzman D, Adachi JD, Papaioannou A, Josse R, Kovacs CS, Olszynski WP, Towheed T, Hanley DA, Kaiser SM, Prior J, Jamal S, Kreiger N, Brown JP, Johansson H, Oden A, McCloskey E, Kanis JA, Leslie WD, CaMos Research G (2011) Fracture prediction and calibration of a Canadian FRAX(R) tool: a population-based report from CaMos. Osteoporos Int 22(3):829–837. https://doi.org/10.1007/s00198-010-1465-1

Fried LP, Tangen CM, Walston J, Newman AB, Hirsch C, Gottdiener J, Seeman T, Tracy R, Kop WJ, Burke G, McBurnie MA, Cardiovascular Health Study Collaborative Research G (2001) Frailty in older adults: evidence for a phenotype. J Gerontol A Biol Sci Med Sci 56(3):M146–M156

Gillespie LD, Robertson MC, Gillespie WJ, Sherrington C, Gates S, Clemson LM, Lamb SE (2012) Interventions for preventing falls in older people living in the community. Cochrane Database Syst Rev 9:CD007146. https://doi.org/10.1002/14651858.CD007146.pub3

Godden S, Pollock AM (2001) The use of acute hospital services by elderly residents of nursing and residential care homes. Health Soc Care Community 9(6):367–374

Gonzalez-Montalvo JI, Alarcon T, Gotor P, Queipo R, Velasco R, Hoyos R, Pardo A, Otero A (2016) Prevalence of sarcopenia in acute hip fracture patients and its influence on short-term clinical outcome. Geriatr Gerontol Int 16(9):1021–1027. https://doi.org/10.1111/ggi.12590

Gossiel F, Finigan J, Jacques R, Reid D, Felsenberg D, Roux C, Glueer C, Eastell R (2014) Establishing reference intervals for bone turnover markers in healthy postmenopausal women in a nonfasting state. Bonekey Rep 3:573. https://doi.org/10.1038/bonekey.2014.68

Grant AM, Avenell A, Campbell MK, McDonald AM, MacLennan GS, McPherson GC, Anderson FH, Cooper C, Francis RM, Donaldson C, Gillespie WJ, Robinson CM, Torgerson DJ, Wallace

WA, Group RT (2005) Oral vitamin D3 and calcium for secondary prevention of low-trauma fractures in elderly people (Randomised Evaluation of Calcium Or vitamin D, RECORD): a randomised placebo-controlled trial. Lancet 365(9471):1621–1628. https://doi.org/10.1016/S0140-6736(05)63013-9

Haasum Y, Fastbom J, Fratiglioni L, Johnell K (2012) Undertreatment of osteoporosis in persons with dementia? A population-based study. Osteoporos Int 23(3):1061–1068. https://doi.org/10.1007/s00198-011-1636-8

von Haehling S, Morley JE, Anker SD (2010) An overview of sarcopenia: facts and numbers on prevalence and clinical impact. J Cachexia Sarcopenia Muscle 1(2):129–133. https://doi.org/10.1007/s13539-010-0014-2

Heilmeier U, Hackl M, Skalicky S, Weilner S, Schroeder F, Vierlinger K, Patsch JM, Baum T, Oberbauer E, Lobach I, Burghardt AJ, Schwartz AV, Grillari J, Link TM (2016) Serum miRNA signatures are indicative of skeletal fractures in postmenopausal women with and without type 2 diabetes and influence osteogenic and adipogenic differentiation of adipose tissue-derived mesenchymal stem cells in vitro. J Bone Miner Res 31(12):2173–2192. https://doi.org/10.1002/jbmr.2897

Hippisley-Cox J, Coupland C (2009) Predicting risk of osteoporotic fracture in men and women in England and Wales: prospective derivation and validation of QFractureScores. BMJ 339:b4229. https://doi.org/10.1136/bmj.b4229

Hippisley-Cox J, Coupland C (2012) Derivation and validation of updated QFracture algorithm to predict risk of osteoporotic fracture in primary care in the United Kingdom: prospective open cohort study. BMJ 344:e3427. https://doi.org/10.1136/bmj.e3427

Holroyd C, Harvey N, Dennison E, Cooper C (2012) Epigenetic influences in the developmental origins of osteoporosis. Osteoporos Int 23(2):401–410. https://doi.org/10.1007/s00198-011-1671-5

Honda H, Qureshi AR, Axelsson J, Heimburger O, Suliman ME, Barany P, Stenvinkel P, Lindholm B (2007) Obese sarcopenia in patients with end-stage renal disease is associated with inflammation and increased mortality. Am J Clin Nutr 86(3):633–638

Johansen A, Evans RJ, Stone MD, Richmond PW, Lo SV, Woodhouse KW (1997) Fracture incidence in England and Wales: a study based on the population of Cardiff. Injury 28(9–10):655–660

Johansen A, Boulton C, Hertz K, Ellis M, Burgon V, Rai S, Wakeman R (2017) The National Hip Fracture Database (NHFD) – using a national clinical audit to raise standards of nursing care. Int J Orthop Trauma Nurs 26:3–6. https://doi.org/10.1016/j.ijotn.2017.01.001

Jones D, Song X, Mitnitski A, Rockwood K (2005) Evaluation of a frailty index based on a comprehensive geriatric assessment in a population based study of elderly Canadians. Aging Clin Exp Res 17(6):465–471

Kanis JA, McCloskey E, Johansson H, Oden A, Leslie WD (2012) FRAX((R)) with and without bone mineral density. Calcif Tissue Int 90(1):1–13. https://doi.org/10.1007/s00223-011-9544-7

Kannus P, Parkkari J, Sievanen H, Heinonen A, Vuori I, Jarvinen M (1996) Epidemiology of hip fractures. Bone 18(1 Suppl):57S–63S

Kaufman DW, Kelly JP, Rosenberg L, Anderson TE, Mitchell AA (2002) Recent patterns of medication use in the ambulatory adult population of the United States: the Slone survey. JAMA: J Am Med Assoc 287(3):337–344

Kemper A, Ng T, Duma S (2006) The biomechanical response of human bone: the influence of bone volume and mineral density. Biomed Sci Instrum 42:284–289

Khaw KT, Stewart AW, Waayer D, Lawes CMM, Toop L, Camargo CA Jr, Scragg R (2017) Effect of monthly high-dose vitamin D supplementation on falls and non-vertebral fractures: secondary and post-hoc outcomes from the randomised, double-blind, placebo-controlled ViDA trial. Lancet Diabetes Endocrinol 5(6):438–447. https://doi.org/10.1016/S2213-8587(17)30103-1

Kim DH, Brown RT, Ding EL, Kiel DP, Berry SD (2011) Dementia medications and risk of falls, syncope, and related adverse events: meta-analysis of randomized controlled trials. J Am Geriatr Soc 59(6):1019–1031. https://doi.org/10.1111/j.1532-5415.2011.03450.x

Landi F, Liperoti R, Russo A, Giovannini S, Tosato M, Capoluongo E, Bernabei R, Onder G (2012) Sarcopenia as a risk factor for falls in elderly individuals: results from the ilSIRENTE study. Clin Nutr 31(5):652–658. https://doi.org/10.1016/j.clnu.2012.02.007

Law M, Withers H, Morris J, Anderson F (2006) Vitamin D supplementation and the prevention of fractures and falls: results of a randomised trial in elderly people in residential accommodation. Age Ageing 35(5):482–486. https://doi.org/10.1093/ageing/afj080

Lekamwasam S, Adachi JD, Agnusdei D, Bilezikian J, Boonen S, Borgstrom F, Cooper C, Diez Perez A, Eastell R, Hofbauer LC, Kanis JA, Langdahl BL, Lesnyak O, Lorenc R, McCloskey E, Messina OD, Napoli N, Obermayer-Pietsch B, Ralston SH, Sambrook PN, Silverman S, Sosa M, Stepan J, Suppan G, Wahl DA, Compston JE (2012) A framework for the development of guidelines for the management of glucocorticoid-induced osteoporosis. Osteoporos Int. https://doi.org/10.1007/s00198-012-1958-1

Leslie WD, Lix LM, Morin SN, Johansson H, Oden A, McCloskey EV, Kanis JA (2016) Adjusting hip fracture probability in men and women using hip axis length: the Manitoba bone density database. J Clin Densitom 19(3):326–331. https://doi.org/10.1016/j.jocd.2015.07.004

Levine ME, Crimmins EM (2012) The impact of insulin resistance and inflammation on the association between sarcopenic obesity and physical functioning. Obesity (Silver Spring). https://doi.org/10.1038/oby.2012.20

Lobo A, Launer LJ, Fratiglioni L, Andersen K, Di Carlo A, Breteler MM, Copeland JR, Dartigues JF, Jagger C, Martinez-Lage J, Soininen H, Hofman A (2000) Prevalence of dementia and major subtypes in Europe: a collaborative study of population-based cohorts. Neurologic Diseases in the Elderly Research Group. Neurology 54(11 Suppl 5):S4–S9

Macdonald HM, Reid IR, Gamble GD, Fraser WD, Tang JC, Wood AD (2018) 25-hydroxyvitamin D threshold for the effects of vitamin D supplements on bone density secondary analysis of a randomized controlled trial. J Bone Miner Res. https://doi.org/10.1002/jbmr.3442

Malkov S, Cawthon PM, Peters KW, Cauley JA, Murphy RA, Visser M, Wilson JP, Harris T, Satterfield S, Cummings S, Shepherd JA, Health ABCS (2015) Hip fractures risk in older men and women associated with DXA-derived measures of thigh subcutaneous fat thickness, cross-sectional muscle area, and muscle density. J Bone Miner Res 30(8):1414–1421. https://doi.org/10.1002/jbmr.2469

Marshall D, Johnell O, Wedel H (1996) Meta-analysis of how well measures of bone mineral density predict occurrence of osteoporotic fractures. BMJ 312(7041):1254–1259

Mayne D, Stout NR, Aspray TJ (2010) Diabetes, falls and fractures. Age Ageing 39(5):522–525. https://doi.org/10.1093/ageing/afq081

Melton LJ 3rd (1993) Hip fractures: a worldwide problem today and tomorrow. Bone 14(Suppl 1):S1–S8

Mitchell R, Harvey L, Brodaty H, Draper B, Close J (2017) One-year mortality after hip fracture in older individuals: the effects of delirium and dementia. Arch Gerontol Geriatr 72:135–141. https://doi.org/10.1016/j.archger.2017.06.006

Moe S, Drueke T, Cunningham J, Goodman W, Martin K, Olgaard K, Ott S, Sprague S, Lameire N, Eknoyan G (2006) Definition, evaluation, and classification of renal osteodystrophy: a position statement from Kidney Disease: Improving Global Outcomes (KDIGO). Kidney Int 69(11):1945–1953. https://doi.org/10.1038/sj.ki.5000414

Morley JE, Vellas B, van Kan GA, Anker SD, Bauer JM, Bernabei R, Cesari M, Chumlea WC, Doehner W, Evans J, Fried LP, Guralnik JM, Katz PR, Malmstrom TK, McCarter RJ, Gutierrez Robledo LM, Rockwood K, von Haehling S, Vandewoude MF, Walston J (2013) Frailty consensus: a call to action. J Am Med Dir Assoc 14(6):392–397. https://doi.org/10.1016/j.jamda.2013.03.022

Morris JA, Tsai PC, Joehanes R, Zheng J, Trajanoska K, Soerensen M, Forgetta V, Castillo-Fernandez JE, Frost M, Spector TD, Christensen K, Christiansen L, Rivadeneira F, Tobias JH, Evans DM, Kiel DP, Hsu YH, Richards JB, Bell JT (2017) Epigenome-wide association of DNA methylation in whole blood with bone mineral density. J Bone Miner Res 32(8):1644–1650. https://doi.org/10.1002/jbmr.3148

Muir SW, Gopaul K, Montero Odasso MM (2012) The role of cognitive impairment in fall risk among older adults: a systematic review and meta-analysis. Age Ageing 41(3):299–308. https://doi.org/10.1093/ageing/afs012

Murad MH, Elamin KB, Abu Elnour NO, Elamin MB, Alkatib AA, Fatourechi MM, Almandoz JP, Mullan RJ, Lane MA, Liu H, Erwin PJ, Hensrud DD, Montori VM (2011) Clinical review: the effect of vitamin D on falls: a systematic review and meta-analysis. J Clin Endocrinol Metab 96(10):2997–3006. https://doi.org/10.1210/jc.2011-1193

National Institute for Health and Care Excellence (2012) Osteoporosis: assessing the risk of fragility fracture. NICE, London

Neuburger J, Currie C, Wakeman R, Georghiou T, Boulton C, Johansen A, Tsang C, Wilson H, Cromwell DA, Jan van der M (2018) Safe working in a 7-day service. Experience of hip fracture care as documented by the UK National Hip Fracture Database. Age Ageing. https://doi.org/10.1093/ageing/afy074

NIH (2001) NIH consensus development panel on osteoporosis prevention, diagnosis, and therapy, March 7–29, 2000: highlights of the conference. South Med J 94(6):569–573

Njeh CF, Kuo CW, Langton CM, Atrah HI, Boivin CM (1997) Prediction of human femoral bone strength using ultrasound velocity and BMD: an in vitro study. Osteoporos Int 7(5):471–477

Pasco JA, Lane SE, Brennan-Olsen SL, Holloway KL, Timney EN, Bucki-Smith G, Morse AG, Dobbins AG, Williams LJ, Hyde NK, Kotowicz MA (2015) The epidemiology of incident fracture from cradle to senescence. Calcif Tissue Int 97(6):568–576. https://doi.org/10.1007/s00223-015-0053-y

Peel NM (2011) Epidemiology of falls in older age. Can J Aging 30(1):7–19. https://doi.org/10.1017/S071498081000070X

Reid IR, Mason B, Horne A, Ames R, Reid HE, Bava U, Bolland MJ, Gamble GD (2006) Randomized controlled trial of calcium in healthy older women. Am J Med 119(9):777–785. https://doi.org/10.1016/j.amjmed.2006.02.038

Reid IR, Bolland MJ, Grey A (2014) Effects of vitamin D supplements on bone mineral density: a systematic review and meta-analysis. Lancet 383(9912):146–155. https://doi.org/10.1016/S0140-6736(13)61647-5

Roberts C, Steer T, Maplethorpe N, Cox L, Meadows S, Nicholson S, Page P, Swan G (2018) National diet and nutrition survey: results from years 7 and 8 (combined) of the Rolling Programme (2014/2015 to 2015/2016). Public Health England and the Food Standards Agency, London

Rockwood K, Mitnitski A (2007) Frailty in relation to the accumulation of deficits. J Gerontol A Biol Sci Med Sci 62(7):722–727

Rudang R, Zoulakis M, Sundh D, Brisby H, Diez-Perez A, Johansson L, Mellstrom D, Darelid A, Lorentzon M (2016) Bone material strength is associated with areal BMD but not with prevalent fractures in older women. Osteoporos Int 27(4):1585–1592. https://doi.org/10.1007/s00198-015-3419-0

Sanders KM, Stuart AL, Williamson EJ, Simpson JA, Kotowicz MA, Young D, Nicholson GC (2010) Annual high-dose oral vitamin D and falls and fractures in older women: a randomized controlled trial. JAMA 303(18):1815–1822. https://doi.org/10.1001/jama.2010.594

Scientific Advisory Committee on Nutrition (2016) Vitamin D and health. Scientific Advisory Committee on Nutrition, London

Singh NA, Quine S, Clemson LM, Williams EJ, Williamson DA, Stavrinos TM, Grady JN, Perry TJ, Lloyd BD, Smith EU, Singh MA (2012) Effects of high-intensity progressive resistance training and targeted multidisciplinary treatment of frailty on mortality and nursing home admissions after hip fracture: a randomized controlled trial. J Am Med Dir Assoc 13(1):24–30. https://doi.org/10.1016/j.jamda.2011.08.005

Smith H, Anderson F, Raphael H, Maslin P, Crozier S, Cooper C (2007) Effect of annual intramuscular vitamin D on fracture risk in elderly men and women – a population-based, randomized, double-blind, placebo-controlled trial. Rheumatology (Oxford) 46(12):1852–1857. https://doi.org/10.1093/rheumatology/kem240

Smith LM, Gallagher JC, Suiter C (2017) Medium doses of daily vitamin D decrease falls and higher doses of daily vitamin D3 increase falls: a randomized clinical trial. J Steroid Biochem Mol Biol. https://doi.org/10.1016/j.jsbmb.2017.03.015

Steiniche T (1995) Bone histomorphometry in the pathophysiological evaluation of primary and secondary osteoporosis and various treatment modalities. APMIS Suppl 51:1–44

Tang BM, Eslick GD, Nowson C, Smith C, Bensoussan A (2007) Use of calcium or calcium in combination with vitamin D supplementation to prevent fractures and bone loss in people aged 50 years and older: a meta-analysis. Lancet 370(9588):657–666. https://doi.org/10.1016/S0140-6736(07)61342-7

Van Pottelbergh G, Vaes B, Jadoul M, Mathei C, Wallemacq P, Degryse JM (2012) The prevalence and detection of chronic kidney disease (CKD)-related metabolic complications as a function of estimated glomerular filtration rate in the oldest old. Arch Gerontol Geriatr 54(3):e419–e425. https://doi.org/10.1016/j.archger.2011.12.010

van Staa TP, Dennison EM, Leufkens HG, Cooper C (2001) Epidemiology of fractures in England and Wales. Bone 29(6):517–522

Vilaca T, Gossiel F, Eastell R (2017) Bone turnover markers: use in fracture prediction. J Clin Densitom 20(3):346–352. https://doi.org/10.1016/j.jocd.2017.06.020

Vilayphiou N, Boutroy S, Sornay-Rendu E, Van Rietbergen B, Chapurlat R (2016) Age-related changes in bone strength from HR-pQCT derived microarchitectural parameters with an emphasis on the role of cortical porosity. Bone 83:233–240. https://doi.org/10.1016/j.bone.2015.10.012

Warensjo E, Byberg L, Melhus H, Gedeborg R, Mallmin H, Wolk A, Michaelsson K (2011) Dietary calcium intake and risk of fracture and osteoporosis: prospective longitudinal cohort study. BMJ 342:d1473. https://doi.org/10.1136/bmj.d1473

World Health Organization A, Life Course U, World Health Organization A, Life Course U (2008) WHO global report on falls prevention in older age

Yu B, Wang CY (2016) Osteoporosis: the result of an 'aged' bone microenvironment. Trends Mol Med 22(8):641–644. https://doi.org/10.1016/j.molmed.2016.06.002

Chapter 17
Neurovascular Ageing and Age-Related Diseases

Raj N. Kalaria and Yoshiki Hase

Abstract Proper functioning of the brain is dependent on integrity of the cerebral vasculature. During ageing, a number of factors including aortic or arterial stiffness, autonomic dysregulation, neurovascular uncoupling and blood-brain barrier (BBB) damage will define the dynamics of brain blood flow and local perfusion. The nature and extent of ageing-related cerebrovascular changes, the degree of involvement of the heart and extracranial vessels and the consequent location of tissue pathology may vary considerably. Atheromatous disease retarding flow is a common vascular insult, which increases exponentially with increasing age. Arteriolosclerosis characterized as a prominent feature of small vessel disease is one of the first changes to occur during the natural history of cerebrovascular pathology. At the capillary level, the cerebral endothelium, which forms the BBB undergoes changes including reduced cytoplasm, fewer mitochondria, loss of tight junctions and thickened basement membranes with collagenosis. Astrocyte end-feet protecting the BBB retract as part of the clasmatodendrotic response whereas pericyte coverage is altered. The consequences of these microvascular changes are lacunar infarcts, cortical and subcortical microinfarcts, microbleeds and diffuse white matter disease, which involves myelin loss and axonal abnormalities. The deeper structures are particularly vulnerable because of the relatively reduced density of the microvascular network formed by perforating and penetrating end arteries. Ultimately, the integrity of both the neurovascular and gliovascular units is compromised such that there is an overall synergistic effect reflecting on ageing associated cerebral perfusion and permeability. More than one protagonist appears to be involved in ageing-related cognitive dysfunction characteristically associated with the neurocognitive disorders.

Keywords Ageing · Alzheimer's disease · Cerebrovascular degeneration · Cognitive impairment · Infarction · Stroke · Small vessel disease

R. N. Kalaria (✉) · Y. Hase
Institute of Neuroscience, Newcastle University, Campus for Ageing & Vitality,
Newcastle upon Tyne, UK
e-mail: raj.kalaria@ncl.ac.uk; Yoshiki.Hase@ncl.ac.uk

© Springer Nature Singapore Pte Ltd. 2019
J. R. Harris, V. I. Korolchuk (eds.), *Biochemistry and Cell Biology of Ageing:*
Part II Clinical Science, Subcellular Biochemistry 91,
https://doi.org/10.1007/978-981-13-3681-2_17

Introduction

Old age coincides with increased frequencies in both cardiovascular and cerebro-vascular diseases, which affect brain function. Pathological changes within blood vessels of the brain or the cardiovascular system are common in ageing. They both may independently or inter-dependently give rise to brain atrophy, strokes with large and small infarctions, cognitive impairment and dementia in the elderly. However, the functional consequences manifest in altered cerebral blood flow, auto-regulation, neurovascular coupling and endothelial permeability (Toth et al. 2017; Nagata et al. 2016).

Cerebrovascular disease (CVD) involves arteriosclerotic changes in the cerebral or systemic vasculature, and often both, that likely begin during a considerable period prior to the manifestation of an overt stroke-like or CVD accident (Fig. 17.1). While early or subtle changes may not be recognised clinically, they are evident radiologically as white matter (WM) and silent lesions, mostly in form of lacunar infarcts. For example, 20% of healthy elderly people will bear magnetic resonance

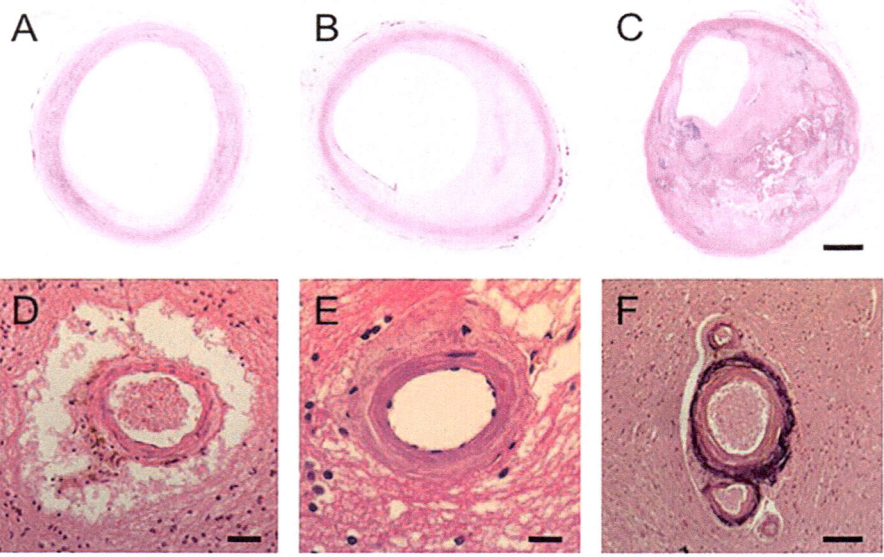

Fig. 17.1 Pathological changes in large artery and small vessel diseases during ageing. (**a–c**) Cross section images of internal carotid arteries showing gradual mild (**a**), moderate (**b**) to severe stenosis (**c**) by accumulation of atheroma. In (**c**), fatty streaks and cholesterol crystals are discerned. The sections were from >75 years old individuals. (**d–f**) Arterioles in the white matter and grey matter showing graded arteriolosclerosis. (**d**) Vascular wall thickening and perivascular spacing is noted. (**e**) Shows almost complete hyalinosis, which is characterised by collagen fibrils. (**f**) The vessel wall is infiltrated with calcium deposition often found in the basal ganglia. Occlusion of these vessels lead to lacunar infarcts and microinfarcts. Magnification Bar: C = 1 cm, D = 50 μm, F = 100 μm

imaging (MRI)-defined silent brain infarcts and up to 50% of these are detected in ageing cohorts (Vermeer et al. 2007). Hypertension-linked small-vessel disease (SVD) is the main risk factor for these infarcts, which may be associated with subtle deficits in physical and cognitive function that commonly go unnoticed, particularly in older age. In recent years, the staging of SVD derived from radiological measures (Wardlaw et al. 2013) refers to an intracranial disorder involving pathological changes within and at the surfaces of brain microvessels including perforating arteries and arterioles, capillaries and venules. While SVD can coexist with atherosclerosis involving large extracranial vessels and cardioembolic disease (Li et al. 2015), it incorporates tissue injury in both cortical and subcortical grey and white matter. In addition, breach of the blood-brain barrier (BBB) can occur in selective areas of the ageing brain. Using advanced dynamic contrast-enhanced MRI sequences with high spatial and temporal resolutions, a study showed that the BBB breaks down in the ageing hippocampus as an early event. The BBB breach was worse in ageing individuals with mild cognitive impairment (Montagne et al. 2015).

Cerebrovascular changes may be conveniently grouped under large and small vessel domains. Large vessel and cardiac embolic events involving atherosclerosis, plaque rupture, intraplaque hemorrhage, thrombotic occlusion, and embolism, dissection and dolichoectasia may lead to macroinfarcts (Kalaria et al. 2015a). Accompanying haemodynamic events cause borderzone or watershed lesions (~5 mm) as wedge shaped regions of pallor and rarefaction extending into the WM. In Western communities, 80% of the strokes involve ischaemic infarcts whereas about 20% attributed to haemorrhages, which most often occur as intracerebral bleeds from small arteries or larger vessels as subarachnoid haemorrhages. Systematic investigation on the natural history and staging of CVD indicates that vessel wall modifications such as arteriolosclerosis or deposition of amyloid like proteins were the most common and earliest changes during ageing. These were followed by perivascular spacing with lacunar and regional microinfarcts occurring as consequent but independent processes. The regional progression of the changes were frontal > temporal lobe $\geq$ basal ganglia. In terms of dementia, vascular dementia (VaD) patients have the highest burden of vascular pathology whereas Alzheimer's disease (AD) subjects have the second and those with Dementia with Lewy Bodies as the last but all much greater than ageing controls (Deramecourt et al. 2012).

In this chapter, we focus on ageing related macro- and microvascular changes and common diseases that result from the alterations in the vasculature. We discuss ageing-related changes in the cellular components of arteries, arterioles, capillaries and veins and how these alter the BBB. We also discuss the consequent tissue changes, which contribute to cerebrovascular disease and lead to cognitive impairment and dementia.

Neurovascular Ageing and Brain Perfusion

On the whole, different studies using various brain imaging together with radio-tracer techniques suggest cerebral blood flow (CBF) declines with age (Nagata et al. 2016; Ambarki et al. 2015; Zhang et al. 2017). In comparison to younger subjects, some areas of the brains e.g. parietal cortex of older individuals are more vulnerable and it has been difficult to assess the white matter usually due to low signal. Several factors may play a role in this decline (Nagata et al. 2016). Ageing-related reduction in cardiac output may impair CBF regulation (Tarumi and Zhang 2018). CBF may also be affected by both ageing-related arterial stiffness and exposure to vascular risk factors such as hypertension and hyperlipidaemia. Consistent with this, increased aortic stiffness has been associated with reduced wave reflection at the interface between the aorta and the carotid arteries, transmission of excessive flow pulsatility into the brain, microvascular structural damage in both the grey matter and WM and reduced cognitive function (Mitchell et al. 2011; Nagata et al. 2016). Adequate blood to the brain is normally delivered by cerebral arteries of sufficient luminal calibre, lined by non-activated but 'alert' endothelial cells on robust intima supported by structurally and functionally intact tunica media and adventitia. Many disease processes thicken the arterial wall, particularly the intima, and narrow the lumen and weaken the vessel wall, making it more susceptible to dilatation and rupture. Such disorders frequently also damage the endothelium or produce factors that activate platelets and the clotting cascade and cause thrombosis. This leads to atheromatous disease with consequent degrees of occlusion.

Risk Factors and Neurovascular Disease

Risk factors for brain vascular disease are by and large the same as those for athero-sclerotic and arteriosclerotic diseases in the cardiovascular system. Hypertension, diabetes mellitus, obesity, metabolic syndrome, dyslipidaemia, in addition to ageing are traditionally considered as major risk factors for strokes and cerebral SVD. All these vascular morbidities may occur singly but often in combination frequently during ageing. It is difficult to define which particular risk factor might be primarily responsible for ageing-related vascular changes and the degree and extent of conse-quent brain damage.

Hypertensive Angiopathy

Hypertension is not in itself a disease of blood vessels but the deleterious effects of high blood pressure are mediated by structural changes in smaller arteries. Experimental studies have demonstrated that chronic hypertension shifts the autoregulatory limits towards higher arterial pressure values (Nagata et al. 2016; Toth et al. 2017). This protective response allows maintenance of constant CBF even at increased arterial pressure and prevents the undesired effects of increased systemic pressure to the capillary network (Toth et al. 2017). Symptoms of cerebral hypoperfusion develop when the mean arterial blood pressure falls to ~40% of baseline levels. In hypertensive patients, such a reduction is reached at a correspondingly higher level of arterial pressure than in normotensive individuals. However, sustained hypertension during ageing can have two main consequences. First, it aggravates atherosclerotic changes in both extracranial and intracranial larger arteries. For this to happen, it has been shown experimentally that a critical level of circulating lipoproteins is a prerequisite. For example, in spontaneously hypertensive rats a lipid-rich diet was needed to induce atherosclerosis (Horie 1977). In addition, hypertension promotes atherosclerosis to extend more distally into the intracranial compartment to affect smaller arteries and arterioles (~2 mm in diameter). The leptomeningeal arteries over the convexities are spared in normotensives with atherosclerosis, whereas in hypertensive patients they appear as hardened non-collapsed yellowish blood vessels. Lesions similar to those in large vessel atherosclerosis are present in walls of arterioles down to 100 μm in diameter. Second, the persistent high blood pressure leads to focal disruption of capillaries, which are the seat of the BBB. Immunocytochemical and electron microscopic studies reveal that the brightly eosinophilic 'fibrinoid' material in thickened blood vessel walls contains extracellular deposits of proteinaceous material of different plasma proteins. At these sites, the basement membrane under the damaged or regenerated endothelial cells is thickened or reduplicated. The vascular smooth muscle cells or myocytes degenerate and are replaced by collagenous connective tissue (Kalaria et al. 2015a).

Cerebral Amyloid Angiopathy

Cerebral amyloid angiopathy (CAA) is characterised by the deposition of amyloid protein in intracranial cerebral arteries (Table 17.1). The arteries already vulnerable to effects of ageing invariably accumulate amyloid fibrils in addition to other proteins with the eventual loss of myocytes, particularly within the medial-adventitial layers (Kawai et al. 1993). CAA is never found in the venous system but the basement membranes of capillaries often bear micro amyloid deposits (Kalaria et al. 1996). This observation lead to the classification of two types of CAA, Type I representing arteries and Type II associated with capillaries. CAA is most commonly found in AD (Attems et al. 2011; Love et al. 2015; Grinberg and Thal 2010; Charidimou et al. 2012) but it also evident in ageing people in the absence of significant AD pathology (Cohen et al. 1997). In a study of surgical biopsies exhibiting cerebral and cerebellar infarction, CAA was more common in subjects with infarction than in age-matched controls with non-vascular lesions (Cadavid et al. 2000). There is also an association between severe CAA and cerebrovascular lesions coexisting with AD, including lacunar infarcts, microinfarcts and haemorrhages (Ellis

Table 17.1 Types of changes in cerebral blood vessels during ageing

Primary vascular disease	Common disease type	Vascular distribution	Predominant tissue changes	Associated conditions or syndromes
Atherosclerotic disease	Cardiac and carotid atherosclerosis	Aorta, carotid, intracranial-MCA branches	Cortical and territorial infarcts; WML	Large vessel dementia or multi-infarct dementia
		Aorta, coronary	Infarcts, laminar necrosis, rarefaction	Hypoperfusive dementia
Embolic disease	Cardioembolism	Intracranial arteries, MCA	Large and small infarcts	Multi-infarct dementia (MID)
Arteriolosclerosis	Cerebral small vessel disease	Perforating and penetrating arteries, lenticulostriate arteries	Cortical infarcts, lacunar infarcts/ lacunes, microinfarcts, WML	Small vessel dementia; subcortical ischaemic vascular dementia; strategic infarct dementia
	Hypertensive vasculopathy			Hypertensive encephalopathy with impairment; strategic infarct dementia
Amyloid angiopathy	Amyloid β, prion protein, cystatin C, transthyretin, gelsolin)	Leptomeninges, intracerebral arteries	Cortical microinfarcts, lacunar infarcts, WML	Mild and major VCI

Data summarised and updated from several source references (Filosto et al. 2007; Caplan 2008; Ferro et al. 2010; Kalaria et al. 2015b; Kalaria 2016). Several disorders may also occur with other co-morbidities such as coronary artery disease, congestive heart failure, hypertension, diabetes, hyperlipidaemia, hypercoagulability, renal disease, atrial fibrillation and valvular heart disease. VCI is determined as Mild or Major forms when two or more cognitive domains are affected per harmonisation guidelines or per VCD (Hachinski et al. 2006a; Sachdev et al. 2014; Skrobot et al. 2016)

CAA cerebral amyloid angiopathy, *ICH* intracerebral haemorrhage, *MCA* middle cerebral artery, *SVD* small vessel disease, *VaD* vascular dementia, *VCD* vascular cognitive disorder, *VCI* vascular cognitive impairment, *WML* white matter lesion

et al. 1996; Olichney et al. 1995; Okamoto et al. 2012). Furthermore, there is evidence to suggest CAA is related to WM changes but not exclusively in the oldest old (Tanskanen et al. 2013). CAA is an independent substrate for cognitive impairment and contributes to cognitive dysfunction (Pfeifer et al. 2002; Arvanitakis et al. 2011b). Tissue microstructural damage caused by CAA prior to pre-intracerebral haemorrhage (ICH) is independently associated with cognitive dysfunction (Viswanathan et al. 2008).

Large Artery Disease

Atherosclerosis

Atherosclerosis is likely the most common cause of large artery disease (Fig. 17.1). It is a systemic disease, which increases exponentially with age (Fig. 17.2). It can affect the entire arterial network although marked regional variations exist with segmental location of lesions. The carotid arteries are common extracranial sites of atheromas, which are qualitatively similar to those in the aorta and coronary arteries (Jashari et al. 2013; Kalaria et al. 2015a). Atherosclerosis generally develops slowly and apparently begins at younger ages in Western communities (Fig. 17.2). The risk factors are in common with stroke, including hypertension, diabetes mellitus, dyslipidaemia and cigarette smoking. High serum lipids and blood pressure appear to be associated with a high prevalence of atherosclerosis in both the extracranial and intracranial arteries, whereas low or normal lipids and high blood pressure levels are associated mainly with intracranial and intracerebral arterial disease.

Besides the internal carotid artery, the initial segment of the middle cerebral artery, the distal portion of the vertebral arteries and the basilar artery are recurrent

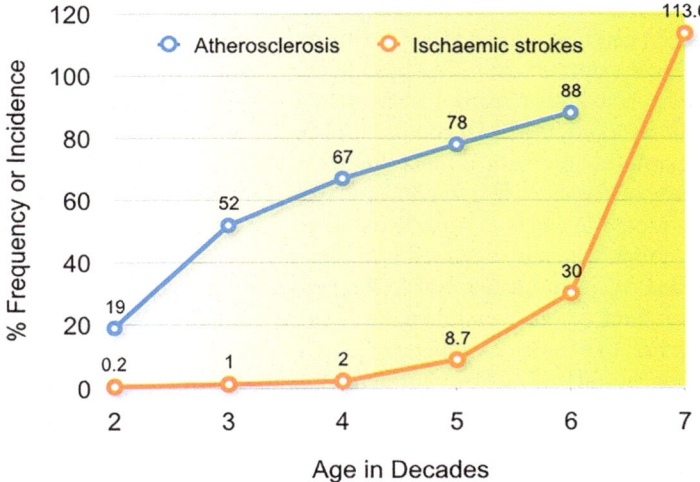

Fig. 17.2 Atherosclerosis and Strokes in relation to Age. Plots show increment in percentages of atherosclerosis and age-standardised stroke incidence per 10,000 person-years by each decade in ageing individuals. These plots illustrate that vascular changes such as atherosclerosis and likely arteriolosclerosis precede overt neurological symptoms including strokes, cognitive impairment and dementia by a number of years. Atherosclerosis is represented in coronary arteries in a North American sample whereas all cause ischaemic strokes are represent in large samples from high income countries including North America and western Europe. Prevalence of severe carotid stenosis increases with prevalence of coronary stenosis. Coronary atherosclerosis appears more common in stroke patients. (Data for the plots derived from Tuzcu et al. 2001; Jashari et al. 2013; Krishnamurthi et al. 2013)

sites for atheromas (Chen et al. 2008; Ogata et al. 2008). Atheromas may also develop in smaller segments of the main cerebral arteries in the very elderly (>80 years) (Kalaria et al. 2012). Atherosclerosis of the anterior cerebral arteries can further block small anterior perforating arteries (Roher et al. 2011). In accord with the American Heart Association criteria, atherosclerotic lesions from type I to type III are clinically silent whereas those from stage IV to VI may progressively obstruct the lumen of medium size arteries to the point of producing a clinical event (Stary 2000).

Pathogenesis of Atherosclerosis

Recent advances indicate the complexities of the pathogenesis of atherosclerosis in arteries (Libby et al. 2011). A key feature is accumulation of lipids in the arterial intima, which is initiated by dysfunction of the underlying arterial endothelium. This leads to a series of interdependent cellular and molecular processes, including modification of lipids, migration and proliferation of smooth muscle and inflammatory cells, production of pro-inflammatory mediators, and possibly also invasion by microorganisms (Libby et al. 2011). Epidemiological and experimental evidence suggests as in coronary arteries, atherosclerotic lesions in walls of large and medium-sized cerebral arteries develop due to increased dietary lipids, particularly cholesterol and saturated fats. The lipid hypothesis postulates that atherosclerosis commences when high plasma levels of low density lipoprotein (LDL) lead to excessive accumulation of LDL-cholesterol in the arterial intima. With increasing age, LDL particles are retained much longer in the arterial intima than in other tissues because of the lack of lymphatic vessels in the intima although not in the media. The proteoglycan matrix further slows down the movement of LDL particles across the intima, which is separated from the media by a poorly permeable internal elastic lamina. As a result, intimal LDL increases some tenfold compared to the extracellular fluid of other extrahepatic tissues. Intimal cells also do not express LDL receptors like other extrahepatic tissues, which retards LDL uptake with increased concentrations of LDL within the intima. The retained LDL particles may be modified by oxidation and peroxidation, as well as proteolysis and aggregation, leading to accumulation of lipid droplets. Cholesterol derived from modified LDL particles and stored in macrophages may be returned to the circulation by high density lipoprotein (HDL), but there is often impairment of this initial stage of reverse cholesterol transport owing to low HDL levels during ageing.

A further prerequisite of atherogenesis is disruption of the endothelial barrier function resulting from injury to endothelial cells by hypertension, dyslipidaemia, free radicals and toxic substances such as those in cigarette smoke. The damaged endothelium is also an excessive outlet of lipids into the intima. Chemical modification of LDL particles is essential for the initiation of the cascade that leads to pathological accumulation of lipids in the intima. Modified LDL particles are also a potent chemoattractant for circulating monocytes and impede the migration of macrophages from the intima to the circulation. The monocytes collecting within the intima and, to a lesser extent, myocytes take up and deposit modified LDL-cholesterol much more effectively than native LDL by exploiting specific scavenger receptors. These cells are transformed into intimal foam cells, clusters of which form the initial lesion visible to the naked eye as yellowish fatty streaks. The efflux of excess cholesterol from cells is mediated by the transmembrane protein adenosine triphosphate (ATP)-binding cassette transporter A1 (ABCA1). Reduced expression of ABCA1 in carotid plaques appears a key factor in the pathogenesis of atherosclerotic lesions.

Fatty streaks may appear in the carotid arteries during the first decade and are themselves innocuous to the circulation. Their appearance does not always lead to more severe atherosclerosis. However, fatty streaks develop into more advanced lesions, as atherosclerotic plaques or atheromas, over decades. The plaques are fibrous if the collagenous connective tissue component predominates over the lipids. Plaques are usually located in specific sites, such as on the outer aspects of the bifurcations of arteries. In these sites, the intima is thickened and laminar blood flow is affected. Inflammatory cells enter the plaques from the circulation through interactions between various adhesion molecules and their ligands on the surface of endothelial cells and leukocytes. Inflammatory and endothelial cells secrete multiple growth factors, such as platelet-derived, epidermal, fibroblast and transforming growth factors, and cytokines, such as interleukins, tumour necrosis factor and leukotrienes, which may recruit additional cells, including myocytes from the media, into these lesions. They also induce the transformation of myocytes from contractile cells into cells that actively synthesize extracellular proteins. The myocytes alone are responsible for the production of extracellular matrix components (mainly collagens I and III) of the fibrous cap underneath the intact endothelium. Beneath the cap, clusters of myocytes, foam cells and lymphocytes collect, together with a central core of necrotic cell debris and extracellular lipids, frequently including cholesterol crystals. As long as the fibrous cap remains intact, the plaque is stable.

Although they may already be advanced and occlude the lumen of the affected artery to a considerable extent, they do not usually reduce blood flow sufficiently to cause neurological symptoms. Even 5% of the cross-sectional area of the lumen may be sufficient to maintain blood flow at the level needed for normal brain function. The turbulence caused by the stenosis, could exacerbate endothelial damage and lead to the development of complicated plaques, which are unstable with the size of the lipid core increasing and fibrous cap becoming thinner. Thinning of the fibrous caps leads to plaque instability, a process modulated by inflammatory cells, which are found at the sites of ulceration or rupture of an atheromatous plaque. Plaques with thin caps become unstable, prone to ulceration and thrombosis. A critical event in atherosclerosis is damage to the endothelium, which is invariably affected in complicated plaques in carotid artery occlusion. Such disruption exposes tissue factors that activate coagulation, and thrombus forms over the plaque, causing narrowing of the lumen and increased risk of embolism. The high frequency of embolic strokes highlights the danger posed by the development of thromboemboli. An embolus leads to abrupt occlusion of the vascular lumen, whereas a local thrombotic process is usually slow and may allow time for collateral channels to develop.

Cerebral Small Vessel Disease

Besides age, hypertension and diabetes mellitus are the main risk factors for SVD. The importance of these risk factors for clinical ischaemic lacunar stroke has been questioned but their contribution to the structural pathology of SVD is generally accepted (Bailey et al. 2012). The term small vessels is applied in the CNS to perforating arteries and arterioles, with diameters 40–900 μm, which emerge from the leptomeningeal arteries, enter the brain parenchyma from the surface of the brain and extend a variable depth into the parenchyma (Fig. 17.1). The progressive age-related arteriolosclerotic changes with continued segmental loss of myocytes in the smaller intracerebral end-arteries and arterioles (Lammie et al. 1997) likely promote loss of elasticity to dilate and constrict in response to variations of systemic blood pressure or auto-regulation, which in turn causes fluctuations in blood flow response and tissue perfusion. The deeper structures and WM would be most vulnerable because the vessels are end arteries almost devoid of anastomoses. Small vessel pathology also likely leads to oedema and damage of the BBB with chronic leakage of fluid and macromolecules in the WM (Wardlaw 2010; Giwa et al. 2012; Ho and Garcia 2000). In the very old, there is often evidence of remote haemorrhage in form of perivascular hemosiderin. Hypertension linked SVD would further promote decreased production and increased degradation of nitric oxide resulting in endothelial dysfunction (Kalaria 2012). However, the two most common cerebrovascular pathologies associated with SVD are lacunar or small deep infarcts (cystic lesions generally <1 cm) and primary non-traumatic ICH. The consequences of multiple lacunar infarcts and incidental microinfarcts occurring in both the deep grey matter and WM structures is subcortical ischaemic VaD.

The four major structural pathologies described in penetrating small arteries in SVD are atherosclerosis, fibrinoid necrosis (lipohyalinosis), arteriolosclerosis and microaneurysms (Deramecourt et al. 2012). Originally reported (Fisher 1982) as segmental arterial disorganization and later described as lipohyalinosis, and small vessel atherosclerosis are the two most common causes of lacunar infarcts. Fibrinoid necrosis as an acute form of lipohyalinosis was more associated with ICH. SVD also entails expansion of the perivascular space and pallor of adjacent perivascular myelin, with associated astrocytic gliosis.

Intracranial Atherosclerosis

Small vessel atherosclerosis affects vessels 200–900 μm in diameter (Kalaria et al. 2012). Plaques may be found in proximal segments of penetrating arteries (microatheroma), at the junction of the branching and parent arteries (junctional atheroma) and in the parent vessel overlying the branch origin. The pathogenesis of atherosclerosis in the small vessels does not differ substantially from that in larger vessels.

The atheroma in small arteries also contains CD68-positive macrophages (Kalaria et al. 2012). Atheroma in small vessels causes complete occlusion at an earlier stage of plaque evolution.

Fibrinoid Necrosis and Lipohyalinosis

The mural disease in SVD, affecting mainly arterioles 40–300 μm in diameter described as lipohyalinosis is closely linked to longstanding hypertension and increases the risk of rupture of small arteries with consequent ICH (Jackson and Sudlow 2005). In the early stages, the walls are thickened by eosinophilic fibrinoid material, which appears to be composed mainly of plasma proteins, with abundant fibrin, formed by the leakage of the BBB together with remnants of myocytes. At this fibrinoid-necrosis stage of SVD, the BBB is disrupted and the affected vessels are prone to rupture. With time, the fibrinoid material is replaced by collagen produced by fibroblasts and the arteriolar walls become more homogeneous and less structured. Lipids are usually only a minor component in these lesions, and hyalinosis refers to acellular fibrosis. It has been recommended that the term 'lipohyalinosis' should be abandoned and replaced by descriptions based on appropriate stains: fibrinoid change if histological or immunocytochemical stains verify the presence of fibrin, and fibrosis if collagen is the main constituent of thickened arterial walls. The homogeneous eosinophilia in haematoxylin- and eosin (H&E) -stained sections may result from either fibrinoid change (necrosis) or collagenous fibrosis (hyalinosis). These two are probably consecutive changes and can be readily distinguished using special stains but appear deceptively similar with H&E.

Arteriolosclerosis

Arteriolosclerosis describes non-fibrinoid hyaline thickening of 40–150 μm-diameter arteriolar vessels. There is clearly morphological overlap with the late stage of Fisher's lipohyalinosis. The frequency of arteriolosclerosis rises with age. The thickened walls narrow the lumen and increase the sclerotic index (SI = 1-[internal diameter/external diameter]), a measure devised to indicate severity of degenerative fibrous thickening of the tunica media (Lammie et al. 1997). Healthy intracerebral arteries are relatively thin-walled with a wide lumen in relation to the wall thickness: their SI range is below 0.4 and in young people even below 0.3. Arteriolosclerosis tends to be associated with ischaemic white matter disease and VaD rather than lacunar infarcts. The thickened fibrotic arteries seldom rupture.

Microaneurysms

Miliary micro-aneurysms arise in the context of hypertension, at weakened sites in vessel walls. They resemble small sacs, 0.3- to 2-mm across, arising from parent arteries/arterioles 100–300 μm in diameter. The walls of the aneurysms are composed of hyaline connective tissue, damaged myocytes and elastica interna. Rupture of micro-aneurysms typically produces globular haemorrhages; if 'healed' by thrombosis and fibrosis, these are transformed into fibrocollagenous balls. Alkaline phosphatase histochemistry and high-resolution micro-radiography has shown the great majority of 'micro-aneurysms' to be complex tortuosities. These were most common at the interface between the grey and white matter and their number increased with age, but hypertension had no effect on their prevalence.

Venous Collagenosis

Venous collagenosis is mostly seen in older brains and increases in tandem with white matter disease (Lin et al. 2017). There is gradual thickening of the walls of periventricular veins and venules. Walls of the veins composed of collagen types I and III, may be dilated or in advanced cases, severely stenosed or occluded but generally lack inflammatory cells. It is proposed that collagenosis initially dilates the veins, making them macroscopically visible and causing venous insufficiency with consequent vessel leakage and vasogenic oedema. While WM changes focus on the arterial system, narrowing and, in many cases, occlusion of veins and venules by collagenous thickening of the vessel walls is readily seen adjacent to the caudal epithelium. Thickening of the walls of periventricular veins and venules by collagen (collagenosis) increases with age, and perivenous collagenosis increases in brains with leukoaraiosis (Brown et al. 2002b). The presence of apoptotic cells in WM adjacent to areas of leukoaraiosis suggests these lesions are dynamic, with progressive cell loss and expansion (Brown et al. 2002b). Vascular stenosis caused by collagenosis may induce chronic ischemia or oedema in the deep WM leading to capillary loss and more widespread effects on the brain (Brown and Thore 2011).

Cerebral Capillaries and the Blood Brain Barrier (BBB)

Smaller intra-parenchymal vessels or microvessels, also consisting of precapillaries and capillaries lacking myocytes, control permeability of water, glucose and nutrients. Capillaries are the anatomical representative of the BBB and comprise and 'epithelial type' of the single-cell layer of endothelial cells supported by a robust basement membrane (basal lamina). The endothelial cells are connected end to end with intervening tight junctions to form a specialised sieve. This unique endothelium (Sweeney et al. 2018), which isolates the brain neuropil from the systemic circulation, is endowed with a host of transporters and receptors that are important in maintaining a stable internal milieu. It is also under direct neuronal control mediated by transmitter-specific receptors. Previous autoradiographic evidence suggests endothelial cell turnover is relatively low after maturation and during ageing. They behave almost like post-mitotic cells in the ageing brain (Kalaria 1996). Properties of the endothelium have been studied using the isolated cerebral microvessel preparations from various species (Kalaria and Harik 1988). Ageing related changes in the cellular and molecular components of brain capillaries as prominent features of the neuro- and gliovascular units have significant impact on functions of the BBB (Table 17.2).

A number of studies have described ageing-related morphological changes in the brain microvasculature consisting of the capillary bed (Kalaria 1996). These include convolutes, tortuosity, looping and twisting of capillaries and arterioles. The presence of "strings", "streamers" and abnormal endothelial processes are also apparent. The capillary endothelium could also be attenuated or absent and a reduction in the number of interendothelial junction clefts. The most consistent change recorded is the thickening of the basement membrane, which may reduplicated and accompanied by inclusions of flocculent material or vacuoles. Some studies have also recorded the thinning of the capillary endothelium in aged rats, monkeys and humans (Kalaria 1996). The thickness of the basement membrane of capillaries in frontal and occipital cortices is increased in ageing subjects with or without diabetes mellitus (Johnson et al. 1982). It is plausible that the thinning of the endothelium and increases in basement membrane account for the lack of detectable change in vessel diameter reported in some studies (Burke et al. 2014). Differential loss of the endothelium during ageing is consistent with impairment of functional proteins, as indicated by reductions in specific transport processes, such as those for glucose and choline (Mooradian 1988). Between the sixth and seventh decades, aged subjects revealed increased capillary diameter, volume and total length, as well as decreased surface area (Kalaria 1996), but these changes were also evident in hypertensive subjects within the same age range (Hunziker et al. 1979).

Using biopsy tissue, Stewart et al. (1987) also reported that capillary walls, especially those of the white matter, became thinner with ageing, and this change was attributed to the thinning of endothelial cytoplasm with reduce mitochondrial density and loss of pericytes. Pericytes are intriguing cellular elements associated with vasculature, but currently, specific markers to identify them are lacking (Sweeney et al. 2016). Pericytes were reported to be decreased in the cortex in ageing (Stewart et al. 1987). It was concluded that the loss of pericytes may indicate that the BBB in the elderly is less able to compensate for transient leaks in older subjects (Stewart et al. 1987). In tandem with these cellular changes, we recently reported that astrocytes undergo clasmatodendrosis and retraction of their end-feet during ageing (Chen et al. 2016). The consequences of this would affect water permeability at the BBB resulting in tissue oedema.

Several studies have recorded effects of ageing on biochemical features associated with the cerebral microvasculature (Sweeney et al. 2018). Collectively, these

Table 17.2 Effects of ageing on components of the neurovascular and gliovascular units

Structure	Cellular component(s)	Types of cellular or organelle change(s)	Key protein(s)	Functional effect(s)
Neurovascular Unit (grey matter)	Endothelial cells	↓ cytoplasm, mitochondria, tight junctions	GLUT1, Na+/K+ ATPase, ICAM1, VEGF	Selective functional impairment of BBB functions; angiogensis
		↑ basement membrane; ↑ collagen fibrils	COL4, proteoglycans, laminin, fibrinogen	Thickening of BL, ↑ MMPs
	Astrocytes	↓ end-feet	Aquaporin; GFAP	Lack of control of water transport; reduced reactive cell responses
	Pericytes	↓ cellular coverage (more evident in WM)	TGFRβ1, BMP4,	Decreased pericyte functions impacting on permeability
	Nerve terminals	↓ afferents local and projection (brainstem) neurons	ACh, NA	Altered excitatory and inhibitory influences
Gliovascular Unit (white matter)[a]	Oligodendrocytes	↑ cell numbers, oligodendrocyte precursor cells	GST-γ	Attempted for remyelination; modification of inflammatory responses
	Microglia	↑ hypertrophic cell numbers	CD68, Iba1, TREM2	Increased perivascular macrophage activity

Data summarised and updated from several source references (Kalaria 1996; Kalaria et al. 2015a; Kalaria 2016)

ACh acetylcholine, *COL4* collagen IV, *BBB* blood-brain barrier, *BL* basal lamina, *BMP4* bone marrow protein 4, *GFAP* glial acidic fibrillary protein; *GLUT1* glucose transporter 1, *GST-γ* glutathione S-transferase-gamma, *ICAM1* intracellular adhesion molecule, *MMPs* metallomatrix proteins, *NA* noradrenaline, *VEGF* vascular endothelial growth factor

[a]Endothelial cells, astrocytes, pericytes and nerve endings not listed but in general changes similar to that in grey matter

studies also involving in vitro assays of isolated microvascular fractions suggest ageing-related decline in synthesis and composition of several 'vascular proteins, which contribute to alterations in the mechanical and functional properties of both resistance and perfusion function-related cerebral vessels (Kalaria 1996). Selective biochemical changes are also evident in the cerebral microvasculature of AD and Down's syndrome (DS) subjects (Sweeney et al. 2018). We reported the density of the glucose transporter 1 (GLUT1) declines with age and is markedly reduced in

AD compared to age-matched cognitively intact ageing controls (Kalaria 1992; Kawai et al. 1990). These original observations corroborate the early cerebral microvascular degeneration in GLUT1 deficient mice (Winkler et al. 2015). The transferrin receptor (TfR) is increased in the cerebral microvasculature in AD subjects with advanced disease. This increase may represent neovascularization, particularly if TfR is a marker of maturing cells (Kalaria et al. 1992). This is consistent with the absolute increase in vessel length, which may be induced by recurrent hypoxic episodes during ageing (Burke et al. 2014). Ageing-related mechanisms that potentially produce oxidative damage in the suggest increased nitric oxide synthase protein and nitric oxide production in capillaries from AD subjects compared with age-matched normal controls (Dorheim et al. 1994).

Consequences of Neurovascular Changes

Macroinfarcts or Large infarction

Macroinfarcts largely occur because of blockage in more proximal arterial territories by emboli or thrombi originating from the heart or the carotid and rarely vertebral arteries. The breaking up an old atheroma from the carotid artery is a typical example causing such occlusion. In routine post-mortem examination of cerebrovascular disease cases, visible infarcts may range from 5 to 50 mm. Acute lobar infarcts involving blockage in the very proximal parts middle cerebral artery may also be rarely evident. There is greater predilection for cortical infarcts when there is more severe stenosis of the carotid arteries.

Lacunar Infarcts/Lacunes

Lacunar infarcts increase with age. They represent small foci of ischaemic necrosis resulting from narrowing or occlusion of perforating arteries branching directly from larger cerebral arteries (Fisher 1982). Lacunar infarcts are frequently multiple and bilateral and often coexist with other vascular lesions such as large infarcts or diffuse WM changes. Whether single or multiple, they may be asymptomatic or symptomatic, depending on their location and the volume of normal brain tissue lost. Perivascular oedema and thickening, inflammation and disintegration of the arteriolar wall are common, whereas vessel occlusion is rare (Bailey et al. 2012). Severe cribriform change and greater numbers of subcortical microvascular infarcts both lacunar and microinfarcts are associated with declining cognitive function (Smallwood et al. 2012; Kalaria 2012; Skrobot et al. 2016).

Microinfarcts

It is not clear how microinfarcts occur. Age-related changes in blood rheology or hemodynamics including severe hypotension and atherosclerosis appear to play a major role in the genesis of cortical watershed microinfarcts. Microvascular infarcts (lacunar infarcts and microinfarcts) also predict poor prognosis in the elderly (Ballard et al. 2000; Vinters et al. 2000; Brown et al. 2002a). Microinfarcts, visible only upon microscopy, are described as attenuated lesions of indistinct nature occurring in both cortical and subcortical regions and of up to 500 μm diameter. These foci exhibit usually a small vessel, pallor, perivascular neuronal loss, axonal damage (WM) and gliosis. They are estimated to occur in thousands (Westover et al. 2013) and typically occur in about 30% of cerebrovascular disease sample which comes to post-mortem (Arvanitakis et al. 2011a). Microinfarcts may have predilection in certain brain regions and could not only be influenced by the type of WM change but also the nature of vascular pathology (Ince et al. 2017). Cortical microinfarcts are increased in the presence of CAA (Okamoto et al. 2012; Haglund et al. 2006) and commonly found in tandem with CAA in the occipital cortex (Kovari et al. 2017).

Microhaemorrhages

Both radiological cerebral microbleeds and foci of haemosiderin containing single crystalloids or larger perivascular aggregates are found in brains of cognitively intact older subjects including those diagnosed with dementia, including VaD and AD. Cerebral microhaemorrhages or microbleeds detected by T2* weighted MRI are small, dot like hypotense abnormalities, and have been associated with extravasated haemosiderin derived from erythrocytes, lipohyalinosis and CAA (Fazekas et al. 1999). They are likely a surrogate marker of SVD evident on MRI along with lacunes and WM changes (Van der Flier and Cordonnier 2012). Microbleeds are mainly thought to result from hypertensive vasculopathy, but the frequent co-occurrence of lobar microbleeds suggests that neurodegenerative pathology or CAA is also of importance (Werring et al. 2011). Cerebral microhaemorrhages may occur in the absence of large intracerebral haemorrhages.

Cerebral microbleeds detected by MR imaging are a surrogate for ischaemic SVD rather than exclusively haemorrhagic diathesis (Janaway et al. 2014). Greater putamen haemosiderin was significantly associated with indices of small vessel ischaemia including microinfarcts, arteriolosclerosis and perivascular spacing and with lacunes in any brain region but not large artery disease or other measures of neurodegenerative pathology.

Vascular Cognitive Impairment During Ageing

Vascular cognitive impairment (VCI) was constructed to empower an unique label for all ageing-related conditions of vascular origin or impaired brain perfusion (O'Brien et al. 2003; Hachinski et al. 2006a; Gorelick et al. 2011). During further refining, the description vascular cognitive disorder (Sachdev 1999) was devised to incorporate a continuum comprising cognitive disorders of vascular aetiology with diverse pathologies and clinical manifestations. However, for practical purposes the categories of mild and major vascular cognitive disorders were introduced (Association 2013). The major neurocognitive disorder classification, meant to describe frank dementia as a substitute for VaD fits better with patients, and more adapted to neurodegenerative cognitive disorders for which memory impairment is not predominant but comprises substantial frontal lobe changes (Sachdev et al. 2014). More refinement towards a standardised diagnosis of VCI was achieved through a Delphi analysis (Skrobot et al. 2017a, b). This led to the use of 'Mild' and 'Major' subdivisions of the severity of impairment aligning with the revised terminology in DSM-5. Major forms of VCI or VaD are now proposed to comprise four main subtypes including subcortical ischaemic VaD or SIVaD, multi-infarct dementia or cortical dementia, post-stroke dementia (PSD) and mixed dementias, which could be subdivided according to respective neurodegenerative pathologies. Cognitive impairment or dementia following stroke is relatively common (Leys et al. 2005; Pendlebury and Rothwell 2009; Mok et al. 2017). Incident dementia after stroke or PSD may develop within 3 months or after a stabilisation period of a year or longer after stroke injury (Allan et al. 2012; Bejot et al. 2011; Pohjasvaara et al. 1997). However, PSD can have a complex aetiology with varying combinations of large artery disease and SVD as well as non-vascular pathology. At least 75% of PSD cases fulfilling relevant clinical guidelines for VCI (Hachinski et al. 2006b) are pathologically confirmed as VaD with little age-related AD pathology (Allan et al. 2012). Furthermore, the presence of any age-related hippocampal neurodegenerative changes did not differentiate demented from non-demented post-stroke subjects (Akinyemi et al. 2017).

Summary

Neurovascular changes impact on brain health during ageing in a significant way. Atherosclerosis and arteriolosclerosis in both extracranial and intracranial locations contribute parenchymal loss. Large vessel stenosis predominantly leads to cortical infarcts whereas small vessel may predispose to lacunes, microinfarcts and diffuse white matter disease. The white matter is most vulnerable and is often seen as one of the earliest to be affected. Specific ageing-related changes in the cellular components of the neurovascular and gliovascular units promote early disruption in the functions of the BBB prior to development of overt pathology. The degrees and duration of the resultant pathologies correlate with cognitive impairment.

Acknowledgments We thank Yumi Yamamoto, Arthur Oakley and Janet Slade, previous members of the Neurovascular Research Group for providing continued technical support.

Funding Sources Our work has been supported by the RCUK Newcastle Centre for Brain Ageing and Vitality, Medical Research Council (UK), Alzheimer's Research UK, the Dunhill Medical Trust, UK and the Newcastle National Institute for Health Research Biomedical Research Centre in Ageing and Age Related Diseases, Newcastle upon Tyne Hospitals National Health Service Foundation Trust.

Competing Interests We declare no competing interests.

References

Akinyemi RO, Allan L, Oakley A, Kalaria R (2017) Hippocampal neurodegenerative pathology in post-stroke dementia compared to other dementias and ageing controls. Front Neurosci 11:717. https://doi.org/10.3389/fnins.2017.00717

Allan LM, Rowan EN, Firbank MJ, Thomas AJ, Parry SW, Polvikoski TM, O'Brien JT, Kalaria RN (2012) Long term incidence of dementia, predictors of mortality and pathological diagnosis in older stroke survivors. Brain J Neurol 134(Pt 12):3716–3727. https://doi.org/10.1093/brain/awr273

Ambarki K, Wahlin A, Zarrinkoob L, Wirestam R, Petr J, Malm J, Eklund A (2015) Accuracy of parenchymal cerebral blood flow measurements using pseudocontinuous arterial spin-labeling in healthy volunteers. AJNR Am J Neuroradiol 36(10):1816–1821. https://doi.org/10.3174/ajnr.A4367

Arvanitakis Z, Leurgans SE, Barnes LL, Bennett DA, Schneider JA (2011a) Microinfarct pathology, dementia, and cognitive systems. Stroke 42(3):722–727. https://doi.org/10.1161/STROKEAHA.110.595082

Arvanitakis Z, Leurgans SE, Wang Z, Wilson RS, Bennett DA, Schneider JA (2011b) Cerebral amyloid angiopathy pathology and cognitive domains in older persons. Ann Neurol 69(2):320–327. https://doi.org/10.1002/ana.22112

Association AP (2013) Diagnostic and statistical manual of mental disorders, vol 5. APA, USA

Attems J, Jellinger K, Thal DR, Van Nostrand W (2011) Review: sporadic cerebral amyloid angiopathy. Neuropathol Appl Neurobiol 37(1):75–93. https://doi.org/10.1111/j.1365-2990.2010.01137.x

Bailey EL, Smith C, Sudlow CL, Wardlaw JM (2012) Pathology of lacunar ischemic stroke in humans – a systematic review. Brain Pathol 22(5):583–591. https://doi.org/10.1111/j.1750-3639.2012.00575.x

Ballard C, McKeith I, O'Brien J, Kalaria R, Jaros E, Ince P, Perry R (2000) Neuropathological substrates of dementia and depression in vascular dementia, with a particular focus on cases with small infarct volumes. Dement Geriatr Cogn Disord 11(2):59–65

Bejot Y, Aboa-Eboule C, Durier J, Rouaud O, Jacquin A, Ponavoy E, Richard D, Moreau T, Giroud M (2011) Prevalence of early dementia after first-ever stroke: a 24-year population-based study. Stroke; J Cereb Circ 42(3):607–612. https://doi.org/10.1161/STROKEAHA.110.595553

Brown WR, Thore CR (2011) Review: cerebral microvascular pathology in ageing and neurodegeneration. Neuropathol Appl Neurobiol 37(1):56–74. https://doi.org/10.1111/j.1365-2990.2010.01139.x

Brown WR, Moody DM, Challa VR, Thore CR, Anstrom JA (2002a) Apoptosis in leukoaraiosis lesions. J Neurol Sci 203–204:169–171

Brown WR, Moody DM, Challa VR, Thore CR, Anstrom JA (2002b) Venous collagenosis and arteriolar tortuosity in leukoaraiosis. J Neurol Sci 203–204:159–163

Burke MJC, Nelson L, Slade JY, Oakley AE, Khundakar AA, Kalaria RN (2014) Morphometry of the hippocampal microvasculature in post-stroke and age- related dementias. Neuropathol Appl Neurobiol 40(3):284–295. https://doi.org/10.1111/nan.12085

Cadavid D, Mena H, Koeller K, Frommelt RA (2000) Cerebral beta amyloid angiopathy is a risk factor for cerebral ischemic infarction. A case control study in human brain biopsies. J Neuropathol Exp Neurol 59(9):768–773

Caplan LR (2008) Uncommon causes of stroke, 2nd edn. Cambridge University Press, Cambridge

Charidimou A, Gang Q, Werring DJ (2012) Sporadic cerebral amyloid angiopathy revisited: recent insights into pathophysiology and clinical spectrum. J Neurol Neurosurg Psychiatry 83(2):124–137. https://doi.org/10.1136/jnnp-2011-301308

Chen XY, Wong KS, Lam WW, Zhao HL, Ng HK (2008) Middle cerebral artery atherosclerosis: histological comparison between plaques associated with and not associated with infarct in a postmortem study. Cerebrovasc Dis 25(1–2):74–80. https://doi.org/10.1159/000111525

Chen A, Akinyemi RO, Hase Y, Firbank MJ, Ndung'u MN, Foster V, Craggs LJ, Washida K, Okamoto Y, Thomas AJ, Polvikoski TM, Allan LM, Oakley AE, O'Brien JT, Horsburgh K, Ihara M, Kalaria RN (2016) Frontal white matter hyperintensities, clasmatodendrosis and gliovascular abnormalities in ageing and post-stroke dementia. Brain 139(Pt 1):242–258. https://doi.org/10.1093/brain/awv328

Cohen DL, Hedera P, Premkumar DR, Friedland RP, Kalaria RN (1997) Amyloid-beta protein angiopathies masquerading as Alzheimer's disease? Ann N Y Acad Sci 826:390–395

Deramecourt V, Slade JY, Oakley AE, Perry RH, Ince PG, Maurage CA, Kalaria RN (2012) Staging and natural history of cerebrovascular pathology in dementia. Neurology 78(14):1043–1050. https://doi.org/10.1212/WNL.0b013e31824e8e7f

Dorheim MA, Tracey WR, Pollock JS, Grammas P (1994) Nitric oxide synthase activity is elevated in brain microvessels in Alzheimer's disease. Biochem Biophys Res Commun 205(1):659–665. https://doi.org/10.1006/bbrc.1994.2716

Ellis RJ, Olichney JM, Thal LJ, Mirra SS, Morris JC, Beekly D, Heyman A (1996) Cerebral amyloid angiopathy in the brains of patients with Alzheimer's disease: the CERAD experience, Part XV. Neurology 46(6):1592–1596

Fazekas F, Kleinert R, Roob G, Kleinert G, Kapeller P, Schmidt R, Hartung HP (1999) Histopathologic analysis of foci of signal loss on gradient-echo T2*-weighted MR images in patients with spontaneous intracerebral hemorrhage: evidence of microangiopathy-related microbleeds. AJNR Am J Neuroradiol 20(4):637–642

Ferro JM, Massaro AR, Mas JL (2010) Aetiological diagnosis of ischaemic stroke in young adults. Lancet Neurol 9(11):1085–1096. https://doi.org/10.1016/S1474-4422(10)70251-9

Filosto M, Tomelleri G, Tonin P, Scarpelli M, Vattemi G, Rizzuto N, Padovani A, Simonati A (2007) Neuropathology of mitochondrial diseases. Biosci Rep 27(1–3):23–30. https://doi.org/10.1007/s10540-007-9034-3

Fisher CM (1982) Lacunar strokes and infarcts: a review. Neurology 32(8):871–876

Giwa MO, Williams J, Elderfield K, Jiwa NS, Bridges LR, Kalaria RN, Markus HS, Esiri MM, Hainsworth AH (2012) Neuropathologic evidence of endothelial changes in cerebral small vessel disease. Neurology 78(3):167–174. https://doi.org/10.1212/WNL.0b013e3182407968

Gorelick PB, Scuteri A, Black SE, Decarli C, Greenberg SM, Iadecola C, Launer LJ, Laurent S, Lopez OL, Nyenhuis D, Petersen RC, Schneider JA, Tzourio C, Arnett DK, Bennett DA, Chui HC, Higashida RT, Lindquist R, Nilsson PM, Roman GC, Sellke FW, Seshadri S, American Heart Association Stroke Council CoE, Prevention CoCNCoCR, Intervention, Council on Cardiovascular S, Anesthesia (2011) Vascular contributions to cognitive impairment and dementia: a statement for healthcare professionals from the american heart association/american stroke association. Stroke 42(9):2672–2713. https://doi.org/10.1161/STR.0b013e3182299496

Grinberg LT, Thal DR (2010) Vascular pathology in the aged human brain. Acta Neuropathol 119(3):277–290. https://doi.org/10.1007/s00401-010-0652-7

Hachinski V, Iadecola C, Petersen RC, Breteler MM, Nyenhuis DL, Black SE, Powers WJ, DeCarli C, Merino JG, Kalaria RN, Vinters HV, Holtzman DM, Rosenberg GA, Dichgans M, Marler JR, Leblanc GG (2006a) National institute of neurological disorders and stroke-canadian stroke network vascular cognitive impairment harmonization standards. Stroke 37(9):2220–2241

Hachinski V, Iadecola C, Petersen RC, Breteler MM, Nyenhuis DL, Black SE, Powers WJ, DeCarli C, Merino JG, Kalaria RN, Vinters HV, Holtzman DM, Rosenberg GA, Wallin A, Dichgans M, Marler JR, Leblanc GG (2006b) National institute of neurological disorders and stroke-canadian stroke network vascular cognitive impairment harmonization standards. Stroke 37(9):2220–2241. https://doi.org/10.1161/01.STR.0000237236.88823.47

Haglund M, Passant U, Sjobeck M, Ghebremedhin E, Englund E (2006) Cerebral amyloid angiopathy and cortical microinfarcts as putative substrates of vascular dementia. Int J Geriatr Psychiatry 21(7):681–687. https://doi.org/10.1002/gps.1550

Ho KL, Garcia JH (2000) Neuropathology of the small blood vessels in selected disease of the cerebral white matter. In: Pantoni L, Inzitari D, Wallin A (eds) The matter of white matter. Current issues in neurodegenerative diseases. Academic Pharmaceutical Productions, Utrecht, pp 247–273

Horie R (1977) Studies on stroke in relation to cerebrovascular atherogenesis in stroke-prone spontaneously hypertensive rats (SHRSP). Nihon Geka Hokan 46(3):191–213

Hunziker O, Abdel'Al S, Schulz U (1979) The aging human cerebral cortex: a stereological characterization of changes in the capillary net. J Gerontol 34(3):345–350

Ince PG, Minett T, Forster G, Brayne C, Wharton SB, Medical Research Council Cognitive F, Ageing Neuropathology S (2017) Microinfarcts in an older population-representative brain donor cohort (MRC CFAS): prevalence, relation to dementia and mobility, and implications for the evaluation of cerebral small vessel disease. Neuropathol Appl Neurobiol 43(5):409–418. https://doi.org/10.1111/nan.12363

Jackson C, Sudlow C (2005) Comparing risks of death and recurrent vascular events between lacunar and non-lacunar infarction. Brain 128(Pt 11):2507–2517. https://doi.org/10.1093/brain/awh636

Janaway BM, Simpson JE, Hoggard N, Highley JR, Forster G, Drew D, Gebril OH, Matthews FE, Brayne C, Wharton SB, Ince PG, Function MRCC, Ageing Neuropathology S (2014) Brain haemosiderin in older people: pathological evidence for an ischaemic origin of magnetic resonance imaging (MRI) microbleeds. Neuropathol Appl Neurobiol 40(3):258–269. https://doi.org/10.1111/nan.12062

Jashari F, Ibrahimi P, Nicoll R, Bajraktari G, Wester P, Henein MY (2013) Coronary and carotid atherosclerosis: similarities and differences. Atherosclerosis 227(2):193–200. https://doi.org/10.1016/j.atherosclerosis.2012.11.008

Johnson PC, Brendel K, Meezan E (1982) Thickened cerebral cortical capillary basement membranes in diabetics. Arch Pathol Lab Med 106(5):214–217

Kalaria RN (1992) The blood-brain-barrier and cerebral microcirculation in Alzheimer-disease. Cerebrovas Brain Met 4(3):226–260

Kalaria RN (1996) Cerebral vessels in ageing and Alzheimer's disease. Pharmacol Ther 72(3):193–214

Kalaria RN (2012) Cerebrovascular disease and mechanisms of cognitive impairment: evidence from clinicopathological studies in humans. Stroke 43(9):2526–2534. https://doi.org/10.1161/STROKEAHA.112.655803

Kalaria RN (2016) Neuropathological diagnosis of vascular cognitive impairment and vascular dementia with implications for Alzheimer's disease. Acta Neuropathol 131(5):659–685. https://doi.org/10.1007/s00401-016-1571-z

Kalaria RN, Harik SI (1988) Adenosine receptors and the nucleoside transporter in human-brain vasculature. J Cerebr Blood F Met 8(1):32–39. https://doi.org/10.1038/Jcbfm.1988.5

Kalaria RN, Sromek SM, Grahovac I, Harik SI (1992) Transferrin receptors of rat and human brain and cerebral microvessels and their status in Alzheimers-disease. Brain Res 585(1–2):87–93. https://doi.org/10.1016/0006-8993(92)91193-I

Kalaria RN, Premkumar DR, Pax AB, Cohen DL, Lieberburg I (1996) Production and increased detection of amyloid beta protein and amyloidogenic fragments in brain microvessels, meningeal vessels and choroid plexus in Alzheimer's disease. Brain Res Mol Brain Res 35(1–2):58–68

Kalaria RN, Perry RH, O'Brien J, Jaros E (2012) Atheromatous disease in small intracerebral vessels, microinfarcts and dementia. Neuropathol Appl Neurobiol. https://doi.org/10.1111/j.1365-2990.2012.01264.x

Kalaria R, Ferrer I, Love S (2015a) Vascular disease, hypoxia and related conditions. In: Love S, Arie P, Ironside J, Budka H (eds) Greenfield's neuropathology, vol 1, 9th edn. CRC Press, London, pp 59–209

Kalaria RN, Ferrer I, Love S (2015b) Vascular disease, hypoxia and related conditions. In: Love S, Perry A, Ironside J, Budka H (eds) Greenfield's neuropathology, vol 1, 9th edn. CRC Press, London, pp 59–209

Kawai M, Kalaria RN, Harik SI, Perry G (1990) The relationship of amyloid plaques to cerebral capillaries in Alzheimers-disease. Am J Pathol 137(6):1435–1446

Kawai M, Kalaria RN, Cras P, Siedlak SL, Velasco ME, Shelton ER, Chan HW, Greenberg BD, Perry G (1993) Degeneration of vascular muscle-cells in cerebral amyloid angiopathy of Alzheimer-disease. Brain Res 623(1):142–146. https://doi.org/10.1016/0006-8993(93)90021-E

Kovari E, Herrmann FR, Gold G, Hof PR, Charidimou A (2017) Association of cortical microinfarcts and cerebral small vessel pathology in the ageing brain. Neuropathol Appl Neurobiol 43(6):505–513. https://doi.org/10.1111/nan.12366

Krishnamurthi RV, Feigin VL, Forouzanfar MH, Mensah GA, Connor M, Bennett DA, Moran AE, Sacco RL, Anderson LM, Truelsen T, O'Donnell M, Venketasubramanian N, Barker-Collo S, Lawes CM, Wang W, Shinohara Y, Witt E, Ezzati M, Naghavi M, Murray C, Global Burden of Diseases IRFS, Group GBDSE (2013) Global and regional burden of first-ever ischaemic and haemorrhagic stroke during 1990–2010: findings from the global burden of disease study 2010. Lancet Glob Health 1(5):e259–e281. https://doi.org/10.1016/S2214-109X(13)70089-5

Lammie GA, Brannan F, Slattery J, Warlow C (1997) Nonhypertensive cerebral small-vessel disease. An autopsy study. Stroke 28(11):2222–2229

Leys D, Henon H, Mackowiak-Cordoliani MA, Pasquier F (2005) Poststroke dementia. Lancet Neurol 4(11):752–759

Li L, Yiin GS, Geraghty OC, Schulz UG, Kuker W, Mehta Z, Rothwell PM, Oxford Vascular S (2015) Incidence, outcome, risk factors, and long-term prognosis of cryptogenic transient ischaemic attack and ischaemic stroke: a population-based study. Lancet Neurol 14(9):903–913. https://doi.org/10.1016/S1474-4422(15)00132-5

Libby P, Ridker PM, Hansson GK (2011) Progress and challenges in translating the biology of atherosclerosis. Nature 473(7347):317–325. https://doi.org/10.1038/nature10146

Lin J, Wang D, Lan L, Fan Y (2017) Multiple factors involved in the pathogenesis of white matter lesions. Biomed Res Int 2017:9372050. https://doi.org/10.1155/2017/9372050

Love S, Chalmers K, Ince P, Esiri M, Attems J, Kalaria R, Jellinger K, Yamada M, McCarron M, Minett T, Matthews F, Greenberg S, Mann D, Kehoe PG (2015) Erratum: development, appraisal, validation and implementation of a consensus protocol for the assessment of cerebral amyloid angiopathy in post-mortem brain tissue. Am J Neurodegener Dis 4(2):49

Mitchell GF, van Buchem MA, Sigurdsson S, Gotal JD, Jonsdottir MK, Kjartansson O, Garcia M, Aspelund T, Harris TB, Gudnason V, Launer LJ (2011) Arterial stiffness, pressure and flow pulsatility and brain structure and function: the age, gene/environment susceptibility – Reykjavik study. Brain 134(Pt 11):3398–3407. https://doi.org/10.1093/brain/awr253

Mok VC, Lam BY, Wong A, Ko H, Markus HS, Wong LK (2017) Early-onset and delayed-onset poststroke dementia – revisiting the mechanisms. Nat Rev Neurol 13(3):148–159. https://doi.org/10.1038/nrneurol.2017.16

Montagne A, Barnes SR, Sweeney MD, Halliday MR, Sagare AP, Zhao Z, Toga AW, Jacobs RE, Liu CY, Amezcua L, Harrington MG, Chui HC, Law M, Zlokovic BV (2015) Blood-brain barrier breakdown in the aging human hippocampus. Neuron 85(2):296–302. https://doi.org/10.1016/j.neuron.2014.12.032

Mooradian AD (1988) Effect of aging on the blood-brain barrier. Neurobiol Aging 9(1):31–39

Nagata K, Yamazaki T, Takano D, Maeda T, Fujimaki Y, Nakase T, Sato Y (2016) Cerebral circulation in aging. Ageing Res Rev 30:49–60. https://doi.org/10.1016/j.arr.2016.06.001

O'Brien JT, Erkinjuntti T, Reisberg B, Roman G, Sawada T, Pantoni L, Bowler JV, Ballard C, DeCarli C, Gorelick PB, Rockwood K, Burns A, Gauthier S, DeKosky ST (2003) Vascular cognitive impairment. Lancet Neurol 2(2):89–98

Ogata J, Yutani C, Otsubo R, Yamanishi H, Naritomi H, Yamaguchi T, Minematsu K (2008) Heart and vessel pathology underlying brain infarction in 142 stroke patients. Ann Neurol 63(6):770–781

Okamoto Y, Yamamoto T, Kalaria RN, Senzaki H, Maki T, Hase Y, Kitamura A, Washida K, Yamada M, Ito H, Tomimoto H, Takahashi R, Ihara M (2012) Cerebral hypoperfusion accelerates cerebral amyloid angiopathy and promotes cortical microinfarcts. Acta Neuropathol 123(3):381–394. https://doi.org/10.1007/s00401-011-0925-9

Olichney JM, Hansen LA, Hofstetter CR, Grundman M, Katzman R, Thal LJ (1995) Cerebral infarction in Alzheimer's disease is associated with severe amyloid angiopathy and hypertension. Arch Neurol 52(7):702–708

Pendlebury ST, Rothwell PM (2009) Prevalence, incidence, and factors associated with pre-stroke and post-stroke dementia: a systematic review and meta-analysis. Lancet Neurol 8(11):1006–1018. S1474-4422(09)70236-4 [pii]. https://doi.org/10.1016/S1474-4422(09)70236-4

Pfeifer LA, White LR, Ross GW, Petrovitch H, Launer LJ (2002) Cerebral amyloid angiopathy and cognitive function: the HAAS autopsy study. Neurology 58(11):1629–1634

Pohjasvaara T, Erkinjuntti T, Vataja R, Kaste M (1997) Dementia three months after stroke. Baseline frequency and effect of different definitions of dementia in the Helsinki Stroke Aging Memory study (SAM) cohort. Stroke; J Cereb Circ 28(4):785–792

Roher AE, Tyas SL, Maarouf CL, Daugs ID, Kokjohn TA, Emmerling MR, Garami Z, Belohlavek M, Sabbagh MN, Sue LI, Beach TG (2011) Intracranial atherosclerosis as a contributing factor to Alzheimer's disease dementia. Alzheimers Dement 7(4):436–444. S1552-5260(10)02421-0 [pii]. https://doi.org/10.1016/j.jalz.2010.08.228

Sachdev P (1999) Vascular cognitive disorder. Int J Geriatr Psychiatry 14(5):402–403

Sachdev P, Kalaria R, O'Brien J, Skoog I, Alladi S, Black SE, Blacker D, Blazer DG, Chen C, Chui H, Ganguli M, Jellinger K, Jeste DV, Pasquier F, Paulsen J, Prins N, Rockwood K, Roman G, Scheltens P, Internationlal Society for Vascular B, Cognitive D (2014) Diagnostic criteria for vascular cognitive disorders: a VASCOG statement. Alzheimer Dis Assoc Disord 28(3):206–218. https://doi.org/10.1097/WAD.0000000000000034

Skrobot OA, Attems J, Esiri M, Hortobagyi T, Ironside JW, Kalaria RN, King A, Lammie GA, Mann D, Neal J, Ben-Shlomo Y, Kehoe PG, Love S (2016) Vascular cognitive impairment neuropathology guidelines (VCING): the contribution of cerebrovascular pathology to cognitive impairment. Brain 139(11):2957–2969. https://doi.org/10.1093/brain/aww214

Skrobot OA, Black SE, Chen C, DeCarli C, Erkinjuntti T, Ford GA, Kalaria RN, O'Brien J, Pantoni L, Pasquier F, Roman GC, Wallin A, Sachdev P, Skoog I, group V, Ben-Shlomo Y, Passmore AP, Love S, Kehoe PG (2017a) Progress toward standardized diagnosis of vascular cognitive impairment: guidelines from the vascular impairment of cognition classification consensus study. Alzheimers Dement. https://doi.org/10.1016/j.jalz.2017.09.007

Skrobot OA, O'Brien J, Black S, Chen C, DeCarli C, Erkinjuntti T, Ford GA, Kalaria RN, Pantoni L, Pasquier F, Roman GC, Wallin A, Sachdev P, Skoog I, group V, Ben-Shlomo Y, Passmore AP, Love S, Kehoe PG (2017b) The vascular impairment of cognition classification consensus study. Alzheimers Dement 13(6):624–633. https://doi.org/10.1016/j.jalz.2016.10.007

Smallwood A, Oulhaj A, Joachim C, Christie S, Sloan C, Smith AD, Esiri M (2012) Cerebral subcortical small vessel disease and its relation to cognition in elderly subjects: a pathological study in the Oxford Project to Investigate Memory and Ageing (OPTIMA) cohort. Neuropathol Appl Neurobiol 38(4):337–343. https://doi.org/10.1111/j.1365-2990.2011.01221.x

Stary HC (2000) Natural history and histological classification of atherosclerotic lesions: an update. Arterioscler Thromb Vasc Biol 20(5):1177–1178

Stewart PA, Magliocco M, Hayakawa K, Farrell CL, Del Maestro RF, Girvin J, Kaufmann JC, Vinters HV, Gilbert J (1987) A quantitative analysis of blood-brain barrier ultrastructure in the aging human. Microvasc Res 33(2):270–282

Sweeney MD, Ayyadurai S, Zlokovic BV (2016) Pericytes of the neurovascular unit: key functions and signaling pathways. Nat Neurosci 19(6):771–783. https://doi.org/10.1038/nn.4288

Sweeney MD, Sagare AP, Zlokovic BV (2018) Blood-brain barrier breakdown in Alzheimer disease and other neurodegenerative disorders. Nat Rev Neurol 14(3):133–150. https://doi.org/10.1038/nrneurol.2017.188

Tanskanen M, Kalaria RN, Notkola IL, Makela M, Polvikoski T, Myllykangas L, Sulkava R, Kalimo H, Paetau A, Scheltens P, Barkhof F, van Straaten E, Erkinjuntti T (2013) Relationships between white matter hyperintensities, cerebral amyloid angiopathy and dementia in a population-based sample of the oldest old. Curr Alzheimer Res 10(10):1090–1097

Tarumi T, Zhang R (2018) Cerebral blood flow in normal aging adults: cardiovascular determinants, clinical implications, and aerobic fitness. J Neurochem 144(5):595–608. https://doi.org/10.1111/jnc.14234

Toth P, Tarantini S, Csiszar A, Ungvari Z (2017) Functional vascular contributions to cognitive impairment and dementia: mechanisms and consequences of cerebral autoregulatory dysfunction, endothelial impairment, and neurovascular uncoupling in aging. Am J Phys Heart Circ Phys 312(1):H1–H20. https://doi.org/10.1152/ajpheart.00581.2016

Tuzcu EM, Kapadia SR, Tutar E, Ziada KM, Hobbs RE, McCarthy PM, Young JB, Nissen SE (2001) High prevalence of coronary atherosclerosis in asymptomatic teenagers and young adults: evidence from intravascular ultrasound. Circulation 103(22):2705–2710

Van der Flier WM, Cordonnier C (2012) Microbleeds in vascular dementia: clinical aspects. Exp Gerontol 47(11):853–857. https://doi.org/10.1016/j.exger.2012.07.007

Vermeer SE, Longstreth WT Jr, Koudstaal PJ (2007) Silent brain infarcts: a systematic review. Lancet Neurol 6(7):611–619. https://doi.org/10.1016/S1474-4422(07)70170-9

Vinters HV, Ellis WG, Zarow C, Zaias BW, Jagust WJ, Mack WJ, Chui HC (2000) Neuropathologic substrates of ischemic vascular dementia. J Neuropathol Exp Neurol 59(11):931–945

Viswanathan A, Patel P, Rahman R, Nandigam RN, Kinnecom C, Bracoud L, Rosand J, Chabriat H, Greenberg SM, Smith EE (2008) Tissue microstructural changes are independently associated with cognitive impairment in cerebral amyloid angiopathy. Stroke; J Cereb Circ 39(7):1988–1992. https://doi.org/10.1161/STROKEAHA.107.509091

Wardlaw JM (2010) Blood-brain barrier and cerebral small vessel disease. J Neurol Sci 299(1–2):66–71

Wardlaw JM, Smith C, Dichgans M (2013) Mechanisms of sporadic cerebral small vessel disease: insights from neuroimaging. Lancet Neurol 12(5):483–497. https://doi.org/10.1016/S1474-4422(13)70060-7

Werring DJ, Gregoire SM, Cipolotti L (2011) Cerebral microbleeds and vascular cognitive impairment. J Neurol Sci 299(1–2):131–135. S0022-510X(10)00401-6 [pii]. https://doi.org/10.1016/j.jns.2010.08.034

Westover MB, Bianchi MT, Yang C, Schneider JA, Greenberg SM (2013) Estimating cerebral microinfarct burden from autopsy samples. Neurology 80(15):1365–1369. https://doi.org/10.1212/WNL.0b013e31828c2f52

Winkler EA, Nishida Y, Sagare AP, Rege SV, Bell RD, Perlmutter D, Sengillo JD, Hillman S, Kong P, Nelson AR, Sullivan JS, Zhao Z, Meiselman HJ, Wendy RB, Soto J, Abel ED, Makshanoff J, Zuniga E, De Vivo DC, Zlokovic BV (2015) GLUT1 reductions exacerbate Alzheimer's disease vasculo-neuronal dysfunction and degeneration. Nat Neurosci 18(4):521–530. https://doi.org/10.1038/nn.3966

Zhang N, Gordon ML, Goldberg TE (2017) Cerebral blood flow measured by arterial spin labeling MRI at resting state in normal aging and Alzheimer's disease. Neurosci Biobehav Rev 72:168–175. https://doi.org/10.1016/j.neubiorev.2016.11.023

Printed by Printforce, the Netherlands